SOCIAL SCIENCES

Applied Calculus

for the Managerial, Life, and Social Sciences

FIFTH EDITION

S. T. Tan

Stonehill College

BROOKS/COLE

THOMSON LEARNING

Australia • Canada • Mexico • Singapore • Spain • United Kingdom • United States

Sponsoring Editor: *Curt Hinrichs*
Assistant Editor: *Ann Day*
Editorial Assistant: *Suzannah Alexander*
Marketing Manager: *Karin Sandberg*
Marketing Assistant: *Darcie D. Pool*
Print/Media Buyer: *Barbara Britton*
Advertising Project Manager: *Brian Chaffee*
Production Service: *Cecile Joyner, The Cooper Company*
Text Designer: *Delgado Design, Inc.*
Photo Researcher: *Terri Wright*
Copy Editor: *Betty Duncan*
Cover Designer: *Lisa Henry*

Cover Illustration: *Judith Harkness*
Cover Printing: *Phoenix Color Corp.*
Compositor: *TechBooks*
Printer: *Quebecor/World–Versailles*
Photo Credits: **Page 3:** David Young-Wolff/PhotoEdit
59: Terry Powell/The Photographer's Window
183: Robert J. Western **266:** NASA **277:** PhotoDisc
373: Robert J. Western **439:** PhotoDisc **541:** Roy Corral/
CORBIS **601:** Terje Rakke/The Image Bank **692:** Harold
Sund/The Image Bank **728:** The Photographer's Window
772: Kay Chernusl/The Image Bank **856:** Barrie Rokeach/The
Image Bank

For permission to use material from this work, contact us by
Web: http://www.thomsonrights.com
Fax: 1-800-730-2215
Phone: 1-800-730-2214

Wadsworth/Thomson Learning
10 Davis Drive
Belmont, CA 94002-3098
USA

For more information about our products, contact us:
Thomson Learning Academic Resource Center
1-800-423-0563
http://www.brookscole.com

International Headquarters
Thomson Learning
International Division
290 Harbor Drive, 2nd Floor
Stamford, CT 06902-7477
USA

UK/Europe/Middle East/South Africa
Thomson Learning
Berkshire House
168-173 High Holborn
London WC1V 7AA
United Kingdom

Asia
Thomson Learning
60 Albert Street, #15-01
Albert Complex
Singapore 189969

Canada
Nelson Thomson Learning
1120 Birchmount Road
Toronto, Ontario M1K 5G4
Canada

Library of Congress Cataloging-in-Publication Data

Tan, Soo Tang.
　　Applied calculus for the managerial, life, and social
　　　　sciences / S. T. Tan.—5th ed.
　　　　p.　　cm.
　　Includes index.
　　Rev. ed. of: College mathematics. 4th ed. © 1999.
　　ISBN 0-534-37843-9
　　　　1. Calculus.　I. Tan, Soo Tang. Applied calculus.
　　II. Title.

QA303.T14　2002
515—dc21　　　　　　　　　　　　2001025725

CONTENTS

* Sections marked with an asterisk are not prerequisites for later material.

The first two sections of this chapter contain a brief review of algebra. We then introduce the Cartesian coordinate system, which allows us to represent points in the plane in terms of ordered pairs of real numbers. This in turn enables us to compute the distance between two points algebraically. This chapter also covers straight lines. The slope of a straight line plays an important role in the study of calculus.

What sales figure can be predicted for next year? In Example 10, page 46, you will see how the manager of a local sporting goods store used sales figures from the previous years to predict the sales level for next year.

1.1 Precalculus Review I

Sections 1.1 and 1.2 review some of the basic concepts and techniques of algebra that are essential in the study of calculus. The material in this review will help you work through the examples and exercises in this book. You can read through this material now and do the exercises in areas where you feel a little "rusty," or you can review the material on an as-needed basis as you study the text. We begin our review with a discussion of real numbers.

THE REAL NUMBER LINE

The real number system is made up of the set of real numbers together with the usual operations of addition, subtraction, multiplication, and division.

Real numbers may be represented geometrically by points on a line. Such a line is called the **real number,** or **coordinate, line** and can be constructed as follows. Arbitrarily select a point on a straight line to represent the number zero. This point is called the **origin.** If the line is horizontal, then a point at a convenient distance to the right of the origin is chosen to represent the number 1. This determines the scale for the number line. Each positive real number lies at an appropriate distance to the right of the origin, and each negative real number lies at an appropriate distance to the left of the origin (Figure 1.1).

FIGURE 1.1
The real number line

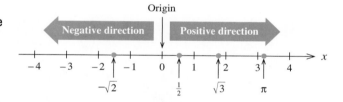

A *one-to-one correspondence* is set up between the set of all real numbers and the set of points on the number line; that is, exactly one point on the line is associated with each real number. Conversely, exactly one real number is associated with each point on the line. The real number that is associated with a point on the real number line is called the **coordinate** of that point.

INTERVALS

Throughout this book, we will often restrict our attention to certain subsets of the set of real numbers. For example, if x denotes the number of cars rolling off a plant assembly line each day, then x must be nonnegative—that is, $x \geq 0$. Further, suppose management decides that the daily production must not exceed 200 cars. Then, x must satisfy the inequality $0 \leq x \leq 200$.

More generally, we will be interested in the following subsets of real numbers: open intervals, closed intervals, and half-open intervals. The set of all real numbers that lie *strictly* between two fixed numbers a and b is called an **open interval** (a, b). It consists of all real numbers x that satisfy the inequalities $a < x < b$, and it is called "open" because neither of its end points is included in the interval. A **closed interval** contains *both* of its end points. Thus, the set of all real numbers x that satisfy the inequalities $a \leq x \leq b$ is the closed interval $[a, b]$. Notice that square brackets are used to indicate that the end points are included in this interval. **Half-open intervals** contain only *one* of their end points. Thus, the interval $[a, b)$ is the set of all real numbers x that satisfy $a \leq x < b$, whereas the interval $(a, b]$ is described by the inequalities $a < x \leq b$. Examples of these **finite intervals** are illustrated in Table 1.1.

Table 1.1   Finite Intervals

Interval		Graph	Example	
Open	(a, b)	(open circles at a and b)	$(-2, 1)$	$-3\ -2\ -1\ 0\ 1\ 2\ 3$
Closed	$[a, b]$	(closed at a and b)	$[-1, 2]$	$-1\ 0\ 1\ 2$
Half-open	$(a, b]$	(open at a, closed at b)	$(\frac{1}{2}, 3]$	$0\ \frac{1}{2}\ 1\ 2\ 3$
Half-open	$[a, b)$	(closed at a, open at b)	$[-\frac{1}{2}, 3)$	$-\frac{1}{2}\ 0\ 1\ 2\ 3$

In addition to finite intervals, we will encounter **infinite intervals.** Examples of infinite intervals are the half lines (a, ∞), $[a, \infty)$, $(-\infty, a)$, and $(-\infty, a]$ defined by the set of all real numbers that satisfy $x > a$, $x \geq a$, $x < a$, and $x \leq a$, respectively. The symbol ∞, called *infinity,* is not a real number. It is used here only for notational purposes in conjunction with the definition of infinite intervals. The notation $(-\infty, \infty)$ is used for the set of all real numbers x since, by definition, the inequalities $-\infty < x < \infty$ hold for any real number x. Infinite intervals are illustrated in Table 1.2.

Table 1.2   Infinite Intervals

Interval	Graph	Example	
(a, ∞)	(open at a, extends right)	$(2, \infty)$	$0\ 1\ 2$
$[a, \infty)$	(closed at a, extends right)	$[-1, \infty)$	$-1\ 0$
$(-\infty, a)$	(extends left, open at a)	$(-\infty, 1)$	$0\ 1$
$(-\infty, a]$	(extends left, closed at a)	$(-\infty, -\frac{1}{2}]$	$-\frac{1}{2}\ 0\ 1\ 2$

PROPERTIES OF INEQUALITIES

In practical applications, intervals are often found by solving one or more inequalities involving a variable. In such situations, the following properties may be used to advantage.

Properties of Inequalities

If a, b, and c are any real numbers, then

		Example
Property 1	If $a < b$ and $b < c$, then $a < c$.	$2 < 3$ and $3 < 8$, so $2 < 8$
Property 2	If $a < b$, then $a + c < b + c$.	$-5 < -3$, so $-5 + 2 < -3 + 2$; that is, $-3 < -1$
Property 3	If $a < b$ and $c > 0$, then $ac < bc$.	$-5 < -3$, and since $2 > 0$, we have $(-5)(2) < (-3)(2)$; that is, $-10 < -6$
Property 4	If $a < b$ and $c < 0$, then $ac > bc$.	$-2 < 4$, and since $-3 < 0$, we have $(-2)(-3) > (4)(-3)$; that is, $6 > -12$

Similar properties hold if each inequality sign, $<$, between a and b is replaced by $\geq$, $>$, or $\leq$.

A real number is a *solution of an inequality* involving a variable if a true statement is obtained when the variable is replaced by that number. The set of all real numbers satisfying the inequality is called the *solution set*.

EXAMPLE 1   Find the set of real numbers that satisfy $-1 \leq 2x - 5 < 7$.

SOLUTION ✔   Add 5 to each member of the given double inequality, obtaining

$$4 \leq 2x < 12$$

Next, multiply each member of the resulting double inequality by 1/2, yielding

$$2 \leq x < 6$$

Thus, the solution is the set of all values of x lying in the interval $[2, 6)$.

■ ■ ■ ■

To add or subtract two or more algebraic expressions, first remove the parentheses and then combine like terms. The resulting expression is written in order of decreasing degree from left to right.

EXAMPLE 1

a. $(2x^4 + 3x^3 + 4x + 6) - (3x^4 + 9x^3 + 3x^2)$
$= 2x^4 + 3x^3 + 4x + 6 - 3x^4 - 9x^3 - 3x^2$ (Remove parentheses.)

$= 2x^4 - 3x^4 + 3x^3 - 9x^3 - 3x^2 + 4x + 6$
$= -x^4 - 6x^3 - 3x^2 + 4x + 6$ (Combine like terms.)

b. $2t^3 - \{t^2 - [t - (2t - 1)] + 4\}$
$= 2t^3 - \{t^2 - [t - 2t + 1] + 4\}$
$= 2t^3 - \{t^2 - [-t + 1] + 4\}$ (Remove parentheses and combine like terms within brackets.)
$= 2t^3 - \{t^2 + t - 1 + 4\}$ (Remove brackets.)
$= 2t^3 - \{t^2 + t + 3\}$ (Combine like terms within the braces.)
$= 2t^3 - t^2 - t - 3$ (Remove braces.)   ■ ■ ■ ■

An algebraic expression is said to be **simplified** if none of its terms are similar. Observe that when the algebraic expression in Example 1b was simplified, the innermost grouping symbols were removed first; that is, the parentheses ( ) were removed first, the brackets [ ] second, and the braces { } third.

When algebraic expressions are multiplied, each term of one algebraic expression is multiplied by each term of the other. The resulting algebraic expression is then simplified.

EXAMPLE 2

Perform the indicated operations:

a. $(x^2 + 1)(3x^2 + 10x + 3)$ **b.** $(e^t + e^{-t})e^t - e^t(e^t - e^{-t})$

SOLUTION ✔

a. $(x^2 + 1)(3x^2 + 10x + 3) = x^2(3x^2 + 10x + 3) + 1(3x^2 + 10x + 3)$
$= 3x^4 + 10x^3 + 3x^2 + 3x^2 + 10x + 3$
$= 3x^4 + 10x^3 + 6x^2 + 10x + 3$

b. $(e^t + e^{-t})e^t - e^t(e^t - e^{-t}) = e^{2t} + e^0 - e^{2t} + e^0$
$= e^{2t} - e^{2t} + e^0 + e^0$
$= 1 + 1$ (Recall that $e^0 = 1$.)
$= 2$ ■ ■ ■ ■

Certain product formulas that are frequently used in algebraic computations are given in Table 1.5.

---

* The symbol ℝ indicates that these examples were selected from the calculus portion of the text in order to help you review the algebraic computations you will *actually* be using in calculus.

Table 1.5

Formula	Example
$(a + b)^2 = a^2 + 2ab + b^2$	$(2x + 3y)^2 = (2x)^2 + 2(2x)(3y) + (3y)^2$
	$= 4x^2 + 12xy + 9y^2$
$(a - b)^2 = a^2 - 2ab + b^2$	$(4x - 2y)^2 = (4x)^2 - 2(4x)(2y) + (2y)^2$
	$= 16x^2 - 16xy + 4y^2$
$(a + b)(a - b) = a^2 - b^2$	$(2x + y)(2x - y) = (2x)^2 - (y)^2$
	$= 4x^2 - y^2$

FACTORING

Factoring is the process of expressing an algebraic expression as a product of other algebraic expressions. For example, by applying the distributive property, we may write

$$3x^2 - x = x(3x - 1)$$

The first step in factoring an algebraic expression is to check to see whether it contains any common terms. If it does, the greatest common term is then factored out. For example, the common factor of the algebraic expression $2a^2x + 4ax + 6a$ is $2a$, because

$$2a^2x + 4ax + 6a = 2a \cdot ax + 2a \cdot 2x + 2a \cdot 3 = 2a(ax + 2x + 3)$$

 EXAMPLE 3   Factor out the greatest common factor in each of the following expressions:

a. $-0.3t^2 + 3t$ **b.** $2x^{3/2} - 3x^{1/2}$ **c.** $2ye^{xy^2} + 2xy^3e^{xy^2}$

d. $4x(x + 1)^{1/2} - 2x^2 \left(\dfrac{1}{2}\right)(x + 1)^{-1/2}$

SOLUTION ✔   **a.** $-0.3t^2 + 3t = -0.3t(t - 10)$
b. $2x^{3/2} - 3x^{1/2} = x^{1/2}(2x - 3)$
c. $2ye^{xy^2} + 2xy^3e^{xy^2} = 2ye^{xy^2}(1 + xy^2)$

d. $4x(x + 1)^{1/2} - 2x^2 \left(\dfrac{1}{2}\right)(x + 1)^{-1/2} = 4x(x + 1)^{1/2} - x^2(x + 1)^{-1/2}$

$$= x(x + 1)^{-1/2}[4(x + 1)^{1/2}(x + 1)^{1/2} - x]$$
$$= x(x + 1)^{-1/2}[4(x + 1) - x]$$
$$= x(x + 1)^{-1/2}(4x + 4 - x) = x(x + 1)^{-1/2}(3x + 4)$$

Here we select $(x + 1)^{-1/2}$ as the common factor because it is "contained" in each algebraic term. In particular, observe that

$$(x + 1)^{-1/2}(x + 1)^{1/2}(x + 1)^{1/2} = (x + 1)^{1/2} \qquad ■■■■$$

Sometimes an algebraic expression may be factored by regrouping and rearranging its terms and then factoring out a common term. This technique is illustrated in Example 4.

EXAMPLE 4 Factor:

a. $2ax + 2ay + bx + by$ **b.** $3x\sqrt{y} - 4 - 2\sqrt{y} + 6x$

SOLUTION ✔ **a.** First, factor the common term $2a$ from the first two terms and the common term b from the last two terms. Thus,

$$2ax + 2ay + bx + by = 2a(x + y) + b(x + y)$$

Since $(x + y)$ is common to both terms of the polynomial, we may factor it out. Hence,

$$2a(x + y) + b(x + y) = (x + y)(2a + b)$$

b. $3x\sqrt{y} - 4 - 2\sqrt{y} + 6x = 3x\sqrt{y} - 2\sqrt{y} + 6x - 4$
$$= \sqrt{y}(3x - 2) + 2(3x - 2)$$
$$= (3x - 2)(\sqrt{y} + 2)$$

The first step in factoring a polynomial is to find the common factors. The next step is to express the polynomial as the product of a constant and/or one or more prime polynomials.

Certain product formulas that are useful in factoring binomials and trinomials are listed in Table 1.6.

Table 1.6

Formula	Example
Difference of two squares	
$x^2 - y^2 = (x + y)(x - y)$	$x^2 - 36 = (x + 6)(x - 6)$
	$8x^2 - 2y^2 = 2(4x^2 - y^2)$
	$\quad = 2(2x + y)(2x - y)$
	$9 - a^6 = (3 + a^3)(3 - a^3)$
Perfect-square trinomial	
$x^2 + 2xy + y^2 = (x + y)^2$	$x^2 + 8x + 16 = (x + 4)^2$
$x^2 - 2xy + y^2 = (x - y)^2$	$4x^2 - 4xy + y^2 = (2x - y)^2$
Sum of two cubes	
$x^3 + y^3 = (x + y)(x^2 - xy + y^2)$	$z^3 + 27 = z^3 + (3)^3$
	$\quad = (z + 3)(z^2 - 3z + 9)$
Difference of two cubes	
$x^3 - y^3 = (x - y)(x^2 + xy + y^2)$	$8x^3 - y^6 = (2x)^3 - (y^2)^3$
	$\quad = (2x - y^2)(4x^2 + 2xy^2 + y^4)$

The factors of the second-degree polynomial with integral coefficients

$$px^2 + qx + r$$

are $(ax + b)(cx + d)$, where $ac = p$, $ad + bc = q$, and $bd = r$. Since only a limited number of choices are possible, we use a trial-and-error method to factor polynomials having this form.

For example, to factor $x^2 - 2x - 3$, we first observe that the only possible first-degree terms are

$$(x \qquad)(x \qquad) \qquad \text{(Since the coefficient of } x^2 \text{ is 1)}$$

Next, we observe that the product of the constant terms is (-3). This gives us the following possible factors:

$$(x - 1)(x + 3)$$
$$(x + 1)(x - 3)$$

Looking once again at the polynomial $x^2 - 2x - 3$, we see that the coefficient of x is -2. Checking to see which set of factors yields -2 for the coefficient of x, we find that

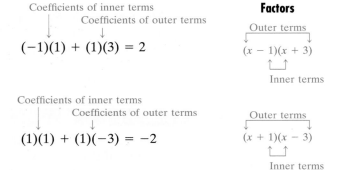

and we conclude that the correct factorization is

$$x^2 - 2x - 3 = (x + 1)(x - 3)$$

With practice, you will soon find that you can perform many of these steps mentally and the need to write out each step will be eliminated.

 EXAMPLE 5   Factor:

a. $3x^2 + 4x - 4$ **b.** $3x^2 - 6x - 24$

    a. Using trial and error, we find that the correct factorization is

$$3x^2 + 4x - 4 = (3x - 2)(x + 2)$$

b. Since each term has the common factor 3, we have

$$3x^2 - 6x - 24 = 3(x^2 - 2x - 8)$$

Using the trial-and-error method of factorization, we find that

$$x^2 - 2x - 8 = (x - 4)(x + 2)$$

Thus, we have

$$3x^2 - 6x - 24 = 3(x - 4)(x + 2)$$ ■ ■ ■ ■

ROOTS OF POLYNOMIAL EQUATIONS

A polynomial equation of degree n in the variable x is an equation of the form

$$a_n x^n + a_{n-1} x^{n-1} + \cdots + a_0 = 0$$

where n is a nonnegative integer and $a_0, a_1, \ldots, a_n$ are real numbers with $a_n \neq 0$. For example, the equation

$$-2x^5 + 8x^3 - 6x^2 + 3x + 1 = 0$$

is a polynomial equation of degree 5 in x.

   The **roots of a polynomial equation** are precisely the values of x that satisfy the given equation.* One way of finding the roots of a polynomial equation is to first factor the polynomial and then solve the resulting equation. For example, the polynomial equation

$$x^3 - 3x^2 + 2x = 0$$

may be rewritten in the form

$$x(x^2 - 3x + 2) = 0 \quad \text{or} \quad x(x - 1)(x - 2) = 0$$

Since the product of two real numbers can be equal to zero if and only if one (or both) of the factors is equal to zero, we have

$$x = 0, \quad x - 1 = 0, \quad \text{or} \quad x - 2 = 0$$

from which we see that the desired roots are $x = 0$, 1, and 2.

THE QUADRATIC FORMULA

In general, the problem of finding the roots of a polynomial equation is a difficult one. But the roots of a quadratic equation (a polynomial equation of degree 2) are easily found either by factoring or by using the following quadratic formula.

Quadratic Formula

> The solutions of the equation $ax^2 + bx + c = 0$ $(a \neq 0)$ are given by
>
> $$x = \frac{-b \pm \sqrt{b^2 - 4ac}}{2a}$$

---

* In this book, we are interested only in the *real* roots of an equation.

EXAMPLE 6

Solve each of the following quadratic equations:

a. $2x^2 + 5x - 12 = 0$ **b.** $x^2 = -3x + 8$

SOLUTION ✔

a. The equation is in standard form, with $a = 2$, $b = 5$, and $c = -12$. Using the quadratic formula, we find

$$x = \frac{-b \pm \sqrt{b^2 - 4ac}}{2a} = \frac{-5 \pm \sqrt{5^2 - 4(2)(-12)}}{2(2)}$$

$$= \frac{-5 \pm \sqrt{121}}{4} = \frac{-5 \pm 11}{4}$$

$$= -4 \text{ or } \frac{3}{2}$$

This equation can also be solved by factoring. Thus,

$$2x^2 + 5x - 12 = (2x - 3)(x + 4) = 0$$

from which we see that the desired roots are $x = 3/2$ or $x = -4$, as obtained earlier.

b. We first rewrite the given equation in the standard form $x^2 + 3x - 8 = 0$, from which we see that $a = 1$, $b = 3$, and $c = -8$. Using the quadratic formula, we find

$$x = \frac{-b \pm \sqrt{b^2 - 4ac}}{2a} = \frac{-3 \pm \sqrt{3^2 - 4(1)(-8)}}{2(1)}$$

$$= \frac{-3 \pm \sqrt{41}}{2}$$

That is, the solutions are

$$\frac{-3 + \sqrt{41}}{2} \approx 1.7 \quad \text{and} \quad \frac{-3 - \sqrt{41}}{2} \approx -4.7$$

In this case, the quadratic formula proves quite handy!   ■ ■ ■ ■

RATIONAL EXPRESSIONS

Quotients of polynomials are called **rational expressions.** Examples of rational expressions are

$$\frac{6x - 1}{2x + 3}, \quad \frac{3x^2y^3 - 2xy}{4x}, \quad \frac{2}{5ab}$$

Since rational expressions are quotients in which the variables represent real numbers, the properties of the real numbers apply to rational expressions as well, and operations with rational fractions are performed in the same manner as operations with arithmetic fractions. For example, using the properties of the real number system, we may write

$$\frac{ac}{bc} = \frac{a}{b} \cdot \frac{c}{c} = \frac{a}{b} \cdot 1 = \frac{a}{b}$$

where a, b, and c are any real numbers and b and c are not zero.

Similarly, using the same properties of real numbers, we may write

$$\frac{(x + 2)(x - 3)}{(x - 2)(x - 3)} = \frac{x + 2}{x - 2} \qquad (x \neq 2, 3)$$

after "canceling" the common factors.

An example of incorrect cancellation is

$$\frac{\cancel{3} + 4x}{\cancel{3}} \neq 1 + 4x$$

Instead, we need to write

$$\frac{3 + 4x}{3} = \frac{3}{3} + \frac{4x}{3} = 1 + \frac{4x}{3}$$

A rational expression is simplified, or in lowest terms, when the numerator and denominator have no common factors other than 1 and -1 and the expression contains no negative exponents.

EXAMPLE 7   Simplify the following expressions:

a. $\dfrac{x^2 + 2x - 3}{x^2 + 4x + 3}$ **b.** $\dfrac{[(t^2 + 4)(2t - 4) - (t^2 - 4t + 4)(2t)]}{(t^2 + 4)^2}$

SOLUTION ✔

a. $\dfrac{x^2 + 2x - 3}{x^2 + 4x + 3} = \dfrac{(x + 3)(x - 1)}{(x + 3)(x + 1)} = \dfrac{x - 1}{x + 1}$

b. $\dfrac{[(t^2 + 4)(2t - 4) - (t^2 - 4t + 4)(2t)]}{(t^2 + 4)^2}$

$$= \frac{2t^3 - 4t^2 + 8t - 16 - 2t^3 + 8t^2 - 8t}{(t^2 + 4)^2} \qquad \text{(Carry out the indicated multiplication.)}$$

$$= \frac{4t^2 - 16}{(t^2 + 4)^2} \qquad \text{(Combine like terms.)}$$

$$= \frac{4(t^2 - 4)}{(t^2 + 4)^2} \qquad \text{(Factor.)} \qquad ■ ■ ■ ■$$

The operations of multiplication and division are performed with algebraic fractions in the same manner as with arithmetic fractions (Table 1.7).

Table 1.7

Operation	Example
If P, Q, R, and S are polynomials, then	
Multiplication	
$\dfrac{P}{Q} \cdot \dfrac{R}{S} = \dfrac{PR}{QS} \qquad (Q, S \neq 0)$	$\dfrac{2x}{y} \cdot \dfrac{(x + 1)}{(y - 1)} = \dfrac{2x(x + 1)}{y(y - 1)} = \dfrac{2x^2 + 2x}{y^2 - y}$
Division	
$\dfrac{P}{Q} \div \dfrac{R}{S} = \dfrac{P}{Q} \cdot \dfrac{S}{R} = \dfrac{PS}{QR} \qquad (Q, R, S \neq 0)$	$\dfrac{x^2 + 3}{y} \div \dfrac{y^2 + 1}{x} = \dfrac{x^2 + 3}{y} \cdot \dfrac{x}{y^2 + 1} = \dfrac{x^3 + 3x}{y^3 + y}$

When rational expressions are multiplied and divided, the resulting expressions should be simplified.

EXAMPLE 8

Perform the indicated operations and simplify:

$$\frac{2x-8}{x+2} \cdot \frac{x^2+4x+4}{x^2-16}$$

SOLUTION ✔

$$\frac{2x-8}{x+2} \cdot \frac{x^2+4x+4}{x^2-16}$$

$$=\frac{2(x-4)}{x+2} \cdot \frac{(x+2)^2}{(x+4)(x-4)}$$

$$=\frac{2(x-4)(x+2)(x+2)}{(x+2)(x+4)(x-4)} \qquad \text{[Cancel the common factors } (x+2)(x-4).\text{]}$$

$$=\frac{2(x+2)}{x+4}$$

■ ■ ■ ■

For rational expressions, the operations of addition and subtraction are performed by finding a common denominator of the fractions and then adding or subtracting the fractions. Table 1.8 shows the rules for fractions with equal denominators.

Table 1.8

Operation	Example
If P, Q, and R are polynomials, then	
Addition	
$\dfrac{P}{R} + \dfrac{Q}{R} = \dfrac{P+Q}{R}$ $(R \neq 0)$	$\dfrac{2x}{x+2} + \dfrac{6x}{x+2} = \dfrac{2x+6x}{x+2} = \dfrac{8x}{x+2}$
Subtraction	
$\dfrac{P}{R} - \dfrac{Q}{R} = \dfrac{P-Q}{R}$ $(R \neq 0)$	$\dfrac{3y}{y-x} - \dfrac{y}{y-x} = \dfrac{3y-y}{y-x} = \dfrac{2y}{y-x}$

To add or subtract fractions that have different denominators, first find a common denominator, preferably the least common denominator (LCD). Then carry out the indicated operations following the procedure described in Table 1.8.

To find the LCD of two or more rational expressions:

1. *Find the prime factors of each denominator.*
2. *Form the product of the different prime factors that occur in the denominators.* Each prime factor in this product should be raised to the highest power of that factor appearing in the denominators.

$$\frac{x}{2+y} \neq \frac{x}{2} + \frac{x}{y}$$

EXAMPLE **9**   Simplify:

a. $\dfrac{2x}{x^2+1} + \dfrac{6(3x^2)}{x^3+2}$ b. $\dfrac{1}{x+h} - \dfrac{1}{x}$

SOLUTION ✔ a. $\dfrac{2x}{x^2+1} + \dfrac{6(3x^2)}{x^3+2}$

$$= \frac{2x(x^3+2) + 6(3x^2)(x^2+1)}{(x^2+1)(x^3+2)}$$ [LCD $= (x^2+1)(x^3+2)$]

$$= \frac{2x^4 + 4x + 18x^4 + 18x^2}{(x^2+1)(x^3+2)}$$

$$= \frac{20x^4 + 18x^2 + 4x}{(x^2+1)(x^3+2)}$$

$$= \frac{2x(10x^3 + 9x + 2)}{(x^2+1)(x^3+2)}$$

b. $\dfrac{1}{x+h} - \dfrac{1}{x} = \dfrac{1}{x+h} \cdot \dfrac{x}{x} - \dfrac{1}{x} \cdot \dfrac{x+h}{x+h}$ [LCD $= (x)(x+h)$]

$$= \frac{x}{x(x+h)} - \frac{x+h}{x(x+h)}$$

$$= \frac{x-x-h}{x(x+h)}$$

$$= \frac{-h}{x(x+h)}$$ ■ ■ ■ ■

OTHER ALGEBRAIC FRACTIONS

The techniques used to simplify rational expressions may also be used to simplify algebraic fractions in which the numerator and denominator are not polynomials, as illustrated in Example 10.

EXAMPLE **10**   Simplify:

a. $\dfrac{1 + \dfrac{1}{x+1}}{x - \dfrac{4}{x}}$ b. $\dfrac{x^{-1} + y^{-1}}{x^{-2} - y^{-2}}$

SOLUTION ✔

a.
$$\frac{1 + \dfrac{1}{x+1}}{x - \dfrac{4}{x}} = \frac{1 \cdot \dfrac{x+1}{x+1} + \dfrac{1}{x+1}}{x \cdot \dfrac{x}{x} - \dfrac{4}{x}} = \frac{\dfrac{x+1+1}{x+1}}{\dfrac{x^2-4}{x}}$$

$$= \frac{x+2}{x+1} \cdot \frac{x}{x^2-4} = \frac{x+2}{x+1} \cdot \frac{x}{(x+2)(x-2)}$$

$$= \frac{x}{(x+1)(x-2)}$$

b.
$$\frac{x^{-1} + y^{-1}}{x^{-2} - y^{-2}} = \frac{\dfrac{1}{x} + \dfrac{1}{y}}{\dfrac{1}{x^2} - \dfrac{1}{y^2}} = \frac{\dfrac{y+x}{xy}}{\dfrac{y^2-x^2}{x^2y^2}} \qquad \left(x^{-n} = \frac{1}{x^n}\right)$$

$$= \frac{y+x}{xy} \cdot \frac{x^2y^2}{y^2 - x^2} = \frac{y+x}{xy} \cdot \frac{(xy)^2}{(y+x)(y-x)}$$

$$= \frac{xy}{y-x}$$ ■■■■

EXAMPLE 11   Perform the given operations and simplify:

a. $\dfrac{x^2(2x^2+1)^{1/2}}{x-1} \cdot \dfrac{4x^3 - 6x^2 + x - 2}{x(x-1)(2x^2+1)}$ **b.** $\dfrac{12x^2}{\sqrt{2x^2+3}} + 6\sqrt{2x^2+3}$

SOLUTION ✔

a. $\dfrac{x^2(2x^2+1)^{1/2}}{x-1} \cdot \dfrac{4x^3 - 6x^2 + x - 2}{x(x-1)(2x^2+1)} = \dfrac{x(4x^3 - 6x^2 + x - 2)}{(x-1)^2(2x^2+1)^{1-1/2}}$

$$= \frac{x(4x^3 - 6x^2 + x - 2)}{(x-1)^2(2x^2+1)^{1/2}}$$

b. $\dfrac{12x^2}{\sqrt{2x^2+3}} + 6\sqrt{2x^2+3} = \dfrac{12x^2}{(2x^2+3)^{1/2}} + 6(2x^2+3)^{1/2}$

$$= \frac{12x^2 + 6(2x^2+3)^{1/2}(2x^2+3)^{1/2}}{(2x^2+3)^{1/2}}$$

$$= \frac{12x^2 + 6(2x^2+3)}{(2x^2+3)^{1/2}}$$

$$= \frac{24x^2 + 18}{(2x^2+3)^{1/2}} = \frac{6(4x^2+3)}{\sqrt{2x^2+3}}$$ ■■■■

RATIONALIZING ALGEBRAIC FRACTIONS

When the denominator of an algebraic fraction contains sums or differences involving radicals, we may **rationalize the denominator**—that is, transform the fraction into an equivalent one with a denominator that does not contain radicals. In doing so, we make use of the fact that

$$(\sqrt{a} + \sqrt{b})(\sqrt{a} - \sqrt{b}) = (\sqrt{a})^2 - (\sqrt{b})^2$$
$$= a - b$$

This procedure is illustrated in Example 12.

EXAMPLE 12   Rationalize the denominator: $\dfrac{1}{1 + \sqrt{x}}$.

SOLUTION ✔   Upon multiplying the numerator and the denominator by $(1 - \sqrt{x})$, we obtain

$$\frac{1}{1 + \sqrt{x}} = \frac{1}{1 + \sqrt{x}} \cdot \frac{1 - \sqrt{x}}{1 - \sqrt{x}}$$
$$= \frac{1 - \sqrt{x}}{1 - (\sqrt{x})^2}$$
$$= \frac{1 - \sqrt{x}}{1 - x}$$

■ ■ ■ ■

In other situations, it may be necessary to rationalize the numerator of an algebraic expression. In calculus, for example, one encounters the following problem.

EXAMPLE 13   Rationalize the numerator: $\dfrac{\sqrt{1 + h} - 1}{h}$.

SOLUTION ✔

$$\frac{\sqrt{1 + h} - 1}{h} = \frac{\sqrt{1 + h} - 1}{h} \cdot \frac{\sqrt{1 + h} + 1}{\sqrt{1 + h} + 1}$$
$$= \frac{(\sqrt{1 + h})^2 - (1)^2}{h(\sqrt{1 + h} + 1)}$$
$$= \frac{1 + h - 1}{h(\sqrt{1 + h} + 1)} \qquad [(\sqrt{1 + h})^2 = \sqrt{1 + h} \cdot \sqrt{1 + h}$$
$$= 1 + h]$$
$$= \frac{h}{h(\sqrt{1 + h} + 1)}$$
$$= \frac{1}{\sqrt{1 + h} + 1}$$

■ ■ ■ ■

1.2 Exercises

In Exercises 1–22, perform the indicated operations and simplify each expression.

1. $(7x^2 - 2x + 5) + (2x^2 + 5x - 4)$

2. $(3x^2 + 5xy + 2y) + (4 - 3xy - 2x^2)$

3. $(5y^2 - 2y + 1) - (y^2 - 3y - 7)$

4. $3(2a - b) - 4(b - 2a)$

5. $x - \{2x - [-x - (1 - x)]\}$

6. $3x^2 - \{x^2 + 1 - x[x - (2x - 1)]\} + 2$

7. $\left(\dfrac{1}{3} - 1 + e\right) - \left(-\dfrac{1}{3} - 1 + e^{-1}\right)$

8. $-\dfrac{3}{4}y - \dfrac{1}{4}x + 100 + \dfrac{1}{2}x + \dfrac{1}{4}y - 120$

9. $3\sqrt{8} + 8 - 2\sqrt{y} + \dfrac{1}{2}\sqrt{x} - \dfrac{3}{4}\sqrt{y}$

10. $\dfrac{8}{9}x^2 + \dfrac{2}{3}x + \dfrac{16}{3}x^2 - \dfrac{16}{3}x - 2x + 2$

11. $(x + 8)(x - 2)$ **12.** $(5x + 2)(3x - 4)$

13. $(a + 5)^2$ **14.** $(3a - 4b)^2$

15. $(x + 2y)^2$ **16.** $(6 - 3x)^2$

17. $(2x + y)(2x - y)$ **18.** $(3x + 2)(2 - 3x)$

19. $(x^2 - 1)(2x) - x^2(2x)$

20. $(x^{1/2} + 1)\left(\dfrac{1}{2}x^{-1/2}\right) - (x^{1/2} - 1)\left(\dfrac{1}{2}x^{-1/2}\right)$

21. $2(t + \sqrt{t})^2 - 2t^2$ **22.** $2x^2 + (-x + 1)^2$

In Exercises 23–30, factor out the greatest common factor from each expression.

23. $4x^5 - 12x^4 - 6x^3$

24. $4x^2y^2z - 2x^5y^2 + 6x^3y^2z^2$

25. $7a^4 - 42a^2b^2 + 49a^3b$

26. $3x^{2/3} - 2x^{1/3}$ **27.** $e^{-x} - xe^{-x}$

28. $2ye^{xy^2} + 2xy^3e^{xy^2}$ **29.** $2x^{-5/2} - \dfrac{3}{2}x^{-3/2}$

30. $\dfrac{1}{2}\left(\dfrac{2}{3}u^{3/2} - 2u^{1/2}\right)$

In Exercises 31–44, factor each expression.

31. $6ac + 3bc - 4ad - 2bd$

32. $3x^3 - x^2 + 3x - 1$

33. $4a^2 - b^2$ **34.** $12x^2 - 3y^2$

35. $10 - 14x - 12x^2$ **36.** $x^2 - 2x - 15$

37. $3x^2 - 6x - 24$ **38.** $3x^2 - 4x - 4$

39. $12x^2 - 2x - 30$ **40.** $(x + y)^2 - 1$

41. $9x^2 - 16y^2$ **42.** $8a^2 - 2ab - 6b^2$

43. $x^6 + 125$ **44.** $x^3 - 27$

In Exercises 45–52, perform the indicated operations and simplify each expression.

45. $(x^2 + y^2)x - xy(2y)$ **46.** $2kr(R - r) - kr^2$

47. $2(x - 1)(2x + 2)^3[4(x - 1) + (2x + 2)]$

48. $5x^2(3x^2 + 1)^4(6x) + (3x^2 + 1)^5(2x)$

49. $4(x - 1)^2(2x + 2)^3(2) + (2x + 2)^4(2)(x - 1)$

50. $(x^2 + 1)(4x^3 - 3x^2 + 2x) - (x^4 - x^3 + x^2)(2x)$

51. $(x^2 + 2)^2[5(x^2 + 2)^2 - 3](2x)$

52. $(x^2 - 4)(x^2 + 4)(2x + 8) - (x^2 + 8x - 4)(4x^3)$

In Exercises 53–58, find the real roots of each equation by factoring.

53. $x^2 + x - 12 = 0$ **54.** $3x^2 - x - 4 = 0$

55. $4t^2 + 2t - 2 = 0$ **56.** $-6x^2 + x + 12 = 0$

57. $\dfrac{1}{4}x^2 - x + 1 = 0$ **58.** $\dfrac{1}{2}a^2 + a - 12 = 0$

In Exercises 59–64, solve the equation by using the quadratic formula.

59. $4x^2 + 5x - 6 = 0$ **60.** $3x^2 - 4x + 1 = 0$

61. $8x^2 - 8x - 3 = 0$ **62.** $x^2 - 6x + 6 = 0$

63. $2x^2 + 4x - 3 = 0$ **64.** $2x^2 + 7x - 15 = 0$

In Exercises 65–70, simplify the expression.

65. $\dfrac{x^2 + x - 2}{x^2 - 4}$

66. $\dfrac{2a^2 - 3ab - 9b^2}{2ab^2 + 3b^3}$

67. $\dfrac{12t^2 + 12t + 3}{4t^2 - 1}$

68. $\dfrac{x^3 + 2x^2 - 3x}{-2x^2 - x + 3}$

69. $\dfrac{(4x - 1)(3) - (3x + 1)(4)}{(4x - 1)^2}$

70. $\dfrac{(1 + x^2)^2(2) - 2x(2)(1 + x^2)(2x)}{(1 + x^2)^4}$

In Exercises 71–88, perform the indicated operations and simplify each expression.

71. $\dfrac{2a^2 - 2b^2}{b - a} \cdot \dfrac{4a + 4b}{a^2 + 2ab + b^2}$

72. $\dfrac{x^2 - 6x + 9}{x^2 - x - 6} \cdot \dfrac{3x + 6}{2x^2 - 7x + 3}$

73. $\dfrac{3x^2 + 2x - 1}{2x + 6} \div \dfrac{x^2 - 1}{x^2 + 2x - 3}$

74. $\dfrac{3x^2 - 4xy - 4y^2}{x^2y} \div \dfrac{(2y - x)^2}{x^3y}$

75. $\dfrac{58}{3(3t + 2)} + \dfrac{1}{3}$

76. $\dfrac{a + 1}{3a} + \dfrac{b - 2}{5b}$

77. $\dfrac{2x}{2x - 1} - \dfrac{3x}{2x + 5}$

78. $\dfrac{-xe^x}{x + 1} + e^x$

79. $\dfrac{4}{x^2 - 9} - \dfrac{5}{x^2 - 6x + 9}$

80. $\dfrac{x}{1 - x} + \dfrac{2x + 3}{x^2 - 1}$

81. $\dfrac{1 + \dfrac{1}{x}}{1 - \dfrac{1}{x}}$

82. $\dfrac{\dfrac{1}{x} + \dfrac{1}{y}}{1 - \dfrac{1}{xy}}$

83. $\dfrac{4x^2}{2\sqrt{2x^2 + 7}} + \sqrt{2x^2 + 7}$

84. $6(2x + 1)^2\sqrt{x^2 + x} + \dfrac{(2x + 1)^4}{2\sqrt{x^2 + x}}$

85. $\dfrac{2x(x + 1)^{-1/2} - (x + 1)^{1/2}}{x^2}$

86. $\dfrac{(x^2 + 1)^{1/2} - 2x^2(x^2 + 1)^{-1/2}}{1 - x^2}$

87. $\dfrac{(2x + 1)^{1/2} - (x + 2)(2x + 1)^{-1/2}}{2x + 1}$

88. $\dfrac{2(2x - 3)^{1/3} - (x - 1)(2x - 3)^{-2/3}}{(2x - 3)^{2/3}}$

In Exercises 89–94, rationalize the denominator of each expression.

89. $\dfrac{1}{\sqrt{3} - 1}$

90. $\dfrac{1}{\sqrt{x} + 5}$

91. $\dfrac{1}{\sqrt{x} - \sqrt{y}}$

92. $\dfrac{a}{1 - \sqrt{a}}$

93. $\dfrac{\sqrt{a} + \sqrt{b}}{\sqrt{a} - \sqrt{b}}$

94. $\dfrac{2\sqrt{a} + \sqrt{b}}{2\sqrt{a} - \sqrt{b}}$

In Exercises 95–100, rationalize the numerator of each expression.

95. $\dfrac{\sqrt{x}}{3}$

96. $\dfrac{\sqrt[3]{y}}{x}$

97. $\dfrac{1 - \sqrt{3}}{3}$

98. $\dfrac{\sqrt{x} - 1}{x}$

99. $\dfrac{1 + \sqrt{x + 2}}{\sqrt{x + 2}}$

100. $\dfrac{\sqrt{x + 3} - \sqrt{x}}{3}$

In Exercises 101–104, determine whether the statement is true or false. If it is true, explain why it is true. If it is false, give an example to show why it is false.

101. If $b^2 - 4ac > 0$, then $ax^2 + bx + c = 0$, $a \neq 0$, has two real roots.

102. If $b^2 - 4ac < 0$, then $ax^2 + bx + c = 0$, $a \neq 0$, has no real roots.

103. $\dfrac{a}{b + c} = \left(\dfrac{a}{b} + \dfrac{a}{c}\right)$

104. $\sqrt{(a + b)(b - a)} = \sqrt{b^2 - a^2}$ for all real numbers a and b.

1.3 The Cartesian Coordinate System

THE CARTESIAN COORDINATE SYSTEM

FIGURE 1.3

The Cartesian coordinate system

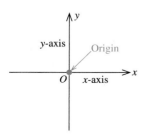

In Section 1.1 we saw how a one-to-one correspondence between the set of real numbers and the points on a straight line leads to a coordinate system on a line (a one-dimensional space).

A similar representation for points in a plane (a two-dimensional space) is realized through the Cartesian coordinate system, which is constructed as follows: Take two perpendicular lines, one of which is normally chosen to be horizontal. These lines intersect at a point O, called the **origin** (Figure 1.3). The horizontal line is called the **x-axis,** and the vertical line is called the **y-axis.** A number scale is set up along the x-axis, with the positive numbers lying to the right of the origin and the negative numbers lying to the left of it. Similarly, a number scale is set up along the y-axis, with the positive numbers lying above the origin and the negative numbers lying below it.

The number scales on the two axes need not be the same. Indeed, in many applications different quantities are represented by x and y. For example, x may represent the number of typewriters sold and y the total revenue resulting from the sales. In such cases it is often desirable to choose different number scales to represent the different quantities. Note, however, that the zeros of both number scales coincide at the origin of the two-dimensional coordinate system.

FIGURE 1.4

An ordered pair (x, y)

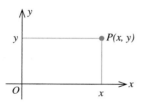

A point in the plane can now be represented uniquely in this coordinate system by an ordered pair of numbers—that is, a pair (x, y), where x is the first number and y the second. To see this, let P be any point in the plane (Figure 1.4). Draw perpendiculars from P to the x-axis and y-axis, respectively. Then the number x is precisely the number that corresponds to the point on the x-axis at which the perpendicular through P hits the x-axis. Similarly, y is the number that corresponds to the point on the y-axis at which the perpendicular through P crosses the y-axis.

Conversely, given an ordered pair (x, y) with x as the first number and y the second, a point P in the plane is uniquely determined as follows: Locate the point on the x-axis represented by the number x and draw a line through that point parallel to the y-axis. Next, locate the point on the y-axis represented by the number y and draw a line through that point parallel to the x-axis. The point of intersection of these two lines is the point P (see Figure 1.4).

In the ordered pair (x, y), x is called the **abscissa,** or **x-coordinate,** y is called the **ordinate,** or **y-coordinate,** and x and y together are referred to as the **coordinates** of the point P.

Letting (a, b) denote the point P with x-coordinate a and y-coordinate b, the points $A = (2, 3)$, $B = (-2, 3)$, $C = (-2, -3)$, $D = (2, -3)$, $E = (3, 2)$, $F = (4, 0)$, and $G = (0, -5)$ are plotted in Figure 1.5. The fact that, in general, $(x, y) \neq (y, x)$ is clearly illustrated by points A and E.

The axes divide the plane into four quadrants. Quadrant I consists of the points (x, y) that satisfy $x > 0$ and $y > 0$; Quadrant II, the points (x, y), where

FIGURE 1.5
Several points in the Cartesian plane

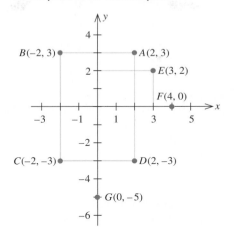

FIGURE 1.6
The four quadrants in the Cartesian plane

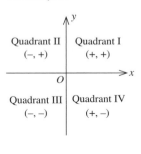

FIGURE 1.7
The distance d between the points (x_1, y_1) and (x_2, y_2)

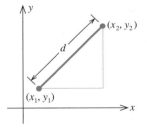

$x < 0$ and $y > 0$; Quadrant III, the points (x, y), where $x < 0$ and $y < 0$; and Quadrant IV, the points (x, y), where $x > 0$ and $y < 0$ (Figure 1.6).

THE DISTANCE FORMULA

One immediate benefit that arises from using the Cartesian coordinate system is that the distance between any two points in the plane may be expressed solely in terms of their coordinates. Suppose, for example, (x_1, y_1) and (x_2, y_2) are any two points in the plane (Figure 1.7). Then the distance between these two points can be computed using the following formula.

Distance Formula

> The distance d between two points $P_1(x_1, y_1)$ and $P_2(x_2, y_2)$ in the plane is given by
> $$d = \sqrt{(x_2 - x_1)^2 + (y_2 - y_1)^2} \qquad \textbf{(1)}$$

For a proof of this result, see Exercise 46, page 36.
     In what follows, we give several applications of the distance formula.

 EXAMPLE 1   Find the distance between the points $(-4, 3)$ and $(2, 6)$.

SOLUTION ✔   Let $P_1(-4, 3)$ and $P_2(2, 6)$ be points in the plane. Then, we have

$$x_1 = -4, \qquad y_1 = 3, \qquad x_2 = 2, \qquad y_2 = 6$$

Using Formula (1), we have

$$d = \sqrt{[2 - (-4)]^2 + (6 - 3)^2}$$
$$= \sqrt{6^2 + 3^2}$$
$$= \sqrt{45} = 3\sqrt{5}$$

■ ■ ■ ■

> **Group Discussion**
> Refer to Example 1. Suppose we label the point $(2, 6)$ as P_1 and the point $(-4, 3)$ as P_2. (1) Show that the distance d between the two points is the same as that obtained earlier. (2) Prove that, in general, the distance d in formula (1) is independent of the way we label the two points.

EXAMPLE 2

Let $P(x, y)$ denote a point lying on the circle with radius r and center $C(h, k)$ (Figure 1.8). Find a relationship between x and y.

FIGURE 1.8
A circle with radius r and center $C(h, k)$

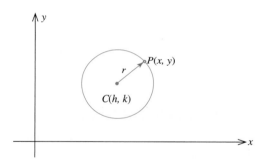

SOLUTION ✔

By the definition of a circle, the distance between $C(h, k)$ and $P(x, y)$ is r. Using Formula (1), we have

$$\sqrt{(x - h)^2 + (y - k)^2} = r$$

which, upon squaring both sides, gives the equation

$$(x - h)^2 + (y - k)^2 = r^2$$

that must be satisfied by the variables x and y.

■ ■ ■ ■

A summary of the result obtained in Example 2 follows.

Equation of a Circle

An equation of the circle with center $C(h, k)$ and radius r is given by

$$(x - h)^2 + (y - k)^2 = r^2 \tag{2}$$

EXAMPLE **3**

Find an equation of the circle with:

a. Radius 2 and center $(-1, 3)$.
b. Radius 3 and center located at the origin.

SOLUTION ✔

a. We use Formula (2) with $r = 2$, $h = -1$, and $k = 3$, obtaining

$$[x - (-1)]^2 + (y - 3)^2 = 2^2 \quad \text{or} \quad (x + 1)^2 + (y - 3)^2 = 4$$

(Figure 1.9a).

b. Using Formula (2) with $r = 3$ and $h = k = 0$, we obtain

$$x^2 + y^2 = 3^2 \quad \text{or} \quad x^2 + y^2 = 9$$

(Figure 1.9b).

■ ■ ■ ■

FIGURE 1.9

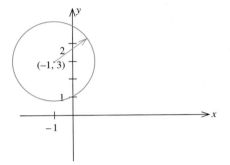

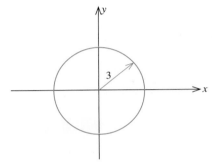

(a) The circle with radius 2 and center $(-1, 3)$ **(b)** The circle with radius 3 and center $(0, 0)$

 Group Discussion
1. Use the distance formula to help you describe the set of points in the xy-plane satisfying each of the following inequalities.
a. $(x - h)^2 + (y - k)^2 \leq r^2$ **b.** $(x - h)^2 + (y - k)^2 < r^2$
c. $(x - h)^2 + (y - k)^2 \geq r^2$ **d.** $(x - h)^2 + (y - k)^2 > r^2$
2. Consider the equation $x^2 + y^2 = 4$.
a. Show that $y = \pm\sqrt{4 - x^2}$.
b. Describe the set of points (x, y) in the xy-plane satisfying the following equations:
(i) $y = \sqrt{4 - x^2}$
(ii) $y = -\sqrt{4 - x^2}$

APPLICATION

EXAMPLE **4**

In the following diagram (Figure 1.10), S represents the position of a power relay station located on a straight coastal highway, and M shows the location of a marine biology experimental station on an island. A cable is to be laid

FIGURE 1.10
Cable connecting relay station S to experimental station M

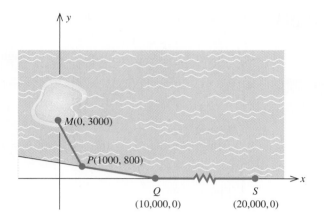

connecting the relay station with the experimental station. If the cost of running the cable on land is $2 per running foot and the cost of running the cable underwater is $6 per running foot, find the total cost for laying the cable.

SOLUTION ✔   The length of cable required on land is given by the distance from P to Q plus the distance from Q to S. The distance is

$$\sqrt{(10,000 - 1000)^2 + (0 - 800)^2} + \sqrt{(20,000 - 10,000)^2 + (0 - 0)^2}$$
$$= \sqrt{9000^2 + 800^2} + 10,000$$
$$= \sqrt{81,640,000} + 10,000$$
$$\approx 19,035.49$$

or approximately 19,035.49 feet. Next, we see that the length of cable required underwater is given by the distance from M to P. This distance is

$$\sqrt{(0 - 1000)^2 + (3000 - 800)^2} = \sqrt{1000^2 + 2200^2}$$
$$= \sqrt{5,840,000}$$
$$\approx 2416.61$$

or approximately 2416.61 feet. Therefore, the total cost for laying the cable is

$$2(19,035.49) + 6(2416.61) = 52,570.64$$

or approximately $52,571.   ■ ■ ■ ■

Group Discussion

In the Cartesian coordinate system, the two axes are perpendicular to each other. Consider a coordinate system in which the x- and y-axes are noncollinear (that is, the axes do not lie along a straight line) and are not perpendicular to each other (see the accompanying figure).

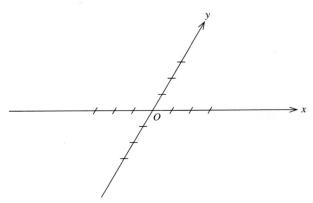

1. Describe how a point is represented in this coordinate system by an ordered pair (x, y) of real numbers. Conversely, show how an ordered pair (x, y) of real numbers uniquely determines a point in the plane.

2. Suppose you want to find a formula for the distance between two points $P_1(x_1, y_1)$ and $P_2(x_2, y_2)$ in the plane. What is the advantage that the Cartesian coordinate system has over the coordinate system under consideration? Comment on your answer.

SELF-CHECK EXERCISES 1.3

1. a. Plot the points $A(4, -2)$, $B(2, 3)$, and $C(-3, 1)$.
   b. Find the distance between the points A and B; between B and C; between A and C.
   c. Use the Pythagorean theorem to show that the triangle with vertices A, B, and C is a right triangle.

2. The figure at the top of page 34 shows the location of cities A, B, and C. Suppose a pilot wishes to fly from city A to city C but must make a mandatory stopover in city B. If the single-engine light plane has a range of 650 miles, can she make the trip without refueling in city B?

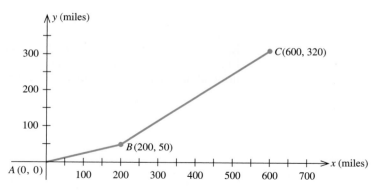

Solutions to Self-Check Exercises 1.3 can be found on page 36.

1.3 Exercises

In Exercises 1–6, refer to the following figure and determine the coordinates of each point and the quadrant in which it is located.

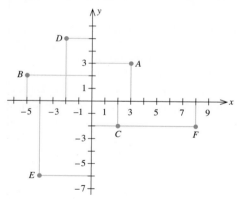

1. A   **2.** B   **3.** C
4. D   **5.** E   **6.** F

In Exercises 7–12, refer to the following figure.

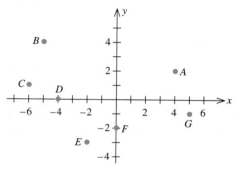

7. Which point has coordinates $(4, 2)$?

8. What are the coordinates of point B?

9. Which points have negative y-coordinates?

10. Which point has a negative x-coordinate and a negative y-coordinate?

11. Which point has an x-coordinate that is equal to zero?

12. Which point has a y-coordinate that is equal to zero?

In Exercises 13–20, sketch a set of coordinate axes and plot each point.

13. $(-2, 5)$ **14.** $(1, 3)$

15. $(3, -1)$ **16.** $(3, -4)$

17. $(8, -7/2)$ **18.** $(-5/2, 3/2)$

19. $(4.5, -4.5)$ **20.** $(1.2, -3.4)$

In Exercises 21–24, find the distance between the given points.

21. $(1, 3)$ and $(4, 7)$ **22.** $(1, 0)$ and $(4, 4)$

23. $(-1, 3)$ and $(4, 9)$ **24.** $(-2, 1)$ and $(10, 6)$

25. Find the coordinates of the points that are 10 units away from the origin and have a y-coordinate equal to -6.

26. Find the coordinates of the points that are 5 units away from the origin and have an x-coordinate equal to 3.

27. Show that the points $(3, 4)$, $(-3, 7)$, $(-6, 1)$, and $(0, -2)$ form the vertices of a square.

28. Show that the triangle with vertices $(-5, 2)$, $(-2, 5)$, and $(5, -2)$ is a right triangle.

In Exercises 29–34, find an equation of the circle that satisfies the given conditions.

29. Radius 5 and center $(2, -3)$

30. Radius 3 and center $(-2, -4)$

31. Radius 5 and center at the origin

32. Center at the origin and passes through $(2, 3)$

33. Center $(2, -3)$ and passes through $(5, 2)$

34. Center $(-a, a)$ and radius $2a$

35. DISTANCE TRAVELED A grand tour of four cities begins at city A and makes successive stops at cities B, C, and D before returning to city A. If the cities are located as shown in the following figure, find the total distance covered on the tour.

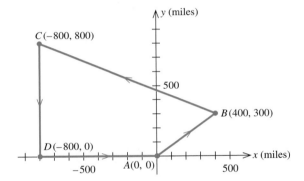

36. DELIVERY CHARGES A furniture store offers free setup and delivery services to all points within a 25-mi radius of its warehouse distribution center. If you live 20 mi east and 14 mi south of the warehouse, will you incur a delivery charge? Justify your answer.

37. TRAVEL TIME Towns A, B, C, and D are located as shown in the following figure. Two highways link town A to

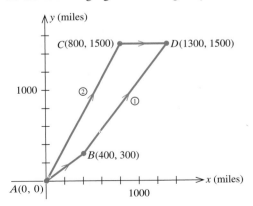

town D. Route 1 runs from town A to town D via town B, and route 2 runs from town A to town D via town C. If a salesman wishes to drive from town A to town D and traffic conditions are such that he could expect to average the same speed on either route, which highway should he take in order to arrive in the shortest time?

38. MINIMIZING SHIPPING COSTS Refer to the figure for Exercise 37. Suppose a fleet of 100 automobiles are to be shipped from an assembly plant in town A to town D. They may be shipped either by freight train along Route 1 at a cost of 11 cents/mile per automobile or by truck along Route 2 at a cost of $10\frac{1}{2}$ cents/mile per automobile. Which means of transportation minimizes the shipping cost? What is the net savings?

39. CONSUMER DECISIONS Ivan wishes to determine which antenna he should purchase for his home. The TV store has supplied him with the following information:

Range in Miles			
VHF	**UHF**	**Model**	**Price**
30	20	A	$40
45	35	B	$50
60	40	C	$60
75	55	D	$70

Ivan wishes to receive channel 17 (VHF), which is located 25 mi east and 35 mi north of his home, and channel 38 (UHF), which is located 20 mi south and 32 mi west of his home. Which model will allow him to receive both channels at the least cost? (Assume that the terrain between Ivan's home and both broadcasting stations is flat.)

40. CALCULATING THE COST OF LAYING CABLE In the following diagram, S represents the position of a power relay station located on a coastal highway, and M shows the location of a marine biology experimental station on an

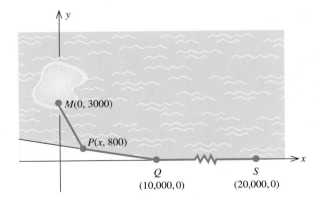

island. A cable is to be laid connecting the relay station with the experimental station. If the cost of running the cable on land is $2/running foot and the cost of running cable under water is $6/running foot, find an expression in terms of x that gives the total cost for laying the cable. Use this expression to find the total cost when $x = 900$; when $x = 1000$.

41. Two ships leave port at the same time. Ship A sails north at a speed of 20 mph while ship B sails east at a speed of 30 mph.
a. Find an expression in terms of the time t (in hours) giving the distance between the two ships.
b. Using the expression obtained in part (a), find the distance between the two ships 2 hr after leaving port.

In Exercises 42–45, determine whether the statement is true or false. If it is true, explain why it is true. If it is false, give an example to show why it is false.

42. The point $(-a, b)$ is symmetric to the point (a, b) with respect to the y-axis.

43. The point $(-a, -b)$ is symmetric to the point (a, b) with respect to the origin.

44. If the distance between the points $P_1(a, b)$ and $P_2(c, d)$ is D, then the distance between the points $P_1(a, b)$ and $P_3(kc, kd)$, $(k \neq 0)$, is given by $|k|D$.

45. The circle with equation $kx^2 + ky^2 = a^2$ lies inside the circle with equation $x^2 + y^2 = a^2$, provided $k > 1$.

46. Let (x_1, y_1) and (x_2, y_2) be two points lying in the xy-plane. Show that the distance between the two points is given by

$$d = \sqrt{(x_2 - x_1)^2 + (y_2 - y_1)^2}$$

Hint: Refer to the accompanying figure and use the Pythagorean theorem.

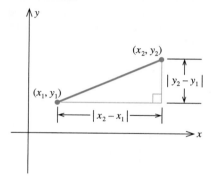

SOLUTIONS TO SELF-CHECK EXERCISES 1.3

1. a. The points are plotted in the following figure:

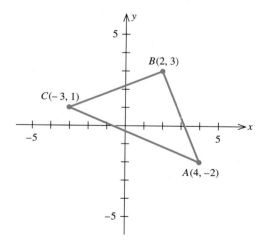

b. The distance between A and B is

$$d(A, B) = \sqrt{(2 - 4)^2 + [3 - (-2)]^2}$$
$$= \sqrt{(-2)^2 + 5^2} = \sqrt{4 + 25} = \sqrt{29}$$

The distance between B and C is

$$d(B, C) = \sqrt{(-3 - 2)^2 + (1 - 3)^2}$$
$$= \sqrt{(-5)^2 + (-2)^2} = \sqrt{25 + 4} = \sqrt{29}$$

The distance between A and C is

$$d(A, C) = \sqrt{(-3 - 4)^2 + [1 - (-2)]^2}$$
$$= \sqrt{(-7)^2 + 3^2} = \sqrt{49 + 9} = \sqrt{58}$$

c. We will show that

$$[d(A, C)]^2 = [d(A, B)]^2 + [d(B, C)]^2$$

From part (b), we see that $[d(A, B)]^2 = 29$, $[d(B, C)]^2 = 29$, and $[d(A, C)]^2 = 58$, and the desired result follows.

2. The distance between city A and city B is

$$d(A, B) = \sqrt{200^2 + 50^2} \approx 206$$

or 206 mi. The distance between city B and city C is

$$d(B, C) = \sqrt{[600 - 200]^2 + [320 - 50]^2}$$
$$= \sqrt{400^2 + 270^2} \approx 483$$

or 483 mi. Therefore, the total distance the pilot would have to cover is 689 mi, so she must refuel in city B.

1.4 Straight Lines

FIGURE 1.11
Linear depreciation of an asset

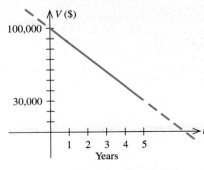

In computing income tax, business firms are allowed by law to depreciate certain assets such as buildings, machines, furniture, automobiles, and so on, over a period of time. Linear depreciation, or the straight-line method, is often used for this purpose. The graph of the straight line shown in Figure 1.11 describes the book value V of a computer that has an initial value of $100,000 and that is being depreciated linearly over 5 years with a scrap value of $30,000. Note that only the solid portion of the straight line is of interest here.

The book value of the computer at the end of year t, where t lies between 0 and 5, can be read directly from the graph. But there is one shortcoming in this approach: The result depends on how accurately you draw and read the graph. A better and more accurate method is based on finding an *algebraic* representation of the depreciation line.

Slope of a Line

To see how a straight line in the xy-plane may be described algebraically, we need to first recall certain properties of straight lines. Let L denote the unique straight line that passes through the two distinct points (x_1, y_1) and (x_2, y_2). If $x_1 = x_2$, then L is a vertical line, and the slope is undefined (Figure 1.12).

FIGURE 1.12
m is undefined

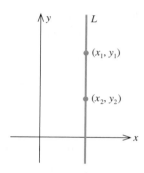

If $x_1 \ne x_2$, we define the slope of L as follows:

Slope of a Nonvertical Line

If (x_1, y_1) and (x_2, y_2) are any two distinct points on a nonvertical line L, then the slope m of L is given by

$$m = \frac{\Delta y}{\Delta x} = \frac{y_2 - y_1}{x_2 - x_1} \qquad (3)$$

See Figure 1.13.

FIGURE 1.13

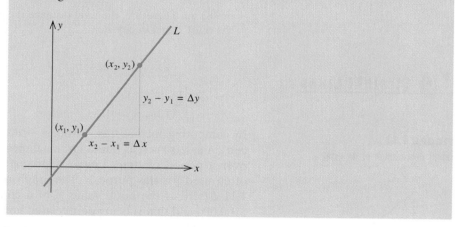

Observe that the slope of a straight line is a constant whenever it is defined. The number $\Delta y = y_2 - y_1$ (Δy is read "delta y") is a measure of the vertical change in y, and $\Delta x = x_2 - x_1$ is a measure of the horizontal change in x, as shown in Figure 1.13. From this figure we can see that the slope m of a straight line L is a measure of the *rate of change of y with respect to x*.

Figure 1.14a shows a straight line L_1 with slope 2. Observe that L_1 has the property that a unit increase in x results in a 2-unit increase in y. To see

FIGURE 1.14

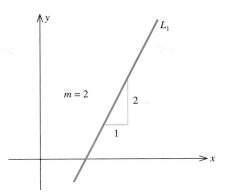

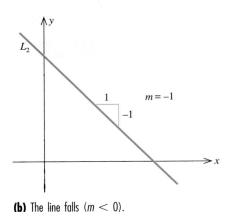

(a) The line rises ($m > 0$).

(b) The line falls ($m < 0$).

FIGURE 1.15
A family of straight lines

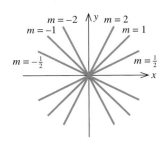

this, let $\Delta x = 1$ in Formula (3) so that $m = \Delta y$. Since $m = 2$, we conclude that $\Delta y = 2$. Similarly, Figure 1.14b shows a line L_2 with slope -1. Observe that a straight line with positive slope slants upward from left to right (y increases as x increases), whereas a line with negative slope slants downward from left to right (y decreases as x increases). Finally, Figure 1.15 shows a family of straight lines passing through the origin with indicated slopes.

EXAMPLE 1   Sketch the straight line that passes through the point $(-2, 5)$ and has slope $-4/3$.

SOLUTION ✔   First, plot the point $(-2, 5)$ (Figure 1.16). Next, recall that a slope of $-4/3$ indicates that an increase of 1 unit in the x-direction produces a *decrease* of $4/3$ units in the y-direction, or equivalently, a 3-unit increase in the x-direction produces a $3(4/3)$, or 4-unit, decrease in the y-direction. Using this information, we plot the point $(1, 1)$ and draw the line through the two points.

FIGURE 1.16
L has slope $-4/3$ and passes through $(-2, 5)$.

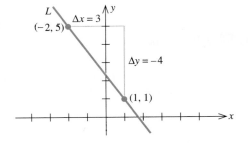

Group Discussion
Show that the slope of a nonvertical line is independent of the two distinct points $P_1(x_1, y_1)$ and $P_2(x_2, y_2)$ used to compute it.

Hint: Suppose we pick two other distinct points, $P_3(x_3, y_3)$ and $P_4(x_4, y_4)$ lying on L. Draw a picture and use similar triangles to demonstrate that using P_3 and P_4 gives the same value as that obtained using P_1 and P_2.

EXAMPLE **2**   Find the slope m of the line that passes through the points $(-1, 1)$ and $(5, 3)$.

SOLUTION ✔   Choose (x_1, y_1) to be the point $(-1, 1)$ and (x_2, y_2) to be the point $(5, 3)$. Then, with $x_1 = -1$, $y_1 = 1$, $x_2 = 5$, and $y_2 = 3$, we find

$$m = \frac{y_2 - y_1}{x_2 - x_1} = \frac{3 - 1}{5 - (-1)} = \frac{1}{3} \qquad \text{[Using (3)]}$$

(Figure 1.17). Try to verify that the result obtained would have been the same had we chosen the point $(-1, 1)$ to be (x_2, y_2) and the point $(5, 3)$ to be (x_1, y_1).

FIGURE 1.17

L passes through $(5, 3)$ and $(-1, 1)$.

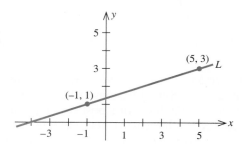

EXAMPLE **3**   Find the slope of the line that passes through the points $(-2, 5)$ and $(3, 5)$.

SOLUTION ✔   The slope of the required line is given by

FIGURE 1.18

The slope of the horizontal line L is 0.

$$m = \frac{5 - 5}{3 - (-2)} = \frac{0}{5} = 0$$

(Figure 1.18).

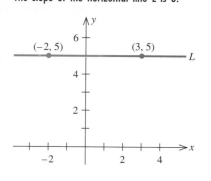

REMARK In general, the slope of a horizontal line is zero. ■ ■ ■

We can use the slope of a straight line to determine whether a line is parallel to another line.

Parallel Lines

> Two distinct lines are **parallel** if and only if their slopes are equal or their slopes are undefined.

EXAMPLE **4**   Let L_1 be a line that passes through the points $(-2, 9)$ and $(1, 3)$ and let L_2 be the line that passes through the points $(-4, 10)$ and $(3, -4)$. Determine whether L_1 and L_2 are parallel.

SOLUTION ✔ The slope m_1 of L_1 is given by

$$m_1 = \frac{3 - 9}{1 - (-2)} = -2$$

The slope m_2 of L_2 is given by

$$m_2 = \frac{-4 - 10}{3 - (-4)} = -2$$

Since $m_1 = m_2$, the lines L_1 and L_2 are in fact parallel (Figure 1.19). ■ ■ ■ ■

FIGURE 1.19
L_1 and L_2 have the same slope and hence are parallel.

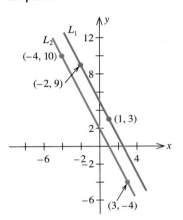

Equations of Lines

We will now show that every straight line lying in the xy-plane may be represented by an equation involving the variables x and y. One immediate benefit of this is that problems involving straight lines may be solved algebraically.

Let L be a straight line parallel to the y-axis (perpendicular to the x-axis) (Figure 1.20). Then, L crosses the x-axis at some point $(a, 0)$ with the x-coordinate given by $x = a$, where a is some real number. Any other point on L has the form $(a, \bar{y})$, where $\bar{y}$ is an appropriate number. Therefore, the vertical line L is described by the sole condition

$$x = a$$

and this is, accordingly, the equation of L. For example, the equation $x = -2$ represents a vertical line 2 units to the left of the y-axis, and the equation $x = 3$ represents a vertical line 3 units to the right of the y-axis (Figure 1.21).

Next, suppose L is a nonvertical line so that it has a well-defined slope m. Suppose (x_1, y_1) is a fixed point lying on L and (x, y) is a variable point on L distinct from (x_1, y_1) (Figure 1.22).

FIGURE 1.20
The vertical line $x = a$

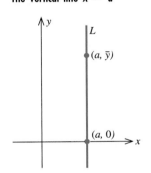

FIGURE 1.21
The vertical lines $x = -2$ and $x = 3$

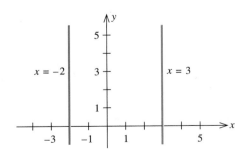

FIGURE 1.22
L passes through (x_1, y_1) and has slope m.

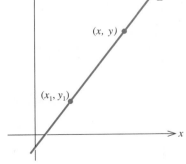

Using Formula (3) with the point $(x_2, y_2) = (x, y)$, we find that the slope of L is given by

$$m = \frac{y - y_1}{x - x_1}$$

Upon multiplying both sides of the equation by $x - x_1$, we obtain Formula (4).

Point-Slope Form

> An equation of the line that has slope m and passes through the point (x_1, y_1) is given by
>
> $$y - y_1 = m(x - x_1) \qquad \textbf{(4)}$$

Equation (4) is called the **point-slope form of the equation of a line** since it utilizes a given point (x_1, y_1) on a line and the slope m of the line.

EXAMPLE 5 Find an equation of the line that passes through the point $(1, 3)$ and has slope 2.

SOLUTION ✔ Using the point-slope form of the equation of a line with the point $(1, 3)$ and $m = 2$, we obtain

$$y - 3 = 2(x - 1) \qquad [(y - y_1) = m(x - x_1)]$$

which, when simplified, becomes

$$2x - y + 1 = 0$$

(Figure 1.23).   ■ ■ ■ ■

FIGURE 1.23
L passes through $(1, 3)$ and has slope 2.

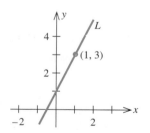

EXAMPLE 6 Find an equation of the line that passes through the points $(-3, 2)$ and $(4, -1)$.

SOLUTION ✔ The slope of the line is given by

$$m = \frac{-1 - 2}{4 - (-3)} = -\frac{3}{7}$$

Using the point-slope form of the equation of a line with the point $(4, -1)$ and the slope $m = -3/7$, we have

$$y + 1 = -\frac{3}{7}(x - 4) \qquad [(y - y_1) = m(x - x_1)]$$

$$7y + 7 = -3x + 12$$

$$3x + 7y - 5 = 0$$

(Figure 1.24).

FIGURE 1.24
L passes through $(-3, 2)$ and $(4, -1)$.

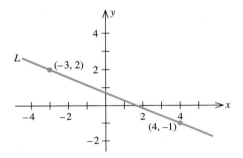

■ ■ ■ ■

Group Discussion

Consider the slope-intercept form of a straight line $y = mx + b$. Describe the family of straight lines obtained by keeping:

1. The value of m fixed and allowing the value of b to vary.
2. The value of b fixed and allowing the value of m to vary.

We can use the slope of a straight line to determine whether a line is perpendicular to another line.

Perpendicular Lines

If L_1 and L_2 are two distinct nonvertical lines that have slopes m_1 and m_2, respectively, then L_1 is **perpendicular** to L_2 (written $L_1 \perp L_2$) if and only if

$$m_1 = -\frac{1}{m_2}$$

If the line L_1 is vertical (so that its slope is undefined), then L_1 is perpendicular to another line, L_2, if and only if L_2 is horizontal (so that its slope is zero). For a proof of these results, see Exercise 83, page 52.

EXAMPLE 7   Find an equation of the line that passes through the point (3, 1) and is perpendicular to the line of Example 5.

SOLUTION ✔   Since the slope of the line in Example 5 is 2, the slope of the required line is given by $m = -1/2$, the negative reciprocal of 2. Using the point-slope form of the equation of a line, we obtain

$$y - 1 = -\frac{1}{2}(x - 3) \qquad [(y - y_1) = m(x - x_1)]$$

$$2y - 2 = -x + 3$$

$$x + 2y - 5 = 0$$

(See Figure 1.25).

FIGURE 1.25
L_2 is perpendicular to L_1 and passes through (3, 1).

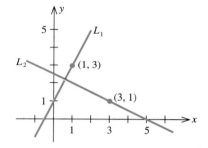

FIGURE 1.26
The line L has x-intercept a and y-intercept b.

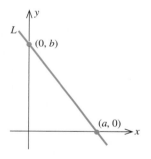

A straight line L that is neither horizontal nor vertical cuts the x-axis and the y-axis at, say, points $(a, 0)$ and $(0, b)$, respectively (Figure 1.26). The numbers a and b are called the **x-intercept** and **y-intercept,** respectively, of L.

Exploring with Technology

1. Use a graphing utility to plot the straight lines L_1 and L_2 with equations $2x + y - 5 = 0$ and $41x + 20y - 11 = 0$ on the same set of axes using the standard viewing rectangle.
   a. Can you tell if the lines L_1 and L_2 are parallel to each other?
   b. Verify your observations by computing the slopes of L_1 and L_2 algebraically.
2. Use a graphing utility to plot the straight lines L_1 and L_2 with equations $x + 2y - 5 = 0$ and $5x - y + 5 = 0$ on the same set of axes using the standard viewing rectangle.
   a. Can you tell if the lines L_1 and L_2 are perpendicular to each other?
   b. Verify your observation by computing the slopes of L_1 and L_2 algebraically.

Now, let L be a line with slope m and y-intercept b. Using Formula (4), the point-slope form of the equation of a line, with the point $(0, b)$ and slope m, we have

$$y - b = m(x - 0)$$
$$y = mx + b$$

Slope-Intercept Form

The equation of the line that has slope m and intersects the y-axis at the point $(0, b)$ is given by

$$y = mx + b \qquad\qquad (5)$$

EXAMPLE 8   Find an equation of the line that has slope 3 and y-intercept -4.

SOLUTION ✔   Using Equation (5) with $m = 3$ and $b = -4$, we obtain the required equation

$$y = 3x - 4 \qquad\qquad ■ ■ ■ ■$$

EXAMPLE 9   Determine the slope and y-intercept of the line whose equation is $3x - 4y = 8$.

SOLUTION ✔   Rewrite the given equation in the slope-intercept form and obtain

$$y = \frac{3}{4}x - 2$$

Comparing this result with Equation (5), we find $m = 3/4$ and $b = -2$, and we conclude that the slope and y-intercept of the given line are 3/4 and -2, respectively.   ■ ■ ■ ■

Exploring with Technology

1. Use a graphing utility to plot the straight lines with equations $y = -2x + 3$, $y = -x + 3$, $y = x + 3$, and $y = 2.5x + 3$ on the same set of axes using the standard viewing rectangle. What effect does changing the coefficient m of x in the equation $y = mx + b$ have on its graph?
2. Use a graphing utility to plot the straight lines with equations $y = 2x - 2$, $y = 2x - 1$, $y = 2x$, $y = 2x + 1$, and $y = 2x + 4$ on the same set of axes using the standard viewing rectangle. What effect does changing the constant b in the equation $y = mx + b$ have on its graph?
3. Describe in words the effect of changing both m and b in the equation $y = mx + b$.

APPLICATIONS

EXAMPLE 10

The sales manager of a local sporting goods store plotted sales versus time for the last 5 years and found the points to lie approximately along a straight line (Figure 1.27). By using the points corresponding to the first and fifth years, find an equation of the trend line. What sales figure can be predicted for the sixth year?

SOLUTION ✔

Using Formula (3) with the points (1, 20) and (5, 60), we find that the slope of the required line is given by

$$m = \frac{60 - 20}{5 - 1} = 10$$

FIGURE 1.27
Sales of a sporting goods store

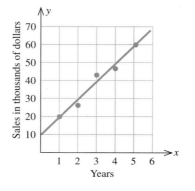

Next, using the point-slope form of the equation of a line with the point (1, 20) and $m = 10$, we obtain

$$y - 20 = 10(x - 1) \qquad [(y - y_1) = m(x - x_1)]$$
$$y = 10x + 10$$

as the required equation.

The sales figure for the sixth year is obtained by letting $x = 6$ in the last equation, giving

$$y = 70$$

or $70,000. ▪▪▪▪

EXAMPLE 11

Suppose an art object purchased for $50,000 is expected to appreciate in value at a constant rate of $5000 per year for the next 5 years. Use Formula (5) to write an equation predicting the value of the art object in the next several years. What will its value be 3 years from the date of purchase?

SOLUTION ✔

Let x denote the time (in years) that has elapsed since the date the object was purchased and let y denote the object's value (in dollars). Then, $y = 50,000$ when $x = 0$. Furthermore, the slope of the required equation is given by $m = 5000$, since each unit increase in x (1 year) implies an increase of 5000 units (dollars) in y. Using (5) with $m = 5000$ and $b = 50,000$, we obtain

$$y = 5000x + 50,000 \qquad (y = mx + b)$$

Three years from the date of purchase, the value of the object will be given by

$$y = 5000(3) + 50,000$$

or $65,000. ▪▪▪▪

> **Group Discussion**
> Refer to Example 11. Can the equation predicting the value of the art object be used to predict long-term growth?

GENERAL EQUATION OF A LINE

We have considered several forms of the equation of a straight line in the plane. These different forms of the equation are equivalent to each other. In fact, each is a special case of the following equation.

General Form of a Linear Equation

The equation

$$Ax + By + C = 0 \qquad (6)$$

where A, B, and C are constants and A and B are not both zero, is called the general form of a linear equation in the variables x and y.

We will now state (without proof) an important result concerning the algebraic representation of straight lines in the plane.

THEOREM 1

An equation of a straight line is a linear equation; conversely, every linear equation represents a straight line.

This result justifies the use of the adjective *linear* describing Equation (6).

EXAMPLE 12   Sketch the straight line represented by the equation

$$3x - 4y - 12 = 0$$

SOLUTION ✔   Since every straight line is uniquely determined by two distinct points, we need find only two such points through which the line passes in order to sketch it. For convenience let us compute the x- and y-intercepts. Setting $y = 0$, we find $x = 4$; thus, the x-intercept is 4. Setting $x = 0$ gives $y = -3$, and the y-intercept is -3. A sketch of the line appears in Figure 1.28.

FIGURE 1.28
To sketch $3x - 4y - 12 = 0$, first find the x-intercept, 4, and the y-intercept, -3.

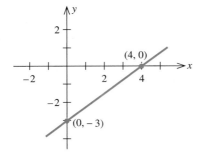

Following is a summary of the common forms of the equations of straight lines discussed in this section.

Equations of Straight Lines

Vertical line:	$x = a$
Horizontal line:	$y = b$
Point-slope form:	$y - y_1 = m(x - x_1)$
Slope-intercept form:	$y = mx + b$
General form:	$Ax + By + C = 0$

1. Determine the number a so that the line passing through the points $(a, 2)$ and $(3, 6)$ is parallel to a line with slope 4.

2. Find an equation of the line that passes through the point $(3, -1)$ and is perpendicular to a line with slope $-1/2$.

3. Does the point $(3, -3)$ lie on the line with equation $2x - 3y - 12 = 0$? Sketch the graph of the line.

4. The percentage of people over age 65 who have high school diplomas is summarized in the following table:

Year, x	1960	1965	1970	1975	1980	1985	1990
Percentage with Diplomas, y	20	25	30	36	42	47	52

Source: The World Almanac

a. Plot the percentage of people over age 65 who have high school diplomas (y) versus the year (x).
b. Draw the straight line L through the points (1960, 20) and (1990, 52).
c. Find an equation of the line L.
d. Assuming the trend continued, estimate the percentage of people over age 65 who had high school diplomas by the year 2000.

Solutions to Self-Check Exercises 1.4 can be found on page 53.

1.4 Exercises

In Exercises 1–6, match the statement with one of the graphs (a)–(f).

1. The slope of the line is zero.

2. The slope of the line is undefined.

3. The slope of the line is positive, and its y-intercept is positive.

4. The slope of the line is positive, and its y-intercept is negative.

5. The slope of the line is negative, and its x-intercept is negative.

6. The slope of the line is negative, and its x-intercept is positive.

a.

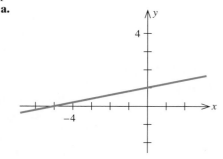

b.

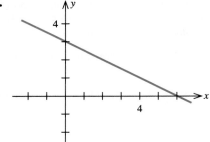

c.

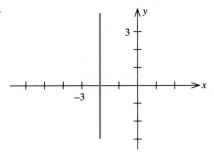

d.

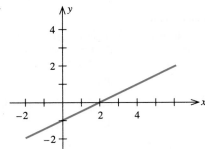

e.

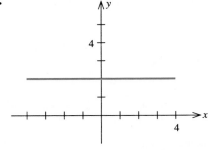

f.

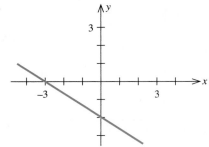

In Exercises 7–10, find the slope of the line shown in each figure.

7.

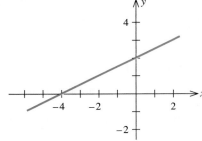

8.

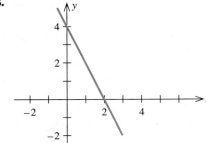

9.

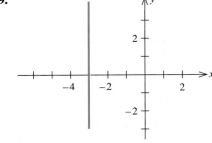

10.

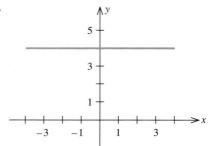

In Exercises 11–16, find the slope of the line that passes through each pair of points.

11. (4, 3) and (5, 8)

12. (4, 5) and (3, 8)

13. (−2, 3) and (4, 8)

14. (−2, −2) and (4, −4)

15. (a, b) and (c, d)

16. (−a + 1, b − 1) and (a + 1, −b)

17. Given the equation $y = 4x - 3$, answer the following questions:
a. If x increases by 1 unit, what is the corresponding change in y?
b. If x decreases by 2 units, what is the corresponding change in y?

18. Given the equation $2x + 3y = 4$, answer the following questions:
a. Is the slope of the line described by this equation positive or negative?
b. As x increases in value, does y increase or decrease?
c. If x decreases by 2 units, what is the corresponding change in y?

In Exercises 19 and 20, determine whether the line through each pair of points is parallel.

19. $A(1, -2)$, $B(-3, -10)$ and $C(1, 5)$, $D(-1, 1)$

20. $A(2, 3)$, $B(2, -2)$ and $C(-2, 4)$, $D(-2, 5)$

In Exercises 21 and 22, determine whether the line through each pair of points is perpendicular.

21. $A(-2, 5)$, $B(4, 2)$ and $C(-1, -2)$, $D(3, 6)$

22. $A(2, 0)$, $B(1, -2)$ and $C(4, 2)$, $D(-8, 4)$

23. If the line passing through the points $(1, a)$ and $(4, -2)$ is parallel to the line passing through the points $(2, 8)$ and $(-7, a + 4)$, what is the value of a?

24. If the line passing through the points $(a, 1)$ and $(5, 8)$ is parallel to the line passing through the points $(4, 9)$ and $(a + 2, 1)$, what is the value of a?

25. Find an equation of the horizontal line that passes through $(-4, -3)$.

26. Find an equation of the vertical line that passes through $(0, 5)$.

In Exercises 27–30, find an equation of the line that passes through the point and has the indicated slope m.

27. $(3, -4)$; $m = 2$

28. $(2, 4)$; $m = -1$

29. $(-3, 2)$; $m = 0$

30. $(1, 2)$; $m = -1/2$

In Exercises 31–34, find an equation of the line that passes through the given points.

31. $(2, 4)$ and $(3, 7)$

32. $(2, 1)$ and $(2, 5)$

33. $(1, 2)$ and $(-3, -2)$

34. $(-1, -2)$ and $(3, -4)$

In Exercises 35–38, find an equation of the line that has slope m and y-intercept b.

35. $m = 3$; $b = 4$

36. $m = -2$; $b = -1$

37. $m = 0$; $b = 5$

38. $m = -1/2$; $b = 3/4$

In Exercises 39–44, write the equation in the slope-intercept form and then find the slope and y-intercept of the corresponding line.

39. $x - 2y = 0$

40. $y - 2 = 0$

41. $2x - 3y - 9 = 0$

42. $3x - 4y + 8 = 0$

43. $2x + 4y = 14$

44. $5x + 8y - 24 = 0$

45. Find an equation of the line that passes through the point $(-2, 2)$ and is parallel to the line $2x - 4y - 8 = 0$.

46. Find an equation of the line that passes through the point $(2, 4)$ and is perpendicular to the line $3x + 4y - 22 = 0$.

In Exercises 47–52, find an equation of the line that satisfies the given condition.

47. The line parallel to the x-axis and 6 units below it

48. The line passing through the origin and parallel to the line joining the points $(2, 4)$ and $(4, 7)$

49. The line passing through the point (a, b) with slope equal to zero

50. The line passing through $(-3, 4)$ and parallel to the x-axis

51. The line passing through $(-5, -4)$ and parallel to the line joining $(-3, 2)$ and $(6, 8)$

52. The line passing through (a, b) with undefined slope

53. Given that the point $P(-3, 5)$ lies on the line $kx + 3y + 9 = 0$, find k.

54. Given that the point $P(2, -3)$ lies on the line $-2x + ky + 10 = 0$, find k.

In Exercises 55–60, sketch the straight line defined by the given linear equation by finding the x- and y-intercepts.
Hint: See Example 12, page 47.

55. $3x - 2y + 6 = 0$ **56.** $2x - 5y + 10 = 0$

57. $x + 2y - 4 = 0$ **58.** $2x + 3y - 15 = 0$

59. $y + 5 = 0$ **60.** $-2x - 8y + 24 = 0$

61. Show that an equation of a line through the points $(a, 0)$ and $(0, b)$ with $a \neq 0$ and $b \neq 0$ can be written in the form

$$\frac{x}{a} + \frac{y}{b} = 1$$

(Recall that the numbers a and b are the x- and y-intercepts, respectively, of the line. This form of an equation of a line is called the **intercept form.**)

In Exercises 62–65, use the results of Exercise 61 to find an equation of a line with the given x- and y-intercepts.

62. x-intercept 3; y-intercept 4

63. x-intercept -2; y-intercept -4

64. x-intercept $-\frac{1}{2}$; y-intercept $\frac{3}{4}$

65. x-intercept 4; y-intercept $-\frac{1}{2}$

In Exercises 66 and 67, determine whether the given points lie on a straight line.

66. $A(-1, 7)$, $B(2, -2)$, and $C(5, -9)$

67. $A(-2, 1)$, $B(1, 7)$, and $C(4, 13)$

68. SOCIAL SECURITY CONTRIBUTIONS For wages less than the maximum taxable wage base, Social Security contributions by employees are 7.65% of the employee's wages.

a. Find an equation that expresses the relationship between the wages earned (x) and the Social Security taxes paid (y) by an employee who earns less than the maximum taxable wage base.
b. For each additional dollar that an employee earns, by how much is his or her Social Security contribution increased? (Assume that the employee's wages are less than the maximum taxable wage base.)
c. What Social Security contributions will an employee who earns \$35,000 (which is less than the maximum taxable wage base) be required to make?

69. COLLEGE ADMISSIONS Using data compiled by the Admissions Office at Faber University, college admissions officers estimate that 55% of the students who are offered admission to the freshman class at the university will actually enroll.
a. Find an equation that expresses the relationship between the number of students who actually enroll (y) and the number of students who are offered admission to the university (x).
b. If the desired freshman class size for the upcoming academic year is 1100 students, how many students should be admitted?

70. WEIGHT OF WHALES The equation $W = 3.51L - 192$, expressing the relationship between the length L (in feet) and the expected weight W (in British tons) of adult blue whales, was adopted in the late 1960s by the International Whaling Commission.
a. What is the expected weight of an 80-ft blue whale?
b. Sketch the straight line that represents the equation.

71. THE NARROWING GENDER GAP Since the founding of the Equal Employment Opportunity Commission and the passage of equal-pay laws, the gulf between men's and women's earnings has continued to close gradually. At the beginning of 1990 ($t = 0$), women's wages were 68% of men's wages. However, women's wages were projected to be 80% of men's wages by the beginning of the year 2000 ($t = 10$). If this gap between women's and men's wages continued to narrow *linearly*, what percentage of men's wages were women's wages expected to be at the beginning of 2002?
Source: Journal of Economic Perspectives

72. IDEAL HEIGHTS AND WEIGHTS FOR WOMEN The Venus Health Club for Women provides its members with the following table, which gives the average desirable weight (in pounds) for women of a certain height (in inches):

Height, x	60	63	66	69	72
Weight, y	108	118	129	140	152

a. Plot the weight (y) versus the height (x).

b. Draw a straight line L through the points corresponding to heights of 5 ft and 6 ft.

c. Derive an equation of the line L.

d. Using the equation of part (c), estimate the average desirable weight for a woman who is 5 ft 5 in. tall.

73. **COST OF A COMMODITY** A manufacturer obtained the following data relating the cost y (in dollars) to the number of units (x) of a commodity produced:

No. of Units Produced, x	0	20	40	60	80	100
Cost, y	200	208	222	230	242	250

a. Plot the cost (y) versus the quantity produced (x).

b. Draw the straight line through the points $(0, 200)$ and $(100, 250)$.

c. Derive an equation of the straight line of part (b).

d. Taking this equation to be an approximation of the relationship between the cost and the level of production, estimate the cost of producing 54 units of the commodity.

74. **DIGITAL TV SERVICES** The percentage of homes with digital TV services, which stood at 5% at the beginning of 1999 ($t = 0$) is projected to grow linearly so that at the beginning of 2003 ($t = 4$) the percentage of such homes is projected to be 25%.

a. Derive an equation of the line passing through the points $A(0, 5)$ and $B(4, 25)$.

b. Plot the line with the equation found in part (a).

c. Using the equation found in part (a), find the percentage of homes with digital TV services at the beginning of 2001.

Source: Paul Kagan Associates

75. **SALES GROWTH** Metro Department Store's annual sales (in millions of dollars) during the past 5 yr were:

Annual Sales, y	5.8	6.2	7.2	8.4	9.0
Year, x	1	2	3	4	5

a. Plot the annual sales (y) versus the year (x).

b. Draw a straight line L through the points corresponding to the first and fifth years.

c. Derive an equation of the line L.

d. Using the equation found in part (c), estimate Metro's annual sales 4 yr from now ($x = 9$).

In Exercises 76–80, determine whether the statement is true or false. If it is true, explain why it is true. If it is false, give an example to show why it is false.

76. Suppose the slope of a line L is $-1/2$ and P is a given point on L. If Q is the point on L lying 4 units to the left of P, then Q is situated 2 units above P.

77. The line with equation $Ax + By + C = 0$, $(B \neq 0)$, and the line with equation $ax + by + c = 0$, $(b \neq 0)$, are parallel if $Ab - aB = 0$.

78. If the slope of the line L_1 is positive, then the slope of a line L_2 perpendicular to L_1 may be positive or negative.

79. The lines with equations $ax + by + c_1 = 0$ and $bx - ay + c_z = 0$, where $a \neq 0$ and $b \neq 0$, are perpendicular to each other.

80. If L is the line with equation $Ax + By + C = 0$, where $A \neq 0$, then L crosses the x-axis at the point $\left(-\dfrac{C}{A}, 0\right)$.

81. Is there a difference between the statements "The slope of a straight line is zero" and "The slope of a straight line does not exist (is not defined)"? Explain your answer.

82. Show that two distinct lines with equations $a_1 x + b_1 y + c_1 = 0$ and $a_2 x + b_2 y + c_2 = 0$, respectively, are parallel if and only if $a_1 b_2 - b_1 a_2 = 0$.

Hint: Write each equation in the slope-intercept form and compare.

83. Prove that if a line L_1 with slope m_1 is perpendicular to a line L_2 with slope m_2, then $m_1 m_2 = -1$.

Hint: Refer to the following figure. Show that $m_1 = b$ and $m_2 = c$. Next, apply the Pythagorean theorem to triangles OAC, OCB, and OBA to show that $1 = -bc$.

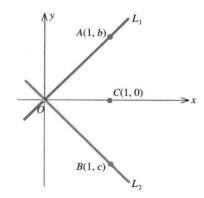

Solutions to Self-Check Exercises 1.4

1. The slope of the line that passes through the points $(a, 2)$ and $(3, 6)$ is

$$m = \frac{6 - 2}{3 - a}$$

$$= \frac{4}{3 - a}$$

Since this line is parallel to a line with slope 4, m must be equal to 4; that is,

$$\frac{4}{3 - a} = 4$$

or, upon multiplying both sides of the equation by $3 - a$,

$$4 = 4(3 - a)$$
$$4 = 12 - 4a$$
$$4a = 8$$
$$a = 2$$

2. Since the required line L is perpendicular to a line with slope $-1/2$, the slope of L is

$$-\frac{1}{-1/2} = 2$$

Next, using the point-slope form of the equation of a line, we have

$$y - (-1) = 2(x - 3)$$
$$y + 1 = 2x - 6$$
$$y = 2x - 7$$

3. Substituting $x = 3$ and $y = -3$ into the left-hand side of the given equation, we find

$$2(3) - 3(-3) - 12 = 3$$

which is not equal to zero (the right-hand side). Therefore, $(3, -3)$ does not lie on the line with equation $2x - 3y - 12 = 0$.

Setting $x = 0$, we find $y = -4$, the y-intercept. Next, setting $y = 0$ gives $x = 6$, the x-intercept. We now draw the line passing through the points $(0, -4)$ and $(6, 0)$ as shown.

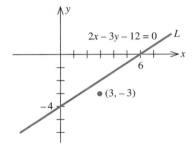

4. a. and **b.** See the accompanying figure.

c. The slope of L is

$$m = \frac{52 - 20}{1990 - 1960} = \frac{32}{30} = \frac{16}{15}$$

Using the point-slope form of the equation of a line with the point $(1960, 20)$, we find

$$y - 20 = \frac{16}{15}(x - 1960) = \frac{16}{15}x - \frac{(16)(1960)}{15}$$

$$y = \frac{16}{15}x - \frac{6272}{3} + 20$$

$$= \frac{16}{15}x - \frac{6212}{3}$$

d. To estimate the percentage of people over age 65 who had high school diplomas by the year 2000, let $x = 2000$ in the equation obtained in part (c). Thus, the required estimate is

$$y = \frac{16}{15}(2000) - \frac{6212}{3} \approx 62.67$$

or approximately 63%.

Group projects for this chapter can be found at the Brooks/Cole Web site:
http://www.brookscole.com/product/0534378439

CHAPTER 1 Summary of Principal Formulas and Terms

Formulas

1. Quadratic formula $\qquad\qquad x = \dfrac{-b \pm \sqrt{b^2 - 4ac}}{2a}$

2. Distance between two points $\qquad d = \sqrt{(x_2 - x_1)^2 + (y_2 - y_1)^2}$

3. Slope of a line $\qquad\qquad\qquad m = \dfrac{y_2 - y_1}{x_2 - x_1}$

4. Equation of a vertical line $\qquad x = a$

5. Equation of a horizontal line $\qquad y = b$

6. Point-slope form of the equation of a line $\qquad y - y_1 = m(x - x_1)$

7. Slope-intercept form of the equation of a line $\qquad y = mx + b$

8. General equation of a line $\qquad Ax + By + C = 0$

Terms

real number (coordinate) line

open interval

closed interval

half-open interval

finite interval

infinite interval

absolute value

triangle inequality

polynomial

roots of a polynomial equation

Cartesian coordinate system

ordered pair

parallel lines

perpendicular lines

CHAPTER 1 REVIEW EXERCISES

In Exercises 1–4, find the values of *x* that satisfy the inequality (inequalities).

1. $-x + 3 \le 2x + 9$

2. $-2 \le 3x + 1 \le 7$

3. $x - 3 > 2$ or $x + 3 < -1$

4. $2x^2 > 50$

In Exercises 5–8, evaluate the expression.

5. $|-5 + 7| + |-2|$

6. $\left| \dfrac{5 - 12}{-4 - 3} \right|$

7. $|2\pi - 6| - \pi$

8. $|\sqrt{3} - 4| + |4 - 2\sqrt{3}|$

In Exercises 9–14, evaluate the expression.

9. $\left(\dfrac{9}{4} \right)^{3/2}$

10. $\dfrac{5^6}{5^4}$

11. $(3 \cdot 4)^{-2}$

12. $(-8)^{5/3}$

13. $\dfrac{(3 \cdot 2^{-3})(4 \cdot 3^5)}{2 \cdot 9^3}$

14. $\dfrac{3\sqrt[3]{54}}{\sqrt[3]{18}}$

In Exercises 15–19, simplify the expression.

15. $\dfrac{4(x^2 + y)^3}{x^2 + y}$

16. $\dfrac{a^6 b^{-5}}{(a^3 b^{-2})^{-3}}$

17. $\dfrac{\sqrt[4]{16x^5yz}}{\sqrt[4]{81xyz^5}}$ $(x, y, \text{and } z \text{ positive})$

18. $(2x^3)(-3x^{-2})\left(\dfrac{1}{6} x^{-1/2} \right)$

19. $\left(\dfrac{3xy^2}{4x^3y} \right)^{-2} \left(\dfrac{3xy^3}{2x^2} \right)^3$

In Exercises 20–23, factor the expression.

20. $-2\pi^2 r^3 + 100\pi r^2$

21. $2v^3 w + 2vw^3 + 2u^2 vw$

22. $16 - x^2$

23. $12t^3 - 6t^2 - 18t$

In Exercises 24–27, solve the equation by factoring.

24. $8x^2 + 2x - 3 = 0$

25. $-6x^2 - 10x + 4 = 0$

26. $-x^3 - 2x^2 + 3x = 0$

27. $2x^4 + x^2 = 1$

In Exercises 28 and 29, use the quadratic formula to solve the quadratic equation.

28. $x^2 - 2x - 5 = 0$

29. $2x^2 + 8x + 7 = 0$

In Exercises 30–33, perform the indicated operations and simplify the expression.

30. $\dfrac{(t + 6)(60) - (60t + 180)}{(t + 6)^2}$

31. $\dfrac{6x}{2(3x^2 + 2)} + \dfrac{1}{4(x + 2)}$

32. $\dfrac{2}{3}\left(\dfrac{4x}{2x^2 - 1} \right) + 3\left(\dfrac{3}{3x - 1} \right)$

33. $\dfrac{-2x}{\sqrt{x + 1}} + 4\sqrt{x + 1}$

34. Rationalize the numerator:

$$\frac{\sqrt{x} - 1}{x - 1}$$

35. Rationalize the denominator:

$$\frac{\sqrt{x} - 1}{2\sqrt{x}}$$

In Exercises 36 and 37, find the distance between the two points.

36. $(-2, -3)$ and $(1, -7)$

37. $\left(\dfrac{1}{2}, \sqrt{3} \right)$ and $\left(-\dfrac{1}{2}, 2\sqrt{3} \right)$

In Exercises 38–43, find an equation of the line *L* that passes through the point (−2, 4) and satisfies the condition.

38. L is a vertical line.

39. L is a horizontal line.

40. L passes through the point $\left(3, \dfrac{7}{2} \right)$.

41. The x-intercept of L is 3.

42. L is parallel to the line $5x - 2y = 6$.

43. L is perpendicular to the line $4x + 3y = 6$.

44. Find an equation of the straight line that passes through the point $(2, 3)$ and is parallel to the line with equation $3x + 4y - 8 = 0$.

45. Find an equation of the straight line that passes through the point $(-1, 3)$ and is parallel to the line passing through the points $(-3, 4)$ and $(2, 1)$.

46. Find an equation of the line that passes through the point $(-2, -4)$ and is perpendicular to the line with equation $2x - 3y - 24 = 0$.

47. Sketch the graph of the equation $3x - 4y = 24$.

48. Sketch the graph of the line that passes through the point $(3, 2)$ and has slope $-2/3$.

49. Find the minimum cost C (in dollars) given that

$$2(1.5C + 80) \leq 2(2.5C - 20)$$

50. Find the maximum revenue R (in dollars) given that

$$12(2R - 320) \leq 4(3R + 240)$$

 Additional study hints and sample chapter tests can be found at the Brooks/Cole Web site: http://www.brookscole.com/product/0534378439

FUNCTIONS, LIMITS, AND THE DERIVATIVE

2

In this chapter we define a *function*, a special relationship between two variables. The concept of a function enables us to describe many relationships that exist in applications. We also begin the study of differential calculus. Historically, differential calculus was developed in response to the problem of finding the tangent line to an arbitrary curve. But it quickly became apparent that solving this problem provided mathematicians with a method for solving many practical problems involving the rate of change of one quantity with respect to another. The basic tool used in differential calculus is the *derivative* of a function. The concept of the derivative is based, in turn, on a more fundamental notion—that of the *limit* of a function.

How does the change in the demand for a certain make of tires affect the unit price of the tires? The management of the Titan Tire Company has determined the demand function that relates the unit price of its Super Titan tires to the quantity demanded. In Example 7, page 165, you will see how this function can be used to compute the rate of change of the unit price of the Super Titan tires with respect to the quantity demanded.

2.1 Functions and Their Graphs

FUNCTIONS

A manufacturer would like to know how his company's profit is related to its production level; a biologist would like to know how the size of the population of a certain culture of bacteria will change over time; a psychologist would like to know the relationship between the learning time of an individual and the length of a vocabulary list; and a chemist would like to know how the initial speed of a chemical reaction is related to the amount of substrate used. In each instance we are concerned with the same question: How does one quantity depend upon another? The relationship between two quantities is conveniently described in mathematics by using the concept of a function.

Function

> A **function** is a rule that assigns to each element in a set A one and only one element in a set B.

The set A is called the **domain** of the function. It is customary to denote a function by a letter of the alphabet, such as the letter f. If x is an element in the domain of a function f, then the element in B that f associates with x is written $f(x)$ (read "f of x") and is called the value of f at x. The set comprising all the values assumed by $y = f(x)$ as x takes on all possible values in its domain is called the **range** of the function f.

We can think of a function f as a machine. The domain is the set of inputs (raw material) for the machine, the rule describes how the input is to be processed, and the value(s) of the function are the outputs of the machine (Figure 2.1).

We can also think of a function f as a mapping in which an element x in the domain of f is mapped onto a unique element $f(x)$ in B (Figure 2.2).

FIGURE 2.1
A function machine

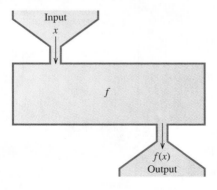

FIGURE 2.2
The function f viewed as a mapping

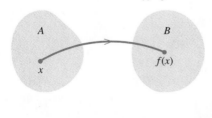

REMARKS

1. It is important to understand that the output $f(x)$ associated with an input x is unique. To appreciate the importance of this uniqueness property, consider a rule that associates with each item x in a department store its selling price y. Then, each x must correspond to *one and only one y*. Notice, however, that different x's may be associated with the same y. In the context of the present example, this says that different items may have the same price.

2. Although the sets A and B that appear in the definition of a function may be quite arbitrary, in this book they will denote sets of real numbers.

■ ■ ■

An example of a function may be taken from the familiar relationship between the area of a circle and its radius. Letting x and y denote the radius and area of a circle, respectively, we have, from elementary geometry,

$$y = \pi x^2 \tag{1}$$

Equation (1) defines y as a function of x since for each admissible value of x (that is, for each nonnegative number representing the radius of a certain circle) there corresponds precisely one number $y = \pi x^2$ that gives the area of the circle. The rule defining this "area function" may be written as

$$f(x) = \pi x^2 \tag{2}$$

To compute the area of a circle of radius 5 inches, we simply replace x in Equation (2) with the number 5. Thus, the area of the circle is

$$f(5) = \pi 5^2 = 25\pi$$

or 25π square inches

In general, to evaluate a function at a specific value of x, we replace x with that value, as illustrated in Examples 1 and 2.

EXAMPLE 1   Let the function f be defined by the rule $f(x) = 2x^2 - x + 1$. Compute:

a. $f(1)$ **b.** $f(-2)$ **c.** $f(a)$ **d.** $f(a + h)$

SOLUTION ✔

a. $f(1) = 2(1)^2 - (1) + 1 = 2 - 1 + 1 = 2$
b. $f(-2) = 2(-2)^2 - (-2) + 1 = 8 + 2 + 1 = 11$
c. $f(a) = 2(a)^2 - (a) + 1 = 2a^2 - a + 1$
d. $f(a + h) = 2(a + h)^2 - (a + h) + 1 = 2a^2 + 4ah + 2h^2 - a - h + 1$

■ ■ ■ ■

EXAMPLE 2   The Thermo-Master Company manufactures an indoor–outdoor thermometer at its Mexican subsidiary. Management estimates that the profit (in dollars) realizable by Thermo-Master in the manufacture and sale of x thermometers per week is

$$P(x) = -0.001x^2 + 8x - 5000$$

b. $f(3) = \dfrac{\sqrt{3+1}}{3} = \dfrac{\sqrt{4}}{3} = \dfrac{2}{3}$

c. $f(a+h) = \dfrac{\sqrt{(a+h)+1}}{a+h} = \dfrac{\sqrt{a+h+1}}{a+h}$

2. a. For t in the subdomain $[0, 6]$, the rule for f is given by $f(t) = 6t + 17$. The equation $y = 6t + 17$ is a linear equation, so that portion of the graph of f is the line segment joining the points $(0, 17)$ and $(6, 53)$. Next, in the subdomain $(6, 20]$, the rule for f is given by $f(t) = 15.98(t - 6)^{1/4} + 53$. Using a calculator, we construct the following table of values of $f(t)$ for selected values of t.

t	6	8	10	12	14	16	18	20
$f(t)$	53	72	75.6	78	79.9	81.4	82.7	83.9

We have included $t = 6$ in the table, although it does not lie in the subdomain of the function under consideration, in order to help us obtain a better sketch of that portion of the graph of f in the subdomain $(6, 20]$. The graph of f is as follows:

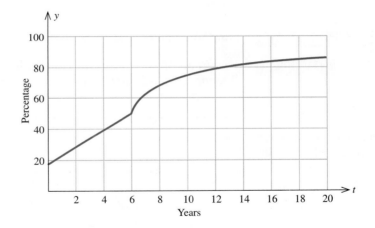

b. The percentage of all self-serve gas sales at the beginning of 1978 is found by evaluating f at $t = 4$. Since this point lies in the interval $[0, 6]$, we use the rule $f(t) = 6t + 17$ and find

$$f(4) = 6(4) + 17$$

giving 41% as the required figure. The percentage of all self-serve gas sales at the beginning of 1994 is given by

$$f(20) = 15.98(20 - 6)^{1/4} + 53$$

or approximately 83.9%.

3. A point (x, y) lies on the graph of the function f if and only if the coordinates satisfy the equation $y = f(x)$. Now,

$$f(4) = \sqrt{2(4) + 1} + 2 = \sqrt{9} + 2 = 5 \neq 6$$

and we conclude that the given point does not lie on the graph of f.

GRAPHING A FUNCTION

Most of the graphs of functions in this book can be plotted with the help of a graphing utility. Furthermore, a graphing utility can be used to analyze the nature of a function. However, the amount and accuracy of the information obtained using a graphing utility depend on the experience and sophistication of the user. As you progress through this book, you will see that the more knowledge of calculus you gain, the more effective the graphing utility will prove as a tool in problem solving.

FINDING A SUITABLE VIEWING RECTANGLE

The first step in plotting the graph of a function with a graphing utility is to select a suitable viewing rectangle. We usually do this by experimenting. For example, you might first plot the graph using the *standard viewing rectangle* $[-10, 10]$ by $[-10, 10]$. If necessary, you then might adjust the viewing rectangle by enlarging it or reducing it to obtain a sufficiently complete view of the graph or at least the portion of the graph that is of interest.

EXAMPLE 1 Plot the graph of $f(x) = 2x^2 - 4x - 5$ in the standard viewing rectangle.

SOLUTION ✔ The graph of f, shown in Figure T1, is a parabola. From our previous work (Example 6, Section 2.1), we know that the figure does give a good view of the graph.

FIGURE T1
The graph of $y = 2x^2 - 4x - 5$ in the viewing rectangle $[-10, 10]$ $\times [-10, 10]$

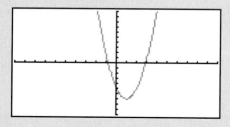

EXAMPLE 2 Let $f(x) = x^3(x - 3)^4$.

a. Plot the graph of f in the standard viewing rectangle.
b. Plot the graph of f in the rectangle $[-1, 5] \times [-40, 40]$.

SOLUTION ✔ **a.** The graph of f in the standard viewing rectangle is shown in Figure T2a. Since the graph does not appear to be complete, we need to adjust the viewing rectangle.

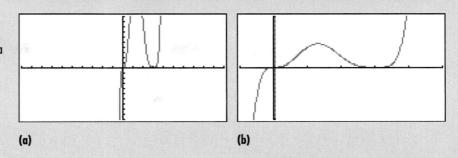

(a)   (b)

b. The graph of f in the rectangle $[-1, 5] \times [-40, 40]$, shown in Figure T2b, is an improvement over the previous graph. (Later we will be able to show that the figure does in fact give a rather complete view of the graph of f.)

■ ■ ■ ■

EVALUATING A FUNCTION

A graphing utility can be used to find the value of a function with minimal effort, as the next example shows.

EXAMPLE 3   Let $f(x) = x^3 - 4x^2 + 4x + 2$.

a. Plot the graph of f in the standard viewing rectangle.
b. Find $f(3)$ and verify your result by direct computation.
c. Find $f(4.215)$.

SOLUTION ✔   **a.** The graph of f is shown in Figure T3.

FIGURE T3
The graph of
$f(x) = x^3 - 4x^2 + 4x + 2$
in the standard viewing rectangle

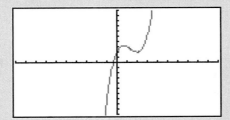

b. Using the evaluation function of the graphing utility and the value 3 for x, we find $y = 5$. This result is verified by computing

$$f(3) = 3^3 - 4(3^2) + 4(3) + 2 = 27 - 36 + 12 + 2 = 5$$

c. Using the evaluation function of the graphing utility and the value 4.215 for x, we find $y = 22.679738375$. Thus, $f(4.215) = 22.679738375$. The efficacy of the graphing utility is clearly demonstrated here! ■■■■

EXAMPLE **4**

The anticipated rise in the number of Alzheimer's patients in the United States is given by

$$f(t) = -0.0277t^4 + 0.3346t^3 - 1.1261t^2 + 1.7575t + 3.7745 \qquad (0 \le t \le 6)$$

where $f(t)$ is measured in millions and t is measured in decades, with $t = 0$ corresponding to the beginning of 1990.

a. Use a graphing utility to plot the graph of f in the viewing rectangle $[0, 7] \times [0, 12]$.
b. What was the anticipated number of Alzheimer's patients in the United States at the beginning of the year 2000 ($t = 1$)? At the beginning of 2030 ($t = 4$)?

Source: Alzheimer's Association

SOLUTION ✔   **a.** The graph of f in the viewing rectangle $[0, 7] \times [0, 12]$ is shown in Figure T4.

FIGURE T4
The graph of f in the viewing rectangle $[0, 7] \times [0, 12]$

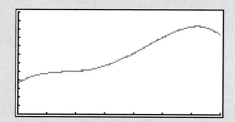

b. Using the evaluation function of the graphing utility and the value 1 for x, we see that the anticipated number of Alzheimer's patients at the beginning of the year 2000 was

$$f(1) = 4.7128$$

or approximately 4.7 million. The anticipated number of Alzheimer's patients at the beginning of 2030 is given by

$$f(4) = 7.1101$$

or approximately 7.1 million. ■■■■

In Exercises 1–8, plot the graph of the function f in the standard viewing window.

1. $f(x) = 2x^2 - 16x + 29$ 2. $f(x) = -x^2 - 10x - 20$

3. $f(x) = x^3 - 2x^2 + x - 2$

4. $f(x) = -2.01x^3 + 1.21x^2 - 0.78x + 1$

5. $f(x) = 0.2x^4 - 2.1x^2 + 1$

6. $f(x) = -0.4x^4 + 1.2x - 1.2$

7. $f(x) = 2x\sqrt{x^2 + 1}$ 8. $f(x) = \dfrac{\sqrt{x} + 1}{\sqrt{x} - 1}$

In Exercises 9–20, plot the graph of the function f in (a) the standard viewing window and (b) the indicated window.

9. $f(x) = 2x^2 - 32x + 125$; $[5, 15] \times [-5, 10]$

10. $f(x) = x^2 + 20x + 95$; $[-20, 10] \times [-15, -5]$

11. $f(x) = x^3 - 20x^2 + 8x - 10$; $[-20, 20] \times [-1200, 100]$

12. $f(x) = -2x^3 + 10x^2 - 15x - 5$; $[-10, 10] \times [-100, 100]$

13. $f(x) = x^4 - 2x^2 + 8$; $[-2, 2] \times [6, 10]$

14. $f(x) = x^4 - 2x^3$; $[-1, 3] \times [-2, 2]$

15. $f(x) = x + \dfrac{1}{x}$; $[-1, 3] \times [-5, 5]$

16. $f(x) = \dfrac{4}{x^2 - 8}$; $[-5, 5] \times [-5, 5]$

17. $f(x) = 2 - \dfrac{1}{x^2 + 1}$; $[-3, 3] \times [0, 3]$

18. $f(x) = x - 2\sqrt{x}$; $[0, 20] \times [-2, 10]$

19. $f(x) = x\sqrt{4 - x^2}$; $[-3, 3] \times [-2, 2]$

20. $f(x) = \dfrac{\sqrt{x} - 1}{x}$; $[0, 50] \times [-0.25, 0.25]$

In Exercises 21–30, plot the graph of the function f in an appropriate viewing window. (Note: The answer is not unique.)

21. $f(x) = x^2 - 4x + 16$ 22. $f(x) = -x^2 + 2x - 11$

23. $f(x) = 2x^3 - 10x^2 + 5x - 10$

24. $f(x) = -x^3 + 5x^2 - 14x + 20$

25. $f(x) = 2x^4 - 3x^3 + 5x^2 - 20x + 40$

26. $f(x) = -2x^4 + 5x^2 - 4$

27. $f(x) = \dfrac{x^3}{x^3 + 1}$ 28. $f(x) = \dfrac{2x^4 - 3x}{x^2 - 1}$

29. $f(x) = 0.2\sqrt{x} - 0.3x^3$ 30. $f(x) = \sqrt{x}(2x - 1)^3$

In Exercises 31–34, use the evaluation function of your graphing utility to find the value of f at the given value of x and verify your result by direct computation.

31. $f(x) = -3x^3 + 5x^2 - 2x + 8$; $x = -1$

32. $f(x) = 2x^4 - 3x^3 + 2x^2 + x - 5$; $x = 2$

33. $f(x) = \dfrac{x^4 - 3x^2}{x - 2}$; $x = 1$ 34. $f(x) = \dfrac{\sqrt{x^2 - 1}}{3x + 4}$; $x = 2$

In Exercises 35–42, use the evaluation function of your graphing utility to find the value of f at the indicated value of x. Express your answer accurate to four decimal places.

35. $f(x) = 3x^3 - 2x^2 + x - 4$; $x = 2.145$

36. $f(x) = 2x^3 + 5x^2 + 3x + 1$; $x = -0.27$

37. $f(x) = 5x^4 - 2x^2 + 8x - 3$; $x = 1.28$

38. $f(x) = 4x^4 - 3x^3 + 1$; $x = -2.42$

39. $f(x) = \dfrac{2x^3 - 3x + 1}{3x - 2}$; $x = 2.41$

40. $f(x) = \dfrac{2x + 5}{3x^2 - 4x + 1}$; $x = -1.72$

41. $f(x) = \sqrt{2x^2 + 1} + \sqrt{3x^2 - 1}$; $x = 0.62$

42. $f(x) = 2x(3x^3 + 5)^{1/3}$; $x = -6.24$

43. **Manufacturing Capacity** Data obtained from the Federal Reserve show that the annual increase in manufacturing capacity between 1988 and 1994 is given by

extreme to an interpretation of data associated with the problem on the other. The model for the accumulated amount of a fixed bank account mentioned earlier may be derived theoretically (see Chapter 5). Later we will see how linear equations (models) can be constructed from a given set of data points.

In calculus we are concerned primarily with how one (dependent) variable depends on one or more (independent) variables. Consequently, most of our mathematical models will involve functions of one or more variables.* Once a function has been constructed to describe a specific real-world problem, a host of questions pertaining to the problem may be answered by analyzing the function (mathematical model). For example, if we have a function that gives the population of a certain culture of bacteria at any time t, then we can determine how fast the population is increasing or decreasing at any time t, and so on. Conversely, if we have a model that gives the rate of change of the cost of producing a certain item as a function of the level of production and if we know the fixed cost incurred in producing this item, then we can find the total cost incurred in producing a certain number of those items.

Before going on, let us look at two mathematical models. The first one is used to estimate spending by business on computer security, and the second is used to project the growth of the number of people enrolled in health maintenance organizations (HMOs). These models are derived from data using the least-squares technique. In the Using Technology section on page 105, you can see how mathematical models are constructed from raw data.

EXAMPLE 1

The estimated spending (in billions of dollars) by businesses on computer security equipment and services from 1987 to 1993 is given in the following table. The figures include spending for protection against computer criminals who steal, erase, or alter data, along with protection against fires, electrical failures, and natural disasters.

Year	1987	1988	1989	1990	1991	1992	1993
Spending	0.49	0.59	0.66	0.73	0.81	0.93	1.02

A mathematical model approximating the amount of spending over the period in question is given by

$$S(t) = 0.0864t + 0.4879$$

where t is measured in years, with $t = 0$ corresponding to 1987.

a. Sketch the graph of the function S and the given data on the same set of axes.
b. Assuming that this trend continued, what was the spending by business on computer security equipment and services in 1995 ($t = 8$).
c. What is the rate of increase of the annual expenditure over the period in question?

Source: Frost & Sullivan, Inc.

---

* Functions of more than one variable will be studied later.

FIGURE 2.14
Estimated spending by businesses on computer security equipment and services

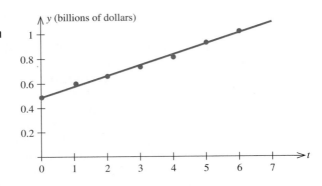

FIGURE 2.14
Estimated spending by businesses on computer security equipment and services

SOLUTION ✔   **a.** The graph of S is shown in Figure 2.14.
b. The estimated spending in 1995 is

$$S(8) = 0.0864(8) + 0.4879$$
$$\approx 1.1791$$

or approximately $1.18 billion.

c. The function S is linear, and so we conclude that the annual increase in the expenditure is given by the slope of the straight line represented by S, which is approximately $0.09 billion per year.   ■ ■ ■ ■

    EXAMPLE 2   The number of people (in millions) enrolled in HMOs from 1988 to 1998 is given in the following table.

Year	1988	1990	1992	1994	1996	1998
No. of People	32.7	36.5	41.4	51.1	66.5	78.0

A mathematical model approximating the number of people, $N(t)$, enrolled in HMOs during this period is

$$N(t) = -0.0258t^3 + 0.7465t^2 - 0.3491t + 33.1444 \qquad (0 \leq t \leq 10)$$

where t is measured in years and $t = 0$ corresponds to 1988.

a. Sketch the graph of the function N to see how the model compares with the actual data.
b. Assume that this trend continues and use the model to predict how many people will be enrolled in HMOs at the beginning of 2002.

SOLUTION ✔   **a.** The graph of the function N is shown in Figure 2.15.
b. The number of people that will be enrolled in HMOs at the beginning of 2002 is given by

$$N(14) = -0.0258(14)^3 + 0.7465(14)^2 - 0.3491(14) + 33.1444$$
$$= 103.7758$$

or approximately 103.8 million people.

FIGURE 2.15
The graph of $y = N(t)$ approximates the number of people enrolled in HMOs from 1988 to 1998.

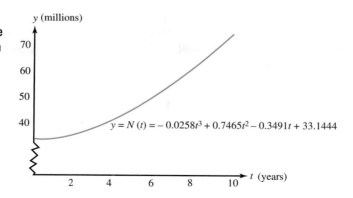

We will discuss several mathematical models from the field of economics later in this section, but first we review some important functions that are the basis for many mathematical models.

POLYNOMIAL FUNCTIONS

We begin by recalling a special class of functions, polynomial functions.

Polynomial Function

> A **polynomial function** of degree n is a function of the form
>
> $$f(x) = a_0 x^n + a_1 x^{n-1} + \cdots + a_{n-1} x + a_n \qquad (a_0 \neq 0)$$
>
> where $a_0, a_1, \ldots, a_n$ are constants and n is a nonnegative integer.

For example, the functions

$$f(x) = 4x^5 - 3x^4 + x^2 - x + 8$$
$$g(x) = 0.001x^3 - 2x^2 + 20x + 400$$

are polynomial functions of degrees 5 and 3, respectively. Observe that a polynomial function is defined everywhere so that it has domain $(-\infty, \infty)$.

A polynomial function of degree 1 $(n = 1)$

$$f(x) = a_0 x + a_1 \qquad (a_0 \neq 0)$$

is the equation of a straight line in the slope-intercept form with slope $m = a_0$ and y-intercept $b = a_1$ (see Section 1.4). For this reason, a polynomial function of degree 1 is called a **linear function.** For example, the linear function $f(x) = 2x + 3$ may be written as a linear equation in x and y—namely, $y = 2x + 3$ or $2x - y + 3 = 0$. Conversely, the linear equation $2x - 3y + 4 = 0$ can be solved for y in terms of x to yield the linear function $y = f(x) = \frac{2}{3}x + \frac{4}{3}$.

A polynomial function of degree 2 is referred to as a **quadratic function.** A polynomial function of degree 3 is called a **cubic function,** and so on. The mathematical model in Example 2 involves a cubic function.

RATIONAL AND POWER FUNCTIONS

Another important class of functions is rational functions. A **rational function** is simply the quotient of two polynomials. Examples of rational functions are

$$F(x) = \frac{3x^3 + x^2 - x + 1}{x - 2}$$

$$G(x) = \frac{x^2 + 1}{x^2 - 1}$$

In general, a rational function has the form

$$R(x) = \frac{f(x)}{g(x)}$$

where $f(x)$ and $g(x)$ are polynomial functions. Since division by zero is not allowed, we conclude that the domain of a rational function is the set of all real numbers except the zeros of g—that is, the roots of the equation $g(x) = 0$. Thus, the domain of the function F is the set of all numbers except $x = 2$, whereas the domain of the function G is the set of all numbers except those that satisfy $x^2 - 1 = 0$ or $x = \pm 1$.

Functions of the form

$$f(x) = x^r$$

where r is any real number, are called **power functions.** We encountered examples of power functions earlier in our work. For example, the functions

$$f(x) = \sqrt{x} = x^{1/2} \qquad \text{and} \qquad g(x) = \frac{1}{x^2} = x^{-2}$$

are power functions.

Many of the functions we will encounter later will involve combinations of the functions introduced here. For example, the following functions may be viewed as suitable combinations of such functions:

$$f(x) = \sqrt{\frac{1 - x^2}{1 + x^2}}$$

$$g(x) = \sqrt{x^2 - 3x + 4}$$

$$h(x) = (1 + 2x)^{1/2} + \frac{1}{(x^2 + 2)^{3/2}}$$

As with polynomials of degree 3 or greater, analyzing the properties of these functions is facilitated by using the tools of calculus, to be developed later.

EXAMPLE 3   A study of driving costs based on 1992 model compact (six-cylinder) cars found that the average cost (car payments, gas, insurance, upkeep, and depreciation), measured in cents per mile, is approximated by the function

$$C(x) = \frac{2095}{x^{2.2}} + 20.08$$

where x (in thousands) denotes the number of miles the car is driven in 1 year. Using this model, estimate the average cost of driving a compact car 10,000 miles a year and 20,000 miles a year.
Source: Runzheimer International Study

SOLUTION ✔   The average cost of driving a compact car 10,000 miles a year is given by

$$C(10) = \frac{2095}{10^{2.2}} + 20.08$$

$$\approx 33.3$$

or approximately 33 cents per mile. The average cost of driving 20,000 miles a year is given by

$$C(20) = \frac{2095}{20^{2.2}} + 20.08$$

$$\approx 23.0$$

or approximately 23 cents per mile.   ■ ■ ■ ■

SOME ECONOMIC MODELS

In the remainder of this section, we look at some economic models.

In a free market economy, consumer demand for a particular commodity depends on the commodity's unit price. A **demand equation** expresses the relationship between the unit price and the quantity demanded. The graph of the demand equation is called a **demand curve.** In general, the quantity demanded of a commodity decreases as the commodity's unit price increases, and vice versa. Accordingly, a **demand function** defined by $p = f(x)$, where p measures the unit price and x measures the number of units of the commodity in question, is generally characterized as a decreasing function of x; that is, $p = f(x)$ decreases as x increases. Since both x and p assume only nonnegative values, the demand curve is that part of the graph of $f(x)$ that lies in the first quadrant (Figure 2.16).

In a competitive market a relationship also exists between the unit price of a commodity and the commodity's availability in the market. In general, an increase in the commodity's unit price induces the producer to increase the supply of the commodity. Conversely, a decrease in the unit price generally leads to a drop in the supply. The equation that expresses the relation between

FIGURE 2.16
A demand curve

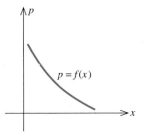

FIGURE 2.17
A supply curve

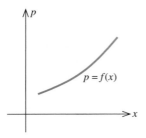

FIGURE 2.18
Market equilibrium corresponds to (x_0, p_0), the point at which the supply and demand curves intersect.

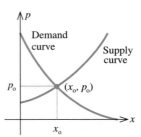

the unit price and the quantity supplied is called a **supply equation,** and its graph is called a **supply curve.** A supply function defined by $p = f(x)$ is generally characterized as an increasing function of x; that is, $p = f(x)$ increases as x increases. Since both x and p assume only nonnegative values, the supply curve is that part of the graph of $f(x)$ that lies in the first quadrant (Figure 2.17).

Under pure competition, the price of a commodity will eventually settle at a level dictated by the following condition: that the supply of the commodity be equal to the demand for it. If the price is too high, the consumer will not buy, and if the price is too low, the supplier will not produce. **Market equilibrium** prevails when the quantity produced is equal to the quantity demanded. The quantity produced at market equilibrium is called the **equilibrium quantity,** and the corresponding price is called the **equilibrium price.**

Market equilibrium corresponds to the point at which the demand curve and the supply curve intersect. In Figure 2.18 x_0 represents the equilibrium quantity and p_0 the equilibrium price. The point (x_0, p_0) lies on the supply curve and therefore satisfies the supply equation. At the same time it also lies on the demand curve and therefore satisfies the demand equation. Thus, to find the point (x_0, p_0), and hence the equilibrium quantity and price, we solve the demand and supply equations simultaneously for x and p. For meaningful solutions, x and p must both be positive.

EXAMPLE 4

The demand function for a certain brand of videocassette is given by

$$p = d(x) = -0.01x^2 - 0.2x + 8$$

and the corresponding supply function is given by

$$p = s(x) = 0.01x^2 + 0.1x + 3$$

where p is expressed in dollars and x is measured in units of a thousand. Find the equilibrium quantity and price.

SOLUTION ✔

We solve the following system of equations:

$$p = -0.01x^2 - 0.2x + 8$$
$$p = 0.01x^2 + 0.1x + 3$$

FIGURE 2.19
The supply curve and the demand curve intersect at the point (10, 5).

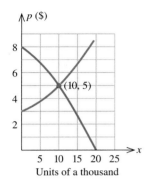

Substituting the first equation into the second yields

$$-0.01x^2 - 0.2x + 8 = 0.01x^2 + 0.1x + 3$$

which is equivalent to

$$0.02x^2 + 0.3x - 5 = 0$$
$$2x^2 + 30x - 500 = 0$$
$$x^2 + 15x - 250 = 0$$
$$(x + 25)(x - 10) = 0$$

Thus, $x = -25$ or $x = 10$. Since x must be nonnegative, the root $x = -25$ is rejected. Therefore, the equilibrium quantity is 10,000 videocassettes. The equilibrium price is given by

$$p = 0.01(10)^2 + 0.1(10) + 3 = 5$$

or $5 per videocassette (Figure 2.19).   ■ ■ ■ ■

Exploring with Technology

1. **a.** Use a graphing utility to plot the straight lines L_1 and L_2 with equations $y = 2x - 1$ and $y = 2.1x + 3$, respectively, on the same set of axes using the standard viewing rectangle. Do the lines appear to intersect?
   b. Plot the straight lines L_1 and L_2 using the viewing rectangle $[-100, 100] \times [-100, 100]$. Do the lines appear to intersect? Can you find the point of intersection using TRACE and ZOOM? Using the "intersection" function of your graphing utility?
   c. Find the point of intersection of L_1 and L_2 algebraically.
   d. Comment on the effectiveness of the methods of solutions in parts (b) and (c).
2. **a.** Use a graphing utility to plot the straight lines L_1 and L_2 with equations $y = 3x - 2$ and $y = -2x + 3$, respectively, on the same set of axes using the standard viewing rectangle. Then use TRACE and ZOOM to find the point of intersection of L_1 and L_2. Repeat using the "intersection" function of your graphing utility.
   b. Find the point of intersection of L_1 and L_2 algebraically.
   c. Comment on the effectiveness of the methods.

CONSTRUCTING MATHEMATICAL MODELS

We close this section by showing how some mathematical models can be constructed using elementary geometric and algebraic arguments.

EXAMPLE 5   The owner of the Rancho Los Feliz has 3000 yards of fencing material with which to enclose a rectangular piece of grazing land along the straight portion of a river. Fencing is not required along the river. Letting x denote the width of the rectangle, find a function f in the variable x giving the area of the grazing land if she uses all of the fencing material (Figure 2.20).

FIGURE 2.20
The rectangular grazing land has width x and length y.

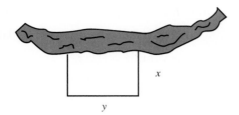

SOLUTION ✔ The area of the rectangular grazing land is $A = xy$. Next, observe that the amount of fencing is $2x + y$ and this must be equal to 3000 since all the fencing material is used; that is,

$$2x + y = 3000$$

From the equation we see that $y = 3000 - 2x$. Substituting this value of y into the expression for A gives

$$A = xy = x(3000 - 2x) = 3000x - 2x^2$$

Finally, observe that both x and y must be nonnegative since they represent the width and length of a rectangle, respectively. Thus, $x \geq 0$ and $y \geq 0$. But the latter is equivalent to $3000 - 2x \geq 0$, or $x \leq 1500$. So the required function is $f(x) = 3000x - 2x^2$ with domain $0 \leq x \leq 1500$. ■■■■

REMARK Observe that if we view the function $f(x) = 3000x - 2x^2$ strictly as a mathematical entity, then its domain is the set of all real numbers. But physical consideration dictates that its domain should be restricted to the interval $[0, 1500]$. ■■■

EXAMPLE 6 If exactly 200 people sign up for a charter flight, the Leisure World Travel Agency charges $300 per person. However, if more than 200 people sign up for the flight (assume this is the case), then each fare is reduced by $1 for each additional person. Letting x denote the number of passengers above 200, find a function giving the revenue realized by the company.

SOLUTION ✔ If there are x passengers above 200, then the number of passengers signing up for the flight is $200 + x$. Furthermore, the fare will be $\$(300 - x)$ per passenger. Therefore, the revenue will be

$$R = (200 + x)(300 - x) \qquad \text{(Number of passengers times the fare}$$
$$= -x^2 + 100x + 60{,}000 \qquad \text{per passenger)}$$

Clearly, x must be nonnegative, and $300 - x \geq 0$, or $x \leq 300$. So the required function is $f(x) = -x^2 + 100x + 60{,}000$ with domain $[0, 300]$. ■■■■

SELF-CHECK EXERCISES 2.3

1. Thomas Young has suggested the following rule for calculating the dosage of medicine for children from ages 1 to 12 years. If a denotes the adult dosage (in milligrams) and t is the age of the child (in years), then the child's dosage is given by

$$D(t) = \frac{at}{t + 12}$$

   If the adult dose of a substance is 500 mg, how much should a 4-yr-old child receive?

2. The demand function for Mrs. Baker's cookies is given by

$$d(x) = -\frac{2}{15}x + 4$$

   where $d(x)$ is the wholesale price in dollars per pound and x is the quantity demanded each week, measured in thousands of pounds. The supply function for the cookies is given by

$$s(x) = \frac{1}{75}x^2 + \frac{1}{10}x + \frac{3}{2}$$

   where $s(x)$ is the wholesale price in dollars per pound and x is the quantity, in thousands of pounds, that will be made available in the market per week by the supplier.

   a. Sketch the graphs of the functions d and s.
   b. Find the equilibrium quantity and price.

   Solutions to Self-Check Exercises 2.3 can be found on page 108.

2.3 Exercises

In Exercises 1–8, determine whether the equation defines y as a linear function of x. If so, write it in the form $y = mx + b$.

1. $2x + 3y = 6$

2. $-2x + 4y = 7$

3. $x = 2y - 4$

4. $2x = 3y + 8$

5. $2x - 4y + 9 = 0$

6. $3x - 6y + 7 = 0$

7. $2x^2 - 8y + 4 = 0$

8. $3\sqrt{x} + 4y = 0$

In Exercises 9–14, determine whether the given function is a polynomial function, a rational function, or some other function. State the degree of each polynomial function.

9. $f(x) = 3x^6 - 2x^2 + 1$

10. $f(x) = \dfrac{x^2 - 9}{x - 3}$

11. $G(x) = 2(x^2 - 3)^3$

12. $H(x) = 2x^{-3} + 5x^{-2} + 6$

13. $f(t) = 2t^2 + 3\sqrt{t}$

14. $f(r) = \dfrac{6r}{(r^3 - 8)}$

15. Find the constants m and b in the linear function $f(x) = mx + b$ so that $f(0) = 2$ and $f(3) = -1$.

16. Find the constants m and b in the linear function $f(x) = mx + b$ so that $f(2) = 4$ and the straight line represented by f has slope -1.

17. A manufacturer has a monthly fixed cost of $40,000 and a production cost of $8 for each unit produced. The product sells for $12/unit.
   a. What is the cost function?
   b. What is the revenue function?
   c. What is the profit function?
   d. Compute the profit (loss) corresponding to production levels of 8000 and 12,000 units.

18. A manufacturer has a monthly fixed cost of $100,000 and a production cost of $14 for each unit produced. The product sells for $20/unit.
   a. What is the cost function?
   b. What is the revenue function?
   c. What is the profit function?
   d. Compute the profit (loss) corresponding to production levels of 12,000 and 20,000 units.

19. DISPOSABLE INCOME Economists define the *disposable annual income* for an individual by the equation $D = (1 - r)T$, where T is the individual's total income and r is the net rate at which he or she is taxed. What is the disposable income for an individual whose income is $40,000 and whose net tax rate is 28%?

20. DRUG DOSAGES A method sometimes used by pediatricians to calculate the dosage of medicine for children is based on the child's surface area. If a denotes the adult dosage (in milligrams) and S is the surface area of the child (in square meters), then the child's dosage is given by

$$D(S) = \frac{Sa}{1.7}$$

If the adult dose of a substance is 500 mg, how much should a child whose surface area is 0.4 m² receive?

21. COWLING'S RULE Cowling's rule is a method for calculating pediatric drug dosages. If a denotes the adult dosage (in milligrams) and t is the age of the child (in years), then the child's dosage is given by

$$D(t) = \left(\frac{t+1}{24}\right)a$$

If the adult dose of a substance is 500 mg, how much should a 4-yr-old child receive?

22. WORKER EFFICIENCY An efficiency study showed that the average worker at Delphi Electronics assembled cordless telephones at the rate of

$$f(t) = -\frac{3}{2}t^2 + 6t + 10 \qquad (0 \le t \le 4)$$

phones/hour, t hr after starting work during the morning shift. At what rate does the average worker assemble telephones 2 hr after starting work?

23. REVENUE FUNCTIONS The revenue (in dollars) realized by Apollo, Inc., from the sale of its inkjet printers is given by

$$R(x) = -0.1x^2 + 500x$$

where x denotes the number of units manufactured per month. What is Apollo's revenue when 1000 units are produced?

24. EFFECT OF ADVERTISING ON SALES The quarterly profit of Cunningham Realty depends on the amount of money x spent on advertising per quarter according to the rule

$$P(x) = -\frac{1}{8}x^2 + 7x + 30 \qquad (0 \le x \le 50)$$

where $P(x)$ and x are measured in thousands of dollars. What is Cunningham's profit when its quarterly advertising budget is $28,000?

25. E-MAIL USAGE The number of international e-mailings per day (in millions) is approximated by the function

$$f(t) = 38.57t^2 - 24.29t + 79.14 \qquad (0 \le t \le 4)$$

where t is measured in years with $t = 0$ corresponding to the beginning of 1998.
   a. Sketch the graph of f.
   b. How many international e-mailings per day were there at the beginning of the year 2000?
   Source: Pioneer Consulting

26. DOCUMENT MANAGEMENT The size (measured in millions of dollars) of the document-management business is described by the function

$$f(t) = 0.22t^2 + 1.4t + 3.77 \qquad (0 \le t \le 6)$$

where t is measured in years with $t = 0$ corresponding to the beginning of 1996.
   a. Sketch the graph of f.
   b. What was the size of the document-management business at the beginning of the year 2000?
   Source: Sun Trust Equitable Securities

27. REACTION OF A FROG TO A DRUG Experiments conducted by A. J. Clark suggest that the response $R(x)$ of a frog's heart muscle to the injection of x units of acetylcholine (as a percentage of the maximum possible effect of the drug) may be approximated by the rational function

$$R(x) = \frac{100x}{b + x} \qquad (x \ge 0)$$

where b is a positive constant that depends on the particular frog.
   a. If a concentration of 40 units of acetylcholine produces a response of 50% for a certain frog, find the "response function" for this frog.
   b. Using the model found in part (a), find the response of the frog's heart muscle when 60 units of acetylcholine are administered.

28. FORECASTING SALES The annual sales of the Crimson Drug Store are expected to be given by

$$S(t) = 2.3 + 0.4t$$

million dollars t yr from now, whereas the annual sales of the Cambridge Drug Store are expected to be given by

$$S(t) = 1.2 + 0.6t$$

million dollars t yr from now. When will the annual sales of the Cambridge Drug Store first surpass the annual sales of the Crimson Drug Store?

29. **CRICKET CHIRPING AND TEMPERATURE** Entomologists have discovered that a linear relationship exists between the number of chirps of crickets of a certain species and the air temperature. When the temperature is 70°F, the crickets chirp at the rate of 120 times/minute, and when the temperature is 80°F, they chirp at the rate of 160 times/minute.
   a. Find an equation giving the relationship between the air temperature T and the number of chirps per minute, N, of the crickets.
   b. Find N as a function of T and use this formula to determine the rate at which the crickets chirp when the temperature is 102°F.

30. **LINEAR DEPRECIATION** In computing income tax, business firms are allowed by law to depreciate certain assets such as buildings, machines, furniture, automobiles, and so on, over a period of time. The linear depreciation, or straight-line method, is often used for this purpose. Suppose an asset has an initial value of $\$C$ and is to be depreciated linearly over n years with a scrap value of $\$S$. Show that the book value of the asset at any time t ($0 \le t \le n$) is given by the linear function

$$V(t) = C - \frac{(C - S)}{n}t$$

   Hint: Find an equation of the straight line that passes through the points $(0, C)$ and (n, S). Then rewrite the equation in the slope-intercept form.

31. **LINEAR DEPRECIATION** Using the linear depreciation model of Exercise 30, find the book value of a printing machine at the end of the second year if its initial value is $\$100,000$ and it is depreciated linearly over 5 years with a scrap value of $\$30,000$.

32. **PRICE OF IVORY** According to the World Wildlife Fund, a group in the forefront of the fight against illegal ivory trade, the price of ivory (in dollars per kilo) compiled from a variety of legal and black market sources is approximated by the function

$$f(t) = \begin{cases} 8.37t + 7.44 & \text{if } 0 \le t \le 8 \\ 2.84t + 51.68 & \text{if } 8 < t \le 30 \end{cases}$$

where t is measured in years and $t = 0$ corresponds to the beginning of 1970.

a. Sketch the graph of the function f.
b. What was the price of ivory at the beginning of 1970? At the beginning of 1990?

33. **SALES OF DIGITAL TVs** The number of homes with digital TVs is expected to grow according to the function

$$f(t) = 0.1714t^2 + 0.6657t + 0.7143 \qquad (0 \le t \le 6)$$

where t is measured in years with $t = 0$ corresponding to the beginning of the year 2000 and $f(t)$ is measured in millions of homes.
   a. How many homes had digital TVs at the beginning of the year 2000?
   b. How many homes will have digital TVs at the beginning of 2005?
   Source: Consumer Electronics Manufacturers Association

34. **SENIOR CITIZENS' HEALTH CARE** According to a study conducted for the Senate Select Committee on Aging, the out-of-pocket cost to senior citizens for health care, $f(t)$ (as a percentage of income), in year t where $t = 0$ corresponds to 1977, is given by

$$y = \begin{cases} \frac{2}{7}t + 12 & \text{if } 0 \le t \le 7 \\ t + 7 & \text{if } 7 < t \le 10 \\ \frac{3}{5}t + 11 & \text{if } 10 < t < 20 \end{cases}$$

   a. Sketch the graph of f.
   b. What was the out-of-pocket cost to senior citizens for health care in 1982? In 1992?
   Source: Senate Select Committee on Aging, AARP

35. **PRICE OF AUTOMOBILE PARTS** For years, automobile manufacturers had a monopoly on the replacement-parts market, particularly for sheet metal parts such as fenders, doors, and hoods, the parts most often damaged in a crash. Beginning in the late 1970s, however, competition appeared on the scene. In a report conducted by an insurance company to study the effects of the competition, the price of an OEM (original equipment manufacturer) fender for a particular 1983 model car was found to be

$$f(t) = \frac{110}{\frac{1}{2}t + 1} \qquad (0 \le t \le 2)$$

where $f(t)$ is measured in dollars and t is in years. Over the same period of time, the price of a non-OEM fender for the car was found to be

$$g(t) = 26\left(\frac{1}{4}t^2 - 1\right)^2 + 52 \qquad (0 \le t \le 2)$$

where $g(t)$ is also measured in dollars. Find a function $h(t)$ that gives the difference in price between an OEM fender and a non-OEM fender. Compute $h(0)$, $h(1)$, and $h(2)$. What does the result of your computation seem to say about the price gap between OEM and non-OEM fenders over the 2 yr?

For the demand equations in Exercises 36–39, where x represents the quantity demanded in units of a thousand and p is the unit price in dollars, (a) sketch the demand curve and (b) determine the quantity demanded when the unit price is set at \$p.

36. $p = -x^2 + 36$; $p = 11$ **37.** $p = -x^2 + 16$; $p = 7$

38. $p = \sqrt{9 - x^2}$; $p = 2$ **39.** $p = \sqrt{18 - x^2}$; $p = 3$

For the supply equations in Exercises 40–43, where x is the quantity supplied in units of a thousand and p is the unit price in dollars, (a) sketch the supply curve and (b) determine the price at which the supplier will make 2000 units of the commodity available in the market.

40. $p = 2x^2 + 18$ **41.** $p = x^2 + 16x + 40$

42. $p = x^3 + x + 10$ **43.** $p = x^3 + 2x + 3$

44. DEMAND FOR CLOCK RADIOS   In the accompanying figure, L_1 is the demand curve for the model A clock radios manufactured by Ace Radio, Inc., and L_2 is the demand curve for their model B clock radios. Which line has the greater slope? Interpret your results.

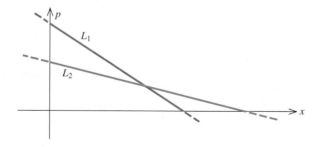

45. SUPPLY OF CLOCK RADIOS   In the accompanying figure, L_1 is the supply curve for the model A clock radios manufactured by Ace Radio, Inc., and L_2 is the supply curve for their model B clock radios. Which line has the greater slope? Interpret your results.

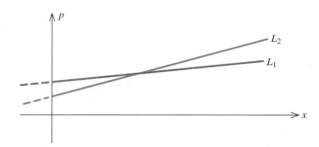

46. DEMAND FOR SMOKE ALARMS   The demand function for the Sentinel smoke alarm is given by

$$p = \frac{30}{0.02x^2 + 1} \qquad (0 \le x \le 10)$$

where x (measured in units of a thousand) is the quantity demanded per week and p is the unit price in dollars. Sketch the graph of the demand function. What is the unit price that corresponds to a quantity demanded of 10,000 units?

47. DEMAND FOR COMMODITIES   Assume that the demand function for a certain commodity has the form

$$p = \sqrt{-ax^2 + b} \qquad (a \ge 0, b \ge 0)$$

where x is the quantity demanded, measured in units of a thousand, and p is the unit price in dollars. Suppose the quantity demanded is 6000 ($x = 6$) when the unit price is \$8 and 8000 ($x = 8$) when the unit price is \$6. Determine the demand equation. What is the quantity demanded when the unit price is set at \$7.50?

48. SUPPLY FUNCTIONS   The supply function for the Luminar desk lamp is given by

$$p = 0.1x^2 + 0.5x + 15$$

where x is the quantity supplied (in thousands) and p is the unit price in dollars. Sketch the graph of the supply function. What unit price will induce the supplier to make 5000 lamps available in the marketplace?

49. SUPPLY FUNCTIONS   Suppliers of transistor radios will market 10,000 units when the unit price is \$20 and 62,500 units when the unit price is \$35. Determine the supply function if it is known to have the form

$$p = a\sqrt{x} + b \qquad (a \ge 0, b \ge 0)$$

where x is the quantity supplied and p is the unit price in dollars. Sketch the graph of the supply function. What unit price will induce the supplier to make 40,000 transistor radios available in the marketplace?

50. Suppose the demand and supply equations for a certain commodity are given by $p = ax + b$ and $p = cx + d$, respectively, where $a < 0$, $c > 0$, and $b > d > 0$ (see the accompanying figure).
a. Find the equilibrium quantity and equilibrium price in terms of a, b, c, and d.
b. Use part (a) to determine what happens to the market equilibrium if c is increased while a, b, and d remain fixed. Interpret your answer in economic terms.
c. Use part (a) to determine what happens to the market equilibrium if b is decreased while a, c, and d remain fixed. Interpret your answer in economic terms.

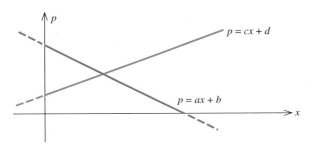

For each pair of supply and demand equations in Exercises 51–54, where x represents the quantity demanded in units of a thousand and p the unit price in dollars, find the equilibrium quantity and the equilibrium price.

51. $p = -2x^2 + 80$ and $p = 15x + 30$

52. $p = -x^2 - 2x + 100$ and $p = 8x + 25$

53. $11p + 3x - 66 = 0$ and $2p^2 + p - x = 10$

54. $p = 60 - 2x^2$ and $p = x^2 + 9x + 30$

55. MARKET EQUILIBRIUM   The weekly demand and supply functions for Sportsman 5×7 tents are given by

$$p = -0.1x^2 - \ x + 40$$
$$p = \ \ 0.1x^2 + 2x + 20$$

respectively, where p is measured in dollars and x is measured in units of a hundred. Find the equilibrium quantity and price.

56. MARKET EQUILIBRIUM   The management of the Titan Tire Company has determined that the weekly demand and supply functions for their Super Titan tires are given by

$$p = 144 - \ x^2$$
$$p = \ \ 48 + \frac{1}{2}x^2$$

respectively, where p is measured in dollars and x is measured in units of a thousand. Find the equilibrium quantity and price.

57. WALKING VERSUS RUNNING   The oxygen consumption (in milliliter/pound/minute) for a person walking at x mph is approximated by the function

$$f(x) = \frac{5}{3}x^2 + \frac{5}{3}x + 10 \qquad (0 \le x \le 9)$$

where the oxygen consumption for a runner at x mph is approximated by the function

$$g(x) = 11x + 10 \qquad (4 \le x \le 9)$$

a. Sketch the graphs of f and g.
b. At what speed is the oxygen consumption the same for a walker as it is for a runner? What is the level of oxygen consumption at that speed?
c. What happens to the oxygen consumption of the walker and the runner at speeds beyond that found in part (b)?
Source: Exercise Physiology, by William McArdley, Frank Katch, and Victor Katch

58. ENCLOSING AN AREA   Patricia wishes to have a rectangular-shaped garden in her backyard. She has 80 ft of fencing material with which to enclose her garden. Letting x denote the width of the garden, find a function f in the variable x giving the area of the garden. What is its domain?

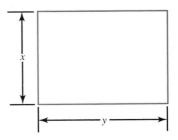

59. ENCLOSING AN AREA   Patricia's neighbor, Juanita, also wishes to have a rectangular-shaped garden in her backyard. But Juanita wants her garden to have an area of 250 ft². Letting x denote the width of the garden, find a function f in the variable x giving the length of the fencing material required to construct the garden. What is the domain of the function?
Hint: Refer to the figure for Exercise 58. The amount of fencing material required is equal to the perimeter of the rectangle, which is twice the width plus twice the length of the rectangle.

(continued on p. 108)

FINDING THE POINTS OF INTERSECTION OF TWO GRAPHS AND MODELING

A graphing utility can be used to find the point(s) of intersection of the graphs of two functions.

EXAMPLE 1

Find the points of intersection of the graphs of

$$f(x) = 0.3x^2 - 1.4x - 3 \quad \text{and} \quad g(x) = -0.4x^2 + 0.8x + 6.4$$

SOLUTION ✔

The graphs of both f and g in the standard viewing rectangle are shown in Figure T1. Using **TRACE** and **ZOOM** or the function for finding the points of intersection of two graphs on your graphing utility, we find the point(s) of intersection, accurate to four decimal places, to be $(-2.4158, 2.1329)$ and $(5.5587, -1.5125)$.

FIGURE T1

The graphs of f and g in the standard viewing window

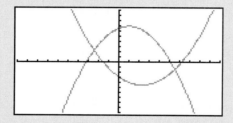

EXAMPLE 2

Consider the demand and supply functions

$$p = d(x) = -0.01x^2 - 0.2x + 8 \quad \text{and} \quad p = s(x) = 0.01x^2 + 0.1x + 3$$

of Example 4 in Section 2.3.

a. Plot the graphs of d and s in the viewing rectangle $[0, 15] \times [0, 10]$.
b. Verify that the equilibrium point is $(10, 5)$, as obtained in Example 4.

SOLUTION ✔

a. The graphs of d and s are shown in Figure T2.

FIGURE T2

The graphs of d and s in the window $[0, 15] \times [0, 10]$

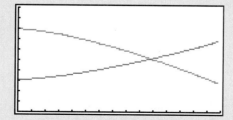

b. Using **TRACE** and **ZOOM** or the function for finding the point of intersection of two graphs, we see that $x = 10$ and $y = 5$, so the equilibrium point is $(10, 5)$, as obtained before.

Constructing Mathematical Models from Raw Data

A graphing utility can sometimes be used to construct mathematical models from sets of data. For example, if the points corresponding to the given data are scattered about a straight line, then one uses **LINR** (linear regression) from the **STAT CALC** (statistical calculation) menu of the graphing utility to obtain a function (model) that approximates the data at hand. If the points seem to be scattered along a parabola (the graph of a quadratic function), then one uses **P2REG** (second-order polynomial regression), and so on.

Details for using the items in the STAT CALC menu can be found at the Web site:
http://www.brookscole.com/product/0534378439

EXAMPLE 3

Indian Gaming Industry

The following data gives the estimated gross revenues (in billions of dollars) from the Indian gaming industries from 1990 ($t = 0$) to 1997 ($t = 7$).

Year	0	1	2	3	4	5	6	7
Revenue	0.5	0.7	1.6	2.6	3.4	4.8	5.6	6.8

a. Use a graphing utility to find a polynomial function f of degree 4 that models the data.
b. Plot the graph of the function f, using the viewing rectangle $[0, 8] \times [0, 10]$.
c. Use the function evaluation capability of the graphing utility to compute $f(0), f(1), \ldots, f(7)$ and compare these values with the original data.

Source: Christiansen/Cummings Associates

SOLUTION ✔

a. Choosing **P4REG** (fourth-order polynomial regression) from the **STAT CALC** (statistical calculations) menu of a graphing utility, we find

$$f(t) = 0.00379t^4 - 0.06616t^3 + 0.41667t^2 - 0.07291t + 0.48333$$

b. The graph of f is shown in Figure T3.

FIGURE T3
The graph of f in the viewing rectangle $[0, 8] \times [0, 10]$

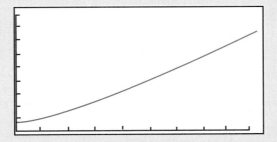

c. The required values, which compare favorably with the given data, follow:

t	0	1	2	3	4	5	6	7
$f(t)$	0.5	0.8	1.5	2.5	3.6	4.6	5.7	6.8

Exercises

In Exercises 1–6, find the points of intersection of the graphs of the functions. Express your answer accurate to four decimal places.

1. $f(x) = 1.2x + 3.8; g(x) = -0.4x^2 + 1.2x + 7.5$

2. $f(x) = 0.2x^2 - 1.3x - 3; g(x) = -1.3x + 2.8$

3. $f(x) = 0.3x^2 - 1.7x - 3.2; g(x) = -0.4x^2 + 0.9x + 6.7$

4. $f(x) = -0.3x^2 + 0.6x + 3.2; g(x) = 0.2x^2 - 1.2x - 4.8$

5. $f(x) = 0.3x^3 - 1.8x^2 + 2.1x - 2; g(x) = 2.1x - 4.2$

6. $f(x) = -0.2x^3 + 1.2x^2 - 1.2x + 2; g(x) = -0.2x^2 + 0.8x + 2.1$

7. The monthly demand and supply functions for a certain brand of wall clock are given by

$$p = -0.2x^2 - 1.2x + 50$$
$$p = 0.1x^2 + 3.2x + 25$$

respectively, where p is measured in dollars and x is measured in units of a hundred.

a. Plot the graphs of both functions in an appropriate viewing rectangle.

b. Find the equilibrium quantity and price.

8. The quantity demanded x (in units of a hundred) of Mikado miniature cameras per week is related to the unit price p (in dollars) by

$$p = -0.2x^2 + 80$$

The quantity x (in units of a hundred) that the supplier is willing to make available in the market is related to the unit price p (in dollars) by

$$p = 0.1x^2 + x + 40$$

a. Plot the graphs of both functions in an appropriate viewing rectangle.

b. Find the equilibrium quantity and price.

In Exercises 9–14, use the STAT CALC menu of a graphing utility to construct a mathematical model associated with given data.

9. SALES OF DIGITAL SIGNAL PROCESSORS The projected sales (in billions of dollars) of digital signal processors (DSPs) follow:

Year	1997	1998	1999	2000	2001	2002
Sales	3.1	4	5	6.2	8	10

a. Use **P2REG** to find a second-degree polynomial regression model for the data. Let $t = 0$ correspond to 1997.

b. Plot the graph of the function f found in part (a), using the viewing rectangle $[0, 5] \times [0, 12]$.

c. Compute the values of $f(t)$ for $t = 0, 1, 2, 3, 4$, and 5. How does your model compare with the given data?

Source: A. G. Edwards & Sons, Inc.

10. PRISON POPULATION The following data gives the past, present, and projected U.S. prison population (in millions) from 1980 through 2005.

Year	1980	1985	1990	1995	2000	2005
Population	0.52	0.77	1.18	1.64	2.23	3.20

a. Letting $t = 0$ correspond to the beginning of 1980 and supposing t is measured in 5-yr intervals, use **P2REG** to find a second-degree polynomial regression model based on the given data.

b. Plot the graph of the function f found in part (a), using the viewing rectangle $[0, 5] \times [0, 3.5]$.

c. Compute $f(0), f(1), f(2), f(3), f(4)$, and $f(5)$. Compare these values with the given data.

11. DIGITAL TV SHIPMENTS The estimated number of digital TV shipments between the year 2000 and 2006 (in millions of units) is given in the following table:

Year	2000	2001	2002	2003	2004	2005	2006
Units Shipped	0.63	1.43	2.57	4.1	6	8.1	10

a. Use **P3REG** to find a third-degree polynomial regression model for the data. Let $t = 0$ correspond to the year 2000.
b. Plot the graph of the function f found in part (a), using the viewing rectangle $[0, 6] \times [0, 11]$.
c. Compute the values of $f(t)$ for $t = 0, 1, 2, 3, 4, 5,$ and 6.

Source: Consumer Electronics Manufacturers Association

12. **ON-LINE SHOPPING** The following data gives the revenue per year (in billions of dollars) from Internet shopping.

Year	1997	1998	1999	2000	2001
Revenue	2.4	5	8	12	17.4

a. Use **P3REG** to find a third-degree polynomial regression model for the data. Let $t = 0$ correspond to 1997.
b. Plot the graph of the function f found in part (a), using the viewing rectangle $[0, 4] \times [0, 20]$.
c. Compare the values of f at $t = 0, 1, 2, 3,$ and 4 with the given data.

Source: Forrester Research, Inc.

13. **CABLE AD REVENUE** The past and projected revenues (in billions of dollars) from cable advertisement for the years 1995 through the year 2000 follow:

Year	1995	1996	1997	1998	1999	2000
Ad Revenue	5.1	6.6	8.1	9.4	11.1	13.7

a. Use **P3REG** to find a third-degree polynomial regression model for the data. Let $t = 0$ correspond to 1995.

b. Plot the graph of the function f found in part (a), using the viewing rectangle $[0, 6] \times [0, 14]$.
c. Compare the values of f at $t = 1, 2, 3, 4,$ and 5 with the given data.

Source: National Cable Television Association

14. **ON-LINE SPENDING** The following data gives the worldwide spending and projected spending (in billions of dollars) on the Web from 1997 through 2002.

Year	1997	1998	1999	2000	2001	2002
Spending	5.0	10.5	20.5	37.5	60	95

a. Choose **P4REG** to find a fourth-degree polynomial regression model for the data. Let $t = 0$ correspond to 1997.
b. Plot the graph of the function f found in part (a), using the viewing rectangle $[0, 5] \times [0, 100]$.

Source: International Data Corporation

15. **MARIJUANA ARRESTS** The number of arrests (in thousands) for marijuana sales and possession in New York City from 1992 through 1997 is given below.

Year	1992	1993	1994	1995	1996	1997
No. of Arrests	5.0	5.8	8.8	11.7	18.5	27.5

a. Use **P4REG** to find a fourth-degree polynomial regression model for the data. Let $t = 0$ correspond to 1992.
b. Plot the graph of the function f found in part (a), using the viewing rectangle $[0, 5] \times [0, 30]$.
c. Compare the values of f at $t = 0, 1, 2, 3, 4,$ and 5 with the given data.

Source: New York State Division of Criminal Justice Services

are shown in Figure 2.27a–b. Observe that the graphs are identical except when $x = 2$. The function g is defined for all values of x and, in particular, its value at $x = 2$ is $g(2) = 4(2 + 2) = 16$. Thus, the point $(2, 16)$ is on the graph of g. However, the function f is not defined at $x = 2$. Since $f(x) = g(x)$ for all values of x except $x = 2$, it follows that the graph of f must look exactly like the graph of g, with the exception that the point $(2, 16)$ is missing from the graph of f. This illustrates graphically why we can evaluate the limit of f by evaluating the limit of the "equivalent" function g.

FIGURE 2.27
The graphs of $f(x)$ and $g(x)$ are identical except at the point $(2, 16)$.

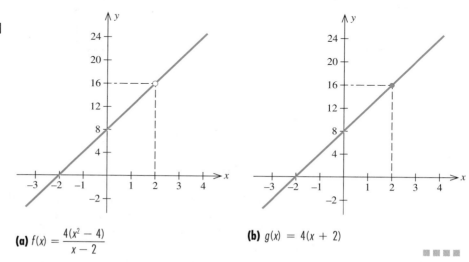

(a) $f(x) = \dfrac{4(x^2 - 4)}{x - 2}$

(b) $g(x) = 4(x + 2)$

REMARK Notice that the limit in Example 5 is the same limit that we evaluated earlier when we discussed the instantaneous velocity of a maglev at a specified time. ■ ■ ■

Exploring with Technology

1. Use a graphing utility to plot the graph of

$$f(x) = \frac{4(x^2 - 4)}{x - 2}$$

in the viewing rectangle $[0, 3] \times [0, 20]$. Then use **ZOOM** and **TRACE** to find

$$\lim_{x \to 2} \frac{4(x^2 - 4)}{x - 2}$$

2. Use a graphing utility to plot the graph of $g(x) = 4(x + 2)$ in the viewing rectangle $[0, 3] \times [0, 20]$. Then use **ZOOM** and **TRACE** to find $\lim_{x \to 2} 4(x + 2)$. What happens to the y-value when x takes on the value 2? Explain.

3. Can you distinguish between the graphs of f and g?

4. Reconcile your results with those of Example 5.

EXAMPLE 6   Evaluate:

$$\lim_{h \to 0} \frac{\sqrt{1+h} - 1}{h}$$

SOLUTION ✔   Letting h approach zero, we obtain the indeterminate form 0/0. Next, we rationalize the numerator of the quotient (see page 25) by multiplying both the numerator and the denominator by the expression $(\sqrt{1+h} + 1)$, obtaining

$$\frac{\sqrt{1+h} - 1}{h} = \frac{(\sqrt{1+h} - 1)(\sqrt{1+h} + 1)}{h(\sqrt{1+h} + 1)}$$

$$= \frac{1 + h - 1}{h(\sqrt{1+h} + 1)} \qquad [(\sqrt{a} - \sqrt{b})(\sqrt{a} + \sqrt{b}) = a - b]$$

$$= \frac{h}{h(\sqrt{1+h} + 1)}$$

$$= \frac{1}{\sqrt{1+h} + 1}$$

Therefore,

$$\lim_{h \to 0} \frac{\sqrt{1+h} - 1}{h} = \lim_{h \to 0} \frac{1}{\sqrt{1+h} + 1}$$

$$= \frac{1}{\sqrt{1} + 1} = \frac{1}{2}$$ ■ ■ ■ ■

Exploring with Technology

1. Use a graphing utility to plot the graph of

$$g(x) = \frac{\sqrt{1+x} - 1}{x}$$

in the viewing rectangle $[-1, 2] \times [0, 1]$. Then use **ZOOM** and **TRACE** to find

$$\lim_{x \to 0} \frac{\sqrt{1+x} - 1}{x}$$

by observing the values of $g(x)$ as x approaches 0 from the left and from the right.
2. Use a graphing utility to plot the graph of

$$f(x) = \frac{1}{\sqrt{1+x} + 1}$$

in the viewing rectangle $[-1, 2] \times [0, 1]$. Then use **ZOOM** and **TRACE** to find

$$\lim_{x \to 0} \frac{1}{\sqrt{1+x} + 1}$$

What happens to the y-value when x takes on the value 0? Explain.
3. Can you distinguish between the graphs of f and g?
4. Reconcile your results with those of Example 6.

LIMITS AT INFINITY

Up to now we have studied the limit of a function as x approaches a (finite) number a. There are occasions, however, when we want to know whether $f(x)$ approaches a unique number as x increases without bound. Consider, for example, the function P, giving the number of fruit flies (*Drosophila*) in a container under controlled laboratory conditions, as a function of a time t. The graph of P is shown in Figure 2.28. You can see from the graph of P that, as t increases without bound (gets larger and larger), $P(t)$ approaches the number 400. This number, called the *carrying capacity* of the environment, is determined by the amount of living space and food available, as well as other environmental factors.

FIGURE 2.28
The graph of $P(t)$ gives the population of fruit flies in a laboratory experiment.

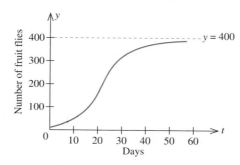

As another example, suppose we are given the function

$$f(x) = \frac{2x^2}{1 + x^2}$$

and we want to determine what happens to $f(x)$ as x gets larger and larger. Picking the sequence of numbers 1, 2, 5, 10, 100, and 1000 and computing the corresponding values of $f(x)$, we obtain the following table of values:

x	1	2	5	10	100	1000
$f(x)$	1	1.6	1.92	1.98	1.9998	1.999998

From the table, we see that as x gets larger and larger, $f(x)$ gets closer and closer to 2. The graph of the function f shown in Figure 2.29 confirms this observation. We call the line $y = 2$ a **horizontal asymptote.***

FIGURE 2.29

The graph of $y = \dfrac{2x^2}{1 + x^2}$ has a horizontal asymptote at $y = 2$.

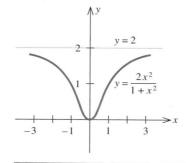

* We will discuss asymptotes in greater detail in Section 4.3.

In this situation we say that the limit of the function

$$f(x) = \frac{2x^2}{1 + x^2}$$

as x increases without bound is 2, written

$$\lim_{x \to \infty} \frac{2x^2}{1 + x^2} = 2$$

In the general case, the following definition is applicable:

Limit of a Function at Infinity

The function f has the limit L as x increases without bound (or, as x approaches infinity), written

$$\lim_{x \to \infty} f(x) = L$$

if $f(x)$ can be made arbitrarily close to L by taking x large enough.
Similarly, the function f has the limit M as x decreases without bound (or as x approaches negative infinity), written

$$\lim_{x \to -\infty} f(x) = M$$

if $f(x)$ can be made arbitrarily close to M by taking x to be negative and sufficiently large in absolute value.

EXAMPLE 7   Let f and g be the functions

$$f(x) = \begin{cases} -1 & \text{if } x < 0 \\ 1 & \text{if } x \geq 0 \end{cases} \quad \text{and} \quad g(x) = \frac{1}{x^2}$$

Evaluate:

a. $\lim\limits_{x \to \infty} f(x)$ and $\lim\limits_{x \to -\infty} f(x)$ **b.** $\lim\limits_{x \to \infty} g(x)$ and $\lim\limits_{x \to -\infty} g(x)$

SOLUTION ✔   The graphs of $f(x)$ and $g(x)$ are shown in Figure 2.30. Referring to the graphs of the respective functions, we see that

a. $\lim\limits_{x \to \infty} f(x) = 1$ and $\lim\limits_{x \to -\infty} f(x) = -1$

b. $\lim\limits_{x \to \infty} \frac{1}{x^2} = 0$ and $\lim\limits_{x \to -\infty} \frac{1}{x^2} = 0$

FIGURE 2.30

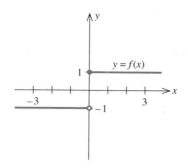

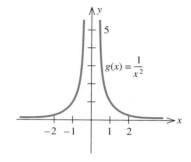

(a) $\lim\limits_{x \to \infty} f(x) = 1$ and $\lim\limits_{x \to -\infty} f(x) = -1$ **(b)** $\lim\limits_{x \to \infty} g(x) = 0$ and $\lim\limits_{x \to -\infty} g(x) = 0$

All the properties of limits listed in Theorem 1 are valid when a is replaced by ∞ or $-\infty$. In addition, we have the following property for the limit at infinity.

THEOREM 2

For all $n > 0$,

$$\lim_{x \to \infty} \frac{1}{x^n} = 0 \quad \text{and} \quad \lim_{x \to -\infty} \frac{1}{x^n} = 0$$

provided that $\dfrac{1}{x^n}$ is defined.

Exploring with Technology

1. Use a graphing utility to plot the graphs of

$$y_1 = \frac{1}{x^{0.5}}, \quad y_2 = \frac{1}{x}, \quad \text{and} \quad y_3 = \frac{1}{x^{1.5}}$$

in the viewing rectangle $[0, 200] \times [0, 0.5]$. What can you say about $\lim\limits_{x \to \infty} \dfrac{1}{x^n}$ if $n = 0.5$, $n = 1$, and $n = 1.5$? Are these results predicted by Theorem 2?

2. Use a graphing utility to plot the graphs of

$$y_1 = \frac{1}{x} \quad \text{and} \quad y_2 = \frac{1}{x^{5/3}}$$

in the viewing rectangle $[-50, 0] \times [-0.5, 0]$. What can you say about $\lim\limits_{x \to -\infty} \dfrac{1}{x^n}$ if $n = 1$ and $n = \dfrac{5}{3}$?

Are these results predicted by Theorem 2?

Hint: To graph y_2, write it in the form $y2 = 1/(x^{(1/3)})^5$.

In evaluating the limit at infinity of a rational function, the following technique is often used: *Divide the numerator and denominator of the expression by x^n, where n is the highest power present in the denominator of the expression.*

EXAMPLE 8   Evaluate:

$$\lim_{x \to \infty} \frac{x^2 - x + 3}{2x^3 + 1}$$

SOLUTION ✔   Since the limits of both the numerator and the denominator do not exist as x approaches infinity, the property pertaining to the limit of a quotient (Property 5) is not applicable. Let us divide the numerator and denominator of the rational expression by x^3, obtaining

$$\lim_{x \to \infty} \frac{x^2 - x + 3}{2x^3 + 1} = \lim_{x \to \infty} \frac{\dfrac{1}{x} - \dfrac{1}{x^2} + \dfrac{3}{x^3}}{2 + \dfrac{1}{x^3}}$$

$$= \frac{0 - 0 + 0}{2 + 0} = \frac{0}{2} \qquad \text{(Using Theorem 2)}$$

$$= 0$$

■ ■ ■ ■

EXAMPLE 9   Let

$$f(x) = \frac{3x^2 + 8x - 4}{2x^2 + 4x - 5}$$

Compute $\lim_{x \to \infty} f(x)$ if it exists.

SOLUTION ✔   Again, we see that Property 5 is not applicable. Dividing the numerator and the denominator by x^2, we obtain

$$\lim_{x \to \infty} \frac{3x^2 + 8x - 4}{2x^2 + 4x - 5} = \lim_{x \to \infty} \frac{3 + \dfrac{8}{x} - \dfrac{4}{x^2}}{2 + \dfrac{4}{x} - \dfrac{5}{x^2}}$$

$$= \frac{\lim\limits_{x \to \infty} 3 + 8 \lim\limits_{x \to \infty} \dfrac{1}{x} - 4 \lim\limits_{x \to \infty} \dfrac{1}{x^2}}{\lim\limits_{x \to \infty} 2 + 4 \lim\limits_{x \to \infty} \dfrac{1}{x} - 5 \lim\limits_{x \to \infty} \dfrac{1}{x^2}}$$

$$= \frac{3 + 0 - 0}{2 + 0 - 0}$$

$$= \frac{3}{2} \qquad \text{(Using Theorem 2)}$$

■ ■ ■ ■

EXAMPLE 10   Let $f(x) = \dfrac{2x^3 - 3x^2 + 1}{x^2 + 2x + 4}$ and evaluate:

a. $\lim\limits_{x \to \infty} f(x)$ **b.** $\lim\limits_{x \to -\infty} f(x)$

SOLUTION ✔   **a.** Dividing the numerator and the denominator of the rational expression by x^2, we obtain

$$\lim_{x \to \infty} \frac{2x^3 - 3x^2 + 1}{x^2 + 2x + 4} = \lim_{x \to \infty} \frac{2x - 3 + \dfrac{1}{x^2}}{1 + \dfrac{2}{x} + \dfrac{4}{x^2}}$$

Since the numerator becomes arbitrarily large whereas the denominator approaches 1 as x approaches infinity, we see that the quotient $f(x)$ gets larger and larger as x approaches infinity. In other words, the limit does not exist. We indicate this by writing

$$\lim_{x \to \infty} \frac{2x^3 - 3x^2 + 1}{x^2 + 2x + 4} = \infty$$

b. Once again, dividing both the numerator and the denominator by x^2, we obtain

$$\lim_{x \to -\infty} \frac{2x^3 - 3x^2 + 1}{x^2 + 2x + 4} = \lim_{x \to -\infty} \frac{2x - 3 + \dfrac{1}{x^2}}{1 + \dfrac{2}{x} + \dfrac{4}{x^2}}$$

In this case the numerator becomes arbitrarily large in magnitude but negative in sign, whereas the denominator approaches 1 as x approaches negative infinity. Therefore, the quotient $f(x)$ decreases without bound, and the limit does not exist. We indicate this by writing

$$\lim_{x \to -\infty} \frac{2x^3 - 3x^2 + 1}{x^2 + 2x + 4} = -\infty$$

■ ■ ■ ■

Example 11 gives an application of the concept of the limit of a function at infinity.

EXAMPLE 11   The Custom Office Company makes a line of executive desks. It is estimated that the total cost of making x Senior Executive Model desks is $C(x) = 100x + 200{,}000$ dollars per year, so that the average cost of making x desks is given by

$$\overline{C}(x) = \frac{C(x)}{x} = \frac{100x + 200{,}000}{x} = 100 + \frac{200{,}000}{x}$$

dollars per desk. Evaluate $\lim\limits_{x \to \infty} \overline{C}(x)$ and interpret your results.

SOLUTION ✓

$$\lim_{x \to \infty} \overline{C}(x) = \lim_{x \to \infty} \left(100 + \frac{200,000}{x} \right)$$

$$= \lim_{x \to \infty} 100 + \lim_{x \to \infty} \frac{200,000}{x} = 100$$

A sketch of the graph of the function $\overline{C}(x)$ appears in Figure 2.31. The result we obtained is fully expected if we consider its economic implications. Note that as the level of production increases, the fixed cost per desk produced, represented by the term $(200,000/x)$, drops steadily. The average cost should approach a constant unit cost of production—$100 in this case.

FIGURE 2.31
As the level of production increases, the average cost approaches $100 per desk.

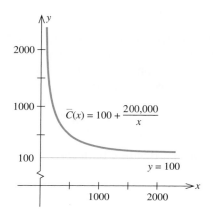

$$\overline{C}(x) = 100 + \frac{200,000}{x}$$

$y = 100$

Group Discussion

Consider the graph of the function f depicted in the following figure:

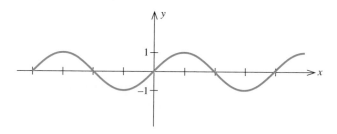

It has the property that the curve oscillates between $y = -1$ and $y = 1$ indefinitely in either direction.

1. Explain why $\lim_{x \to -\infty} f(x)$ and $\lim_{x \to \infty} f(x)$ do not exist.

2. Compare this function with those of Example 10. More specifically, discuss the different ways each function fails to have a limit at infinity or minus infinity.

1. Find the indicated limit if it exists.

a. $\lim\limits_{x\to 3}\dfrac{\sqrt{x^2+7}+\sqrt{3x-5}}{x+2}$

b. $\lim\limits_{x\to -1}\dfrac{x^2-x-2}{2x^2-x-3}$

2. The average cost per disc (in dollars) incurred by the Herald Record Company in pressing x compact audio discs is given by the average cost function

$$\overline{C}(x)=1.8+\frac{3000}{x}$$

Evaluate $\lim\limits_{x\to\infty}\overline{C}(x)$ and interpret your result.

Solutions to Self-Check Exercises 2.4 can be found on page 135.

2.4 Exercises

In Exercises 1–8, use the graph of the given function f to determine $\lim\limits_{x\to a} f(x)$ at the indicated value of a, if it exists.

1.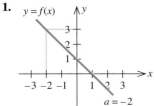

2.

3.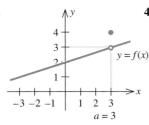

4.

5.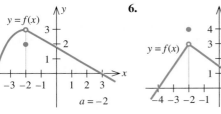

6.

7.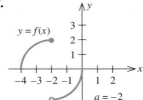
$a=-2$

8.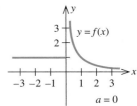
$a=0$

In Exercises 9–16, complete the table by computing $f(x)$ at the given values of x. Use these results to estimate the indicated limit (if it exists).

9. $f(x)=x^2+1;\ \lim\limits_{x\to 2} f(x)$

x	1.9	1.99	1.999	2.001	2.01	2.1
$f(x)$						

10. $f(x)=2x^2-1;\ \lim\limits_{x\to 1} f(x)$

x	0.9	0.99	0.999	1.001	1.01	1.1
$f(x)$						

11. $f(x) = \dfrac{|x|}{x}$; $\lim\limits_{x\to 0} f(x)$

x	-0.1	-0.01	-0.001	0.001	0.01	0.1
$f(x)$						

12. $f(x) = \dfrac{|x-1|}{x-1}$; $\lim\limits_{x\to 1} f(x)$

x	0.9	0.99	0.999	1.001	1.01	1.1
$f(x)$						

13. $f(x) = \dfrac{1}{(x-1)^2}$; $\lim\limits_{x\to 1} f(x)$

x	0.9	0.99	0.999	1.001	1.01	1.1
$f(x)$						

14. $f(x) = \dfrac{1}{x-2}$; $\lim\limits_{x\to 2} f(x)$

x	1.9	1.99	1.999	2.001	2.01	2.1
$f(x)$						

15. $f(x) = \dfrac{x^2 + x - 2}{x - 1}$; $\lim\limits_{x\to 1} f(x)$

x	0.9	0.99	0.999	1.001	1.01	1.1
$f(x)$						

16. $f(x) = \dfrac{x - 1}{x - 1}$; $\lim\limits_{x\to 1} f(x)$

x	0.9	0.99	0.999	1.001	1.01	1.1
$f(x)$						

In Exercises 17–22, sketch the graph of the function f and evaluate $\lim\limits_{x\to a} f(x)$, if it exists, for the given values of a.

17. $f(x) = \begin{cases} x - 1 & \text{if } x \le 0 \\ -1 & \text{if } x > 0 \end{cases}$ $(a = 0)$

18. $f(x) = \begin{cases} x - 1 & \text{if } x \le 3 \\ -2x + 8 & \text{if } x > 3 \end{cases}$ $(a = 3)$

19. $f(x) = \begin{cases} x & \text{if } x < 1 \\ 0 & \text{if } x = 1 \\ -x + 2 & \text{if } x > 1 \end{cases}$ $(a = 1)$

20. $f(x) = \begin{cases} -2x + 4 & \text{if } x < 1 \\ 4 & \text{if } x = 1 \\ x^2 + 1 & \text{if } x > 1 \end{cases}$ $(a = 1)$

21. $f(x) = \begin{cases} |x| & \text{if } x \ne 0 \\ 1 & \text{if } x = 0 \end{cases}$ $(a = 0)$

22. $f(x) = \begin{cases} |x - 1| & \text{if } x \ne 1 \\ 0 & \text{if } x = 1 \end{cases}$ $(a = 1)$

In Exercises 23–40, find the indicated limit.

23. $\lim\limits_{x\to 2} 3$

24. $\lim\limits_{x\to -2} -3$

25. $\lim\limits_{x\to 3} x$

26. $\lim\limits_{x\to -2} -3x$

27. $\lim\limits_{x\to 1}(1 - 2x^2)$

28. $\lim\limits_{t\to 3}(4t^2 - 2t + 1)$

29. $\lim\limits_{x\to 1}(2x^3 - 3x^2 + x + 2)$

30. $\lim\limits_{x\to 0}(4x^5 - 20x^2 + 2x + 1)$

31. $\lim\limits_{s\to 0}(2s^2 - 1)(2s + 4)$

32. $\lim\limits_{x\to 2}(x^2 + 1)(x^2 - 4)$

33. $\lim\limits_{x\to 2} \dfrac{2x + 1}{x + 2}$

34. $\lim\limits_{x\to 1} \dfrac{x^3 + 1}{2x^3 + 2}$

35. $\lim\limits_{x\to 2} \sqrt{x + 2}$

36. $\lim\limits_{x\to -2} \sqrt[3]{5x + 2}$

37. $\lim\limits_{x\to -3} \sqrt{2x^4 + x^2}$

38. $\lim\limits_{x\to 2} \sqrt{\dfrac{2x^3 + 4}{x^2 + 1}}$

39. $\lim\limits_{x\to -1} \dfrac{\sqrt{x^2 + 8}}{2x + 4}$

40. $\lim\limits_{x\to 3} \dfrac{x\sqrt{x^2 + 7}}{2x - \sqrt{2x + 3}}$

In Exercises 41–48, find the indicated limit given that $\lim\limits_{x\to a} f(x) = 3$ and $\lim\limits_{x\to a} g(x) = 4$.

41. $\lim\limits_{x\to a}[f(x) - g(x)]$

42. $\lim\limits_{x\to a} 2f(x)$

43. $\lim\limits_{x\to a}[2f(x) - 3g(x)]$

44. $\lim\limits_{x\to a}[f(x)g(x)]$

45. $\lim\limits_{x\to a} \sqrt{g(x)}$

46. $\lim\limits_{x\to a} \sqrt[3]{5f(x) + 3g(x)}$

47. $\lim\limits_{x\to a} \dfrac{2f(x) - g(x)}{f(x)g(x)}$

48. $\lim\limits_{x\to a} \dfrac{g(x) - f(x)}{f(x) + \sqrt{g(x)}}$

In Exercises 49–62, find the indicated limit, if it exists.

49. $\lim\limits_{x\to 1} \dfrac{x^2 - 1}{x - 1}$

50. $\lim\limits_{x\to -2} \dfrac{x^2 - 4}{x + 2}$

51. $\lim\limits_{x\to 0} \dfrac{x^2 - x}{x}$

52. $\lim\limits_{x\to 0} \dfrac{2x^2 - 3x}{x}$

53. $\lim\limits_{x\to -5} \dfrac{x^2 - 25}{x + 5}$

54. $\lim\limits_{b\to -3} \dfrac{b + 1}{b + 3}$

55. $\lim\limits_{x\to 1} \dfrac{x}{x - 1}$

56. $\lim\limits_{x\to 2} \dfrac{x + 2}{x - 2}$

57. $\lim\limits_{x\to -2} \dfrac{x^2 - x - 6}{x^2 + x - 2}$

58. $\lim\limits_{z\to 2} \dfrac{z^3 - 8}{z - 2}$

59. $\lim\limits_{x\to 1} \dfrac{\sqrt{x} - 1}{x - 1}$

Hint: Multiply by $\dfrac{\sqrt{x} + 1}{\sqrt{x} + 1}$.

60. $\lim\limits_{x\to 4} \dfrac{x - 4}{\sqrt{x} - 2}$

Hint: See Exercise 59.

61. $\lim\limits_{x\to 1} \dfrac{x - 1}{x^3 + x^2 - 2x}$

62. $\lim\limits_{x\to -2} \dfrac{4 - x^2}{2x^2 + x^3}$

In Exercises 63–68, use the graph of the function f to determine $\lim\limits_{x\to\infty} f(x)$ and $\lim\limits_{x\to-\infty} f(x)$, if they exist.

63.
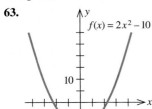
$f(x) = 2x^2 - 10$

64.
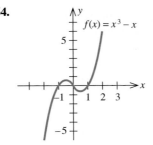
$f(x) = x^3 - x$

65.
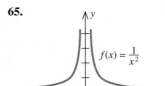
$f(x) = \dfrac{1}{x^2}$

66.
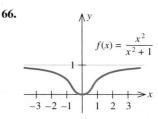
$f(x) = \dfrac{x^2}{x^2 + 1}$

67.
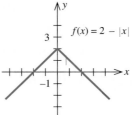
$f(x) = 2 - |x|$

68.
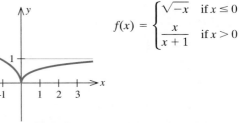
$f(x) = \begin{cases} \sqrt{-x} & \text{if } x \le 0 \\ \dfrac{x}{x + 1} & \text{if } x > 0 \end{cases}$

In Exercises 69–72, complete the table by computing $f(x)$ at the given values of x. Use the results to guess at the indicated limits, if they exist.

69. $f(x) = \dfrac{1}{x^2 + 1}$; $\lim\limits_{x\to\infty} f(x)$ and $\lim\limits_{x\to-\infty} f(x)$

x	1	10	100	1000
$f(x)$				

x	-1	-10	-100	-1000
$f(x)$				

70. $f(x) = \dfrac{2x}{x + 1}$; $\lim\limits_{x\to\infty} f(x)$ and $\lim\limits_{x\to-\infty} f(x)$

x	1	10	100	1000
$f(x)$				

x	-5	-10	-100	-1000
$f(x)$				

(continued on p. 133)

Finding the Limit of a Function

A graphing utility can be used to help us find the limit of a function, if it exists, as illustrated in the following examples.

EXAMPLE 1   Let $f(x) = \dfrac{x^3 - 1}{x - 1}$.

a. Plot the graph of f in the viewing rectangle $[-2, 2] \times [0, 4]$.

b. Use ZOOM to find $\lim\limits_{x \to 1} \dfrac{x^3 - 1}{x - 1}$.

c. Verify your result by evaluating the limit algebraically.

SOLUTION ✔   **a.** The graph of f in the viewing rectangle $[-2, 2] \times [0, 4]$ is shown in Figure T1.

FIGURE T1

The graph of $f(x) = \dfrac{x^3 - 1}{x - 1}$ in the viewing rectangle $[-2, 2] \times [0, 4]$

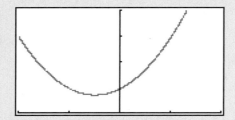

b. Using ZOOM-IN repeatedly, we see that the y-value approaches 3 as the x-value approaches 1. We conclude, accordingly, that

$$\lim_{x \to 1} \frac{x^3 - 1}{x - 1} = 3$$

c. We compute

$$\lim_{x \to 1} \frac{x^3 - 1}{x - 1} = \lim_{x \to 1} \frac{(x - 1)(x^2 + x + 1)}{x - 1}$$
$$= \lim_{x \to 1} (x^2 + x + 1) = 3$$

▪ ▪ ▪ ▪

REMARK   If you attempt to find the limit in Example 1 by using the evaluation function of your graphing utility to find the value of $f(x)$ when $x = 1$, you will see that the graphing utility does not display the y-value. This happens because the point $x = 1$ is not in the domain of f. ▪ ▪ ▪

EXAMPLE **2**

Use **zoom** to find $\lim_{x \to 0} (1 + x)^{1/x}$.

SOLUTION ✔

We first plot the graph of $f(x) = (1 + x)^{1/x}$ in a suitable viewing rectangle. Figure T2 shows a plot of f in the rectangle $[-1, 1] \times [0, 4]$. Using **zoom-in** repeatedly, we see that $\lim_{x \to 0} (1 + x)^{1/x} \approx 2.71828$.

FIGURE T2
The graph of $f(x) = (1 + x)^{1/x}$ in the viewing rectangle $[-1, 1] \times [0, 4]$

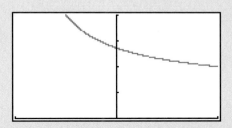

The limit of $f(x) = (1 + x)^{1/x}$ as x approaches zero, denoted by the letter e, plays a very important role in the study of mathematics and its applications (see Section 5.6). Thus,

$$\lim_{x \to 0} (1 + x)^{1/x} = e$$

where, as we have just seen, $e \approx 2.71828$. ■ ■ ■ ■

EXAMPLE 3

When organic waste is dumped into a pond, the oxidation process that takes place reduces the pond's oxygen content. However, given time, nature will restore the oxygen content to its natural level. Suppose that the oxygen content t days after the organic waste has been dumped into the pond is given by

$$f(t) = 100 \left(\frac{t^2 + 10t + 100}{t^2 + 20t + 100} \right)$$

percent of its normal level.

a. Plot the graph of f in the viewing rectangle $[0, 200] \times [70, 100]$.
b. What can you say about $f(t)$ when t is very large?
c. Verify your observation in part (b) by evaluating $\lim_{t \to \infty} f(t)$.

SOLUTION ✔

a. The graph of f is shown in Figure T3.
b. From the graph of f it appears that $f(t)$ approaches 100 steadily as t gets

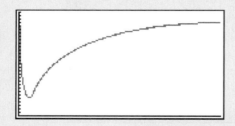

FIGURE T3
The graph of f in the viewing rectangle $[0, 200] \times [70, 100]$

larger and larger. This observation tells us that eventually the oxygen content of the pond will be restored to its natural level.

c. To verify the observation made in part (b), we compute

$$\lim_{t \to \infty} f(t) = \lim_{t \to \infty} 100 \left(\frac{t^2 + 10t + 100}{t^2 + 20t + 100} \right)$$

$$= 100 \lim_{t \to \infty} \left(\frac{1 + \dfrac{10}{t} + \dfrac{100}{t^2}}{1 + \dfrac{20}{t} + \dfrac{100}{t^2}} \right) = 100$$

Exercises

In Exercises 1–10, use a graphing utility to find the indicated limit by first plotting the graph of the function in a suitable viewing rectangle and then using the ZOOM-IN feature of the calculator.

1. $\lim\limits_{x \to 1} \dfrac{2x^3 - 2x^2 + 3x - 3}{x - 1}$

2. $\lim\limits_{x \to -2} \dfrac{2x^3 + 3x^2 - x + 2}{x + 2}$

3. $\lim\limits_{x \to -1} \dfrac{x^3 + 1}{x + 1}$

4. $\lim\limits_{x \to -1} \dfrac{x^4 - 1}{x - 1}$

5. $\lim\limits_{x \to 1} \dfrac{x^3 - x^2 - x + 1}{x^3 - 3x + 2}$

6. $\lim\limits_{x \to 2} \dfrac{x^3 + 2x^2 - 16}{2x^3 - x^2 + 2x - 16}$

7. $\lim\limits_{x \to 0} \dfrac{\sqrt{x + 1} - 1}{x}$

8. $\lim\limits_{x \to 0} \dfrac{(x + 4)^{3/2} - 8}{x}$

9. $\lim\limits_{x \to 0} (1 + 2x)^{1/x}$

10. $\lim\limits_{x \to 0} \dfrac{2^x - 1}{x}$

11. Use a graphing utility to show that $\lim\limits_{x \to 3} \dfrac{2}{x - 3}$ does not exist.

12. Use a graphing utility to show that $\lim\limits_{x \to 2} \dfrac{x^3 - 2x + 1}{x - 2}$ does not exist.

13. **CITY PLANNING** A major developer is building a 5000-acre complex of homes, offices, stores, schools, and churches in the rural community of Marlboro. As a result of this development, the planners have estimated that Marlboro's population (in thousands) t yr from now will be given by

$$P(t) = \frac{25t^2 + 125t + 200}{t^2 + 5t + 40}$$

a. Plot the graph of P in the viewing rectangle $[0, 50] \times [0, 30]$.

b. What will be the population of Marlboro in the long run?
Hint: Find $\lim\limits_{t \to \infty} P(t)$.

71. $f(x) = 3x^3 - x^2 + 10$; $\lim\limits_{x \to \infty} f(x)$ and $\lim\limits_{x \to -\infty} f(x)$

x	1	5	10	100	1000
$f(x)$					

x	-1	-5	-10	-100	-1000
$f(x)$					

72. $f(x) = \dfrac{|x|}{x}$; $\lim\limits_{x \to \infty} f(x)$ and $\lim\limits_{x \to -\infty} f(x)$

x	1	10	100	-1	-10	-100
$f(x)$						

In Exercises 73–80, find the indicated limits, if they exist.

73. $\lim\limits_{x \to \infty} \dfrac{3x + 2}{x - 5}$

74. $\lim\limits_{x \to -\infty} \dfrac{4x^2 - 1}{x + 2}$

75. $\lim\limits_{x \to -\infty} \dfrac{3x^3 + x^2 + 1}{x^3 + 1}$

76. $\lim\limits_{x \to \infty} \dfrac{2x^2 + 3x + 1}{x^4 - x^2}$

77. $\lim\limits_{x \to -\infty} \dfrac{x^4 + 1}{x^3 - 1}$

78. $\lim\limits_{x \to \infty} \dfrac{4x^4 - 3x^2 + 1}{2x^4 + x^3 + x^2 + x + 1}$

79. $\lim\limits_{x \to \infty} \dfrac{x^5 - x^3 + x - 1}{x^6 + 2x^2 + 1}$

80. $\lim\limits_{x \to \infty} \dfrac{2x^2 - 1}{x^3 + x^2 + 1}$

81. Toxic Waste A city's main well was recently found to be contaminated with trichloroethylene, a cancer-causing chemical, as a result of an abandoned chemical dump leaching chemicals into the water. A proposal submitted to city council members indicates that the cost, measured in millions of dollars, of removing x percent of the toxic pollutant is given by

$$C(x) = \frac{0.5x}{100 - x} \qquad (0 < x < 100)$$

a. Find the cost of removing 50%, 60%, 70%, 80%, 90%, and 95% of the pollutant.

b. Evaluate

$$\lim_{x \to 100} \frac{0.5x}{100 - x}$$

and interpret your result.

82. A Doomsday Situation The population of a certain breed of rabbits introduced into an isolated island is given by

$$P(t) = \frac{72}{9 - t} \qquad (0 < t < 9)$$

where t is measured in months.
a. Find the number of rabbits present in the island initially.
b. Show that the population of rabbits is increasing without bound.
c. Sketch the graph of the function P.
(*Comment:* This phenomenon is referred to as a *doomsday situation*.)

83. Average Cost The average cost per disc in dollars incurred by the Herald Record Company in pressing x video discs is given by the average cost function

$$\overline{C}(x) = 2.2 + \frac{2500}{x}$$

Evaluate $\lim\limits_{x \to \infty} \overline{C}(x)$ and interpret your result.

84. Concentration of a Drug in the Bloodstream The concentration of a certain drug in a patient's bloodstream t hr after injection is given by

$$C(t) = \frac{0.2t}{t^2 + 1}$$

mg/cm³. Evaluate $\lim\limits_{t \to \infty} C(t)$ and interpret your result.

85. Box Office Receipts The total worldwide box office receipts for a long-running blockbuster movie are approximated by the function

$$T(x) = \frac{120x^2}{x^2 + 4}$$

where $T(x)$ is measured in millions of dollars and x is the number of months since the movie's release.
a. What are the total box office receipts after the first month? The second month? The third month?
b. What will the movie gross in the long run?

86. POPULATION GROWTH   A major corporation is building a 4325-acre complex of homes, offices, stores, schools, and churches in the rural community of Glen Cove. As a result of this development, the planners have estimated that Glen Cove's population (in thousands) t yr from now will be given by

$$P(t) = \frac{25t^2 + 125t + 200}{t^2 + 5t + 40}$$

a. What is the current population of Glen Cove?
b. What will the population be in the long run?

87. DRIVING COSTS   A study of driving costs of 1992 model subcompact (four-cylinder) cars found that the average cost (car payments, gas, insurance, upkeep, and depreciation), measured in cents per mile, is approximated by the function

$$C(x) = \frac{2010}{x^{2.2}} + 17.80$$

where x denotes the number of miles (in thousands) the car is driven in a year.
a. What is the average cost of driving a subcompact car 5000 mi/yr? 10,000 mi/yr? 15,000 mi/yr? 20,000 mi/yr? 25,000 mi/yr?
b. Use part (a) to help sketch the graph of the function C.
c. What happens to the average cost as the number of miles driven increases without bound?

In Exercises 89–94, determine whether the statement is true or false. If it is true, explain why it is true. If it is false, give an example to show why it is false.

89. If $\lim\limits_{x \to a} f(x)$ exists, then f is defined at $x = a$.

90. If $\lim\limits_{x \to 0} f(x) = 4$ and $\lim\limits_{x \to 0} g(x) = 0$, then $\lim\limits_{x \to 0} f(x)g(x) = 0$.

91. If $\lim\limits_{x \to 2} f(x) = 3$ and $\lim\limits_{x \to 2} g(x) = 0$, then $\lim\limits_{x \to 2} [f(x)]/[g(x)]$ does not exist.

92. If $\lim\limits_{x \to 3} f(x) = 0$ and $\lim\limits_{x \to 3} g(x) = 0$, then $\lim\limits_{x \to 3} [f(x)]/[g(x)]$ does not exist.

93. $\lim\limits_{x \to 2} \left(\dfrac{x}{x + 1} + \dfrac{3}{x - 1} \right) = \lim\limits_{x \to 2} \dfrac{x}{x + 1} + \lim\limits_{x \to 2} \dfrac{3}{x - 1}$

94. $\lim\limits_{x \to 1} \left(\dfrac{2x}{x - 1} - \dfrac{2}{x - 1} \right) = \lim\limits_{x \to 1} \dfrac{2x}{x - 1} - \lim\limits_{x \to 1} \dfrac{2}{x - 1}$

95. SPEED OF A CHEMICAL REACTION   Certain proteins, known as enzymes, serve as catalysts for chemical reactions in living things. In 1913 Leonor Michaelis and L. M. Menten discovered the following formula giving the initial speed V (in moles/liter/second) at which the reaction begins in terms of the amount of substrate x (the substance being acted upon, measured in moles/liter):

$$V = \frac{ax}{x + b}$$

where a and b are positive constants. Evaluate

$$\lim_{x \to \infty} \frac{ax}{x + b}$$

and interpret your result.

96. Show by means of an example that $\lim\limits_{x \to a} [f(x) + g(x)]$ may exist even though neither $\lim\limits_{x \to a} f(x)$ nor $\lim\limits_{x \to a} g(x)$ exists. Does this example contradict Theorem 1?

97. Show by means of an example that $\lim\limits_{x \to a} [f(x)g(x)]$ may exist even though neither $\lim\limits_{x \to a} f(x)$ nor $\lim\limits_{x \to a} g(x)$ exists. Does this example contradict Theorem 1?

THEOREM 5

Existence of Zeros of a Continuous Function

If f is a continuous function on a closed interval $[a, b]$, and if $f(a)$ and $f(b)$ have opposite signs, then there is at least one solution of the equation $f(x) = 0$ in the interval (a, b) (Figure 2.41).

FIGURE 2.41

If $f(a)$ and $f(b)$ have opposite signs, there must be at least one number c $(a < c < b)$ such that $f(c) = 0$.

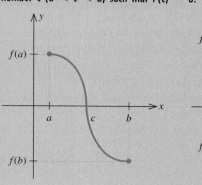
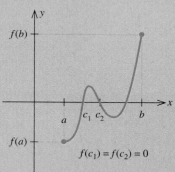

$f(c_1) = f(c_2) = 0$

Geometrically, this property states that if the graph of a continuous function goes from above the x-axis to below the x-axis, or vice versa, it must *cross* the x-axis. This is not necessarily true if the function is discontinuous (Figure 2.42).

FIGURE 2.42

$f(a) < 0$ and $f(b) > 0$, but the graph of f does not cross the x-axis between a and b because f is discontinuous.

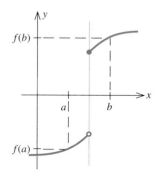

EXAMPLE 5

Let $f(x) = x^3 + x + 1$.

a. Show that f is continuous for all values of x.
b. Compute $f(-1)$ and $f(1)$ and use the results to deduce that there must be at least one point $x = c$, where c lies in the interval $(-1, 1)$ and $f(c) = 0$.

SOLUTION ✔

a. The function f is a polynomial function of degree 3 and is therefore continuous everywhere.

b. $f(-1) = (-1)^3 + (-1) + 1 = -1$
$\quad\;\; f(1) = 1^3 + 1 + 1 = 3$

Since $f(-1)$ and $f(1)$ have opposite signs, Theorem 5 tells us that there must be at least one point $x = c$ with $-1 < c < 1$ such that $f(c) = 0$.   ■ ■ ■ ■

The next example shows how the intermediate value theorem can be used to help us find the zero of a function.

EXAMPLE 6

Let $f(x) = x^3 + x - 1$. Since f is a polynomial function, it is continuous everywhere. Observe that $f(0) = -1$ and $f(1) = 1$ so that Theorem 5 guarantees the existence of at least one root of the equation $f(x) = 0$ in $(0, 1)$.*

We can locate the root more precisely by using Theorem 5 once again as follows: Evaluate $f(x)$ at the midpoint of $[0, 1]$, obtaining

$$f(0.5) = -0.375$$

Because $f(0.5) < 0$ and $f(1) > 0$, Theorem 5 now tells us that the root must lie in $(0.5, 1)$.

Repeat the process: Evaluate $f(x)$ at the midpoint of $[0.5, 1]$, which is

$$\frac{0.5 + 1}{2} = 0.75$$

Thus,

$$f(0.75) = 0.1719$$

Because $f(0.5) < 0$ and $f(0.75) > 0$, Theorem 5 tells us that the root is in $(0.5, 0.75)$. This process can be continued. Table 2.3 summarizes the results of our computations through nine steps.

From Table 2.3 we see that the root is approximately 0.68, accurate to two decimal places. By continuing the process through a sufficient number of steps, we can obtain as accurate an approximation to the root as we please.

■ ■ ■ ■

Table 2.3

Step	Root of $f(x) = 0$ Lies In
1	(0, 1)
2	(0.5, 1)
3	(0.5, 0.75)
4	(0.625, 0.75)
5	(0.625, 0.6875)
6	(0.65625, 0.6875)
7	(0.671875, 0.6875)
8	(0.6796875, 0.6875)
9	(0.6796875, 0.6835937)

REMARK The process of finding the root of $f(x) = 0$ used in Example 6 is called the **method of bisection.** It is crude but effective.   ■ ■ ■

SELF-CHECK EXERCISES 2.5

1. Evaluate $\lim_{x \to -1^-} f(x)$ and $\lim_{x \to -1^+} f(x)$, where

$$f(x) = \begin{cases} 1 & \text{if } x < -1 \\ 1 + \sqrt{x + 1} & \text{if } x \geq -1 \end{cases}$$

Does $\lim_{x \to -1} f(x)$ exist?

---

* It can be shown that f has precisely one zero in $(0, 1)$ (see Exercise 97, Section 4.1.).

2. Determine the values of x for which the given function is discontinuous. At each point of discontinuity, indicate which condition(s) for continuity are violated. Sketch the graph of the function.

a. $f(x) = \begin{cases} -x^2 + 1 & \text{if } x \le 1 \\ x - 1 & \text{if } x > 1 \end{cases}$

b. $g(x) = \begin{cases} -x + 1 & \text{if } x < -1 \\ 2 & \text{if } -1 < x \le 1 \\ -x + 3 & \text{if } x > 1 \end{cases}$

Solutions to Self-Check Exercises 2.5 can be found on page 154.

2.5 Exercises

In Exercises 1–8, use the graph of the function f to find $\lim\limits_{x \to a^-} f(x)$, $\lim\limits_{x \to a^+} f(x)$, and $\lim\limits_{x \to a} f(x)$ at the indicated value of a, if the limit exists.

1.
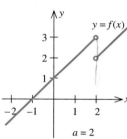
$a = 2$

2.
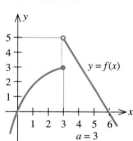
$a = 3$

3.
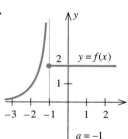
$a = -1$

4.
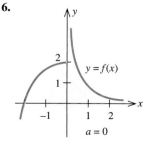
$a = 1$

5.
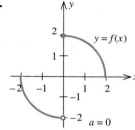
$a = 1$

6.
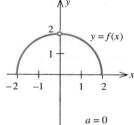
$a = 0$

7.

$a = 0$

8.

$a = 0$

In Exercises 9–14, refer to the graph of the function f and determine whether each statement is true or false.

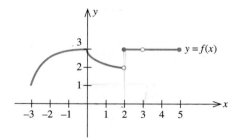

9. $\lim\limits_{x \to -3^+} f(x) = 1$

10. $\lim\limits_{x \to 0} f(x) = f(0)$

11. $\lim\limits_{x \to 2^-} f(x) = 2$

12. $\lim\limits_{x \to 2^+} f(x) = 3$

13. $\lim\limits_{x \to 3} f(x)$ does not exist.

14. $\lim\limits_{x \to 5^-} f(x) = 3$

In Exercises 15–20, refer to the graph of the function *f* and determine whether each statement is true or false.

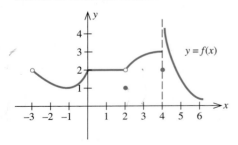

15. $\lim\limits_{x \to -3^+} f(x) = 2$

16. $\lim\limits_{x \to 0} f(x) = 2$

17. $\lim\limits_{x \to 2} f(x) = 1$

18. $\lim\limits_{x \to 4^-} f(x) = 3$

19. $\lim\limits_{x \to 4^+} f(x)$ does not exist.

20. $\lim\limits_{x \to 4} f(x) = 2$

In Exercises 21–42, find the indicated one-sided limit, if it exists.

21. $\lim\limits_{x \to 1^+}(2x + 4)$

22. $\lim\limits_{x \to 1^-}(3x - 4)$

23. $\lim\limits_{x \to 2^-}\dfrac{x - 3}{x + 2}$

24. $\lim\limits_{x \to 1^+}\dfrac{x + 2}{x + 1}$

25. $\lim\limits_{x \to 0^+}\dfrac{1}{x}$

26. $\lim\limits_{x \to 0^-}\dfrac{1}{x}$

27. $\lim\limits_{x \to 0^+}\dfrac{x - 1}{x^2 + 1}$

28. $\lim\limits_{x \to 2^+}\dfrac{x + 1}{x^2 - 2x + 3}$

29. $\lim\limits_{x \to 0^+} \sqrt{x}$

30. $\lim\limits_{x \to 2^+} 2\sqrt{x - 2}$

31. $\lim\limits_{x \to -2^+} (2x + \sqrt{2 + x})$

32. $\lim\limits_{x \to -5^+} x(1 + \sqrt{5 + x})$

33. $\lim\limits_{x \to 1^-}\dfrac{1 + x}{1 - x}$

34. $\lim\limits_{x \to 1^+}\dfrac{1 + x}{1 - x}$

35. $\lim\limits_{x \to 2^-}\dfrac{x^2 - 4}{x - 2}$

36. $\lim\limits_{x \to -3^+}\dfrac{\sqrt{x + 3}}{x^2 + 1}$

37. $\lim\limits_{x \to 3^+}\dfrac{x^2 - 9}{x + 3}$

38. $\lim\limits_{x \to -2^-}\dfrac{\sqrt[3]{x + 10}}{2x^2 + 1}$

39. $\lim\limits_{x \to 0^+} f(x)$ and $\lim\limits_{x \to 0^-} f(x)$, where

$$f(x) = \begin{cases} 2x & \text{if } x < 0 \\ x^2 & \text{if } x \geq 0 \end{cases}$$

40. $\lim\limits_{x \to 0^+} f(x)$ and $\lim\limits_{x \to 0^-} f(x)$, where

$$f(x) = \begin{cases} -x + 1 & \text{if } x \leq 0 \\ 2x + 3 & \text{if } x > 0 \end{cases}$$

41. $\lim\limits_{x \to 1^+} f(x)$ and $\lim\limits_{x \to 1^-} f(x)$, where

$$f(x) = \begin{cases} \sqrt{x + 3} & \text{if } x \geq 1 \\ 2 + \sqrt{x} & \text{if } x < 1 \end{cases}$$

42. $\lim\limits_{x \to 1^+} f(x)$ and $\lim\limits_{x \to 1^-} f(x)$, where

$$f(x) = \begin{cases} x + 2\sqrt{x - 1} & \text{if } x \geq 1 \\ 1 - \sqrt{1 - x} & \text{if } x < 1 \end{cases}$$

In Exercises 43–50, determine the values of *x*, if any, at which each function is discontinuous. At each point of discontinuity, state the condition(s) for continuity that are violated.

43.

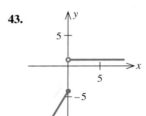

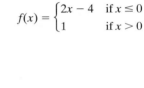

$$f(x) = \begin{cases} 2x - 4 & \text{if } x \leq 0 \\ 1 & \text{if } x > 0 \end{cases}$$

44.

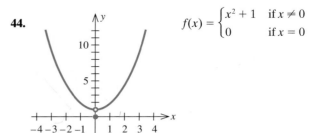

$$f(x) = \begin{cases} x^2 + 1 & \text{if } x \neq 0 \\ 0 & \text{if } x = 0 \end{cases}$$

45.

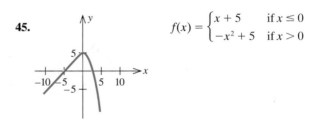

$$f(x) = \begin{cases} x + 5 & \text{if } x \leq 0 \\ -x^2 + 5 & \text{if } x > 0 \end{cases}$$

46.

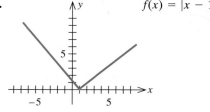

$f(x) = |x - 1|$

47.

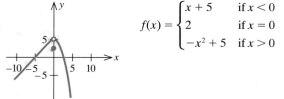

$f(x) = \begin{cases} x + 5 & \text{if } x < 0 \\ 2 & \text{if } x = 0 \\ -x^2 + 5 & \text{if } x > 0 \end{cases}$

48.

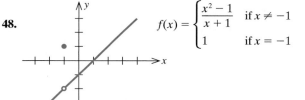

$f(x) = \begin{cases} \dfrac{x^2 - 1}{x + 1} & \text{if } x \neq -1 \\ 1 & \text{if } x = -1 \end{cases}$

49.

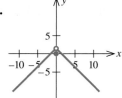

$f(x) = \begin{cases} -|x| + 1 & \text{if } x \neq 0 \\ 0 & \text{if } x = 0 \end{cases}$

50.

$f(x) = \begin{cases} \dfrac{1}{x^2} & \text{if } x \neq 0 \\ 1 & \text{if } x = 0 \end{cases}$

In Exercises 51–66, find the values of x for which each function is continuous.

51. $f(x) = 2x^2 + x - 1$

52. $f(x) = x^3 - 2x^2 + x - 1$

53. $f(x) = \dfrac{2}{x^2 + 1}$

54. $f(x) = \dfrac{x}{2x^2 + 1}$

55. $f(x) = \dfrac{2}{2x - 1}$

56. $f(x) = \dfrac{x + 1}{x - 1}$

57. $f(x) = \dfrac{2x + 1}{x^2 + x - 2}$

58. $f(x) = \dfrac{x - 1}{x^2 + 2x - 3}$

59. $f(x) = \begin{cases} x & \text{if } x \leq 1 \\ 2x - 1 & \text{if } x > 1 \end{cases}$

60. $f(x) = \begin{cases} -x + 1 & \text{if } x \leq -1 \\ x + 1 & \text{if } x > -1 \end{cases}$

61. $f(x) = \begin{cases} -2x + 1 & \text{if } x < 0 \\ x^2 + 1 & \text{if } x \geq 0 \end{cases}$

62. $f(x) = \begin{cases} x + 1 & \text{if } x \leq 1 \\ -x^2 + 1 & \text{if } x > 1 \end{cases}$

63. $f(x) = \begin{cases} \dfrac{x^2 - 1}{x - 1} & \text{if } x \neq 1 \\ 2 & \text{if } x = 1 \end{cases}$

64. $f(x) = \begin{cases} \dfrac{x^2 - 4}{x + 2} & \text{if } x \neq -2 \\ 1 & \text{if } x = -2 \end{cases}$

65. $f(x) = |x + 1|$

66. $f(x) = \dfrac{|x - 1|}{x - 1}$

In Exercises 67–70, determine all values of x at which the function is discontinuous.

67. $f(x) = \dfrac{2x}{x^2 - 1}$

68. $f(x) = \dfrac{1}{(x - 1)(x - 2)}$

69. $f(x) = \dfrac{x^2 - 2x}{x^2 - 3x + 2}$

70. $f(x) = \dfrac{x^2 - 3x + 2}{x^2 - 2x}$

71. THE POSTAGE FUNCTION   The graph of the "postage function"

$$f(x) = \begin{cases} 34 & \text{if } 0 < x \leq 1 \\ 55 & \text{if } 1 < x \leq 2 \\ \vdots \\ 265 & \text{if } 11 < x \leq 12 \end{cases}$$

where x denotes the weight of a parcel in ounces and $f(x)$ the postage in cents, is shown in the figure on page 148. Determine the values of x for which f is discontinuous.

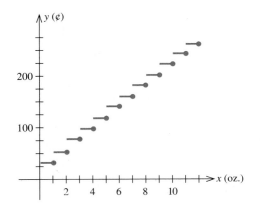

72. INVENTORY CONTROL As part of an optimal inventory policy, the manager of an office supply company orders 500 reams of photocopy paper every 20 days. The accompanying graph shows the *actual* inventory level of paper in an office supply store during the first 60 business days of 2001. Determine the values of t for which the "inventory function" is discontinuous and give an interpretation of the graph.

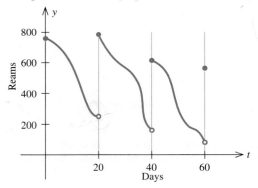

73. LEARNING CURVES The following graph describes the progress Michael made in solving a problem correctly during a mathematics quiz. Here, y denotes the percentage of work completed, and x is measured in minutes. Give an interpretation of the graph.

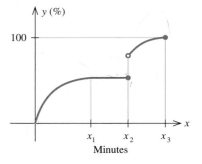

74. AILING FINANCIAL INSTITUTIONS The Franklin Savings and Loan Company acquired two ailing financial institutions in 1992. One of them was acquired at time $t = T_1$, and the other was acquired at time $t = T_2$ ($t = 0$ corresponds to the beginning of 1992). The following graph shows the total amount of money on deposit with Franklin. Explain the significance of the discontinuities of the function at T_1 and T_2.

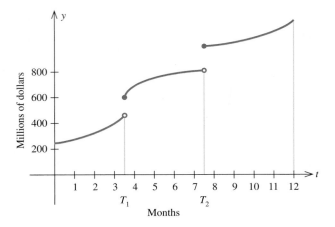

75. ENERGY CONSUMPTION The following graph shows the amount of home heating oil remaining in a 200-gallon tank over a 120-day period ($t = 0$ corresponds to October 1). Explain why the function is discontinuous at $t = 40$, $t = 70$, $t = 95$, and $t = 110$.

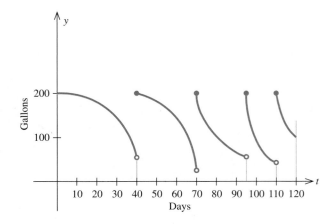

76. PRIME INTEREST RATE The function P, whose graph follows, gives the prime rate (the interest rate banks charge their best corporate customers) as a function of time for the first 32 wk in 1989. Determine the values of t for which P is discontinuous and interpret your results.

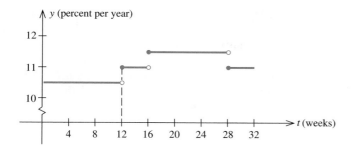

77. ADMINISTRATION OF AN INTRAVENOUS SOLUTION A dextrose solution is being administered to a patient intravenously. The 1-liter (L) bottle holding the solution is removed and replaced by another as soon as the contents drop to approximately 5% of the initial (1-L) amount. The rate of discharge is constant, and it takes 6 hr to discharge 95% of the contents of a full bottle. Draw a graph showing the amount of dextrose solution in a bottle in the IV system over a 24-hr period, assuming that we started with a full bottle.

78. COMMISSIONS The base salary of a salesman working on commission is $12,000. For each $50,000 of sales beyond $100,000, he is paid a $1000 commission. Sketch a graph showing his earnings as a function of the level of his sales x. Determine the values of x for which the function f is discontinuous.

79. PARKING FEES The fee charged per car in a downtown parking lot is $1 for the first half hour and $.50 for each additional half hour or part thereof, subject to a maximum of $5. Derive a function f relating the parking fee to the length of time a car is left in the lot. Sketch the graph of f and determine the values of x for which the function f is discontinuous.

80. COMMODITY PRICES The function that gives the cost of a certain commodity is defined by

$$C(x) = \begin{cases} 5x & \text{if } 0 < x < 10 \\ 4x & \text{if } 10 \le x < 30 \\ 3.5x & \text{if } 30 \le x < 60 \\ 3.25x & \text{if } x \ge 60 \end{cases}$$

where x is the number of pounds of a certain commodity sold and $C(x)$ is measured in dollars. Sketch the graph of the function C and determine the values of x for which the function C is discontinuous.

81. ENERGY EXPENDED BY A FISH Suppose that a fish swimming a distance of L ft at a speed of v ft/sec relative to the water and against a current flowing at the rate of

u ft/sec ($u < v$) expends a total energy given by

$$E(v) = \frac{aLv^3}{v - u}$$

where E is measured in foot-pounds (ft-lb) and a is a constant.
a. Evaluate $\lim\limits_{v \to u^+} E(v)$ and interpret your result.
b. Evaluate $\lim\limits_{v \to \infty} E(v)$ and interpret your result.

82. Let

$$f(x) = \begin{cases} x + 2 & \text{if } x \le 1 \\ kx^2 & \text{if } x > 1 \end{cases}$$

Find the value of k that will make f continuous on $(-\infty, \infty)$.

83. Let

$$f(x) = \begin{cases} \dfrac{x^2 - 4}{x + 2} & \text{if } x \ne -2 \\ k & \text{if } x = -2 \end{cases}$$

For what value of k will f be continuous on $(-\infty, \infty)$?

84. a. Suppose f is continuous at a and g is discontinuous at a. Is the sum $f + g$ discontinuous at a? Explain.
b. Suppose f and g are both discontinuous at a. Is the sum $f + g$ necessarily discontinuous at a? Explain.

85. a. Suppose f is continuous at a and g is discontinuous at a. Is the product fg necessarily discontinuous at a? Explain.
b. Suppose f and g are both discontinuous at a. Is the product fg necessarily discontinuous at a? Explain.

In Exercises 86–89, (a) show that the function f is continuous for all values of x in the interval $[a, b]$ and (b) prove that f must have at least one zero in the interval (a, b) by showing that $f(a)$ and $f(b)$ have opposite signs.

86. $f(x) = x^2 - 6x + 8;\ a = 1,\ b = 3$

87. $f(x) = x^3 - 2x^2 + 3x + 2;\ a = -1,\ b = 1$

88. $f(x) = 2x^3 - 3x^2 - 36x + 14;\ a = 0,\ b = 1$

89. $f(x) = 2x^{5/3} - 5x^{4/3};\ a = 14,\ b = 16$

(continued on p. 154)

FINDING THE POINTS OF DISCONTINUITY OF A FUNCTION

You can very often recognize the points of discontinuity of a function f by examining its graph. For example, Figure T1 shows the graph of $f(x) = x/(x^2 - 1)$ obtained using a graphing utility. It is evident that f is discontinuous at $x = -1$ and $x = 1$. This observation is also borne out by the fact that both these points are not in the domain of f.

FIGURE T1

The graph of $f(x) = \dfrac{x}{x^2 - 1}$ in the viewing rectangle $[-4, 4] \times [-10, 10]$

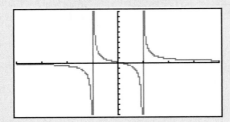

Consider the function

$$g(x) = \frac{2x^3 + x^2 - 7x - 6}{x^2 - x - 2}$$

Using a graphing utility we obtain the graph of g shown in Figure T2. An examination of this graph does not reveal any points of discontinuity. However, if we factor both the numerator and the denominator of the rational expression, we see that

$$g(x) = \frac{(x + 1)(x - 2)(2x + 3)}{(x + 1)(x - 2)}$$

$$= 2x + 3$$

provided $x \neq -1$ and $x \neq 2$, so that its graph in fact looks like that shown in Figure T3.

FIGURE T2

The graph of $g(x) = \dfrac{2x^3 + x^2 - 7x - 6}{x^2 - x - 2}$ in the standard viewing rectangle

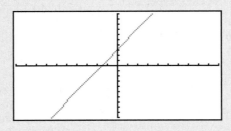

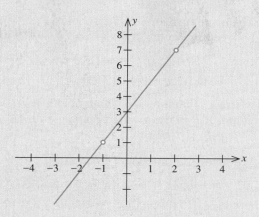

This example shows the limitation of the graphing utility and reminds us of the importance of studying functions analytically!

GRAPHING FUNCTIONS DEFINED PIECEWISE

The following example illustrates how to plot the graphs of functions defined in a piecewise manner on a graphing utility.

EXAMPLE **1**

Plot the graph of

$$f(x) = \begin{cases} x + 1 & \text{if } x \le 1 \\ \dfrac{2}{x} & \text{if } x > 1 \end{cases}$$

SOLUTION ✔

We enter the function

$$y1 = (x + 1)(x \le 1) + (2/x)(x > 1)$$

The graph of the function in the viewing rectangle $[-5, 5] \times [-2, 4]$ is shown in Figure T4.

FIGURE **T4**
The graph of f in the viewing rectangle
$[-5, 5] \times [-2, 4]$

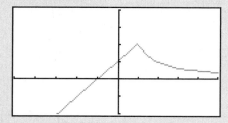

EXAMPLE **2**

The percentage of U.S. households, $P(t)$, watching television during weekdays between the hours of 4 P.M. and 4 A.M. is given by

$$P(t) = \begin{cases} 0.01354t^4 - 0.49375t^3 + 2.58333t^2 + 3.8t + 31.60704 & \text{if } 0 \le t \le 8 \\ 1.35t^2 - 33.05t + 208 & \text{if } 8 < t \le 12 \end{cases}$$

where t is measured in hours, with $t = 0$ corresponding to 4 P.M. Plot the graph of P in the viewing rectangle $[0, 12] \times [0, 80]$.
Source: A. C. Nielsen Co.

SOLUTION ✔   We enter the function

$$y2 = (0.01354x^4 - 0.49375x^3 + 2.58333x^2 + 3.8x + 31.60704)(x \ge 0)(x \le 8)$$
$$+ (1.35x^2 - 33.05x + 208)(x > 8)(x \le 12)$$

The graph of P is shown in Figure T5.

FIGURE T5
The graph of P in the viewing rectangle $[0, 12] \times [0, 80]$

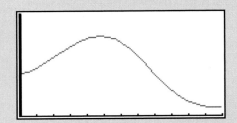

Exercises

In Exercises 1–10, use a graphing utility to plot the graph of *f* and to spot the points of discontinuity of *f*. Then use analytical means to verify your observation and find all points of discontinuity.

1. $f(x) = \dfrac{2}{x^2 - x}$

2. $f(x) = \dfrac{2x + 1}{x^2 + x - 2}$

3. $f(x) = \dfrac{\sqrt{x}}{x^2 - x - 2}$

4. $f(x) = \dfrac{3}{\sqrt{x}(x + 1)}$

5. $f(x) = \dfrac{6x^3 + x^2 - 2x}{2x^2 - x}$

6. $f(x) = \dfrac{2x^3 - x^2 - 13x - 6}{2x^2 - 5x - 3}$

7. $f(x) = \dfrac{2x^4 - 3x^3 - 2x^2}{2x^2 - 3x - 2}$

8. $f(x) = \dfrac{6x^4 - x^3 + 5x^2 - 1}{6x^2 - x - 1}$

9. $f(x) = \dfrac{x^3 + x^2 - 2x}{x^4 + 2x^3 - x - 2}$

Hint: $x^4 + 2x^3 - x - 2 = (x^3 - 1)(x + 2)$

10. $f(x) = \dfrac{x^3 - x}{x^{4/3} - x + x^{1/3} - 1}$

Hint: $x^{4/3} - x + x^{1/3} - 1 = (x^{1/3} - 1)(x + 1)$
Can you explain why part of the graph is missing?

In Exercises 11–14, use a graphing utility to plot the graph of *f* in the indicated viewing rectangle.

11. $f(x) = \begin{cases} -1 & \text{if } x \le 1 \\ x + 1 & \text{if } x > 1; [-5, 5] \times [-2, 8] \end{cases}$

12. $f(x) = \begin{cases} \frac{1}{3}x^2 - 2x & \text{if } x \le 3 \\ -x + 6 & \text{if } x > 3; [0, 7] \times [-5, 5] \end{cases}$

13. $f(x) = \begin{cases} 2 & \text{if } x \le 0 \\ \sqrt{4 - x^2} & \text{if } x > 0; [-2, 2] \times [-4, 4] \end{cases}$

14. $f(x) = \begin{cases} -x^2 + x + 2 & \text{if } x \le 1 \\ 2x^3 - x^2 - 4 & \text{if } x > 1; [-4, 4] \times [-5, 5] \end{cases}$

15. **FLIGHT PATH OF A PLANE** The function

$$f(x) = \begin{cases} 0 & \text{if } 0 \le x < 1 \\ -0.00411523x^3 + 0.0679012x^2 & \\ \quad - 0.123457x + 0.0596708 & \text{if } 1 \le x < 10 \\ 1.5 & \text{if } 10 \le x \le 100 \end{cases}$$

where both x and $f(x)$ are measured in units of 1000 ft, describes the flight path of a plane taking off from the origin and climbing to an altitude of 15,000 ft. Plot the graph of f to visualize the trajectory of the plane.

16. **HOME SHOPPING INDUSTRY** According to industry sources, revenue from the home shopping industry for the years since its inception may be approximated by the function

$$R(t) = \begin{cases} -0.03t^3 + 0.25t^2 - 0.12t & \text{if } 0 \le t \le 3 \\ 0.57t - 0.63 & \text{if } 3 < t \le 11 \end{cases}$$

where $R(t)$ measures the revenue in billions of dollars and t is measured in years, with $t = 0$ corresponding to the beginning of 1984. Plot the graph of R.

Source: Paul Kagan Associates

Table 2.4 (continued)

x stands for	y stands for	$\dfrac{f(a+h)-f(a)}{h}$ measures the	$\displaystyle\lim_{h\to 0}\dfrac{f(a+h)-f(a)}{h}$ measures the
Time	**Volume of sales** at time x	Average rate of change in the volume of sales over the time interval $[a, a + h]$	Instantaneous rate of change in the volume of sales at time $x = a$
Time	**Population** of *Drosophila* (fruit flies) at time x	Average rate of growth of the fruit fly population over the time interval $[a, a + h]$	Instantaneous rate of change of the fruit fly population at time $x = a$
Temperature in a chemical reaction	**Amount of product formed in the chemical reaction** when the temperature is x degrees	Average rate of formation of chemical product over the temperature range $[a, a + h]$	Instantaneous rate of formation of chemical product when the temperature is a degrees

DIFFERENTIABILITY AND CONTINUITY

In practical applications, one encounters functions that fail to be **differentiable**—that is, do not have a derivative at certain values in the domain of the function f. It can be shown that a continuous function f fails to be differentiable at a point $x = a$ when the graph of f makes an abrupt change of direction at that point. We call such a point a "corner." A function also fails to be differentiable at a point where the tangent line is vertical since the slope of a vertical line is undefined. These cases are illustrated in Figure 2.54.

FIGURE 2.54

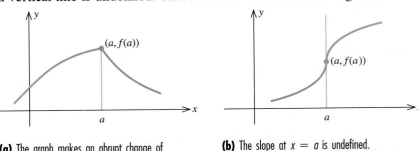

(a) The graph makes an abrupt change of direction at $x = a$.

(b) The slope at $x = a$ is undefined.

The next example illustrates a function that is not differentiable at a point.

EXAMPLE 8   Mary works at the B&O department store, where, on a weekday, she is paid $6 per hour for the first 8 hours and $9 per hour for overtime. The function

$$f(x) = \begin{cases} 6x & \text{if } 0 \le x \le 8 \\ 9x - 24 & \text{if } 8 < x \end{cases}$$

gives Mary's earnings on a weekday in which she worked x hours. Sketch the graph of the function f and explain why it is not differentiable at $x = 8$.

SOLUTION ✔   The graph of f is shown in Figure 2.55. Observe that the graph of f has a corner at $x = 8$ and consequently is not differentiable at $x = 8$.

FIGURE 2.55
The function f is not differentiable at (8, 48).

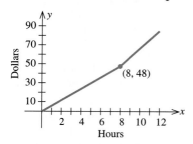

We close this section by mentioning the connection between the continuity and the differentiability of a function at a given value $x = a$ in the domain of f. By reexamining the function of Example 8, it becomes clear that f is continuous everywhere and, in particular, when $x = 8$. This shows that in general the continuity of a function at a point $x = a$ does not necessarily imply the differentiability of the function at that point. The converse, however, is true: If a function f is differentiable at a point $x = a$, then it is continuous there.

Exploring with Technology

1. Use a graphing utility to plot the graph of $f(x) = x^{1/3}$ in the viewing rectangle $[-2, 2] \times [-2, 2]$.
2. Use the graphing utility to draw the tangent line to the graph of f at the point $(0, 0)$. Can you explain why the process breaks down?

Differentiability and Continuity

If a function is differentiable at $x = a$, then it is continuous at $x = a$.

For a proof of this result, see Exercise 59, page 174.

EXAMPLE 9   Figure 2.56 depicts a portion of the graph of a function. Explain why the function fails to be differentiable at each of the points $x = a, b, c, d, e, f,$ and g.

FIGURE 2.56
The graph of this function is not differentiable at the points a–g.

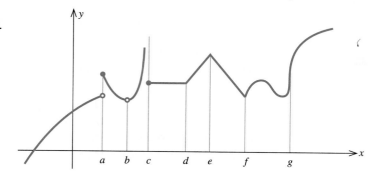

SOLUTION ✔ The function fails to be differentiable at the points $x = a$, b, and c because it is discontinuous at each of these points. The derivative of the function does not exist at $x = d$, e, and f because it has a kink at each of these points. Finally, the function is not differentiable at $x = g$ because the tangent line is vertical at that point.   ■ ■ ■ ■

> **Group Discussion**
> Suppose a function f is differentiable at $x = a$. Can there be two tangent lines to the graphs of f at the point $(a, f(a))$? Explain your answer.

SELF-CHECK EXERCISES 2.6

1. Let $f(x) = -x^2 - 2x + 3$.
   a. Find the derivative f' of f, using the definition of the derivative.
   b. Find the slope of the tangent line to the graph of f at the point $(0, 3)$.
   c. Find the rate of change of f when $x = 0$.
   d. Find an equation of the tangent line to the graph of f at the point $(0, 3)$.
   e. Sketch the graph of f and the tangent line to the curve at the point $(0, 3)$.

2. The losses (in millions of dollars) due to bad loans extended chiefly in agriculture, real estate, shipping, and energy by the Franklin Bank are estimated to be

$$A = f(t) = -t^2 + 10t + 30 \qquad (0 \leq t \leq 10)$$

   where t is the time in years ($t = 0$ corresponds to the beginning of 1994). How fast were the losses mounting at the beginning of 1997? At the beginning of 1999? How fast will the losses be mounting at the beginning of 2001? Interpret your results.

Solutions to Self-Check Exercises 2.6 can be found on page 174.

2.6 Exercises

1. **AVERAGE WEIGHT OF AN INFANT** The following graph shows the weight measurements of the average infant from the time of birth ($t = 0$) through age 2 ($t = 24$). By computing the slopes of the respective tangent lines, estimate the rate of change of the average infant's weight when $t = 3$ and when $t = 18$. What is the average rate of change in the average infant's weight over the first year of life?

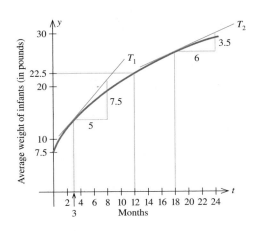

2. **FORESTRY** The following graph shows the volume of wood produced in a single-species forest. Here $f(t)$ is

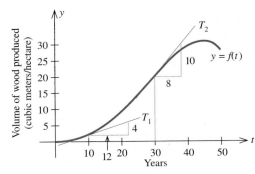

measured in cubic meters per hectare and t is measured in years. By computing the slopes of the respective tangent lines, estimate the rate at which the wood grown is changing at the beginning of year 10 and at the beginning of year 30.

Source: The Random House Encyclopedia

3. **TV-VIEWING PATTERNS** The following graph shows the percentage of U.S. households watching television during a 24-hr period on a weekday ($t = 0$ corresponds to 6 A.M.). By computing the slopes of the respective tangent lines, estimate the rate of change of the percentage of households watching television at 4 P.M. and 11 P.M.

Source: A. C. Nielsen Company

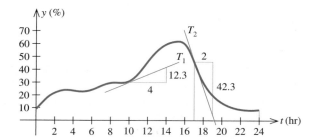

4. **CROP YIELD** Productivity and yield of cultivated crops are often reduced by insect pests. The following graph shows the relationship between the yield of a certain crop, $f(x)$, as a function of the density of aphids x. (Aphids are small insects that suck plant juices.) Here, $f(x)$ is measured in kilograms/4000 square meters, and x is measured in hundreds of aphids per bean stem. By computing the slopes of the respective tangent lines, estimate the rate of change of the crop yield with respect to the density of aphids when that density is 200 aphids/bean stem and when it is 800 aphids/bean stem.

Source: The Random House Encyclopedia

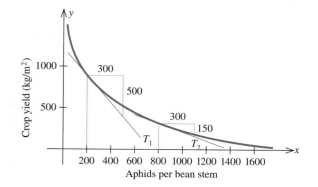

5. The position of car A and car B, starting out side by side and traveling along a straight road, is given by $s = f(t)$ and $s = g(t)$, respectively, where s is measured in feet and t is measured in seconds (see the accompanying figure).

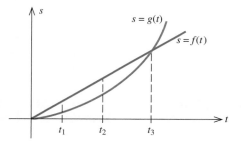

a. Which car is traveling faster at t_1?
b. What can you say about the speed of the cars at t_2?
Hint: Compare tangent lines.
c. Which car is traveling faster at t_3?
d. What can you say about the positions of the cars at t_3?

6. The velocity of car A and car B, starting out side by side and traveling along a straight road, is given by $v = f(t)$ and $v = g(t)$, respectively, where v is measured in feet/second and t is measured in seconds (see the accompanying figure).

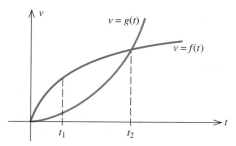

a. What can you say about the velocity and acceleration of the two cars at t_1? (Acceleration is the rate of change of velocity.)
b. What can you say about the velocity and acceleration of the two cars at t_2?

7. **EFFECT OF A BACTERICIDE ON BACTERIA** In the following figure, $f(t)$ gives the population P_1 of a certain bacteria culture at time t after a portion of bactericide A was introduced into the population at $t = 0$. The graph of g gives the population P_2 of a similar bacteria culture at time t after a portion of bactericide B was introduced into the population at $t = 0$.

a. Which population is decreasing faster at t_1?
b. Which population is decreasing faster at t_2?

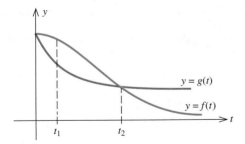

c. Which bactericide is more effective in reducing the population of bacteria in the short run? In the long run?

8. MARKET SHARE The following figure shows the devastating effect the opening of a new discount department store had on an established department store in a small town. The revenue of the discount store at time t (in months) is given by $f(t)$ million dollars, whereas the revenue of the established department store at time t is given by $g(t)$ million dollars. Answer the following questions by giving the value of t at which the specified event took place.

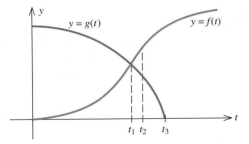

a. The revenue of the established department store is decreasing at the slowest rate.
b. The revenue of the established department store is decreasing at the fastest rate.
c. The revenue of the discount store first overtakes that of the established store.
d. The revenue of the discount store is increasing at the fastest rate.

In Exercises 9–16, use the four-step process to find the slope of the tangent line to the graph of the given function at any point.

9. $f(x) = 13$ **10.** $f(x) = -6$

11. $f(x) = 2x + 7$ **12.** $f(x) = 8 - 4x$

13. $f(x) = 3x^2$ **14.** $f(x) = -\dfrac{1}{2}x^2$

15. $f(x) = -x^2 + 3x$ **16.** $f(x) = 2x^2 + 5x$

In Exercises 17–22, find the slope of the tangent line to the graph of each function at the given point and determine an equation of the tangent line.

17. $f(x) = 2x + 7$ at $(2, 11)$

18. $f(x) = -3x + 4$ at $(-1, 7)$

19. $f(x) = 3x^2$ at $(1, 3)$

20. $f(x) = 3x - x^2$ at $(-2, -10)$

21. $f(x) = -\dfrac{1}{x}$ at $\left(3, -\dfrac{1}{3}\right)$

22. $f(x) = \dfrac{3}{2x}$ at $\left(1, \dfrac{3}{2}\right)$

23. Let $f(x) = 2x^2 + 1$.
   a. Find the derivative f' of f.
   b. Find an equation of the tangent line to the curve at the point $(1, 3)$.
   c. Sketch the graph of f.

24. Let $f(x) = x^2 + 6x$.
   a. Find the derivative f' of f.
   b. Find the point on the graph of f where the tangent line to the curve is horizontal.
   Hint: Find the value of x for which $f'(x) = 0$.
   c. Sketch the graph of f and the tangent line to the curve at the point found in part (b).

25. Let $f(x) = x^2 - 2x + 1$.
   a. Find the derivative f' of f.
   b. Find the point on the graph of f where the tangent line to the curve is horizontal.
   c. Sketch the graph of f and the tangent line to the curve at the point found in part (b).
   d. What is the rate of change of f at this point?

26. Let $f(x) = \dfrac{1}{x - 1}$.
   a. Find the derivative f' of f.
   b. Find an equation of the tangent line to the curve at the point $(-1, -\frac{1}{2})$.
   c. Sketch the graph of f.

27. Let $y = f(x) = x^2 + x$.
   a. Find the average rate of change of y with respect to x in the interval from $x = 2$ to $x = 3$, from $x = 2$ to $x = 2.5$, and from $x = 2$ to $x = 2.1$.
   b. Find the (instantaneous) rate of change of y at $x = 2$.
   c. Compare the results obtained in part (a) with that of part (b).

28. Let $y = f(x) = x^2 - 4x$.
  a. Find the average rate of change of y with respect to x in the interval from $x = 3$ to $x = 4$, from $x = 3$ to $x = 3.5$, and from $x = 3$ to $x = 3.1$.
  b. Find the (instantaneous) rate of change of y at $x = 3$.
  c. Compare the results obtained in part (a) with that of part (b).

29. VELOCITY OF A CAR Suppose the distance s (in feet) covered by a car moving along a straight road t sec after starting from rest is given by the function $f(t) = 2t^2 + 48t$.
  a. Calculate the average velocity of the car over the time intervals [20, 21], [20, 20.1], and [20, 20.01].
  b. Calculate the (instantaneous) velocity of the car when $t = 20$.
  c. Compare the results of part (a) with that of part (b).

30. VELOCITY OF A BALL THROWN INTO THE AIR A ball is thrown straight up with an initial velocity of 128 ft/sec, so that its height (in feet) after t sec is given by $s(t) = 128t - 16t^2$.
  a. What is the average velocity of the ball over the time intervals [2, 3], [2, 2.5], and [2, 2.1]?
  b. What is the instantaneous velocity at time $t = 2$?
  c. What is the instantaneous velocity at time $t = 5$? Is the ball rising or falling at this time?
  d. When will the ball hit the ground?

31. During the construction of a high-rise building, a worker accidentally dropped his portable electric screwdriver from a height of 400 ft. After t sec, the screwdriver had fallen a distance of $s = 16t^2$ ft.
  a. How long did it take the screwdriver to reach the ground?
  b. What was the average velocity of the screwdriver between the time it was dropped and the time it hit the ground?
  c. What was the velocity of the screwdriver at the time it hit the ground?

32. A hot air balloon rises vertically from the ground so that its height after t sec is $h = \frac{1}{2}t^2 + \frac{1}{2}t$ ft ($0 \le t \le 60$).
  a. What is the height of the balloon at the end of 40 sec?
  b. What is the average velocity of the balloon between $t = 0$ and $t = 40$?
  c. What is the velocity of the balloon at the end of 40 sec?

33. At a temperature of 20°C, the volume V (in liters) of 1.33 g of O_2 is related to its pressure p (in atmospheres) by the formula $V = 1/p$.

  a. What is the average rate of change of V with respect to p as p increases from $p = 2$ to $p = 3$?
  b. What is the rate of change of V with respect to p when $p = 2$?

34. COST OF PRODUCING SURFBOARDS The total cost $C(x)$ (in dollars) incurred by the Aloha Company in manufacturing x surfboards a day is given by

$$C(x) = -10x^2 + 300x + 130 \qquad (0 \le x \le 15)$$

  a. Find $C'(x)$.
  b. What is the rate of change of the total cost when the level of production is ten surfboards a day?
  c. What is the average cost Aloha incurs in manufacturing ten surfboards a day?

35. EFFECT OF ADVERTISING ON PROFIT The quarterly profit (in thousands of dollars) of Cunningham Realty is given by

$$P(x) = -\frac{1}{3}x^2 + 7x + 30 \qquad (0 \le x \le 50)$$

where x (in thousands of dollars) is the amount of money Cunningham spends on advertising per quarter.
  a. Find $P'(x)$.
  b. What is the rate of change of Cunningham's quarterly profit if the amount it spends on advertising is $10,000/quarter ($x = 10$) and $30,000/quarter ($x = 30$)?

36. DEMAND FOR TENTS The demand function for the Sportsman 5×7 tents is given by

$$p = f(x) = -0.1x^2 - x + 40$$

where p is measured in dollars and x is measured in units of a thousand.
  a. Find the average rate of change in the unit price of a tent if the quantity demanded is between 5000 and 5050 tents; between 5000 and 5010 tents.
  b. What is the rate of change of the unit price if the quantity demanded is 5000?

37. A COUNTRY'S GDP The gross domestic product (GDP) of a certain country is projected to be

$$N(t) = t^2 + 2t + 50 \qquad (0 \le t \le 5)$$

billion dollars t yr from now. What will be the rate of change of the country's GDP 2 yr and 4 yr from now?

38. GROWTH OF BACTERIA Under a set of controlled laboratory conditions, the size of the population of a certain bacteria culture at time t (in minutes) is described by the function

$$P = f(t) = 3t^2 + 2t + 1$$

Find the rate of population growth at $t = 10$ min.

In Exercises 39–43, let x and $f(x)$ represent the given quantities. Fix $x = a$ and let h be a small positive number. Give an interpretation of the quantities

$$\frac{f(a+h) - f(a)}{h} \quad \textbf{and} \quad \lim_{h \to 0} \frac{f(a+h) - f(a)}{h}$$

39. x denotes time, and $f(x)$ denotes the population of seals at time x.

40. x denotes time, and $f(x)$ denotes the prime interest rate at time x.

41. x denotes time, and $f(x)$ denotes a country's industrial production.

42. x denotes the level of production of a certain commodity, and $f(x)$ denotes the total cost incurred in producing x units of the commodity.

43. x denotes altitude, and $f(x)$ denotes atmospheric pressure.

In each of Exercises 44–49, the graph of a function is shown. For each function, state whether or not (a) $f(x)$ has a limit at $x = a$, (b) $f(x)$ is continuous at $x = a$, and (c) $f(x)$ is differentiable at $x = a$. Justify your answers.

44.

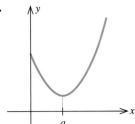

45.

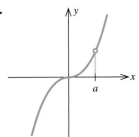

46.

47.

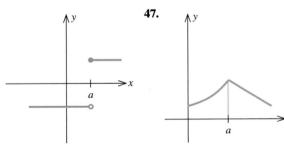

48.

49.

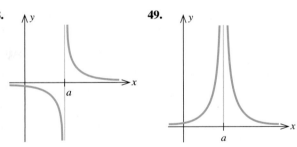

50. The distance s (in feet) covered by a motorcycle traveling in a straight line and starting from rest in t sec is given by the function

$$s(t) = -0.1t^3 + 2t^2 + 24t$$

Calculate the motorcycle's average velocity over the time interval $[2, 2 + h]$ for $h = 1, 0.1, 0.01, 0.001, 0.0001$, and 0.00001 and use your results to guess at the motorcycle's instantaneous velocity at $t = 2$.

51. The daily total cost $C(x)$ incurred by Trappee and Sons, Inc., for producing x cases of Texa-Pep hot sauce is given by

$$C(x) = 0.000002x^3 + 5x + 400$$

Calculate

$$\frac{C(100 + h) - C(100)}{h}$$

for $h = 1, 0.1, 0.01, 0.001$, and 0.0001 and use your results to estimate the rate of change of the total cost function when the level of production is 100 cases/day.

In Exercises 52 and 53, determine whether the statement is true or false. If it is true, explain why it is true. If it is false, give an example to show why it is false.

52. If f is continuous at $x = a$, then f is differentiable at $x = a$.

53. If f is continuous at $x = a$ and g is differentiable at $x = a$, then $\lim_{x \to a} f(x)g(x) = f(a)g(a)$.

54. Sketch the graph of the function $f(x) = |x + 1|$ and show that the function does not have a derivative at $x = -1$.

55. Sketch the graph of the function $f(x) = 1/(x - 1)$ and show that the function does not have a derivative at $x = 1$.

56. Let

$$f(x) = \begin{cases} x^2 & \text{if } x \leq 1 \\ ax + b & \text{if } x > 1 \end{cases}$$

Find the values of a and b so that f is continuous and has a derivative at $x = 1$. Sketch the graph of f.

57. Sketch the graph of the function $f(x) = x^{2/3}$. Is the function continuous at $x = 0$? Does $f'(0)$ exist? Why, or why not?

58. Prove that the derivative of the function $f(x) = |x|$ for $x \neq 0$ is given by

$$f'(x) = \begin{cases} 1 & \text{if } x > 0 \\ -1 & \text{if } x < 0 \end{cases}$$

Hint: Recall the definition of the absolute value of a number.

59. Show that if a function f is differentiable at a point $x = a$, then f must be continuous at that point.

Hint: Write

$$f(x) - f(a) = \left[\frac{f(x) - f(a)}{x - a} \right] (x - a)$$

Use the product rule for limits and the definition of the derivative to show that

$$\lim_{x \to a} [f(x) - f(a)] = 0$$

SOLUTIONS TO SELF-CHECK EXERCISES 2.6

1. a. $f'(x) = \lim_{h \to 0} \dfrac{f(x + h) - f(x)}{h}$

$$= \lim_{h \to 0} \frac{[-(x + h)^2 - 2(x + h) + 3] - (-x^2 - 2x + 3)}{h}$$

$$= \lim_{h \to 0} \frac{-x^2 - 2xh - h^2 - 2x - 2h + 3 + x^2 + 2x - 3}{h}$$

$$= \lim_{h \to 0} \frac{h(-2x - h - 2)}{h}$$

$$= \lim_{h \to 0} (-2x - h - 2) = -2x - 2$$

b. From the result of part (a), we see that the slope of the tangent line to the graph of f at any point $(x, f(x))$ is given by

$$f'(x) = -2x - 2$$

In particular, the slope of the tangent line to the graph of f at $(0, 3)$ is

$$f'(0) = -2$$

c. The rate of change of f when $x = 0$ is given by $f'(0) = -2$, or -2 units/unit change in x.

d. Using the result from part (b), we see that an equation of the required tangent line is

$$y - 3 = -2(x - 0)$$
$$y = -2x + 3$$

e.

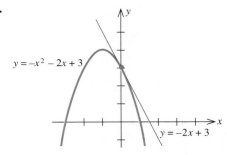

$y = -x^2 - 2x + 3$

$y = -2x + 3$

2. The rate of change of the losses at any time t is given by

$$f'(t) = \lim_{h \to 0} \frac{f(t+h) - f(t)}{h}$$

$$= \lim_{h \to 0} \frac{[-(t+h)^2 + 10(t+h) + 30] - (-t^2 + 10t + 30)}{h}$$

$$= \lim_{h \to 0} \frac{-t^2 - 2th - h^2 + 10t + 10h + 30 + t^2 - 10t - 30}{h}$$

$$= \lim_{h \to 0} \frac{h(-2t - h + 10)}{h}$$

$$= \lim_{h \to 0} (-2t - h + 10)$$

$$= -2t + 10$$

Therefore, the rate of change of the losses suffered by the bank at the beginning of 1997 ($t = 3$) was

$$f'(3) = -2(3) + 10 = 4$$

That is, the losses were increasing at the rate of $4 million/year. At the beginning of 1999 ($t = 5$),

$$f'(5) = -2(5) + 10 = 0$$

and we see that the growth in losses due to bad loans was zero at this point. At the beginning of 2001 ($t = 7$),

$$f'(7) = -2(7) + 10 = -4$$

and we conclude that the losses will be decreasing at the rate of $4 million/year.

Group projects for this chapter can be found at the Brooks/Cole Web site:
http://www.brookscole.com/product/0534378439

GRAPHING A FUNCTION AND ITS TANGENT LINES

We can use a graphing utility to plot the graph of a function f and the tangent line at any point on the graph.

EXAMPLE 1 Let $f(x) = x^2 - 4x$.

a. Find an equation of the tangent line to the graph of f at the point $(3, -3)$.
b. Plot both the graph of f and the tangent line found in part (a) on the same set of axes.

SOLUTION ✔ **a.** The slope of the tangent line at any point on the graph of f is given by $f'(x)$. But from Example 4 (page 162) we find $f'(x) = 2x - 4$. Using this result, we see that the slope of the required tangent line is

$$f'(3) = 2(3) - 4 = 2$$

Finally, using the point-slope form of the equation of a line, we find that an equation of the tangent line is

$$y - (-3) = 2(x - 3)$$
$$y + 3 = 2x - 6$$
$$y = 2x - 9.$$

b. The graph of f in the standard viewing rectangle and the tangent line of interest are shown in Figure T1.

FIGURE T1
The graph of $f(x) = x^2 - 4x$ and the tangent line $y = 2x - 9$ in the standard viewing rectangle

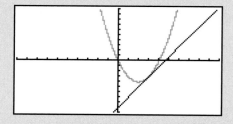

REMARK Some graphing utilities will draw both the graph of a function f and the tangent line to the graph of f at a specified point when the function and the specified value of x are entered.

FINDING THE DERIVATIVE OF A FUNCTION AT A GIVEN POINT

The numerical derivative operation of a graphing utility can be used to give an approximate value of the derivative of a function for a given value of x.

EXAMPLE 2

Let $f(x) = \sqrt{x}$.

a. Use the numerical derivative operation of a graphing utility to find the derivative of f at (4, 2).
b. Find an equation of the tangent line to the graph of f at (4, 2).
c. Plot the graph of f and the tangent line on the same set of axes.

SOLUTION ✔

a. Using the numerical derivative operation of a graphing utility, we find that

$$f'(4) = \frac{1}{4}$$

b. An equation of the required tangent line is

$$y - 2 = \frac{1}{4}(x - 4)$$

$$y = \frac{1}{4}x + 1$$

c. The graph of f and the tangent line in the viewing rectangle $[0, 15] \times [0, 4]$ is shown in Figure T2.

FIGURE T2
The graph of $f(x) = \sqrt{x}$ and the tangent line $y = \frac{1}{4}x + 1$ in the viewing rectangle $[0, 15] \times [0, 4]$

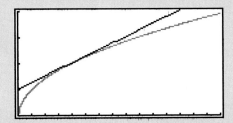

■ ■ ■ ■

In Exercises 1–10, (a) find an equation of the tangent line to the graph of *f* at the indicated point and (b) use a graphing utility to plot the graph of *f* and the tangent line on the same set of axes. Use a suitable viewing rectangle.

1. $f(x) = 4x - 3;\ (2, 5)$

2. $f(x) = -2x + 5;\ (1, 3)$

3. $f(x) = 2x^2 + x;\ (-2, 6)$

4. $f(x) = -x^2 + 2x;\ (1, 1)$

5. $f(x) = 2x^2 + x - 3;\ (2, 7)$

6. $f(x) = -3x^2 + 2x - 1;\ (1, -2)$

7. $f(x) = x + \dfrac{1}{x};\ (1, 2)$

8. $f(x) = x - \dfrac{1}{x};\ (1, 0)$

9. $f(x) = \sqrt{x};\ (4, 2)$

10. $f(x) = \dfrac{1}{\sqrt{x}};\ \left(4, \dfrac{1}{2}\right)$

In Exercises 11–20, (a) use the numerical derivative operation of a graphing utility to find the derivative of *f* for the given value of *x* (to two desired places of accuracy), (b) find an equation of the tangent line to the graph of *f* at the indicated point, and (c) plot the graph of *f* and the tangent line on the same set of axes. Use a suitable viewing rectangle.

11. $f(x) = x^3 + x + 1;\ x = 1;\ (1, 3)$

12. $f(x) = -2x^3 + 3x^2 + 2;\ x = -1;\ (-1, 7)$

13. $f(x) = x^4 - 3x^2 + 1;\ x = 2;\ (2, 5)$

14. $f(x) = -x^4 + 3x + 1;\ x = 1;\ (1, 3)$

15. $f(x) = x - \sqrt{x};\ x = 4;\ (4, 2)$

16. $f(x) = x^{3/2} - x;\ x = 4;\ (4, 4)$

17. $f(x) = \dfrac{1}{x+1};\ x = 1;\ \left(1, \dfrac{1}{2}\right)$

18. $f(x) = \dfrac{x}{x+1};\ x = 3;\ \left(3, \dfrac{3}{4}\right)$

19. $f(x) = x\sqrt{x^2 + 1};\ x = 2;\ (2, 2\sqrt{5})$

20. $f(x) = \dfrac{x}{\sqrt{x^2 + 1}};\ x = 1;\ \left(1, \dfrac{\sqrt{2}}{2}\right)$

CHAPTER 2 Summary of Principal Formulas and Terms

Formulas

1. Average rate of change of f over $[x, x + h]$
or
Slope of the secant line to the graph of f through $(x, f(x))$ and $(x + h, f(x + h))$
or
Difference quotient

$$\frac{f(x + h) - f(x)}{h}$$

2. Instantaneous rate of change of f at $(x, f(x))$
or
Slope of tangent line to the graph of f at $(x, f(x))$ at x
or
Derivative of f

$$\lim_{h \to 0} \frac{f(x + h) - f(x)}{h}$$

Terms

function	power function
domain	demand function
range	supply function
independent variable	market equilibrium
dependent variable	equilibrium quantity
ordered pairs	equilibrium price
graph of a function	limit of a function
graph of an equation	indeterminate form
vertical-line test	limit of a function at infinity
composite function	right-hand limit of a function
polynomial function	left-hand limit of a function
linear function	continuity of a function at a point
quadratic function	secant line
cubic function	tangent line to the graph of f
rational function	differentiable function

CHAPTER 2 REVIEW EXERCISES

1. Find the domain of each function:

a. $f(x) = \sqrt{9 - x}$ **b.** $f(x) = \dfrac{x + 3}{2x^2 - x - 3}$

2. Let $f(x) = 3x^2 + 5x - 2$. Find:
a. $f(-2)$ **b.** $f(a + 2)$
c. $f(2a)$ **d.** $f(a + h)$

SELF-CHECK EXERCISES 3.1

1. Find the derivative of each of the following functions using the rules of differentiation.
   a. $f(x) = 1.5x^2 + 2x^{1.5}$
   b. $g(x) = 2\sqrt{x} + \dfrac{3}{\sqrt{x}}$

2. Let $f(x) = 2x^3 - 3x^2 + 2x - 1$.
   a. Compute $f'(x)$.
   b. What is the slope of the tangent line to the graph of f when $x = 2$?
   c. What is the rate of change of the function f at $x = 2$?

3. A certain country's gross domestic product (GDP) (in millions of dollars) is described by the function

$$G(t) = -2t^3 + 45t^2 + 20t + 6000 \qquad (0 \le t \le 11)$$

where $t = 0$ corresponds to the beginning of 1990.
   a. At what rate was the GDP changing at the beginning of 1995? At the beginning of 1997? At the beginning of 2000?
   b. What was the average rate of growth of the GDP over the period 1995–2000?

Solutions to Self-Check Exercises 3.1 can be found on page 197.

3.1 Exercises

In Exercises 1–34, find the derivative of the function _f_ by using the rules of differentiation.

1. $f(x) = -3$

2. $f(x) = 365$

3. $f(x) = x^5$

4. $f(x) = x^7$

5. $f(x) = x^{2.1}$

6. $f(x) = x^{0.8}$

7. $f(x) = 3x^2$

8. $f(x) = -2x^3$

9. $f(r) = \pi r^2$

10. $f(r) = \frac{4}{3}\pi r^3$

11. $f(x) = 9x^{1/3}$

12. $f(x) = \frac{5}{4}x^{4/5}$

13. $f(x) = 3\sqrt{x}$

14. $f(u) = \dfrac{2}{\sqrt{u}}$

15. $f(x) = 7x^{-12}$

16. $f(x) = 0.3x^{-1.2}$

17. $f(x) = 5x^2 - 3x + 7$

18. $f(x) = x^3 - 3x^2 + 1$

19. $f(x) = -x^3 + 2x^2 - 6$

20. $f(x) = x^4 - 2x^2 + 5$

21. $f(x) = 0.03x^2 - 0.4x + 10$

22. $f(x) = 0.002x^3 - 0.05x^2 + 0.1x - 20$

23. $f(x) = \dfrac{x^3 - 4x^2 + 3}{x}$

24. $f(x) = \dfrac{x^3 + 2x^2 + x - 1}{x}$

25. $f(x) = 4x^4 - 3x^{5/2} + 2$

26. $f(x) = 5x^{4/3} - \dfrac{2}{3}x^{3/2} + x^2 - 3x + 1$

27. $f(x) = 3x^{-1} + 4x^{-2}$

28. $f(x) = -\dfrac{1}{3}(x^{-3} - x^6)$

29. $f(t) = \dfrac{4}{t^4} - \dfrac{3}{t^3} + \dfrac{2}{t}$

30. $f(x) = \dfrac{5}{x^3} - \dfrac{2}{x^2} - \dfrac{1}{x} + 200$

31. $f(x) = 2x - 5\sqrt{x}$

32. $f(t) = 2t^2 + \sqrt{t^3}$

33. $f(x) = \dfrac{2}{x^2} - \dfrac{3}{x^{1/3}}$

34. $f(x) = \dfrac{3}{x^3} + \dfrac{4}{\sqrt{x}} + 1$

35. Let $f(x) = 2x^3 - 4x$. Find:
   a. $f'(-2)$ **b.** $f'(0)$ **c.** $f'(2)$

36. Let $f(x) = 4x^{5/4} + 2x^{3/2} + x$. Find:
   a. $f'(0)$ **b.** $f'(16)$

(continued on p. 194)

FINDING THE RATE OF CHANGE OF A FUNCTION

We can use the numerical derivative operation of a graphing utility to obtain the value of the derivative at a given value of x. Since the derivative of a function $f(x)$ measures the rate of change of the function with respect to x, the numerical derivative operation can be used to answer questions pertaining to the rate of change of one quantity y with respect to another quantity x, where $y = f(x)$, for a specific value of x.

EXAMPLE 1

Let $y = 3t^3 + 2\sqrt{t}$.

a. Use the numerical derivative operation of a graphing utility to find how fast y is changing with respect to t when $t = 1$.
b. Verify the result of part (a), using the rules of differentiation of this section.

SOLUTION ✔

a. Write $f(t) = 3t^3 + 2\sqrt{t}$. Using the numerical derivative operation of a graphing utility, we find that the rate of change of y with respect to t when $t = 1$ is given by $f'(1) = 10$.

b. Here, $f(t) = 3t^3 + 2t^{1/2}$ and

$$f'(t) = 9t^2 + 2\left(\frac{1}{2}t^{-1/2}\right) = 9t^2 + \frac{1}{\sqrt{t}}$$

Using this result, we see that when $t = 1$, y is changing at the rate of

$$f'(1) = 9(1^2) + \frac{1}{\sqrt{1}} = 10$$

units per unit change in t, as obtained earlier. ■ ■ ■ ■

EXAMPLE 2

According to the U.S. Department of Energy and the Shell Development Company, a typical car's fuel economy depends on the speed it is driven and is approximated by the function

$$f(x) = 0.00000310315x^4 - 0.000455174x^3$$
$$+ 0.00287869x^2 + 1.25986x \qquad (0 \le x \le 75)$$

where x is measured in mph and $f(x)$ is measured in miles per gallon (mpg).

a. Use a graphing utility to graph the function f on the interval $[0, 75]$.
b. Find the rate of change of f when $x = 20$ and $x = 50$.
c. Interpret your results.
Source: U.S. Department of Energy and the Shell Development Company

FIGURE T1

SOLUTION ✔

a. The result is shown in Figure T1.
b. Using the numerical derivative operation of a graphing utility, we see that $f'(20) = 0.9280996$. The rate of change of f when $x = 50$ is given by $f'(50) = -0.314501$.
c. The results of part (b) tell us that when a typical car is being driven at 20 mph, its fuel economy increases at the rate of approximately 0.9 mpg per 1 mph increase in its speed. At a speed of 50 mph, its fuel economy decreases at the rate of approximately 0.3 mpg per 1 mph increase in its speed. ■ ■ ■ ■

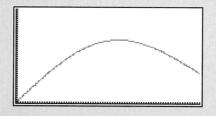

In Exercises 1–6, use the numerical derivative operation of a graphing utility to find the rate of change of f(x) at the given value of x. Give your answer accurate to four decimal places.

1. $f(x) = 4x^5 - 3x^3 + 2x^2 + 1; x = 0.5$

2. $f(x) = -x^5 + 4x^2 + 3; x = 0.4$

3. $f(x) = x - 2\sqrt{x}; x = 3$

4. $f(x) = \dfrac{\sqrt{x} - 1}{x}; x = 2$

5. $f(x) = x^{1/2} - x^{1/3}; x = 1.2$

6. $f(x) = 2x^{5/4} + x; x = 2$

7. CARBON MONOXIDE IN THE ATMOSPHERE The projected average global atmosphere concentration of carbon monoxide is approximated by the function

$$f(t) = 0.881443t^4 - 1.45533t^3 + 0.695876t^2$$
$$+ 2.87801t + 293 \qquad (0 \le t \le 4)$$

where t is measured in 40-yr intervals, with $t = 0$ corresponding to the beginning of 1860 and $f(t)$ is measured in parts per million by volume.
a. Use a graphing utility to plot the graph of f in the viewing rectangle $[0, 4] \times [280, 400]$.
b. Use a graphing utility to estimate how fast the projected average global atmospheric concentration of carbon monoxide was changing at the beginning of the year 1900 ($t = 1$) and at the beginning of 2000 ($t = 3.5$).
Source: "Beyond the Limits," Meadows et al.

8. GROWTH OF HMOs Based on data compiled by the Group Health Association of America, the number of people receiving their care in an HMO (Health Maintenance Organization) from the beginning of 1984 through 1994 is approximated by the function

$$f(t) = 0.0514t^3 - 0.853t^2 + 6.8147t$$
$$+ 15.6524 \qquad (0 \le t \le 11)$$

where $f(t)$ gives the number of people in millions and t is measured in years, with $t = 0$ corresponding to the beginning of 1984.
a. Use a graphing utility to plot the graph of f in the viewing window $[0, 12] \times [0, 80]$.
b. How fast was the number of people receiving their care in an HMO growing at the beginning of 1992?
Source: Group Health Association of America

9. HOME SALES According to the Greater Boston Real Estate Board—Multiple Listing Service, the average number of days a single-family home remains for sale from listing to accepted offer is approximated by the function

$$f(t) = 0.0171911t^4 - 0.662121t^3 + 6.18083t^2$$
$$- 8.97086t + 53.3357 \qquad (0 \le t \le 10)$$

where t is measured in years, with $t = 0$ corresponding to the beginning of 1984.
a. Use a graphing utility to plot the graph of f in the viewing rectangle $[0, 12] \times [0, 120]$.
b. How fast was the average number of days a single-family home remained for sale from listing to accepted offer changing at the beginning of 1984 ($t = 0$)? At the beginning of 1988 ($t = 4$)?
Source: Greater Boston Real Estate Board—Multiple Listing Service

10. SPREAD OF HIV The estimated number of children newly infected with HIV through mother-to-child contact worldwide is given by

$$f(t) = -0.2083t^3 + 3.0357t^2 + 44.0476t$$
$$+ 200.2857 \qquad (0 \le t \le 12)$$

where $f(t)$ is measured in thousands and t is measured in years with $t = 0$ corresponding to the beginning of 1990.
a. Use a graphing utility to plot the graph of f in the viewing rectangle $[0, 12] \times [0, 800]$.
b. How fast was the estimated number of children newly infected with HIV through mother-to-child contact worldwide increasing at the beginning of the year 2000?
Source: United Nations

11. MANUFACTURING CAPACITY Data obtained from the Federal Reserve shows that the annual change in manufacturing capacity between 1988 and 1994 is given by

$$f(t) = 0.0388889t^3 - 0.283333t^2$$
$$+ 0.477778t + 2.04286 \qquad (0 \le t \le 6)$$

where $f(t)$ is a percentage and t is measured in years, with $t = 0$ corresponding to the beginning of 1988.
a. Use a graphing utility to plot the graph of f in the viewing rectangle $[0, 8] \times [0, 4]$.
b. How fast was $f(t)$ changing at the beginning of 1990 ($t = 2$)? At the beginning of 1992 ($t = 4$)?
Source: Federal Reserve

In Exercises 37–40, find the given limit by evaluating the derivative of a suitable function at an appropriate point.

Hint: Look at the definition of the derivative.

37. $\lim\limits_{h \to 0} \dfrac{(1 + h)^3 - 1}{h}$

38. $\lim\limits_{x \to 1} \dfrac{x^5 - 1}{x - 1}$

Hint: Let $h = x - 1$.

39. $\lim\limits_{h \to 0} \dfrac{3(2 + h)^2 - (2 + h) - 10}{h}$

40. $\lim\limits_{t \to 0} \dfrac{1 - (1 + t)^2}{t(1 + t)^2}$

In Exercises 41–44, find the slope and an equation of the tangent line to the graph of the function f at the specified point.

41. $f(x) = 2x^2 - 3x + 4;\ (2, 6)$

42. $f(x) = -\dfrac{5}{3}x^2 + 2x + 2;\ \left(-1, -\dfrac{5}{3}\right)$

43. $f(x) = x^4 - 3x^3 + 2x^2 - x + 1;\ (1, 0)$

44. $f(x) = \sqrt{x} + \dfrac{1}{\sqrt{x}};\ \left(4, \dfrac{5}{2}\right)$

45. Let $f(x) = x^3$.
 a. Find the point on the graph of f where the tangent line is horizontal.
 b. Sketch the graph of f and draw the horizontal tangent line.

46. Let $f(x) = x^3 - 4x^2$. Find the point(s) on the graph of f where the tangent line is horizontal.

47. Let $f(x) = x^3 + 1$.
 a. Find the point(s) on the graph of f where the slope of the tangent line is equal to 12.
 b. Find the equation(s) of the tangent line(s) of part (a).
 c. Sketch the graph of f showing the tangent line(s).

48. Let $f(x) = \frac{2}{3}x^3 + x^2 - 12x + 6$. Find the values of x for which:
 a. $f'(x) = -12$ **b.** $f'(x) = 0$
 c. $f'(x) = 12$

49. Let $f(x) = \frac{1}{4}x^4 - \frac{1}{3}x^3 - x^2$. Find the point(s) on the graph of f where the slope of the tangent line is equal to:
 a. $-2x$ **b.** 0 **c.** $10x$

50. A straight line perpendicular to and passing through the point of tangency of the tangent line is called the *normal* to the curve. Find an equation of the tangent line and the normal to the curve $y = x^3 - 3x + 1$ at the point $(2, 3)$.

51. GROWTH OF A CANCEROUS TUMOR   The volume of a spherical cancer tumor is given by the function

$$V(r) = \frac{4}{3}\pi r^3$$

where r is the radius of the tumor in centimeters. Find the rate of change in the volume of the tumor when:

 a. $r = \dfrac{2}{3}$ cm   **b.** $r = \dfrac{5}{4}$ cm

52. VELOCITY OF BLOOD IN AN ARTERY   The velocity (in centimeters per second) of blood r centimeters from the central axis of an artery is given by

$$v(r) = k(R^2 - r^2)$$

where k is a constant and R is the radius of the artery (see the accompanying figure). Suppose that $k = 1000$ and $R = 0.2$ cm. Find $v(0.1)$ and $v'(0.1)$ and interpret your results.

Blood vessel

53. EFFECT OF STOPPING ON AVERAGE SPEED   According to data from a study by General Motors, the average speed of your trip A (in mph) is related to the number of stops per mile you make on the trip x by the equation

$$A = \frac{26.5}{x^{0.45}}$$

Compute dA/dx for $x = 0.25$ and $x = 2$ and interpret your results.
Source: General Motors

54. WORKER EFFICIENCY   An efficiency study conducted for the Elektra Electronics Company showed that the number of "Space Commander" walkie-talkies assembled by the average worker t hr after starting work at 8 A.M. is given by

$$N(t) = -t^3 + 6t^2 + 15t$$

 a. Find the rate at which the average worker will be assembling walkie-talkies t hr after starting work.
 b. At what rate will the average worker be assembling walkie-talkies at 10 A.M.? At 11 A.M.?
 c. How many walkie-talkies will the average worker assemble between 10 A.M. and 11 A.M.?

55. **CONSUMER PRICE INDEX** An economy's consumer price index (CPI) is described by the function

$$I(t) = -0.2t^3 + 3t^2 + 100 \qquad (0 \le t \le 10)$$

where $t = 0$ corresponds to 1990.
   a. At what rate was the CPI changing in 1995? In 1997? In 2000?
   b. What was the average rate of increase in the CPI over the period from 1995 to 2000?

56. **EFFECT OF ADVERTISING ON SALES** The relationship between the amount of money x that the Cannon Precision Instruments Corporation spends on advertising and the company's total sales $S(x)$ is given by the function

$$S(x) = -0.002x^3 + 0.6x^2 + x + 500 \qquad (0 \le x \le 200)$$

where x is measured in thousands of dollars. Find the rate of change of the sales with respect to the amount of money spent on advertising. Are Cannon's total sales increasing at a faster rate when the amount of money spent on advertising is (a) $100,000 or (b) $150,000?

57. **POPULATION GROWTH** A study prepared for a Sunbelt town's chamber of commerce projected that the town's population in the next 3 yr will grow according to the rule

$$P(t) = 50,000 + 30t^{3/2} + 20t$$

where $P(t)$ denotes the population t months from now. How fast will the population be increasing 9 mo and 16 mo from now?

58. **CURBING POPULATION GROWTH** Five years ago, the government of a Pacific island state launched an extensive propaganda campaign toward curbing the country's population growth. According to the Census Department, the population (measured in thousands of people) for the following 4 yr was

$$P(t) = -\frac{1}{3}t^3 + 64t + 3000$$

where t is measured in years and $t = 0$ at the start of the campaign. Find the rate of change of the population at the end of years 1, 2, 3, and 4. Was the plan working?

59. **CONSERVATION OF SPECIES** A certain species of turtle faces extinction because dealers collect truckloads of turtle eggs to be sold as aphrodisiacs. After severe conservation measures are implemented, it is hoped that the turtle population will grow according to the rule

$$N(t) = 2t^3 + 3t^2 - 4t + 1000 \qquad (0 \le t \le 10)$$

where $N(t)$ denotes the population at the end of year t.

Find the rate of growth of the turtle population when $t = 2$ and $t = 8$. What will be the population 10 yr after the conservation measures are implemented?

60. **FLIGHT OF A ROCKET** The altitude (in feet) of a rocket t sec into flight is given by

$$s = f(t) = -2t^3 + 114t^2 + 480t + 1 \qquad (t \ge 0)$$

   a. Find an expression v for the rocket's velocity at any time t.
   b. Compute the rocket's velocity when $t = 0, 20, 40$, and 60. Interpret your results.
   c. Using the results from the solution to part (b), find the maximum altitude attained by the rocket.
   Hint: At its highest point, the velocity of the rocket is zero.

61. **STOPPING DISTANCE OF A RACING CAR** During a test by the editors of an auto magazine, the stopping distance s (in feet) of the MacPherson X-2 racing car conformed to the rule

$$s = f(t) = 120t - 15t^2 \qquad (t \ge 0)$$

where t was the time (in seconds) after the brakes were applied.
   a. Find an expression for the car's velocity v at any time t.
   b. What was the car's velocity when the brakes were first applied?
   c. What was the car's stopping distance for that particular test?
   Hint: The stopping time is found by setting $v = 0$.

62. **DEMAND FUNCTIONS** The demand function for the Luminar desk lamp is given by

$$p = f(x) = -0.1x^2 - 0.4x + 35$$

where x is the quantity demanded (measured in thousands) and p is the unit price in dollars.
   a. Find $f'(x)$.
   b. What is the rate of change of the unit price when the quantity demanded is 10,000 units ($x = 10$)? What is the unit price at that level of demand?

63. **INCREASE IN TEMPORARY WORKERS** According to the Labor Department, the number of temporary workers (in millions) is estimated to be

$$N(t) = 0.025t^2 + 0.255t + 1.505 \qquad (0 \le t \le 5)$$

where t is measured in years, with $t = 0$ corresponding to 1991.
   a. How many temporary workers were there at the beginning of 1994?
   b. How fast was the number of temporary workers growing at the beginning of 1994?
   Source: Labor Department

SELF-CHECK EXERCISES 3.2

1. Find the derivative of $f(x) = \dfrac{2x+1}{x^2-1}$.

2. What is the slope of the tangent line to the graph of
$$f(x) = (x^2 + 1)(2x^3 - 3x^2 + 1)$$
   at the point $(2, 25)$? How fast is the function f changing when $x = 2$?

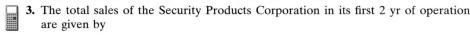

3. The total sales of the Security Products Corporation in its first 2 yr of operation are given by
$$S = f(t) = \frac{0.3t^3}{1 + 0.4t^2} \qquad (0 \le t \le 2)$$
   where S is measured in millions of dollars and $t = 0$ corresponds to the date Security Products began operations. How fast were the sales increasing at the beginning of the company's second year of operation?

Solutions to Self-Check Exercises 3.2 can be found on page 210.

3.2 Exercises

In Exercises 1–30, find the derivative of the given function.

1. $f(x) = 2x(x^2 + 1)$

2. $f(x) = 3x^2(x - 1)$

3. $f(t) = (t - 1)(2t + 1)$

4. $f(x) = (2x + 3)(3x - 4)$

5. $f(x) = (3x + 1)(x^2 - 2)$

6. $f(x) = (x + 1)(2x^2 - 3x + 1)$

7. $f(x) = (x^3 - 1)(x + 1)$

8. $f(x) = (x^3 - 12x)(3x^2 + 2x)$

9. $f(w) = (w^3 - w^2 + w - 1)(w^2 + 2)$

10. $f(x) = \dfrac{1}{5}x^5 + (x^2 + 1)(x^2 - x - 1) + 28$

11. $f(x) = (5x^2 + 1)(2\sqrt{x} - 1)$

12. $f(t) = (1 + \sqrt{t})(2t^2 - 3)$

13. $f(x) = (x^2 - 5x + 2)\left(x - \dfrac{2}{x}\right)$

14. $f(x) = (x^3 + 2x + 1)\left(2 + \dfrac{1}{x^2}\right)$

15. $f(x) = \dfrac{1}{x - 2}$

16. $g(x) = \dfrac{3}{2x + 4}$

17. $f(x) = \dfrac{x - 1}{2x + 1}$

18. $f(t) = \dfrac{1 - 2t}{1 + 3t}$

19. $f(x) = \dfrac{1}{x^2 + 1}$

20. $f(u) = \dfrac{u}{u^2 + 1}$

21. $f(s) = \dfrac{s^2 - 4}{s + 1}$

22. $f(x) = \dfrac{x^3 - 2}{x^2 + 1}$

23. $f(x) = \dfrac{\sqrt{x}}{x^2 + 1}$

24. $f(x) = \dfrac{x^2 + 1}{\sqrt{x}}$

25. $f(x) = \dfrac{x^2 + 2}{x^2 + x + 1}$

26. $f(x) = \dfrac{x + 1}{2x^2 + 2x + 3}$

27. $f(x) = \dfrac{(x + 1)(x^2 + 1)}{x - 2}$

28. $f(x) = (3x^2 - 1)\left(x^2 - \dfrac{1}{x}\right)$

29. $f(x) = \dfrac{x}{x^2 - 4} - \dfrac{x - 1}{x^2 + 4}$

30. $f(x) = \dfrac{x + \sqrt{3x}}{3x - 1}$

In Exercises 31–34, suppose f and g are functions that are differentiable at $x = 1$ and that $f(1) = 2$, $f'(1) = -1$, $g(1) = -2$, $g'(1) = 3$. Find the value of $h'(1)$.

31. $h(x) = f(x)g(x)$

32. $h(x) = (x^2 + 1)g(x)$

33. $h(x) = \dfrac{xf(x)}{x + g(x)}$

34. $h(x) = \dfrac{f(x)g(x)}{f(x) - g(x)}$

In Exercises 35–38, find the derivative of each of the given functions and evaluate $f'(x)$ at the given value of x.

35. $f(x) = (2x - 1)(x^2 + 3); x = 1$

36. $f(x) = \dfrac{2x + 1}{2x - 1}; x = 2$

37. $f(x) = \dfrac{x}{x^4 - 2x^2 - 1}; x = -1$

38. $f(x) = (\sqrt{x} + 2x)(x^{3/2} - x); x = 4$

In Exercises 39–42, find the slope and an equation of the tangent line to the graph of the function f at the specified point.

39. $f(x) = (x^3 + 1)(x^2 - 2); (2, 18)$

40. $f(x) = \dfrac{x^2}{x + 1}; \left(2, \dfrac{4}{3}\right)$

41. $f(x) = \dfrac{x + 1}{x^2 + 1}; (1, 1)$

42. $f(x) = \dfrac{1 + 2x^{1/2}}{1 + x^{3/2}}; \left(4, \dfrac{5}{9}\right)$

43. Find an equation of the tangent line to the graph of the function $f(x) = (x^3 + 1)(3x^2 - 4x + 2)$ at the point $(1, 2)$.

44. Find an equation of the tangent line to the graph of the function $f(x) = \dfrac{3x}{x^2 - 2}$ at the point $(2, 3)$.

45. Let $f(x) = (x^2 + 1)(2 - x)$. Find the point(s) on the graph of f where the tangent line is horizontal.

46. Let $f(x) = \dfrac{x}{x^2 + 1}$. Find the point(s) on the graph of f where the tangent line is horizontal.

47. Find the point(s) on the graph of the function $f(x) = (x^2 + 6)(x - 5)$ where the slope of the tangent line is equal to -2.

48. Find the point(s) on the graph of the function $f(x) = \dfrac{x + 1}{x - 1}$ where the slope of the tangent line is equal to $-1/2$.

49. A straight line perpendicular to and passing through the point of tangency of the tangent line is called the *normal* to the curve. Find the equation of the tangent line and the normal to the curve $y = \dfrac{1}{1 + x^2}$ at the point $(1, \tfrac{1}{2})$.

50. CONCENTRATION OF A DRUG IN THE BLOODSTREAM   The concentration of a certain drug in a patient's bloodstream t hr after injection is given by

$$C(t) = \frac{0.2t}{t^2 + 1}$$

a. Find the rate at which the concentration of the drug is changing with respect to time.
b. How fast is the concentration changing $\tfrac{1}{2}$ hr, 1 hr, and 2 hr after the injection?

51. COST OF REMOVING TOXIC WASTE   A city's main well was recently found to be contaminated with trichloroethylene, a cancer-causing chemical, as a result of an abandoned chemical dump leaching chemicals into the water. A proposal submitted to the city's council members indicates that the cost, measured in millions of dollars, of removing x percent of the toxic pollutant is given by

$$C(x) = \frac{0.5x}{100 - x}$$

Find $C'(80)$, $C'(90)$, $C'(95)$, and $C'(99)$ and interpret your results.

52. DRUG DOSAGES   Thomas Young has suggested the following rule for calculating the dosage of medicine for children 1 to 12 yr old. If a denotes the adult dosage (in milligrams) and if t is the child's age (in years), then the child's dosage is given by

$$D(t) = \frac{at}{t + 12}$$

Suppose that the adult dosage of a substance is 500 mg. Find an expression that gives the rate of change of a child's dosage with respect to the child's age. What is the rate of change of a child's dosage with respect to his or her age for a 6-yr-old child? For a 10-yr-old child?

53. EFFECT OF BACTERICIDE   The number of bacteria $N(t)$ in a certain culture t min after an experimental bactericide is introduced obeys the rule

$$N(t) = \frac{10{,}000}{1 + t^2} + 2000$$

Find the rate of change of the number of bacteria in the culture 1 minute after and 2 minutes after the bactericide is introduced. What is the population of the bacteria in the culture 1 min and 2 min after the bactericide is introduced?

54. DEMAND FUNCTIONS   The demand function for the Sicard wristwatch is given by

$$d(x) = \frac{50}{0.01x^2 + 1} \qquad (0 \le x \le 20)$$

where x (measured in units of a thousand) is the quantity demanded per week and $d(x)$ is the unit price in dollars.

a. Find $d'(x)$.

b. Find $d'(5)$, $d'(10)$, and $d'(15)$ and interpret your results.

55. LEARNING CURVES From experience, the Emory Secretarial School knows that the average student taking Advanced Typing will progress according to the rule

$$N(t) = \frac{60t + 180}{t + 6} \qquad (t \geq 0)$$

where $N(t)$ measures the number of words per minute the student can type after t wk in the course.

a. Find an expression for $N'(t)$.

b. Compute $N'(t)$ for $t = 1$, 3, 4, and 7 and interpret your results.

c. Sketch the graph of the function N. Does it confirm the results obtained in part (b)?

d. What will be the average student's typing speed at the end of the 12-wk course?

56. BOX OFFICE RECEIPTS The total worldwide box office receipts for a long-running movie are approximated by the function

$$T(x) = \frac{120x^2}{x^2 + 4}$$

where $T(x)$ is measured in millions of dollars and x is the number of years since the movie's release. How fast are the total receipts changing 1 yr, 3 yr, and 5 yr after its release?

57. FORMALDEHYDE LEVELS A study on formaldehyde levels in 900 homes indicates that emissions of various chemicals can decrease over time. The formaldehyde level (parts per million) in an average home in the study is given by

$$f(t) = \frac{0.055t + 0.26}{t + 2} \qquad (0 \leq t \leq 12)$$

where t is the age of the house in years. How fast is the formaldehyde level of the average house dropping when it is new? At the beginning of its fourth year?

Source: Bonneville Power Administration

58. POPULATION GROWTH A major corporation is building a 4325-acre complex of homes, offices, stores, schools, and churches in the rural community of Glen Cove. As a result of this development, the planners have estimated that Glen Cove's population (in thousands) t yr from now will be given by

$$P(t) = \frac{25t^2 + 125t + 200}{t^2 + 5t + 40}$$

a. Find the rate at which Glen Cove's population is changing with respect to time.

b. What will be the population after 10 yr? At what rate will the population be increasing when $t = 10$?

In Exercises 59–62, determine whether the statement is true or false. If it is true, explain why it is true. If it is false, give an example to show why it is false.

59. If f and g are differentiable, then

$$\frac{d}{dx}[f(x)g(x)] = f'(x)g'(x).$$

60. If f is differentiable, then

$$\frac{d}{dx}[xf(x)] = f(x) + xf'(x).$$

61. If f is differentiable, then

$$\frac{d}{dx}\left[\frac{f(x)}{x^2}\right] = \frac{f'(x)}{2x}.$$

62. If f, g, and h are differentiable, then

$$\frac{d}{dx}\left[\frac{f(x)g(x)}{h(x)}\right]$$
$$= \frac{f'(x)g(x)h(x) + f(x)g'(x)h(x) - f(x)g(x)h'(x)}{[h(x)]^2}.$$

63. Extend the product rule for differentiation to the following case involving the product of three differentiable functions: Let $h(x) = u(x)v(x)w(x)$ and show that $h'(x) = u(x)v(x)w'(x) + u(x)v'(x)w(x) + u'(x)v(x)w(x)$.

Hint: Let $f(x) = u(x)v(x)$, $g(x) = w(x)$, and $h(x) = f(x)g(x)$, and apply the product rule to the function h.

64. Prove the quotient rule for differentiation (Rule 6).

Hint: Verify the following steps:

a. $\dfrac{k(x+h) - k(x)}{h} = \dfrac{f(x+h)g(x) - f(x)g(x+h)}{hg(x+h)g(x)}$

b. By adding $[-f(x)g(x) + f(x)g(x)]$ to the numerator and simplifying,

$$\frac{k(x+h) - k(x)}{h} = \frac{1}{g(x+h)g(x)}$$
$$\times \left\{ \left[\frac{f(x+h) - f(x)}{h}\right] \cdot g(x) \right.$$
$$\left. - \left[\frac{g(x+h) - g(x)}{h}\right] \cdot f(x) \right\}$$

c. $k'(x) = \lim\limits_{h \to 0} \dfrac{k(x+h) - k(x)}{h}$

$$= \frac{g(x)f'(x) - f(x)g'(x)}{[g(x)]^2}$$

THE PRODUCT AND QUOTIENT RULES

EXAMPLE 1

Let $f(x) = (2\sqrt{x} + 0.5x)\left(0.3x^3 + 2x - \dfrac{0.3}{x}\right)$. Find $f'(0.2)$.

SOLUTION ✔

Using the numerical derivative operation of a graphing utility, we find

$$f'(0.2) = 6.47974948127$$
　　　　■ ■ ■ ■

EXAMPLE 2

Importance of Time in Treating Heart Attacks

According to the American Heart Association, the treatment benefit for heart attacks depends on the time to treatment and is described by the function

$$f(t) = \frac{-16.94t + 203.28}{t + 2.0328} \qquad (0 \le t \le 12)$$

where t is measured in hours and $f(t)$ is a percentage.

a. Use a graphing utility to graph the function f using the viewing rectangle [0, 13] × [0, 120].
b. Use a graphing utility to find the derivative of f when $t = 0$ and $t = 2$.
c. Interpret the results obtained in part (b).
Source: American Heart Association

SOLUTION ✔

a. The graph of f is shown in Figure T1.

FIGURE T1

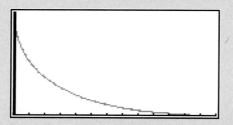

b. Using the numerical derivative operation of a graphing utility, we find

$$f'(0) \approx -57.5266$$

$$f'(2) \approx -14.6165$$

c. The results of part (b) show that the treatment benefit drops off at the rate of 58% per hour at the time when the heart attack first occurs and falls off at the rate of 15% per hour when the time to treatment is 2 hours. Thus, it is extremely urgent that a patient suffering a heart attack receive medical attention as soon as possible. ■ ■ ■ ■

Exercises

In Exercises 1–6, use the numerical derivative operation of a graphing utility to find the rate of change of $f(x)$ at the given value of x. Give your answer accurate to four decimal places.

1. $f(x) = (2x^2 + 1)(x^3 + 3x + 4); x = -0.5$

2. $f(x) = (\sqrt{x} + 1)(2x^2 + x - 3); x = 1.5$

3. $f(x) = \dfrac{\sqrt{x} - 1}{\sqrt{x} + 1}; x = 3$

4. $f(x) = \dfrac{\sqrt{x}(x^2 + 4)}{x^3 + 1}; x = 4$

5. $f(x) = \dfrac{\sqrt{x}(1 + x^{-1})}{x + 1}; x = 1$

6. $f(x) = \dfrac{x^2(2 + \sqrt{x})}{1 + \sqrt{x}}; x = 1$

7. NEW CONSTRUCTION JOBS The president of a major housing construction company claims that the number of construction jobs created in the next t months is given by

$$f(t) = 1.42 \left(\frac{7t^2 + 140t + 700}{3t^2 + 80t + 550} \right)$$

where $f(t)$ is measured in millions of jobs per year. At what rate will construction jobs be created 1 yr from now, assuming her projection is correct?

8. POPULATION GROWTH A major corporation is building a 4325-acre complex of homes, offices, stores, schools, and churches in the rural community of Glen Cove. As a result of this development, the planners have estimated that Glen Cove's population (in thousands) t yr from now will be given by

$$P(t) = \frac{25t^2 + 125t + 200}{t^2 + 5t + 40}$$

a. What will be the population 10 yr from now?
b. At what rate will the population be increasing 10 yr from now?

SOLUTIONS TO SELF-CHECK EXERCISES 3.2

1. We use the quotient rule to obtain

$$f'(x) = \frac{(x^2 - 1)\dfrac{d}{dx}(2x + 1) - (2x + 1)\dfrac{d}{dx}(x^2 - 1)}{(x^2 - 1)^2}$$

$$= \frac{(x^2 - 1)(2) - (2x + 1)(2x)}{(x^2 - 1)^2}$$

$$= \frac{2x^2 - 2 - 4x^2 - 2x}{(x^2 - 1)^2}$$

$$= \frac{-2x^2 - 2x - 2}{(x^2 - 1)^2}$$

$$= -\frac{2(x^2 + x + 1)}{(x^2 - 1)^2}$$

2. The slope of the tangent line to the graph of f at any point is given by

$$f'(x) = (x^2 + 1)\frac{d}{dx}(2x^3 - 3x^2 + 1)$$

$$+ (2x^3 - 3x^2 + 1)\frac{d}{dx}(x^2 + 1)$$

$$= (x^2 + 1)(6x^2 - 6x) + (2x^3 - 3x^2 + 1)(2x)$$

In particular, the slope of the tangent line to the graph of f when $x = 2$ is

$$f'(2) = (2^2 + 1)[6(2^2) - 6(2)]$$

$$+ [2(2^3) - 3(2^2) + 1][2(2)]$$

$$= 60 + 20 = 80$$

Note that it is not necessary to simplify the expression for $f'(x)$ since we are required only to evaluate the expression at $x = 2$. We also conclude, from this result, that the function f is changing at the rate of 80 units/unit change in x when $x = 2$.

3. The rate at which the company's total sales are changing at any time t is given by

$$S'(t) = \frac{(1 + 0.4t^2)\dfrac{d}{dt}(0.3t^3) - (0.3t^3)\dfrac{d}{dt}(1 + 0.4t^2)}{(1 + 0.4t^2)^2}$$

$$= \frac{(1 + 0.4t^2)(0.9t^2) - (0.3t^3)(0.8t)}{(1 + 0.4t^2)^2}$$

Therefore, at the beginning of the second year of operation, Security Products' sales were increasing at the rate of

$$S'(1) = \frac{(1 + 0.4)(0.9) - (0.3)(0.8)}{(1 + 0.4)^2} = 0.520408$$

or \$520,408/year.

3.3 The Chain Rule

This section introduces another rule of differentiation called the **chain rule.** When used in conjunction with the rules of differentiation developed in the last two sections, the chain rule enables us to greatly enlarge the class of functions we are able to differentiate.

THE CHAIN RULE

Consider the function $h(x) = (x^2 + x + 1)^2$. If we were to compute $h'(x)$ using only the rules of differentiation from the previous sections, then our approach might be to expand $h(x)$. Thus,

$$h(x) = (x^2 + x + 1)^2 = (x^2 + x + 1)(x^2 + x + 1)$$
$$= x^4 + 2x^3 + 3x^2 + 2x + 1$$

from which we find

$$h'(x) = 4x^3 + 6x^2 + 6x + 2$$

But what about the function $H(x) = (x^2 + x + 1)^{100}$? The same technique may be used to find the derivative of the function H, but the amount of work involved in this case would be prodigious! Consider, also, the function $G(x) = \sqrt{x^2 + 1}$. For each of the two functions H and G, the rules of differentiation of the previous sections cannot be applied directly to compute the derivatives H' and G'.

Observe that both H and G are **composite functions;** that is, each is composed of, or built up from, simpler functions. For example, the function H is composed of the two simpler functions $f(x) = x^2 + x + 1$ and $g(x) = x^{100}$ as follows:

$$H(x) = g[f(x)] = [f(x)]^{100}$$
$$= (x^2 + x + 1)^{100}$$

In a similar manner, we see that the function G is composed of the two simpler functions $f(x) = x^2 + 1$ and $g(x) = \sqrt{x}$. Thus,

$$G(x) = g[f(x)] = \sqrt{f(x)}$$
$$= \sqrt{x^2 + 1}$$

As a first step toward finding the derivative h' of a composite function $h = g \circ f$ defined by $h(x) = g[f(x)]$, we write

$$u = f(x) \qquad \text{and} \qquad y = g[f(x)] = g(u)$$

The dependency of h on g and f is illustrated in Figure 3.5. Since u is a function of x, we may compute the derivative of u with respect to x, if f is a differentiable function, obtaining $du/dx = f'(x)$. Next, if g is a differentiable function of u, we may compute the derivative of g with respect to u, obtaining $dy/du = g'(u)$. Now, since the function h is composed of the function g and the function

In Exercises 47–52, find $\dfrac{dy}{du}$, $\dfrac{du}{dx}$, and $\dfrac{dy}{dx}$.

47. $y = u^{4/3}$ and $u = 3x^2 - 1$

48. $y = \sqrt{u}$ and $u = 7x - 2x^2$

49. $y = u^{-2/3}$ and $u = 2x^3 - x + 1$

50. $y = 2u^2 + 1$ and $u = x^2 + 1$

51. $y = \sqrt{u} + \dfrac{1}{\sqrt{u}}$ and $u = x^3 - x$

52. $y = \dfrac{1}{u}$ and $u = \sqrt{x} + 1$

53. Suppose $F(x) = g(f(x))$ and $f(2) = 3$, $f'(2) = -3$, $g(3) = 5$, and $g'(3) = 4$. find $F'(2)$.

54. Suppose $h = f \circ g$. Find $h'(0)$ given that $f(0) = 6$, $f'(5) = -2$, $g(0) = 5$, and $g'(0) = 3$.

55. Suppose $F(x) = f(x^2 + 1)$. Find $F'(1)$ if $f'(2) = 3$.

56. Let $F(x) = f(f(x))$. Does it follow that $F'(x) = [f'(x)]^2$? Hint: Let $f(x) = x^2$.

57. Suppose $h = g \circ f$. Does it follow that $h' = g' \circ f'$? Hint: Let $f(x) = x$ and $g(x) = x^2$.

58. Suppose $h = f \circ g$. Show that $h' = (f' \circ g)g'$.

In Exercises 59–62, find an equation of the tangent line to the graph of the function at the given point.

59. $f(x) = (1 - x)(x^2 - 1)^2$; $(2, -9)$

60. $f(x) = \left(\dfrac{x + 1}{x - 1}\right)^2$; $(3, 4)$

61. $f(x) = x\sqrt{2x^2 + 7}$; $(3, 15)$

62. $f(x) = \dfrac{8}{\sqrt{x^2 + 6x}}$; $(2, 2)$

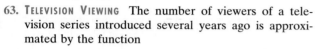 **63.** TELEVISION VIEWING The number of viewers of a television series introduced several years ago is approximated by the function

$$N(x) = (60 + 2x)^{2/3} \qquad (1 \le x \le 26)$$

where $N(x)$ (measured in millions) denotes the number of weekly viewers of the series in the xth week. Find the rate of increase of the weekly audience at the end of week 2 and at the end of week 12. How many viewers were there in week 2 and in week 24?

64. MALE LIFE EXPECTANCY Suppose the life expectancy of a male at birth in a certain country is described by the function

$$f(t) = 46.9(1 + 1.09t)^{0.1} \qquad (0 \le t \le 150)$$

where t is measured in years and $t = 0$ corresponds to the beginning of 1900. How long can a male born at the beginning of the year 2000 in that country expect to live? What is the rate of change of the life expectancy of a male born in that country at the beginning of the year 2000?

65. CONCENTRATION OF CARBON MONOXIDE IN THE AIR According to a joint study conducted by Oxnard's Environmental Management Department and a state government agency, the concentration of carbon monoxide in the air due to automobile exhaust t yr from now is given by

$$C(t) = 0.01(0.2t^2 + 4t + 64)^{2/3}$$

parts per million.
a. Find the rate at which the level of carbon monoxide is changing with respect to time.
b. Find the rate at which the level of carbon monoxide will be changing 5 yr from now.

66. CONTINUING EDUCATION ENROLLMENT The registrar of Kellogg University estimates that the total student enrollment in the Continuing Education division will be given by

$$N(t) = -\dfrac{20{,}000}{\sqrt{1 + 0.2t}} + 21{,}000$$

where $N(t)$ denotes the number of students enrolled in the division t yr from now. Find an expression for $N'(t)$. How fast is the student enrollment increasing currently? How fast will it be increasing 5 yr from now?

67. AIR POLLUTION According to the South Coast Air Quality Management District, the level of nitrogen dioxide, a brown gas that impairs breathing, present in the atmosphere on a certain May day in downtown Los Angeles is approximated by

$$A(t) = 0.03t^3(t - 7)^4 + 60.2 \qquad (0 \le t \le 7)$$

where $A(t)$ is measured in pollutant standard index and t is measured in hours, with $t = 0$ corresponding to 7 A.M.
a. Find $A'(t)$.
b. Find $A'(1)$, $A'(3)$, and $A'(4)$ and interpret your results.

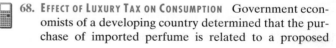 **68.** EFFECT OF LUXURY TAX ON CONSUMPTION Government economists of a developing country determined that the purchase of imported perfume is related to a proposed

"luxury tax" by the formula

$$N(x) = \sqrt{10{,}000 - 40x - 0.02x^2} \qquad (0 \le x \le 200)$$

where $N(x)$ measures the percentage of normal consumption of perfume when a "luxury tax" of x percent is imposed on it. Find the rate of change of $N(x)$ for taxes of 10%, 100%, and 150%.

69. PULSE RATE OF AN ATHLETE The pulse rate (the number of heartbeats per minute) of a long-distance runner t seconds after leaving the starting line is given by

$$P(t) = \frac{300\sqrt{\tfrac{1}{2}t^2 + 2t + 25}}{t + 25} \qquad (t \ge 0)$$

Compute $P'(t)$. How fast is the athlete's pulse rate increasing 10 sec, 60 sec, and 2 min into the run? What is her pulse rate 2 min into the run?

70. THURSTONE LEARNING MODEL Psychologist L. L. Thurstone suggested the following relationship between learning time T and the length of a list n:

$$T = f(n) = An\sqrt{n - b}$$

where A and b are constants that depend on the person and the task.
a. Compute dT/dn and interpret your result.
b. For a certain person and a certain task, suppose $A = 4$ and $b = 4$. Compute $f'(13)$ and $f'(29)$ and interpret your results.

71. OIL SPILLS In calm waters, the oil spilling from the ruptured hull of a grounded tanker spreads in all directions. Assuming that the area polluted is a circle and that its radius is increasing at a rate of 2 ft/sec, determine how fast the area is increasing when the radius of the circle is 40 ft.

72. ARTERIOSCLEROSIS Refer to Example 7, page 217. Suppose the radius of an individual's artery is 1 cm and the thickness of the plaque (in centimeters) t yr from now is given by

$$h = g(t) = \frac{0.5t^2}{t^2 + 10} \qquad (0 \le t \le 10)$$

How fast will the arterial opening be decreasing 5 yr from now?

73. TRAFFIC FLOW Opened in the late 1950s, the Central Artery in downtown Boston was designed to move 75,000 vehicles a day. The number of vehicles moved per day is approximated by the function

$$x = f(t) = 6.25t^2 + 19.75t + 74.75 \qquad (0 \le t \le 5)$$

where x is measured in thousands and t in decades, with $t = 0$ corresponding to the beginning of 1959. Suppose

the average speed of traffic flow in mph is given by

$$S = g(x) = -0.00075x^2 + 67.5 \qquad (75 \le x \le 350)$$

where x has the same meaning as before. What was the rate of change of the average speed of traffic flow at the beginning of 1999? What was the average speed of traffic flow at that time?

74. HOTEL OCCUPANCY RATES The occupancy rate of the all-suite Wonderland Hotel, located near an amusement park, is given by the function

$$r(t) = \frac{10}{81}t^3 - \frac{10}{3}t^2 + \frac{200}{9}t + 60 \qquad (0 \le t \le 12)$$

where t is measured in months and $t = 0$ corresponds to the beginning of January. Management has estimated that the monthly revenue (in thousands of dollars per month) is approximated by the function

$$R(r) = -\frac{3}{5000}r^3 + \frac{9}{50}r^2 \qquad (0 \le r \le 100)$$

where r is the occupancy rate.
a. Find an expression that gives the rate of change of Wonderland's occupancy rate with respect to time.
b. Find an expression that gives the rate of change of Wonderland's monthly revenue with respect to the occupancy rate.
c. What is the rate of change of Wonderland's monthly revenue with respect to time at the beginning of January? At the beginning of June?
Hint: Use the chain rule to find $R'(r(0))r'(0)$ and $R'(r(6))r'(6)$.

75. EFFECT OF HOUSING STARTS ON JOBS The president of a major housing construction firm claims that the number of construction jobs created is given by

$$N(x) = 1.42x$$

where x denotes the number of housing starts. Suppose the number of housing starts in the next t mo is expected to be

$$x(t) = \frac{7t^2 + 140t + 700}{3t^2 + 80t + 550}$$

million units/year. Find an expression that gives the rate at which the number of construction jobs will be created t mo from now. At what rate will construction jobs be created 1 yr from now?

76. DEMAND FOR PCs The quantity demanded per month, x, of a certain make of personal computer (PC) is related to the average unit price, p (in dollars), of PCs by the equation

$$x = f(p) = \frac{100}{9}\sqrt{810{,}000 - p^2}$$

It is estimated that t mo from now, the average price of a PC will be given by

$$p(t) = \frac{400}{1 + \frac{1}{8}\sqrt{t}} + 200 \qquad (0 \le t \le 60)$$

dollars. Find the rate at which the quantity demanded per month of the PCs will be changing 16 mo from now.

77. CRUISE SHIP BOOKINGS The management of Cruise World, operators of Caribbean luxury cruises, expects that the percentage of young adults booking passage on their cruises in the years ahead will rise dramatically. They have constructed the following model, which gives the percentage of young adult passengers in year t:

$$p = f(t) = 50\left(\frac{t^2 + 2t + 4}{t^2 + 4t + 8}\right) \qquad (0 \le t \le 5)$$

Young adults normally pick shorter cruises and generally spend less on their passage. The following model gives an approximation of the average amount of money R (in dollars) spent per passenger on a cruise when the percentage of young adults is p:

$$R(p) = 1000\left(\frac{p + 4}{p + 2}\right)$$

Find the rate at which the price of the average passage will be changing 2 yr from now.

In Exercises 78–81, determine whether the statement is true or false. If it is true, explain why it is true. If it is false, give an example to show why it is false.

78. If f and g are differentiable and $h = f \circ g$, then $h'(x) = f'[g(x)]g'(x)$.

79. If f is differentiable and c is a constant, then

$$\frac{d}{dx}[f(cx)] = cf'(cx).$$

80. If f is differentiable, then

$$\frac{d}{dx}\sqrt{f(x)} = \frac{f'(x)}{2\sqrt{f(x)}}$$

81. If f is differentiable, then

$$\frac{d}{dx}\left[f\left(\frac{1}{x}\right)\right] = f'\left(\frac{1}{x}\right)$$

82. In Section 3.1 we proved that

$$\frac{d}{dx}(x^n) = nx^{n-1}$$

for the special case when $n = 2$. Use the chain rule to show that

$$\frac{d}{dx}(x^{1/n}) = \frac{1}{n}x^{1/n-1}$$

for any nonzero integer n, assuming that $f(x) = x^{1/n}$ is differentiable.

Hint: Let $f(x) = x^{1/n}$ so that $[f(x)]^n = x$. Differentiate both sides with respect to x.

83. With the aid of Exercise 82, prove that

$$\frac{d}{dx}(x^r) = rx^{r-1}$$

for any rational number r.

Hint: Let $r = m/n$, where m and n are integers with $n \ne 0$, and write $x^r = (x^m)^{1/n}$.

SOLUTIONS TO SELF-CHECK EXERCISES 3.3

1. Rewriting, we have

$$f(x) = -(2x^2 - 1)^{-1/2}$$

Using the general power rule, we find

$$f'(x) = -\frac{d}{dx}(2x^2 - 1)^{-1/2}$$

$$= -\left(-\frac{1}{2}\right)(2x^2 - 1)^{-3/2}\frac{d}{dx}(2x^2 - 1)$$

$$= \frac{1}{2}(2x^2 - 1)^{-3/2}(4x)$$

$$= \frac{2x}{(2x^2 - 1)^{3/2}}$$

2. a. The life expectancy at birth of a female born at the beginning of 1980 is given by

$$g(80) = 50.02[1 + 1.09(80)]^{0.1} \approx 78.29$$

or approximately 78 yr. Similarly, the life expectancy at birth of a female born at the beginning of the year 2000 is given by

$$g(100) = 50.02[1 + 1.09(100)]^{0.1} \approx 80.04$$

or approximately 80 yr.

b. The rate of change of the life expectancy at birth of a female born at any time t is given by $g'(t)$. Using the general power rule, we have

$$g'(t) = 50.02 \frac{d}{dt}(1 + 1.09t)^{0.1}$$

$$= (50.02)(0.1)(1 + 1.09t)^{-0.9} \frac{d}{dt}(1 + 1.09t)$$

$$= (50.02)(0.1)(1.09)(1 + 1.09t)^{-0.9}$$

$$= 5.45218(1 + 1.09t)^{-0.9}$$

$$= \frac{5.45218}{(1 + 1.09t)^{0.9}}$$

3.4 Marginal Functions in Economics

Marginal analysis is the study of the rate of change of economic quantities. For example, an economist is not merely concerned with the value of an economy's gross domestic product (GDP) at a given time but is equally concerned with the rate at which it is growing or declining. In the same vein, a manufacturer is not only interested in the total cost corresponding to a certain level of production of a commodity but is also interested in the rate of change of the total cost with respect to the level of production, and so on. Let's begin with an example to explain the meaning of the adjective *marginal*, as used by economists.

COST FUNCTIONS

 EXAMPLE 1 Suppose the total cost in dollars incurred per week by the Polaraire Company for manufacturing x refrigerators is given by the total cost function

$$C(x) = 8000 + 200x - 0.2x^2 \qquad (0 \le x \le 400)$$

a. What is the actual cost incurred for manufacturing the 251st refrigerator?
b. Find the rate of change of the total cost function with respect to x when $x = 250$.
c. Compare the results obtained in parts (a) and (b).

SOLUTION ✔

a. The actual cost incurred in producing the 251st refrigerator is the difference between the total cost incurred in producing the first 251 refrigerators and the total cost of producing the first 250 refrigerators:

$$C(251) - C(250) = [8000 + 200(251) - 0.2(251)^2]$$
$$- [8000 + 200(250) - 0.2(250)^2]$$
$$= 45,599.8 - 45,500$$
$$= 99.80$$

or $99.80.

b. The rate of change of the total cost function C with respect to x is given by the derivative of C—that is, $C'(x) = 200 - 0.4x$. Thus, when the level of production is 250 refrigerators, the rate of change of the total cost with respect to x is given by

$$C'(250) = 200 - 0.4(250)$$
$$= 100$$

or $100.

c. From the solution to part (a), we know that the actual cost for producing the 251st refrigerator is $99.80. This answer is very closely approximated by the answer to part (b), $100. To see why this is so, observe that the difference $C(251) - C(250)$ may be written in the form

$$\frac{C(251) - C(250)}{1} = \frac{C(250 + 1) - C(250)}{1} = \frac{C(250 + h) - C(250)}{h}$$

where $h = 1$. In other words, the difference $C(251) - C(250)$ is precisely the average rate of change of the total cost function C over the interval $[250, 251]$, or, equivalently, the slope of the secant line through the points $(250, 45,500)$ and $(251, 45,599.8)$. However, the number $C'(250) = 100$ is the instantaneous rate of change of the total cost function C at $x = 250$, or, equivalently, the slope of the tangent line to the graph of C at $x = 250$.

Now when h is small, the average rate of change of the function C is a good approximation to the instantaneous rate of change of the function C, or, equivalently, the slope of the secant line through the points in question is a good approximation to the slope of the tangent line through the point in question. Thus, we may expect

$$C(251) - C(250) = \frac{C(251) - C(250)}{1} = \frac{C(250 + h) - C(250)}{h}$$
$$\approx \lim_{h \to 0} \frac{C(250 + h) - C(250)}{h} = C'(250)$$

which is precisely the case in this example.   ▨ ▨ ▨ ▨

The actual cost incurred in producing an additional unit of a certain commodity given that a plant is already at a certain level of operation is called the **marginal cost.** Knowing this cost is very important to management in their decision-making processes. As we saw in Example 1, the marginal cost is approximated by the rate of change of the total cost function evaluated at the appropriate point. For this reason, economists have defined the **marginal cost function** to be the derivative of the corresponding total cost function. In other words, if C is a total cost function, then the marginal cost function is defined to be its derivative C'. Thus, the adjective *marginal* is synonymous with *derivative of.*

EXAMPLE 2   A subsidiary of the Elektra Electronics Company manufactures a programmable pocket calculator. Management determined that the daily total cost of producing these calculators (in dollars) is given by

$$C(x) = 0.0001x^3 - 0.08x^2 + 40x + 5000$$

where x stands for the number of calculators produced.

a. Find the marginal cost function.
b. What is the marginal cost when $x = 200, 300, 400,$ and 600?
c. Interpret your results.

SOLUTION ✔   **a.** The marginal cost function C' is given by the derivative of the total cost function C. Thus,

$$C'(x) = 0.0003x^2 - 0.16x + 40$$

b. The marginal cost when $x = 200, 300, 400,$ and 600 is given by

$$C'(200) = 0.0003(200)^2 - 0.16(200) + 40 = 20$$
$$C'(300) = 0.0003(300)^2 - 0.16(300) + 40 = 19$$
$$C'(400) = 0.0003(400)^2 - 0.16(400) + 40 = 24$$
$$C'(600) = 0.0003(600)^2 - 0.16(600) + 40 = 52$$

or $20, $19, $24, and $52, respectively.

c. From the results of part (b), we see that Elektra's actual cost for producing the 201st calculator is approximately $20. The actual cost incurred for producing one additional calculator when the level of production is already 300 calculators is approximately $19, and so on. Observe that when the level of production is already 600 units, the actual cost of producing one additional unit is approximately $52. The higher cost for producing this additional unit when the level of production is 600 units may be the result of several factors, among them excessive costs incurred because of overtime or higher maintenance, production breakdown caused by greater stress and strain on the equipment, and so on. The graph of the total cost function appears in Figure 3.7.

■ ■ ■ ■

FIGURE 3.7
The cost of producing x calculators is given by $C(x)$.

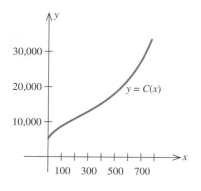

AVERAGE COST FUNCTIONS

Let's now introduce another marginal concept closely related to the marginal cost. Let $C(x)$ denote the total cost incurred in producing x units of a certain commodity. Then the **average cost** of producing x units of the commodity is obtained by dividing the total production cost by the number of units produced. This leads to the following definition.

Average Cost Function

Suppose $C(x)$ is a total cost function. Then the **average cost function**, denoted by $\overline{C}(x)$ (read "C bar of x"), is

$$\frac{C(x)}{x} \tag{4}$$

The derivative $\overline{C}'(x)$ of the average cost function, called the **marginal average cost function,** measures the rate of change of the average cost function with respect to the number of units produced.

EXAMPLE 3 The total cost of producing x units of a certain commodity is given by

$$C(x) = 400 + 20x$$

dollars.

a. Find the average cost function $\overline{C}$.
b. Find the marginal average cost function $\overline{C}'$.
c. Interpret the results obtained in parts (a) and (b).

SOLUTION ✔ **a.** The average cost function is given by

$$\overline{C}(x) = \frac{C(x)}{x} = \frac{400 + 20x}{x}$$

$$= 20 + \frac{400}{x}$$

b. The marginal average cost function is

$$\overline{C}'(x) = -\frac{400}{x^2}$$

c. Since the marginal average cost function is negative for all admissible values of x, the rate of change of the average cost function is negative for all $x > 0$; that is, $\overline{C}(x)$ decreases as x increases. However, the graph of $\overline{C}$ always lies

FIGURE 3.8
As the level of production increases, the average cost approaches $20.

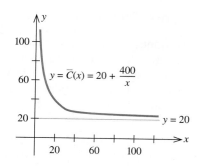

above the horizontal line $y = 20$, but it approaches the line since

$$\lim_{x \to \infty} \overline{C}(x) = \lim_{x \to \infty}\left(20 + \frac{400}{x}\right) = 20$$

A sketch of the graph of the function $\overline{C}(x)$ appears in Figure 3.8. This result is fully expected if we consider the economic implications. Note that as the level of production increases, the fixed cost per unit of production, represented by the term $(400/x)$, drops steadily. The average cost approaches the constant unit of production, which is $20 in this case. ■ ■ ■ ■

EXAMPLE 4   Once again consider the subsidiary of the Elektra Electronics Company. The daily total cost for producing its programmable calculators is given by

$$C(x) = 0.0001x^3 - 0.08x^2 + 40x + 5000$$

dollars, where x stands for the number of calculators produced (see Example 2).

a. Find the average cost function $\overline{C}$.
b. Find the marginal average cost function $\overline{C}'$. Compute $\overline{C}'(500)$.
c. Sketch the graph of the function $\overline{C}$ and interpret the results obtained in parts (a) and (b).

SOLUTION ✓   **a.** The average cost function is given by

$$\overline{C}(x) = \frac{C(x)}{x} = 0.0001x^2 - 0.08x + 40 + \frac{5000}{x}$$

b. The marginal average cost function is given by

$$\overline{C}'(x) = 0.0002x - 0.08 - \frac{5000}{x^2}$$

Also,

$$\overline{C}'(500) = 0.0002(500) - 0.08 - \frac{5000}{(500)^2} = 0$$

c. To sketch the graph of the function $\overline{C}$, observe that if x is a small positive number, then $\overline{C}(x) > 0$. Furthermore, $\overline{C}(x)$ becomes arbitrarily large as x approaches zero from the right, since the term $(5000/x)$ becomes arbitrarily

FIGURE 3.9
The average cost reaches a minimum of $35 when 500 calculators are produced.

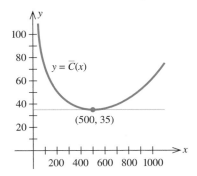

large as x approaches zero. Next, the result $\overline{C}'(500) = 0$ obtained in part (b) tells us that the tangent line to the graph of the function $\overline{C}$ is horizontal at the point (500, 35) on the graph. Finally, plotting the points on the graph corresponding to, say, $x = 100, 200, 300, \ldots , 900$, we obtain the sketch in Figure 3.9. As expected, the average cost drops as the level of production increases. But in this case, as opposed to the case in Example 3, the average cost reaches a minimum value of $35, corresponding to a production level of 500, and *increases* thereafter.

This phenomenon is typical in situations where the marginal cost increases from some point on as production increases, as in Example 2. This situation is in contrast to that of Example 3, in which the marginal cost remains constant at any level of production.   ■ ■ ■ ■

Exploring with Technology

Refer to Example 4.

1. Use a graphing utility to plot the graph of the average cost function

$$\overline{C}(x) = 0.0001x^2 - 0.08x + 40 + \frac{5000}{x}$$

using the viewing rectangle $[0, 1000] \times [0, 100]$. Then, using **ZOOM** and **TRACE**, show that the lowest point on the graph of $\overline{C}$ is (500, 35).
2. Draw the tangent line to the graph of $\overline{C}$ at (500, 35). What is its slope? Is this expected?
3. Plot the graph of the marginal average cost function

$$\overline{C}'(x) = 0.0002x - 0.08 - \frac{5000}{x^2}$$

using the viewing rectangle $[0, 2000] \times [-1, 1]$. Then use **ZOOM** and **TRACE** to show that the zero of the function $\overline{C}'$ occurs at $x = 500$. Verify this result using the root-finding capability of your graphing utility. Is this result compatible with that obtained in part (2)? Explain your answer.

REVENUE FUNCTIONS

Another marginal concept, the *marginal revenue function,* is associated with the revenue function R, given by

$$R(x) = px \tag{5}$$

where x is the number of units sold of a certain commodity and p is the unit selling price. In general, however, the unit selling price p of a commodity is related to the quantity x of the commodity demanded. This relationship, $p = f(x)$, is called a *demand equation* (see Section 2.3). Solving the demand equation for p in terms of x, we obtain the unit price function f, given by

$$p = f(x)$$

Thus, the revenue function R is given by

$$R(x) = px = xf(x)$$

where f is the unit price function. The derivative R' of the function R, called the **marginal revenue function**, measures the rate of change of the revenue function.

EXAMPLE 5 Suppose the relationship between the unit price p in dollars and the quantity demanded x of the Acrosonic model F loudspeaker system is given by the equation

$$p = -0.02x + 400 \qquad (0 \le x \le 20{,}000)$$

a. Find the revenue function R.
b. Find the marginal revenue function R'.
c. Compute $R'(2000)$ and interpret your result.

SOLUTION ✓ **a.** The revenue function R is given by

$$
\begin{aligned}
R(x) &= px \\
&= x(-0.02x + 400) \\
&= -0.02x^2 + 400x \qquad (0 \le x \le 20{,}000)
\end{aligned}
$$

b. The marginal revenue function R' is given by

$$R'(x) = -0.04x + 400$$

c.

$$R'(2000) = -0.04(2000) + 400 = 320$$

Thus, the actual revenue to be realized from the sale of the 2001st loudspeaker system is approximately $320.   ■ ■ ■ ■

PROFIT FUNCTIONS

Our final example of a marginal function involves the profit function. The profit function P is given by

$$P(x) = R(x) - C(x) \qquad \text{(6)}$$

where R and C are the revenue and cost functions and x is the number of units of a commodity produced and sold. The **marginal profit function** $P'(x)$ measures the rate of change of the profit function P and provides us with a good approximation of the actual profit or loss realized from the sale of the $(x + 1)$st unit of the commodity (assuming the xth unit has been sold).

EXAMPLE 6

Refer to Example 5. Suppose the cost of producing x units of the Acrosonic model F loudspeaker is

$$C(x) = 100x + 200{,}000 \text{ dollars}$$

a. Find the profit function P.
b. Find the marginal profit function P'.
c. Compute $P'(2000)$ and interpret your result.
d. Sketch the graph of the profit function P.

SOLUTION ✔

a. From the solution to Example 5(a), we have

$$R(x) = -0.02x^2 + 400x$$

Thus, the required profit function P is given by

$$
\begin{aligned}
P(x) &= R(x) - C(x) \\
&= (-0.02x^2 + 400x) - (100x + 200{,}000) \\
&= -0.02x^2 + 300x - 200{,}000
\end{aligned}
$$

b. The marginal profit function P' is given by

$$P'(x) = -0.04x + 300$$

c.
$$P'(2000) = -0.04(2000) + 300 = 220$$

Thus, the actual profit realized from the sale of the 2001st loudspeaker system is approximately $220.
d. The graph of the profit function P appears in Figure 3.10.   ■ ■ ■ ■

FIGURE 3.10
The total profit made when x loudspeakers are produced is given by $P(x)$.

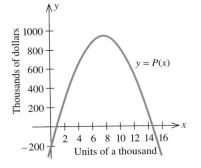

ELASTICITY OF DEMAND

Finally, let us use the marginal concepts introduced in this section to derive an important criterion used by economists to analyze the demand function: *elasticity of demand*.

In what follows, it will be convenient to write the demand function f in the form $x = f(p)$; that is, we will think of the quantity demanded of a certain commodity as a function of its unit price. Since the quantity demanded of a commodity usually decreases as its unit price increases, the function f is typically a decreasing function of p (Figure 3.11a).

FIGURE 3.11

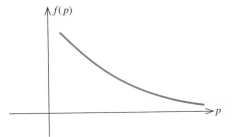

(a) A demand function

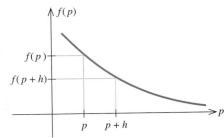

(b) $f(p + h)$ is the quantity demanded when the unit price increases from p to $p + h$ dollars.

Suppose the unit price of a commodity is increased by h dollars from p dollars to $(p + h)$ dollars (Figure 3.11b). Then the quantity demanded drops from $f(p)$ units to $f(p + h)$ units, a change of $[f(p + h) - f(p)]$ units. The percentage change in the unit price is

$$\frac{h}{p}(100) \qquad \left(\frac{\text{Change in unit price}}{\text{Price } p}\right)(100)$$

and the corresponding percentage change in the quantity demanded is

$$100\left[\frac{f(p + h) - f(p)}{f(p)}\right] \qquad \left(\frac{\text{Change in quantity demanded}}{\text{Quantity demanded at price } p}\right)(100)$$

Now, one good way to measure the effect that a percentage change in price has on the percentage change in the quantity demanded is to look at the ratio of the latter to the former. We find

$$\frac{\text{Percentage change in the quantity demanded}}{\text{Percentage change in the unit price}} = \frac{100\left[\dfrac{f(p + h) - f(p)}{f(p)}\right]}{100\left(\dfrac{h}{p}\right)}$$

$$= \frac{\dfrac{f(p + h) - f(p)}{h}}{\dfrac{f(p)}{p}}$$

If f is differentiable at p, then

$$\frac{f(p + h) - f(p)}{h} \approx f'(p)$$

when h is small. Therefore, if h is small, then the ratio is approximately equal to

$$\frac{f'(p)}{\dfrac{f(p)}{p}} = \frac{pf'(p)}{f(p)}$$

Economists call the negative of this quantity the *elasticity of demand.*

Elasticity of Demand

> If f is a differentiable demand function defined by $x = f(p)$, then the **elasticity of demand** at price p is given by
>
> $$E(p) = -\frac{pf'(p)}{f(p)} \tag{7}$$

REMARK It will be shown later (Section 4.1) that if f is decreasing on an interval, then $f'(p) < 0$ for p in that interval. In light of this, we see that since both p and $f(p)$ are positive, the quantity $\dfrac{pf'(p)}{f(p)}$ is negative. Because economists would rather work with a positive value, the elasticity of demand $E(p)$ is defined to be the negative of this quantity.   ■ ■ ■

EXAMPLE 7

Consider the demand equation

$$p = -0.02x + 400 \qquad (0 \le x \le 20{,}000)$$

which describes the relationship between the unit price in dollars and the quantity demanded x of the Acrosonic model F loudspeaker systems.

a. Find the elasticity of demand $E(p)$.
b. Compute $E(100)$ and interpret your result.
c. Compute $E(300)$ and interpret your result.

SOLUTION ✔

a. Solving the given demand equation for x in terms of p, we find

$$x = f(p) = -50p + 20{,}000$$

from which we see that

$$f'(p) = -50$$

Therefore,

$$E(p) = -\frac{pf'(p)}{f(p)} = -\frac{p(-50)}{-50p + 20{,}000}$$

$$= \frac{p}{400 - p}$$

b. $E(100) = \dfrac{100}{400 - 100} = \dfrac{1}{3}$, which is the elasticity of demand when $p = 100$.
To interpret this result, recall that $E(100)$ is the negative of the ratio of the percentage change in the quantity demanded to the percentage change in the unit price when $p = 100$. Therefore, our result tells us that when the unit price p is set at \$100 per speaker, an increase of 1% in the unit price will cause an increase of approximately 0.33% in the quantity demanded.

c. $E(300) = \dfrac{300}{400 - 300} = 3$, which is the elasticity of demand when $p = 300$. It tells us that when the unit price is set at \$300 per speaker, an increase of 1% in the unit price will cause a decrease of approximately 3% in the quantity demanded.   ■ ■ ■ ■

Economists often use the following terminology to describe demand in terms of elasticity.

Elasticity of Demand

The demand is said to be **elastic** if $E(p) > 1$.
The demand is said to be **unitary** if $E(p) = 1$.
The demand is said to be **inelastic** if $E(p) < 1$.

As an illustration, our computations in Example 7 revealed that demand for Acrosonic loudspeakers is elastic when $p = 300$ but inelastic when $p = 100$. These computations confirm that when demand is elastic, a small percentage change in the unit price will result in a greater percentage change in the quantity demanded; and when demand is inelastic, a small percentage change in the unit price will cause a smaller percentage change in the quantity demanded. Finally, when demand is unitary, a small percentage change in the unit price will result in the same percentage change in the quantity demanded.

We can describe the way revenue responds to changes in the unit price using the notion of elasticity. If the quantity demanded of a certain commodity is related to its unit price by the equation $x = f(p)$, then the revenue realized through the sale of x units of the commodity at a price of p dollars each is

$$R(p) = px = pf(p)$$

The rate of change of the revenue with respect to the unit price p is given by

$$R'(p) = f(p) + pf'(p)$$
$$= f(p)\left[1 + \frac{pf'(p)}{f(p)}\right]$$
$$= f(p)[1 - E(p)]$$

Now, suppose demand is elastic when the unit price is set at a dollars. Then $E(a) > 1$, and so $1 - E(a) < 0$. Since $f(p)$ is positive for all values of p, we see that

$$R'(a) = f(a)[1 - E(a)] < 0$$

and so $R(p)$ is decreasing at $p = a$. This implies that a small increase in the unit price when $p = a$ results in a decrease in the revenue, whereas a small decrease in the unit price will result in an increase in the revenue. Similarly, you can show that if the demand is inelastic when the unit price is set at a dollars, then a small increase in the unit price will cause the revenue to increase, and a small decrease in the unit price will cause the revenue to decrease. Finally, if the demand is unitary when the unit price is set at a dollars, then $E(a) = 1$ and $R'(a) = 0$. This implies that a small increase or decrease in the unit price will not result in a change in the revenue. The following statements summarize this discussion.

---

1. If the demand is elastic at p $(E(p) > 1)$, then an increase in the unit price will cause the revenue to decrease, whereas a decrease in the unit price will cause the revenue to increase.

2. If the demand is inelastic at p $(E(p) < 1)$, then an increase in the unit price will cause the revenue to increase, and a decrease in the unit price will cause the revenue to decrease.

3. If the demand is unitary at p $(E(p) = 1)$, then an increase in the unit price will cause the revenue to stay about the same.

These results are illustrated in Figure 3.12.

FIGURE 3.12
The revenue is increasing on an interval where the demand is inelastic, decreasing on an interval where the demand is elastic, and stationary at the point where the demand is unitary.

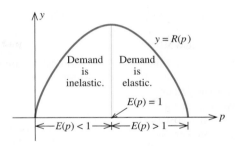

REMARK As an aid to remembering this, note the following:

1. If demand is elastic, then the change in revenue and the change in the unit price move in opposite directions.
2. If demand is inelastic, then they move in the same direction.   ■ ■ ■

EXAMPLE 8   Refer to Example 7.

a. Is demand elastic, unitary, or inelastic when $p = 100$? When $p = 300$?
b. If the price is \$100, will raising the unit price slightly cause the revenue to increase or decrease?

SOLUTION ✔   **a.** From the results of Example 7, we see that $E(100) = \frac{1}{3} < 1$ and $E(300) = 3 > 1$. We conclude accordingly that demand is inelastic when $p = 100$ and elastic when $p = 300$.
b. Since demand is inelastic when $p = 100$, raising the unit price slightly will cause the revenue to increase.   ■ ■ ■ ■

SELF-CHECK EXERCISES 3.4

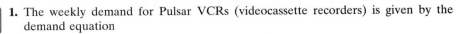

1. The weekly demand for Pulsar VCRs (videocassette recorders) is given by the demand equation

$$p = -0.02x + 300 \qquad (0 \le x \le 15{,}000)$$

where p denotes the wholesale unit price in dollars and x denotes the quantity demanded. The weekly total cost function associated with manufacturing these VCRs is

$$C(x) = 0.000003x^3 - 0.04x^2 + 200x + 70{,}000 \text{ dollars}$$

a. Find the revenue function R and the profit function P.
b. Find the marginal cost function C', the marginal revenue function R', and the marginal profit function P'.
c. Find the marginal average cost function $\overline{C}'$.
d. Compute $C'(3000)$, $R'(3000)$, and $P'(3000)$ and interpret your results.

2. Refer to the preceding exercise. Determine whether the demand is elastic, unitary, or inelastic when $p = 100$ and when $p = 200$.

Solutions to Self-Check Exercises 3.4 can be found on page 239.

3.4 Exercises

1. **PRODUCTION COSTS** The graph of a typical total cost function $C(x)$ associated with the manufacture of x units of a certain commodity is shown in the following figure.
   a. Explain why the function C is always increasing.
   b. As the level of production x increases, the cost per unit drops so that $C(x)$ increases but at a slower pace. However, a level of production is soon reached at which the cost per unit begins to increase dramatically (due to a shortage of raw material, overtime, breakdown of machinery due to excessive stress and strain) so that $C(x)$ continues to increase at a faster pace. Use the graph of C to find the approximate level of production x_0 where this occurs.

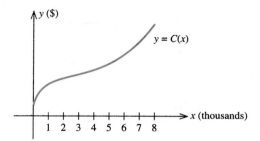

A calculator is recommended for Exercises 2–33.

2. **MARGINAL COST** The total weekly cost (in dollars) incurred by the Lincoln Record Company in pressing x long-playing records is

   $$C(x) = 2000 + 2x - 0.0001x^2 \qquad (0 \le x \le 6000)$$

   a. What is the actual cost incurred in producing the 1001st and the 2001st record?
   b. What is the marginal cost when $x = 1000$ and 2000?

3. **MARGINAL COST** A division of Ditton Industries manufactures the Futura model microwave oven. The daily cost (in dollars) of producing these microwave ovens is

   $$C(x) = 0.0002x^3 - 0.06x^2 + 120x + 5000$$

   where x stands for the number of units produced.
   a. What is the actual cost incurred in manufacturing the 101st oven? The 201st oven? The 301st oven?
   b. What is the marginal cost when $x = 100, 200$, and 300?

4. **MARGINAL AVERAGE COST** The Custom Office Company makes a line of executive desks. It is estimated that the total cost for making x units of their Senior Executive Model is

   $$C(x) = 100x + 200,000$$

   dollars per year.
   a. Find the average cost function $\overline{C}$.
   b. Find the marginal average cost function $\overline{C}'$.
   c. What happens to $\overline{C}(x)$ when x is very large? Interpret your results.

5. **MARGINAL AVERAGE COST** The management of the ThermoMaster Company, whose Mexican subsidiary manufactures an indoor–outdoor thermometer, has estimated that the total weekly cost (in dollars) for producing x thermometers is

   $$C(x) = 5000 + 2x$$

   a. Find the average cost function $\overline{C}$.
   b. Find the marginal average cost function $\overline{C}'$.
   c. Interpret your results.

6. Find the average cost function $\overline{C}$ and the marginal average cost function $\overline{C}'$ associated with the total cost function C of Exercise 2.

7. Find the average cost function $\overline{C}$ and the marginal average cost function $\overline{C}'$ associated with the total cost function C of Exercise 3.

8. **MARGINAL REVENUE** The Williams Commuter Air Service realizes a monthly revenue of

   $$R(x) = 8000x - 100x^2$$

   dollars when the price charged per passenger is x dollars.
   a. Find the marginal revenue R'.
   b. Compute $R'(39)$, $R'(40)$, and $R'(41)$. What do your results imply?

9. **MARGINAL REVENUE** The management of the Acrosonic Company plans to market the Electro-Stat, an electrostatic speaker system. The marketing department has determined that the demand for these speakers is

   $$p = -0.04x + 800 \qquad (0 \le x \le 20,000)$$

   where p denotes the speaker's unit price (in dollars) and x denotes the quantity demanded.
   a. Find the revenue function R.
   b. Find the marginal revenue function R'.
   c. Compute $R'(5000)$ and interpret your result.

10. MARGINAL PROFIT Lynbrook West, an apartment complex, has 100 two-bedroom units. The monthly profit (in dollars) realized from renting x apartments is

$$P(x) = -10x^2 + 1760x - 50,000$$

a. What is the actual profit realized from renting the 51st unit, assuming that 50 units have already been rented?
b. Compute the marginal profit when $x = 50$ and compare your results with that obtained in part (a).

11. MARGINAL PROFIT Refer to Exercise 9. Acrosonic's production department estimates that the total cost (in dollars) incurred in manufacturing x Electro-Stat speaker systems in the first year of production will be

$$C(x) = 200x + 300,000$$

a. Find the profit function P.
b. Find the marginal profit function P'.
c. Compute $P'(5000)$ and $P'(8000)$.
d. Sketch the graph of the profit function and interpret your results.

12. MARGINAL COST, REVENUE, AND PROFIT The weekly demand for the Pulsar 25 color console television is

$$p = 600 - 0.05x \qquad (0 \le x \le 12,000)$$

where p denotes the wholesale unit price in dollars and x denotes the quantity demanded. The weekly total cost function associated with manufacturing the Pulsar 25 is given by

$$C(x) = 0.000002x^3 - 0.03x^2 + 400x + 80,000$$

where $C(x)$ denotes the total cost incurred in producing x sets.
a. Find the revenue function R and the profit function P.
b. Find the marginal cost function C', the marginal revenue function R', and the marginal profit function P'.
c. Compute $C'(2000)$, $R'(2000)$, and $P'(2000)$ and interpret your results.
d. Sketch the graphs of the functions C, R, and P and interpret parts (b) and (c) using the graphs obtained.

13. MARGINAL COST, REVENUE, AND PROFIT The Pulsar Corporation also manufactures a series of 19-inch color television sets. The quantity x of these sets demanded each week is related to the wholesale unit price p by the equation

$$p = -0.006x + 180$$

The weekly total cost incurred by Pulsar for producing x sets is

$$C(x) = 0.000002x^3 - 0.02x^2 + 120x + 60,000$$

dollars. Answer the questions in Exercise 12 for these data.

14. MARGINAL AVERAGE COST Find the average cost function $\overline{C}$ associated with the total cost function C of Exercise 12.
a. What is the marginal average cost function $\overline{C}'$?
b. Compute $\overline{C}'(5,000)$ and $\overline{C}'(10,000)$ and interpret your results.
c. Sketch the graph of $\overline{C}$.

15. MARGINAL AVERAGE COST Find the average cost function $\overline{C}$ associated with the total cost function C of Exercise 13.
a. What is the marginal average cost function $\overline{C}'$?
b. Compute $\overline{C}'(5,000)$ and $\overline{C}'(10,000)$ and interpret your results.

16. MARGINAL REVENUE The quantity of Sicard wristwatches demanded per month is related to the unit price by the equation

$$p = \frac{50}{0.01x^2 + 1} \qquad (0 \le x \le 20)$$

where p is measured in dollars and x in units of a thousand.
a. Find the revenue function R.
b. Find the marginal revenue function R'.
c. Compute $R'(2)$ and interpret your result.

17. MARGINAL PROPENSITY TO CONSUME The consumption function of the U.S. economy for 1929 to 1941 is

$$C(x) = 0.712x + 95.05$$

where $C(x)$ is the personal consumption expenditure and x is the personal income, both measured in billions of dollars. Find the rate of change of consumption with respect to income, dC/dx. This quantity is called the *marginal propensity to consume*.

18. MARGINAL PROPENSITY TO CONSUME Refer to Exercise 17. Suppose a certain economy's consumption function is

$$C(x) = 0.873x^{1.1} + 20.34$$

where $C(x)$ and x are measured in billions of dollars. Find the marginal propensity to consume when $x = 10$.

19. MARGINAL PROPENSITY TO SAVE   Suppose $C(x)$ measures an economy's personal consumption expenditure and x the personal income, both in billions of dollars. Then,

$$S(x) = x - C(x) \qquad \text{(Income minus consumption)}$$

measures the economy's savings corresponding to an income of x billion dollars. Show that

$$\frac{dS}{dx} = 1 - \frac{dC}{dx}$$

The quantity dS/dx is called the *marginal propensity to save*.

20. Refer to Exercise 19. For the consumption function of Exercise 17, find the marginal propensity to save.

21. Refer to Exercise 19. For the consumption function of Exercise 18, find the marginal propensity to save when $x = 10$.

For each demand equation in Exercises 22–27, compute the elasticity of demand and determine whether the demand is elastic, unitary, or inelastic at the indicated price.

22. $x = -\dfrac{3}{2}p + 9;\ p = 2$

23. $x = -\dfrac{5}{4}p + 20;\ p = 10$

24. $x + \dfrac{1}{3}p - 20 = 0;\ p = 30$

25. $0.4x + p - 20 = 0;\ p = 10$

26. $p = 144 - x^2;\ p = 96$

27. $p = 169 - x^2;\ p = 29$

28. ELASTICITY OF DEMAND   The management of the Titan Tire Company has determined that the quantity demanded x of their Super Titan tires per week is related to the unit price p by the equation

$$x = \sqrt{144 - p}$$

where p is measured in dollars and x in units of a thousand.
a. Compute the elasticity of demand when $p = 63$, 96, and 108.
b. Interpret the results obtained in part (a).
c. Is the demand elastic, unitary, or inelastic when $p = 63$, 96, and 108?

29. ELASTICITY OF DEMAND   The demand equation for the Roland portable hair dryer is given by

$$x = \frac{1}{5}(225 - p^2) \qquad (0 \le p \le 15)$$

where x (measured in units of a hundred) is the quantity demanded per week and p is the unit price in dollars.
a. Is the demand elastic or inelastic when $p = 8$ and when $p = 10$?
b. When is the demand unitary?
Hint: Solve $E(p) = 1$ for p.
c. If the unit price is lowered slightly from \$10, will the revenue increase or decrease?
d. If the unit price is increased slightly from \$8, will the revenue increase or decrease?

30. ELASTICITY OF DEMAND   The quantity demanded per week x (in units of a hundred) of the Mikado miniature camera is related to the unit price p (in dollars) by the demand equation

$$x = \sqrt{400 - 5p} \qquad (0 \le p \le 80)$$

a. Is the demand elastic or inelastic when $p = 40$? When $p = 60$?
b. When is the demand unitary?
c. If the unit price is lowered slightly from \$60, will the revenue increase or decrease?
d. If the unit price is increased slightly from \$40, will the revenue increase or decrease?

31. ELASTICITY OF DEMAND   The proprietor of the Showplace, a video club, has estimated that the rental price p (in dollars) of prerecorded videocassette tapes is related to the quantity x rented per week by the demand equation

$$x = \frac{2}{3}\sqrt{36 - p^2} \qquad (0 \le p \le 6)$$

Currently, the rental price is \$2/tape.
a. Is the demand elastic or inelastic at this rental price?
b. If the rental price is increased, will the revenue increase or decrease?

32. ELASTICITY OF DEMAND   The demand function for a certain make of exercise bicycle sold exclusively through cable television is

$$p = \sqrt{9 - 0.02x} \qquad (0 \le x \le 450)$$

where p is the unit price in hundreds of dollars and x is the quantity demanded per week. Compute the elasticity of demand and determine the range of prices corresponding to inelastic, unitary, and elastic demand.
Hint: Solve the equation $E(p) = 1$.

33. ELASTICITY OF DEMAND The demand equation for the Sicard wristwatch is given by

$$x = 10 \sqrt{\frac{50 - p}{p}} \qquad (0 < p \leq 50)$$

where x (measured in units of a thousand) is the quantity demanded per week and p is the unit price in dollars. Compute the elasticity of demand and determine the range of prices corresponding to inelastic, unitary, and elastic demand.

In Exercises 34 and 35, determine whether the statement is true or false. If it is true, explain why it is true. If it is false, give an example to show why it is false.

34. If C is a differentiable total cost function, then the marginal average cost function is

$$\overline{C}'(x) = \frac{xC'(x) - C(x)}{x^2}$$

35. If the marginal profit function is positive at $x = a$, then it makes sense to decrease the level of production.

SOLUTIONS TO SELF-CHECK EXERCISES 3.4

1. **a.** $R(x) = px$
   $$= x(-0.02x + 300)$$
   $$= -0.02x^2 + 300x \qquad (0 \leq x \leq 15,000)$$

   $P(x) = R(x) - C(x)$
   $$= -0.02x^2 + 300x$$
   $$\quad - (0.000003x^3 - 0.04x^2 + 200x + 70,000)$$
   $$= -0.000003x^3 + 0.02x^2 + 100x - 70,000$$

   b. $C'(x) = 0.000009x^2 - 0.08x + 200$
   $R'(x) = -0.04x + 300$
   $P'(x) = -0.000009x^2 + 0.04x + 100$

   c. The average cost function is

   $$\overline{C}(x) = \frac{C(x)}{x}$$
   $$= \frac{0.000003x^3 - 0.04x^2 + 200x + 70,000}{x}$$
   $$= 0.000003x^2 - 0.04x + 200 + \frac{70,000}{x}$$

   Therefore, the marginal average cost function is

   $$\overline{C}'(x) = 0.000006x - 0.04 - \frac{70,000}{x^2}$$

   d. Using the results from part (b), we find

   $$C'(3000) = 0.000009(3000)^2 - 0.08(3000) + 200$$
   $$= 41$$

That is, when the level of production is already 3000 VCRs, the actual cost of producing one additional VCR is approximately \$41. Next,

$$R'(3000) = -0.04(3000) + 300 = 180$$

That is, the actual revenue to be realized from selling the 3001st VCR is approximately \$180. Finally,

$$P'(3000) = -0.000009(3000)^2 + 0.04(3000) + 100$$
$$= 139$$

That is, the actual profit realized from selling the 3001st VCR is approximately \$139.

2. We first solve the given demand equation for x in terms of p, obtaining

$$x = f(p) = -50p + 15{,}000$$
$$f'(p) = -50$$

Therefore,

$$E(p) = -\frac{pf'(p)}{f(p)} = -\frac{p}{-50p + 15{,}000}(-50)$$
$$= \frac{p}{300 - p} \qquad (0 \le p < 300)$$

Next, we compute

$$E(100) = \frac{100}{300 - 100} = \frac{1}{2} < 1$$

and we conclude that demand is inelastic when $p = 100$. Also,

$$E(200) = \frac{200}{300 - 200} = 2 > 1$$

and we see that demand is elastic when $p = 200$.

3.5 Higher-Order Derivatives

HIGHER-ORDER DERIVATIVES

The derivative f' of a function f is also a function. As such, the differentiability of f' may be considered. Thus, the function f' has a derivative f'' at a point x in the domain of f' if the limit of the quotient

$$\frac{f'(x + h) - f'(x)}{h}$$

exists as h approaches zero. In other words, it is the derivative of the first derivative.

The function f'' obtained in this manner is called the **second derivative** of the function f, just as the derivative f' of f is often called the first derivative of f. Continuing in this fashion, we are led to considering the third, fourth, and higher-order derivatives of f whenever they exist. Notations for the first, second, third, and, in general, nth derivatives of a function f at a point x are

$$f'(x), \ f''(x), \ f'''(x), \ \dots, \ f^{(n)}(x)$$

or
$$D^1 f(x), \ D^2 f(x), \ D^3 f(x), \ \dots, \ D^n f(x)$$

If f is written in the form $y = f(x)$, then the notations for its derivatives are

$$y', \ y'', \ y''', \ \dots, \ y^{(n)}$$

$$\frac{dy}{dx}, \ \frac{d^2 y}{dx^2}, \ \frac{d^3 y}{dx^3}, \ \dots, \ \frac{d^n y}{dx^n}$$

or
$$D^1 y, \ D^2 y, \ D^3 y, \ \dots, \ D^n y$$

respectively.

EXAMPLE 1   Find the derivatives of all orders of the polynomial function

$$f(x) = x^5 - 3x^4 + 4x^3 - 2x^2 + x - 8$$

SOLUTION ✔   We have

$$f'(x) = 5x^4 - 12x^3 + 12x^2 - 4x + 1$$

$$f''(x) = \frac{d}{dx} f'(x) = 20x^3 - 36x^2 + 24x - 4$$

$$f'''(x) = \frac{d}{dx} f''(x) = 60x^2 - 72x + 24$$

$$f^{(4)}(x) = \frac{d}{dx} f'''(x) = 120x - 72$$

$$f^{(5)}(x) = \frac{d}{dx} f^{(4)}(x) = 120$$

and, in general,

$$f^{(n)}(x) = 0 \qquad \text{(for } n > 5\text{)}$$ ■ ■ ■ ■

EXAMPLE 2   Find the third derivative of the function f defined by $y = x^{2/3}$. What is its domain?

SOLUTION ✔   We have

$$y' = \frac{2}{3} x^{-1/3}$$

$$y'' = \left(\frac{2}{3}\right)\left(-\frac{1}{3}\right) x^{-4/3} = -\frac{2}{9} x^{-4/3}$$

so the required derivative is

$$y''' = \left(-\frac{2}{9}\right)\left(-\frac{4}{3}\right) x^{-7/3} = \frac{8}{27} x^{-7/3} = \frac{8}{27 x^{7/3}}$$

FIGURE 3.13
The graph of the function $y = x^{2/3}$

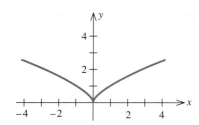

The common domain of the functions f', f'', and f''' is the set of all real numbers except $x = 0$. The domain of $y = x^{2/3}$ is the set of all real numbers. The graph of the function $y = x^{2/3}$ appears in Figure 3.13.   ■ ■ ■ ■

REMARK Always simplify an expression before differentiating it to obtain the next order derivative.   ■ ■ ■

EXAMPLE 3   Find the second derivative of the function $y = (2x^2 + 3)^{3/2}$.

SOLUTION ✔   We have, using the general power rule,

$$y' = \frac{3}{2}(2x^2 + 3)^{1/2}(4x) = 6x(2x^2 + 3)^{1/2}$$

Next, using the product rule and then the chain rule, we find

$$y'' = (6x) \cdot \frac{d}{dx}(2x^2 + 3)^{1/2} + \left[\frac{d}{dx}(6x)\right](2x^2 + 3)^{1/2}$$

$$= (6x)\left(\frac{1}{2}\right)(2x^2 + 3)^{-1/2}(4x) + 6(2x^2 + 3)^{1/2}$$

$$= 12x^2(2x^2 + 3)^{-1/2} + 6(2x^2 + 3)^{1/2}$$

$$= 6(2x^2 + 3)^{-1/2}[2x^2 + (2x^2 + 3)]$$

$$= \frac{6(4x^2 + 3)}{\sqrt{2x^2 + 3}}$$ ■ ■ ■ ■

APPLICATIONS

Just as the derivative of a function f at a point x measures the rate of change of the function f at that point, the second derivative of f (the derivative of f') measures the rate of change of the derivative f' of the function f. The third derivative of the function f, f''', measures the rate of change of f'', and so on.

In Chapter 4 we will discuss applications involving the geometric interpretation of the second derivative of a function. The following example gives an interpretation of the second derivative in a familiar role.

EXAMPLE **4**

Refer to the example on page 111. The distance s (in feet) covered by a maglev moving along a straight track t seconds after starting from rest is given by the function $s = 4t^2$ ($0 \leq t \leq 10$). What is the maglev's acceleration at the end of 30 seconds?

SOLUTION ✔

The velocity of the maglev t seconds from rest is given by

$$v = \frac{ds}{dt} = \frac{d}{dt}(4t^2) = 8t$$

The acceleration of the maglev t seconds from rest is given by the rate of change of the velocity of t—that is,

$$a = \frac{d}{dt}v = \frac{d}{dt}\left(\frac{ds}{dt}\right) = \frac{d^2s}{dt^2} = \frac{d}{dt}(8t) = 8$$

or 8 feet per second per second, normally abbreviated, 8 ft/sec^2.   ■ ■ ■ ■

EXAMPLE **5**

A ball is thrown straight up into the air from the roof of a building. The height of the ball as measured from the ground is given by

$$s = -16t^2 + 24t + 120$$

where s is measured in feet and t in seconds. Find the velocity and acceleration of the ball 3 seconds after it is thrown into the air.

SOLUTION ✔

The velocity v and acceleration a of the ball at any time t are given by

$$v = \frac{ds}{dt} = \frac{d}{dt}(-16t^2 + 24t + 120) = -32t + 24$$

and

$$a = \frac{d^2s}{dt^2} = \frac{d}{dt}\left(\frac{ds}{dt}\right) = \frac{d}{dt}(-32t + 24) = -32$$

Therefore, the velocity of the ball 3 seconds after it is thrown into the air is

$$v = -32(3) + 24 = -72$$

That is, the ball is falling downward at a speed of 72 ft/sec. The acceleration of the ball is 32 ft/sec^2 downward at any time during the motion.   ■ ■ ■ ■

FIGURE 3.14
The CPI of a certain economy from year a to year b is given by $I(t)$.

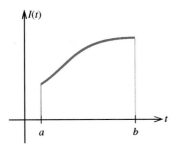

Another interpretation of the second derivative of a function—this time from the field of economics—follows. Suppose the consumer price index (CPI) of an economy between the years a and b is described by the function $I(t)$ ($a \leq t \leq b$) (Figure 3.14). Then, the first derivative of I, $I'(t)$, gives the rate of inflation of the economy at any time t. The second derivative of I, $I''(t)$, gives the *rate of change of the inflation rate* at any time t. Thus, when the economist or politician claims that "inflation is slowing," what he or she is saying is that the rate of inflation is decreasing. Mathematically, this is equivalent to noting that the second derivative $I''(t)$ is negative at the time t under consideration. Observe that $I'(t)$ could be positive at a time when $I''(t)$ is negative (see Example 6). Thus, one may not draw the conclusion from the aforementioned quote that prices of goods and services are about to drop!

EXAMPLE 6   An economy's CPI is described by the function

$$I(t) = -0.2t^3 + 3t^2 + 100 \qquad (0 \le t \le 9)$$

where $t = 0$ corresponds to the year 1991. Compute $I'(6)$ and $I''(6)$ and use these results to show that even though the CPI was rising at the beginning of 1997, "inflation was moderating" at that time.

SOLUTION ✔   We find

$$I'(t) = -0.6t^2 + 6t \qquad \text{and} \qquad I''(t) = -1.2t + 6$$

FIGURE 3.15

The CPI of an economy is given by $I(t)$.

Thus,

$$I'(6) = -0.6(6)^2 + 6(6) = 14.4$$

$$I''(6) = -1.2(6) + 6 = -1.2$$

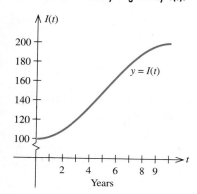

Our computations reveal that at the beginning of 1997 ($t = 6$), the CPI was increasing at the rate of 14.4 points per year, whereas the rate of the inflation rate was decreasing by 1.2 points per year. Thus, inflation was moderating at that time (Figure 3.15). In Section 4.2, we will see that relief actually began in early 1996.   ■ ■ ■ ■

SELF-CHECK EXERCISES 3.5

1. Find the third derivative of

$$f(x) = 2x^5 - 3x^3 + x^2 - 6x + 10$$

2. Let

$$f(x) = \frac{1}{1 + x}$$

Find $f'(x)$, $f''(x)$, and $f'''(x)$.

3. A certain species of turtle faces extinction because dealers collect truckloads of turtle eggs to be sold as aphrodisiacs. After severe conservation measures are implemented, it is hoped that the turtle population will grow according to the rule

$$N(t) = 2t^3 + 3t^2 - 4t + 1000 \qquad (0 \le t \le 10)$$

where $N(t)$ denotes the population at the end of year t. Compute $N''(2)$ and $N''(8)$ and interpret your results.

Solutions to Self-Check Exercises 3.5 can be found on page 248.

3.5 Exercises

In Exercises 1–20, find the first and second derivatives of the given function.

1. $f(x) = 4x^2 - 2x + 1$

2. $f(x) = -0.2x^2 + 0.3x + 4$

3. $f(x) = 2x^3 - 3x^2 + 1$

4. $g(x) = -3x^3 + 24x^2 + 6x - 64$

5. $h(t) = t^4 - 2t^3 + 6t^2 - 3t + 10$

6. $f(x) = x^5 - x^4 + x^3 - x^2 + x - 1$

7. $f(x) = (x^2 + 2)^5$ **8.** $g(t) = t^2(3t + 1)^4$

9. $g(t) = (2t^2 - 1)^2(3t^2)$

10. $h(x) = (x^2 + 1)^2(x - 1)$

11. $f(x) = (2x^2 + 2)^{7/2}$

12. $h(w) = (w^2 + 2w + 4)^{5/2}$

13. $f(x) = x(x^2 + 1)^2$ **14.** $g(u) = u(2u - 1)^3$

15. $f(x) = \dfrac{x}{2x + 1}$ **16.** $g(t) = \dfrac{t^2}{t - 1}$

17. $f(s) = \dfrac{s - 1}{s + 1}$ **18.** $f(u) = \dfrac{u}{u^2 + 1}$

19. $f(u) = \sqrt{4 - 3u}$ **20.** $f(x) = \sqrt{2x - 1}$

In Exercises 21–28, find the third derivative of the given function.

21. $f(x) = 3x^4 - 4x^3$

22. $f(x) = 3x^5 - 6x^4 + 2x^2 - 8x + 12$

23. $f(x) = \dfrac{1}{x}$ **24.** $f(x) = \dfrac{2}{x^2}$

25. $g(s) = \sqrt{3s - 2}$ **26.** $g(t) = \sqrt{2t + 3}$

27. $f(x) = (2x - 3)^4$ **28.** $g(t) = (\tfrac{1}{2}t^2 - 1)^5$

29. ACCELERATION OF A FALLING OBJECT   During the construction of an office building, a hammer is accidentally dropped from a height of 256 ft. The distance the hammer falls in t sec is $s = 16t^2$. What is the hammer's velocity when it strikes the ground? What is its acceleration?

30. ACCELERATION OF A CAR   The distance s (in feet) covered by a car t sec after starting from rest is given by

$$s = -t^3 + 8t^2 + 20t \qquad (0 \le t \le 6)$$

Find a general expression for the car's acceleration at any time $t (0 \le t \le 6)$. Show that the car is decelerating $2\frac{2}{3}$ sec after starting from rest.

31. CRIME RATES   The number of major crimes committed in Bronxville between 1988 and 1995 is approximated by the function

$$N(t) = -0.1t^3 + 1.5t^2 + 100 \qquad (0 \le t \le 7)$$

where $N(t)$ denotes the number of crimes committed in year t and $t = 0$ corresponds to the year 1988. Enraged by the dramatic increase in the crime rate, Bronxville's citizens, with the help of the local police, organized "Neighborhood Crime Watch" groups in early 1992 to combat this menace.
a. Verify that the crime rate was increasing from 1988 through 1995.
Hint: Compute $N'(0), N'(1), \ldots, N'(7)$.
b. Show that the Neighborhood Crime Watch program was working by computing $N''(4)$, $N''(5)$, $N''(6)$, and $N''(7)$.

32. GDP OF A DEVELOPING COUNTRY   A developing country's gross domestic product (GDP) from 1992 to 2000 is approximated by the function

$$G(t) = -0.2t^3 + 2.4t^2 + 60 \qquad (0 \le t \le 8)$$

where $G(t)$ is measured in billions of dollars and $t = 0$ corresponds to the year 1992.
a. Compute $G'(0), G'(1), \ldots, G'(8)$.
b. Compute $G''(0), G''(1), \ldots, G''(8)$.
c. Using the results obtained in parts (a) and (b), show that after a spectacular growth rate in the early years, the growth of the GDP cooled off.

33. TEST FLIGHT OF A VTOL   In a test flight of the McCord Terrier, McCord Aviation's experimental VTOL (vertical takeoff and landing) aircraft, it was determined that t sec after lift-off, when the craft was operated in the
(continued on p. 248)

FINDING THE SECOND DERIVATIVE OF A FUNCTION AT A GIVEN POINT

Some graphing utilities have the capability of numerically computing the second derivative of a function at a point. If your graphing utility has this capability, use it to work through the examples and exercises of this section.

EXAMPLE 1

Use the (second) numerical derivative operation of a graphing utility to find the second derivative of $f(x) = \sqrt{x}$ when $x = 4$.

SOLUTION ✔

Using the (second) numerical derivative operation of a graphing utility, we find

$$f''(4) = \text{der2} \, (x\char`^.5, x, 4) = -0.03125 \qquad \blacksquare\blacksquare\blacksquare\blacksquare$$

EXAMPLE 2

The anticipated rise in Alzheimer's patients in the United States is given by

$$f(t) = -0.02765t^4 + 0.3346t^3 - 1.1261t^2$$
$$+ \, 1.7575t + 3.7745 \qquad (0 \le t \le 6)$$

where $f(t)$ is measured in millions and t is measured in decades, with $t = 0$ corresponding to the beginning of 1990.

a. How fast is the number of Alzheimer's patients in the United States anticipated to be changing at the beginning of 2030?
b. How fast is the rate of change of the number of Alzheimer's patients in the United States anticipated to be changing at the beginning of 2030?
c. Plot the graph of f in the viewing rectangle $[0, 7] \times [0, 12]$.
Source: Alzheimer's Association

SOLUTION ✔

a. Using the numerical derivative operation of a graphing utility, we find that the number of Alzheimer's patients at the beginning of 2030 can be anticipated to be changing at the rate of

$$f'(4) = 1.7311$$

That is, the number is increasing at the rate of approximately 1.7 million patients per decade.
b. Using the (second) numerical derivative operation of a graphing utility, we find that

$$f''(4) = 0.4694$$

That is, the rate of change of the number of Alzheimer's patients is increasing at the rate of approximately 0.5 million patients per decade per decade.

c. The graph is shown in Figure T1.

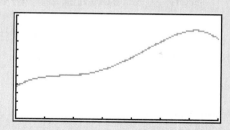

Exercises

In Exercises 1–8, find the value of the second derivative of *f* at the given value of *x*. Express your answer correct to four decimal places.

1. $f(x) = 2x^3 - 3x^2 + 1$; $x = -1$

2. $f(x) = 2.5x^5 - 3x^3 + 1.5x + 4$; $x = 2.1$

3. $f(x) = 2.1x^{3.1} - 4.2x^{1.7} + 4.2$; $x = 1.4$

4. $f(x) = 1.7x^{4.2} - 3.2x^{1.3} + 4.2x - 3.2$; $x = 2.2$

5. $f(x) = \dfrac{x^2 + 2x - 5}{x^3 + 1}$; $x = 2.1$

6. $f(x) = \dfrac{x^3 + x + 2}{2x^2 - 5x + 4}$; $x = 1.2$

7. $f(x) = \dfrac{x^{1/2} + 2x^{3/2} + 1}{2x^{1/2} + 3}$; $x = 0.5$

8. $f(x) = \dfrac{\sqrt{x} - 1}{2x + \sqrt{x} + 4}$; $x = 2.3$

9. RATE OF BANK FAILURES   The Federal Deposit Insurance Corporation (FDIC) estimates that the rate at which banks were failing between 1982 and 1994 is given by

$$f(t) = -0.063447t^4 - 1.953283t^3 + 14.632576t^2$$
$$- 6.684704t + 47.458874 \qquad (0 \le t \le 12)$$

where $f(t)$ is measured in the number of banks per year and t is measured in years, with $t = 0$ corresponding to the beginning of 1982. Compute $f''(6)$ and interpret your results.
Source: Federal Deposit Insurance Corporation

10. MULTIMEDIA SALES   According to the Electronic Industries Association, sales in the multimedia market (hardware and software) are expected to be

$$S(t) = -0.0094t^4 + 0.1204t^3 - 0.0868t^2$$
$$+ 0.0195t + 3.3325 \qquad (0 \le t \le 10)$$

where $S(t)$ is measured in billions of dollars and t is measured in years, with $t = 0$ corresponding to 1990. Compute $S''(7)$ and interpret your results.
Source: Electronics Industries Association

vertical takeoff mode, its altitude (in feet) was

$$h(t) = \frac{1}{16}t^4 - t^3 + 4t^2 \qquad (0 \le t \le 8)$$

a. Find an expression for the craft's velocity at time t.
b. Find the craft's velocity when $t = 0$ (the initial velocity), $t = 4$, and $t = 8$.
c. Find an expression for the craft's acceleration at time t.
d. Find the craft's acceleration when $t = 0$, 4, and 8.
e. Find the craft's height when $t = 0$, 4, and 8.

34. U.S. CENSUS   According to the U.S. Census Bureau, the number of Americans aged 45 to 54 will be approximately

$$N(t) = -0.00233t^4 + 0.00633t^3 - 0.05417t^2$$
$$+ 1.3467t + 25$$

million people in year t, where $t = 0$ corresponds to the beginning of 1990. Compute $N'(10)$ and $N''(10)$ and interpret your results.
Source: U.S. Census Bureau

35. AIR PURIFICATION   During testing of a certain brand of air purifier, it was determined that the amount of smoke remaining t min after the start of the test was

$$A(t) = -0.00006t^5 + 0.00468t^4 - 0.1316t^3$$
$$+ 1.915t^2 - 17.63t + 100$$

percent of the original amount. Compute $A'(10)$ and $A''(10)$ and interpret your results.
Source: Consumer Reports

In Exercises 36–39, determine whether the statement is true or false. If it is true, explain why it is true. If it is false, give an example to show why it is false.

36. If the second derivative of f exists at $x = a$, then $f''(a) = [f'(a)]^2$.

37. If $h = fg$ where f and g have second-order derivatives, then

$$h''(x) = f''(x)g(x) + 2f'(x)g'(x) + f(x)g''(x)$$

38. If $f(x)$ is a polynomial function of degree n, then $f^{(n+1)}(x) = 0$.

39. Suppose $P(t)$ represents the population of bacteria at time t and suppose $P'(t) > 0$ and $P''(t) < 0$; then the population is increasing at time t but at a decreasing rate.

40. Let f be the function defined by the rule $f(x) = x^{7/3}$. Show that f has first- and second-order derivatives at all points x, in particular at $x = 0$. Show also that the third derivative of f does *not* exist at $x = 0$.

41. Construct a function f that has derivatives of order up through and including n at a point a but fails to have the $(n + 1)$st derivative there.
Hint: See Exercise 40.

42. Show that a polynomial function has derivatives of all orders.
Hint: Let $P(x) = a_0 x^n + a_1 x^{n-1} + a_2 x^{n-2} + \cdots + a_n$ be a polynomial of degree n, where n is a positive integer and a_0, $a_1, \ldots, a_n$ are constants with $a_0 \ne 0$. Compute $P'(x), P''(x), \ldots$.

SOLUTIONS TO SELF-CHECK EXERCISES 3.5

1. $f'(x) = 10x^4 - 9x^2 + 2x - 6$
$f''(x) = 40x^3 - 18x + 2$
$f'''(x) = 120x^2 - 18$

2. We write $f(x) = (1 + x)^{-1}$ and use the general power rule, obtaining

$$f'(x) = (-1)(1 + x)^{-2}\frac{d}{dx}(1 + x) = -(1 + x)^{-2}(1)$$

$$= -(1 + x)^{-2} = -\frac{1}{(1 + x)^2}$$

Continuing, we find

$$f''(x) = -(-2)(1+x)^{-3}$$

$$= 2(1+x)^{-3} = \frac{2}{(1+x)^3}$$

$$f'''(x) = 2(-3)(1+x)^{-4}$$

$$= -6(1+x)^{-4}$$

$$= -\frac{6}{(1+x)^4}$$

3. $N'(t) = 6t^2 + 6t - 4$

   $N''(t) = 12t + 6 = 6(2t + 1)$

Therefore, $N''(2) = 30$ and $N''(8) = 102$. The results of our computations reveal that at the end of year 2, the *rate* of growth of the turtle population is increasing at the rate of 30 turtles/year/year. At the end of year 8, the rate is increasing at the rate of 102 turtles/year/year. Clearly, the conservation measures are paying off handsomely.

3.6 Implicit Differentiation and Related Rates

DIFFERENTIATING IMPLICITLY

Up to now we have dealt with functions expressed in the form $y = f(x)$; that is, the dependent variable y is expressed *explicitly* in terms of the independent variable x. However, not all functions are expressed in this form. Consider, for example, the equation

$$x^2 y + y - x^2 + 1 = 0 \qquad \textbf{(8)}$$

This equation does express y *implicitly* as a function of x. In fact, solving (8) for y in terms of x, we obtain

$$(x^2 + 1)y = x^2 - 1 \qquad \text{(Implicit equation)}$$

$$y = f(x) = \frac{x^2 - 1}{x^2 + 1} \qquad \text{(Explicit equation)}$$

which gives an explicit representation of f.

Next, consider the equation

$$y^4 - y^3 - y + 2x^3 - x = 8$$

When certain restrictions are placed on x and y, this equation defines y as a function of x. But in this instance, we would be hard pressed to find y explicitly in terms of x. The following question arises naturally: How does one go about computing dy/dx in this case?

As it turns out, thanks to the chain rule, a method *does* exist for computing the derivative of a function directly from the implicit equation defining the function. This method is called **implicit differentiation** and is demonstrated in the next several examples.

3.7 Differentials

The Millers are planning to buy a house in the near future and estimate that they will need a 30-year fixed-rate mortgage for $120,000. If the interest rate increases from the present rate of 9% per year to 9.4% per year between now and the time the Millers decide to secure the loan, approximately how much more per month will their mortgage be? (You will be asked to answer this question in Exercise 44, page 272.)

Questions such as this, in which one wishes to *estimate* the change in the dependent variable (monthly mortgage payment) corresponding to a small change in the independent variable (interest rate per year), occur in many real-life applications. For example:

■ An economist would like to know how a small increase in a country's capital expenditure will affect the country's gross domestic output.

■ A sociologist would like to know how a small increase in the amount of capital investment in a housing project will affect the crime rate.

■ A businesswoman would like to know how raising a product's unit price by a small amount will affect her profit.

■ A bacteriologist would like to know how a small increase in the amount of a bactericide will affect a population of bacteria.

To calculate these changes and estimate their effects, we use the *differential* of a function, a concept that will be introduced shortly.

INCREMENTS

Let x denote a variable quantity and suppose x changes from x_1 to x_2. This change in x is called the **increment in x** and is denoted by the symbol Δx (read "delta x"). Thus,

$$\Delta x = x_2 - x_1 \qquad \text{(Final value minus initial value)} \qquad \textbf{(9)}$$

EXAMPLE 1   Find the increment in x:

a. As x changes from 3 to 3.2   **b.** As x changes from 3 to 2.7

SOLUTION ✔   **a.** Here, $x_1 = 3$ and $x_2 = 3.2$, so

$$\Delta x = x_2 - x_1 = 3.2 - 3 = 0.2$$

b. Here, $x_1 = 3$ and $x_2 = 2.7$. Therefore,

$$\Delta x = x_2 - x_1 = 2.7 - 3 = -0.3 \qquad \blacksquare\,\blacksquare\,\blacksquare\,\blacksquare$$

Observe that Δx plays the same role that h played in Section 2.4.

Now, suppose two quantities, x and y, are related by an equation $y = f(x)$, where f is a function. If x changes from x to $x + \Delta x$, then the

JOHN DECKER

TITLE: Mortgage Counselor
INSTITUTION: A major bank

John Decker stresses that he and his colleagues "strive to grant loans. That's our job." But before allowing someone to file a formal application, Decker takes the person over several "prequalification hurdles" to gauge his or her ability to handle a mortgage.

To start, Decker relies on a two-tiered, debt-to-income ratio to see whether a person has sufficient gross monthly income to make payments. Under the first tier, the proposed monthly payment (principal and interest, property tax, homeowner's insurance, and, when applicable, a condo fee) cannot exceed 28% of an individual's gross monthly income. If Decker's initial calculations are positive, he then must determine the individual's ability to meet monthly mortgage payments while also repaying other debts, such as car and student loans, credit cards, alimony, and so on. These combined payments cannot exceed 36% percent of gross monthly income. A typical person with an $800 mortgage obligation and $550 in other payments would have to earn $3750 per month to clear these first two hurdles.

Contrary to popular belief, bankers *want* to lend money. "The idea is to grant mortgages," says Decker, which contribute substantially to a bank's profitability. But lending money means making sensible decisions about how much to lend as well as a person's ability to repay the loan.

Using a loan-to-value formula, banks might lend 80% of a property's value. In such cases, the applicant puts 20% down to make the purchase. Or the bank may decide to lend up to 95% or as little as 75% of the appraised value.

Understandably, banks don't like to see bankruptcies, late payments, or liens on a personal credit history. Decker notes, however, that even this hurdle doesn't necessarily mean failure in securing a mortgage. He works closely with each individual to overcome any stigma that might prompt a rejection.

Once Decker has put together a successful mortgage application, it is reviewed internally. Then, even though a mortgage is granted, it might come through at a slightly higher interest rate—10.4% instead of the expected 10% rate. Differentials would be used to compute the change in monthly payments, for although this might seem like a small change, on a large mortgage it can be enough of a variable to affect the new customer's ability to make monthly payments.

Using the debt-to-income ratio, Decker plugs the new variable into his formulas to determine whether a problem exists. These formulas used to compute the monthly payments involve the use of exponential functions. On a $100,000, 30-year, fixed-rate mortgage, payments will increase about $30 per month. Decker notes that such a small increase doesn't usually pose a significant problem.

If a customer can't make the new payment, however, Decker explores alternatives until he finds a solution. For Decker, it comes down to this: "If there is any possible way to give a loan, we're going to make it work."

Decker's job is simple: to lend money to people who want to buy a home. The hard part is deciding whether applicants qualify for one of the bank's 40 different mortgage plans. Sifting through income figures, current indebtedness, and credit history helps Decker determine which individuals make the best mortgage candidates.

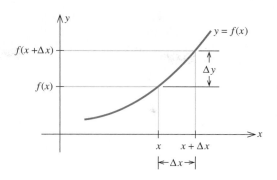

corresponding change in y is called the **increment in y**. It is denoted by Δy and is defined in Figure 3.18 by

$$\Delta y = f(x + \Delta x) - f(x) \tag{10}$$

 EXAMPLE 2   Let $y = x^3$. Find Δx and Δy:

a. When x changes from 2 to 2.01   **b.** When x changes from 2 to 1.98

SOLUTION ✔   Let $f(x) = x^3$.

a. Here, $\Delta x = 2.01 - 2 = 0.01$. Next,

$$\Delta y = f(x + \Delta x) - f(x) = f(2.01) - f(2)$$
$$= (2.01)^3 - 2^3 = 8.120601 - 8 = 0.120601$$

b. Here, $\Delta x = 1.98 - 2 = -0.02$. Next,

$$\Delta y = f(x + \Delta x) - f(x) = f(1.98) - f(2)$$
$$= (1.98)^3 - 2^3 = 7.762392 - 8 = -0.237608 \qquad ■■■■$$

DIFFERENTIALS

We can obtain a relatively quick and simple way of approximating Δy, the change in y due to a small change Δx, by examining the graph of the function f shown in Figure 3.19.

FIGURE 3.19
If Δx is small, dy is a good approximation of Δy.

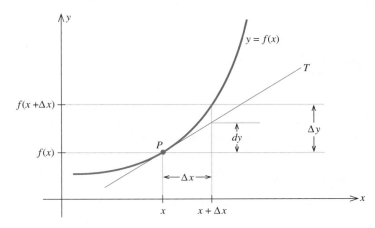

Observe that near the point of tangency P, the tangent line T is close to the graph of f. Therefore, if Δx is small, then dy is a good approximation of Δy. We can find an expression for dy as follows: Notice that the slope of T is given by

$$\frac{dy}{\Delta x} \quad \text{(Rise divided by run)}$$

However, the slope of T is given by $f'(x)$. Therefore, we have

$$\frac{dy}{\Delta x} = f'(x)$$

or $dy = f'(x)\Delta x$. Thus, we have the approximation

$$\Delta y \approx dy = f'(x)\Delta x$$

in terms of the derivative of f at x. The quantity dy is called the *differential of y*.

The Differential

Let $y = f(x)$ define a differentiable function of x. Then,

1. The **differential** dx of the independent variable x is $dx = \Delta x$.
2. The **differential** dy of the dependent variable y is

$$dy = f'(x)\Delta x = f'(x)dx \qquad \textbf{(11)}$$

REMARKS

1. For the independent variable x: There is no difference between Δx and dx—both measure the change in x from x to $x + \Delta x$.
2. For the dependent variable y: Δy measures the *actual* change in y as x changes from x to $x + \Delta x$, whereas dy measures the *approximate* change in y corresponding to the same change in x.
3. The differential dy depends on both x and dx, but for fixed x, dy is a linear function of dx. ■ ■ ■

EXAMPLE 3   Let $y = x^3$.

a. Find the differential dy of y.
b. Use dy to approximate Δy when x changes from 2 to 2.01.
c. Use dy to approximate Δy when x changes from 2 to 1.98.
d. Compare the results of part (b) with those of Example 2.

SOLUTION ✔   **a.** Let $f(x) = x^3$. Then,

$$dy = f'(x)\ dx = 3x^2 dx$$

b. Here, $x = 2$ and $dx = 2.01 - 2 = 0.01$. Therefore,

$$dy = 3x^2\, dx = 3(2)^2(0.01) = 0.12$$

c. Here, $x = 2$ and $dx = 1.98 - 2 = -0.02$. Therefore,

$$dy = 3x^2\, dx = 3(2)^2(-0.02) = -0.24$$

d. As you can see, both approximations 0.12 and -0.24 are quite close to the actual changes of Δy obtained in Example 2: 0.120601 and -0.237608.

■ ■ ■ ■

Observe how much easier it is to find an approximation to the exact change in a function with the help of the differential, rather than calculating the exact change in the function itself. In the following examples, we take advantage of this fact.

EXAMPLE 4   Approximate the value of $\sqrt{26.5}$ using differentials. Verify your result using the $\boxed{\sqrt{}}$ key on your calculator.

SOLUTION ✔   Since we want to compute the square root of a number, let's consider the function $y = f(x) = \sqrt{x}$. Since 25 is the number nearest 26.5 whose square root is readily recognized, let's take $x = 25$. We want to know the change in y, Δy, as x changes from $x = 25$ to $x = 26.5$, an increase of $\Delta x = 1.5$ units. Using Equation (11), we find

$$\Delta y \approx dy = f'(x)\Delta x$$

$$= \left[\frac{1}{2\sqrt{x}}\bigg|_{x=25}\right] \cdot (1.5) = \left(\frac{1}{10}\right)(1.5) = 0.15$$

Therefore,

$$\sqrt{26.5} - \sqrt{25} = \Delta y \approx 0.15$$
$$\sqrt{26.5} \approx \sqrt{25} + 0.15 = 5.15$$

The exact value of $\sqrt{26.5}$, rounded off to five decimal places, is 5.14782. Thus, the error incurred in the approximation is 0.00218.   ■ ■ ■ ■

APPLICATIONS

EXAMPLE 5   The total cost incurred in operating a certain type of truck on a 500-mile trip, traveling at an average speed of v mph, is estimated to be

$$C(v) = 125 + v + \frac{4500}{v}$$

dollars. Find the approximate change in the total operating cost when the average speed is increased from 55 mph to 58 mph.

SOLUTION ✔ With $v = 55$ and $\Delta v = dv = 3$, we find

$$\Delta C \approx dC = C'(v)dv = \left(1 - \frac{4500}{v^2}\right)\bigg|_{v=55} \cdot 3$$

$$= \left(1 - \frac{4500}{3025}\right)(3) \approx -1.46$$

so the total operating cost is found to decrease by $1.46. This might explain why so many independent truckers often exceed the 55 mph speed limit.

■ ■ ■ ■

EXAMPLE 6 The relationship between the amount of money x spent by Cannon Precision Instruments on advertising and Cannon's total sales $S(x)$ is given by the function

$$S(x) = -0.002x^3 + 0.6x^2 + x + 500 \qquad (0 \le x \le 200)$$

where x is measured in thousands of dollars. Use differentials to estimate the change in Cannon's total sales if advertising expenditures are increased from $100,000 ($x = 100$) to $105,000 ($x = 105$).

SOLUTION ✔ The required change in sales is given by

$$\Delta S \approx dS = S'(100)dx$$
$$= -0.006x^2 + 1.2x + 1|_{x=100} \cdot (5) \qquad (dx = 105 - 100 = 5)$$
$$= (-60 + 120 + 1)(5) = 305$$

—that is, an increase of $305,000.

■ ■ ■ ■

EXAMPLE 7

The Rings of Neptune

a. A ring has an inner radius of r units and an outer radius of R units, where $(R - r)$ is small in comparison to r (Figure 3.20a). Use differentials to estimate the area of the ring.

b. Recent observations, including those of Voyager I and II, showed that Neptune's ring system is considerably more complex than had been believed. For one thing, it is made up of a large number of distinguishable rings rather than one continuous great ring as previously thought (Figure 3.20b). The

FIGURE 3.20

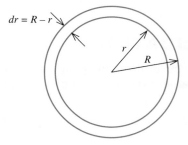

(a) The area of the ring is the circumference of the inner circle times the thickness.

(b) Neptune and its rings.

outermost ring, 1989N1R, has an inner radius of approximately 62,900 kilometers (measured from the center of the planet), and a radial width of approximately 50 kilometers. Using these data, estimate the area of the ring.

SOLUTION ✔ **a.** Using the fact that the area of a circle of radius x is $A = f(x) = \pi x^2$, we find

$$\pi R^2 - \pi r^2 = f(R) - f(r)$$
$$= \Delta A \qquad \text{(Remember, } \Delta A = \text{change in } f \text{ when}$$
$$\approx dA \qquad \qquad x \text{ changes from } x = r \text{ to } x = R.)$$
$$= f'(r)dr$$

where $dr = R - r$. So, we see that the area of the ring is approximately $2\pi r(R - r)$ square units. In words, the area of the ring is approximately equal to

Circumference of the inner circle × Thickness of the ring

b. Applying the results of part (a) with $r = 62{,}900$ and $dr = 50$, we find that the area of the ring is approximately $2\pi(62{,}900)(50)$, or 19,760,618 square kilometers, which is roughly 4% of Earth's surface.   ■ ■ ■ ■

Before looking at the next example, we need to familiarize ourselves with some terminology. If a quantity with exact value q is measured or calculated with an error of Δq, then the quantity $\Delta q/q$ is called the *relative error* in the measurement or calculation of q. If the quantity $\Delta q/q$ is expressed as a percentage, it is then called the *percentage error*. Because Δq is approximated by dq, we normally approximate the relative error $\Delta q/q$ by dq/q.

EXAMPLE **8** Suppose the radius of a ball-bearing is measured to be 0.5 inch, with a maximum error of ±0.0002 inch. Then, the relative error in r is

$$\frac{dr}{r} = \frac{\pm0.0002}{0.5} = \pm0.0004$$

and the percentage error is $\pm0.04\%$.   ■ ■ ■ ■

EXAMPLE **9** Suppose the side of a cube is measured with a maximum percentage error of 2%. Use differentials to estimate the maximum percentage error in the calculated volume of the cube.

SOLUTION ✔ Suppose the side of the cube is x, so its volume is

$$V = x^3$$

We are given that $\left|\dfrac{dx}{x}\right| \leq 0.02$. Now,

$$dV = 3x^2 dx$$

and so

$$\frac{dV}{V} = \frac{3x^2 dx}{x^3} = 3\frac{dx}{x}$$

Therefore,

$$\left|\frac{dV}{V}\right| = 3\left|\frac{dx}{x}\right| \le 3(0.02) = 0.06$$

and we see that the maximum percentage error in the measurement of the volume of the cube is 6%. ■■■■

Finally, we want to point out that if at some point in reading this section you have a sense of déjà vu, do not be surprised, because the notion of the differential was first used in Section 3.4 (see Example 1). There we took $\Delta x = 1$ since we were interested in finding the marginal cost when the level of production was increased from $x = 250$ to $x = 251$. If we had used differentials, we would have found

$$C(251) - C(250) \approx C'(250)dx$$

so that taking $dx = \Delta x = 1$, we have $C(251) - C(250) \approx C'(250)$, which agrees with the result obtained in Example 1. Thus, in Section 3.4, we touched upon the notion of the differential, albeit in the special case in which $dx = 1$.

SELF-CHECK EXERCISES 3.7

1. Find the differential of $f(x) = \sqrt{x} + 1$.

2. A certain country's government economists have determined that the demand equation for corn in that country is given by

$$p = f(x) = \frac{125}{x^2 + 1}$$

where p is expressed in dollars per bushel and x, the quantity demanded per year, is measured in billions of bushels. The economists are forecasting a harvest of 6 billion bushels for the year. If the actual production of corn were 6.2 billion bushels for the year instead, what would be the approximate drop in the predicted price of corn per bushel?

Solutions to Self-Check Exercises 3.7 can be found on page 273.

3.7 Exercises

In Exercises 1–14, find the differential of the given function.

1. $f(x) = 2x^2$

2. $f(x) = 3x^2 + 1$

3. $f(x) = x^3 - x$

4. $f(x) = 2x^3 + x$

5. $f(x) = \sqrt{x + 1}$

6. $f(x) = \dfrac{3}{\sqrt{x}}$

7. $f(x) = 2x^{3/2} + x^{1/2}$

8. $f(x) = 3x^{5/6} + 7x^{2/3}$

9. $f(x) = x + \dfrac{2}{x}$

10. $f(x) = \dfrac{3}{x - 1}$

11. $f(x) = \dfrac{x - 1}{x^2 + 1}$

12. $f(x) = \dfrac{2x^2 + 1}{x + 1}$

13. $f(x) = \sqrt{3x^2 - x}$

14. $f(x) = (2x^2 + 3)^{1/3}$

15. Let f be a function defined by

$$y = f(x) = x^2 - 1$$

a. Find the differential of f.

b. Use your result from part (a) to find the approximate change in y if x changes from 1 to 1.02.

c. Find the actual change in y if x changes from 1 to 1.02 and compare your result with that obtained in part (b).

16. Let f be a function defined by

$$y = f(x) = 3x^2 - 2x + 6$$

a. Find the differential of f.

b. Use your result from part (a) to find the approximate change in y if x changes from 2 to 1.97.

c. Find the actual change in y if x changes from 2 to 1.97 and compare your result with that obtained in part (b).

17. Let f be a function defined by

$$y = f(x) = \frac{1}{x}$$

a. Find the differential of f.

b. Use your result from part (a) to find the approximate change in y if x changes from -1 to -0.95.

c. Find the actual change in y if x changes from -1 to -0.95 and compare your result with that obtained in part (b).

18. Let f be a function defined by

$$y = f(x) = \sqrt{2x + 1}$$

a. Find the differential of f.

b. Use your result from part (a) to find the approximate change in y if x changes from 4 to 4.1.

c. Find the actual change in y if x changes from 4 to 4.1 and compare your result with that obtained in part (b).

In Exercises 19–26, use differentials to approximate the given quantity.

19. $\sqrt{10}$ **20.** $\sqrt{17}$ **21.** $\sqrt{49.5}$

22. $\sqrt{99.7}$ **23.** $\sqrt[3]{7.8}$ **24.** $\sqrt[4]{81.6}$

25. $\sqrt{0.089}$ **26.** $\sqrt[3]{0.00096}$

27. Use a differential to approximate $\sqrt{4.02} + \dfrac{1}{\sqrt{4.02}}$.

Hint: Let $f(x) = \sqrt{x} + \dfrac{1}{\sqrt{x}}$ and compute dy with $x = 4$ and $dx = 0.02$.

28. Use a differential to approximate $\dfrac{2(4.98)}{(4.98)^2 + 1}$.

Hint: Study the hint for Exercise 27.

▢ **A calculator is recommended for the remainder of this exercise set.**

29. ERROR ESTIMATION   The length of each edge of a cube is 12 cm, with a possible error in measurement of 0.02 cm. Use differentials to estimate the error that might occur when the volume of the cube is calculated.

30. ESTIMATING THE AMOUNT OF PAINT REQUIRED   A coat of paint of thickness 0.05 cm is to be applied uniformly to the faces of a cube of edge 30 cm. Use differentials to find the approximate amount of paint required for the job.

31. ERROR ESTIMATION   A hemisphere-shaped dome of radius 60 ft is to be coated with a layer of rust-proofer before painting. Use differentials to estimate the amount of rust-proofer needed if the coat is to be 0.01 in. thick. Hint: The volume of a hemisphere of radius r is $V = \frac{2}{3}\pi r^3$.

32. GROWTH OF A CANCEROUS TUMOR   The volume of a spherical cancer tumor is given by

$$V(r) = \frac{4}{3}\pi r^3$$

If the radius of a tumor is estimated at 1.1 cm, with a maximum error in measurement of 0.005 cm, determine the error that might occur when the volume of the tumor is calculated.

33. UNCLOGGING ARTERIES   Research done in the 1930s by the French physiologist Jean Poiseuille showed that the resistance R of a blood vessel of length l and radius r is $R = kl/r^4$, where k is a constant. Suppose a dose of the drug TPA increases r by 10%. How will this affect the resistance R? Assume that l is constant.

34. GROSS DOMESTIC PRODUCT   An economist has determined that a certain country's gross domestic product (GDP) is approximated by the function $f(x) = 640x^{1/5}$, where $f(x)$ is measured in billions of dollars and x is the capital outlay in billions of dollars. Use differentials to estimate the change in the country's GDP if the country's capital expenditure changes from $243 billion to $248 billion.

35. LEARNING CURVES   The length of time (in seconds) a certain individual takes to learn a list of n items is approximated by

$$f(n) = 4n\sqrt{n - 4}$$

Use differentials to approximate the additional time it takes the individual to learn the items on a list when n is increased from 85 to 90 items.

(continued on p. 272)

FINDING THE DIFFERENTIAL
OF A FUNCTION

The calculation of the differential of f at a given value of x involves the evaluation of the derivative of f at that point and can be facilitated through the use of the numerical derivative function.

EXAMPLE 1

Use dy to approximate Δy if $y = x^2(2x^2 + x + 1)^{2/3}$ and x changes from 2 to 1.98.

SOLUTION ✔

Let $f(x) = x^2(2x^2 + x + 1)^{2/3}$. Since $dx = 1.98 - 2 = -0.02$, we find the required approximation to be

$$dy = f'(2) \cdot (-0.02)$$

But using the numerical derivative operation, we find

$$f'(2) = 30.5758132855$$

and so

$$dy = (-0.02)(30.5758132855) = -0.611516266 \qquad \blacksquare\blacksquare\blacksquare\blacksquare$$

EXAMPLE 2

The Meyers are considering the purchase of a house in the near future and estimate that they will need a loan of $120,000. Based on a 30-year conventional mortgage with an interest rate of r per year, their monthly repayment will be

$$P = \frac{10{,}000r}{1 - \left(1 + \dfrac{r}{12}\right)^{-360}}$$

dollars. If the interest rate increases from the present rate of 10%/year to 10.2% per year between now and the time the Meyers decide to secure the loan, approximately how much more per month will their mortgage payment be?

SOLUTION ✔

Let's write

$$P = f(r) = \frac{10{,}000r}{1 - \left(1 + \dfrac{r}{12}\right)^{-360}}$$

Then the increase in the mortgage payment will be approximately

$$dP = f'(0.1)dr = f'(0.1)(0.002) \qquad \text{(Since } dr = 0.102 - 0.1)$$
$$= (8867.59947979)(0.002) \approx 17.7352 \qquad \text{(Using the numerical derivative operation)}$$

or approximately $17.74 per month. $\qquad \blacksquare\blacksquare\blacksquare\blacksquare$

In Exercises 1–6, use *dy* to approximate Δy for the function $y = f(x)$ when x changes from $x = a$ to $x = b$.

1. $f(x) = 0.21x^7 - 3.22x^4 + 5.43x^2 + 1.42x + 12.42$; $a = 3$, $b = 3.01$

2. $f(x) = \dfrac{0.2x^2 + 3.1}{1.2x + 1.3}$; $a = 2$, $b = 1.96$

3. $f(x) = \sqrt{2.2x^2 + 1.3x + 4}$; $a = 1$, $b = 1.03$

4. $f(x) = x\sqrt{2x^3 - x + 4}$; $a = 2$, $b = 1.98$

5. $f(x) = \dfrac{\sqrt{x^2 + 4}}{x - 1}$; $a = 4$, $b = 4.1$

6. $f(x) = 2.1x^2 + \dfrac{3}{\sqrt{x}} + 5$; $a = 3$, $b = 2.95$

7. CALCULATING MORTGAGE PAYMENTS Refer to Example 2. How much more per month will the Meyers' mortgage payment be if the interest rate increases from 10% to 10.3%/year? To 10.4%/year? To 10.5%/year?

8. ESTIMATING THE AREA OF A RING OF NEPTUNE The ring 1989N2R of the planet Neptune has an inner radius of approximately 53,200 km (measured from the center of the planet) and a radial width of 15 km. Use differentials to estimate the area of the ring.

9. EFFECT OF PRICE INCREASE ON QUANTITY DEMANDED The quantity demanded per week of the Alpha Sports Watch, x (in thousands), is related to its unit price of p dollars by the equation

$$x = f(p) = 10\sqrt{\frac{50 - p}{p}} \qquad (0 \le p \le 50)$$

Use differentials to find the decrease in the quantity of the watches demanded per week if the unit price is increased from \$40 to \$42.

36. **EFFECT OF ADVERTISING ON PROFITS** The relationship between Cunningham Realty's quarterly profits, $P(x)$, and the amount of money x spent on advertising per quarter is described by the function

$$P(x) = -\frac{1}{8}x^2 + 7x + 30 \qquad (0 \leq x \leq 50)$$

where both $P(x)$ and x are measured in thousands of dollars. Use differentials to estimate the increase in profits when advertising expenditure each quarter is increased from $24,000 to $26,000.

37. **EFFECT OF MORTGAGE RATES ON HOUSING STARTS** A study prepared for the National Association of Realtors estimates that the number of housing starts per year over the next 5 yr will be

$$N(r) = \frac{7}{1 + 0.02r^2}$$

million units, where r (percent) is the mortgage rate. Use differentials to estimate the decrease in the number of housing starts when the mortgage rate is increased from 12% to 12.5%.

38. **SUPPLY-PRICE** The supply equation for a certain brand of transistor radio is given by

$$p = s(x) = 0.3 \sqrt{x} + 10$$

where x is the quantity supplied and p is the unit price in dollars. Use differentials to approximate the change in price when the quantity supplied is increased from 10,000 units to 10,500 units.

39. **DEMAND-PRICE** The demand function for the Sentinel smoke alarm is given by

$$p = d(x) = \frac{30}{0.02x^2 + 1}$$

where x is the quantity demanded (in units of a thousand) and p is the unit price in dollars. Use differentials to estimate the change in the price p when the quantity demanded changes from 5000 to 5500 units/wk.

40. **SURFACE AREA OF AN ANIMAL** Animal physiologists use the formula

$$S = kW^{2/3}$$

to calculate an animal's surface area (in square meters) from its weight W (in kilograms), where k is a constant that depends on the animal under consideration. Sup-

pose a physiologist calculates the surface area of a horse ($k = 0.1$). If the horse's weight is estimated at 300 kg, with a maximum error in measurement of 0.6 kg, determine the percentage error in the calculation of the horse's surface area.

41. **FORECASTING PROFITS** The management of Trappee and Sons, Inc., forecast that they will sell 200,000 cases of their Texa-Pep hot sauce next year. Their annual profit is described by

$$P(x) = -0.000032x^3 + 6x - 100$$

thousand dollars, where x is measured in thousands of cases. If the maximum error in the forecast is 15%, determine the corresponding error in Trappee's profits.

42. **FORECASTING COMMODITY PRICES** A certain country's government economists have determined that the demand equation for soybeans in that country is given by

$$p = f(x) = \frac{55}{2x^2 + 1}$$

where p is expressed in dollars per bushel and x, the quantity demanded per year, is measured in billions of bushels. The economists are forecasting a harvest of 1.8 billion bushels for the year, with a maximum error of 15% in their forecast. Determine the corresponding maximum error in the predicted price per bushel of soybeans.

43. **CRIME STUDIES** A sociologist has found that the number of serious crimes in a certain city per year is described by the function

$$N(x) = \frac{500(400 + 20x)^{1/2}}{(5 + 0.2x)^2}$$

where x (in cents per dollar deposited) is the level of reinvestment in the area in conventional mortgages by the city's ten largest banks. Use differentials to estimate the change in the number of crimes if the level of reinvestment changes from 20 cents/dollar deposited to 22 cents/dollar deposited.

44. **FINANCING A HOME** The Millers are planning to buy a home in the near future and estimate that they will need a 30-yr fixed-rate mortgage for $120,000. Their monthly payment P (in dollars) can be computed using the formula

$$P = \frac{10,000r}{1 - \left(1 + \dfrac{r}{12}\right)^{-360}}$$

where r is the interest rate per year.

a. Find the differential of P.

b. If the interest rate increases from the present rate of 9%/year to 9.2%/year between now and the time the Millers decide to secure the loan, approximately how much more will their monthly mortgage payment be? How much more will it be if the interest rate increases to 9.3%/year? To 9.4%/year? To 9.5%/year?

In Exercises 45 and 46, determine whether the statement is true or false. If it is true, explain why it is true. If it is false, give an example to show why it is false.

45. If $y = ax + b$ where a and b are constants, then $\Delta y = dy$.

46. If $A = f(x)$, then the percentage change in A is

$$\frac{100f'(x)}{f(x)} dx$$

SOLUTIONS TO SELF-CHECK EXERCISES 3.7

1. We find

$$f'(x) = \frac{1}{2}x^{-1/2} = \frac{1}{2\sqrt{x}}$$

Therefore, the required differential of f is

$$dy = \frac{1}{2\sqrt{x}} dx$$

2. We first compute the differential

$$dp = -\frac{250x}{(x^2 + 1)^2} dx$$

Next, using Equation (11) with $x = 6$ and $dx = 0.2$, we find

$$\Delta p \approx dp = -\frac{250(6)}{(36 + 1)^2}(0.2) = -0.22$$

or a drop in price of 22 cents/bushel.

CHAPTER 3 Summary of Principal Formulas and Terms

Formulas

1. Derivative of a constant	$\dfrac{d}{dx}(c) = 0$, c a constant
2. Power rule	$\dfrac{d}{dx}(x^n) = nx^{n-1}$
3. Constant multiple rule	$\dfrac{d}{dx}(cu) = c\dfrac{du}{dx}$, c a constant
4. Sum rule	$\dfrac{d}{dx}(u \pm v) = \dfrac{du}{dx} \pm \dfrac{dv}{dx}$
5. Product rule	$\dfrac{d}{dx}(uv) = u\dfrac{dv}{dx} + v\dfrac{du}{dx}$

6. Quotient rule

$$\frac{d}{dx}\left(\frac{u}{v}\right) = \frac{v\frac{du}{dx} - u\frac{dv}{dx}}{v^2}$$

7. Chain rule

$$\frac{dy}{dx} = \frac{dy}{du} \cdot \frac{du}{dx}$$

8. General power rule

$$\frac{d}{dx}(u^n) = nu^{n-1}\frac{du}{dx}$$

9. Average cost function

$$\overline{C}(x) = \frac{C(x)}{x}$$

10. Revenue function

$$R(x) = px$$

11. Profit function

$$P(x) = R(x) - C(x)$$

12. Elasticity of demand

$$E(p) = -\frac{pf'(p)}{f(p)}$$

13. Differential of y

$$dy = f'(x)dx$$

Terms

marginal cost function	unitary demand
marginal average cost function	inelastic demand
marginal revenue function	second derivative of f
marginal profit function	implicit differentiation
elastic demand	related rates

CHAPTER 3 REVIEW EXERCISES

In Exercises 1–30, find the derivative of the given function.

1. $f(x) = 3x^5 - 2x^4 + 3x^2 - 2x + 1$

2. $f(x) = 4x^6 + 2x^4 + 3x^2 - 2$

3. $g(x) = -2x^{-3} + 3x^{-1} + 2$

4. $f(t) = 2t^2 - 3t^3 - t^{-1/2}$

5. $g(t) = 2t^{-1/2} + 4t^{-3/2} + 2$

6. $h(x) = x^2 + \dfrac{2}{x}$

7. $f(t) = t + \dfrac{2}{t} + \dfrac{3}{t^2}$

8. $g(s) = 2s^2 - \dfrac{4}{s} + \dfrac{2}{\sqrt{s}}$

9. $h(x) = x^2 - \dfrac{2}{x^{3/2}}$

10. $f(x) = \dfrac{x+1}{2x-1}$

11. $g(t) = \dfrac{t^2}{2t^2 + 1}$

12. $h(t) = \dfrac{\sqrt{t}}{\sqrt{t} + 1}$

13. $f(x) = \dfrac{\sqrt{x} - 1}{\sqrt{x} + 1}$

14. $f(t) = \dfrac{t}{2t^2 + 1}$

15. $f(x) = \dfrac{x^2(x^2 + 1)}{x^2 - 1}$

16. $f(x) = (2x^2 + x)^3$

17. $f(x) = (3x^3 - 2)^8$

18. $h(x) = (\sqrt{x} + 2)^5$

19. $f(t) = \sqrt{2t^2 + 1}$

20. $g(t) = \sqrt[3]{1 - 2t^3}$

21. $s(t) = (3t^2 - 2t + 5)^{-2}$

22. $f(x) = (2x^3 - 3x^2 + 1)^{-3/2}$

23. $h(x) = \left(x + \dfrac{1}{x}\right)^2$

24. $h(x) = \dfrac{1 + x}{(2x^2 + 1)^2}$

25. $h(t) = (t^2 + t)^4(2t^2)$

26. $f(x) = (2x + 1)^3(x^2 + x)^2$

27. $g(x) = \sqrt{x}(x^2 - 1)^3$

28. $f(x) = \dfrac{x}{\sqrt{x^3 + 2}}$

29. $h(x) = \dfrac{\sqrt{3x + 2}}{4x - 3}$

30. $f(t) = \dfrac{\sqrt{2t + 1}}{(t + 1)^3}$

In Exercises 31–36, find the second derivative of the given function.

31. $f(x) = 2x^4 - 3x^3 + 2x^2 + x + 4$

32. $g(x) = \sqrt{x} + \dfrac{1}{\sqrt{x}}$

33. $h(t) = \dfrac{t}{t^2 + 4}$

34. $f(x) = (x^3 + x + 1)^2$

35. $f(x) = \sqrt{2x^2 + 1}$

36. $f(t) = t(t^2 + 1)^3$

In Exercises 37–42, find *dy/dx* by implicit differentiation.

37. $6x^2 - 3y^2 = 9$

38. $2x^3 - 3xy = 4$

39. $y^3 + 3x^2 = 3y$

40. $x^2 + 2x^2 y^2 + y^2 = 10$

41. $x^2 - 4xy - y^2 = 12$

42. $3x^2 y - 4xy + x - 2y = 6$

43. Let $f(x) = 2x^3 - 3x^2 - 16x + 3$.
  a. Find the points on the graph of f at which the slope of the tangent line is equal to -4.
  b. Find the equation(s) of the tangent line(s) of part (a).

44. Let $f(x) = \frac{1}{3}x^3 + \frac{1}{2}x^2 - 4x + 1$.
  a. Find the points on the graph of f at which the slope of the tangent line is equal to -2.
  b. Find the equation(s) of the tangent line(s) of part (a).

45. Find an equation of the tangent line to the graph of $y = \sqrt{4 - x^2}$ at the point $(1, \sqrt{3})$.

46. Find an equation of the tangent line to the graph of $y = x(x + 1)^5$ at the point $(1, 32)$.

47. Find the third derivative of the function

$$f(x) = \dfrac{1}{2x - 1}$$

What is its domain?

48. WORLDWIDE NETWORKED PCS   The number of worldwide networked PCs (in millions) is given by

$$N(t) = 3.136t^2 + 3.954t + 116.468 \qquad (0 \le t \le 9)$$

where t is measured in years, with $t = 0$ corresponding to the beginning of 1991.
  a. How many worldwide networked PCs were there at the beginning of 1997?
  b. How fast was the number of worldwide networked PCs changing at the beginning of 1997?

49. The number of subscribers to CNC Cable Television in the town of Randolph is approximated by the function

$$N(x) = 1000(1 + 2x)^{1/2} \qquad (1 \le x \le 30)$$

where $N(x)$ denotes the number of subscribers to the service in the xth week. Find the rate of increase in the number of subscribers at the end of the 12th week.

50. The total weekly cost in dollars incurred by the Herald Record Company in pressing x video discs is given by the total cost function

$$C(x) = 2500 + 2.2x \qquad (0 \le x \le 8000)$$

  a. What is the marginal cost when $x = 1000$? When $x = 2000$?
  b. Find the average cost function $\overline{C}$ and the marginal average cost function $\overline{C}'$.
  c. Using the results from part (b), show that the average cost incurred by Herald in pressing a video disc approaches \$2.20/disc when the level of production is high enough.

51. The marketing department of Telecon Corporation has determined that the demand for their cordless phones obeys the relationship

$$p = -0.02x + 600 \qquad (0 \le x \le 30{,}000)$$

where p denotes the phone's unit price (in dollars) and x denotes the quantity demanded.
  a. Find the revenue function R.
  b. Find the marginal revenue function R'.
  c. Compute $R'(10{,}000)$ and interpret your result.

52. The weekly demand for the Lectro-Copy photocopying machine is given by the demand equation

$$p = 2000 - 0.04x \qquad (0 \le x \le 50{,}000)$$

where p denotes the wholesale unit price in dollars and x denotes the quantity demanded. The weekly total cost function for manufacturing these copiers is given by

$$C(x) = 0.000002x^3 - 0.02x^2 + 1000x + 120{,}000$$

where $C(x)$ denotes the total cost incurred in producing x units.
  a. Find the revenue function R, the profit function P, and the average cost function $\overline{C}$.
  b. Find the marginal cost function C', the marginal revenue function R', the marginal profit function P', and the marginal average cost function $\overline{C}'$.
  c. Compute $C'(3000)$, $R'(3000)$, and $P'(3000)$.
  d. Compute $\overline{C}'(5000)$ and $\overline{C}'(8000)$ and interpret your results.

APPLICATIONS OF THE DERIVATIVE

4

This chapter further explores the power of the derivative, which we use to help analyze the properties of functions. The information obtained can then be used to accurately sketch graphs of functions. We also see how the derivative is used in solving a large class of optimization problems, including finding what level of production will yield a maximum profit for a company, finding what level of production will result in minimal cost to a company, finding the maximum height attained by a rocket, finding the maximum velocity at which air is expelled when a person coughs, and a host of other problems.

What is the maximum altitude and the maximum velocity attained by the rocket? In Example 7, page 348, you will see how the techniques of calculus can be used to help answer these questions.

Exploring with Technology

Refer to Example 8.

1. Use a graphing utility to plot the graphs of $f(x) = x + 1/x$ and its derivative function $f'(x) = 1 - 1/x^2$, using the viewing rectangle $[-4, 4] \times [-8, 8]$.
2. By studying the graph of f', determine the critical points of f. Next, note the sign of $f'(x)$ immediately to the left and to the right of each critical point. What can you conclude about each critical point? Are your conclusions borne out by the graph of f?

APPLICATIONS

EXAMPLE 9   The profit function of the Acrosonic Company is given by

$$P(x) = -0.02x^2 + 300x - 200{,}000$$

dollars, where x is the number of Acrosonic model F loudspeaker systems produced. Find where the function P is increasing and where it is decreasing.

SOLUTION ✔   The derivative P' of the function P is

$$P'(x) = -0.04x + 300 = -0.04(x - 7500)$$

Thus, $P'(x) = 0$ when $x = 7500$. Furthermore, $P'(x) > 0$ for x in the interval $(0, 7500)$, and $P'(x) < 0$ for x in the interval $(7500, \infty)$. This means that the profit function P is increasing on $(0, 7500)$ and decreasing on $(7500, \infty)$ (Figure 4.24).

FIGURE 4.24
The profit function is increasing on $(0, 7500)$ and decreasing on $(7500, \infty)$.

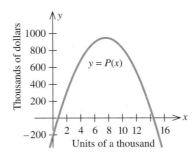

■ ■ ■ ■

EXAMPLE 10   The number of major crimes committed in the city of Bronxville from 1993 to 2000 is approximated by the function

$$N(t) = -0.1t^3 + 1.5t^2 + 100 \qquad (0 \le t \le 7)$$

where $N(t)$ denotes the number of crimes committed in year t and $t = 0$ corresponds to the beginning of 1993. Find where the function N is increasing and where it is decreasing.

SOLUTION ✔   The derivative N' of the function N is

$$N'(t) = -0.3t^2 + 3t = -0.3t(t - 10)$$

Since $N'(t) > 0$ for t in the interval $(0, 7)$, the function N is increasing throughout that interval (Figure 4.25).

FIGURE 4.25
The number of crimes, $N(t)$, is increasing over the 7-year interval.

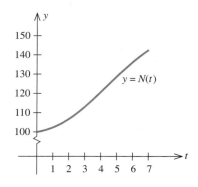

■ ■ ■ ■

SELF-CHECK EXERCISES 4.1

1. Find the intervals where the function $f(x) = \frac{2}{3}x^3 - x^2 - 12x + 3$ is increasing and the intervals where it is decreasing.

2. Find the relative extrema of $f(x) = \dfrac{x^2}{1 - x^2}$.

Solutions to Self-Check Exercises 4.1 can be found on page 301.

4.1 Exercises

In Exercises 1–8, you are given the graph of a function f. Determine the intervals where f is increasing, constant, or decreasing.

1.
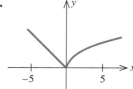

2.
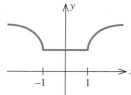

3.
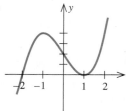

4.
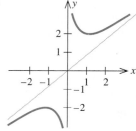

5.
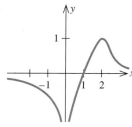

6.
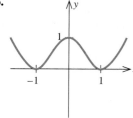

7.
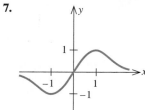

8.
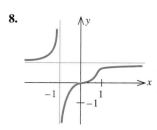

9. THE BOSTON MARATHON The graph of the function *f* shown in the accompanying figure gives the elevation of that part of the Boston Marathon course that includes the notorious Heartbreak Hill. Determine the intervals (stretches of the course) where the function *f* is increasing (the runner is laboring), where it is constant (the runner is taking a breather), and where it is decreasing (the runner is coasting).

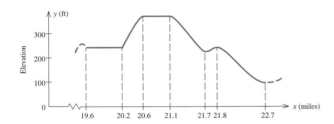

In Exercises 10–35, find the interval(s) where each function is increasing and the interval(s) where it is decreasing.

10. $f(x) = 4 - 5x$

11. $f(x) = 3x + 5$

12. $f(x) = 2x^2 + x + 1$

13. $f(x) = x^2 - 3x$

14. $f(x) = x^3 - 3x^2$

15. $g(x) = x - x^3$

16. $f(x) = x^3 - 3x + 4$

17. $g(x) = x^3 + 3x^2 + 1$

18. $f(x) = \frac{2}{3}x^3 - 2x^2 - 6x - 2$

19. $f(x) = \frac{1}{3}x^3 - 3x^2 + 9x + 20$

20. $g(x) = x^4 - 2x^2 + 4$

21. $h(x) = x^4 - 4x^3 + 10$

22. $h(x) = \dfrac{1}{2x + 3}$

23. $f(x) = \dfrac{1}{x - 2}$

24. $g(t) = \dfrac{2t}{t^2 + 1}$

25. $h(t) = \dfrac{t}{t - 1}$

26. $f(x) = x^{2/3} + 5$

27. $f(x) = x^{3/5}$

28. $f(x) = (x - 5)^{2/3}$

29. $f(x) = \sqrt{x + 1}$

30. $g(x) = x\sqrt{x + 1}$

31. $f(x) = \sqrt{16 - x^2}$

32. $h(x) = \dfrac{x^2}{x - 1}$

33. $f(x) = \dfrac{x^2 - 1}{x}$

34. $g(x) = \dfrac{x}{(x + 1)^2}$

35. $f(x) = \dfrac{1}{(x - 1)^2}$

In Exercises 36–43, you are given the graph of a function *f*. Determine the relative maxima and relative minima, if any.

36.

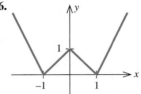

37.

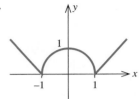

38.

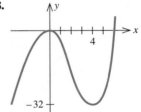

39.

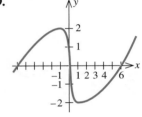

40.

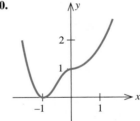

41.

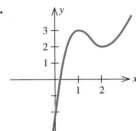

42.

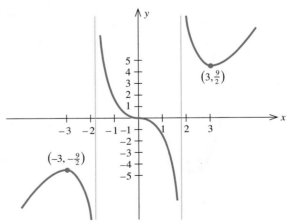

43.

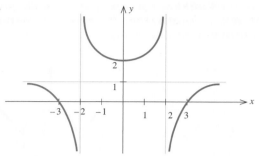

In Exercises 44–47, match the graph of the function with the graph of its derivative in (a)–(d).

44.

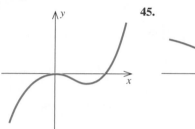

45.

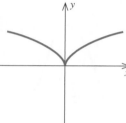

46.

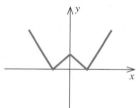

47.

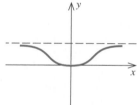

(a)

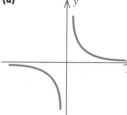

(b)

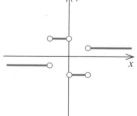

(c)

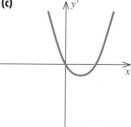

(d)

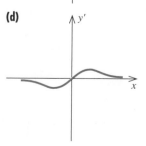

In Exercises 48–71, find the relative maxima and relative minima, if any, of each function.

48. $g(x) = x^2 + 3x + 8$ **49.** $f(x) = x^2 - 4x$

50. $h(t) = -t^2 + 6t + 6$ **51.** $f(x) = \dfrac{1}{2} x^2 - 2x + 4$

52. $f(x) = x^{5/3}$ **53.** $f(x) = x^{2/3} + 2$

54. $f(x) = x^3 - 3x + 6$ **55.** $g(x) = x^3 - 3x^2 + 4$

56. $f(x) = \dfrac{1}{2} x^4 - x^2$

57. $F(x) = \dfrac{1}{3} x^3 - x^2 - 3x + 4$

58. $h(x) = \dfrac{1}{2} x^4 - 3x^2 + 4x - 8$

59. $g(x) = x^4 - 4x^3 + 8$

60. $F(t) = 3t^5 - 20t^3 + 20$

61. $f(x) = 3x^4 - 2x^3 + 4$

62. $h(x) = \dfrac{x}{x+1}$ **63.** $g(x) = \dfrac{x+1}{x}$

64. $g(x) = 2x^2 + \dfrac{4000}{x} + 10$

65. $f(x) = x + \dfrac{9}{x} + 2$

66. $g(x) = \dfrac{x}{x^2 - 1}$ **67.** $f(x) = \dfrac{x}{1 + x^2}$

68. $g(t) = \dfrac{t^2}{1 + t^2}$ **69.** $f(x) = \dfrac{x^2}{x^2 - 4}$

70. $g(x) = x\sqrt{x - 4}$ **71.** $f(x) = (x - 1)^{2/3}$

72. A stone is thrown straight up from the roof of an 80-ft building. The distance (in feet) of the stone from the ground at any time t (in seconds) is given by

$$h(t) = -16t^2 + 64t + 80$$

When is the stone rising, and when is it falling? If the stone were to miss the building, when would it hit the ground? Sketch the graph of h.
Hint: The stone is on the ground when $h(t) = 0$.

73. PROFIT FUNCTIONS The Mexican subsidiary of the Thermo-Master Company manufactures an indoor–outdoor thermometer. Management estimates that the profit (in dollars) realizable by the company for the manufacture and sale of x units of thermometers per week is

$$P(x) = -0.001x^2 + 8x - 5000$$

Find the intervals where the profit function P is increasing and the intervals where P is decreasing.

74. **FLIGHT OF A ROCKET** The height (in feet) attained by a rocket t sec into flight is given by the function

$$h(t) = -\frac{1}{3}t^3 + 16t^2 + 33t + 10$$

When is the rocket rising, and when is it descending?

75. **ENVIRONMENT OF FORESTS** Following the lead of the National Wildlife Federation, the Department of the Interior of a South American country began to record an index of environmental quality that measured progress and decline in the environmental quality of its forests. The index for the years 1984 through 1994 is approximated by the function

$$I(t) = \frac{1}{3}t^3 - \frac{5}{2}t^2 + 80 \qquad (0 \le t \le 10)$$

where $t = 0$ corresponds to the year 1984. Find the intervals where the function I is increasing and the intervals where it is decreasing. Interpret your results.
Source: World Almanac

76. **AVERAGE SPEED OF A HIGHWAY VEHICLE** The average speed of a vehicle on a stretch of Route 134 between 6 A.M. and 10 A.M. on a typical weekday is approximated by the function

$$f(t) = 20t - 40\sqrt{t} + 50 \qquad (0 \le t \le 4)$$

where $f(t)$ is measured in miles per hour and t is measured in hours, with $t = 0$ corresponding to 6 A.M. Find the interval where f is increasing and the interval where f is decreasing and interpret your results.

77. **AVERAGE COST** The average cost (in dollars) incurred by the Lincoln Record Company per week in pressing x compact discs is given by

$$\overline{C}(x) = -0.0001x + 2 + \frac{2000}{x} \qquad (0 < x \le 6000)$$

Show that $\overline{C}(x)$ is always decreasing over the interval $(0, 6000)$.

78. **AIR POLLUTION** According to the South Coast Air Quality Management District, the level of nitrogen dioxide, a brown gas that impairs breathing, present in the atmosphere on a certain May day in downtown Los Angeles is approximated by

$$A(t) = 0.03t^3(t - 7)^4 + 60.2 \qquad (0 \le t \le 7)$$

where $A(t)$ is measured in pollutant standard index (PSI) and t is measured in hours with $t = 0$ corresponding to 7 A.M. At what time of day is the air pollution increasing, and at what time is it decreasing?

79. **PROJECTED RETIREMENT FUNDS** Based on data from the Central Provident Fund of a certain country (a government agency similar to the Social Security Administration), the estimated cash in the fund in 1995 is given by

$$A(t) = -96.6t^4 + 403.6t^3$$
$$+ 660.9t^2 + 250 \qquad (0 \le t \le 5)$$

where $A(t)$ is measured in billions of dollars and t is measured in decades, with $t = 0$ corresponding to the year 1995. Find the interval where A is increasing and the interval where A is decreasing and interpret your results.
Hint: Use the quadratic formula.

80. **LEARNING CURVES** The Emory Secretarial School finds from experience that the average student taking Advanced Typing will progress according to the rule

$$N(t) = \frac{60t + 180}{t + 6} \qquad (t \ge 0)$$

where $N(t)$ measures the number of words per minute the student can type after t weeks in the course. Compute $N'(t)$ and use this result to show that the function N is increasing on the interval $(0, \infty)$.

81. **DRUG CONCENTRATION IN THE BLOOD** The concentration (in milligrams per cubic centimeter) of a certain drug in a patient's body t hr after injection is given by

$$C(t) = \frac{t^2}{2t^3 + 1} \qquad (0 \le t \le 4)$$

When is the concentration of the drug increasing, and when is it decreasing?

82. **AGE OF DRIVERS IN CRASH FATALITIES** The number of crash fatalities per 100,000 vehicle miles of travel (based on 1994 data) is approximated by the model

$$f(x) = \frac{15}{0.08333x^2 + 1.91667x + 1} \qquad (0 \le x \le 11)$$

where x is the age of the driver in years with $x = 0$ corresponding to age 16. Show that f is decreasing on $(0, 11)$ and interpret your result.
Source: National Highway Traffic Safety Administration

83. **AIR POLLUTION** The amount of nitrogen dioxide, a brown gas that impairs breathing, present in the atmosphere on a certain May day in the city of Long Beach is approximated by

$$A(t) = \frac{136}{1 + 0.25(t - 4.5)^2} + 28 \qquad (0 \le t \le 11)$$

where $A(t)$ is measured in pollutant standard index (PSI) and t is measured in hours, with $t = 0$ corresponding to
(continued on p. 300)

Using the First Derivative to Analyze a Function

A graphing utility is an effective tool for analyzing the properties of functions. This is especially true when we also bring into play the power of calculus, as the following examples show.

EXAMPLE 1

Let $f(x) = 2.4x^4 - 8.2x^3 + 2.7x^2 + 4x + 1$.

a. Use a graphing utility to plot the graph of f.
b. Find the intervals where f is increasing and the intervals where f is decreasing.
c. Find the relative extrema of f.

SOLUTION ✔

a. The graph of f in the viewing rectangle $[-2, 4] \times [-10, 10]$ is shown in Figure T1.

FIGURE T1
The graph of f in the viewing rectangle $[-2, 4] \times [-10, 10]$

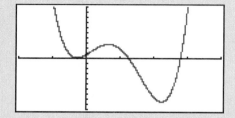

b. We compute

$$f'(x) = 9.6x^3 - 24.6x^2 + 5.4x + 4$$

and observe that f' is continuous everywhere, so the critical points of f occur at values of x where $f'(x) = 0$. To solve this last equation, observe that $f'(x)$ is a *polynomial function* of degree 3. The easiest way to solve the polynomial equation

$$9.6x^3 - 24.6x^2 + 5.4x + 4 = 0$$

is to use the function on a graphing utility for solving polynomial equations. (Not all graphing utilities have this function.) You can also use TRACE and ZOOM, but this will not give the same accuracy without a much greater effort.

We find

$$x_1 \approx 2.22564943249, \qquad x_2 \approx 0.63272944121, \qquad x_3 \approx -0.295878873696$$

Referring to Figure T1, we conclude that f is decreasing on $(-\infty, -0.2959)$ and $(0.6327, 2.2256)$ (correct to four decimal places) and f is increasing on $(-0.2959, 0.6327)$ and $(2.2256, \infty)$.

c. Using the evaluation function of a graphing utility, we find the value of f at each of the critical points found in part (b). Upon referring to Figure T1 once again, we see that $f(x_3) \approx 0.2836$ and $f(x_1) \approx -8.2366$ are relative minimum values of f and $f(x_2) \approx 2.9194$ is a relative maximum value of f.

■ ■ ■ ■

REMARK The equation $f'(x) = 0$ in Example 1 is a polynomial equation, and so it is easily solved using the function for solving polynomial equations. We could also solve the equation using the function for finding the roots of equations, but that would require much more work. For equations that are *not* polynomial equations, however, our only choice is to use the function for finding the roots of equations.

■ ■ ■

If the derivative of a function is difficult to compute or simplify and we do not require great precision in the solution, we can find the relative extrema of the function using a combination of **ZOOM** and **TRACE.** This technique, which does not require the use of the derivative of f, is illustrated in the following example.

EXAMPLE 2

Let $f(x) = x^{1/3}(x^2 + 1)^{-3/2}3^{-x}$.

a. Use a graphing utility to plot the graph of f.*
b. Find the relative extrema of f.

SOLUTION ✔

a. The graph of f in the viewing rectangle $[-4, 2] \times [-2, 1]$ is shown in Figure T2.

FIGURE T2
The graph of f in the viewing rectangle $[-4, 2] \times [-2, 1]$

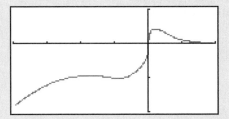

b. From the graph of f in Figure T2, we see that f has relative maxima when $x \approx -2$ and $x \approx 0.25$ and a relative minimum when $x \approx -0.75$. To obtain a better approximation of the first relative maximum, we zoom-in with the

---

* Functions of the form $f(x) = 3^{-x}$ are called exponential functions, and we will study them in greater detail in Chapter 5.

cursor at approximately the point on the graph corresponding to $x \approx -2$. Then, using **TRACE,** we see that a relative maximum occurs when $x \approx -1.76$ with value $y \approx -1.01$. Similarly, we find the other relative maximum where $x \approx 0.20$ with value $y \approx 0.44$. Repeating the procedure, we find the relative minimum at the point where $x \approx -0.86$ and $y \approx -1.07$. ▪▪▪▪

Finally, we comment that if you have access to a computer and software such as Derive, Maple, or Mathematica, then symbolic differentiation will yield the derivative $f'(x)$ of any differentiable function. This software will also solve the equation $f'(x) = 0$ with ease. Thus, the use of a computer will simplify even more greatly the analysis of functions.

Exercises

In Exercises 1–4, use a graphing utility to find (a) the intervals where f is increasing and the intervals where f is decreasing and (b) the relative extrema of f. Express your answers accurate to four decimal places.

1. $f(x) = 3.4x^4 - 6.2x^3 + 1.8x^2 + 3x - 2$

2. $f(x) = 1.8x^4 - 9.1x^3 + 5x - 4$

3. $f(x) = 2x^5 - 5x^3 + 8x^2 - 3x + 2$

4. $f(x) = 3x^5 - 4x^2 + 3x - 1$

In Exercises 5–8, use the ZOOM and TRACE features of a graphing utility to find (a) the intervals where f is increasing and the intervals where f is decreasing and (b) the relative extrema of f. Express your answers accurate to two decimal places.

5. $f(x) = (2x + 1)^{1/3}(x^2 + 1)^{-2/3}$

6. $f(x) = [x^2(x^3 - 1)]^{1/3} + \dfrac{1}{x}$

7. $f(x) = x - \sqrt{1 - x^2}$

8. $f(x) = \dfrac{\sqrt{x}(x^2 - 1)^2}{x - 2}$

9. RATE OF BANK FAILURES The Federal Deposit Insurance Company (FDIC) estimates that the rate at which banks were failing between 1982 and 1994 is given by

$$f(t) = 0.063447t^4 - 1.953283t^3 + 14.632576t^2$$
$$- 6.684704t + 47.458874 \qquad (0 \le t \le 12)$$

where $f(t)$ is measured in the number of banks per year and t is measured in years, with $t = 0$ corresponding to the beginning of 1982.
a. Use a graphing utility to plot the graph of f in the viewing rectangle $[0, 13] \times [0, 220]$.
b. Determine the intervals where f is increasing and where f is decreasing and interpret your result.
c. Find the relative maximum of f and interpret your result.
Source: Federal Deposit Insurance Corporation

10. MANUFACTURING CAPACITY Data obtained from the Federal Reserve show that the annual increase in manufacturing capacity between 1988 and 1994 is given by

$$f(t) = 0.0388889t^3 - 0.283333t^2 + 0.477778t$$
$$+ 2.04286 \qquad (0 \le t \le 6)$$

where $f(t)$ is a percentage and t is measured in years, with $t = 0$ corresponding to the beginning of 1988.

a. Use a graphing utility to plot the graph of f in the viewing rectangle $[0, 8] \times [0, 4]$.

b. Determine the intervals where f is increasing and where f is decreasing and interpret your result.

Source: Federal Reserve

11. **HOME SALES** According to the Greater Boston Real Estate Board—Multiple Listing Service, the average number of days a single-family home remains for sale from listing to accepted offer is approximated by the function

$$f(t) = 0.0171911t^4 - 0.662121t^3 + 6.18083t^2$$
$$- 8.97086t + 53.3357 \qquad (0 \le t \le 10)$$

where t is measured in years, with $t = 0$ corresponding to the beginning of 1984.

a. Use a graphing utility to plot the graph of f in the viewing rectangle $[0, 12] \times [0, 120]$.

b. Use the graph of f to find, approximately, the intervals where f is increasing and the intervals where f is decreasing. What does this result tell us about the sales of single-family homes in the greater Boston area from 1984 through 1994?

Source: Greater Boston Real Estate Board—Multiple Listing Service

12. **MORNING TRAFFIC RUSH** The speed of traffic flow on a certain stretch of Route 123 between 6 A.M. and 10 A.M. on a typical weekday is approximated by the function

$$f(t) = 20t - 40\sqrt{t} + 52 \qquad (0 \le t \le 4)$$

where $f(t)$ is measured in miles per hour and t is measured in hours, with $t = 0$ corresponding to 6 A.M. Find the interval where f is increasing, the interval where f is decreasing, and the relative extrema of f. Interpret your results.

13. **AIR POLLUTION** The amount of nitrogen dioxide, a brown gas that impairs breathing, present in the atmosphere on a certain May day in the city of Long Beach, is approximated by

$$A(t) = \frac{136}{1 + 0.25(t - 4.5)^2} + 28 \qquad (0 \le t \le 11)$$

where $A(t)$ is measured in pollutant standard index (PSI) and t is measured in hours, with $t = 0$ corresponding to 7 A.M. When is the PSI increasing and when is it decreasing? At what time is the PSI highest, and what is its value at that time?

The Second Derivative Test

1. Compute $f'(x)$ and $f''(x)$.
2. Find all the critical points of f at which $f'(x) = 0$.
3. Compute $f''(c)$ for each such critical point c.
   a. If $f''(c) < 0$, then f has a relative maximum at c.
   b. If $f''(c) > 0$, then f has a relative minimum at c.
   c. If $f''(c) = 0$, the test fails; that is, it is inconclusive.

REMARK As stated in step 3c, the second derivative test does not yield a conclusion if $f''(c) = 0$ or if $f''(c)$ does not exist. In other words, $x = c$ may give rise to a relative extremum or an inflection point (see Exercise 91, page 320). In such cases, you should revert to the first derivative test.   ■ ■ ■

EXAMPLE 8

Determine the relative extrema of the function

$$f(x) = x^3 - 3x^2 - 24x + 32$$

using the second derivative test. (See Example 7, Section 4.1.)

SOLUTION ✔

We have

$$f'(x) = 3x^2 - 6x - 24 = 3(x + 2)(x - 4)$$

so $f'(x) = 0$ gives $x = -2$ and $x = 4$, the critical points of f, as in Example 7. Next, we compute

$$f''(x) = 6x - 6 = 6(x - 1)$$

Since

$$f''(-2) = 6(-2 - 1) = -18 < 0$$

the second derivative test implies that $f(-2) = 60$ is a relative maximum of f. Also,

$$f''(4) = 6(4 - 1) = 18 > 0$$

and the second derivative test implies that $f(4) = -48$ is a relative minimum of f, which confirms the results obtained earlier.   ■ ■ ■ ■

Group Discussion
Suppose a function f has the following properties:

1. $f''(x) > 0$ for all x in an interval (a, b).
2. There is a point c between a and b such that $f'(c) = 0$.

What special property can you ascribe to the point $(c, f(c))$? Answer the question if Property 1 is replaced by the property that $f''(x) < 0$ for all x in (a, b).

COMPARING THE FIRST AND SECOND DERIVATIVE TESTS

Notice that both the first derivative test and the second derivative test are used to classify the critical points of f. What are the pros and cons of the two tests? Since the second derivative test is applicable only when f'' exists, it is less versatile than the first derivative test. For example, it cannot be used to locate the relative minimum $f(0) = 0$ of the function $f(x) = x^{2/3}$.

Furthermore, the second derivative test is inconclusive when f'' is equal to zero at a critical point of f, whereas the first derivative test always yields positive conclusions. The second derivative test is also inconvenient to use when f'' is difficult to compute. On the plus side, if f'' is computed easily, then we use the second derivative test since it involves just the evaluation of f'' at the critical point(s) of f. Also, the conclusions of the second derivative test are important in theoretical work.

We close this section by summarizing the different roles played by the first derivative f' and the second derivative f'' of a function f in determining the properties of the graph of f. The first derivative f' tells us where f is increasing and where f is decreasing, whereas the second derivative f'' tells us where f is concave upward and where f is concave downward. These different properties of f are reflected by the signs of f' and f'' in the interval of interest. The following table shows the general characteristics of the function f for various possible combinations of the signs of f' and f'' in the interval (a, b).

Signs of f' and f''	Properties of the Graph of f	General Shape of the Graph of f
$f'(x) > 0$ $f''(x) > 0$	f increasing f concave upward	⌣
$f'(x) > 0$ $f''(x) < 0$	f increasing f concave downward	/
$f'(x) < 0$ $f''(x) > 0$	f decreasing f concave upward	\
$f'(x) < 0$ $f''(x) < 0$	f decreasing f concave downward	⌐

SELF-CHECK EXERCISES 4.2

1. Determine where the function $f(x) = 4x^3 - 3x^2 + 6$ is concave upward and where it is concave downward.

2. Using the second derivative test, if applicable, find the relative extrema of the function $f(x) = 2x^3 - \frac{1}{2}x^2 - 12x - 10$.

3. A certain country's gross domestic product (GDP) (in millions of dollars) in year t is described by the function

$$G(t) = -2t^3 + 45t^2 + 20t + 6000 \qquad (0 \leq t \leq 11)$$

where $t = 0$ corresponds to the beginning of the year 1989. Find the inflection point of the function G and discuss its significance.

Solutions to Self-Check Exercises 4.2 can be found on page 320.

4.2 Exercises

In Exercises 1–8, you are given the graph of a function f. Determine the intervals where f is concave upward and where it is concave downward. Also, find all inflection points of f, if any.

1.

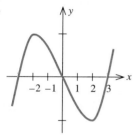

2.

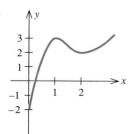

3.

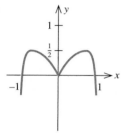

4.

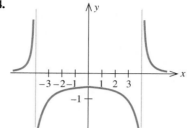

5.

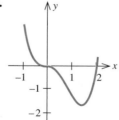

6.

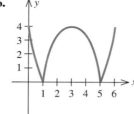

7.

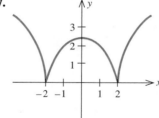

8.

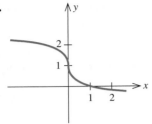

In Exercises 9–12, determine which graph—a, b, or c—is the graph of the function f with the specified properties.

9. $f(2) = 1$, $f'(2) > 0$, and $f''(2) < 0$

(a)

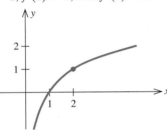

(b)

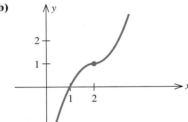

(c)

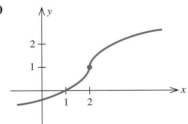

10. $f(1) = 2$, $f'(x) > 0$ on $(-\infty, 1) \cup (1, \infty)$, and $f''(1) = 0$

(a)

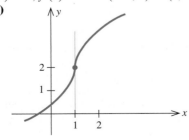

(b)

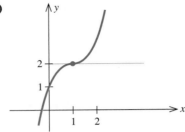

(c)

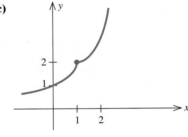

11. $f'(0)$ is undefined, f is decreasing on $(-\infty, 0)$, f is concave downward on $(0, 3)$, and f has an inflection point at $x = 3$.

(a)

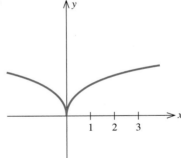

(b)

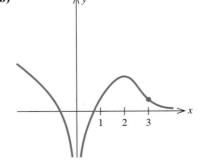

(c)

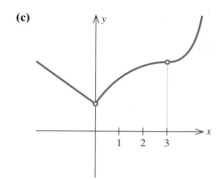

12. f is decreasing on $(-\infty, 2)$ and increasing on $(2, \infty)$, f is concave upward on $(1, \infty)$, and f has inflection points at $x = 0$ and $x = 1$.

(a)

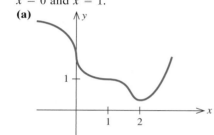

(b)

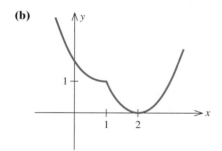

(c)

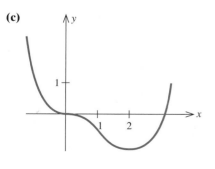

13. EFFECT OF ADVERTISING ON BANK DEPOSITS The following graphs were used by the CEO of the Madison Savings Bank to illustrate what effect a projected promotional campaign would have on its deposits over the next year. The functions D_1 and D_2 give the projected amount of money on deposit with the bank over the next 12 mo with and without the proposed promotional campaign, respectively.
a. Determine the signs of $D_1'(t)$, $D_2'(t)$, $D_1''(t)$, and $D_2''(t)$ on the interval $(0, 12)$.
b. Explain the significance of your results in the context of the problem.

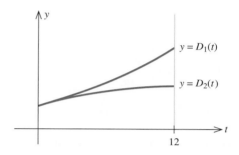

14. ASSEMBLY TIME OF A WORKER In the following graph, $N(t)$ gives the number of transistor radios assembled by the average worker by the tth hour, where $t = 0$ corresponds to 8 A.M. and $0 \le t \le 4$. Explain the significance of the inflection point P shown on the graph.

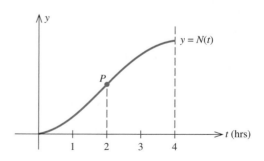

15. WATER POLLUTION When organic waste is dumped into a pond, the oxidation process that takes place reduces the pond's oxygen content. However, given time, nature will restore the oxygen content to its natural level. In the graph on page 316, $P(t)$ gives the oxygen content (as a percentage of its normal level) t days after organic waste has been dumped into the pond. Explain the significance of the inflection point Q.

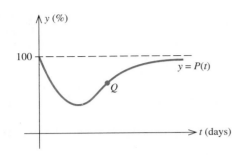

In Exercises 16–21 show that the function is concave upward wherever it is defined.

16. $f(x) = 4x^2 - 12x + 7$

17. $g(x) = x^4 + \dfrac{1}{2}x^2 + 6x + 10$

18. $h(x) = \dfrac{1}{x^2}$ **19.** $f(x) = \dfrac{1}{x^4}$

20. $h(x) = \sqrt{x^2 + 4}$ **21.** $g(x) = -\sqrt{4 - x^2}$

In Exercises 22–43, determine where the function is concave upward and where it is concave downward.

22. $g(x) = -x^2 + 3x + 4$ **23.** $f(x) = 2x^2 - 3x + 4$

24. $g(x) = x^3 - x$ **25.** $f(x) = x^3 - 1$

26. $f(x) = 3x^4 - 6x^3 + x - 8$

27. $f(x) = x^4 - 6x^3 + 2x + 8$

28. $f(x) = \sqrt[3]{x}$ **29.** $f(x) = x^{4/7}$

30. $g(x) = \sqrt{x - 2}$ **31.** $f(x) = \sqrt{4 - x}$

32. $g(x) = \dfrac{x}{x + 1}$ **33.** $f(x) = \dfrac{1}{x - 2}$

34. $g(x) = \dfrac{x}{1 + x^2}$ **35.** $f(x) = \dfrac{1}{2 + x^2}$

36. $f(x) = \dfrac{x + 1}{x - 1}$ **37.** $h(t) = \dfrac{t^2}{t - 1}$

38. $h(r) = -\dfrac{1}{(r - 2)^2}$ **39.** $g(x) = x + \dfrac{1}{x^2}$

40. $f(x) = (x - 2)^{2/3}$ **41.** $g(t) = (2t - 4)^{1/3}$

42. $f(x) = \dfrac{x^2 - 2}{x^3}$ **43.** $f(x) = \dfrac{x^2}{x^2 - 1}$

In Exercises 44–55, find the inflection points, if any, of each function.

44. $g(x) = x^3 - 6x$ **45.** $f(x) = x^3 - 2$

46. $g(x) = 2x^3 - 3x^2 + 18x - 8$

47. $f(x) = 6x^3 - 18x^2 + 12x - 15$

48. $f(x) = x^4 - 2x^3 + 6$ **49.** $f(x) = 3x^4 - 4x^3 + 1$

50. $f(x) = \sqrt[5]{x}$ **51.** $g(t) = \sqrt[3]{t}$

52. $f(x) = (x - 2)^{4/3}$ **53.** $f(x) = (x - 1)^3 + 2$

54. $f(x) = 2 + \dfrac{3}{x}$ **55.** $f(x) = \dfrac{2}{1 + x^2}$

In Exercises 56–73, find the relative extrema, if any, of each function. Use the second derivative test, if applicable.

56. $g(x) = 2x^2 + 3x + 7$ **57.** $f(x) = -x^2 + 2x + 4$

58. $g(x) = x^3 - 6x$ **59.** $f(x) = 2x^3 + 1$

60. $f(x) = 2x^3 + 3x^2 - 12x - 4$

61. $f(x) = \dfrac{1}{3}x^3 - 2x^2 - 5x - 10$

62. $f(t) = 2t + \dfrac{3}{t}$ **63.** $g(t) = t + \dfrac{9}{t}$

64. $f(x) = \dfrac{2x}{x^2 + 1}$ **65.** $f(x) = \dfrac{x}{1 - x}$

66. $g(x) = x^2 + \dfrac{2}{x}$ **67.** $f(t) = t^2 - \dfrac{16}{t}$

68. $g(x) = \dfrac{1}{1 + x^2}$ **69.** $g(s) = \dfrac{s}{1 + s^2}$

70. $f(x) = \dfrac{x^2}{x^2 + 1}$ **71.** $f(x) = \dfrac{x^4}{x - 1}$

72. $f(x) = \dfrac{x^2 + 4}{x^2 - 1}$ **73.** $g(x) = \dfrac{2 - x}{(x + 2)^3}$

74. EFFECT OF BUDGET CUTS ON DRUG-RELATED CRIMES The following graphs were used by a police commissioner to illustrate what effect a budget cut would have on crime in the city. The number $N_1(t)$ gives the projected number

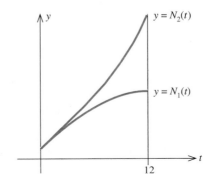

of drug-related crimes in the next 12 mo. The number $N_2(t)$ gives the projected number of drug-related crimes in the same time frame if next year's budget is cut.

a. Explain why $N_1'(t)$ and $N_2'(t)$ are both positive on the interval (0, 12).

b. What are the signs of $N_1''(t)$ and $N_2''(t)$ on the interval (0, 12)?

c. Interpret the results of part (b).

75. **DEMAND FOR RNs** The following graph gives the total number of help-wanted ads for RNs (registered nurses) in 22 cities over the last 12 mo as a function of time t (t measured in months).

a. Explain why $N'(t)$ is positive on the interval (0, 12).

b. Determine the signs of $N''(t)$ on the interval (0, 6) and the interval (6, 12).

c. Interpret the results of part (b).

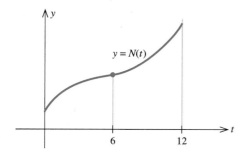

76. In the following figure, water is poured into the vase at a constant rate (in appropriate units), and the water level rises to a height of $f(t)$ units at time t as measured from the base of the vase. The graph of f follows. Explain the shape of the curve in terms of its concavity. What is the significance of the inflection point?

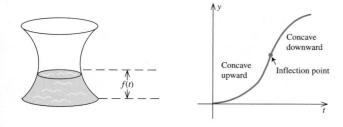

77. In the following figure, water is poured into an urn at a constant rate (in appropriate units), and the water level rises to a height of $f(t)$ units at time t as measured from the base of the urn. Sketch the graph of f and explain its shape, indicating where it is concave upward and concave downward. Indicate the inflection point on the graph and explain its significance.

Hint: Study Exercise 76.

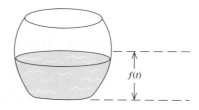

78. **EFFECT OF ADVERTISING ON HOTEL REVENUE** The total annual revenue R of the Miramar Resorts Hotel is related to the amount of money x the hotel spends on advertising its services by the function

$$R(x) = -0.003x^3 + 1.35x^2 + 2x + 8000$$
$$(0 \le x \le 400)$$

where both R and x are measured in thousands of dollars. Find the inflection point of R and discuss its significance.

79. **EFFECT OF ADVERTISING ON SALES** The total sales S of the Cannon Precision Instruments Corporation is related to the amount of money x that Cannon spends on advertising its products by the function

$$S(x) = -0.002x^3 + 0.6x^2 + x + 500$$
$$(0 \le x \le 200)$$

where S and x are measured in thousands of dollars. Find the inflection point of the function S and discuss its significance.

80. **FORECASTING PROFITS** As a result of increasing energy costs, the growth rate of the profit of the 4-yr-old Venice Glassblowing Company has begun to decline. Venice's management, after consulting with energy experts, decides to implement certain energy-conservation measures aimed at cutting energy bills. The general manager reports that, according to his calculations, the growth rate of Venice's profit should be on the increase again within 4 yr. If Venice's profit (in hundreds of dollars) x years from now is given by the function

$$P(x) = x^3 - 9x^2 + 40x + 50 \qquad (0 \le x \le 8)$$

determine whether the general manager's forecast will be accurate.

Hint: Find the inflection point of the function P and study the concavity of P.

81. **WORKER EFFICIENCY** An efficiency study conducted for the Elektra Electronics Company showed that the number of Space Commander walkie-talkies assembled by the average worker t hours after starting work at 8 A.M. is given by

$$N(t) = -t^3 + 6t^2 + 15t \qquad (0 \le t \le 4)$$

At what time during the morning shift is the average worker performing at peak efficiency?

(continued on p. 320)

FINDING THE INFLECTION POINTS
OF A FUNCTION

A graphing utility can be used to find the inflection points of a function and hence the intervals where the graph of the function is concave upward and the intervals where it is concave downward. Some graphing utilities have an operation for finding inflection points directly. If your graphing utility has this capability, use it to work through the example and exercises in this section.

EXAMPLE 1

Let $f(x) = 2.5x^5 - 12.4x^3 + 4.2x^2 - 5.2x + 4$.

a. Use a graphing utility to plot the graph of f.
b. Find the inflection points of f.
c. Find the intervals where f is concave upward and where it is concave downward.

SOLUTION ✔

a. The graph of f using the viewing rectangle $[-3, 3] \times [-25, 60]$ is shown in Figure T1.

FIGURE T1
The graph of f in the viewing rectangle $[-3, 3] \times [-25, 60]$

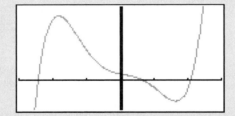

b. From Figure T1 we see that f has three inflection points—one occurring at the point where the x-coordinate is approximately -1, another at the point where $x \approx 0$, and the third at the point where $x \approx 1$. To find the first inflection point, we use the inflection operation, moving the cursor to the point on the graph of f where $x \approx -1$. We obtain the point $(-1.2728, 34.6395)$ (accurate to four decimal places). Next, setting the cursor near the point $x = 0$ yields the inflection point $(0.1139, 3.4440)$. Finally, with the cursor set at $x = 1$, we obtain the third inflection point $(1.1589, -10.4594)$.
c. From the results of part (b), we see that f is concave upward on the intervals $(-1.2728, 0.1139)$ and $(1.1589, \infty)$ and concave downward on $(-\infty, -1.2728)$ and $(0.1139, 1.1589)$.

In Exercises 1–8, use a graphing utility to find (a) the intervals where f is concave upward and the intervals where f is concave downward and (b) the inflection points of f. Express your answers accurate to four decimal places.

1. $f(x) = 1.8x^4 - 4.2x^3 + 2.1x + 2$

2. $f(x) = -2.1x^4 + 3.1x^3 + 2x^2 - x + 1.2$

3. $f(x) = 1.2x^5 - 2x^4 + 3.2x^3 - 4x + 2$

4. $f(x) = -2.1x^5 + 3.2x^3 - 2.2x^2 + 4.2x - 4$

5. $f(x) = x^3(x^2 + 1)^{-1/3}$

6. $f(x) = x^2(x^3 - 1)^3$

7. $f(x) = \dfrac{x^2 - 1}{x^3}$

8. $f(x) = \dfrac{x + 1}{\sqrt{x}}$

9. GROWTH OF HMOs Based on data compiled by the Group Health Association of America, the number of people receiving their care in an HMO (Health Maintenance Organization) from the beginning of 1984 through 1994 is approximated by the function

$$f(t) = 0.0514t^3 - 0.853t^2 + 6.8147t + 15.6524$$
$$(0 < t \le 11)$$

where $f(t)$ gives the number of people in millions and t is measured in years, with $t = 0$ corresponding to the beginning of 1984.
a. Use a graphing utility to plot the graph of f in the viewing rectangle $[0, 12] \times [0, 120]$.
b. Find the points of inflection of f.
c. At what time in the given time interval was the number of people receiving their care at an HMO increasing fastest?
Source: Group Health Association of America

10. MANUFACTURING CAPACITY Data obtained from the Federal Reserve show that the annual increase in manufac-

turing capacity between 1988 and 1994 is given by

$$f(t) = 0.0388889t^3 - 0.283333t^2 + 0.477778t$$
$$+ 2.04286 \qquad (0 \le t \le 6)$$

where $f(t)$ is a percentage and t is measured in years, with $t = 0$ corresponding to the beginning of 1988.
a. Use a graphing utility to plot the graph of f in the viewing rectangle $[0, 8] \times [0, 4]$.
b. Find the point of inflection and interpret your result.
Source: Federal Reserve

11. Time on the Market According to the Greater Boston Real Estate Board—Multiple Listing Service, the average number of days a single-family home remains for sale from listing to accepted offer is approximated by the function

$$f(t) = 0.0171911t^4 - 0.662121t^3 + 6.18083t^2$$
$$- 8.97086t + 53.3357 \qquad (0 \le t \le 10)$$

where t is measured in years, with $t = 0$ corresponding to the beginning of 1984.
a. Use a graphing utility to plot the graph of f in the viewing rectangle $[0, 12] \times [0, 120]$.
b. Find the points of inflection and interpret your result.
Source: Greater Boston Real Estate Board—Multiple Listing Service

12. MULTIMEDIA SALES According to the Electronic Industries Association, sales in the multimedia market (hardware and software) are expected to be

$$S(t) = -0.0094t^4 + 0.1204t^3 - 0.0868t^2$$
$$+ 0.0195t + 3.3325 \qquad (0 \le t \le 10)$$

where $S(t)$ is measured in billions of dollars and t is measured in years, with $t = 0$ corresponding to 1990.
a. Plot the graph of S in the viewing rectangle $[0, 12] \times [0, 25]$.
b. Find the inflection point of S and interpret your result.
Source: Electronic Industries Association

82. COST OF PRODUCING CALCULATORS A subsidiary of Elektra Electronics manufactures programmable calculators. Management determines that the daily cost $C(x)$ (in dollars) of producing these calculators is

$$C(x) = 0.0001x^3 - 0.08x^2 + 40x + 5000$$

where x is the number of calculators produced. Find the inflection point of the function C and interpret your result.

83. FLIGHT OF A ROCKET The altitude (in feet) of a rocket t sec into flight is given by

$$s = f(t) = -t^3 + 54t^2 + 480t + 6$$

Find the point of inflection of the function f and interpret your result. What is the maximum velocity attained by the rocket?

84. AIR POLLUTION The level of ozone, an invisible gas that irritates and impairs breathing, present in the atmosphere on a certain May day in the city of Riverside was approximated by

$$A(t) = 1.0974t^3 - 0.0915t^4 \qquad (0 \le t \le 11)$$

where $A(t)$ is measured in pollutant standard index (PSI) and t is measured in hours, with $t = 0$ corresponding to 7 A.M. Use the second derivative test to show that the function A has a relative maximum at approximately $t = 9$. Interpret your results.

In Exercises 85–87, determine whether the statement is true or false. If it is true, explain why it is true. If it is false, give an example to show why it is false.

85. If the graph of f is concave upward on (a, b) then the graph of $-f$ is concave downward on (a, b).

86. If the graph of f is concave upward on (a, c) and concave downward on (c, b), where $a < c < b$, then f has an inflection point at $x = c$.

87. If $x = c$ is a critical point of f where $a < c < b$ and $f''(x) < 0$ on (a, b), then f has a relative maximum at $x = c$.

88. Show that the quadratic function

$$f(x) = ax^2 + bx + c \qquad (a \ne 0)$$

is concave upward if $a > 0$ and concave downward if $a < 0$. Thus, by examining the sign of the coefficient of x^2, one can tell immediately whether the parabola opens upward or downward.

89. Suppose f has an inflection point at $(a, f(a))$. Must the function f' have a relative extremum at $x = a$? Explain your answer.

90. Show that the cubic function

$$f(x) = ax^3 + bx^2 + cx + d \qquad (a \ne 0)$$

has one and only one inflection point. Find the coordinates of this point.

91. Consider the functions $f(x) = x^3$, $g(x) = x^4$, and $h(x) = -x^4$.
a. Show that $x = 0$ is a critical point of each of the functions f, g, and h.
b. Show that the second derivative of each of the functions f, g, and h equals zero at $x = 0$.
c. Show that f has neither a relative maximum nor a relative minimum at $x = 0$, that g has a relative minimum at $x = 0$, and that h has a relative maximum at $x = 0$.

SOLUTIONS TO SELF-CHECK EXERCISES 4.2

1. We first compute

$$f'(x) = 12x^2 - 6x$$
$$f''(x) = 24x - 6 = 6(4x - 1)$$

Observe that f'' is continuous everywhere and has a zero at $x = \frac{1}{4}$. The sign diagram of f'' is shown in the accompanying figure.

$$- - - - - - 0 + + + + + +$$

$$\begin{array}{ccc} & | & | & \longrightarrow x \\ 0 & & \frac{1}{4} \end{array}$$

From the sign diagram for f'', we see that f is concave upward on $(\frac{1}{4}, \infty)$ and concave downward on $(-\infty, \frac{1}{4})$.

2. First, we find the critical points of f by solving the equation

$$f'(x) = 6x^2 - x - 12 = 0$$

That is,

$$(3x + 4)(2x - 3) = 0$$

giving $x = -\frac{4}{3}$ and $x = \frac{3}{2}$. Next, we compute

$$f''(x) = 12x - 1$$

Since

$$f''\left(-\frac{4}{3}\right) = 12\left(-\frac{4}{3}\right) - 1 = -17 < 0$$

the second derivative test implies that $f\left(-\frac{4}{3}\right) = \frac{10}{27}$ is a relative maximum of f. Also,

$$f''\left(\frac{3}{2}\right) = 12\left(\frac{3}{2}\right) - 1 = 17 > 0$$

and we see that $f(\frac{3}{2}) = -\frac{179}{8}$ is a relative minimum.

3. We compute the second derivative of G. Thus,

$$G'(t) = -6t^2 + 90t + 20$$
$$G''(t) = -12t + 90$$

Now, G'' is continuous everywhere, and $G''(t) = 0$, where $t = \frac{15}{2}$, giving $t = \frac{15}{2}$ as the only candidate for an inflection point of G. Since $G''(t) > 0$ for $t < \frac{15}{2}$ and $G''(t) < 0$ for $t > \frac{15}{2}$, we see that the point $(\frac{15}{2}, \frac{15,675}{2})$ is an inflection point of G. The results of our computations tell us that the country's GDP was increasing most rapidly at the beginning of July 1996.

4.3 Curve Sketching

A REAL-LIFE EXAMPLE

As we have seen on numerous occasions, the graph of a function is a useful aid for visualizing the function's properties. From a practical point of view, the graph of a function also gives, at one glance, a complete summary of all the information captured by the function.

Consider, for example, the graph of the function giving the Dow-Jones Industrial Average (DJIA) on Black Monday, October 19, 1987 (Figure 4.44). Here, $t = 0$ corresponds to 8:30 A.M., when the market was open for business, and $t = 7.5$ corresponds to 4 P.M., the closing time. The following information may be gleaned from studying the graph.

The graph is *decreasing* rapidly from $t = 0$ to $t = 1$, reflecting the sharp drop in the index in the first hour of trading. The point (1, 2047) is a *relative minimum* point of the function, and this turning point coincides with the start

4.3 Exercises

In Exercises 1–10, find the horizontal and vertical asymptotes of the graph.

1.

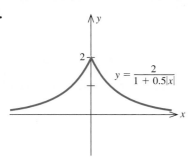

$y = \dfrac{2}{1 + 0.5|x|}$

2.

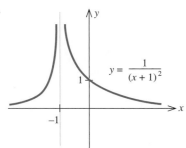

$y = \dfrac{1}{(x + 1)^2}$

3.

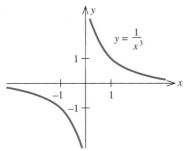

$y = \dfrac{1}{x^3}$

4.

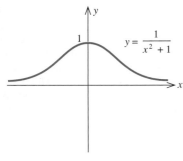

$y = \dfrac{1}{x^2 + 1}$

5.

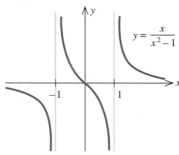

$y = \dfrac{x}{x^2 - 1}$

6.

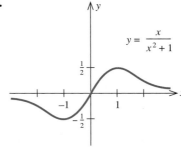

$y = \dfrac{x}{x^2 + 1}$

7.

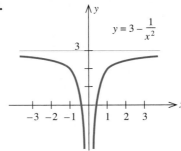

$y = 3 - \dfrac{1}{x^2}$

8.

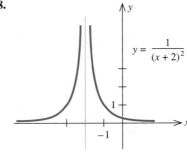

$y = \dfrac{1}{(x + 2)^2}$

9.

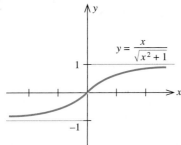

10.

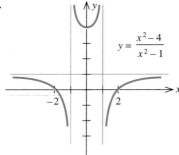

In Exercises 11–28, find the horizontal and vertical asymptotes of the graph of the function. (You need not sketch the graph.)

11. $f(x) = \dfrac{1}{x}$

12. $f(x) = \dfrac{1}{x + 2}$

13. $f(x) = -\dfrac{2}{x^2}$

14. $g(x) = \dfrac{1}{1 + 2x^2}$

15. $f(x) = \dfrac{x - 1}{x + 1}$

16. $g(t) = \dfrac{t + 1}{2t - 1}$

17. $h(x) = x^3 - 3x^2 + x + 1$

18. $g(x) = 2x^3 + x^2 + 1$

19. $f(t) = \dfrac{t^2}{t^2 - 9}$

20. $g(x) = \dfrac{x^3}{x^2 - 4}$

21. $f(x) = \dfrac{3x}{x^2 - x - 6}$

22. $g(x) = \dfrac{2x}{x^2 + x - 2}$

23. $g(t) = 2 + \dfrac{5}{(t - 2)^2}$

24. $f(x) = 1 + \dfrac{2}{x - 3}$

25. $f(x) = \dfrac{x^2 - 2}{x^2 - 4}$

26. $h(x) = \dfrac{2 - x^2}{x^2 + x}$

27. $g(x) = \dfrac{x^3 - x}{x(x + 1)}$

28. $f(x) = \dfrac{x^4 - x^2}{x(x - 1)(x + 2)}$

In Exercises 29 and 30, you are given the graphs of two functions f and g. One function is the derivative function of the other. Identify each of them.

29.

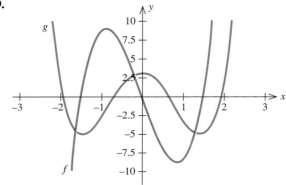

30.

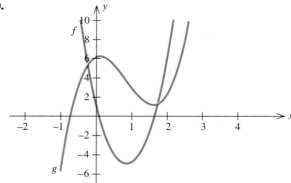

31. TERMINAL VELOCITY A skydiver leaps from the gondola of a hot-air balloon. As she free-falls, air resistance, which is proportional to her velocity, builds up to a point where it balances the force due to gravity. The resulting motion may be described in terms of her velocity as follows: Starting at rest (zero velocity), her velocity increases and approaches a constant velocity, called the *terminal velocity*. Sketch a graph of her velocity v versus time t.

In Exercises 32–35, use the information summarized in the table to sketch the graph of *f*.

32. $f(x) = x^3 - 3x^2 + 1$

Domain	$(-\infty, \infty)$
Intercept	y-intercept: 1
Asymptotes	None
Intervals where f is ↗ and ↘	↗ on $(-\infty, 0) \cup (2, \infty)$; ↘ on $(0, 2)$
Relative extrema	Rel. max. at $(0, 1)$; rel. min. at $(2, -3)$
Concavity	Downward on $(-\infty, 1)$; upward on $(1, \infty)$
Point of inflection	$(1, -1)$

33. $f(x) = \frac{1}{9}(x^4 - 4x^3)$

Domain	$(-\infty, \infty)$
Intercepts	x-intercepts: 0, 4; y-intercept: 0
Asymptotes	None
Intervals where f is ↗ and ↘	↗ on $(3, \infty)$; ↘ on $(-\infty, 0) \cup (0, 3)$
Relative extrema	Rel. min. at $(3, -3)$
Concavity	Downward on $(0, 2)$; upward on $(-\infty, 0) \cup (2, \infty)$
Points of inflection	$(0, 0)$ and $(2, -16/9)$

34. $f(x) = \dfrac{4x - 4}{x^2}$

Domain	$(-\infty, 0) \cup (0, \infty)$
Intercept	x-intercept: 1
Asymptotes	x-axis and y-axis
Intervals where f is ↗ and ↘	↗ on $(0, 2)$; ↘ on $(-\infty, 0) \cup (2, \infty)$
Relative extrema	Rel. max. at $(2, 1)$
Concavity	Downward on $(-\infty, 0) \cup (0, 3)$; upward on $(3, \infty)$
Points of inflection	$(3, 8/9)$

35. $f(x) = x - 3x^{1/3}$

Domain	$(-\infty, \infty)$
Intercepts	x-intercepts: $\pm 3\sqrt{3}$, 0
Asymptotes	None
Intervals where f is ↗ and ↘	↗ on $(-\infty, -1) \cup (1, \infty)$; ↘ on $(-1, 1)$
Relative extrema	Rel. max. at $(-1, 2)$; rel. min. at $(1, -2)$
Concavity	Downward on $(-\infty, 0)$; upward on $(0, \infty)$
Points of inflection	$(0, 0)$

In Exercises 36–59, sketch the graph of the function, using the curve-sketching guide of this section.

36. $f(x) = x^2 - 2x + 3$

37. $g(x) = 4 - 3x - 2x^3$

38. $f(x) = 2x^3 + 1$

39. $h(x) = x^3 - 3x + 1$

40. $f(t) = 2t^3 - 15t^2 + 36t - 20$

41. $f(x) = -2x^3 + 3x^2 + 12x + 2$

42. $f(t) = 3t^4 + 4t^3$

43. $h(x) = \frac{3}{2}x^4 - 2x^3 - 6x^2 + 8$

44. $f(x) = \sqrt{x^2 + 5}$

45. $f(t) = \sqrt{t^2 - 4}$

46. $f(x) = \sqrt[3]{x^2}$

47. $g(x) = \frac{1}{2}x - \sqrt{x}$

48. $f(x) = \dfrac{1}{x + 1}$

49. $g(x) = \dfrac{2}{x - 1}$

50. $g(x) = \dfrac{x}{x - 1}$

51. $h(x) = \dfrac{x + 2}{x - 2}$

52. $g(x) = \dfrac{x}{x^2 - 4}$

53. $f(t) = \dfrac{t^2}{1 + t^2}$

54. $f(x) = \dfrac{x^2 - 9}{x^2 - 4}$

55. $g(t) = -\dfrac{t^2 - 2}{t - 1}$

56. $h(x) = \dfrac{1}{x^2 - x - 2}$

57. $g(t) = \dfrac{t + 1}{t^2 - 2t - 1}$

58. $g(x) = (x + 2)^{3/2} + 1$

59. $h(x) = (x - 1)^{2/3} + 1$

60. COST OF REMOVING TOXIC POLLUTANTS A city's main well was recently found to be contaminated with trichloroethylene (a cancer-causing chemical) as a result of an abandoned chemical dump leaching chemicals into the water. A proposal submitted to the city council indicated that the cost, measured in millions of dollars, of removing $x\%$ of the toxic pollutants is given by

$$C(x) = \frac{0.5x}{100 - x}$$

a. Find the vertical asymptote of $C(x)$.
b. Is it possible to remove 100% of the toxic pollutant from the water?

61. AVERAGE COST OF PRODUCING VIDEO DISCS The average cost per disc (in dollars) incurred by the Herald Record Company in pressing x video discs is given by the average cost function

$$\overline{C}(x) = 2.2 + \frac{2500}{x}$$

a. Find the horizontal asymptote of $\overline{C}(x)$.
b. What is the limiting value of the average cost?

(continued on p. 338)

Using Technology

ANALYZING THE PROPERTIES OF A FUNCTION

One of the main purposes of studying Section 4.3 is to see how the many concepts of calculus come together to paint a picture of a function. The techniques of graphing also play a very practical role. For example, using the techniques of graphing developed in Section 4.3, you can tell if the graph of a function generated by a graphing utility is reasonably complete. Furthermore, these techniques can often reveal details that are missing from a graph.

EXAMPLE 1

Consider the function $f(x) = 2x^3 - 3.5x^2 + x - 10$. A plot of the graph of f in the standard viewing rectangle is shown in Figure T1. Since the domain of f is the interval $(-\infty, \infty)$, we see that Figure T1 does not reveal the part of the graph to the left of the y-axis. This suggests that we enlarge the viewing rectangle accordingly. Figure T2 shows the graph of f in the viewing rectangle $[-10, 10] \times [-20, 10]$.

FIGURE T1

The graph of f in the standard viewing rectangle

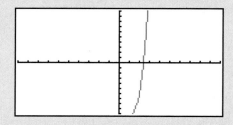

FIGURE T2

The graph of f in the viewing rectangle $[-10, 10] \times [-20, 10]$

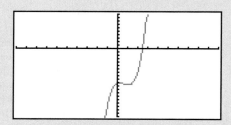

The behavior of f for large values of f [$\lim\limits_{x \to -\infty} f(x) = -\infty$ and $\lim\limits_{x \to \infty} f(x) = \infty$] suggests that this viewing rectangle has captured a sufficiently complete picture of f. Next, an analysis of the first derivative of f,

$$f'(x) = 6x^2 - 7x + 1 = (6x - 1)(x - 1)$$

334

reveals that f has critical values at $x = 1/6$ and $x = 1$. In fact, a sign diagram of f' shows that f has a relative maximum at $x = 1/6$ and a relative minimum at $x = 1$, details that are not revealed in the graph of f shown in Figure T2. To examine this portion of the graph of f, we use, say, the viewing rectangle $[-1, 2] \times [-11, -8]$. The resulting graph of f is shown in Figure T3, which certainly reveals the hitherto missing details! Thus, through an interaction of calculus and a graphing utility, we are able to obtain a good picture of the properties of f.

FIGURE T3
The graph of f in the viewing rectangle $[-1, 2] \times [-11, -8]$

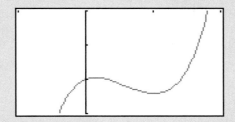

FINDING x-INTERCEPTS

As noted in Section 4.3, it is not always easy to find the x-intercepts of the graph of a function. But this information is very important in applications. By using the function for solving polynomial equations or the function for finding the roots of an equation, we can solve the equation $f(x) = 0$ quite easily and hence yield the x-intercepts of the graph of a function.

EXAMPLE **2**

Let $f(x) = x^3 - 3x^2 + x + 1.5$.

a. Use the function for solving polynomial equations on a graphing utility to find the x-intercepts of the graph of f.
b. Use the function for finding the roots of an equation on a graphing utility to find the x-intercepts of the graph of f.

SOLUTION ✔

a. Observe that f is a polynomial function of degree 3, and so we may use the function for solving polynomial equations to solve the equation $x^3 - 3x^2 + x + 1.5 = 0$ $[f(x) = 0]$. We find that the solutions (x-intercepts) are

$$x_1 \approx -0.525687120865, \qquad x_2 \approx 1.2586520225, \qquad x_3 \approx 2.26703509836$$

FIGURE **T4**
The graph of
$f(x) = x^3 - 3x^2 + x + 1.5$

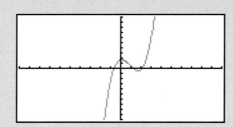

b. Using the graph of f (Figure T4), we see that $x_1 \approx -0.5$, $x_2 \approx 1$, and $x_3 \approx 2$. Using the function for finding the roots of an equation on a graphing utility, and these values of x as initial guesses, we find

$$x_1 \approx -0.5256871209, \qquad x_2 \approx 1.2586520225, \qquad x_3 \approx 2.2670350984$$

REMARK The function for solving polynomial equations on a graphing utility will solve a polynomial equation $f(x) = 0$, where f is a polynomial function. The function for finding the roots of a polynomial, however, will solve equations $f(x) = 0$ even if f is not a polynomial.

EXAMPLE 3

Unless payroll taxes are increased significantly and/or benefits are scaled back drastically, it is a matter of time before the current Social Security system goes broke. Based on data from the Board of Trustees of the Social Security Administration, the assets of the system—the Social Security "trust fund"—may be approximated by

$$f(t) = -0.0129t^4 + 0.3087t^3 + 2.1760t^2 + 62.8466t + 506.2955 \qquad (0 \le t \le 35)$$

where $f(t)$ is measured in millions of dollars and t is measured in years, with $t = 0$ corresponding to 1995.

a. Use a graphing calculator to sketch the graph of f.
b. Based on this model, when can the Social Security system be expected to go broke?
Source: Social Security Administration

SOLUTION ✔
a. The graph of f in the window $[0, 35] \times [-1000, 3500]$ is shown in Figure T5.
b. Using the function for finding the roots on a graphing utility, we find that $y = 0$ when $t \approx 34.1$, and this tells us that the system is expected to go broke around 2029.

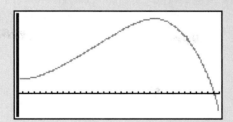

Exercises

In Exercises 1–4, use the method of Example 1 to analyze the function. (*Note:* Your answers will not be unique.)

1. $f(x) = 4x^3 - 4x^2 + x + 10$

2. $f(x) = x^3 + 2x^2 + x - 12$

3. $f(x) = \dfrac{1}{2}x^4 + x^3 + \dfrac{1}{2}x^2 - 10$

4. $f(x) = 2.25x^4 - 4x^3 + 2x^2 + 2$

In Exercises 5–10, find the *x*-intercepts of the graph of *f*. Give your answer accurate to four decimal places.

5. $f(x) = 0.2x^3 - 1.2x^2 + 0.8x + 2.1$

6. $f(x) = -0.5x^3 + 1.7x^2 - 1.2$

7. $f(x) = 0.3x^4 - 1.2x^3 + 0.8x^2 + 1.1x - 2$

8. $f(x) = -0.2x^4 + 0.8x^3 - 2.1x + 1.2$

9. $f(x) = 2x^2 - \sqrt{x + 1} - 3$

10. $f(x) = x - \sqrt{1 - x^2}$

FIGURE 4.65
The maximum altitude of the rocket is
143,655 feet.

FIGURE 4.65
The maximum altitude of the rocket is
143,655 feet.

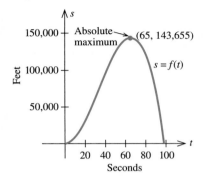

To find the absolute maximum of f, compute

$$f'(t) = -3t^2 + 192t + 195$$
$$= -3(t - 65)(t + 1)$$

and solve the equation $f'(t) = 0$, obtaining $t = -1$ and $t = 65$. Ignore $t = -1$ since it lies outside the interval $[0, T]$. This leaves the critical point $t = 65$ of f. Continuing, we compute

$$f(0) = 5, \qquad f(65) = 143,655, \qquad f(T) = 0$$

and conclude, accordingly, that the absolute maximum value of f is 143,655. Thus, the maximum altitude of the rocket is 143,655 feet, attained 65 seconds into flight. The graph of f is sketched in Figure 4.65.

b. To find the maximum velocity attained by the rocket, find the largest value of the function that describes the rocket's velocity at any time t—namely,

$$v = f'(t) = -3t^2 + 192t + 195 \qquad (t \geq 0)$$

We find the critical point of v by setting $v' = 0$. But

$$v' = -6t + 192$$

and the critical point of v is $t = 32$. Since

$$v'' = -6 < 0$$

FIGURE 4.66
The maximum velocity of the rocket is
3267 feet per second.

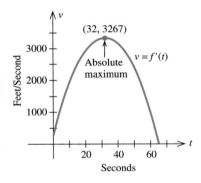

the second derivative test implies that a relative maximum of v occurs at $t = 32$. Our computation has in fact clarified the property of the "velocity curve." Since $v'' < 0$ everywhere, the velocity curve is concave downward everywhere. With this observation, we assert that the relative maximum must in fact be the absolute maximum of v. The maximum velocity of the rocket is given by evaluating v at $t = 32$,

$$f'(32) = -3(32)^2 + 192(32) + 195$$

or 3267 feet per second. The graph of the velocity function v is sketched in Figure 4.66. ■ ■ ■ ■

SELF-CHECK EXERCISES 4.4

1. Let $f(x) = x - 2\sqrt{x}$.
   a. Find the absolute extrema of f on the interval $[0, 9]$.
   b. Find the absolute extrema of f.

2. Find the absolute extrema of $f(x) = 3x^4 + 4x^3 + 1$ on $[-2, 1]$.

3. The operating rate (expressed as a percentage) of factories, mines, and utilities in a certain region of the country on the tth day of the year 2000 is given by the function

$$f(t) = 80 + \frac{1200t}{t^2 + 40,000} \qquad (0 \leq t \leq 250)$$

On which day of the first 250 days of 2000 was the manufacturing capacity operating rate highest?

Solutions to Self-Check Exercises 4.4 can be found on page 356.

4.4 Exercises

In Exercises 1–8, you are given the graph of a function _f_ defined on the indicated interval. Find the absolute maximum and the absolute minimum of _f_, if they exist.

1.

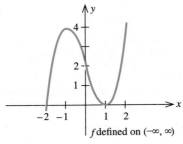

_f_ defined on $(-\infty, \infty)$

2.

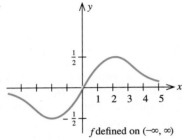

_f_ defined on $(-\infty, \infty)$

3.

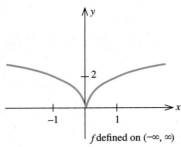

_f_ defined on $(-\infty, \infty)$

4.

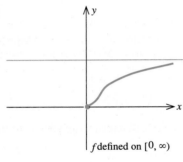

_f_ defined on $[0, \infty)$

5.

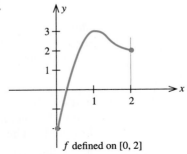

_f_ defined on $[0, 2]$

6.

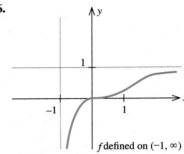

_f_ defined on $(-1, \infty)$

7.

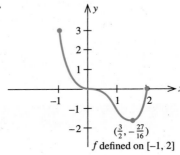

$(\frac{3}{2}, -\frac{27}{16})$

_f_ defined on $[-1, 2]$

8.

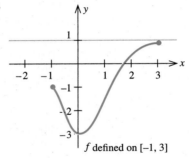

_f_ defined on $[-1, 3]$

In Exercises 9–38, find the absolute maximum value and the absolute minimum value, if any, of the given function.

9. $f(x) = 2x^2 + 3x - 4$

10. $g(x) = -x^2 + 4x + 3$

11. $h(x) = x^{1/3}$

12. $f(x) = x^{2/3}$

13. $f(x) = \dfrac{1}{1 + x^2}$

14. $f(x) = \dfrac{x}{1 + x^2}$

15. $f(x) = x^2 - 2x - 3$ on $[-2, 3]$

16. $g(x) = x^2 - 2x - 3$ on $[0, 4]$

17. $f(x) = -x^2 + 4x + 6$ on $[0, 5]$

18. $f(x) = -x^2 + 4x + 6$ on $[3, 6]$

19. $f(x) = x^3 + 3x^2 - 1$ on $[-3, 2]$

20. $g(x) = x^3 + 3x^2 - 1$ on $[-3, 1]$

21. $g(x) = 3x^4 + 4x^3$ on $[-2, 1]$

22. $f(x) = \dfrac{1}{2}x^4 - \dfrac{2}{3}x^3 - 2x^2 + 3$ on $[-2, 3]$

23. $f(x) = \dfrac{x + 1}{x - 1}$ on $[2, 4]$

24. $g(t) = \dfrac{t}{t - 1}$ on $[2, 4]$

25. $f(x) = 4x + \dfrac{1}{x}$ on $[1, 3]$

26. $f(x) = 9x - \dfrac{1}{x}$ on $[1, 3]$

27. $f(x) = \dfrac{1}{2}x^2 - 2\sqrt{x}$ on $[0, 3]$

28. $g(x) = \dfrac{1}{8}x^2 - 4\sqrt{x}$ on $[0, 9]$

29. $f(x) = \dfrac{1}{x}$ on $(0, \infty)$

30. $g(x) = \dfrac{1}{x + 1}$ on $(0, \infty)$

31. $f(x) = 3x^{2/3} - 2x$ on $[0, 3]$

32. $g(x) = x^2 + 2x^{2/3}$ on $[-2, 2]$

33. $f(x) = x^{2/3}(x^2 - 4)$ on $[-1, 2]$

34. $f(x) = x^{2/3}(x^2 - 4)$ on $[-1, 3]$

35. $f(x) = \dfrac{x}{x^2 + 2}$ on $[-1, 2]$

36. $f(x) = \dfrac{1}{x^2 + 2x + 5}$ on $[-2, 1]$

37. $f(x) = \dfrac{x}{\sqrt{x^2 + 1}}$ on $[-1, 1]$

38. $g(x) = x\sqrt{4 - x^2}$ on $[0, 2]$

39. A stone is thrown straight up from the roof of an 80-ft building. The height (in feet) of the stone at any time t (in seconds), measured from the ground, is given by

$$h(t) = -16t^2 + 64t + 80$$

What is the maximum height the stone reaches?

40. MAXIMIZING PROFITS   Lynbrook West, an apartment complex, has 100 two-bedroom units. The monthly profit (in dollars) realized from renting out x apartments is given by

$$P(x) = -10x^2 + 1760x - 50{,}000$$

How many units should be rented out in order to maximize the monthly rental profit? What is the maximum monthly profit realizable?

41. MAXIMIZING PROFITS   The estimated monthly profit (in dollars) realizable by the Cannon Precision Instruments Corporation for manufacturing and selling x units of its model M1 camera is

$$P(x) = -0.04x^2 + 240x - 10{,}000$$

How many cameras should Cannon produce per month in order to maximize its profits?

42. FLIGHT OF A ROCKET   The altitude (in feet) attained by a model rocket t sec into flight is given by the function

$$h(t) = -\frac{1}{3}t^3 + 4t^2 + 20t + 2$$

Find the maximum altitude attained by the rocket.

43. FEMALE SELF-EMPLOYED WORKFORCE   Based on data obtained from the U.S. Department of Labor, the number of nonfarm, full-time, self-employed women can be approximated by

$$N(t) = 0.81t - 1.14\sqrt{t} + 1.53 \qquad (0 \le t \le 6)$$

where $N(t)$ is measured in millions and t is measured in 5-yr intervals, with $t = 0$ corresponding to the beginning of 1963. Determine the absolute extrema of the function N on the interval $[0, 6]$. Interpret your results.
Source: U.S. Department of Labor

44. MAXIMIZING PROFITS   The management of Trappee and Sons, Inc., producers of the famous Texa-Pep hot sauce, estimate that their profit (in dollars) from the daily production and sale of x cases (each case consisting of 24 bottles) of the hot sauce is given by

$$P(x) = -0.000002x^3 + 6x - 400$$

What is the largest possible profit Trappee can make in 1 day?

45. **MAXIMIZING PROFITS**   The quantity demanded per month of the Walter Serkin recording of Beethoven's *Moonlight Sonata,* manufactured by Phonola Record Industries, is related to the price per compact disc. The equation

$$p = -0.00042x + 6 \qquad (0 \le x \le 12{,}000)$$

where p denotes the unit price in dollars and x is the number of discs demanded, relates the demand to the price. The total monthly cost (in dollars) for pressing and packaging x copies of this classical recording is given by

$$C(x) = 600 + 2x - 0.00002x^2 \qquad (0 \le x \le 20{,}000)$$

How many copies should Phonola produce per month in order to maximize its profits?

Hint: The revenue is $R(x) = px$, and the profit is $P(x) = R(x) - C(x)$.

46. **MAXIMIZING PROFIT**   A manufacturer of tennis rackets finds that the total cost $C(x)$ (in dollars) of manufacturing x rackets/day is given by $C(x) = 400 + 4x + 0.0001x^2$. Each racket can be sold at a price of p dollars, where p is related to x by the demand equation $p = 10 - 0.0004x$. If all rackets that are manufactured can be sold, find the daily level of production that will yield a maximum profit for the manufacturer.

47. **MAXIMIZING PROFIT**   The weekly demand for the Pulsar 25-in. color console television is given by the demand equation

$$p = -0.05x + 600 \qquad (0 \le x \le 12{,}000)$$

where p denotes the wholesale unit price in dollars and x denotes the quantity demanded. The weekly total cost function associated with manufacturing these sets is given by

$$C(x) = 0.000002x^3 - 0.03x^2 + 400x + 80{,}000$$

where $C(x)$ denotes the total cost incurred in producing x sets. Find the level of production that will yield a maximum profit for the manufacturer.

Hint: Use the quadratic formula.

48. **MINIMIZING AVERAGE COSTS**   Suppose the total cost function for manufacturing a certain product is $C(x) = 0.2(0.01x^2 + 120)$ dollars, where x represents the number of units produced. Find the level of production that will minimize the average cost.

49. **MINIMIZING PRODUCTION COSTS**   The total monthly cost (in dollars) incurred by Cannon Precision Instruments Corporation for manufacturing x units of the model M1

camera is given by the function

$$C(x) = 0.0025x^2 + 80x + 10{,}000$$

a. Find the average cost function $\overline{C}$.
b. Find the level of production that results in the smallest average production cost.
c. Find the level of production for which the average cost is equal to the marginal cost.
d. Compare the result of part (c) with that of part (b).

50. **MINIMIZING PRODUCTION COSTS**   The daily total cost (in dollars) incurred by Trappee and Sons, Inc., for producing x cases of Texa-Pep hot sauce is given by the function

$$C(x) = 0.000002x^3 + 5x + 400$$

Using this function, answer the questions posed in Exercise 49.

51. **MAXIMIZING REVENUE**   Suppose the quantity demanded per week of a certain dress is related to the unit price p by the demand equation $p = \sqrt{800 - x}$, where p is in dollars and x is the number of dresses made. How many dresses should be made and sold per week in order to maximize the revenue?

Hint: $R(x) = px$.

52. **MAXIMIZING REVENUE**   The quantity demanded per month of the Sicard wristwatch is related to the unit price by the equation

$$p = \frac{50}{0.01x^2 + 1} \qquad (0 \le x \le 20)$$

where p is measured in dollars and x is measured in units of a thousand. How many watches must be sold to yield a maximum revenue?

53. **OXYGEN CONTENT OF A POND**   When organic waste is dumped into a pond, the oxidation process that takes place reduces the pond's oxygen content. However, given time, nature will restore the oxygen content to its natural level. Suppose the oxygen content t days after organic waste has been dumped into the pond is given by

$$f(t) = 100 \left[\frac{t^2 - 4t + 4}{t^2 + 4} \right] \qquad (0 \le t < \infty)$$

percent of its normal level.
a. When is the level of oxygen content lowest?
b. When is the rate of oxygen regeneration greatest?

54. **AIR POLLUTION**   The amount of nitrogen dioxide, a brown gas that impairs breathing, present in the atmosphere

on a certain May day in the city of Long Beach is approximated by

$$A(t) = \frac{136}{1 + 0.25(t - 4.5)^2} + 28 \qquad (0 \le t \le 11)$$

where $A(t)$ is measured in pollutant standard index (PSI) and t is measured in hours, with $t = 0$ corresponding to 7 A.M. Determine the time of day when the pollution is at its highest level.

55. **MAXIMIZING REVENUE** The average revenue is defined as the function

$$\overline{R}(x) = \frac{R(x)}{x} \qquad (x > 0)$$

Prove that if a revenue function $R(x)$ is concave downward $[R''(x) < 0]$, then the level of sales that will result in the largest average revenue occurs when $\overline{R}(x) = R'(x)$.

56. **VELOCITY OF BLOOD** According to a law discovered by the nineteenth-century physician Jean Louis Marie Poiseuille, the velocity (in centimeters per second) of blood r cm from the central axis of an artery is given by

$$v(r) = k(R^2 - r^2)$$

where k is a constant and R is the radius of the artery. Show that the velocity of blood is greatest along the central axis.

57. **GDP OF A DEVELOPING COUNTRY** A developing country's gross domestic product (GDP) from 1993 to 2001 is approximated by the function

$$G(t) = -0.2t^3 + 2.4t^2 + 60 \qquad (0 \le t \le 8)$$

where $G(t)$ is measured in billions of dollars and $t = 0$ corresponds to the year 1993. Show that the growth rate of the country's GDP was maximal in 1997.

58. **CRIME RATES** The number of major crimes committed in the city of Bronxville between 1987 and 1994 is approximated by the function

$$N(t) = -0.1t^3 + 1.5t^2 + 100 \qquad (0 \le t \le 7)$$

where $N(t)$ denotes the number of crimes committed in year t ($t = 0$ corresponds to the year 1987). Enraged by the dramatic increase in the crime rate, the citizens of Bronxville, with the help of the local police, organized "Neighborhood Crime Watch" groups in early 1991 to combat this menace. Show that the growth in the crime rate was maximal in 1992, giving credence to the claim that the Neighborhood Crime Watch program was working.

59. **SOCIAL SECURITY SURPLUS** Based on data from the Social Security Administration, the estimated cash in the Social Security retirement and disability trust funds may be approximated by

$$f(t) = -0.0129t^4 + 0.3087t^3 + 2.1760t^2 + 62.8466t$$
$$+ 506.2955 \qquad (0 \le t \le 35)$$

where $f(t)$ is measured in billions of dollars and t is measured in years, with $t = 0$ corresponding to the year 1995. Show that the Social Security surplus will be at its highest level at approximately the middle of the year 2018.

Hint: Show that $t = 23.6811$ is an approximate critical point of $f'(t)$.

Source: Social Security Administration

60. **ENERGY EXPENDED BY A FISH** It has been conjectured that a fish swimming a distance of L ft at a speed of v ft/sec relative to the water and against a current flowing at the rate of u ft/sec ($u < v$) expends a total energy given by

$$E(v) = \frac{aLv^3}{v - u}$$

where E is measured in foot-pounds (ft-lb) and a is a constant. Find the speed v at which the fish must swim in order to minimize the total energy expended. (*Note:* This result has been verified by biologists.)

61. **REACTION TO A DRUG** The strength of a human body's reaction R to a dosage D of a certain drug is given by

$$R = D^2 \left(\frac{k}{2} - \frac{D}{3} \right)$$

where k is a positive constant. Show that the maximum reaction is achieved if the dosage is k units.

62. Refer to Exercise 61. Show that the rate of change in the reaction R with respect to the dosage D is maximal if $D = k/2$.

In Exercises 63–66, determine whether the statement is true or false. If it is true, explain why it is true. If it is false, give an example to show why it is false.

63. If f is defined on a closed interval $[a, b]$, then f has an absolute maximum value.

64. If f is continuous on an open interval (a, b), then f does not have an absolute minimum value.

65. If f is not continuous on the closed interval $[a, b]$, then f cannot have an absolute maximum value.

(continued on p. 356)

FINDING THE ABSOLUTE EXTREMA OF A FUNCTION

Some graphing utilities have a function for finding the absolute maximum and the absolute minimum values of a continuous function on a closed interval. If your graphing utility has this capability, use it to work through the example and exercises of this section.

EXAMPLE 1   Let $f(x) = \dfrac{2x + 4}{(x^2 + 1)^{3/2}}$.

a. Use a graphing utility to plot the graph of f in the viewing rectangle $[-3, 3] \times [-1, 5]$.

b. Find the absolute maximum and absolute minimum values of f on the interval $[-3, 3]$. Express your answers accurate to four decimal places.

SOLUTION ✔   **a.** The graph of f is shown in Figure T1.

FIGURE T1
The graph of f in the viewing rectangle $[-3, 3] \times [-1, 5]$

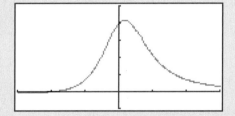

b. Using the function on a graphing utility for finding the absolute minimum value of a continuous function on a closed interval, we find the absolute minimum value of f to be -0.0632. Similarly, using the function for finding the absolute maximum value, we find the absolute maximum value to be 4.1593.

■ ■ ■ ■

REMARK   Some graphing utilities will enable you to find the absolute minimum and absolute maximum values of a continuous function on a closed interval without having to graph the function.   ■ ■ ■

In Exercises 1–6, use a graphing utility to find the absolute maximum and the absolute minimum values of f in the given interval using the method of Example 1. Express your answers accurate to four decimal places.

1. $f(x) = 3x^4 - 4.2x^3 + 6.1x - 2; [-2, 3]$

2. $f(x) = 2.1x^4 - 3.2x^3 + 4.1x^2 + 3x - 4; [-1, 2]$

3. $f(x) = \dfrac{2x^3 - 3x^2 + 1}{x^2 + 2x - 8}; [-3, 1]$

4. $f(x) = \sqrt{x}(x^3 - 4)^2; [0.5, 1]$

5. $f(x) = \dfrac{x^3 - 1}{x^2}; [1, 3]$

6. $f(x) = \dfrac{x^3 - x^2 + 1}{x - 2}; [1, 3]$

7. RATE OF BANK FAILURES The Federal Deposit Insurance Company (FDIC) estimates that the rate at which banks were failing between 1982 and 1994 is given by

$$f(t) = 0.063447t^4 - 1.953283t^3 + 14.632576t^2$$
$$- 6.684704t + 47.458874 \qquad (0 \le t \le 12)$$

where $f(t)$ is the number of banks per year and t is measured in years, with $t = 0$ corresponding to the beginning of 1982.
a. Use a graphing utility to plot the graph of f in the viewing rectangle $[0, 12] \times [0, 220]$.
b. What is the highest rate of bank failures during the period in question?
Source: Federal Deposit Insurance Corporation

8. BOSTON'S DAILY TEMPERATURE FOR 1993 The average daily temperature in Boston for 1993 is given by

$$f(t) = 0.0434841t^4 - 1.18523t^3 + 9.38548t^2$$
$$- 17.7553t + 38.9272 \qquad (0 \le t \le 12)$$

where $f(t)$ is measured in degrees Fahrenheit and t is in months, with $t = 0$ corresponding to the beginning of 1993. Find the absolute extrema for f and interpret your results.
Source: Robert Lautzenheiser, Climatologist

9. TIME ON THE MARKET According to the Greater Boston Real Estate Board—Multiple Listing Service, the average number of days a single-family home remains for sale from listing to accepted offer is approximated by the function

$$f(t) = 0.0171911t^4 - 0.662121t^3 + 6.18083t^2$$
$$- 8.97086t + 53.3357 \qquad (0 \le t \le 10)$$

where t is measured in years, with $t = 0$ corresponding to the beginning of 1984.
a. Use a graphing utility to plot the graph of f in the viewing rectangle $[0, 12] \times [0, 120]$.
b. Find the absolute maximum value and the absolute minimum value of f in the interval $[0, 12]$. Interpret your results.
Source: Greater Boston Real Estate Board—Multiple Listing Service

10. WHY SSS BENEFITS MAY EXCEED PAYROLL TAXES Unless payroll taxes are increased significantly and/or benefits are scaled back drastically, it is a matter of time before the current Social Security system goes broke. Based on data from the Board of Trustees of the Social Security Administration, the assets of the system—the Social Security "trust fund"—may be approximated by

$$f(t) = -0.0129t^4 + 0.3087t^3 + 2.1760t^2$$
$$+ 62.8466t + 506.2955 \qquad (0 \le t \le 35)$$

where $f(t)$ is measured in millions of dollars and t is measured in years, with $t = 0$ corresponding to 1995.
a. Use a graphing calculator to sketch the graph of f.
b. Based on this model, when will the Social Security system start to pay out more benefits than it gets in payroll taxes?
Source: Social Security Administration

66. If $f''(x) < 0$ on (a, b) and $f'(c) = 0$ where $a < c < b$, then $f(c)$ is the absolute maximum value of f on $[a, b]$.

67. Let f be a constant function—that is, let $f(x) = c$, where c is some real number. Show that every point $x = a$ is an absolute maximum and, at the same time, an absolute minimum of f.

68. Show that a polynomial function defined on the interval $(-\infty, \infty)$ cannot have both an absolute maximum and an absolute minimum unless it is a constant function.

69. One condition that must be satisfied before Theorem 3

(page 342) is applicable is that the function f must be continuous on the closed interval $[a, b]$. Define a function f on the closed interval $[-1, 1]$ by

$$f(x) = \begin{cases} \dfrac{1}{x} & \text{if } x \in [-1, 1] \qquad (x \neq 0) \\ 0 & \text{if } x = 0 \end{cases}$$

a. Show that f is not continuous at $x = 0$.
b. Show that $f(x)$ does not attain an absolute maximum or an absolute minimum on the interval $[-1, 1]$.
c. Confirm your results by sketching the function f.

Solutions to Self-Check Exercises 4.4

1. a. The function f is continuous in its domain and differentiable in the interval $(0, 9)$. The derivative of f is

$$f'(x) = 1 - x^{-1/2} = \frac{x^{1/2} - 1}{x^{1/2}}$$

and it is equal to zero when $x = 1$. Evaluating $f(x)$ at the end points $x = 0$ and $x = 9$ and at the critical point $x = 1$ of f, we have

$$f(0) = 0, \qquad f(1) = -1, \qquad f(9) = 3$$

From these results, we see that -1 is the absolute minimum value of f and 3 is the absolute maximum value of f.

b. In this case, the domain of f is the interval $[0, \infty)$, which is not closed. Therefore, we resort to the graphic method. Using the techniques of graphing, we sketch in the accompanying figure the graph of f.

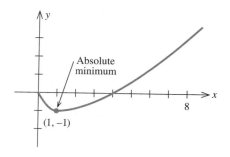

The graph of f shows that -1 is the absolute minimum value of f, but f has no absolute maximum since $f(x)$ increases without bound as x increases without bound.

2. The function f is continuous on the interval $[-2, 1]$. It is also differentiable on the open interval $(-2, 1)$. The derivative of f is

$$f'(x) = 12x^3 + 12x^2 = 12x^2(x + 1)$$

and it is continuous on $(-2, 1)$. Setting $f'(x) = 0$ gives $x = -1$ and $x = 0$ as critical points of f. Evaluating $f(x)$ at these critical points of f as well as at the end points of the interval $[-2, 1]$, we obtain

$$f(-2) = 17, \qquad f(-1) = 0, \qquad f(0) = 1, \qquad f(1) = 8$$

From these results, we see that 0 is the absolute minimum value of f and 17 is the absolute maximum value of f.

3. The problem is solved by finding the absolute maximum of the function f on $[0, 250]$. Differentiating $f(t)$, we obtain

$$f'(t) = \frac{(t^2 + 40{,}000)(1200) - 1200t(2t)}{(t^2 + 40{,}000)^2}$$

$$= \frac{-1200(t^2 - 40{,}000)}{(t^2 + 40{,}000)^2}$$

Upon setting $f'(t) = 0$ and solving the resulting equation, we obtain $t = -200$ or 200. Since -200 lies outside the interval $[0, 250]$, we are interested only in the critical point $t = 200$ of f. Evaluating $f(t)$ at $t = 0$, $t = 200$, and $t = 250$, we find

$$f(0) = 80, \qquad f(200) = 83, \qquad f(250) = 82.93$$

We conclude that the manufacturing capacity operating rate was the highest on the 200th day of 2000—that is, a little past the middle of July 2000.

4.5 Optimization II

Section 4.4 outlined how to find the solution to certain optimization problems in which the objective function is given. In this section we consider problems in which we are required to first find the appropriate function to be optimized. The following guidelines will be useful for solving these problems.

Guidelines for Solving Optimization Problems

1. Assign a letter to each variable mentioned in the problem. If appropriate, draw and label a figure.
2. Find an expression for the quantity to be optimized.
3. Use the conditions given in the problem to write the quantity to be optimized as a function f of *one* variable. Note any restrictions to be placed on the domain of f from physical considerations of the problem.
4. Optimize the function f over its domain using the methods of Section 4.4.

REMARK In carrying out step 4, remember that if the function f to be optimized is continuous on a closed interval, then the absolute maximum and absolute minimum of f are, respectively, the largest and smallest values of $f(x)$ on the set composed of the critical points of f and the end points of the interval. If the domain of f is not a closed interval, then we resort to the graphic method. ■■■

18. **DESIGNING A GRAIN SILO** A grain silo has the shape of a right circular cylinder surmounted by a hemisphere (see the accompanying figure). If the silo is to have a capacity of 504π ft^3, find the radius and height of the silo that requires the least amount of material to construct.

Hint: The volume of the silo is $\pi r^2 h + \frac{2}{3}\pi r^3$, and the surface area (including the floor) is $\pi(3r^2 + 2rh)$.

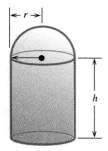

19. **MINIMIZING CONSTRUCTION COSTS** In the following diagram, S represents the position of a power relay station located on a straight coast, and E shows the location of a marine biology experimental station on an island. A cable is to be laid connecting the relay station with the experimental station. If the cost of running the cable on land is $1/running foot and the cost of running the cable under water is $3/running foot, locate the point P that will result in a minimum cost (solve for x).

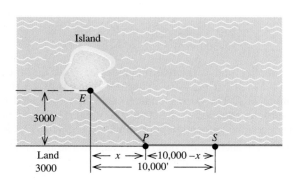

20. **FLIGHTS OF BIRDS** During daylight hours, some birds fly more slowly over water than over land because some of their energy is expended in overcoming the downdrafts of air over open bodies of water. Suppose a bird that flies at a constant speed of 4 mph over water and 6 mph over land starts its journey at the point E on an island and ends at its nest N on the shore of the mainland, as shown in the accompanying figure. Find the location of the point P that allows the bird to complete its journey in the minimum time (solve for x).

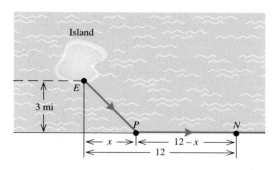

21. **OPTIMAL SPEED OF A TRUCK** A truck gets $400/x$ mpg when driven at a constant speed of x mph (between 50 and 70 mph). If the price of fuel is $1/gallon and the driver is paid $8/hour, at what speed between 50 and 70 mph is it most economical to drive?

22. **INVENTORY CONTROL AND PLANNING** The demand for motorcycle tires imported by the Dixie Import-Export Company is 40,000/year and may be assumed to be uniform throughout the year. The cost of ordering a shipment of tires is $400, and the cost of storing each tire for a year is $2. Determine how many tires should be in each shipment if the ordering and storage costs are to be minimized. (Assume that each shipment arrives just as the previous one has been sold.)

23. **INVENTORY CONTROL AND PLANNING** The McDuff Preserves Company expects to bottle and sell 2,000,000 32-oz jars of jam. The company orders its containers from the Consolidated Bottle Company. The cost of ordering a shipment of bottles is $200, and the cost of storing each empty bottle for a year is $.40. How many orders should McDuff place per year and how many bottles should be in each shipment if the ordering and storage costs are to be minimized? (Assume that each shipment of bottles is used up before the next shipment arrives.)

24. **INVENTORY CONTROL AND PLANNING** The Neilsen Cookie Company sells its assorted butter cookies in containers that have a net content of 1 lb. The estimated demand for the cookies is 1,000,000 units. The setup cost for each production run is $500, and the manufacturing cost is $.50 for each container of cookies. The cost of storing each container of cookies over the year is $.40. Assuming uniformity of demand throughout the year and instantaneous production, how many containers of cookies should Neilsen produce per production run in order to minimize the production cost?

Hint: Following the method of Example 5, show that the total production cost is given by the function

$$C(x) = \frac{500,000,000}{x} + 0.2x + 500,000$$

Then minimize the function C on the interval $(0, 1,000,000)$.

SOLUTIONS TO SELF-CHECK EXERCISES 4.5

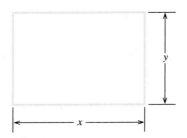

1. Let x and y (measured in feet) denote the length and width of the rectangular garden. Since the area is to be 300 ft², we have

$$xy = 300$$

Next, the amount of fencing to be used is given by the perimeter, and this quantity is to be minimized. Thus, we want to minimize

$$2x + 2y$$

or, since $y = 300/x$ (obtained by solving for y in the first equation), we see that the expression to be minimized is

$$f(x) = 2x + 2\left(\frac{300}{x}\right)$$

$$= 2x + \frac{600}{x}$$

for positive values of x. Now,

$$f'(x) = 2 - \frac{600}{x^2}$$

Setting $f'(x) = 0$ yields $x = -\sqrt{300}$ or $x = \sqrt{300}$. We consider only the critical point $x = \sqrt{300}$ since $-\sqrt{300}$ lies outside the interval $(0, \infty)$. We then compute

$$f''(x) = \frac{1200}{x^3}$$

Since

$$f''(\sqrt{300}) > 0$$

the second derivative test implies that a relative minimum of f occurs at $x = \sqrt{300}$. In fact, since $f''(x) > 0$ for all x in $(0, \infty)$, we conclude that $x = \sqrt{300}$ gives rise to the absolute minimum of f. The corresponding value of y, obtained by substituting this value of x into the equation $xy = 300$, is $y = \sqrt{300}$. Therefore, the required dimensions of the vegetable garden are approximately 17.3 ft × 17.3 ft.

2. Let x denote the number of tires in each production run. Then, the average number of tires in storage is $x/2$, so the storage cost incurred by the company is $2(x/2)$, or x dollars. Next, since the company needs to manufacture 1,000,000 tires for the year in order to meet the demand, the number of production runs is $1,000,000/x$. This gives setup costs amounting to

$$4000\left(\frac{1,000,000}{x}\right) = \frac{4,000,000,000}{x}$$

dollars for the year. The total manufacturing cost is \$20,000,000. Thus, the total yearly cost incurred by the company is given by

$$C(x) = x + \frac{4,000,000,000}{x} + 20,000,000$$

Differentiating $C(x)$, we find

$$C'(x) = 1 - \frac{4{,}000{,}000{,}000}{x^2}$$

Setting $C'(x) = 0$ gives $x = 63{,}246$ as the critical point in the interval $(0, 1{,}000{,}000)$. Next, we find

$$C''(x) = \frac{8{,}000{,}000{,}000}{x^3}$$

Since $C''(x) > 0$ for all $x > 0$, we see that C is concave upward for all $x > 0$. Furthermore, $C''(63{,}246) > 0$ implies that $x = 63{,}246$ gives rise to a relative minimum of C (by the second derivative test). Since C is always concave upward for $x > 0$, $x = 63{,}246$ gives the absolute minimum of C. Therefore, the company should manufacture 63,246 tires in each production run.

CHAPTER 4 Summary of Principal Terms

Terms

increasing function   concave downward
decreasing function   inflection point
relative maximum   second derivative test
relative minimum   vertical asymptote
relative extrema   horizontal asymptote
critical point   absolute extrema
first derivative test   absolute maximum value
concave upward   absolute minimum value

CHAPTER 4 REVIEW EXERCISES

In Exercises 1–10, (a) find the intervals where the given function f is increasing and where it is decreasing, (b) find the relative extrema of f, (c) find the intervals where f is concave upward and where it is concave downward, and (d) find the inflection points, if any, of f.

1. $f(x) = \frac{1}{3}x^3 - x^2 + x - 6$

2. $f(x) = (x - 2)^3$

3. $f(x) = x^4 - 2x^2$

4. $f(x) = x + \frac{4}{x}$

5. $f(x) = \frac{x^2}{x - 1}$

6. $f(x) = \sqrt{x - 1}$

7. $f(x) = (1 - x)^{1/3}$

8. $f(x) = x\sqrt{x - 1}$

9. $f(x) = \frac{2x}{x + 1}$

10. $f(x) = \frac{-1}{1 + x^2}$

In Exercises 11–18, obtain as much information as possible on each of the given functions. Then use this information to sketch the graph of the function.

11. $f(x) = x^2 - 5x + 5$

12. $f(x) = -2x^2 - x + 1$

13. $g(x) = 2x^3 - 6x^2 + 6x + 1$

14. $g(x) = \frac{1}{3}x^3 - x^2 + x - 3$

15. $h(x) = x\sqrt{x-2}$

16. $h(x) = \frac{2x}{1+x^2}$

17. $f(x) = \frac{x-2}{x+2}$

18. $f(x) = x - \frac{1}{x}$

In Exercises 19–22, find the horizontal and vertical asymptotes of the graphs of the given functions. Do not sketch the graphs.

19. $f(x) = \frac{1}{2x+3}$

20. $f(x) = \frac{2x}{x+1}$

21. $f(x) = \frac{5x}{x^2 - 2x - 8}$

22. $f(x) = \frac{x^2 + x}{x(x-1)}$

In Exercises 23–32, find the absolute maximum value and the absolute minimum value, if any, of the given function.

23. $f(x) = 2x^2 + 3x - 2$

24. $g(x) = x^{2/3}$

25. $g(t) = \sqrt{25 - t^2}$

26. $f(x) = \frac{1}{3}x^3 - x^2 + x + 1$ on $[0, 2]$

27. $h(t) = t^3 - 6t^2$ on $[2, 5]$

28. $g(x) = \frac{x}{x^2 + 1}$ on $[0, 5]$

29. $f(x) = x - \frac{1}{x}$ on $[1, 3]$

30. $h(t) = 8t - \frac{1}{t^2}$ on $[1, 3]$

31. $f(s) = s\sqrt{1 - s^2}$ on $[-1, 1]$

32. $f(x) = \frac{x^2}{x-1}$ on $[-1, 3]$

33. Odyssey Travel Agency's monthly profit (in thousands of dollars) depends on the amount of money x (in thousands of dollars) spent on advertising per month according to the rule

$$P(x) = -x^2 + 8x + 20$$

What should Odyssey's monthly advertising budget be in order to maximize its monthly profits?

34. The Department of the Interior of an African country began to record an index of environmental quality to measure progress or decline in the environmental quality of its wildlife. The index for the years 1984 through 1994 is approximated by the function

$$I(t) = \frac{50t^2 + 600}{t^2 + 10} \qquad (0 \le t \le 10)$$

a. Compute $I'(t)$ and show that $I(t)$ is decreasing on the interval $(0, 10)$.
b. Compute $I''(t)$. Study the concavity of the graph of I.
c. Sketch the graph of I.
d. Interpret your results.

35. The weekly demand for video discs manufactured by the Herald Record Company is given by

$$p = -0.0005x^2 + 60$$

where p denotes the unit price in dollars and x denotes the quantity demanded. The weekly total cost function associated with producing these discs is given by

$$C(x) = -0.001x^2 + 18x + 4000$$

where $C(x)$ denotes the total cost incurred in pressing x discs. Find the production level that will yield a maximum profit for the manufacturer.
Hint: Use the quadratic formula.

36. The total monthly cost (in dollars) incurred by the Carlota Music Company in manufacturing x units of its Professional Series guitars is given by the function

$$C(x) = 0.001x^2 + 100x + 4000$$

a. Find the average cost function $\overline{C}$.
b. Determine the production level that will result in the smallest average production cost.

37. The average worker at Wakefield Avionics, Inc., can assemble

$$N(t) = -2t^3 + 12t^2 + 2t \qquad (0 \le t \le 4)$$

ready-to-fly radio-controlled model airplanes t hr into the 8 A.M. to 12 noon morning shift. At what time during this shift is the average worker performing at peak efficiency?

38. You wish to construct a closed rectangular box that has a volume of 4 ft³. The length of the base of the box will be twice as long as its width. The material for the top and bottom of the box costs 30 cents/square foot. The material for the sides of the box costs 20 cents/square foot. Find the dimensions of the least expensive box that can be constructed.

39. The Lehen Vinters Company imports a certain brand of beer. The demand, which may be assumed to be uniform, is 800,000 cases/year. The cost of ordering a shipment of beer is $500, and the cost of storing each case of beer for a year is $2. Determine how many cases of beer should be in each shipment if the ordering and storage costs are to be kept to a minimum. (Assume that each shipment of beer arrives just as the previous one has been sold.)

40. Let

$$f(x) = \begin{cases} x^3 + 1 & \text{if } x \neq 0 \\ 2 & \text{if } x = 0 \end{cases}$$

a. Compute $f'(x)$ and show that it does not change sign as we move across $x = 0$.
b. Show that f has a relative maximum at $x = 0$. Does this contradict the first derivative test? Explain your answer.

5

EXPONENTIAL AND LOGARITHMIC FUNCTIONS

The exponential function is, without doubt, the most important function in mathematics and its applications. After a brief introduction to the exponential function and its *inverse*, the logarithmic function, we learn how to differentiate such functions. This lays the foundation for exploring the many applications involving exponential functions. For example, we look at the role played by exponential functions in computing earned interest on a bank account, in studying the growth of a bacteria population in the laboratory, in studying the way radioactive matter decays, in studying the rate at which a factory worker learns a certain process, and in studying the rate at which a communicable disease is spread over time.

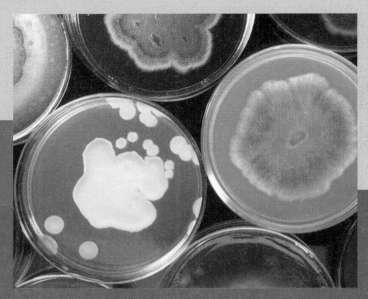

How many bacteria will there be in a culture at the end of a certain period of time? How fast will the bacteria population be growing at the end of that time? Example 1, page 426, answers these questions.

5.2 Logarithmic Functions

LOGARITHMS

You are already familiar with exponential equations of the form

$$b^y = x \qquad (b > 0, b \neq 1)$$

where the variable x is expressed in terms of a real number b and a variable y. But what about solving this same equation for y? You may recall from your study of algebra that the number y is called the **logarithm of x to the base b** and is denoted by **$\log_b x$.** It is the exponent to which the base b must be raised in order to obtain the number x.

**Logarithm of
x to the Base b**

$$y = \log_b x \qquad \text{if and only if} \qquad x = b^y \qquad (x > 0)$$

Observe that the logarithm $\log_b x$ is defined only for positive values of x.

EXAMPLE 1

a. $\log_{10} 100 = 2$ since $100 = 10^2$
b. $\log_5 125 = 3$ since $125 = 5^3$
c. $\log_3 \dfrac{1}{27} = -3$ since $\dfrac{1}{27} = \dfrac{1}{3^3} = 3^{-3}$
d. $\log_{20} 20 = 1$ since $20 = 20^1$ ■ ■ ■ ■

EXAMPLE 2

Solve each of the following equations for x.

a. $\log_3 x = 4$ **b.** $\log_{16} 4 = x$ **c.** $\log_x 8 = 3$

SOLUTION ✔

a. By definition, $\log_3 x = 4$ implies $x = 3^4 = 81$.
b. $\log_{16} 4 = x$ is equivalent to $4 = 16^x = (4^2)^x = 4^{2x}$, or $4^1 = 4^{2x}$, from which we deduce that

$$2x = 1 \qquad (b^m = b^n \Rightarrow m = n)$$
$$x = \frac{1}{2}$$

c. Referring once again to the definition, we see that the equation $\log_x 8 = 3$ is equivalent to

$$8 = (2^3) = x^3$$
$$x = 2 \qquad (a^m = b^m \Rightarrow a = b)$$ ■ ■ ■ ■

The two widely used systems of logarithms are the system of **common logarithms,** which uses the number 10 as the base, and the system of **natural logarithms,** which uses the irrational number $e = 2.71828\ldots$ as the base. Also, it is standard practice to write **log** for $\log_{10}$ and **ln** for $\log_e$.

Logarithmic Notation

$$\log x = \log_{10} x \qquad \text{(Common logarithm)}$$
$$\ln x = \log_e x \qquad \text{(Natural logarithm)}$$

The system of natural logarithms is widely used in theoretical work. Using natural logarithms rather than logarithms to other bases often leads to simpler expressions.

LAWS OF LOGARITHMS

Computations involving logarithms are facilitated by the following **laws of logarithms.**

Laws of Logarithms

If m and n are positive numbers, then

1. $\log_b mn = \log_b m + \log_b n$

2. $\log_b \dfrac{m}{n} = \log_b m - \log_b n$

3. $\log_b m^n = n \log_b m$

4. $\log_b 1 = 0$

5. $\log_b b = 1$

 Do not confuse the expression $\log\ m/n$ (Law 2) with the expression $\log\ m/\log\ n$. For example,

$$\log \frac{100}{10} = \log 100 - \log 10 = 2 - 1 = 1 \neq \frac{\log 100}{\log 10} = \frac{2}{1} = 2$$

You will be asked to prove these laws in Exercises 53–55. Their derivations are based on the definition of a logarithm and the corresponding laws of exponents. The following examples illustrate the properties of logarithms.

EXAMPLE 3
a. $\log(2 \cdot 3) = \log 2 + \log 3$ **b.** $\ln \dfrac{5}{3} = \ln 5 - \ln 3$

c. $\log \sqrt{7} = \log 7^{1/2} = \dfrac{1}{2} \log 7$ **d.** $\log_5 1 = 0$

e. $\log_{45} 45 = 1$ ■ ■ ■ ■

EXAMPLE 4
Given that $\log 2 \approx 0.3010$, $\log 3 \approx 0.4771$, and $\log 5 \approx 0.6990$, use the laws of logarithms to find

a. $\log 15$ **b.** $\log 7.5$ **c.** $\log 81$ **d.** $\log 50$

SOLUTION ✔   **a.** Note that $15 = 3 \cdot 5$, so by Law 1 for logarithms,

$$\begin{aligned}
\log 15 &= \log 3 \cdot 5 \\
&= \log 3 + \log 5 \\
&\approx 0.4771 + 0.6990 \\
&= 1.1761
\end{aligned}$$

b. Observing that $7.5 = 15/2 = (3 \cdot 5)/2$, we apply Laws 1 and 2, obtaining

$$\begin{aligned}
\log 7.5 &= \log \frac{(3)(5)}{2} \\
&= \log 3 + \log 5 - \log 2 \\
&\approx 0.4771 + 0.6990 - 0.3010 \\
&= 0.8751
\end{aligned}$$

c. Since $81 = 3^4$, we apply Law 3 to obtain

$$\begin{aligned}
\log 81 &= \log 3^4 \\
&= 4 \log 3 \\
&\approx 4(0.4771) \\
&= 1.9084
\end{aligned}$$

d. We write $50 = 5 \cdot 10$ and find

$$\begin{aligned}
\log 50 &= \log(5)(10) \\
&= \log 5 + \log 10 \\
&\approx 0.6990 + 1 \qquad \text{(Using Law 5)} \\
&= 1.6990
\end{aligned}$$

■ ■ ■ ■

EXAMPLE 5   Expand and simplify the following expressions:

a. $\log_3 x^2 y^3$ **b.** $\log_2 \dfrac{x^2 + 1}{2^x}$ **c.** $\ln \dfrac{x^2 \sqrt{x^2 - 1}}{e^x}$

SOLUTION ✔   **a.** $\log_3 x^2 y^3 = \log_3 x^2 + \log_3 y^3$ (Law 1)

$\qquad\qquad\quad = 2 \log_3 x + 3 \log_3 y$ (Law 3)

b. $\log_2 \dfrac{x^2 + 1}{2^x} = \log_2(x^2 + 1) - \log_2 2^x$ (Law 2)

$\qquad\qquad\quad = \log_2(x^2 + 1) - x \log_2 2$ (Law 3)

$\qquad\qquad\quad = \log_2(x^2 + 1) - x$ (Law 5)

c. $\ln \dfrac{x^2 \sqrt{x^2 - 1}}{e^x} = \ln \dfrac{x^2(x^2 - 1)^{1/2}}{e^x}$ (Rewriting)

$\qquad\qquad\quad = \ln x^2 + \ln(x^2 - 1)^{1/2} - \ln e^x$ (Laws 1 and 2)

$\qquad\qquad\quad = 2 \ln x + \dfrac{1}{2} \ln(x^2 - 1) - x \ln e$ (Law 3)

$\qquad\qquad\quad = 2 \ln x + \dfrac{1}{2} \ln(x^2 - 1) - x$ (Law 5)

■ ■ ■ ■

LOGARITHMIC FUNCTIONS AND THEIR GRAPHS

The definition of the logarithm implies that if b and n are positive numbers and b is different from 1, then the expression $\log_b n$ is a real number. This enables us to define a logarithmic function as follows:

Logarithmic Function

The function defined by

$$f(x) = \log_b x \qquad (b > 0, b \neq 1)$$

is called the **logarithmic function with base b.** The domain of f is the set of all positive numbers.

One easy way to obtain the graph of the logarithmic function $y = \log_b x$ is to construct a table of values of the logarithm (base b). However, another method—and a more instructive one—is based on exploiting the intimate relationship between logarithmic and exponential functions.

If a point (u, v) lies on the graph of $y = \log_b x$, then

$$v = \log_b u$$

But we can also write this equation in exponential form as

$$u = b^v$$

So the point (v, u) also lies on the graph of the function $y = b^x$. Let's look at the relationship between the points (u, v) and (v, u) and the line $y = x$ (Figure 5.7). If we think of the line $y = x$ as a mirror, then the point (v, u) is the mirror reflection of the point (u, v). Similarly, the point (u, v) is a mirror reflection of the point (v, u). We can take advantage of this relationship to help us draw the graph of logarithmic functions. For example, if we wish to draw the graph of $y = \log_b x$, where $b > 1$, then we need only draw the mirror reflection of the graph of $y = b^x$ with respect to the line $y = x$ (Figure 5.8).

FIGURE 5.7
The points (u, v) and (v, u) are mirror reflections of each other.

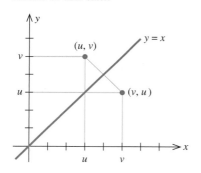

FIGURE 5.8
The graphs of $y = b^x$ and $y = \log_b x$ are mirror reflections of each other.

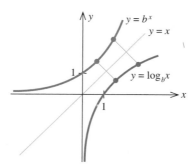

You may discover the following properties of the logarithmic function by taking the reflection of the graph of an appropriate exponential function (Exercises 31 and 32).

Properties of the Logarithmic Function

The logarithmic function $y = \log_b x$ ($b > 0, b \neq 1$) has the following properties:

1. Its domain is $(0, \infty)$.
2. Its range is $(-\infty, \infty)$.
3. Its graph passes through the point $(1, 0)$.
4. It is continuous on $(0, \infty)$.
5. It is increasing on $(0, \infty)$ if $b > 1$ and decreasing on $(0, \infty)$ if $b < 1$.

EXAMPLE 6   Sketch the graph of the function $y = \ln x$.

SOLUTION ✔   We first sketch the graph of $y = e^x$. Then, the required graph is obtained by tracing the mirror reflection of the graph of $y = e^x$ with respect to the line $y = x$ (Figure 5.9).

FIGURE 5.9
The graph of $y = \ln x$ is the mirror reflection of the graph of $y = e^x$.

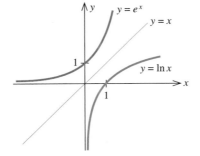

PROPERTIES RELATING THE EXPONENTIAL AND LOGARITHMIC FUNCTIONS

We made use of the relationship that exists between the exponential function $f(x) = e^x$ and the logarithmic function $g(x) = \ln x$ when we sketched the graph of g in Example 6. This relationship is further described by the following properties, which are an immediate consequence of the definition of the logarithm of a number.

Properties Relating e^x and $\ln x$

$$e^{\ln x} = x \qquad (x > 0) \qquad\qquad (2)$$
$$\ln e^x = x \qquad (\text{for any real number } x) \qquad (3)$$

(Try to verify these properties.)

From Properties 2 and 3, we conclude that the composite function

$$(f \circ g)(x) = f[g(x)]$$
$$= e^{\ln x} = x$$
$$(g \circ f)(x) = g[f(x)]$$
$$= \ln e^x = x$$

Thus,

$$f[g(x)] = g[f(x)]$$
$$= x$$

Any two functions f and g that satisfy this relationship are said to be **inverses** of each other. Note that the function f undoes what the function g does, and vice versa, so the composition of the two functions in any order results in the identity function $F(x) = x$.

The relationships expressed in Equations (2) and (3) are useful in solving equations that involve exponentials and logarithms.

Exploring with Technology

You can demonstrate the validity of Properties 5 and 6, which state that the exponential function $f(x) = e^x$ and the logarithmic function $g(x) = \ln x$ are inverses of each other as follows:

1. Sketch the graph of $(f \circ g)(x) = e^{\ln x}$, using the viewing rectangle $[0, 10] \times [0, 10]$. Interpret the result.
2. Sketch the graph of $(g \circ f)(x) = \ln e^x$, using the standard viewing rectangle. Interpret the result.

 **EXAMPLE 7**   Solve the equation $2e^{x+2} = 5$.

SOLUTION ✔   We first divide both sides of the equation by 2 to obtain

$$e^{x+2} = \frac{5}{2} = 2.5$$

Next, taking the natural logarithm of each side of the equation and using Equation (3), we have

$$\ln e^{x+2} = \ln 2.5$$
$$x + 2 = \ln 2.5$$
$$x = -2 + \ln 2.5$$
$$\approx -1.08$$

■ ■ ■ ■

  EXAMPLE 8   Solve the equation $5 \ln x + 3 = 0$.

SOLUTION ✔   Adding -3 to both sides of the equation leads to

$$5 \ln x = -3$$
$$\ln x = -\frac{3}{5} = -0.6$$

and so

$$e^{\ln x} = e^{-0.6}$$

Using equation (2), we conclude that

$$x = e^{-0.6}$$
$$\approx 0.55$$

■ ■ ■ ■

> **Group Discussion**
>
> Consider the equation $y = y_0 b^{kx}$, where y_0 and k are positive constants and $b > 0$, $b \neq 1$. Suppose we want to express y in the form $y = y_0 e^{px}$. Use the laws of logarithms to show that $p = k \ln b$ and hence that $y = y_0 e^{(k \ln b)x}$ is an alternative form of $y = y_0 b^{kx}$ using the base e.

SELF-CHECK EXERCISES 5.2

1. Sketch the graph of $y = 3^x$ and $y = \log_3 x$ on the same set of axes.
2. Solve the equation $3e^{x+1} - 2 = 4$.

Solutions to Self-Check Exercises 5.2 can be found on page 391.

5.2 Exercises

In Exercises 1–10, express the given equation in logarithmic form.

1. $2^6 = 64$

2. $3^5 = 243$

3. $3^{-2} = \dfrac{1}{9}$

4. $5^{-3} = \dfrac{1}{125}$

5. $\left(\dfrac{1}{3}\right)^1 = \dfrac{1}{3}$

6. $\left(\dfrac{1}{2}\right)^{-4} = 16$

7. $32^{3/5} = 8$

8. $81^{3/4} = 27$

9. $10^{-3} = 0.001$

10. $16^{-1/4} = 0.5$

In Exercises 11–16, use the facts that log 3 = 0.4771 and log 4 = 0.6021 to find the value of the given logarithm.

11. $\log 12$

12. $\log \dfrac{3}{4}$

13. $\log 16$

14. $\log \sqrt{3}$

15. $\log 48$

16. $\log \dfrac{1}{300}$

In Exercises 17–26, use the laws of logarithms to simplify the given expression.

17. $\log x(x + 1)^4$

18. $\log x(x^2 + 1)^{-1/2}$

19. $\log \dfrac{\sqrt{x + 1}}{x^2 + 1}$

20. $\ln \dfrac{e^x}{1 + e^x}$

21. $\ln xe^{-x^2}$

22. $\ln x(x + 1)(x + 2)$

23. $\ln \dfrac{x^{1/2}}{x^2\sqrt{1 + x^2}}$

24. $\ln \dfrac{x^2}{\sqrt{x}(1 + x)^2}$

25. $\ln x^x$

26. $\ln x^{x^2+1}$

In Exercises 27–30, sketch the graph of the given equation.

27. $y = \log_3 x$

28. $y = \log_{1/3} x$

29. $y = \ln 2x$

30. $y = \ln \dfrac{1}{2} x$

In Exercises 31 and 32, sketch the graphs of the given equations on the same coordinate axes.

31. $y = 2^x$ and $y = \log_2 x$

32. $y = e^{3x}$ and $y = \ln 3x$

In Exercises 33–42, use logarithms to solve the given equation for t.

33. $e^{0.4t} = 8$

34. $\dfrac{1}{3} e^{-3t} = 0.9$

35. $5e^{-2t} = 6$

36. $4e^{t-1} = 4$

37. $2e^{-0.2t} - 4 = 6$

38. $12 - e^{0.4t} = 3$

39. $\dfrac{50}{1 + 4e^{0.2t}} = 20$

40. $\dfrac{200}{1 + 3e^{-0.3t}} = 100$

41. $A = Be^{-t/2}$

42. $\dfrac{A}{1 + Be^{t/2}} = C$

43. BLOOD PRESSURE A normal child's systolic blood pressure may be approximated by the function

$$p(x) = m(\ln x) + b$$

where $p(x)$ is measured in millimeters of mercury, x is measured in pounds, and m and b are constants. Given that $m = 19.4$ and $b = 18$, determine the systolic blood pressure of a child who weighs 92 lb.

44. MAGNITUDE OF EARTHQUAKES On the Richter scale, the magnitude R of an earthquake is given by the formula

$$R = \log \frac{I}{I_0}$$

where I is the intensity of the earthquake being measured and I_0 is the standard reference intensity.
a. Express the intensity I of an earthquake of magnitude $R = 5$ in terms of the standard intensity I_0.
b. Express the intensity I of an earthquake of magnitude $R = 8$ in terms of the standard intensity I_0. How many times greater is the intensity of an earthquake of magnitude 8 than one of magnitude 5?
c. In modern times the greatest loss of life attributable to an earthquake occurred in eastern China in 1976. Known as the Tangshan earthquake, it registered 8.2 on the Richter scale. How does the intensity of this earthquake compare with the intensity of an earthquake of magnitude $R = 5$?

45. SOUND INTENSITY The relative loudness of a sound D of intensity I is measured in decibels (db), where

$$D = 10 \log \frac{I}{I_0}$$

and I_0 is the standard threshold of audibility.
a. Express the intensity I of a 30-db sound (the sound level of normal conversation) in terms of I_0.
b. Determine how many times greater the intensity of an 80-db sound (rock music) is than that of a 30-db sound.
c. Prolonged noise above 150 db causes immediate and permanent deafness. How does the intensity of a 150-db sound compare with the intensity of an 80-db sound?

46. BAROMETRIC PRESSURE Halley's law states that the barometric pressure (in inches of mercury) at an altitude of x mi above sea level is approximately given by the equation

$$p(x) = 29.92 e^{-0.2x} \qquad (x \geq 0)$$

If the barometric pressure as measured by a hot-air balloonist is 20 in. of mercury, what is the balloonist's altitude?

47. FORENSIC SCIENCE Forensic scientists use the following law to determine the time of death of accident or murder victims. If T denotes the temperature of a body t hr after death, then

$$T = T_0 + (T_1 - T_0)(0.97)^t$$

where T_0 is the air temperature and T_1 is the body temperature at the time of death. John Doe was found murdered at midnight in his house, when the room temperature was 70°F and his body temperature was 80°F. When was he killed? Assume that the normal body temperature is 98.6°F.

In Exercises 48–51, determine whether the statement is true or false. If it is true, explain why it is true. If it is false, give an example to show why it is false.

48. $(\ln x)^3 = 3 \ln x$ for all x in $(0, \infty)$.

49. $\ln a - \ln b = \ln(a - b)$ for all positive real numbers a and b.

50. The function $f(x) = 1/\ln x$ is continuous on $(1, \infty)$.

51. The function $f(x) = \ln |x|$ is continuous for all $x \neq 0$.

52. a. Given that $2^x = e^{kx}$, find k.
b. Show that, in general, if b is a nonnegative real number, then any equation of the form $y = b^x$ may be written in the form $y = e^{kx}$, for some real number k.

53. Use the definition of a logarithm to prove:
a. $\log_b mn = \log_b m + \log_b n$
b. $\log_b \dfrac{m}{n} = \log_b m - \log_b n$

Hint: Let $\log_b m = p$ and $\log_b n = q$. Then, $b^p = m$ and $b^q = n$.

54. Use the definition of a logarithm to prove

$$\log_b m^n = n \log_b m$$

55. Use the definition of a logarithm to prove:
a. $\log_b 1 = 0$
b. $\log_b b = 1$

SOLUTIONS TO SELF-CHECK EXERCISES 5.2

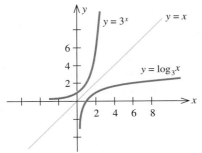

1. First, sketch the graph of $y = 3^x$ with the help of the following table of values:

x	−3	−2	−1	0	1	2	3
$y = 3^x$	1/27	1/9	1/3	0	3	9	27

Next, take the mirror reflection of this graph with respect to the line $y = x$ to obtain the graph of $y = \log_3 x$.

2.
$$3e^{x+1} - 2 = 4$$
$$3e^{x+1} = 6$$
$$e^{x+1} = 2$$
$$\ln e^{x+1} = \ln 2$$
$$(x + 1)\ln e = \ln 2 \quad \text{(Law 3)}$$
$$x + 1 = \ln 2 \quad \text{(Law 5)}$$
$$x = \ln 2 - 1$$
$$\approx -0.3069$$

5.3 Compound Interest

COMPOUND INTEREST

Compound interest is a natural application of the exponential function to the business world. (Albert Einstein called compound interest the greatest invention of mankind.) We begin by recalling that simple interest is interest that is computed only on the original principal. Thus, if I denotes the interest on a principal P (in dollars) at an interest rate of r per year for t years, then we have

$$I = Prt$$

The **accumulated amount** A, the sum of the principal and interest after t years, is given by

$$A = P + I = P + Prt$$
$$= P(1 + rt) \quad \text{(Simple interest formula)} \qquad \textbf{(4)}$$

Frequently, interest earned is periodically added to the principal and thereafter earns interest itself at the same rate. This is called **compound interest.** To find a formula for the accumulated amount, let's consider a numerical example. Suppose $1000 (the principal) is deposited in a bank for a **term** of 3 years, earning interest at the rate of 8% per year (called the **nominal,** or **stated, rate**) compounded annually. Then, using Formula (4) with $P = 1000$, $r = 0.08$, and $t = 1$, we see that the accumulated amount at the end of the first year is

$$A_1 = P(1 + rt)$$
$$= 1000[1 + 0.08(1)] = 1000(1.08) = 1080$$

or $1080.

To find the accumulated amount A_2 at the end of the second year, we use (4) once again, this time with $P = A_1$. (Remember, the principal *and* interest now earn interest over the second year.) We obtain

$$A_2 = P(1 + rt) = A_1(1 + rt)$$
$$= 1000[1 + 0.08(1)][1 + 0.08(1)]$$
$$= 1000[1 + 0.08]^2 = 1000(1.08)^2 \approx 1166.40$$

or approximately $1166.40.

Finally, the accumulated amount A_3 at the end of the third year is found using (4) with $P = A_2$, giving

$$A_3 = P(1 + rt) = A_2(1 + rt)$$
$$= 1000[1 + 0.08(1)]^2[1 + 0.08(1)]$$
$$= 1000[1 + 0.08]^3 = 1000(1.08)^3 \approx 1259.71$$

or approximately $1259.71.

If you reexamine our calculations in this example, you will see that the accumulated amounts at the end of each year have the following form:

First year:	$A_1 = 1000(1 + 0.08)$	or	$A_1 = P(1 + r)$
Second year:	$A_2 = 1000(1 + 0.08)^2$	or	$A_2 = P(1 + r)^2$
Third year:	$A_3 = 1000(1 + 0.08)^3$	or	$A_3 = P(1 + r)^3$

These observations suggest the following general result: If P dollars are invested over a term of t years earning interest at the rate of r per year compounded annually, then the accumulated amount is

$$A = P(1 + r)^t \qquad \textbf{(5)}$$

Formula (5) was derived under the assumption that interest was compounded *annually.* In practice, however, interest is usually compounded more than once a year. The interval of time between successive interest calculations is called the **conversion period.**

If interest at a nominal rate of r per year is compounded m times a year on a principal of P dollars, then the simple interest rate per conversion period is

$$i = \frac{r}{m} \qquad \frac{\text{(Annual interest rate)}}{\text{(Periods per year)}}$$

For example, if the nominal interest rate is 8% per year ($r = 0.08$) and interest is compounded quarterly ($m = 4$), then

$$i = \frac{r}{m} = \frac{0.08}{4} = 0.02$$

or 2% per period.

To find a general formula for the accumulated amount when a principal of P dollars is deposited in a bank for a term of t years and earns interest at the (nominal) rate of r per year compounded m times per year, we proceed as before using Formula (5) repeatedly with the interest rate $i = r/m$. We see that the accumulated amount at the end of each period is as follows:

First period: $A_1 = P(1 + i)$
Second period: $A_2 = A_1(1 + i) = [P(1 + i)](1 + i) = P(1 + i)^2$
Third period: $A_3 = A_2(1 + i) = [P(1 + i)^2](1 + i) = P(1 + i)^3$
$$\vdots \qquad\qquad\qquad \vdots$$
nth period: $A_n = A_{n-1}(1 + i) = [P(1 + i)^{n-1}](1 + i) = P(1 + i)^n$

But there are $n = mt$ periods in t years (number of conversion periods times the term). Therefore, the accumulated amount at the end of t years is given by

$$A = P(1 + i)^n$$

Compound Interest Formula

$$A = P(1 + i)^n \qquad\qquad (6)$$

where $i = r/m$, $n = mt$, and

$A =$ Accumulated amount at the end of n conversion periods
$P =$ Principal
$r =$ Nominal interest rate per year
$m =$ Number of conversion periods per year
$t =$ Term (number of years)

 EXAMPLE 1   Find the accumulated amount after 3 years if $1000 is invested at 8% per year compounded (a) annually, (b) semiannually, (c) quarterly, and (d) monthly.

SOLUTION ✔   **a.** Here, $P = 1000$, $r = 0.08$, and $m = 1$. Thus, $i = r = 0.08$ and $n = 3$, so Formula (6) gives

$$A = 1000(1.08)^3$$
$$= 1259.71$$

or $1259.71.

b. Here, $P = 1000$, $r = 0.08$, and $m = 2$. Thus, $i = 0.08/2 = 0.04$ and $n = (3)(2) = 6$, so that (6) gives

$$A = 1000(1.04)^6$$
$$= 1265.32$$

or $1265.32.

c. In this case, $P = 1000$, $r = 0.08$, and $m = 4$. Thus, $i = 0.08/4 = 0.02$ and $n = (3)(4) = 12$, so (6) gives

$$A = 1000(1.02)^{12}$$
$$= 1268.24$$

or $1268.24.

d. Here, $P = 1000$, $r = 0.08$, and $m = 12$. Thus, $i = 0.08/12 = 0.0067$ and $n = (3)(12) = 36$, so (6) gives

$$A = 1000(1.0067)^{36}$$
$$= 1271.75$$

or $1271.75. These results are summarized in Table 5.2.

Table 5.2				
Nominal Rate, r	Conversion Period	Interest Rate/ Conversion Period	Initial Investment	Accumulated Amount
8%	Annual ($m = 1$)	8%	$1000	$1259.71
8	Semiannual ($m = 2$)	4	1000	1265.32
8	Quarterly ($m = 4$)	2	1000	1268.24
8	Monthly ($m = 12$)	2/3	1000	1271.75

■ ■ ■ ■

EFFECTIVE RATE OF INTEREST

In the last example we saw that the interest actually earned on an investment depends on the frequency with which the interest is compounded. Thus, the stated, or nominal, rate of 8% per year does not reflect the actual rate at which interest is earned. This suggests that we need to find a common basis for comparing interest rates. One such way of comparing interest rates is provided by using the effective rate. The **effective rate** is the *simple* interest rate that would produce the same accumulated amount in 1 year as the nominal rate compounded m times a year. The effective rate is also called the **true rate.**

To derive a relation between the nominal interest rate, r per year compounded m times, and its corresponding effective rate, r_{eff} per year, let's assume an initial investment of P dollars. Then, the accumulated amount after 1 year at a simple interest rate of r_{eff} per year is

$$A = P(1 + r_{\text{eff}})$$

Also, the accumulated amount after 1 year at an interest rate of r per year compounded m times a year is

$$A = P(1 + i)^n = P\left(1 + \frac{r}{m}\right)^m \qquad \text{(Since } i = r/m)$$

Equating the two expressions gives

$$P(1 + r_{\text{eff}}) = P\left(1 + \frac{r}{m}\right)^m$$

$$1 + r_{\text{eff}} = \left(1 + \frac{r}{m}\right)^m \qquad \text{(Dividing both sides by } P)$$

or, upon solving for r_{eff}, we obtain the formula for computing the effective rate of interest:

Effective Rate of Interest Formula

$$r_{\text{eff}} = \left(1 + \frac{r}{m}\right)^m - 1 \qquad\qquad (7)$$

where

$r_{\text{eff}} = $ Effective rate of interest
$r = $ Nominal interest rate per year
$m = $ Number of conversion periods per year

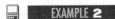 **EXAMPLE 2**

Find the effective rate of interest corresponding to a nominal rate of 8% per year compounded (a) annually, (b) semiannually, (c) quarterly, and (d) monthly.

SOLUTION ✔

a. The effective rate of interest corresponding to a nominal rate of 8% per year compounded annually is of course given by 8% per year. This result is also confirmed by using Formula (7) with $r = 0.08$ and $m = 1$. Thus,

$$r_{\text{eff}} = (1 + 0.08) - 1 = 0.08$$

b. Let $r = 0.08$ and $m = 2$. Then, (7) yields

$$r_{\text{eff}} = \left(1 + \frac{0.08}{2}\right)^2 - 1$$
$$= (1.04)^2 - 1$$
$$= 0.0816$$

so the required effective rate is 8.16% per year.

c. Let $r = 0.08$ and $m = 4$. Then, (7) yields

$$r_{\text{eff}} = \left(1 + \frac{0.08}{4}\right)^4 - 1$$
$$= (1.02)^4 - 1$$
$$= 0.08243$$

so the corresponding effective rate in this case is 8.243% per year.

d. Let $r = 0.08$ and $m = 12$. Then, (7) yields

$$r_{\text{eff}} = \left(1 + \frac{0.08}{12}\right)^{12} - 1$$
$$= (1.0067)^{12} - 1$$
$$= 0.08343$$

so the corresponding effective rate in this case is 8.343% per year.   ■ ■ ■ ■

Now, if the effective rate of interest r_{eff} is known, then the accumulated amount after t years on an investment of P dollars may be more readily computed by using the formula

$$A = P(1 + r_{\text{eff}})^t$$

The 1968 Truth in Lending Act passed by Congress requires that the effective rate of interest be disclosed in all contracts involving interest charges. The passage of this act has benefited consumers because they now have a common basis for comparing the various nominal rates quoted by different financial institutions. Furthermore, knowing the effective rate enables consumers to compute the actual charges involved in a transaction. Thus, if the effective rates of interest found in Example 2 were known, the accumulated values of Example 1, shown in Table 5.3, could have been readily found.

Table 5.3

Nominal Rate	Frequency of Interest Payment	Effective Rate	Initial Investment	Accumulated Amount After 3 Years	
8%	Annually	8%	$1000	$1000(1 + 0.08)^3$	= $1259.71
8	Semiannually	8.16	1000	$1000(1 + 0.0816)^3$	= 1265.32
8	Quarterly	8.243	1000	$1000(1 + 0.08243)^3$	= 1268.23
8	Monthly	8.343	1000	$1000(1 + 0.08343)^3$	= 1271.75

PRESENT VALUE

Let's return to the compound interest Formula (6), which expresses the accumulated amount at the end of n periods when interest at the rate of r is compounded m times a year. The principal P in (6) is often referred to as the **present value,** and the accumulated value A is called the **future value** since it is realized at a future date. In certain instances an investor may wish to determine how much money he should invest now, at a fixed rate of interest, so that he will realize a certain sum at some future date. This problem may be solved by expressing P in terms of A. Thus, from (6) we find

$$P = A(1 + i)^{-n}$$

Here, as before, $i = r/m$, where m is the number of conversion periods per year.

Present Value Formula for Compound Interest

$$P = A(1 + i)^{-n} \qquad\qquad (8)$$

 **EXAMPLE 3**   Find how much money should be deposited in a bank paying interest at the rate of 6% per year compounded monthly so that at the end of 3 years the accumulated amount will be $20,000.

 SOLUTION ✔   Here, $r = 0.06$ and $m = 12$, so $i = 0.06/12 = 0.005$ and $n = (3)(12) = 36$. Thus, the problem is to determine P given that $A = 20,000$. Using Formula (8), we obtain

$$P = 20{,}000(1.005)^{-36}$$
$$\approx 16{,}713$$

or $16,713.   ■ ■ ■ ■

 EXAMPLE 4   Find the present value of $49,158.60 due in 5 years at an interest rate of 10% per year compounded quarterly.

 SOLUTION ✔   Using Formula (8) with $r = 0.1$ and $m = 4$, so that $i = 0.1/4 = 0.025$, $n = (4)(5) = 20$, and $A = 49,158.6$, we obtain

$$P = (49{,}158.6)(1.025)^{-20} \approx 30{,}000$$

or $30,000.   ■ ■ ■ ■

CONTINUOUS COMPOUNDING OF INTEREST

One question that arises naturally in the study of compound interest is: What happens to the accumulated amount over a fixed period of time if the interest is computed more and more frequently?

Intuition suggests that the more often interest is compounded, the larger the accumulated amount will be. This is confirmed by the results of Example 1, where we found that the accumulated amounts did in fact increase when we increased the number of conversion periods per year.

This leads us to another question: Does the accumulated amount approach a limit when the interest is computed more and more frequently over a fixed period of time?

To answer this question, let's look again at the compound interest formula:

$$A = P\left(1 + \frac{r}{m}\right)^{mt} \tag{9}$$

Recall that m is the number of conversion periods per year. So to find an answer to our problem, we should let m approach infinity (get larger and larger) in (9). But first we will rewrite this equation in the form

$$A = P\left[\left(1 + \frac{r}{m}\right)^m\right]^t \qquad \text{[Since } b^{xy} = (b^x)^y]$$

Now, letting $m \to \infty$, we find that

$$\lim_{m \to \infty}\left[P\left(1 + \frac{r}{m}\right)^m\right]^t = P\left[\lim_{m \to \infty}\left(1 + \frac{r}{m}\right)^m\right]^t \qquad \text{(Why?)}$$

Next, upon making the substitution $u = m/r$ and observing that $u \to \infty$ as $m \to \infty$, the foregoing expression reduces to

$$P\left[\lim_{u \to \infty}\left(1 + \frac{1}{u}\right)^{ur}\right]^t = P\left[\lim_{u \to \infty}\left(1 + \frac{1}{u}\right)^u\right]^{rt}$$

But

$$\lim_{u \to \infty}\left(1 + \frac{1}{u}\right)^u = e \qquad \text{[Using (1)]}$$

so

$$\lim_{m \to \infty} P\left[\left(1 + \frac{r}{m}\right)^m\right]^t = Pe^{rt}$$

Our computations tell us that as the frequency with which interest is compounded increases without bound, the accumulated amount approaches Pe^{rt}. In this situation, we say that interest is *compounded continuously*. Let's summarize this important result.

Continuous Compound Interest Formula

$$A = Pe^{rt} \tag{10}$$

where

P = Principal
r = Annual interest rate compounded continuously
t = Time in years
A = Accumulated amount at the end of t years

 **EXAMPLE 5**   Find the accumulated amount after 3 years if $1000 is invested at 8% per year compounded (a) daily (take the number of days in a year to be 365) and (b) continuously.

SOLUTION ✔   **a.** Using Formula (6) with $P = 1000$, $r = 0.08$, $m = 365$, and $n = (365)(3) = 1095$, we find

$$A = 1000 \left(1 + \frac{0.08}{365}\right)^{1095} \approx 1271.22$$

or $1271.22.

b. Here we use Formula (10) with $P = 1000$, $r = 0.08$, and $t = 3$, obtaining

$$A = 1000e^{(0.08)(3)}$$
$$\approx 1271.25$$

or $1271.25.   ■ ■ ■ ■

Exploring with Technology

In the opening paragraph of Section 5.1, we pointed out that the accumulated amount of an account earning interest *compounded continuously* will eventually outgrow by far the accumulated amount of an account earning interest at the same nominal rate but earning simple interest. Illustrate this fact using the following example.

Suppose you deposit $1000 in account I, earning interest at the rate of 10% per year compounded continuously so that the accumulated amount at the end of t years is $A_1(t) = 1000e^{0.1t}$. Suppose you also deposit $1000 in account II, earning simple interest at the rate of 10% per year so that the accumulated amount at the end of t years is $A_2(t) = 1000(1 + 0.1t)$. Use a graphing utility to sketch the graphs of the functions A_1 and A_2 in the viewing rectangle $[0, 20] \times [0, 10{,}000]$ to see the accumulated amounts $A_1(t)$ and $A_2(t)$ over a 20-year period.

Observe that the accumulated amounts corresponding to interest compounded daily and interest compounded continuously differ by very little. The continuous compound interest formula is a very important tool in theoretical work in financial analysis.

If we solve Formula (10) for P, we obtain

$$P = Ae^{-rt} \tag{11}$$

which gives the present value in terms of the future (accumulated) value for the case of continuous compounding.

 EXAMPLE **6**   The Blakely Investment Company owns an office building located in the commercial district of a city. As a result of the continued success of an urban renewal program, local business is enjoying a miniboom. The market value of Blakely's property is

$$V(t) = 300{,}000e^{\sqrt{t}/2}$$

where $V(t)$ is measured in dollars and t is the time in years from the present. If the expected rate of inflation is 9% compounded continuously for the next 10 years, find an expression for the present value $P(t)$ of the market price of the property valid for the next 10 years. Compute $P(7)$, $P(8)$, and $P(9)$, and interpret your results.

SOLUTION ✔   Using Formula (11) with $A = V(t)$ and $r = 0.09$, we find that the present value of the market price of the property t years from now is

$$P(t) = V(t)e^{-0.09t}$$
$$= 300{,}000e^{-0.09t + \sqrt{t}/2} \qquad (0 \le t \le 10)$$

Letting $t = 7$, 8, and 9, respectively, we find that

$$P(7) = 300{,}000e^{-0.09(7) + \sqrt{7}/2} \approx 599{,}837, \text{ or } \$599{,}837$$
$$P(8) = 300{,}000e^{-0.09(8) + \sqrt{8}/2} \approx 600{,}640, \text{ or } \$600{,}640$$
$$P(9) = 300{,}000e^{-0.09(9) + \sqrt{9}/2} \approx 598{,}115, \text{ or } \$598{,}115$$

From the results of these computations, we see that the present value of the property's market price seems to decrease after a certain period of growth. This suggests that there is an optimal time for the owners to sell. Later we will show that the highest present value of the property's market price is $600{,}779$, which occurs at time $t = 7.72$ years.   ▩ ▩ ▩ ▩

Exploring with Technology

The effective rate of interest is given by

$$r_{\text{eff}} = \left(1 + \frac{r}{m}\right)^m - 1$$

where the number of conversion periods per year is m. In Exercise 27 you will be asked to show that the effective rate of interest r_{eff} corresponding to a nominal interest rate r per year compounded continuously is given by

$$\hat{r}_{\text{eff}} = e^r - 1$$

To obtain a visual confirmation of this result, consider the special case where $r = 0.1$ (10% per year).
1. Use a graphing utility to plot the graph of both

$$y_1 = \left(1 + \frac{0.1}{x}\right)^x - 1 \qquad \text{and} \qquad y_2 = e^{0.1} - 1$$

in the viewing rectangle $[0, 3] \times [0, 0.12]$.
2. Does your result seem to imply that

$$\left(1 + \frac{r}{m}\right)^m - 1$$

approaches

$$\hat{r}_{\text{eff}} = e^r - 1$$

as m increases without bound for the special case $r = 0.1$?

SELF-CHECK EXERCISES 5.3

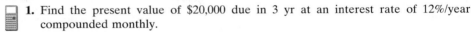 **1.** Find the present value of $20,000 due in 3 yr at an interest rate of 12%/year compounded monthly.

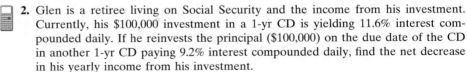 **2.** Glen is a retiree living on Social Security and the income from his investment. Currently, his $100,000 investment in a 1-yr CD is yielding 11.6% interest compounded daily. If he reinvests the principal ($100,000) on the due date of the CD in another 1-yr CD paying 9.2% interest compounded daily, find the net decrease in his yearly income from his investment.

3. a. What is the accumulated amount after 5 yr if $10,000 is invested at 10%/year compounded continuously?
b. Find the present value of $10,000 due in 5 yr at an interest rate of 10%/year compounded continuously.

Solutions to Self-Check Exercises 5.3 can be found on page 404.

5.3 Exercises

 **A calculator is recommended for these exercises.**

In Exercises 1–4, find the accumulated amount A if the principal P is invested at an interest rate of r per year for t years.

1. $P = \$2500$, $r = 7\%$, $t = 10$, compounded semiannually

2. $P = \$12,000$, $r = 8\%$, $t = 10$, compounded quarterly

3. $P = \$150,000$, $r = 10\%$, $t = 4$, compounded monthly

4. $P = \$150,000$, $r = 9\%$, $t = 3$, compounded daily

In Exercises 5 and 6, find the effective rate corresponding to the given nominal rate.

5. a. 10%/year compounded semiannually
b. 9%/year compounded quarterly

6. a. 8%/year compounded monthly
b. 8%/year compounded daily

In Exercises 7 and 8, find the present value of $40,000 due in 4 years at the given rate of interest.

7. a. 8%/year compounded semiannually
b. 8%/year compounded quarterly

8. a. 7%/year compounded monthly
b. 9%/year compounded daily

9. Find the accumulated amount after 4 yr if $5000 is invested at 8%/year compounded continuously.

10. An amount of $25,000 is deposited in a bank that pays interest at the rate of 7%/year, compounded annually. What is the total amount on deposit at the end of 6 yr, assuming there are no deposits or withdrawals during those 6 yr? What is the interest earned in that period of time?

11. HOUSING PRICES The Estradas are planning to buy a house 4 yr from now. Housing experts in their area have estimated that the cost of a home will increase at a rate of 9%/year during that 4-yr period. If this economic prediction holds true, how much can they expect to pay for a house that currently costs $80,000?

12. ENERGY CONSUMPTION A metropolitan utility company in a western city of the United States expects the consumption of electricity to increase by 8%/year during the next decade, due mainly to the expected population increase. If consumption does increase at this rate, find the amount by which the utility company will have to increase its generating capacity in order to meet the area's needs at the end of the decade.

13. PENSION FUNDS The managers of a pension fund have invested $1.5 million in U.S. government certificates of deposit (CDs) that pay interest at the rate of 9.5%/year compounded semiannually over a period of 10 yr. At the end of this period, how much will the investment be worth?

14. **SAVINGS ACCOUNTS** Bernie invested a sum of money 5 yr ago in a savings account, which has since paid interest at the rate of 8%/yr compounded quarterly. His investment is now worth $22,289.22. How much did he originally invest?

15. **LOAN CONSOLIDATION** The proprietors of the Coachmen Inn secured two loans from the Union Bank: one for $8000 due in 3 yr and one for $15,000 due in 6 yr, both at an interest rate of 10%/yr compounded semiannually. The bank agreed to allow the two loans to be consolidated into one loan payable in 5 yr at the same interest rate. How much will the proprietors have to pay the bank at the end of 5 yr?

16. **TAX-DEFERRED ANNUITIES** Kate is in the 28% tax bracket and has $25,000 available for investment during her current tax year. Assume that she remains in the same tax bracket over the next 10 yr and determine the accumulated amount of her investment if she puts the $25,000 into a:
    a. Tax-deferred annuity that pays 12%/year, tax deferred for 10 yr.
    b. Taxable instrument that pays 12%/year for 10 yr.
    Hint: In this case the yield after taxes is 8.64%/year.

17. **CONSUMER PRICE INDEX** At an annual inflation rate of 7.5%, how long will it take the Consumer Price Index (CPI) to double?

18. **INVESTMENT RETURNS** Zoe purchased a house in 1993 for $80,000. In 1999 she sold the house and made a net profit of $28,000. Find the effective annual rate of return on her investment over the 6-yr period.

19. **INVESTMENT RETURNS** Julio purchased 1000 shares of a certain stock for $25,250 (including commissions). He sold the shares 2 yr later and received $32,100 after deducting commissions. Find the effective annual rate of return on his investment over the 2-yr period.

20. **INVESTMENT OPTIONS** Investment A offers a 10% return compounded semiannually, and investment B offers a 9.75% return compounded continuously. Which investment has a higher rate of return over a 4-yr period?

21. **PRESENT VALUE** Find the present value of $59,673 due in 5 yr at an interest rate of 8%/year compounded continuously.

22. **REAL ESTATE INVESTMENTS** A condominium complex was purchased by a group of private investors for $1.4 million and sold 6 yr later for $3.6 million. Find the annual rate of return (compounded continuously) on their investment.

23. **SAVING FOR COLLEGE** Having received a large inheritance, a child's parents wish to establish a trust for the child's college education. If 7 yr from now they need an estimated $70,000, how much should they set aside in trust now, if they invest the money at 10.5% compounded (a) quarterly? (b) Continuously?

24. **EFFECT OF INFLATION ON SALARIES** Omar's current annual salary is $35,000. How much will he need to earn 10 yr from now in order to retain his present purchasing power if the rate of inflation over that period is 6%/year? Assume that inflation is continuously compounded.

25. **PENSIONS** Eleni, who is now 50 years old, is employed by a firm that guarantees her a pension of $40,000/year at age 65. What is the present value of her first year's pension if inflation over the next 15 yr is (a) 6%? (b) 8%? (c) 12%? Assume that inflation is continuously compounded.

26. **REAL ESTATE INVESTMENTS** An investor purchased a piece of waterfront property. Because of the development of a marina in the vicinity, the market value of the property is expected to increase according to the rule

$$V(t) = 80,000e^{\sqrt{t}/2}$$

where $V(t)$ is measured in dollars and t is the time in years from the present. If the rate of inflation is expected to be 9% compounded continuously for the next 8 yr, find an expression for the present value $P(t)$ of the property's market price valid for the next 8 yr. What is $P(t)$ expected to be in 4 yr?

27. Show that the effective rate of interest $\hat{r}_{\text{eff}}$ that corresponds to a nominal interest rate r per year compounded continuously is given by

$$\hat{r}_{\text{eff}} = e^r - 1$$

Hint: From Formula (7) we see that the effective rate $\hat{r}_{\text{eff}}$ corresponding to a nominal interest rate r per year compounded m times a year is given by

$$\hat{r}_{\text{eff}} = \left(1 + \frac{r}{m}\right)^m - 1$$

Let m tend to infinity in this expression.

28. Refer to Exercise 27. Find the effective rate of interest that corresponds to a nominal rate of 10%/year compounded (a) quarterly, (b) monthly, and (c) continuously.

29. **INVESTMENT ANALYSIS** Refer to Exercise 27. Bank A pays interest on deposits at a 7% annual rate compounded quarterly, and Bank B pays interest on deposits at a $7\frac{1}{8}$% annual rate compounded continuously. Which bank has the higher effective rate of interest?

30. **INVESTMENT ANALYSIS** Find the nominal rate of interest that, when compounded monthly, yields an effective rate of interest of 10%/year.
    Hint: Use Equation (7).

31. **INVESTMENT ANALYSIS** Find the nominal rate of interest that, when compounded continuously, yields an effective rate of interest of 10%/year.
    Hint: See Exercise 27.

Portfolio

MISATO NAKAZAKI

TITLE: Assistant Vice President
INSTITUTION: A large investment corporation

In the securities industry, buying and selling stocks and bonds has always required a mastery of concepts and formulas that outsiders find confusing. As a bond seller, Misato Nakazaki routinely uses terms such as *issue, maturity, current yield, callable* and *convertible bonds,* and so on.

These terms, however, are easily defined. When corporations issue bonds, they are borrowing money at a fixed rate of interest. The bonds are scheduled to mature—to be paid back—on a specific date as much as 30 years into the future. Callable bonds allow the issuer to pay off the loans prior to their expected maturity, reducing overall interest payments. In its simplest terms, current yield is the price of a bond multiplied by the interest rate at which the bond is issued. For example, a bond with a face value of $1000 and an interest rate of 10% yields $100 per year in interest payments. When that same bond is resold at a premium on the secondary market for $1200, its current yield nets only an 8.3% rate of return based on the higher purchase price.

Bonds attract investors for many reasons. A key variable is the sensitivity of the bond's price to future changes in interest rates. If investors get locked into a low-paying bond when future bonds pay higher yields, they lose money. Nakazaki stresses that "no one knows for sure what rates will be over time." Employing differentials allows her to calculate interest-rate sensitivity for clients as they ponder purchase decisions.

Computerized formulas, "whose basis is calculus," says Nakazaki, help her factor the endless stream of numbers flowing across her desk.

On a typical day, Nakazaki might be given a bid on "10 million, GMAC, 8.5%, January 2005." Translation: Her customer wants her to buy General Motors Acceptance Corporation bonds with a face value of $10 million and an interest rate of 8.5%, maturing in January 2005.

After she calls her firm's trader to find out the yield on the bond in question, Nakazaki enters the price and other variables, such as the interest rate and date of maturity, and the computer prints out the answers. Nakazaki can then relay to her client the bond's current yield, accrued interest, and so on. In Nakazaki's rapid-fire work environment, such speed is essential. Nakazaki cautions that "computer users have to understand what's behind the formulas." The software "relies on the basics of calculus. If people don't understand the formula, it's useless for them to use the calculations."

With an MBA from New York University, Nakazaki typifies the younger generation of Japanese women who have chosen to succeed in the business world. Since earning her degree, she has sold bonds for a global securities firm in New York City.

Nakazaki's client list reads like a *who's who* of the leading Japanese banks, insurance companies, mutual funds, and corporations. As institutional buyers, her clients purchase large blocks of American corporate bonds and mortgage-backed securities such as Ginnie Maes.

SOLUTIONS TO SELF-CHECK EXERCISES 5.3

1. Using Formula (8) with $r = 0.12$ and $m = 12$, so that

$$i = \frac{0.12}{12} = 0.01, \qquad n = (12)(3) = 36, \qquad A = 20,000$$

we find the required present value to be

$$P = 20,000(1.01)^{-36}$$
$$= 13,978.50$$

or $13,978.50.

2. The accumulated amount of Glen's current investment is found by using Formula (6) with $P = 100,000$, $r = 0.116$, and $m = 360$. Thus,

$$i = \frac{0.116}{360} = 0.0003222 \qquad \text{and} \qquad n = 360$$

so the required accumulated amount is

$$A = 100,000(1.0003222)^{360}$$
$$= 112,296.59$$

or $112,296.59. Next, we compute the accumulated amount of Glen's reinvestment. Once again, using (6) with $P = 100,000$, $r = 0.092$, and $m = 360$ so that

$$i = \frac{0.092}{360} = 0.0002556 \qquad \text{and} \qquad n = 360$$

we find the required accumulated amount in this case to be

$$\overline{A} = 100,000(1.0002556)^{360}$$

or $109,636.95. Therefore, Glen can expect to experience a net decrease in yearly income of

$$112,296.59 - 109,636.95$$

or $2,659.64.

3. a. Using Formula (10) with $P = 10,000$, $r = 0.1$, and $t = 5$, we find that the required accumulated amount is given by

$$A = 10,000e^{(0.1)(5)}$$
$$= 16,487.21$$

or $16,487.21.

b. Using Formula (11) with $A = 10,000$, $r = 0.1$, and $t = 5$, we see that the required present value is given by

$$P = 10,000e^{-(0.1)(5)}$$
$$= 6065.31$$

or $6065.31.

5.4 Differentiation of Exponential Functions

THE DERIVATIVE OF THE EXPONENTIAL FUNCTION

To study the effects of budget deficit-reduction plans at different income levels, it is important to know the income distribution of American families. Based on data from the House Budget Committee, the House Ways and Means Committee, and the U.S. Census Bureau, the graph of f shown in Figure 5.10 gives the number of American families y (in millions) as a function of their annual income x (in thousands of dollars) in 1990.

FIGURE 5.10

The graph of f shows the number of families versus their annual income.

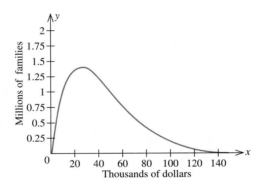

Source: House Budget Committee, House Ways and Means Committee, and U.S. Census Bureau

Observe that the graph of f rises very quickly and then tapers off. From the graph of f, you can see that the bulk of American families earned less than \$100,000 per year. In fact, 95% of U.S. families earned less than \$102,358 per year in 1990. (We will refer to this model again in Using Technology at the end of this section.)

To analyze mathematical models involving exponential and logarithmic functions in greater detail, we need to develop rules for computing the derivative of these functions. We begin by looking at the rule for computing the derivative of the exponential function.

Rule 1: Derivative of the Exponential Function

$$\frac{d}{dx}e^x = e^x$$

Thus, the derivative of the exponential function with base e is equal to the function itself. To demonstrate the validity of this rule, we compute

$$f'(x) = \lim_{h \to 0} \frac{f(x+h) - f(x)}{h}$$

$$= \lim_{h \to 0} \frac{e^{x+h} - e^x}{h}$$

$$= \lim_{h \to 0} \frac{e^x(e^h - 1)}{h} \qquad \text{(Writing } e^{x+h} = e^x e^h \text{ and factoring)}$$

$$= e^x \lim_{h \to 0} \frac{e^h - 1}{h} \qquad \text{(Why?)}$$

To evaluate

$$\lim_{h \to 0} \frac{e^h - 1}{h}$$

let's refer to Table 5.4, which is constructed with the aid of a calculator. From the table, we see that

$$\lim_{h \to 0} \frac{e^h - 1}{h} = 1$$

(Although a rigorous proof of this fact is possible, it is beyond the scope of this book. Also see Example 1, Using Technology, page 414.) Using this result, we conclude that

$$f'(x) = e^x \cdot 1 = e^x$$

as we set out to show.

Table 5.4

h	0.1	0.01	0.001	-0.1	-0.01	-0.001
$\dfrac{e^h - 1}{h}$	1.0517	1.0050	1.0005	0.9516	0.9950	0.9995

EXAMPLE 1   Compute the derivative of each of the following functions:

a. $f(x) = x^2 e^x$ **b.** $g(t) = (e^t + 2)^{3/2}$

SOLUTION ✔   **a.** The product rule gives

$$f'(x) = \frac{d}{dx}(x^2 e^x)$$

$$= x^2 \frac{d}{dx}(e^x) + e^x \frac{d}{dx}(x^2)$$

$$= x^2 e^x + e^x(2x)$$

$$= xe^x(x + 2)$$

To see this, observe that $h(x) = g[f(x)]$, where $g(x) = \ln x$ $(x > 0)$. Since $g'(x) = 1/x$, we have, using the chain rule,

$$h'(x) = g'(f(x))f'(x)$$

$$= \frac{1}{f(x)}f'(x) = \frac{f'(x)}{f(x)}$$

Observe that in the special case $f(x) = x$, $h(x) = \ln x$, so the derivative of h is, by Rule 3, given by $h'(x) = 1/x$.

EXAMPLE 2   Find the derivative of the function $f(x) = \ln(x^2 + 1)$.

SOLUTION ✔   Using Rule 4, we see immediately that

$$f'(x) = \frac{\frac{d}{dx}(x^2 + 1)}{x^2 + 1} = \frac{2x}{x^2 + 1}$$

■■■■

When differentiating functions involving logarithms, the rules of logarithms may be used to advantage, as shown in Examples 3 and 4.

EXAMPLE 3   Differentiate the function $y = \ln[(x^2 + 1)(x^3 + 2)^6]$.

SOLUTION ✔   We first rewrite the given function using the properties of logarithms:

$$y = \ln[(x^2 + 1)(x^3 + 2)^6]$$
$$= \ln(x^2 + 1) + \ln(x^3 + 2)^6 \qquad (\ln mn = \ln m + \ln n)$$
$$= \ln(x^2 + 1) + 6\ln(x^3 + 2) \qquad (\ln m^n = n \ln m)$$

Differentiating and using Rule 4, we obtain

$$y' = \frac{\frac{d}{dx}(x^2 + 1)}{x^2 + 1} + \frac{6\frac{d}{dx}(x^3 + 2)}{x^3 + 2}$$

$$= \frac{2x}{x^2 + 1} + \frac{6(3x^2)}{x^3 + 2} = \frac{2x}{x^2 + 1} + \frac{18x^2}{x^3 + 2}$$

■■■■

Exploring with Technology

Use a graphing utility to plot the graphs of $f(x) = \ln x$; its first derivative function, $f'(x) = 1/x$; and its second derivative function, $f''(x) = -1/x^2$, using the same viewing rectangle $[0, 4] \times [-3, 3]$.

1. Describe the properties of the graph of f revealed by studying the graph of $f'(x)$. What can you say about the rate of increase of f for large values of x?
2. Describe the properties of the graph of f revealed by studying the graph of $f''(x)$. What can you say about the concavity of f for large values of x?

EXAMPLE 4   Find the derivative of the function $g(t) = \ln(t^2 e^{-t^2})$.

SOLUTION ✔   Here again, to save a lot of work, we first simplify the given expression using the properties of logarithms. We have

$$g(t) = \ln(t^2 e^{-t^2})$$
$$= \ln t^2 + \ln e^{-t^2} \qquad (\ln mn = \ln m + \ln n)$$
$$= 2 \ln t - t^2 \qquad (\ln m^n = n \ln m \text{ and } \ln e = 1)$$

Therefore,

$$g'(t) = \frac{2}{t} - 2t = \frac{2(1 - t^2)}{t}$$

■ ■ ■ ■

LOGARITHMIC DIFFERENTIATION

As we saw in the last two examples, the task of finding the derivative of a given function can be made easier by first applying the laws of logarithms to simplify the function. We now illustrate a process called **logarithmic differentiation,** which not only simplifies the calculation of the derivatives of certain functions but also enables us to compute the derivatives of functions we could not otherwise differentiate using the techniques developed thus far.

EXAMPLE 5   Differentiate $y = x(x + 1)(x^2 + 1)$, using logarithmic differentiation.

SOLUTION ✔   First, we take the natural logarithm on both sides of the given equation, obtaining

$$\ln y = \ln x(x + 1)(x^2 + 1)$$

Next, we use the properties of logarithms to rewrite the right-hand side of this equation, obtaining

$$\ln y = \ln x + \ln(x + 1) + \ln(x^2 + 1)$$

If we differentiate both sides of this equation, we have

$$\frac{d}{dx} \ln y = \frac{d}{dx} [\ln x + \ln(x + 1) + \ln(x^2 + 1)]$$
$$= \frac{1}{x} + \frac{1}{x + 1} + \frac{2x}{x^2 + 1} \qquad \text{(Using Rule 4)}$$

To evaluate the expression on the left-hand side, note that y is a function of x. Therefore, writing $y = f(x)$ to remind us of this fact, we have

$$\frac{d}{dx}\ln y = \frac{d}{dx}\ln[f(x)] \qquad \text{[Writing } y = f(x)]$$

$$= \frac{f'(x)}{f(x)} \qquad \text{(Using Rule 4)}$$

$$= \frac{y'}{y} \qquad \text{[Returning to using } y \text{ instead of } f(x)]$$

Therefore, we have

$$\frac{y'}{y} = \frac{1}{x} + \frac{1}{x+1} + \frac{2x}{x^2+1}$$

Finally, solving for y', we have

$$y' = y\left(\frac{1}{x} + \frac{1}{x+1} + \frac{2x}{x^2+1}\right)$$

$$= x(x+1)(x^2+1)\left(\frac{1}{x} + \frac{1}{x+1} + \frac{2x}{x^2+1}\right) \qquad \blacksquare\blacksquare\blacksquare\blacksquare$$

Before considering other examples, let's summarize the important steps involved in logarithmic differentiation.

Finding $\dfrac{dy}{dx}$ by Logarithmic Differentiation

1. Take the natural logarithm on both sides of the equation and use the properties of logarithms to write any "complicated expression" as a sum of simpler terms.

2. Differentiate both sides of the equation with respect to x.

3. Solve the resulting equation for $\dfrac{dy}{dx}$.

EXAMPLE 6   Differentiate $y = x^2(x-1)(x^2+4)^3$.

SOLUTION ✔   Taking the natural logarithm on both sides of the given equation and using the laws of logarithms, we obtain

$$\ln y = \ln x^2(x-1)(x^2+4)^3$$
$$= \ln x^2 + \ln(x-1) + \ln(x^2+4)^3$$
$$= 2\ln x + \ln(x-1) + 3\ln(x^2+4)$$

Differentiating both sides of the equation with respect to x, we have

$$\frac{d}{dx}\ln y = \frac{y'}{y} = \frac{2}{x} + \frac{1}{x-1} + 3\cdot\frac{2x}{x^2+4}$$

Finally, solving for y', we have

$$y' = y\left(\frac{2}{x} + \frac{1}{x-1} + \frac{6x}{x^2+4}\right)$$

$$= x^2(x-1)(x^2+4)^3\left(\frac{2}{x} + \frac{1}{x-1} + \frac{6x}{x^2+4}\right)$$ ■■■■

EXAMPLE 7   Find the derivative of $f(x) = x^x (x > 0)$.

SOLUTION ✔   A word of caution! This function is neither a power function nor an exponential function. Taking the natural logarithm on both sides of the equation gives

$$\ln f(x) = \ln x^x = x \ln x$$

Differentiating both sides of the equation with respect to x, we obtain

$$\frac{f'(x)}{f(x)} = x\frac{d}{dx}\ln x + (\ln x)\frac{d}{dx}x$$

$$= x\left(\frac{1}{x}\right) + \ln x$$

$$= 1 + \ln x$$

Therefore,

$$f'(x) = f(x)(1 + \ln x) = x^x(1 + \ln x)$$ ■■■■

Exploring with Technology

Refer to Example 7.

1. Use a graphing utility to plot the graph of $f(x) = x^x$, using the viewing rectangle $[0, 2] \times [0, 2]$. Then use **ZOOM** and **TRACE** to show that

$$\lim_{x \to 0^+} f(x) = 1$$

2. Use the results of part 1 and Example 7 to show that $\lim_{x \to 0^+} f'(x) = -\infty$. Justify your answer.

SELF-CHECK EXERCISES 5.5

1. Find an equation of the tangent line to the graph of $f(x) = x \ln(2x + 3)$ at the point $(-1, 0)$.

2. Use logarithmic differentiation to compute y', given $y = (2x + 1)^3(3x + 4)^5$.

Solutions to Self-Check Exercises 5.5 can be found on page 424.

5.5 Exercises

In Exercises 1–32, find the derivative of the function.

1. $f(x) = 5 \ln x$

2. $f(x) = \ln 5x$

3. $f(x) = \ln(x + 1)$

4. $g(x) = \ln(2x + 1)$

5. $f(x) = \ln x^8$

6. $h(t) = 2 \ln t^5$

7. $f(x) = \ln \sqrt{x}$

8. $f(x) = \ln(\sqrt{x} + 1)$

9. $f(x) = \ln \dfrac{1}{x^2}$

10. $f(x) = \ln \dfrac{1}{2x^3}$

11. $f(x) = \ln(4x^2 - 6x + 3)$

12. $f(x) = \ln(3x^2 - 2x + 1)$

13. $f(x) = \ln \dfrac{2x}{x + 1}$

14. $f(x) = \ln \dfrac{x + 1}{x - 1}$

15. $f(x) = x^2 \ln x$

16. $f(x) = 3x^2 \ln 2x$

17. $f(x) = \dfrac{2 \ln x}{x}$

18. $f(x) = \dfrac{3 \ln x}{x^2}$

19. $f(u) = \ln(u - 2)^3$

20. $f(x) = \ln(x^3 - 3)^4$

21. $f(x) = \sqrt{\ln x}$

22. $f(x) = \sqrt{\ln x + x}$

23. $f(x) = (\ln x)^3$

24. $f(x) = 2(\ln x)^{3/2}$

25. $f(x) = \ln(x^3 + 1)$

26. $f(x) = \ln\sqrt{x^2 - 4}$

27. $f(x) = e^x \ln x$

28. $f(x) = e^x \ln\sqrt{x + 3}$

29. $f(t) = e^{2t} \ln(t + 1)$

30. $g(t) = t^2 \ln(e^{2t} + 1)$

31. $f(x) = \dfrac{\ln x}{x}$

32. $g(t) = \dfrac{t}{\ln t}$

In Exercises 33–36, find the second derivative of the function.

33. $f(x) = \ln 2x$

34. $f(x) = \ln(x + 5)$

35. $f(x) = \ln(x^2 + 2)$

36. $f(x) = (\ln x)^2$

In Exercises 37–46, use logarithmic differentiation to find the derivative of the function.

37. $y = (x + 1)^2(x + 2)^3$

38. $y = (3x + 2)^4(5x - 1)^2$

39. $y = (x - 1)^2(x + 1)^3(x + 3)^4$

40. $y = \sqrt{3x + 5}(2x - 3)^4$

41. $y = \dfrac{(2x^2 - 1)^5}{\sqrt{x + 1}}$

42. $y = \dfrac{\sqrt{4 + 3x^2}}{\sqrt[3]{x^2 + 1}}$

43. $y = 3^x$

44. $y = x^{x+2}$

45. $y = (x^2 + 1)^x$

46. $y = x^{\ln x}$

47. Find an equation of the tangent line to the graph of $y = x \ln x$ at the point $(1, 0)$.

48. Find an equation of the tangent line to the graph of $y = \ln x^2$ at the point $(2, \ln 4)$.

49. Determine the intervals where the function $f(x) = \ln x^2$ is increasing and where it is decreasing.

50. Determine the intervals where the function $f(x) = \dfrac{\ln x}{x}$ is increasing and where it is decreasing.

51. Determine the intervals of concavity for the function $f(x) = x^2 + \ln x^2$.

52. Determine the intervals of concavity for the function $f(x) = \dfrac{\ln x}{x}$.

53. Find the inflection points of the function $f(x) = \ln(x^2 + 1)$.

54. Find the inflection points of the function $f(x) = x^2 \ln x$.

55. Find the absolute extrema of the function $f(x) = x - \ln x$ on $[\frac{1}{2}, 3]$.

56. Find the absolute extrema of the function $g(x) = \dfrac{x}{\ln x}$ on $[2, \infty)$.

In Exercises 57 and 58, use the guidelines on page 327 to sketch the graph of the given function.

57. $f(x) = \ln(x - 1)$

58. $f(x) = 2x - \ln x$

In Exercises 59 and 60, determine whether the statement is true or false. If it is true, explain why it is true. If it is false, give an example to show why it is false.

59. If $f(x) = \ln 5$, then $f'(x) = 1/5$.

60. If $f(x) = \ln a^x$, then $f'(x) = \ln a$.

61. Prove that $\dfrac{d}{dx} \ln |x| = \dfrac{1}{x}$ $(x \neq 0)$ for the case $x < 0$.

62. Use the definition of the derivative to show that

$$\lim_{x \to 0} \frac{\ln(x + 1)}{x} = 1$$

SOLUTIONS TO SELF-CHECK EXERCISES 5.5

1. The slope of the tangent line to the graph of f at any point $(x, f(x))$ lying on the graph of f is given by $f'(x)$. Using the product rule, we find

$$f'(x) = \frac{d}{dx}[x \ln(2x + 3)]$$

$$= x \frac{d}{dx} \ln(2x + 3) + \ln(2x + 3) \cdot \frac{d}{dx}(x)$$

$$= x \left(\frac{2}{2x + 3}\right) + \ln(2x + 3) \cdot 1$$

$$= \frac{2x}{2x + 3} + \ln(2x + 3)$$

In particular, the slope of the tangent line to the graph of f at the point $(-1, 0)$ is

$$f'(-1) = \frac{-2}{-2 + 3} + \ln 1 = -2$$

Therefore, using the point-slope form of the equation of a line, we see that a required equation is

$$y - 0 = -2(x + 1)$$
$$y = -2x - 2$$

2. Taking the logarithm on both sides of the equation gives

$$\ln y = \ln(2x + 1)^3(3x + 4)^5$$
$$= \ln(2x + 1)^3 + \ln(3x + 4)^5$$
$$= 3 \ln(2x + 1) + 5 \ln(3x + 4)$$

Differentiating both sides of the equation with respect to x, keeping in mind that y is a function of x, we obtain

$$\frac{d}{dx}(\ln y) = \frac{y'}{y} = 3 \cdot \frac{2}{2x + 1} + 5 \cdot \frac{3}{3x + 4}$$

$$= 3\left[\frac{2}{2x + 1} + \frac{5}{3x + 4}\right]$$

$$= \left(\frac{6}{2x + 1} + \frac{15}{3x + 4}\right)$$

and

$$y' = (2x + 1)^3(3x + 4)^5 \cdot \left(\frac{6}{2x + 1} + \frac{15}{3x + 4}\right)$$

5.6 Exponential Functions as Mathematical Models

EXPONENTIAL GROWTH

FIGURE 5.13
Exponential growth

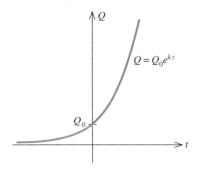

Many problems arising from practical situations can be described mathematically in terms of exponential functions or functions closely related to the exponential function. In this section we look at some applications involving exponential functions from the fields of the life and social sciences.

In Section 5.1 we saw that the exponential function $f(x) = b^x$ is an increasing function when $b > 1$. In particular, the function $f(x) = e^x$ shares this property. From this result one may deduce that the function $Q(t) = Q_0 e^{kt}$, where Q_0 and k are positive constants, has the following properties:

1. $Q(0) = Q_0$
2. $Q(t)$ increases "rapidly" without bound as t increases without bound (Figure 5.13).

Property 1 follows from the computation

$$Q(0) = Q_0 e^0 = Q_0$$

Next, to study the rate of change of the function $Q(t)$, we differentiate it with respect to t, obtaining

$$Q'(t) = \frac{d}{dt}(Q_0 e^{kt})$$

$$= Q_0 \frac{d}{dt}(e^{kt})$$

$$= kQ_0 e^{kt}$$

$$= kQ(t) \qquad \qquad \textbf{(12)}$$

Since $Q(t) > 0$ (because Q_0 is assumed to be positive) and $k > 0$, we see that $Q'(t) > 0$ and so $Q(t)$ is an increasing function of t. Our computation has in fact shed more light on an important property of the function $Q(t)$. Equation (12) says that the rate of increase of the function $Q(t)$ is proportional to the amount $Q(t)$ of the quantity present at time t. The implication is that as $Q(t)$ increases, so does the *rate of increase* of $Q(t)$, resulting in a very rapid increase in $Q(t)$ as t increases without bound.

Thus, the exponential function

$$Q(t) = Q_0 e^{kt} \qquad (0 \leq t < \infty) \qquad \textbf{(13)}$$

provides us with a mathematical model of a quantity $Q(t)$ that is initially present in the amount of $Q(0) = Q_0$ and whose rate of growth at any time t is directly proportional to the amount of the quantity present at time t. Such a quantity is said to exhibit **exponential growth,** and the constant k is called the **growth constant.** Interest earned on a fixed deposit when compounded continuously exhibits exponential growth. Other examples of exponential growth follow.

Under ideal laboratory conditions, the number of bacteria in a culture grows in accordance with the law $Q(t) = Q_0 e^{kt}$, where Q_0 denotes the number of bacteria initially present in the culture, k is some constant determined by the strain of bacteria under consideration, and t is the elapsed time measured in hours. Suppose 10,000 bacteria are present initially in the culture and 60,000 present 2 hours later.

a. How many bacteria will there be in the culture at the end of 4 hours?
b. What is the rate of growth of the population after 4 hours?

SOLUTION ✔

a. We are given that $Q(0) = Q_0 = 10{,}000$, so $Q(t) = 10{,}000 e^{kt}$. Next, the fact that 60,000 bacteria are present 2 hours later translates into $Q(2) = 60{,}000$. Thus,

$$60{,}000 = 10{,}000 e^{2k}$$

$$e^{2k} = 6$$

Taking the natural logarithm on both sides of the equation, we obtain

$$\ln e^{2k} = \ln 6$$

$$2k = \ln 6 \qquad \text{(Since } \ln e = 1\text{)}$$

$$k \approx 0.8959$$

Thus, the number of bacteria present at any time t is given by

$$Q(t) = 10{,}000 e^{0.8959t}$$

In particular, the number of bacteria present in the culture at the end of 4 hours is given by

$$Q(4) = 10{,}000 e^{0.8959(4)}$$
$$= 360{,}029$$

b. The rate of growth of the bacteria population at any time t is given by

$$Q'(t) = kQ(t)$$

Thus, using the result from part (a), we find that the rate at which the population is growing at the end of 4 hours is

$$Q'(4) = kQ(4)$$
$$\approx (0.8959)(360{,}029)$$
$$\approx 322{,}550$$

or approximately 322,550 bacteria per hour.   ■ ■ ■ ■

EXPONENTIAL DECAY

In contrast to exponential growth, a quantity exhibits **exponential decay** if it decreases at a rate that is directly proportional to its size. Such a quantity may be described by the exponential function

$$Q(t) = Q_0 e^{-kt} \qquad [t \in [0, \infty)] \qquad \textbf{(14)}$$

FIGURE 5.14
Exponential decay

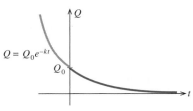

where the positive constant Q_0 measures the amount present initially ($t = 0$) and k is some suitable positive number, called the **decay constant.** The choice of this number is determined by the nature of the substance under consideration. The graph of this function is sketched in Figure 5.14.

To verify the properties ascribed to the function $Q(t)$, we simply compute

$$Q(0) = Q_0 e^0 = Q_0$$

$$Q'(t) = \frac{d}{dt}(Q_0 e^{-kt})$$

$$= Q_0 \frac{d}{dt}(e^{-kt})$$

$$= -kQ_0 e^{-kt} = -kQ(t)$$

EXAMPLE 2   Radioactive substances decay exponentially. For example, the amount of radium present at any time t obeys the law $Q(t) = Q_0 e^{-kt}$, where Q_0 is the initial amount present and k is a suitable positive constant. The **half-life of a radioactive substance** is the time required for a given amount to be reduced by one-half. Now, it is known that the half-life of radium is approximately 1600 years. Suppose initially there are 200 milligrams of pure radium. Find the amount left after t years. What is the amount left after 800 years?

SOLUTION ✔   The initial amount of radium present is 200 milligrams, so $Q(0) = Q_0 = 200$. Thus, $Q(t) = 200e^{-kt}$. Next, the datum concerning the half-life of radium implies that $Q(1600) = 100$, and this gives

$$100 = 200e^{-1600k}$$

$$e^{-1600k} = \frac{1}{2}$$

Taking the natural logarithm on both sides of this equation yields

$$-1600k \ln e = \ln \frac{1}{2}$$

$$-1600k = \ln \frac{1}{2} \qquad (\ln e = 1)$$

$$k = -\frac{1}{1600} \ln \left(\frac{1}{2}\right) = 0.0004332$$

Therefore, the amount of radium left after t years is

$$Q(t) = 200e^{-0.0004332t}$$

In particular, the amount of radium left after 800 years is

$$Q(800) = 200e^{-0.0004332(800)} \approx 141.42$$

or approximately 141 milligrams.

EXAMPLE 3

Carbon 14, a radioactive isotope of carbon, has a half-life of 5770 years. What is its decay constant?

SOLUTION ✔

We have $Q(t) = Q_0e^{-kt}$. Since the half-life of the element is 5770 years, half of the substance is left at the end of that period. That is,

$$Q(5770) = Q_0e^{-5770k} = \frac{1}{2}Q_0$$

$$e^{-5770k} = \frac{1}{2}$$

Taking the natural logarithm on both sides of this equation, we have

$$\ln e^{-5770k} = \ln \frac{1}{2}$$

$$-5770k = -0.693147$$

$$k \approx 0.00012$$

Carbon-14 dating is a well-known method used by anthropologists to establish the age of animal and plant fossils. This method assumes that the proportion of carbon 14 (C-14) present in the atmosphere has remained constant over the past 50,000 years. Professor Willard Libby, recipient of the Nobel Prize in chemistry in 1960, proposed this theory.

The amount of C-14 in the tissues of a living plant or animal is constant. However, when an organism dies, it stops absorbing new quantities of C-14, and the amount of C-14 in the remains diminishes because of the natural decay of the radioactive substance. Thus, the approximate age of a plant or animal fossil can be determined by measuring the amount of C-14 present in the remains.

EXAMPLE 4

A skull from an archeological site has one-tenth the amount of C-14 that it originally contained. Determine the approximate age of the skull.

SOLUTION ✔

Here,

$$Q(t) = Q_0e^{-kt}$$
$$= Q_0e^{-0.00012t}$$

where Q_0 is the amount of C-14 present originally and k, the decay constant, is equal to 0.00012 (see Example 3). Since $Q(t) = (1/10)Q_0$, we have

$$\frac{1}{10}Q_0 = Q_0 e^{-0.00012t}$$

$$\ln\frac{1}{10} = -0.00012t \qquad \text{(Taking the natural logarithm on both sides)}$$

$$t = \frac{\ln\dfrac{1}{10}}{-0.00012}$$

$$\approx 19{,}200$$

or approximately 19,200 years.

■ ■ ■ ■

LEARNING CURVES

The next example shows how the exponential function may be applied to describe certain types of learning processes. Consider the function

$$Q(t) = C - Ae^{-kt}$$

where C, A, and k are positive constants. To sketch the graph of the function Q, observe that its y-intercept is given by $Q(0) = C - A$. Next, we compute

$$Q'(t) = kAe^{-kt}$$

Since both k and A are positive, we see that $Q'(t) > 0$ for all values of t. Thus, $Q(t)$ is an increasing function of t. Also,

$$\lim_{t\to\infty} Q(t) = \lim_{t\to\infty}(C - Ae^{-kt})$$

$$= \lim_{t\to\infty} C - \lim_{t\to\infty} Ae^{-kt}$$

$$= C$$

so $y = C$ is a horizontal asymptote of Q. Thus, $Q(t)$ increases and approaches the number C as t increases without bound. The graph of the function Q is shown in Figure 5.15, where that part of the graph corresponding to the negative values of t is drawn with a gray line since, in practice, one normally restricts the domain of the function to the interval $[0, \infty)$.

Observe that $Q(t)$ $(t > 0)$ increases rather rapidly initially but that the rate of increase slows down considerably after a while. To see this, we compute

$$\lim_{t\to\infty} Q'(t) = \lim_{t\to\infty} kAe^{-kt} = 0$$

This behavior of the graph of the function Q closely resembles the learning pattern experienced by workers engaged in highly repetitive work. For example, the productivity of an assembly-line worker increases very rapidly in the early stages of the training period. This productivity increase is a direct result

FIGURE 5.15
A learning curve

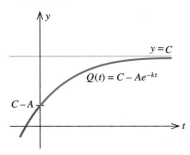

of the worker's training and accumulated experience. But the rate of increase of productivity slows as time goes by, and the worker's productivity level approaches some fixed level due to the limitations of the worker and the machine. Because of this characteristic, the graph of the function $Q(t) = C - Ae^{-kt}$ is often called a **learning curve.**

 EXAMPLE 5

The Camera Division of the Eastman Optical Company produces a 35-mm single-lens reflex camera. Eastman's training department determines that after completing the basic training program, a new, previously inexperienced employee will be able to assemble

$$Q(t) = 50 - 30e^{-0.5t}$$

model F cameras per day, t months after the employee starts work on the assembly line.

a. How many model F cameras can a new employee assemble per day after basic training?
b. How many model F cameras can an employee with 1 month of experience assemble per day? An employee with 2 months of experience? An employee with 6 months of experience?
c. How many model F cameras can the average experienced employee assemble per day?

SOLUTION ✔
a. The number of model F cameras a new employee can assemble is given by

$$Q(0) = 50 - 30 = 20$$

b. The number of model F cameras that an employee with 1 month of experience, 2 months of experience, and 6 months of experience can assemble per day is given by

$$Q(1) = 50 - 30e^{-0.5} \approx 31.80$$

$$Q(2) = 50 - 30e^{-1} \approx 38.96$$

$$Q(6) = 50 - 30e^{-3} \approx 48.51$$

or approximately 32, 39, and 49, respectively.
c. As t increases without bound, $Q(t)$ approaches 50. Hence, the average experienced employee can ultimately be expected to assemble 50 model F cameras per day.   ■ ■ ■ ■

Other applications of the learning curve are found in models that describe the dissemination of information about a product or the velocity of an object dropped into a viscous medium.

LOGISTIC GROWTH FUNCTIONS

Our last example of an application of exponential functions to the description of natural phenomena involves the **logistic** (also called the **S-shaped,** or **sigmoi-**

FIGURE 5.16

A logistic curve

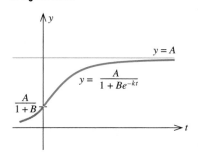

dal) **curve,** which is the graph of the function

$$Q(t) = \frac{A}{1 + Be^{-kt}}$$

where A, B, and k are positive constants. The function Q is called a **logistic growth function,** and the graph of the function Q is sketched in Figure 5.16.

Observe that $Q(t)$ increases rather rapidly for small values of t. In fact, for small values of t, the logistic curve resembles an exponential growth curve. However, the *rate of growth* of $Q(t)$ decreases quite rapidly as t increases and $Q(t)$ approaches the number A as t increases without bound.

Thus, the logistic curve exhibits both the property of rapid growth of the exponential growth curve as well as the "saturation" property of the learning curve. Because of these characteristics, the logistic curve serves as a suitable mathematical model for describing many natural phenomena. For example, if a small number of rabbits were introduced to a tiny island in the South Pacific, the rabbit population might be expected to grow very rapidly at first, but the growth rate would decrease quickly as overcrowding, scarcity of food, and other environmental factors affected it. The population would eventually stabilize at a level compatible with the life-support capacity of the environment. Models describing the spread of rumors and epidemics are other examples of the application of the logistic curve.

 EXAMPLE 6   The number of soldiers at Fort MacArthur who contracted influenza after t days during a flu epidemic is approximated by the exponential model

$$Q(t) = \frac{5000}{1 + 1249e^{-kt}}$$

If 40 soldiers contracted the flu by day 7, find how many soldiers contracted the flu by day 15.

SOLUTION ✔   The given information implies that

$$Q(7) = 40 \quad \text{and} \quad Q(7) = \frac{5000}{1 + 1249e^{-7k}} = 40$$

Thus,

$$40(1 + 1249e^{-7k}) = 5000$$

$$1 + 1249e^{-7k} = \frac{5000}{40} = 125$$

$$e^{-7k} = \frac{124}{1249}$$

$$-7k = \ln\frac{124}{1249}$$

$$k = -\frac{\ln\dfrac{124}{1249}}{7} \approx 0.33$$

Therefore, the number of soldiers who contracted the flu after t days is given by

$$Q(t) = \frac{5000}{1 + 1249e^{-0.33t}}$$

In particular, the number of soldiers who contracted the flu by day 15 is given by

$$Q(15) = \frac{5000}{1 + 1249e^{-15(0.33)}}$$

$$\approx 508$$

or approximately 508 soldiers.   ■ ■ ■ ■

Exploring with Technology

Refer to Example 6.

1. Use a graphing utility to plot the graph of the function Q, using the viewing rectangle $[0, 40] \times [0, 5000]$.
2. Find how long it takes for the first 1000 soldiers to contract the flu.
Hint: Plot the graphs of $y_1 = Q(t)$ and $y_2 = 1000$ and find the point of intersection of the two graphs.

SELF-CHECK EXERCISE 5.6

Suppose that the population (in millions) of a country at any time t grows in accordance with the rule

$$P = \left(P_0 + \frac{I}{k}\right)e^{kt} - \frac{I}{k}$$

where P denotes the population at any time t, k is a constant reflecting the natural growth rate of the population, I is a constant giving the (constant) rate of immigration into the country, and P_0 is the total population of the country at time $t = 0$. The population of the United States in the year 1980 ($t = 0$) was 226.5 million. If the natural growth rate is 0.8% annually ($k = 0.008$) and net immigration is allowed at the rate of half a million people per year ($I = 0.5$) until the end of the century, what is the population of the United States expected to be in the year 2005?

Solutions to Self-Check Exercise 5.6 can be found on page 435.

5.6 Exercises

🖩 **A calculator is recommended for this exercise set.**

1. **EXPONENTIAL GROWTH**   Given that a quantity $Q(t)$ is described by the exponential growth function

$$Q(t) = 400e^{0.05t}$$

where t is measured in minutes, answer the following questions.
a. What is the growth constant?
b. What quantity is present initially?
c. Using a calculator, complete the following table of values:

t	0	10	20	100	1000
Q					

2. **EXPONENTIAL DECAY**   Given that a quantity $Q(t)$ exhibiting exponential decay is described by the function

$$Q(t) = 2000e^{-0.06t}$$

where t is measured in years, answer the following questions.
a. What is the decay constant?
b. What quantity is present initially?
c. Using a calculator, complete the following table of values:

t	0	5	10	20	100
Q					

3. **GROWTH OF BACTERIA**   The growth rate of the bacterium *Escherichia coli*, a common bacterium found in the human intestine, is proportional to its size. Under ideal laboratory conditions, when this bacterium is grown in a nutrient broth medium, the number of cells in a culture doubles approximately every 20 min.
a. If the initial cell population is 100, determine the function $Q(t)$ that expresses the exponential growth of the number of cells of this bacterium as a function of time t (in minutes).
b. How long will it take for a colony of 100 cells to increase to a population of 1 million?
c. If the initial cell population were 1000, how would this alter our model?

4. **WORLD POPULATION**   The world population at the beginning of 1990 was 5.3 billion. Assume that the population continues to grow at its present rate of approximately 2%/year and find the function $Q(t)$ that expresses the world population (in billions) as a function of time t (in years) where $t = 0$ corresponds to the beginning of 1990.
a. Using this function, complete the following table of values and sketch the graph of the function Q.

Year	1990	1995	2000	2005
World Population				

Year	2010	2015	2020	2025
World Population				

b. Find the estimated rate of growth in the year 2005.

5. **WORLD POPULATION**   Refer to Exercise 4.
a. If the world population continues to grow at its present rate of approximately 2%/year, find the length of time t_0 required for the world population to triple in size.
b. Using the time t_0 found in part (a), what would be the world population if the growth rate were reduced to 1.8%?

6. **RESALE VALUE**   A certain piece of machinery was purchased 3 yr ago by the Garland Mills Company for $500,000. Its present resale value is $320,000. Assuming that the machine's resale value decreases exponentially, what will it be 4 yr from now?

7. **ATMOSPHERIC PRESSURE**   If the temperature is constant, then the atmospheric pressure P (in pounds per square inch) varies with the altitude above sea level h in accordance with the law

$$P = p_0 e^{-kh}$$

where p_0 is the atmospheric pressure at sea level and k is a constant. If the atmospheric pressure is 15 lb/in.^{2} at sea level and 12.5 lb/in.^{2} at 4000 ft, find the atmospheric pressure at an altitude of 12,000 ft. How fast is the atmospheric pressure changing with respect to altitude at an altitude of 12,000 ft?

8. **RADIOACTIVE DECAY**   The radioactive element polonium decays according to the law

$$Q(t) = Q_0 \cdot 2^{-(t/140)}$$

where Q_0 is the initial amount and the time t is measured in days. If the amount of polonium left after 280 days is 20 mg, what was the initial amount present?

9. **RADIOACTIVE DECAY** Phosphorus 32 has a half-life of 14.2 days. If 100 g of this substance are present initially, find the amount present after t days. What amount will be left after 7.1 days? How fast is the phosphorus 32 decaying when $t = 7.1$?

10. **NUCLEAR FALLOUT** Strontium 90, a radioactive isotope of strontium, is present in the fallout resulting from nuclear explosions. It is especially hazardous to animal life, including humans, because, upon ingestion of contaminated food, it is absorbed into the bone structure. Its half-life is 27 yr. If the amount of strontium 90 in a certain area is found to be four times the "safe" level, find how much time must elapse before an "acceptable level" is reached.

11. **CARBON-14 DATING** Wood deposits recovered from an archeological site contain 20% of the carbon 14 they originally contained. How long ago did the tree from which the wood was obtained die?

12. **CARBON-14 DATING** Skeletal remains of the so-called "Pittsburgh Man," unearthed in Pennsylvania, had lost 82% of the carbon 14 they originally contained. Determine the approximate age of the bones.

13. **LEARNING CURVES** The American Court Reporting Institute finds that the average student taking Advanced Machine Shorthand, an intensive 20-wk course, progresses according to the function

$$Q(t) = 120(1 - e^{-0.05t}) + 60 \qquad (0 \le t \le 20)$$

where $Q(t)$ measures the number of words (per minute) of dictation that the student can take in machine shorthand after t wk in the course. Sketch the graph of the function Q and answer the following questions.
a. What is the beginning shorthand speed for the average student in this course?
b. What shorthand speed does the average student attain halfway through the course?
c. How many words per minute can the average student take after completing this course?

14. **EFFECT OF ADVERTISING ON SALES** The Metro Department Store found that t wk after the end of a sales promotion the volume of sales was given by a function of the form

$$S(t) = B + Ae^{-kt} \qquad (0 \le t \le 4)$$

where $B = 50,000$ and is equal to the average weekly volume of sales before the promotion. The sales volumes at the end of the first and third weeks were $83,515 and $65,055, respectively. Assume that the sales volume is decreasing exponentially.
a. Find the decay constant k.
b. Find the sales volume at the end of the fourth week.
c. How fast is the sales volume dropping at the end of the fourth week?

15. **DEMAND FOR COMPUTERS** The Universal Instruments Company found that the monthly demand for its new line of Galaxy Home Computers t mo after placing the line on the market was given by

$$D(t) = 2000 - 1500e^{-0.05t} \qquad (t > 0)$$

Graph this function and answer the following questions.
a. What is the demand after 1 mo? After 1 yr? After 2 yr? After 5 yr?
b. At what level is the demand expected to stabilize?
c. Find the rate of growth of the demand after the tenth month.

16. **NEWTON'S LAW OF COOLING** Newton's law of cooling states that the rate at which the temperature of an object changes is proportional to the difference in temperature between the object and that of the surrounding medium. Thus, the temperature $F(t)$ of an object that is greater than the temperature of its surrounding medium is given by

$$F(t) = T + Ae^{-kt}$$

where t is the time expressed in minutes, T is the temperature of the surrounding medium, and A and k are constants. Suppose a cup of instant coffee is prepared with boiling water (212°F) and left to cool on the counter in a room where the temperature is 72°F. If $k = 0.1865$, determine when the coffee will be cool enough to drink (say, 110°F).

17. **SPREAD OF AN EPIDEMIC** During a flu epidemic, the number of children in the Woodbridge Community School System who contracted influenza after t days was given by

$$Q(t) = \frac{1000}{1 + 199e^{-0.8t}}$$

a. How many children were stricken by the flu after the first day?
b. How many children had the flu after 10 days?
c. How many children eventually contracted the disease?

18. **GROWTH OF A FRUIT-FLY POPULATION** On the basis of data collected during an experiment, a biologist found that the growth of the fruit fly (*Drosophila*) with a limited food supply could be approximated by the exponential model

$$N(t) = \frac{400}{1 + 39e^{-0.16t}}$$

where t denotes the number of days since the beginning of the experiment.
a. What was the initial fruit-fly population in the experiment?

b. What was the maximum fruit-fly population that could be expected under this laboratory condition?
c. What was the population of the fruit-fly colony on the 20th day?
d. How fast was the population changing on the 20th day?

19. **PERCENTAGE OF HOUSEHOLDS WITH VCRs** According to estimates by Paul Kroger Associates, the percentage of households that own videocassette recorders (VCRs) is given by

$$P(t) = \frac{68}{1 + 21.67e^{-0.62t}} \qquad (0 \le t \le 12)$$

where t is measured in years, with $t = 0$ corresponding to the beginning of 1985. What percentage of households owned VCRs at the beginning of 1985? At the beginning of 1995?

20. **POPULATION GROWTH IN THE TWENTY-FIRST CENTURY** The U.S. population is approximated by the function

$$P(t) = \frac{616.5}{1 + 4.02e^{-0.5t}}$$

where $P(t)$ is measured in millions of people and t is measured in 30-yr intervals, with $t = 0$ corresponding to 1930. What is the expected population of the United States in 2020 ($t = 3$)?

21. **SPREAD OF A RUMOR** Three hundred students attended the dedication ceremony of a new building on a college campus. The president of the traditionally female college announced a new expansion program, which included plans to make the college coeducational. The number of students who learned of the new program t hr later is given by the function

$$f(t) = \frac{3000}{1 + Be^{-kt}}$$

If 600 students on campus had heard about the new program 2 hr after the ceremony, how many students had heard about the policy after 4 hr? How fast was the rumor spreading 4 hr after the ceremony?

22. **CHEMICAL MIXTURES** Two chemicals react to form another chemical. Suppose the amount of the chemical formed in time t (in hours) is given by

$$x(t) = \frac{15\left[1 - \left(\frac{2}{3}\right)^{3t}\right]}{1 - \frac{1}{4}\left(\frac{2}{3}\right)^{3t}}$$

where $x(t)$ is measured in pounds. How many pounds of the chemical are formed eventually?
Hint: You need to evaluate $\lim_{t \to \infty} x(t)$.

23. **CONCENTRATION OF GLUCOSE IN THE BLOODSTREAM** A glucose solution is administered intravenously into the bloodstream at a constant rate of r mg/hr. As the glucose is being administered, it is converted into other substances and removed from the bloodstream. Suppose the concentration of the glucose solution at time t is given by

$$C(t) = \frac{r}{k} - \left[\left(\frac{r}{k}\right) - C_0\right]e^{-kt}$$

where C_0 is the concentration at time $t = 0$ and k is a constant.
a. Assuming that $C_0 < r/k$, evaluate

$$\lim_{t \to \infty} C(t)$$

and interpret your result.
b. Sketch the graph of the function C.

24. **GOMPERTZ GROWTH CURVE** Consider the function

$$Q(t) = Ce^{-Ae^{-kt}}$$

where $Q(t)$ is the size of a quantity at time t and A, C, and k are positive constants. The graph of this function, called the **Gompertz growth curve,** is used by biologists to describe restricted population growth.
a. Show that the function Q is always increasing.
b. Find the time t at which the growth rate $Q'(t)$ is increasing most rapidly.
Hint: Find the inflection point of Q.
c. Show that $\lim_{t \to \infty} Q(t) = C$ and interpret your result.

SOLUTION TO SELF-CHECK EXERCISE 5.6

We are given that $P_0 = 226.5$, $k = 0.008$, and $I = 0.5$. So

$$P = \left(226.5 + \frac{0.5}{0.008}\right)e^{0.008t} - \frac{0.5}{0.008}$$

$$= 289e^{0.008t} - 62.5$$

Therefore, the population in the year 2005 will be given by

$$P(25) = 289e^{0.2} - 62.5$$
$$\approx 290.5$$

or approximately 290.5 million.

CHAPTER 5 Summary of Principal Formulas and Terms

Formulas

1.	Exponential function with base b	$y = b^x$
2.	The number e	$e = \lim\limits_{m \to \infty} \left(1 + \dfrac{1}{m}\right)^m = 2.71828$
3.	Exponential function with base e	$y = e^x$
4.	Logarithmic function with base b	$y = \log_b x$
5.	Logarithmic function with base e	$y = \ln x$
6.	Inverse properties of $\ln x$ and e	$\ln e^x = x$ and $e^{\ln x} = x$
7.	Compound interest (accumulated amount)	$A = P(1 + i)^n$, where $i = r/m$ and $n = mt$
8.	Effective rate of interest	$r_{\text{eff}} = \left(1 + \dfrac{r}{m}\right)^m - 1$
9.	Compound interest (present value)	$P = A(1 + i)^{-n}$, where $i = r/m$ and $n = mt$
10.	Continuous compound interest	$A = Pe^{rt}$
11.	Derivative of the exponential function	$\dfrac{d}{dx}(e^x) = e^x$
12.	Chain rule for exponential functions	$\dfrac{d}{dx}(e^u) = e^u \dfrac{du}{dx}$
13.	Derivative of the logarithmic function	$\dfrac{d}{dx}\ln|x| = \dfrac{1}{x}$
14.	Chain rule for logarithmic functions	$\dfrac{d}{dx}(\ln u) = \dfrac{1}{u}\dfrac{du}{dx}$

Terms

common logarithm

natural logarithm

logarithmic differentiation

exponential growth

growth constant

exponential decay

decay constant

half-life of a radioactive element

logistic growth function

CHAPTER 5   REVIEW EXERCISES

1. Sketch on the same set of coordinate axes the graphs of the exponential functions defined by the equations.

   a. $y = 2^{-x}$ **b.** $y = \left(\dfrac{1}{2}\right)^x$

In Exercises 2 and 3, express each in logarithmic form.

2. $\left(\dfrac{2}{3}\right)^{-3} = \dfrac{27}{8}$ **3.** $16^{-3/4} = 0.125$

In Exercises 4 and 5, solve each equation for x.

4. $\log_4(2x + 1) = 2$

5. $\ln(x - 1) + \ln 4 = \ln(2x + 4) - \ln 2$

In Exercises 6–8, given that ln 2 = x, ln 3 = y, and ln 5 = z, express each of the given logarithmic values in terms of x, y, and z.

6. $\ln 30$ **7.** $\ln 3.6$ **8.** $\ln 75$

9. Sketch the graph of the function $y = \log_2(x + 3)$.

10. Sketch the graph of the function $y = \log_3(x + 1)$.

In Exercises 11–28, find the derivative of the function.

11. $f(x) = xe^{2x}$

12. $f(t) = \sqrt{t}e^t + t$

13. $g(t) = \sqrt{t}e^{-2t}$

14. $g(x) = e^x\sqrt{1 + x^2}$

15. $y = \dfrac{e^{2x}}{1 + e^{-2x}}$

16. $f(x) = e^{2x^2-1}$

17. $f(x) = xe^{-x^2}$

18. $g(x) = (1 + e^{2x})^{3/2}$

19. $f(x) = x^2e^x + e^x$

20. $g(t) = t \ln t$

21. $f(x) = \ln(e^{x^2} + 1)$

22. $f(x) = \dfrac{x}{\ln x}$

23. $f(x) = \dfrac{\ln x}{x + 1}$

24. $y = (x + 1)e^x$

25. $y = \ln(e^{4x} + 3)$

26. $f(r) = \dfrac{re^r}{1 + r^2}$

27. $f(x) = \dfrac{\ln x}{1 + e^x}$

28. $g(x) = \dfrac{e^{x^2}}{1 + \ln x}$

29. Find the second derivative of the function $y = \ln(3x + 1)$.

30. Find the second derivative of the function $y = x \ln x$.

31. Find $h'(0)$ if $h(x) = g(f(x))$, $g(x) = x + (1/x)$, and $f(x) = e^x$.

32. Find $h'(1)$ if $h(x) = g(f(x))$, $g(x) = \dfrac{x + 1}{x - 1}$, and $f(x) = \ln x$.

33. Use logarithmic differentiation to find the derivative of $f(x) = (2x^3 + 1)(x^2 + 2)^3$.

34. Use logarithmic differentiation to find the derivative of $f(x) = \dfrac{x(x^2 - 2)^2}{(x - 1)}$.

35. Find an equation of the tangent line to the graph of $y = e^{-2x}$ at the point $(1, e^{-2})$.

36. Find an equation of the tangent line to the graph of $y = xe^{-x}$ at the point $(1, e^{-1})$.

37. Sketch the graph of the function $f(x) = xe^{-2x}$.

38. Sketch the graph of the function $f(x) = x^2 - \ln x$.

39. Find the absolute extrema of the function $f(t) = te^{-t}$.

40. Find the absolute extrema of the function

$$g(t) = \frac{\ln t}{t}$$

on $[1, 2]$.

41. A hotel was purchased by a conglomerate for $4.5 million and sold 5 yr later for $8.2 million. Find the annual rate of return (compounded continuously).

42. Find the present value of $119,346 due in 4 yr at an interest rate of 10%/year compounded continuously.

43. A culture of bacteria that initially contained 2000 bacteria has a count of 18,000 bacteria after 2 hr.
   a. Determine the function $Q(t)$ that expresses the exponential growth of the number of cells of this bacterium as a function of time t (in minutes).
   b. Find the number of bacteria present after 4 hr.

44. The radioactive element radium has a half-life of 1600 yr. What is its decay constant?

45. The VCA Television Company found that the monthly demand for its new line of video disc players t mo after placing the players on the market is given by

$$D(t) = 4000 - 3000e^{-0.06t} \qquad (t \geq 0)$$

Graph this function and answer the following questions.
   a. What was the demand after 1 mo? After 1 yr? After 2 yr?
   b. At what level is the demand expected to stabilize?

46. During a flu epidemic, the number of students at a certain university who contracted influenza after t days could be approximated by the exponential model

$$Q(t) = \frac{3000}{1 + 499e^{-kt}}$$

If 90 students contracted the flu by day 10, how many students contracted the flu by day 20?

INTEGRATION

6

Differential calculus is concerned with the problem of finding the rate of change of one quantity with respect to another. In this chapter we begin the study of the other branch of calculus, known as integral calculus. Here we are interested in precisely the opposite problem: If we know the rate of change of one quantity with respect to another, can we find the relationship between the two quantities? The principal tool used in the study of integral calculus is the *antiderivative* of a function, and we develop rules for antidifferentiation, or *integration*, as the process of finding the antiderivative is called. We also show that a link is established between differential and integral calculus—via the fundamental theorem of calculus.

How much will the solar cell panels cost? The head of Soloron Corporation's research and development department has projected that the cost of producing solar cell panels will drop at a certain rate in the next several years. In Example 7, page 460, you will see how this information can be used to predict the cost of solar cell panels in the coming years.

APPLICATIONS

EXAMPLE 11   In a test run of a maglev along a straight elevated monorail track, data obtained from reading its speedometer indicate that the velocity of the maglev at time t can be described by the velocity function

$$v(t) = 8t \qquad (0 \le t \le 30)$$

Find the position function of the maglev. Assume that initially the maglev is located at the origin of a coordinate line.

SOLUTION ✔   Let $s(t)$ denote the position of the maglev at any time t ($0 \le t \le 30$). Then, $s'(t) = v(t)$. So, we have the initial value problem

$$\left.\begin{array}{c} s'(t) = 8t \\ s(0) = 0 \end{array}\right\}$$

Integrating both sides of the differential equation $s'(t) = 8t$, we obtain

$$s(t) = \int s'(t)\, dt = \int 8t\, dt = 4t^2 + C$$

where C is an arbitrary constant. To evaluate C, we use the initial condition $s(0) = 0$ to write

$$s(0) = 4(0) + C = 0 \qquad \text{or} \qquad C = 0$$

Therefore, the required position function is $s(t) = 4t^2$ ($0 \le t \le 30$).

■ ■ ■ ■

EXAMPLE 12   The current circulation of the *Investor's Digest* is 3000 copies per week. The managing editor of the weekly projects a growth rate of

$$4 + 5t^{2/3}$$

copies per week, t weeks from now, for the next 3 years. Based on her projection, what will the circulation of the digest be 125 weeks from now?

SOLUTION ✔   Let $S(t)$ denote the circulation of the digest t weeks from now. Then $S'(t)$ is the rate of change in the circulation in the tth week and is given by

$$S'(t) = 4 + 5t^{2/3}$$

Furthermore, the current circulation of 3000 copies per week translates into the initial condition $S(0) = 3000$. Integrating the differential equation with respect to t gives

$$S(t) = \int S'(t)\, dt = \int (4 + 5t^{2/3})\, dt$$

$$= 4t + 5\left(\frac{t^{5/3}}{\frac{5}{3}}\right) + C = 4t + 3t^{5/3} + C$$

To determine the value of C, we use the condition $S(0) = 3000$ to write

$$S(0) = 4(0) + 3(0) + C = 3000$$

which gives $C = 3000$. Therefore, the circulation of the digest t weeks from now will be

$$S(t) = 4t + 3t^{5/3} + 3000$$

In particular, the circulation 125 weeks from now will be

$$S(125) = 4(125) + 3(125)^{5/3} + 3000 = 12{,}875$$

copies per week.

■■■■

SELF-CHECK EXERCISES 6.1

1. Evaluate $\int \left(\dfrac{1}{\sqrt{x}} - \dfrac{2}{x} + 3e^x \right) dx$.

2. Find the rule for the function f given that (1) the slope of the tangent line to the graph of f at any point $P(x, f(x))$ is given by the expression $3x^2 - 6x + 3$ and (2) the graph of f passes through the point $(2, 9)$.

3. Suppose United Motors' share of the new cars sold in a certain country is changing at the rate of

$$f(t) = -0.01875t^2 + 0.15t - 1.2 \qquad (0 \le t \le 12)$$

percent at year t ($t = 0$ corresponds to the beginning of 1989). The company's market share at the beginning of 1989 was 48.4%. What was United Motors' market share at the beginning of 2001?

Solutions to Self-Check Exercises 6.1 can be found on page 454.

Exercises

In Exercises 1–4, verify directly that F is an antiderivative of f.

1. $F(x) = \dfrac{1}{3}x^3 + 2x^2 - x + 2;\ f(x) = x^2 + 4x - 1$

2. $F(x) = xe^x + \pi;\ f(x) = e^x(1 + x)$

3. $F(x) = \sqrt{2x^2 - 1};\ f(x) = \dfrac{2x}{\sqrt{2x^2 - 1}}$

4. $F(x) = x \ln x - x;\ f(x) = \ln x$

In Exercises 5–8, (a) verify that G is an antiderivative of f, (b) find all antiderivatives of f, and (c) sketch the graphs of a few of the family of antiderivatives found in part (b).

5. $G(x) = 2x;\ f(x) = 2$ 6. $G(x) = 2x^2;\ f(x) = 4x$

7. $G(x) = \dfrac{1}{3}x^3;\ f(x) = x^2$ 8. $G(x) = e^x;\ f(x) = e^x$

In Exercises 9–50, find the indefinite integral.

9. $\displaystyle\int 6\,dx$

10. $\displaystyle\int \sqrt{2}\,dx$

11. $\displaystyle\int x^3\,dx$

12. $\displaystyle\int 2x^5\,dx$

13. $\displaystyle\int x^{-4}\,dx$

14. $\displaystyle\int 3t^{-7}\,dt$

15. $\displaystyle\int x^{2/3}\,dx$

16. $\displaystyle\int 2u^{3/4}\,du$

17. $\displaystyle\int x^{-5/4}\,dx$

18. $\displaystyle\int 3x^{-2/3}\,dx$

19. $\displaystyle\int \dfrac{2}{x^2}\,dx$

20. $\displaystyle\int \dfrac{1}{3x^5}\,dx$

21. $\int \pi\sqrt{t}\, dt$

22. $\int \dfrac{3}{\sqrt{t}}\, dt$

23. $\int (3 - 2x)\, dx$

24. $\int (1 + u + u^2)\, du$

25. $\int (x^2 + x + x^{-3})\, dx$

26. $\int (0.3t^2 + 0.02t + 2)\, dt$

27. $\int 4e^x\, dx$

28. $\int (1 + e^x)\, dx$

29. $\int (1 + x + e^x)\, dx$

30. $\int (2 + x + 2x^2 + e^x)\, dx$

31. $\int \left(4x^3 - \dfrac{2}{x^2} - 1\right) dx$

32. $\int \left(6x^3 + \dfrac{3}{x^2} - x\right) dx$

33. $\int (x^{5/2} + 2x^{3/2} - x)\, dx$

34. $\int (t^{3/2} + 2t^{1/2} - 4t^{-1/2})\, dt$

35. $\int \left(\sqrt{x} + \dfrac{3}{\sqrt{x}}\right) dx$

36. $\int \left(\sqrt[3]{x^2} - \dfrac{1}{x^2}\right) dx$

37. $\int \left(\dfrac{u^3 + 2u^2 - u}{3u}\right) du$

Hint: $\dfrac{u^3 + 2u^2 - u}{3u} = \dfrac{1}{3}u^2 + \dfrac{2}{3}u - \dfrac{1}{3}$

38. $\int \dfrac{x^4 - 1}{x^2}\, dx$

Hint: $\dfrac{x^4 - 1}{x^2} = x^2 - x^{-2}$

39. $\int (2t + 1)(t - 2)\, dt$

40. $\int u^{-2}(1 - u^2 + u^4)\, du$

41. $\int \dfrac{1}{x^2}(x^4 - 2x^2 + 1)\, dx$

42. $\int \sqrt{t}(t^2 + t - 1)\, dt$

43. $\int \dfrac{ds}{(s + 1)^{-2}}$

44. $\int \left(\sqrt{x} + \dfrac{3}{x} - 2e^x\right) dx$

45. $\int (e^t + t^e)\, dt$

46. $\int \left(\dfrac{1}{x^2} - \dfrac{1}{\sqrt[3]{x^2}} + \dfrac{1}{\sqrt{x}}\right) dx$

47. $\int \left(\dfrac{x^3 + x^2 - x + 1}{x^2}\right) dx$

Hint: Simplify the integrand first.

48. $\int \dfrac{t^3 + \sqrt[3]{t}}{t^2}\, dt$

Hint: Simplify the integrand first.

49. $\int \dfrac{(\sqrt{x} - 1)^2}{x^2}\, dx$

Hint: Simplify the integrand first.

50. $\int (x + 1)^2 \left(1 - \dfrac{1}{x}\right) dx$

Hint: Simplify the integrand first.

In Exercises 51–58, find $f(x)$ by solving the initial value problem.

51. $f'(x) = 2x + 1;\ f(1) = 3$

52. $f'(x) = 3x^2 - 6x;\ f(2) = 4$

53. $f'(x) = 3x^2 + 4x - 1;\ f(2) = 9$

54. $f'(x) = \dfrac{1}{\sqrt{x}};\ f(4) = 2$

55. $f'(x) = 1 + \dfrac{1}{x^2};\ f(1) = 2$

56. $f'(x) = e^x - 2x;\ f(0) = 2$

57. $f'(x) = \dfrac{x + 1}{x};\ f(1) = 1$

58. $f'(x) = 1 + e^x + \dfrac{1}{x};\ f(1) = 3 + e$

In Exercises 59–62, find the function f given that the slope of the tangent line to the graph of f at any point $(x, f(x))$ is $f'(x)$ and that the graph of f passes through the given point.

59. $f'(x) = \dfrac{1}{2}x^{-1/2};\ (2, \sqrt{2})$

60. $f'(t) = t^2 - 2t + 3;\ (1, 2)$

61. $f'(x) = e^x + x;\ (0, 3)$ **62.** $f'(x) = \dfrac{2}{x} + 1;\ (1, 2)$

63. VELOCITY OF A CAR   The velocity of a car (in feet/second) t sec after starting from rest is given by the function

$$f(t) = 2\sqrt{t} \qquad (0 \le t \le 30)$$

Find the car's position at any time t.

64. VELOCITY OF A MAGLEV   The velocity (in feet/second) of a maglev is

$$v(t) = 0.2t + 3 \qquad (0 \le t \le 120)$$

At $t = 0$, it is at the station. Find the function giving the position of the maglev at time t, assuming that the motion takes place along a straight stretch of track.

65. COST OF PRODUCING CLOCKS   The Lorimar Watch Company manufactures travel clocks. The daily marginal cost function associated with producing these clocks is

$$C'(x) = 0.000009x^2 - 0.009x + 8$$

where $C'(x)$ is measured in dollars/unit and x denotes the number of units produced. Management has determined that the daily fixed cost incurred in producing these clocks is $120. Find the total cost incurred by Lorimar in producing the first 500 travel clocks/day.

66. **REVENUE FUNCTIONS** The management of the Lorimar Watch Company has determined that the daily marginal revenue function associated with producing and selling their travel clocks is given by

$$R'(x) = -0.009x + 12$$

where x denotes the number of units produced and sold and $R'(x)$ is measured in dollars/unit.
a. Determine the revenue function $R(x)$ associated with producing and selling these clocks.
b. What is the demand equation that relates the whole-sale unit price with the quantity of travel clocks demanded?

67. **PROFIT FUNCTIONS** The Cannon Precision Instruments Corporation makes an automatic electronic flash with Thyrister circuitry. The estimated marginal profit associated with producing and selling these electronic flashes is

$$(-0.004x + 20)$$

dollars/unit/month when the production level is x units per month. Cannon's fixed cost for producing and selling these electronic flashes is $16,000/month. At what level of production does Cannon realize a maximum profit? What is the maximum monthly profit?

68. **COST OF PRODUCING GUITARS** The Carlota Music Company estimates that the marginal cost of manufacturing its Professional Series guitars is

$$C'(x) = 0.002x + 100$$

dollars/month when the level of production is x guitars/month. The fixed costs incurred by Carlota are $4000/month. Find the total monthly cost incurred by Carlota in manufacturing x guitars/month.

69. **QUALITY CONTROL** As part of a quality-control program, the chess sets manufactured by the Jones Brothers Company are subjected to a final inspection before packing. The rate of increase in the number of sets checked per hour by an inspector t hr into the 8 A.M. to 12 noon morning shift is approximately

$$N'(t) = -3t^2 + 12t + 45 \qquad (0 \le t \le 4)$$

a. Find an expression $N(t)$ that approximates the number of sets inspected at the end of t hours.
Hint: $N(0) = 0$.
b. How many sets does the average inspector check during a morning shift?

70. **BALLAST DROPPED FROM A BALLOON** A ballast is dropped from a stationary hot-air balloon that is hovering at an altitude of 400 ft. Its velocity after t sec is $-32t$ ft/sec.
a. Find the height $h(t)$ of the ballast from the ground at time t.
Hint: $h'(t) = -32t$ and $h(0) = 400$.
b. When will the ballast strike the ground?
c. Find the velocity of the ballast when it hits the ground.

Ballast

▭ **A calculator is recommended for Exercises 71–76.**

71. **CABLE TV SUBSCRIBERS** A study conducted by Tele-Cable, Inc., estimates that the number of cable television subscribers will grow at the rate of

$$100 + 210t^{3/4}$$

new subscribers/month t mo from the start date of the service. If 5000 subscribers signed up for the service before the starting date, how many subscribers will there be 16 mo from that date?

72. **AIR POLLUTION** On an average summer day, the level of carbon monoxide (CO) in a city's air is 2 parts per million (ppm). An environmental protection agency's study predicts that, unless more stringent measures are taken to protect the city's atmosphere, the CO concentration present in the air will increase at the rate of

$$0.003t^2 + 0.06t + 0.1$$

ppm/year t yr from now. If no further pollution-control efforts are made, what will be the CO concentration on an average summer day 5 yr from now?

73. **POPULATION GROWTH** The development of Astro World ("The Amusement Park of the Future") on the outskirts of a city will increase the city's population at the rate of

$$4500\sqrt{t} + 1000$$

people/year t yr from the start of construction. The population before construction is 30,000. Determine the projected population 9 yr after construction of the park has begun.

74. **OZONE POLLUTION** The rate of change of the level of ozone, an invisible gas that is an irritant and impairs breathing, present in the atmosphere on a certain May day in the city of Riverside is given by

$$R(t) = 3.2922t^2 - 0.366t^3 \qquad (0 < t < 11)$$

(measured in pollutant standard index per hour). Here, t is measured in hours, with $t = 0$ corresponding to 7 A.M. Find the ozone level $A(t)$ at any time t assuming that at 7 A.M. it is zero.
Hint: $A'(t) = R(t)$ and $A(0) = 0$.

75. **SURFACE AREA OF A HUMAN** Empirical data suggest that the surface area of a 180-cm-tall human body changes at the rate of

$$S'(W) = 0.131773W^{-0.575}$$

square meters/kilogram, where W is the weight of the body in kilograms. If the surface area of a 180-cm-tall human body weighing 70 kg is 1.886277 m², what is the surface area of a human body of the same height weighing 75 kg?

76. **FLIGHT OF A ROCKET** The velocity, in feet/second, of a rocket t sec into vertical flight is given by

$$v(t) = -3t^2 + 192t + 120$$

Find an expression $h(t)$ that gives the rocket's altitude, in feet, t sec after liftoff. What is the altitude of the rocket 30 sec after liftoff?
Hint: $h'(t) = v(t)$; $h(0) = 0$.

77. **BLOOD FLOW IN AN ARTERY** Nineteenth-century physician Jean Louis Marie Poiseuille discovered that the rate of change of the velocity of blood r cm from the central axis of an artery (in centimeters/second/centimeters) is given by

$$a(r) = -kr$$

where k is a constant. If the radius of an artery is R cm, find an expression for the velocity of blood as a function of r (see the accompanying figure).
Hint: $v'(r) = a(r)$ and $v(R) = 0$. (Why?)

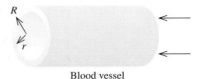

Blood vessel

78. **ACCELERATION OF A CAR** A car traveling along a straight road at 66 ft/sec accelerated to a speed of 88 ft/sec over a distance of 440 ft. What was the acceleration of the car, assuming it was constant?

79. **DECELERATION OF A CAR** What constant deceleration would a car moving along a straight road have to be subjected to if it were brought to rest from a speed of 88 ft/sec in 9 sec? What would be the stopping distance?

80. A tank has a constant cross-sectional area of 50 ft² and an orifice of constant cross-sectional area of $\frac{1}{2}$ ft² located at the bottom of the tank (see the accompanying figure).

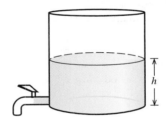

If the tank is filled with water to a height of h ft and allowed to drain, then the height of the water decreases at a rate that is described by the equation

$$\frac{dh}{dt} = -\frac{1}{25}\left(\sqrt{20} - \frac{t}{50}\right) \qquad (0 \le t \le 50\sqrt{20})$$

Find an expression for the height of the water at any time t if its height initially is 20 ft.

81. **LAUNCHING A FIGHTER AIRCRAFT** A fighter aircraft is launched from the deck of a Nimitz-class aircraft carrier with the help of a steam catapult. If the aircraft is to attain a takeoff speed of at least 240 ft/sec after traveling 800 ft along the flight deck, find the minimum acceleration it must be subjected to, assuming it is constant.

82. **BANK DEPOSITS** The Madison Finance Company opened two branches on September 1 ($t = 0$). Branch A is located in an established industrial park, and branch B is located in a fast-growing new development. The net rate at which money was deposited into branch A and branch B in the first 180 business days is given by the graphs of f and g, respectively (see the figure). Which branch has a

larger amount on deposit at the end of 180 business days? Justify your answer.

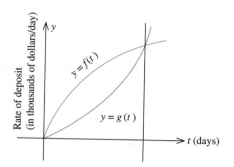

In Exercises 83–86, determine whether the statement is true or false. If it is true, explain why it is true. If it is false, give an example to show why it is false.

83. If F and G are antiderivatives of f on an interval I, then $F(x) = G(x) + C$ on I.

84. If F is an antiderivative of f on an interval I, then $\int f(x)\, dx = F(x)$.

85. If f and g are integrable, then $\int [2f(x) - 3g(x)]\, dx = 2 \int f(x)\, dx - 3 \int g(x)\, dx$.

86. If f and g are integrable, then $\int f(x)g(x)\, dx = [\int f(x)\, dx][\int g(x)\, dx]$.

S OLUTIONS TO S ELF-C HECK E XERCISES **6.1**

1. $\displaystyle\int \left(\frac{1}{\sqrt{x}} - \frac{2}{x} + 3e^x \right) dx = \int \left(x^{-1/2} - \frac{2}{x} + 3e^x \right) dx$

$$= \int x^{-1/2}\, dx - 2 \int \frac{1}{x}\, dx + 3 \int e^x\, dx$$

$$= 2x^{1/2} - 2 \ln|x| + 3e^x + C$$

$$= 2\sqrt{x} - 2 \ln|x| + 3e^x + C$$

2. The slope of the tangent line to the graph of the function f at any point $P(x, f(x))$ is given by the derivative f' of f. Thus, the first condition implies that

$$f'(x) = 3x^2 - 6x + 3$$

which, upon integration, yields

$$f(x) = \int (3x^2 - 6x + 3)\, dx$$

$$= x^3 - 3x^2 + 3x + k$$

where k is the constant of integration.

To evaluate k, we use the initial condition (2), which implies that $f(2) = 9$, or

$$9 = f(2) = 2^3 - 3(2)^2 + 3(2) + k$$

or $k = 7$. Hence, the required rule of definition of the function f is

$$f(x) = x^3 - 3x^2 + 3x + 7$$

3. Let $M(t)$ denote United Motors' market share at year t. Then,

$$M(t) = \int f(t)\, dt$$

$$= \int (-0.01875t^2 + 0.15t - 1.2)\, dt$$

$$= -0.00625t^3 + 0.075t^2 - 1.2t + C$$

To determine the value of C, we use the initial condition $M(0) = 48.4$, obtaining $C = 48.4$. Therefore,

$$M(t) = -0.00625t^3 + 0.075t^2 - 1.2t + 48.4$$

In particular, United Motors' market share of new cars at the beginning of 2001 is given by

$$M(12) = -0.00625(12)^3 + 0.075(12)^2$$
$$-1.2(12) + 48.4 = 34$$

or 34%.

6.2 Integration by Substitution

In Section 6.1 we developed certain rules of integration that are closely related to the corresponding rules of differentiation in Chapters 3 and 5. In this section we introduce a method of integration called the **method of substitution,** which is related to the chain rule for differentiating functions. When used in conjunction with the rules of integration developed earlier, the method of substitution is a powerful tool for integrating a large class of functions.

HOW THE METHOD OF SUBSTITUTION WORKS

Consider the indefinite integral

$$\int 2(2x + 4)^5 \, dx \tag{3}$$

One way of evaluating this integral is to expand the expression $(2x + 4)^5$ and then integrate the resulting integrand term by term. As an alternative approach, let's see if we can simplify the integral by making a change of variable. Write

$$u = 2x + 4$$

with differential

$$du = 2 \, dx$$

If we substitute these quantities into Equation (3), we obtain

$$\int 2(2x + 4)^5 \, dx = \int (2x + 4)^5 (2 \, dx) = \int u^5 \, du$$

$$\underset{\text{Rewriting}}{\uparrow} \qquad\qquad \underset{\begin{cases} u = 2x + 4 \\ du = 2 \, dx \end{cases}}{\uparrow}$$

Now, the last integral involves a power function and is easily evaluated using Rule 2 of Section 6.1. Thus,

$$\int u^5 \, du = \frac{1}{6} u^6 + C$$

Therefore, using this result and replacing u by $2x + 4$, we obtain

$$\int 2(2x + 4)^5 \, dx = \frac{1}{6}(2x + 4)^6 + C$$

We can verify that the foregoing result is indeed correct by computing

$$\frac{d}{dx}\left[\frac{1}{6}(2x + 4)^6\right] = \frac{1}{6} \cdot 6(2x + 4)^5(2) \qquad \text{(Using the chain rule)}$$

$$= 2(2x + 4)^5$$

and observing that the last expression is just the integrand of (3).

THE METHOD OF INTEGRATION BY SUBSTITUTION

To see why the approach used in evaluating the integral in (3) is successful, write

$$f(x) = x^5 \qquad \text{and} \qquad g(x) = 2x + 4$$

Then, $g'(x) = 2 \, dx$. Furthermore, the integrand of (3) is just the composition of f and g. Thus,

$$(f \circ g)(x) = f(g(x))$$
$$= [g(x)]^5 = (2x + 4)^5$$

Therefore, (3) can be written as

$$\int f(g(x))g'(x) \, dx \qquad \qquad \textbf{(4)}$$

Next, let's show that an integral having the form (4) can always be written as

$$\int f(u) \, du \qquad \qquad \textbf{(5)}$$

Suppose F is an antiderivative of f. By the chain rule, we have

$$\frac{d}{dx}[F(g(x))] = F'(g(x))g'(x)$$

Therefore,

$$\int F'(g(x))g'(x) \, dx = F(g(x)) + C$$

Letting $F' = f$ and making the substitution $u = g(x)$, we have

$$\int f(g(x))g'(x) \, dx = F(u) + C = \int F'(u) \, du = \int f(u) \, du$$

as we wished to show. Thus, if the transformed integral is readily evaluated, as is the case with the integral (3), then the method of substitution will prove successful.

Before we look at more examples, let's summarize the steps involved in integration by substitution.

Integration by Substitution

> **Step 1** Let $u = g(x)$, where $g(x)$ is part of the integrand, usually the "inside function" of the composite function $f(g(x))$.
>
> **Step 2** Compute $du = g'(x)\, dx$.
>
> **Step 3** Use the substitution $u = g(x)$ and $du = g'(x)\, dx$ to convert the *entire* integral into one involving *only u*.
>
> **Step 4** Evaluate the resulting integral.
>
> **Step 5** Replace u by $g(x)$ to obtain the final solution as a function of x.

REMARK Sometimes we need to consider different choices of g for the substitution $u = g(x)$ in order to carry out step 3 and/or step 4. ■ ■ ■

EXAMPLE 1 Find $\int 2x(x^2 + 3)^4\, dx$.

SOLUTION ✔

Step 1 Observe that the integrand involves the composite function $(x^2 + 3)^4$ with "inside function" $g(x) = x^2 + 3$. So, we choose $u = x^2 + 3$.

Step 2 Compute $du = 2x\, dx$.

Step 3 Making the substitution $u = x^2 + 3$ and $du = 2x\, dx$, we obtain

$$\int 2x(x^2 + 3)^4\, dx = \int (x^2 + 3)^4(2x\, dx) = \int u^4\, du$$

$$\uparrow$$
$$\text{Rewriting}$$

an integral involving only the variable u.

Step 4 Evaluate

$$\int u^4\, du = \frac{1}{5}u^5 + C$$

Step 5 Replacing u by $x^2 + 3$, we obtain

$$\int 2x(x^2 + 3)^4\, dx = \frac{1}{5}(x^2 + 3)^5 + C$$

■ ■ ■ ■

EXAMPLE 2 Find $\int 3\sqrt{3x + 1}\, dx$.

SOLUTION ✔

Step 1 The integrand involves the composite function $\sqrt{3x + 1}$ with "inside function" $g(x) = 3x + 1$. So, let $u = 3x + 1$.

Step 2 Compute $du = 3\, dx$.

Step 3 Making the substitution $u = 3x + 1$ and $du = 3\, dx$, we obtain

$$\int 3\sqrt{3x + 1}\, dx = \int \sqrt{3x + 1}(3\, dx) = \int \sqrt{u}\, du$$

an integral involving only the variable u.

Step 4   Evaluate

$$\int \sqrt{u} \, du = \int u^{1/2} \, du = \frac{2}{3} u^{3/2} + C$$

Step 5   Replacing u by $3x + 1$, we obtain

$$\int 3\sqrt{3x + 1} \, dx = \frac{2}{3}(3x + 1)^{3/2} + C$$ ■ ■ ■ ■

EXAMPLE 3   Find $\int x^2(x^3 + 1)^{3/2} \, dx$.

SOLUTION ✔   **Step 1**   The integrand contains the composite function $(x^3 + 1)^{3/2}$ with "inside function" $g(x) = x^3 + 1$. So, let $u = x^3 + 1$.

Step 2   Compute $du = 3x^2 \, dx$.

Step 3   Making the substitution $u = x^3 + 1$ and $du = 3x^2 \, dx$, or $x^2 \, dx = \frac{1}{3} \, du$, we obtain

$$\int x^2(x^3 + 1)^{3/2} \, dx = \int (x^3 + 1)^{3/2}(x^2 \, dx)$$

$$= \int u^{3/2} \left(\frac{1}{3} \, du \right) = \frac{1}{3} \int u^{3/2} \, du$$

an integral involving only the variable u.

Step 4   We evaluate

$$\frac{1}{3} \int u^{3/2} \, du = \frac{1}{3} \cdot \frac{2}{5} u^{5/2} + C = \frac{2}{15} u^{5/2} + C$$

Step 5   Replacing u by $x^3 + 1$, we obtain

$$\int x^2(x^3 + 1)^{3/2} \, dx = \frac{2}{15}(x^3 + 1)^{5/2} + C$$ ■ ■ ■ ■

Group Discussion

Let $f(x) = x^2(x^3 + 1)^{3/2}$. Using the result of Example 3, we see that an antiderivative of f is $F(x) = \frac{2}{15}(x^3 + 1)^{5/2}$. However, in terms of u (where $u = x^3 + 1$), an antiderivative of f is $G(u) = \frac{2}{15}u^{5/2}$. Compute $F(2)$. Next, suppose we want to compute $F(2)$ using the function G instead. At what value of u should you evaluate $G(u)$ in order to obtain the desired result? Explain your answer.

In the remaining examples, we drop the practice of labeling the steps involved in evaluating each integral.

EXAMPLE 4   Find $\int e^{-3x}\, dx$.

SOLUTION ✓   Let $u = -3x$ so that $du = -3\, dx$, or $dx = -\frac{1}{3}\, du$. Then,

$$\int e^{-3x}\, dx = \int e^u \left(-\frac{1}{3}\, du \right) = -\frac{1}{3} \int e^u\, du$$

$$= -\frac{1}{3} e^u + C = -\frac{1}{3} e^{-3x} + C$$

EXAMPLE 5   Find $\int \dfrac{x}{3x^2 + 1}\, dx$.

SOLUTION ✓   Let $u = 3x^2 + 1$. Then, $du = 6x\, dx$, or $x\, dx = \frac{1}{6}\, du$. Making the appropriate substitutions, we have

$$\int \frac{x}{3x^2 + 1}\, dx = \int \frac{\frac{1}{6}}{u}\, du$$

$$= \frac{1}{6} \int \frac{1}{u}\, du$$

$$= \frac{1}{6} \ln |u| + C$$

$$= \frac{1}{6} \ln(3x^2 + 1) + C \qquad \text{(Since } 3x^2 + 1 > 0\text{)}$$

EXAMPLE 6   Find $\int \dfrac{(\ln x)^2}{2x}\, dx$.

SOLUTION ✓   Let $u = \ln x$. Then,

$$du = \frac{d}{dx} (\ln x)\, dx = \frac{1}{x}\, dx$$

$$\int \frac{(\ln x)^2}{2x}\, dx = \frac{1}{2} \int \frac{(\ln x)^2}{x}\, dx$$

$$= \frac{1}{2} \int u^2\, du$$

$$= \frac{1}{6} u^3 + C$$

$$= \frac{1}{6} (\ln x)^3 + C$$

> **Group Discussion**
>
> Suppose $\int f(u)\,du = F(u) + C.$
>
> **1.** Show that $\int f(ax + b)\,dx = \dfrac{1}{a} F(ax + b) + C.$
>
> **2.** How can you use this result to facilitate the evaluation of integrals such as $\int (2x + 3)^5\,dx$ and $\int e^{3x-2}\,dx$? Explain your answer.

APPLICATIONS

Examples 7 and 8 show how the method of substitution can be used in practical situations.

EXAMPLE 7 In 1990 the head of the research and development department of the Soloron Corporation claimed that the cost of producing solar cell panels would drop at the rate of

$$\frac{58}{(3t + 2)^2} \qquad (0 \le t \le 10)$$

dollars per peak watt for the next t years, with $t = 0$ corresponding to the beginning of the year 1990. (A peak watt is the power produced at noon on a sunny day.) In 1990 the panels, which are used for photovoltaic power systems, cost \$10 per peak watt. Find an expression giving the cost per peak watt of producing solar cell panels at the beginning of year t. What was the cost at the beginning of 2000?

SOLUTION ✔ Let $C(t)$ denote the cost per peak watt for producing solar cell panels at the beginning of year t. Then,

$$C'(t) = -\frac{58}{(3t + 2)^2}$$

Integrating, we find that

$$C(t) = \int \frac{-58}{(3t + 2)^2}\,dt$$

$$= -58 \int (3t + 2)^{-2}\,dt$$

Let $u = 3t + 2$ so that

$$du = 3\,dt \qquad \text{or} \qquad dt = \frac{1}{3}\,du$$

Then,

$$C(t) = -58 \left(\frac{1}{3}\right) \int u^{-2}\,du$$

$$= -\frac{58}{3}(-1)u^{-1} + k$$

$$= \frac{58}{3(3t + 2)} + k$$

where k is an arbitrary constant. To determine the value of k, note that the cost per peak watt of producing solar cell panels at the beginning of 1990 $(t = 0)$ was 10, or $C(0) = 10$. This gives

$$C(0) = \frac{58}{3(2)} + k = 10$$

or $k = \frac{1}{3}$. Therefore, the required expression is given by

$$C(t) = \frac{58}{3(3t + 2)} + \frac{1}{3}$$

$$= \frac{58 + (3t + 2)}{3(3t + 2)} = \frac{3t + 60}{3(3t + 2)}$$

$$= \frac{t + 20}{3t + 2}$$

The cost per peak watt for producing solar cell panels at the beginning of 2000 is given by

$$C(10) = \frac{10 + 20}{3(10) + 2} \approx 0.94$$

or approximately $.94 per peak watt.   ■ ■ ■ ■

Exploring with Technology

Refer to Example 7.

1. Use a graphing utility to plot the graph of

$$C(t) = \frac{t + 20}{3t + 2}$$

using the viewing rectangle $[0, 10] \times [0, 5]$. Then, use the numerical differentiation capability of the graphing utility to compute $C'(10)$.

2. Plot the graph of

$$C'(t) = -\frac{58}{(3t + 2)^2}$$

using the viewing rectangle $[0, 10] \times [-10, 0]$. Then, use the evaluation capability of the graphing utility to find $C'(10)$. Is this value of $C'(10)$ the same as that obtained in part 1? Explain your answer.

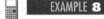   EXAMPLE **8**   A study prepared by the marketing department of the Universal Instruments Company forecasts that, after its new line of Galaxy Home Computers is introduced into the market, sales will grow at the rate of

$$2000 - 1500e^{-0.05t} \qquad (0 \le t \le 60)$$

units per month. Find an expression that gives the total number of computers that will sell t months after they become available on the market. How many computers will Universal sell in the first year they are on the market?

SOLUTION ✔　Let $N(t)$ denote the total number of computers that may be expected to be sold t months after their introduction in the market. Then, the rate of growth of sales is given by $N'(t)$ units per month. Thus,

$$N'(t) = 2000 - 1500e^{-0.05t}$$

so that

$$N(t) = \int (2000 - 1500e^{-0.05t})\,dt$$

$$= \int 2000\,dt - 1500 \int e^{-0.05t}\,dt$$

Upon integrating the second integral by the method of substitution, we obtain

$$N(t) = 2000t + \frac{1500}{0.05}e^{-0.05t} + C \qquad \text{(Let } u = -0.05t,$$
$$\text{then } du = -0.05\,dt.)$$

$$= 2000t + 30{,}000e^{-0.05t} + C$$

To determine the value of C, note that the number of computers sold at the end of month 0 is nil, so $N(0) = 0$. This gives

$$N(0) = 30{,}000 + C = 0 \qquad \text{(Since } e^0 = 1)$$

or $C = -30{,}000$. Therefore, the required expression is given by

$$N(t) = 2000t + 30{,}000e^{-0.05t} - 30{,}000$$
$$= 2000t + 30{,}000(e^{-0.05t} - 1)$$

The number of computers that Universal can expect to sell in the first year is given by

$$N(12) = 2000(12) + 30{,}000(e^{-0.05(12)} - 1)$$
$$= 10{,}464 \text{ units}$$
■ ■ ■ ■

SELF-CHECK EXERCISES 6.2

1. Evaluate $\int \sqrt{2x + 5}\,dx$.

2. Evaluate $\int \dfrac{x^2}{(2x^3 + 1)^{3/2}}\,dx$.

3. Evaluate $\int xe^{2x^2-1}\,dx$.

4. According to a joint study conducted by Oxnard's Environmental Management Department and a state government agency, the concentration of carbon monoxide (CO) in the air due to automobile exhaust is increasing at the rate given by

$$f(t) = \frac{8(0.1t + 1)}{300(0.2t^2 + 4t + 64)^{1/3}}$$

parts per million (ppm) per year t. Currently, the CO concentration due to automobile exhaust is 0.16 ppm. Find an expression giving the CO concentration t yr from now.

Solutions to Self-Check Exercises 6.2 can be found on page 465.

6.2 Exercises

In Exercises 1–50, find the indefinite integral.

1. $\int 4(4x + 3)^4 \, dx$

2. $\int 4x(2x^2 + 1)^7 \, dx$

3. $\int (x^3 - 2x)^2(3x^2 - 2) \, dx$

4. $\int (3x^2 - 2x + 1)(x^3 - x^2 + x)^4 \, dx$

5. $\int \frac{4x}{(2x^2 + 3)^3} \, dx$

6. $\int \frac{3x^2 + 2}{(x^3 + 2x)^2} \, dx$

7. $\int 3t^2 \sqrt{t^3 + 2} \, dt$

8. $\int 3t^2(t^3 + 2)^{3/2} \, dt$

9. $\int (x^2 - 1)^9 x \, dx$

10. $\int x^2(2x^3 + 3)^4 \, dx$

11. $\int \frac{x^4}{1 - x^5} \, dx$

12. $\int \frac{x^2}{\sqrt{x^3 - 1}} \, dx$

13. $\int \frac{2}{x - 2} \, dx$

14. $\int \frac{x^2}{x^3 - 3} \, dx$

15. $\int \frac{0.3x - 0.2}{0.3x^2 - 0.4x + 2} \, dx$

16. $\int \frac{2x^2 + 1}{0.2x^3 + 0.3x} \, dx$

17. $\int \frac{x}{3x^2 - 1} \, dx$

18. $\int \frac{x^2 - 1}{x^3 - 3x + 1} \, dx$

19. $\int e^{-2x} \, dx$

20. $\int e^{-0.02x} \, dx$

21. $\int e^{2-x} \, dx$

22. $\int e^{2t+3} \, dt$

23. $\int xe^{-x^2} \, dx$

24. $\int x^2 e^{x^3-1} \, dx$

25. $\int (e^x - e^{-x}) \, dx$

26. $\int (e^{2x} + e^{-3x}) \, dx$

27. $\int \frac{e^x}{1 + e^x} \, dx$

28. $\int \frac{e^{2x}}{1 + e^{2x}} \, dx$

29. $\int \frac{e^{\sqrt{x}}}{\sqrt{x}} \, dx$

30. $\int \frac{e^{-1/x}}{x^2} \, dx$

31. $\int \frac{e^{3x} + x^2}{(e^{3x} + x^3)^3} \, dx$

32. $\int \frac{e^x - e^{-x}}{(e^x + e^{-x})^{3/2}} \, dx$

33. $\int e^{2x}(e^{2x} + 1)^3 \, dx$

34. $\int e^{-x}(1 + e^{-x}) \, dx$

35. $\int \frac{\ln 5x}{x} \, dx$

36. $\int \frac{(\ln u)^3}{u} \, du$

37. $\int \frac{1}{x \ln x} \, dx$

38. $\int \frac{1}{x(\ln x)^2} \, dx$

39. $\int \frac{\sqrt{\ln x}}{x} \, dx$

40. $\int \frac{(\ln x)^{7/2}}{x} \, dx$

41. $\int \left(xe^{x^2} - \frac{x}{x^2 + 2} \right) dx$

42. $\int \left(xe^{-x^2} + \frac{e^x}{e^x + 3} \right) dx$

43. $\int \frac{x + 1}{\sqrt{x - 1}} \, dx$

Hint: Let $u = \sqrt{x} - 1$.

44. $\int \frac{e^{-u} - 1}{e^{-u} + u} \, du$

Hint: Let $v = e^{-u} + u$.

45. $\int x(x - 1)^5 \, dx$

Hint: $u = x - 1$ implies $x = u + 1$.

46. $\int \frac{t}{t + 1} \, dt$

Hint: $\frac{t}{t + 1} = 1 - \frac{1}{t + 1}$.

47. $\int \frac{1 - \sqrt{x}}{1 + \sqrt{x}} \, dx$

Hint: Let $u = 1 + \sqrt{x}$.

48. $\int \frac{1 + \sqrt{x}}{1 - \sqrt{x}} \, dx$

Hint: Let $u = 1 - \sqrt{x}$.

49. $\int v^2(1 - v)^6 \, dv$

Hint: Let $u = 1 - v$.

50. $\int x^3(x^2 + 1)^{3/2} \, dx$

Hint: Let $u = x^2 + 1$.

In Exercises 51–54, find the function f given that the slope of the tangent line to the graph of f at any point (x, f(x)) is f'(x) and that the graph of f passes through the given point.

51. $f'(x) = 5(2x - 1)^4$; $(1, 3)$

52. $f'(x) = \frac{3x^2}{2\sqrt{x^3 - 1}}$; $(1, 1)$

53. $f'(x) = -2xe^{-x^2+1}$; $(1, 0)$

54. $f'(x) = 1 - \frac{2x}{x^2 + 1}$; $(0, 2)$

55. STUDENT ENROLLMENT The registrar of Kellogg University estimates that the total student enrollment in the Continuing Education division will grow at the rate of

$$N'(t) = 2000(1 + 0.2t)^{-3/2}$$

students/year t yr from now. If the current student enrollment is 1000, find an expression giving the total student enrollment t yr from now. What will be the student enrollment 5 yr from now?

56. **TV VIEWERS: NEWSMAGAZINE SHOWS** The number of viewers of a weekly TV newsmagazine show, introduced in the 1995 season, has been increasing at the rate of

$$3\left(2 + \frac{1}{2}t\right)^{-1/3} \qquad (1 \le t \le 6)$$

million viewers/year in its tth year on the air. The number of viewers of the program during its first year on the air is given by $9(5/2)^{2/3}$ million. Find how many viewers were expected in the 2000 season.

57. **DEMAND: LADIES' BOOTS** The rate of change of the unit price p (in dollars) of Apex ladies' boots is given by

$$p'(x) = \frac{-250x}{(16 + x^2)^{3/2}}$$

where x is the quantity demanded daily in units of a hundred. Find the demand function for these boots if the quantity demanded daily is 300 pairs ($x = 3$) when the unit price is \$50/pair.

■ **A calculator is recommended for the remaining exercises.**

58. **SUPPLY: LADIES' BOOTS** The rate of change of the unit price p (in dollars) of Apex ladies' boots is given by

$$p'(x) = \frac{240x}{(5 - x)^2}$$

where x is the number of pairs that the supplier will make available in the market daily when the unit price is \$p/pair. Find the supply equation for these boots if the quantity the supplier is willing to make available is 200 pairs daily ($x = 2$) when the unit price is \$50/pair.

59. **OIL SPILL** In calm waters the oil spilling from the ruptured hull of a grounded tanker forms an oil slick that is circular in shape. If the radius r of the circle is increasing at the rate of

$$r'(t) = \frac{30}{\sqrt{2t + 4}}$$

feet/minute t min after the rupture occurs, find an expression for the radius at any time t. How large is the polluted area 16 min after the rupture occurred?
Hint: $r(0) = 0$.

60. **LIFE EXPECTANCY OF A FEMALE** Suppose in a certain country the life expectancy at birth of a female is changing at the rate of

$$g'(t) = \frac{5.45218}{(1 + 1.09t)^{0.9}}$$

years/year. Here, t is measured in years, and $t = 0$ corresponds to the beginning of 1900. Find an expression $g(t)$ giving the life expectancy at birth (in years) of a female in that country if the life expectancy at the beginning of 1900 is 50.02 yr. What is the life expectancy at birth of a female born in the year 2000 in that country?

61. **AVERAGE BIRTH HEIGHT OF BOYS** Using data collected at Kaiser Hospital, pediatricians estimate that the average height of male children changes at the rate of

$$h'(t) = \frac{52.8706e^{-0.3277t}}{(1 + 2.449e^{-0.3277t})^2}$$

inches/year, where the child's height $h(t)$ is measured in inches and t, the child's age, is measured in years, with $t = 0$ corresponding to the age at birth. Find an expression $h(t)$ for the average height of a boy at age t if the height at birth of an average child is 19.4 in. What is the height of an average 8-yr-old boy?

62. **LEARNING CURVES** The average student enrolled in the 20-wk Court Reporting I course at the American Institute of Court Reporting progresses according to the rule

$$N'(t) = 6e^{-0.05t} \qquad (0 \le t \le 20)$$

where $N'(t)$ measures the rate of change in the number of words per minute of dictation the student takes in machine shorthand after t wk in the course. Assuming that the average student enrolled in this course begins with a dictation speed of 60 words/minute, find an expression $N(t)$ that gives the dictation speed of the student after t wk in the course.

63. **SALES: LOUDSPEAKERS** In the first year they appeared in the market, 2000 pairs of Acrosonic model F loudspeaker systems were sold. Since then, sales of these loudspeaker systems have been growing at the rate of

$$f'(t) = 2000(3 - 2e^{-t})$$

units/year, where t denotes the number of years these systems have been on the market. Determine the number of systems that were sold in the first 5 yr after their introduction.

64. **AMOUNT OF GLUCOSE IN THE BLOODSTREAM** Suppose a patient is given a continuous intravenous infusion of glucose at a constant rate of r mg/min. Then, the rate at which the

amount of glucose in the bloodstream is changing at time t due to this infusion is given by

$$A'(t) = re^{-at}$$

mg/min, where a is a positive constant associated with the rate at which excess glucose is eliminated from the bloodstream and is dependent on the patient's metabolism rate. Derive an expression for the amount of glucose in the bloodstream at time t.

Hint: $A(0) = 0$.

65. **CONCENTRATION OF A DRUG IN AN ORGAN**   A drug is carried into an organ of volume V cm^3 by a liquid that enters the organ at the rate of a cm^3/sec and leaves it at the rate of b cm^3/sec. The concentration of the drug in the liquid entering the organ is c g/cm^3. If the concentration of the drug in the organ at time t is increasing at the rate of

$$x'(t) = \frac{1}{V}(ac - bx_0)e^{-bt/V}$$

g/cm^3/ sec, and the concentration of the drug in the organ initially is x_0 g/cm^3, show that the concentration of the drug in the organ at time t is given by

$$x(t) = \frac{ac}{b} + \left(x_0 - \frac{ac}{b}\right)e^{-bt/V}$$

SOLUTIONS TO SELF-CHECK EXERCISES 6.2

1. Let $u = 2x + 5$. Then, $du = 2\,dx$, or $dx = \frac{1}{2}\,du$. Making the appropriate substitutions, we have

$$\int \sqrt{2x + 5}\,dx = \int \sqrt{u}\left(\frac{1}{2}\,du\right) = \frac{1}{2}\int u^{1/2}\,du$$

$$= \frac{1}{2}\left(\frac{2}{3}\right)u^{3/2} + C$$

$$= \frac{1}{3}(2x + 5)^{3/2} + C$$

2. Let $u = 2x^3 + 1$, so that $du = 6x^2\,dx$, or $x^2\,dx = \frac{1}{6}\,du$. Making the appropriate substitutions, we have

$$\int \frac{x^2}{(2x^3 + 1)^{3/2}}\,dx = \int \frac{\left(\frac{1}{6}\right)du}{u^{3/2}} = \frac{1}{6}\int u^{-3/2}\,du$$

$$= \left(\frac{1}{6}\right)(-2)u^{-1/2} + C$$

$$= -\frac{1}{3}(2x^3 + 1)^{-1/2} + C$$

$$= -\frac{1}{3\sqrt{2x^3 + 1}} + C$$

3. Let $u = 2x^2 - 1$, so that $du = 4x\,dx$, or $x\,dx = \frac{1}{4}\,du$. Then,

$$\int xe^{2x^2-1}\,dx = \frac{1}{4}\int e^u\,du$$

$$= \frac{1}{4}e^u + C$$

$$= \frac{1}{4}e^{2x^2-1} + C$$

4. Let $C(t)$ denote the CO concentration in the air due to automobile exhaust t yr from now. Then,

$$C'(t) = f(t) = \frac{8(0.1t + 1)}{300(0.2t^2 + 4t + 64)^{1/3}}$$

$$= \frac{8}{300}(0.1t + 1)(0.2t^2 + 4t + 64)^{-1/3}$$

Integrating, we find

$$C(t) = \int \frac{8}{300}(0.1t + 1)(0.2t^2 + 4t + 64)^{-1/3}\, dt$$

$$= \frac{8}{300}\int (0.1t + 1)(0.2t^2 + 4t + 64)^{-1/3}\, dt$$

Let $u = 0.2t^2 + 4t + 64$, so that $du = (0.4t + 4)\, dt = 4(0.1t + 1)\, dt$, or

$$(0.1t + 1)\, dt = \frac{1}{4}\, du$$

Then,

$$C(t) = \frac{8}{300}\left(\frac{1}{4}\right)\int u^{-1/3}\, du$$

$$= \frac{1}{150}\left(\frac{3}{2}u^{2/3}\right) + k$$

$$= 0.01(0.2t^2 + 4t + 64)^{2/3} + k$$

where k is an arbitrary constant. To determine the value of k, we use the condition $C(0) = 0.16$, obtaining

$$C(0) = 0.16 = 0.01(64)^{2/3} + k$$

$$0.16 = 0.16 + k$$

$$k = 0$$

Therefore,

$$C(t) = 0.01(0.2t^2 + 4t + 64)^{2/3}$$

6.3 Area and the Definite Integral

AN INTUITIVE LOOK

Suppose a certain state's annual rate of petroleum consumption over a 4-year period is constant and is given by the function

$$f(t) = 1.2 \qquad (0 \le t \le 4)$$

where t is measured in years and $f(t)$ in millions of barrels per year. Then, the state's total petroleum consumption over the period of time in question is

$$(1.2)(4 - 0) \qquad \text{(Rate of consumption} \cdot \text{time elapsed)}$$

FIGURE 6.5
The total petroleum consumption is given by the area of the rectangular region.

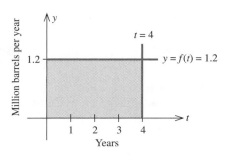

FIGURE 6.6
The daily petroleum consumption is given by the "area" of the shaded region.

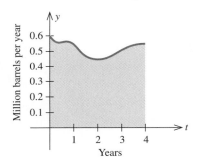

or 4.8 million barrels. If you examine the graph of f shown in Figure 6.5, you will see that this total is just the area of the rectangular region bounded above by the graph of f, below by the t-axis, and to the left and right by the vertical lines $t = 0$ (the y-axis) and $t = 4$, respectively.

Figure 6.6 shows the actual petroleum consumption of a certain New England state over a 4-year period from 1990 ($t = 0$) to 1994 ($t = 4$). Observe that the rate of consumption is not constant; that is, the function f is not a constant function. What is the state's total petroleum consumption over this 4-year period? It seems reasonable to conjecture that it is given by the "area" of the region bounded above by the graph of f, below by the t-axis, and to the left and right by the vertical lines $t = 0$ and $t = 4$, respectively.

This example raises two questions:

1. What is the "area" of the region shown in Figure 6.6?
2. How do we compute this area?

FIGURE 6.7
The area under the graph of f on $[a, b]$

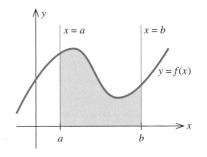

THE AREA PROBLEM

The preceding example touches on the second fundamental problem in calculus: Calculate the area of the region bounded by the graph of a nonnegative function f, the x-axis, and the vertical lines $x = a$ and $x = b$ (Figure 6.7). This area is called the **area under the graph of f** on the interval $[a, b]$, or from a to b.

DEFINING AREA—TWO EXAMPLES

Just as we used the slopes of secant lines (quantities that we could compute) to help us define the slope of the tangent line to a point on the graph of a function, we now adopt a parallel approach and use the areas of rectangles (quantities that we can compute) to help us define the area under the graph of a function. We begin by looking at a specific example.

EXAMPLE 1

Let $f(x) = x^2$ and consider the region R under the graph of f on the interval $[0, 1]$ (Figure 6.8a). To obtain an approximation of the area of R, let's construct four nonoverlapping rectangles as follows: Divide the interval $[0, 1]$ into four subintervals

$$\left[0, \frac{1}{4}\right], \quad \left[\frac{1}{4}, \frac{1}{2}\right], \quad \left[\frac{1}{2}, \frac{3}{4}\right], \quad \left[\frac{3}{4}, 1\right]$$

of equal length $\frac{1}{4}$. Next, construct four rectangles with these subintervals as bases and with heights given by the values of the function at the midpoints

$$\frac{1}{8}, \quad \frac{3}{8}, \quad \frac{5}{8}, \quad \frac{7}{8}$$

of each subinterval. Then, each of these rectangles has width $\frac{1}{4}$ and height

$$f\left(\frac{1}{8}\right), \quad f\left(\frac{3}{8}\right), \quad f\left(\frac{5}{8}\right), \quad f\left(\frac{7}{8}\right)$$

respectively (Figure 6.8b).

FIGURE 6.8
The area of the region under the graph of f on [0, 1] in (a) is approximated by the sum of the areas of the four rectangles in (b).

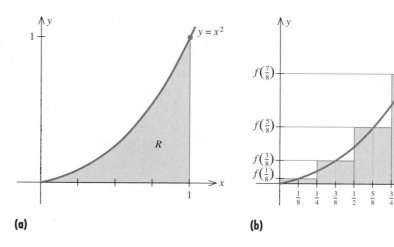

(a)   (b)

If we approximate the area A of S by the sum of the areas of the four rectangles, we obtain

$$A \approx \frac{1}{4} f\left(\frac{1}{8}\right) + \frac{1}{4} f\left(\frac{3}{8}\right) + \frac{1}{4} f\left(\frac{5}{8}\right) + \frac{1}{4} f\left(\frac{7}{8}\right)$$

$$= \frac{1}{4}\left[f\left(\frac{1}{8}\right) + f\left(\frac{3}{8}\right) + f\left(\frac{5}{8}\right) + f\left(\frac{7}{8}\right)\right]$$

$$= \frac{1}{4}\left[\left(\frac{1}{8}\right)^2 + \left(\frac{3}{8}\right)^2 + \left(\frac{5}{8}\right)^2 + \left(\frac{7}{8}\right)^2\right] \qquad \text{[Recall that } f(x) = x^2.]$$

$$= \frac{1}{4}\left(\frac{1}{64} + \frac{9}{64} + \frac{25}{64} + \frac{49}{64}\right) = \frac{21}{64}$$

or approximately 0.328125 square unit.   ■ ■ ■ ■

Following the procedure of Example 1, we can obtain approximations of the area of the region R using any number n of rectangles ($n = 4$ in Example 1). Figure 6.9a shows the approximation of the area A of R using 8 rectangles ($n = 8$), and Figure 6.9b shows the approximation of the area A of R using 16 rectangles.

FIGURE 6.9
As n increases, the number of rectangles increases, and the approximation improves.

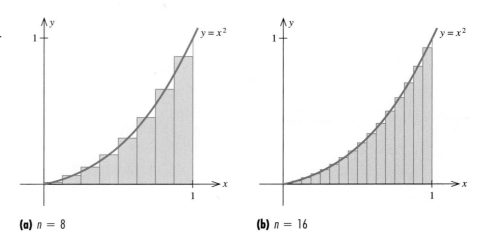

(a) $n = 8$ **(b)** $n = 16$

These figures suggest that the approximations seem to get better as n increases. This is borne out by the results given in Table 6.1, which were obtained using a computer.

Table 6.1							
Number of Rectangles n	4	8	16	32	64	100	200
Approximation of A	0.328125	0.332031	0.333008	0.333252	0.333313	0.333325	0.333331

Our computations seem to suggest that the approximations approach the number $\frac{1}{3}$ as n gets larger and larger. This result suggests that we *define* the area of the region under the graph of $f(x) = x^2$ on the interval $[0, 1]$ to be $\frac{1}{3}$ square unit.

In Example 1 we chose the *midpoint* of each subinterval as the point at which to evaluate $f(x)$ to obtain the height of the approximating rectangle. Let's consider another example, this time choosing the *left end point* of each subinterval.

EXAMPLE 2   Let R be the region under the graph of $f(x) = 16 - x^2$ on the interval $[1, 3]$. Find an approximation of the area A of R using four subintervals of $[1, 3]$ of equal length and picking the left end point of each subinterval to evaluate $f(x)$ to obtain the height of the approximating rectangle.

FIGURE 6.10
The area of R in (a) is approximated by
the sum of the areas of the four rectangles
in (b).

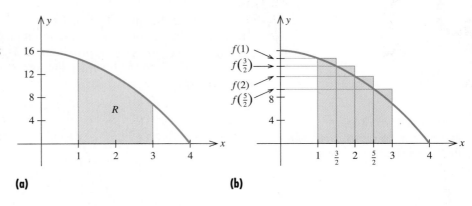

(a)   (b)

SOLUTION ✔   The graph of f is sketched in Figure 6.10a. Since the length of $[1, 3]$ is 2, we see that the length of each subinterval is $\frac{2}{4}$, or $\frac{1}{2}$. Therefore, the four subintervals are

$$\left[1, \frac{3}{2}\right], \qquad \left[\frac{3}{2}, 2\right], \qquad \left[2, \frac{5}{2}\right], \qquad \left[\frac{5}{2}, 3\right]$$

The left end points of these subintervals are 1, $\frac{3}{2}$, 2, and $\frac{5}{2}$, respectively, so the heights of the approximating rectangles are $f(1)$, $f(\frac{3}{2})$, $f(2)$, and $f(\frac{5}{2})$, respectively (Figure 6.10b). Therefore, the required approximation is

$$A \approx \frac{1}{2} f(1) + \frac{1}{2} f\left(\frac{3}{2}\right) + \frac{1}{2} f(2) + \frac{1}{2} f\left(\frac{5}{2}\right)$$

$$= \frac{1}{2}\left[f(1) + f\left(\frac{3}{2}\right) + f(2) + f\left(\frac{5}{2}\right)\right]$$

$$= \frac{1}{2}\left\{[16 - (1)^2] + \left[16 - \left(\frac{3}{2}\right)^2\right]\right.$$

$$\left. + [16 - (2)^2] + \left[16 - \left(\frac{5}{2}\right)^2\right]\right\} \qquad \text{[Recall that } f(x) = 16 - x^2.\text{]}$$

$$= \frac{1}{2}\left(15 + \frac{55}{4} + 12 + \frac{39}{4}\right) = \frac{101}{4}$$

or approximately 25.25 square units.   ■ ■ ■ ■

Table 6.2 shows the approximations of the area A of the region R of Example 2 when n rectangles are used for the approximation and the heights of the approximating rectangles are found by evaluating $f(x)$ at the left end points.

Once again, we see that the approximations seem to approach a unique number as n gets larger and larger—this time the number is $23\frac{1}{3}$. This result

Table 6.2						
Number of Rectangles n	4	10	100	1,000	10,000	50,000	100,000
Approximation of A	25.2500	24.1200	23.4132	23.3413	23.3341	23.3335	23.3334

suggests that we *define* the area of the region under the graph of $f(x) = 16 - x^2$ on the interval $[1, 3]$ to be $23\frac{1}{3}$ square units.

DEFINING AREA—THE GENERAL CASE

Examples 1 and 2 point the way to defining the area A under the graph of an arbitrary but continuous and nonnegative function f on an interval $[a, b]$ (Figure 6.11a).

FIGURE 6.11

The area of the region under the graph of f on $[a, b]$ in (a) is approximated by the sum of the areas of the n rectangles shown in (b).

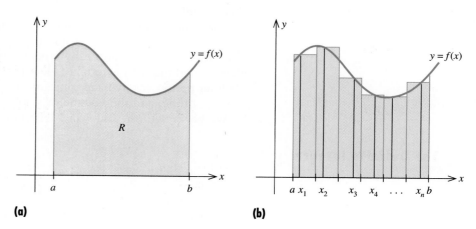

(a)   (b)

Divide the interval $[a, b]$ into n subintervals of equal length $\Delta x = (b - a)/n$. Next, pick n arbitrary points $x_1, x_2, \ldots, x_n$, called *representative points,* from the first, second, $\ldots$, and nth subintervals, respectively (Figure 6.11b). Then, approximating the area A of the region R by the n rectangles of width Δx and heights $f(x_1), f(x_2), \ldots, f(x_n)$, so that the areas of the rectangles are $f(x_1)\Delta x, f(x_2)\Delta x, \ldots, f(x_n)\Delta x$, we have

$$A \approx f(x_1)\Delta x + f(x_2)\Delta x + \cdots + f(x_n)\Delta x$$

The sum on the right-hand side of this expression is called a **Riemann sum** in honor of the German mathematician Bernhard Riemann (1826–1866). Now, as the earlier examples seem to suggest, the Riemann sum will approach a unique number as n becomes arbitrarily large.* We define this number to be the area A of the region R.

The Area Under the Graph of a Function

Let f be a nonnegative continuous function on $[a, b]$. Then, the area of the region under the graph of f is

$$A = \lim_{n \to \infty} [f(x_1) + f(x_2) + \cdots + f(x_n)]\Delta x \qquad (6)$$

where $x_1, x_2, \ldots, x_n$ are arbitrary points in the n subintervals of $[a, b]$ of equal width $\Delta x = (b - a)/n$.

* Even though we chose the representative points to be the midpoints of the subintervals in Example 1 and the left end points in Example 2, it can be shown that each of the respective sums will always approach a unique number as n approaches infinity.

THE DEFINITE INTEGRAL

As we have just seen, the area under the graph of a continuous *nonnegative* function f on an interval $[a, b]$ is defined by the limit of the Riemann sum

$$\lim_{n \to \infty} [f(x_1)\Delta x + f(x_2)\Delta x + \cdots + f(x_n)\Delta x]$$

We now turn our attention to the study of limits of Riemann sums involving functions that are not necessarily nonnegative. Such limits arise in many applications of calculus.

For example, the calculation of the distance covered by a body traveling along a straight line involves evaluating a limit of this form. The computation of the total revenue realized by a company over a certain time period, the calculation of the total amount of electricity consumed in a typical home over a 24-hour period, the average concentration of a drug in a body over a certain interval of time, and the volume of a solid—all involve limits of this type.

We begin with the following definition.

The Definite Integral

Let f be defined on $[a, b]$. If

$$\lim_{n \to \infty} [f(x_1)\Delta x + f(x_2)\Delta x + \cdots + f(x_n)\Delta x]$$

exists for all choices of representative points $x_1, x_2, \ldots, x_n$ in the n subintervals of $[a, b]$ of equal width $\Delta x = (b - a)/n$, then this limit is called the **definite integral of f from a to b** and is denoted by $\int_a^b f(x)\, dx$. Thus,

$$\int_a^b f(x)\, dx = \lim_{n \to \infty} [f(x_1)\Delta x + f(x_2)\Delta x + \cdots + f(x_n)\Delta x] \qquad (7)$$

The number a is the **lower limit of integration**, and the number b is the **upper limit of integration**.

REMARKS

1. If f is nonnegative, then the limit in (7) is the same as the limit in (6); therefore, the definite integral gives the area under the graph of f on $[a, b]$.
2. The limit in (7) is denoted by the integral sign $\int$ because, as we will see later, the definite integral and the antiderivative of a function f are related.
3. It is important to realize that the definite integral $\int_a^b f(x)\, dx$ is a number, whereas the indefinite integral $\int f(x)\, dx$ represents a family of functions (the antiderivatives of f).
4. If the limit in (7) exists, we say that f is **integrable** on the interval $[a, b]$. ■ ■ ■

WHEN IS A FUNCTION INTEGRABLE?

The following theorem, which we state without proof, guarantees that a continuous function is integrable.

Integrability of a Function

Let f be continuous on $[a, b]$. Then, f is integrable on $[a, b]$; that is, the definite integral $\int_a^b f(x)\, dx$ exists.

GEOMETRIC INTERPRETATION OF THE DEFINITE INTEGRAL

If f is nonnegative and integrable on $[a, b]$, then we have the following geometric interpretation of the definite integral $\int_a^b f(x)\, dx$.

Geometric Interpretation of $\int_a^b f(x)\, dx$ for $f(x) \geq 0$ on $[a, b]$

FIGURE 6.12
If $f(x) \geq 0$ on $[a, b]$, then $\int_a^b f(x)\, dx =$ area under the graph of f on $[a, b]$.

If f is nonnegative and continuous on $[a, b]$, then

$$\int_a^b f(x)\, dx \qquad (8)$$

is equal to the area of the region under the graph of f on $[a, b]$ (Figure 6.12).

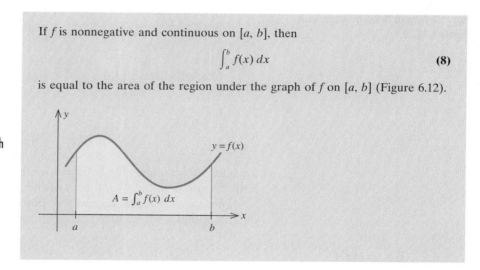

Group Discussion
Suppose f is nonpositive [that is, $f(x) \leq 0$] and continuous on $[a, b]$. Explain why the area of the region below the x-axis and above the graph of f is given by $-\int_a^b f(x)\, dx$.

Next, let's extend our geometric interpretation of the definite integral to include the case where f assumes both positive as well as negative values on $[a, b]$. Consider a typical Riemann sum of the function f,

$$f(x_1)\Delta x + f(x_2)\Delta x + \cdots + f(x_n)\Delta x$$

corresponding to a partition of $[a, b]$ into n subintervals of equal width $(b - a)/n$, where $x_1, x_2, \ldots, x_n$ are representative points in the subintervals. The sum consists of n terms in which a positive term corresponds to the area of a rectangle of height $f(x_k)$ (for some positive integer k) lying above the x-axis and a negative term corresponds to the area of a rectangle of height $-f(x_k)$ lying below the x-axis. (See Figure 6.13, which depicts a situation with $n = 6$).

As n gets larger and larger, the sums of the areas of the rectangles lying above the x-axis seem to give a better and better approximation of the area of the region lying above the x-axis (Figure 6.14). Similarly, the sums of the

FIGURE 6.13
The positive terms in the Riemann sum are associated with the areas of the rectangles that lie above the x-axis, and the negative terms are associated with the areas of those that lie below the x-axis.

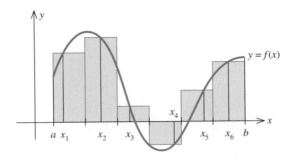

FIGURE 6.14
As n gets larger, the approximations get better. Here, $n = 12$ and we are approximating with twice as many rectangles as in Figure 6.13.

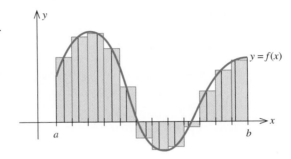

areas of those rectangles lying below the x-axis seem to give a better and better approximation of the area of the region lying below the x-axis.

These observations suggest the following geometric interpretation of the definite integral for an arbitrary continuous function on an interval $[a, b]$.

Geometric Interpretation of $\int_a^b f(x)\, dx$ on $[a, b]$

If f is continuous on $[a, b]$, then

$$\int_a^b f(x)\, dx$$

is equal to the area of the region above $[a, b]$ minus the area of the region below $[a, b]$ (Figure 6.15).

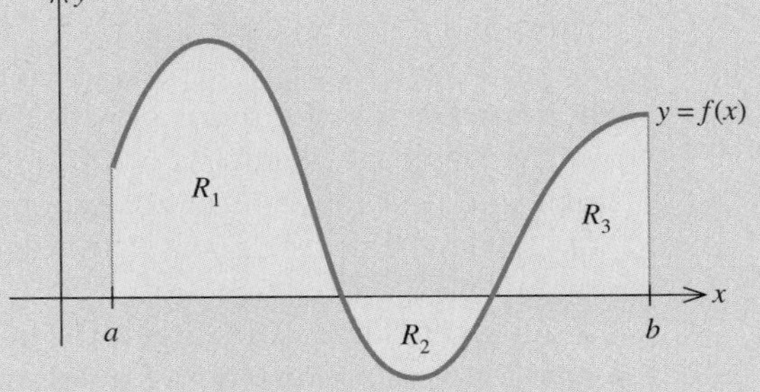

FIGURE 6.15
$\int_a^b f(x)\, dx =$
area of R_1 − area of R_2 + area of R_3

Find an approximation of the area of the region R under the graph of $f(x) = 2x^2 + 1$ on the interval $[0, 3]$, using four subintervals of $[0, 3]$ of equal length and picking the midpoint of each subinterval as a representative point.

The solution to Self-Check Exercise 6.3 can be found on page 477.

6.3 Exercises

In Exercises 1 and 2, find an approximation of the area of the region R under the graph of f by computing the Riemann sum of f corresponding to the partition of the interval into the subintervals shown in the accompanying figures. In each case, use the midpoints of the subintervals as the representative points.

1.

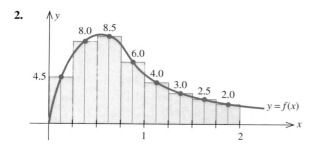

2.

3. Let $f(x) = 3x$.
   a. Sketch the region R under the graph of f on the interval $[0, 2]$ and find its exact area using geometry.
   b. Use a Riemann sum with four subintervals of equal length ($n = 4$) to approximate the area of R. Choose the representative points to be the left end points of the subintervals.

c. Repeat part (b) with eight subintervals of equal length ($n = 8$).
d. Compare the approximations obtained in parts (b) and (c) with the exact area found in part (a). Do the approximations improve with larger n?

4. Repeat Exercise 3, choosing the representative points to be the right end points of the subintervals.

5. Let $f(x) = 4 - 2x$.
   a. Sketch the region R under the graph of f on the interval $[0, 2]$ and find its exact area using geometry.
   b. Use a Riemann sum with five subintervals of equal length ($n = 5$) to approximate the area of R. Choose the representative points to be the left end points of the subintervals.
   c. Repeat part (b) with ten subintervals of equal length ($n = 10$).
   d. Compare the approximations obtained in parts (b) and (c) with the exact area found in part (a). Do the approximations improve with larger n?

6. Repeat Exercise 5, choosing the representative points to be the right end points of the subintervals.

7. Let $f(x) = x^2$ and compute the Riemann sum of f over the interval $[2, 4]$, using:
   a. Two subintervals of equal length ($n = 2$).
   b. Five subintervals of equal length ($n = 5$).
   c. Ten subintervals of equal length ($n = 10$).
   In each case, choose the representative points to be the midpoints of the subintervals.
   d. Can you guess at the area of the region under the graph of f on the interval $[2, 4]$?

8. Repeat Exercise 7, choosing the representative points to be the left end points of the subintervals.

9. Repeat Exercise 7, choosing the representative points to be the right end points of the subintervals.

10. Let $f(x) = x^3$ and compute the Riemann sum of f over the interval $[0, 1]$, using:
   a. Two subintervals of equal length ($n = 2$).
   b. Five subintervals of equal length ($n = 5$).
   c. Ten subintervals of equal length ($n = 10$).
   In each case, choose the representative points to be the midpoints of the subintervals.
   d. Can you guess at the area of the region under the graph of f on the interval $[0, 1]$?

11. Repeat Exercise 10, choosing the representative points to be the left end points of the subintervals.

12. Repeat Exercise 10, choosing the representative points to be the right end points of the subintervals.

In Exercises 13–16, find an approximation of the area of the region R under the graph of the function f on the interval $[a, b]$. In each case, use n subintervals and choose the representative points as indicated.

13. $f(x) = x^2 + 1$; $[0, 2]$; $n = 5$; midpoints

14. $f(x) = 4 - x^2$; $[-1, 2]$; $n = 6$; left end points

15. $f(x) = \dfrac{1}{x}$; $[1, 3]$; $n = 4$; right end points

16. $f(x) = e^x$; $[0, 3]$; $n = 5$; midpoints

17. REAL ESTATE Figure (a) shows a vacant lot with a 100-ft frontage in a development. To estimate its area, we introduce a coordinate system so that the x-axis coincides with the edge of the straight road forming the lower boundary of the property, as shown in Figure (b). Then, thinking of the upper boundary of the property as the graph of a continuous function f over the interval $[0, 100]$, we see that the problem is mathematically equivalent to that of finding the area under the graph of f on $[0, 100]$. To estimate the area of the lot using a Riemann sum, we divide the interval $[0, 100]$ into five equal subintervals of length 20 ft. Then, using surveyor's equipment, we measure the distance from the midpoint of each of these subintervals to the upper boundary of the property. These measurements give the values of $f(x)$ at $x = 10$, 30, 50, 70, and 90. What is the approximate area of the lot?

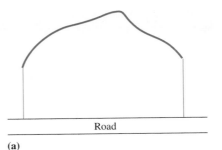

Road

(a)

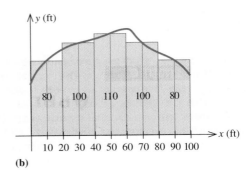

(b)

18. REAL ESTATE Use the technique of Exercise 17 to obtain an estimate of the area of the vacant lot shown in the accompanying figures.

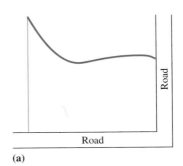

Road

(a)

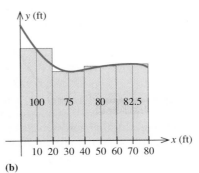

(b)

SOLUTION TO SELF-CHECK EXERCISE 6.3

The length of each subinterval is $\frac{3}{4}$. Therefore, the four subintervals are

$$\left[0, \frac{3}{4}\right], \quad \left[\frac{3}{4}, \frac{3}{2}\right], \quad \left[\frac{3}{2}, \frac{9}{4}\right], \quad \left[\frac{9}{4}, 3\right]$$

The representative points are $\frac{3}{8}$, $\frac{9}{8}$, $\frac{15}{8}$, and $\frac{21}{8}$, respectively. Therefore, the required approximation is

$$\begin{aligned}
A &= \frac{3}{4} f\left(\frac{3}{8}\right) + \frac{3}{4} f\left(\frac{9}{8}\right) + \frac{3}{4} f\left(\frac{15}{8}\right) + \frac{3}{4} f\left(\frac{21}{8}\right) \\
&= \frac{3}{4}\left[f\left(\frac{3}{8}\right) + f\left(\frac{9}{8}\right) + f\left(\frac{15}{8}\right) + f\left(\frac{21}{8}\right)\right] \\
&= \frac{3}{4}\left\{\left[2\left(\frac{3}{8}\right)^2 + 1\right] + \left[2\left(\frac{9}{8}\right)^2 + 1\right] + \left[2\left(\frac{15}{8}\right)^2 + 1\right] + \left[2\left(\frac{21}{8}\right)^2 + 1\right]\right\} \\
&= \frac{3}{4}\left(\frac{41}{32} + \frac{113}{32} + \frac{257}{32} + \frac{473}{32}\right) = \frac{663}{32}
\end{aligned}$$

or approximately 20.72 square units.

6.4 The Fundamental Theorem of Calculus

THE FUNDAMENTAL THEOREM OF CALCULUS

In Section 6.3 we defined the definite integral of an arbitrary continuous function on an interval $[a, b]$ as a limit of Riemann sums. Calculating the value of a definite integral by actually taking the limit of such sums is tedious and in most cases impractical. It is important to realize that the numerical results we obtained in Examples 1 and 2 of Section 6.3 were *approximations* of the respective areas of the regions in question, even though these results enabled us to *conjecture* what the actual areas might be. Fortunately, there is a much better way of finding the exact value of a definite integral.

The following theorem shows how to evaluate the definite integral of a continuous function provided we can find an antiderivative of that function. Because of its importance in establishing the relationship between differentiation and integration, this theorem—discovered independently by Sir Isaac Newton (1642–1727) in England and Gottfried Wilhelm Leibniz (1646–1716) in Germany—is called the **fundamental theorem of calculus.**

THEOREM 2

The Fundamental Theorem of Calculus
Let f be continuous on $[a, b]$. Then,

$$\int_a^b f(x)\, dx = F(b) - F(a) \qquad\qquad (9)$$

where F is any antiderivative of f; that is, $F'(x) = f(x)$.

We will explain why this theorem is true at the end of this section.

When applying the fundamental theorem of calculus, it is convenient to use the notation

$$F(x)\bigg|_a^b = F(b) - F(a)$$

For example, using this notation, Equation (9) is written

$$\int_a^b f(x)\, dx = F(x)\bigg|_a^b = F(b) - F(a)$$

EXAMPLE 1   Let R be the region under the graph of $f(x) = x$ on the interval $[1, 3]$. Use the fundamental theorem of calculus to find the area A of R and verify your result by elementary means.

SOLUTION ✔   The region R is shown in Figure 6.16a. Since f is nonnegative on $[1, 3]$, the area of R is given by the definite integral of f from 1 to 3; that is,

$$A = \int_1^3 x\, dx$$

FIGURE 6.16
The area of R can be computed in two different ways.

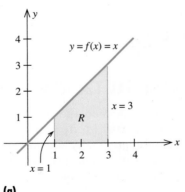

(a)

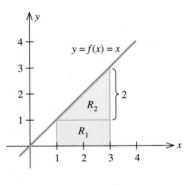

(b)

To evaluate the definite integral, observe that an antiderivative of $f(x) = x$ is $F(x) = \frac{1}{2}x^2 + C$, where C is an arbitrary constant. Therefore, by the fundamental theorem of calculus, we have

$$A = \int_1^3 x\, dx = \frac{1}{2}x^2 + C\, \bigg|_1^3$$

$$= \left(\frac{9}{2} + C\right) - \left(\frac{1}{2} + C\right) = 4 \text{ square units}$$

To verify this result by elementary means, observe that the area A is the area of the rectangle R_1 (width $\times$ height) plus the area of the triangle R_2 ($\frac{1}{2}$ base $\times$ height) (see Figure 6.16b); that is,

$$2(1) + \frac{1}{2}(2)(2) = 2 + 2 = 4$$

which agrees with the result obtained earlier.   ■ ■ ■ ■

Observe that in evaluating the definite integral in Example 1, the constant of integration "dropped out." This is true in general, for if $F(x) + C$ denotes an antiderivative of some function f, then

$$F(x) + C \Big|_a^b = [F(b) + C] - [F(a) + C]$$
$$= F(b) + C - F(a) - C$$
$$= F(b) - F(a)$$

With this fact in mind, we may, in all future computations involving the evaluations of a definite integral, drop the constant of integration from our calculations.

FINDING THE AREA UNDER A CURVE

Having seen how effective the fundamental theorem of calculus is in helping us find the area of simple regions, we now use it to find the area of more complicated regions.

EXAMPLE 2

In Section 6.3 we conjectured that the area of the region R under the graph of $f(x) = x^2$ on the interval $[0, 1]$ was $\frac{1}{3}$ square unit. Use the fundamental theorem of calculus to verify this conjecture.

SOLUTION ✔

The region R is reproduced in Figure 6.17. Observe that f is nonnegative on $[0, 1]$, so the area of R is given by $A = \int_0^1 x^2\, dx$. Since an antide-

FIGURE 6.17
The area of R is $\int_0^1 x^2\, dx = \frac{1}{3}$.

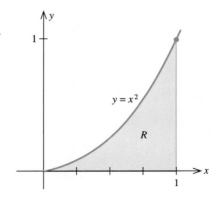

$y = x^2$

R

rivative of $f(x) = x^2$ is $F(x) = \frac{1}{3}x^3$, we see, using the fundamental theorem of calculus, that

$$A = \int_0^1 x^2 \, dx = \frac{1}{3}x^3 \Big|_0^1 = \frac{1}{3}(1) - \frac{1}{3}(0) = \frac{1}{3} \text{ square unit}$$

as we wished to show.   ■ ■ ■ ■

EXAMPLE 3   Find the area of the region R under the graph of $y = x^2 + 1$ from $x = -1$ to $x = 2$.

SOLUTION ✔   The region R under consideration is shown in Figure 6.18. Using the fundamental theorem of calculus, we find that the required area is

$$\int_{-1}^{2} (x^2 + 1) \, dx = \left(\frac{1}{3}x^3 + x\right) \Big|_{-1}^{2}$$

$$= \left[\frac{1}{3}(8) + 2\right] - \left[\frac{1}{3}(-1)^3 + (-1)\right] = 6$$

or 6 square units.

FIGURE 6.18
The area of R is $\int_{-1}^{2} (x^2 + 1) \, dx$.

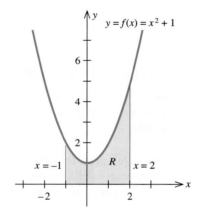

EVALUATING DEFINITE INTEGRALS

In Examples 4 and 5 we use the rules of integration of Section 6.1 to help us evaluate the definite integrals.

EXAMPLE 4   Evaluate $\int_1^3 (3x^2 + e^x) \, dx$.

SOLUTION ✔

$$\int_1^3 (3x^2 + e^x) \, dx = x^3 + e^x \Big|_1^3$$

$$= (27 + e^3) - (1 + e) = 26 + e^3 - e$$ ■ ■ ■ ■

EXAMPLE 5   Evaluate $\int_1^2 \left(\dfrac{1}{x} - \dfrac{1}{x^2} \right) dx$.

SOLUTION ✔

$$\int_1^2 \left(\frac{1}{x} - \frac{1}{x^2} \right) dx = \int_1^2 \left(\frac{1}{x} - x^{-2} \right) dx$$

$$= \ln |x| + \frac{1}{x} \Big|_1^2$$

$$= \left(\ln 2 + \frac{1}{2} \right) - (\ln 1 + 1)$$

$$= \ln 2 - \frac{1}{2} \qquad \text{(Recall, } \ln 1 = 0.)$$

■ ■ ■ ■

Group Discussion

Consider the definite integral $\int_{-1}^1 \dfrac{1}{x^2} \, dx$.

1. Show that a formal application of Equation (9) leads to

$$\int_{-1}^1 \frac{1}{x^2} \, dx = -\frac{1}{x} \Big|_{-1}^1 = -1 - 1 = -2$$

2. Observe that $f(x) = 1/x^2$ is positive at each value of x in $[-1, 1]$ where it is defined. Therefore, one might expect that the definite integral with integrand f has a positive value, if it exists.

3. Explain this apparent contradiction in the result (1) and the observation (2).

APPLICATIONS

EXAMPLE 6   The management of Staedtler Office Equipment has determined that the daily marginal cost function associated with producing battery-operated pencil sharpeners is given by

$$C'(x) = 0.000006x^2 - 0.006x + 4$$

where $C'(x)$ is measured in dollars per unit and x denotes the number of units produced. Management has also determined that the daily fixed cost incurred in producing these pencil sharpeners is $100. Find Staedtler's daily total cost for producing (a) the first 500 units and (b) the 201st through 400th units.

SOLUTION ✔   **a.** Since $C'(x)$ is the marginal cost function, its antiderivative $C(x)$ is the total cost function. The daily fixed cost incurred in producing the pencil sharpeners is $C(0)$ dollars. Since the daily fixed cost is given as $100, we have $C(0) = 100$. We are required to find $C(500)$. Let's compute $C(500) - C(0)$, the net change in the total cost function $C(x)$ over the interval $[0, 500]$. Using the fundamental theorem of calculus, we find

$$C(500) - C(0) = \int_0^{500} C'(x)\, dx$$

$$= \int_0^{500} (0.000006x^2 - 0.006x + 4)\, dx$$

$$= 0.000002x^3 - 0.003x^2 + 4x \Big|_0^{500}$$

$$= [0.000002(500)^3 - 0.003(500)^2 + 4(500)]$$
$$\quad - [0.000002(0)^3 - 0.003(0)^2 + 4(0)]$$

$$= 1500$$

Therefore, $C(500) = 1500 + C(0) = 1500 + 100 = 1600$, so the total cost incurred daily by Staedtler in producing 500 pencil sharpeners is $1600.

b. The daily total cost incurred by Staedtler in producing the 201st through 400th units of battery-operated pencil sharpeners is given by

$$C(400) - C(200) = \int_{200}^{400} C'(x)\, dx$$

$$= \int_{200}^{400} (0.000006x^2 - 0.006x + 4)\, dx$$

$$= 0.000002x^3 - 0.003x^2 + 4x \Big|_{200}^{400}$$

$$= 552$$

or $552.   ■ ■ ■ ■

Since $C'(x)$ is nonnegative for x in the interval $(0, \infty)$, we have the following geometric interpretation of the two definite integrals in Example 6: $\int_0^{500} C'(x)\, dx$ is the area of the region under the graph of the function C' from $x = 0$ to $x = 500$, shown in Figure 6.19a, and $\int_{200}^{400} C'(x)\, dx$ is the area of the region from $x = 200$ to $x = 400$, shown in Figure 6.19b.

FIGURE 6.19

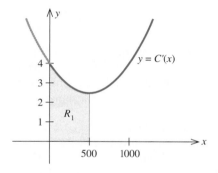

(a) Area of $R_1 = \int_0^{500} C'(x)\, dx$

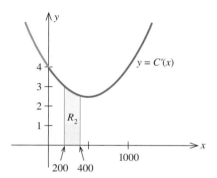

(b) Area of $R_2 = \int_{200}^{400} C'(x)\, dx$

EXAMPLE 7   An efficiency study conducted for the Elektra Electronics Company showed that the rate at which Space Commander walkie-talkies are assembled by the average worker t hours after starting work at 8 A.M. is given by the function

$$f(t) = -3t^2 + 12t + 15 \qquad (0 \le t \le 4)$$

Determine how many walkie-talkies can be assembled by the average worker in the first hour of the morning shift.

SOLUTION ✔   Let $N(t)$ denote the number of walkie-talkies assembled by the average worker t hours after starting work in the morning shift. Then, we have

$$N'(t) = f(t) = -3t^2 + 12t + 15$$

Therefore, the number of units assembled by the average worker in the first hour of the morning shift is

$$N(1) - N(0) = \int_0^1 N'(t)\, dt = \int_0^1 (-3t^2 + 12t + 15)\, dt$$

$$= -t^3 + 6t^2 + 15t \Big|_0^1 = -1 + 6 + 15$$

$$= 20$$

or 20 units.   ■ ■ ■ ■

Exploring with Technology

You can demonstrate graphically that $\int_0^x t\, dt = \frac{1}{2}x^2$ as follows:

1. Plot the graphs of $y1 = \int_0^x t\, dt$ and $y2 = \frac{1}{2}x^2$ on the same set of axes using the viewing rectangle $[-5, 5] \times [0, 10]$.
2. Compare the graphs of $y1$ and $y2$ and draw the desired conclusion.

    EXAMPLE 8   A certain city's rate of electricity consumption is expected to grow exponentially with a growth constant of $k = 0.04$. If the present rate of consumption is 40 million kilowatt-hours (kWh) per year, what should the total production of electricity be over the next 3 years in order to meet the projected demand?

SOLUTION ✔   If $R(t)$ denotes the expected rate of consumption of electricity t years from now, then

$$R(t) = 40e^{0.04t}$$

million kWh per year. Next, if $C(t)$ denotes the expected total consumption of electricity over a period of t years, then

$$C'(t) = R(t)$$

Therefore, the total consumption of electricity expected over the next 3 years is given by

$$\int_0^3 C'(t)\,dt = \int_0^3 40e^{0.04t}\,dt$$

$$= \frac{40}{0.04}\,e^{0.04t}\,\Big|_0^3$$

$$= 1000(e^{0.12} - 1)$$

$$= 127.5$$

or 127.5 million kWh, the amount that must be produced over the next 3 years in order to meet the demand.   ∎ ∎ ∎ ∎

Group Discussion

The definite integral $\int_{-3}^{3} \sqrt{9 - x^2}\,dx$ cannot be evaluated using the fundamental theorem of calculus because the method of this section does not enable us to find an antiderivative of the integrand. But the integral can be evaluated by interpreting it as the area of a certain plane region. What is the region? And what is the value of the integral?

VALIDITY OF THE FUNDAMENTAL THEOREM OF CALCULUS

To demonstrate the plausibility of the fundamental theorem of calculus for the case where f is nonnegative on an interval $[a, b]$, let's define an "area function" A as follows. Let $A(t)$ denote the area of the region R under the graph of $y = f(x)$ from $x = a$ to $x = t$, where $a \le t \le b$ (Figure 6.20).

If h is a small positive number, then $A(t + h)$ is the area of the region under the graph of $y = f(x)$ from $x = a$ to $x = t + h$. Therefore, the difference

$$A(t + h) - A(t)$$

is the area under the graph of $y = f(x)$ from $x = t$ to $x = t + h$ (Figure 6.21).

FIGURE 6.20
$A(t)$ = area under the graph of f from $x = a$ to $x = t$

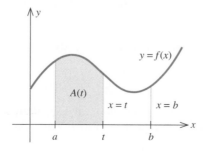

FIGURE 6.21
$A(t + h) - A(t)$ = area under the graph of f from $x = t$ to $x = t + h$

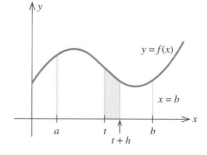

FIGURE 6.22
The area of the rectangle is $h \cdot f(t)$.

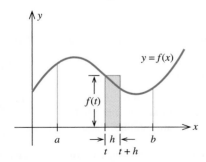

Now, the area of this last region can be approximated by the area of the rectangle of width h and height $f(t)$—that is, by the expression $h \cdot f(t)$ (Figure 6.22). Thus,

$$A(t + h) - A(t) \approx h \cdot f(t)$$

where the approximations improve as h is taken to be smaller and smaller. Dividing both sides of the foregoing relationship by h, we obtain

$$\frac{A(t + h) - A(t)}{h} \approx f(t)$$

Taking the limit as h approaches zero, we find, by the definition of the derivative, that the left-hand side is

$$\lim_{h \to 0} \frac{A(t + h) - A(t)}{h} = A'(t)$$

The right-hand side, which is independent of h, remains constant throughout the limiting process. Because the approximation becomes exact as h approaches zero, we find that

$$A'(t) = f(t)$$

Since the foregoing equation holds for all values of t in the interval $[a, b]$, we have shown that the *area function* A is an antiderivative of the function $f(x)$. By Theorem 1 of Section 6.1, we conclude that $A(x)$ must have the form

$$A(x) = F(x) + C$$

where F is any antiderivative of f and C is an arbitrary constant. To determine the value of C, observe that $A(a) = 0$. This condition implies that

$$A(a) = F(a) + C = 0$$

FIGURE 6.23
The area of R is given by A(b).

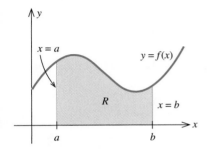

or $C = -F(a)$. Next, since the area of the region R is $A(b)$ (Figure 6.23), we see that the required area is

$$A(b) = F(b) + C$$
$$= F(b) - F(a)$$

Since the area of the region R is

$$\int_a^b f(x)\, dx$$

we have

$$\int_a^b f(x)\, dx = F(b) - F(a)$$

as we set out to show.

SELF-CHECK EXERCISES 6.4

1. Evaluate $\int_0^2 (x + e^x)\, dx$.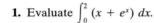

2. The daily marginal profit function associated with producing and selling Texa-Pep hot sauce is

$$P'(x) = -0.000006x^2 + 6$$

where x denotes the number of cases (each case contains 24 bottles) produced and sold daily and $P'(x)$ is measured in dollars/unit. The fixed cost is $400.

   a. What is the total profit realizable from producing and selling 1000 cases of Texa-Pep per day?

   b. What is the additional profit realizable if the production and sale of Texa-Pep is increased from 1000 to 1200 cases/day?

Solutions to Self-Check Exercises 6.4 can be found on page 490.

6.4 Exercises

In Exercises 1–4, find the area of the region under the graph of the function f on the interval $[a, b]$, using the fundamental theorem of calculus. Then verify your result using geometry.

1. $f(x) = 2$; $[1, 4]$ **2.** $f(x) = 4$; $[-1, 2]$

3. $f(x) = 2x$; $[1, 3]$ **4.** $f(x) = -\dfrac{1}{4} x + 1$; $[1, 4]$

In Exercises 5–16, find the area of the region under the graph of the function f on the interval $[a, b]$.

5. $f(x) = 2x + 3$; $[-1, 2]$ **6.** $f(x) = 4x - 1$; $[2, 4]$

7. $f(x) = -x^2 + 4$; $[-1, 2]$

8. $f(x) = 4x - x^2$; $[0, 4]$

9. $f(x) = \dfrac{1}{x}$; $[1, 2]$ **10.** $f(x) = \dfrac{1}{x^2}$; $[2, 4]$

11. $f(x) = \sqrt{x}$; $[1, 9]$ **12.** $f(x) = x^3$; $[1, 3]$

13. $f(x) = 1 - \sqrt[3]{x}$; $[-8, -1]$

14. $f(x) = \dfrac{1}{\sqrt{x}}$; $[1, 9]$

15. $f(x) = e^x$; $[0, 2]$ **16.** $f(x) = e^x - x$; $[1, 2]$

In Exercises 17–40, evaluate the definite integral.

17. $\int_2^4 3\, dx$ **18.** $\int_{-1}^2 -2\, dx$

19. $\int_1^3 (2x + 3)\, dx$ **20.** $\int_{-1}^0 (4 - x)\, dx$

21. $\int_{-1}^3 2x^2\, dx$ **22.** $\int_0^2 8x^3\, dx$

23. $\int_{-2}^2 (x^2 - 1)\, dx$ **24.** $\int_1^4 \sqrt{u}\, du$

25. $\int_1^8 4x^{1/3}\, dx$ **26.** $\int_1^4 2x^{-3/2}\, dx$

27. $\int_0^1 (x^3 - 2x^2 + 1)\, dx$ **28.** $\int_1^2 (t^5 - t^3 + 1)\, dt$

29. $\int_2^4 \dfrac{1}{x}\, dx$ **30.** $\int_1^3 \dfrac{2}{x}\, dx$

31. $\int_0^4 x(x^2 - 1)\, dx$ **32.** $\int_0^2 (x - 4)(x - 1)\, dx$

33. $\int_1^3 (t^2 - t)^2\, dt$ **34.** $\int_{-1}^1 (x^2 - 1)^2\, dx$

35. $\int_{-3}^{-1} \dfrac{1}{x^2}\, dx$ **36.** $\int_1^2 \dfrac{2}{x^3}\, dx$

37. $\int_1^4 \left(\sqrt{x} - \dfrac{1}{\sqrt{x}}\right) dx$ **38.** $\int_0^1 \sqrt{2x}(\sqrt{x} + \sqrt{2})\, dx$

39. $\int_1^4 \dfrac{3x^3 - 2x^2 + 4}{x^2}\, dx$ **40.** $\int_1^2 \left(1 + \dfrac{1}{u} + \dfrac{1}{u^2}\right) du$

🖩 **A calculator is recommended for Exercises 41–47.**

41. **MARGINAL COST** A division of Ditton Industries manufactures a deluxe toaster oven. Management has determined that the daily marginal cost function associated with producing these toaster ovens is given by

$$C'(x) = 0.0003x^2 - 0.12x + 20$$

where $C'(x)$ is measured in dollars/unit and x denotes the number of units produced. Management has also determined that the daily fixed cost incurred in the production is $800.
a. Find the total cost incurred by Ditton in producing the first 300 units of these toaster ovens per day.
b. What is the total cost incurred by Ditton in producing the 201st through 300th units/day?

42. **MARGINAL REVENUE** The management of Ditton Industries has determined that the daily marginal revenue function associated with selling x units of their deluxe toaster ovens is given by

$$R'(x) = -0.1x + 40$$

where $R'(x)$ is measured in dollars/unit.
a. Find the daily total revenue realized from the sale of 200 units of the toaster oven.
b. Find the additional revenue realized when the production (and sales) level is increased from 200 to 300 units.

43. **MARGINAL PROFIT** Refer to Exercise 41. The daily marginal profit function associated with the production and sales of the deluxe toaster ovens is known to be

$$P'(x) = -0.0003x^2 + 0.02x + 20$$

where x denotes the number of units manufactured and sold daily and $P'(x)$ is measured in dollars/unit.
a. Find the total profit realizable from the manufacture and sale of 200 units of the toaster ovens per day.
Hint: $P(200) - P(0) = \int_0^{200} P'(x)\, dx$, $P(0) = -800$.

b. What is the additional daily profit realizable if the production and sale of the toaster ovens are increased from 200 to 220 units/day?

44. **EFFICIENCY STUDIES** Tempco Electronics, a division of Tempco Toys, Inc., manufactures an electronic football game. An efficiency study showed that the rate at which the games are assembled by the average worker t hr after starting work at 8 A.M. is

$$-\frac{3}{2}t^2 + 6t + 20 \qquad (0 \le t \le 4)$$

units/hour.
a. Find the total number of games the average worker can be expected to assemble in the 4-hr morning shift.
b. How many units can the average worker be expected to assemble in the first hour of the morning shift? In the second hour of the morning shift?

45. **SPEEDBOAT RACING** In a recent pretrial run for the world water speed record, the velocity of the *Sea Falcon II* t sec after firing the booster rocket was given by

$$v(t) = -t^2 + 20t + 440 \qquad (0 \le t \le 20)$$

ft/sec. Find the distance covered by the boat over the 20-sec period after the booster rocket was activated.
Hint: The distance is given by $\int_0^{20} v(t)\, dt$.

46. **HAND-HELD COMPUTERS** Annual sales (in millions of units) of hand-held computers are expected to grow in accordance with the function

$$f(t) = 0.18t^2 + 0.16t + 2.64 \qquad (0 \le t \le 6)$$

where t is measured in years, with $t = 0$ corresponding to 1997. How many hand-held computers will be sold over the 6-yr period between the beginning of 1997 and the end of 2002?
Source: Dataquest, Inc.

47. **U.S. CENSUS** According to the U.S. Census Bureau, the number of Americans aged 45 to 54 (which stood at 25 million at the beginning of 1990) grew at the rate of

$$R(t) = 0.00933t^3 + 0.019t^2 - 0.10833t + 1.3467$$

million people/year, t yr from the beginning of 1990. How many Americans aged 45 to 54 were added to the population between 1990 and the year 2000?
Source: U.S. Census Bureau

48. **AIR PURIFICATION** To test air purifiers, the engineers run a purifier in a smoke-filled 10×20-ft room. While conducting a test for a certain brand of air purifier, it was determined that the amount of smoke in the room was decreasing at the rate of

$$R(t) = 0.00032t^4 - 0.01872t^3 + 0.3948t^2$$
$$- 3.83t + 17.63 \qquad (0 \le t \le 20)$$

percent of the (original) amount of the smoke per minute, t min after the start of the test. How much smoke was left in the room 5 min after the start of the test? Ten minutes after the start of the test?
Source: Consumer Reports

(continued on p. 490)

EVALUATING DEFINITE INTEGRALS

Some graphing utilities have an operation for finding the definite integral of a function. If your graphing utility has this capability, use it to work through the example and exercises of this section.

EXAMPLE 1   Use the numerical integral operation of a graphing utility to evaluate

$$\int_{-1}^{2} \frac{2x + 4}{(x^2 + 1)^{3/2}} \, dx$$

SOLUTION ✔   Using the numerical integral operation of a graphing utility, we find

$$\int_{-1}^{2} \frac{2x + 4}{(x^2 + 1)^{3/2}} \, dx \approx 6.92592225992$$

■ ■ ■ ■

Exercises

In Exercises 1–4, use a graphing utility to find the area of the region under the graph of _f_ on the interval [_a, b_]. Express your answer to four decimal places.

1. $f(x) = 0.002x^5 + 0.032x^4 - 0.2x^2 + 2; [-1.1, 2.2]$

2. $f(x) = x\sqrt{x^3 + 1}; [1, 2]$

3. $f(x) = \sqrt{x}e^{-x}; [0, 3]$

4. $f(x) = \dfrac{\ln x}{\sqrt{1 + x^2}}; [1, 2]$

In Exercises 5–10, use a graphing utility to evaluate the definite integral.

5. $\displaystyle\int_{-1.2}^{2.3} (0.2x^4 - 0.32x^3 + 1.2x - 1)\, dx$

6. $\displaystyle\int_{1}^{3} x(x^4 - 1)^{3.2}\, dx$

7. $\displaystyle\int_{0}^{2} \dfrac{3x^3 + 2x^2 + 1}{2x^2 + 3}\, dx$　　**8.** $\displaystyle\int_{1}^{2} \dfrac{\sqrt{x} + 1}{2x^2 + 1}\, dx$

9. $\displaystyle\int_{0}^{2} \dfrac{e^x}{\sqrt{x^2 + 1}}\, dx$　　**10.** $\displaystyle\int_{1}^{3} e^{-x} \ln(x^2 + 1)\, dx$

11. Rework Exercise 47, Exercises 6.4.

12. Rework Exercise 48, Exercises 6.4.

13. THE GLOBAL EPIDEMIC The number of AIDS-related deaths/year in the United States is given by the function

$$f(t) = -53.254t^4 + 673.7t^3 - 2801.07t^2$$
$$+ 8833.379t + 20,000 \qquad (0 \le t \le 9)$$

where $t = 0$ corresponds to the beginning of 1988. Find the total number of AIDS-related deaths in the United States between the beginning of 1988 and the end of 1996.
Source: Centers for Disease Control

14. MARIJUANA ARRESTS The number of arrests for marijuana sales and possession in New York City grew at the rate of approximately

$$f(t) = 0.0125t^4 - 0.01389t^3 + 0.55417t^2$$
$$+ 0.53294t + 4.95238 \qquad (0 \le t \le 5)$$

thousand/year, where t is measured in years, with $t = 0$ corresponding to the beginning of 1992. Find the approximate number of marijuana arrests in the city from the beginning of 1992 to the end of 1997.
Source: State Division of Criminal Justice Services

15. POPULATION GROWTH The population of a certain city is projected to grow at the rate of $18\sqrt{t + 1} \ln \sqrt{t + 1}$ thousand people/year t yr from now. If the current population is 800,000, what will be the population 45 yr from now?

In Exercises 49–51, determine whether the statement is true or false. If it is true, explain why it is true. If it is false, give an example to show why it is false.

49. $\int_{-1}^{1} \frac{1}{x^3} \, dx = -\frac{1}{2x^2} \Big|_{-1}^{1} = -\frac{1}{2} - \left(-\frac{1}{2}\right) = 0$

50. $\int_{0}^{2} (1 - x) \, dx$ gives the area of the region under the graph of $f(x) = 1 - x$ on the interval $[0, 2]$.

51. The total revenue realized in selling the first 5000 units of a product is given by

$$\int_{0}^{500} R'(x) \, dx = R(500) - R(0)$$

where $R(x)$ is the total revenue.

SOLUTIONS TO SELF-CHECK EXERCISES 6.4

1. $\int_{0}^{2} (x + e^x) \, dx = \frac{1}{2}x^2 + e^x \Big|_{0}^{2}$

$$= \left[\frac{1}{2}(2)^2 + e^2\right] - \left[\frac{1}{2}(0) + e^0\right]$$

$$= 2 + e^2 - 1$$

$$= e^2 + 1$$

2. a. We want $P(1000)$. But

$$P(1000) - P(0) = \int_{0}^{1000} P'(x) \, dx = \int_{0}^{1000} (-0.000006x^2 + 6) \, dx$$

$$= -0.000002x^3 + 6x \Big|_{0}^{1000}$$

$$= -0.000002(1000)^3 + 6(1000)$$

$$= 4000$$

So, $P(1000) = 4000 + P(0) = 4000 - 400$, or \$3600/day $[P(0) = -C(0)]$.
b. The additional profit realizable is given by

$$\int_{1000}^{1200} P'(x) \, dx = -0.000002x^3 + 6x \Big|_{1000}^{1200}$$

$$= [-0.000002(1200)^3 + 6(1200)]$$

$$- [-0.000002(1000)^3 + 6(1000)]$$

$$= 3744 - 4000$$

$$= -256$$

That is, the company sustains a loss of \$256/day if production is increased to 1200 cases/day.

6.5 Evaluating Definite Integrals

This section continues our discussion of the applications of the fundamental theorem of calculus.

PROPERTIES OF THE DEFINITE INTEGRAL

Before going on, we list the following useful properties of the definite integral, some of which parallel the rules of integration of Section 6.1.

Properties of the Definite Integral

Let f and g be integrable functions; then,

1. $\displaystyle\int_a^a f(x)\, dx = 0$

2. $\displaystyle\int_a^b f(x)\, dx = -\int_b^a f(x)\, dx$

3. $\displaystyle\int_a^b cf(x)\, dx = c\int_a^b f(x)\, dx \qquad (c, \text{ a constant})$

4. $\displaystyle\int_a^b [f(x) \pm g(x)]\, dx = \int_a^b f(x)\, dx \pm \int_a^b g(x)\, dx$

5. $\displaystyle\int_a^b f(x)\, dx = \int_a^c f(x)\, dx + \int_c^b f(x)\, dx \qquad (a < c < b)$

Property 5 states that if c is a number lying between a and b so that the interval $[a, b]$ is divided into the intervals $[a, c]$ and $[c, b]$, then the integral of f over the interval $[a, b]$ may be expressed as the sum of the integral of f over the interval $[a, c]$ and the integral of f over the interval $[c, b]$.

Property 5 has the following geometric interpretation when f is nonnegative. By definition

$$\int_a^b f(x)\, dx$$

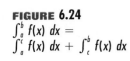

FIGURE 6.24
$\int_a^b f(x)\, dx =$
$\int_a^c f(x)\, dx + \int_c^b f(x)\, dx$

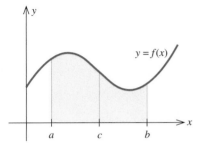

is the area of the region under the graph of $y = f(x)$ from $x = a$ to $x = b$ (Figure 6.24). Similarly, we interpret the definite integrals

$$\int_a^c f(x)\, dx \qquad \text{and} \qquad \int_c^b f(x)\, dx$$

as the areas of the regions under the graph of $y = f(x)$ from $x = a$ to $x = c$ and from $x = c$ to $x = b$, respectively. Since the two regions do not overlap, we see that

$$\int_a^b f(x)\, dx = \int_a^c f(x)\, dx + \int_c^b f(x)\, dx$$

Exercises

In Exercises 1–4, use Property 5 of the properties of the definite integral (page 491) to evaluate the definite integral accurate to six decimal places.

1. $\int_{-1}^{2} f(x)\,dx$, where

$$f(x) = \begin{cases} 2.3x^3 - 3.1x^2 + 2.7x + 3 & \text{if } x < 1 \\ -1.7x^2 + 2.3x + 4.3 & \text{if } x \geq 1 \end{cases}$$

2. $\int_{0}^{3} f(x)\,dx$, where $f(x) = \begin{cases} \dfrac{\sqrt{x}}{1 + x^2} & \text{if } 0 \leq x < 1 \\ 0.5e^{-0.1x^2} & \text{if } x \geq 1 \end{cases}$

3. $\int_{-2}^{2} f(x)\,dx$, where $f(x) = \begin{cases} x^4 - 2x^2 + 4 & \text{if } x < 0 \\ 2\ln(x + e^2) & \text{if } x \geq 0 \end{cases}$

4. $\int_{-2}^{6} f(x)\,dx$, where

$$f(x) = \begin{cases} 2x^3 - 3x^2 + x + 2 & \text{if } \ x < -1 \\ \sqrt{3x + 4} - 5 & \text{if } -1 \leq x \leq 4 \\ x^2 - 3x - 5 & \text{if } \ x > 4 \end{cases}$$

5. **AIDS IN MASSACHUSETTS** The rate of growth (and decline) of the number of AIDS cases diagnosed in Massachusetts from the beginning of 1989 ($t = 0$) through the end of 1997 ($t = 8$) is approximated by the function

$$f(t) = \begin{cases} 69.83333t^2 + 30.16667t + 1000 & \text{if } 0 \leq t < 3 \\ 1719 & \text{if } 3 \leq t < 4 \\ -28.79167t^3 + 491.37500t^2 & \text{if } 4 \leq t \leq 8 \\ \quad - 2985.083333t + 7640 & \end{cases}$$

where $f(t)$ is measured in the number of cases/year. Estimate the total number of AIDS cases diagnosed in Massachusetts from the beginning of 1989 through the end of 1997.

Source: Massachusetts Department of Health

56. Verify by direct computation that

$$\int_1^9 2\sqrt{x}\, dx = 2 \int_1^9 \sqrt{x}\, dx$$

57. Verify by direct computation that

$$\int_0^1 (1 + x - e^x)\, dx = \int_0^1 dx + \int_0^1 x\, dx - \int_0^1 e^x\, dx$$

What properties of the definite integral are demonstrated in this exercise?

58. Verify by direct computation that

$$\int_0^3 (1 + x^3)\, dx = \int_0^1 (1 + x^3)\, dx + \int_1^3 (1 + x^3)\, dx$$

What property of the definite integral is demonstrated here?

59. Verify by direct computation that

$$\int_0^3 (1 + x^3)\, dx$$
$$= \int_0^1 (1 + x^3)\, dx + \int_1^2 (1 + x^3)\, dx + \int_2^3 (1 + x^3)\, dx$$

hence showing that Property 5 may be extended.

60. Evaluate $\int_3^3 (1 + \sqrt{x})e^{-x}\, dx$.

61. Evaluate $\int_3^0 f(x)\, dx$, given that $\int_0^3 f(x)\, dx = 4$.

62. Evaluate $\int_0^3 4f(x)\, dx$, given that $\int_0^3 f(x)\, dx = -1$.

63. Given that $\int_{-1}^2 f(x)\, dx = -2$ and $\int_{-1}^2 g(x)\, dx = 3$, evaluate:

a. $\int_{-1}^2 [2f(x) + g(x)]\, dx$

b. $\int_{-1}^2 [g(x) - f(x)]\, dx$

c. $\int_{-1}^2 [2f(x) - 3g(x)]\, dx$

64. Given that $\int_{-1}^2 f(x)\, dx = 2$ and $\int_0^2 f(x)\, dx = 3$, evaluate:

a. $\int_{-1}^0 f(x)\, dx$

b. $\int_0^2 f(x)\, dx - \int_{-1}^0 f(x)\, dx$

In Exercises 65–69, determine whether the statement is true or false. If it is true, explain why it is true. If it is false, give an example to show why it is false.

65. $\int_2^2 \dfrac{e^x}{\sqrt{1 + x}}\, dx = 0$

66. $\int_1^3 \dfrac{dx}{x - 2} = -\int_3^1 \dfrac{dx}{x - 2}$

67. $\int_0^1 x\sqrt{x + 1}\, dx = \sqrt{x + 1} \int_0^1 x\, dx = \dfrac{1}{2} x^2 \sqrt{x + 1}\, \Big|_0^1 = \dfrac{\sqrt{2}}{2}$

68. If f' is continuous on $[0, 2]$, then $\int_0^2 f'(x)\, dx = f(2) - f(0)$.

69. If f and g are continuous on $[a, b]$ and k is a constant, then

$$\int_a^b [kf(x) + g(x)]\, dx = k \int_a^b f(x)\, dx + \int_a^b g(x)\, dx$$

SOLUTIONS TO SELF-CHECK EXERCISES 6.5

1. Let $u = 2x + 5$. Then, $du = 2\, dx$, or $dx = \frac{1}{2}\, du$. Also, when $x = 0$, $u = 5$, and when $x = 2$, $u = 9$. Therefore,

$$\int_0^2 \sqrt{2x + 5}\, dx = \int_0^2 (2x + 5)^{1/2}\, dx$$
$$= \frac{1}{2} \int_5^9 u^{1/2}\, du$$
$$= \left(\frac{1}{2}\right)\left(\frac{2}{3} u^{3/2}\right)\Big|_5^9$$
$$= \frac{1}{3}[9^{3/2} - 5^{3/2}]$$
$$= \frac{1}{3}(27 - 5\sqrt{5})$$

2. The required average value is given by

$$\frac{1}{2-(-1)} \int_{-1}^{2} (1-x^2)\, dx = \frac{1}{3} \int_{-1}^{2} (1-x^2)\, dx$$

$$= \frac{1}{3}\left(x - \frac{1}{3}x^3\right)\Big|_{-1}^{2}$$

$$= \frac{1}{3}\left[\left(2 - \frac{8}{3}\right) - \left(-1 + \frac{1}{3}\right)\right] = 0$$

3. The average median price of a house over the stated time interval is given by

$$\frac{1}{5-0} \int_{0}^{5} (t^3 - 7t^2 + 17t + 190)\, dt = \frac{1}{5}\left(\frac{1}{4}t^4 - \frac{7}{3}t^3 + \frac{17}{2}t^2 + 190t\right)\Big|_{0}^{5}$$

$$= \frac{1}{5}\left[\frac{1}{4}(5)^4 - \frac{7}{3}(5)^3 + \frac{17}{2}(5)^2 + 190(5)\right]$$

$$= 205.417$$

or \$205,417.

6.6 Area Between Two Curves

Suppose a certain country's petroleum consumption is expected to grow at the rate of $f(t)$ million barrels per year, t years from now, for the next 5 years. Then, the country's total petroleum consumption over the period of time in question is given by the area under the graph of f on the interval $[0, 5]$ (Figure 6.27).

Next, suppose that because of the implementation of certain energy-conservation measures, the rate of growth of petroleum consumption is expected to be $g(t)$ million barrels per year instead. Then, the country's projected total petroleum consumption over the 5-year period is given by the area under the graph of g on the interval $[0, 5]$ (Figure 6.28).

FIGURE 6.27
At a rate of consumption $f(t)$ million barrels per year, the total petroleum consumption is given by the area of the region under the graph of f.

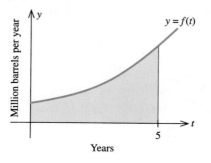

FIGURE 6.28
At a rate of consumption of $g(t)$ million barrels per year, the total petroleum consumption is given by the area of the region under the graph of g.

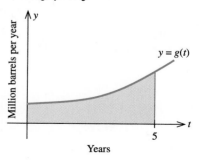

Therefore, the area of the shaded region S lying between the graphs of f and g on the interval $[0, 5]$ (Figure 6.29) gives the amount of petroleum that would be saved over the 5-year period because of the conservation measures.

FIGURE 6.29
The area of S gives the amount of petroleum that would be saved over the 5-year period.

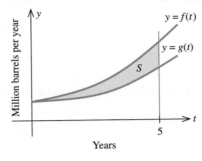

But the area of S is given by

Area under the graph of f on $[a, b]$ − Area under the graph of g on $[a, b]$

$$= \int_0^5 f(t)\, dt - \int_0^5 g(t)\, dt$$

$$= \int_0^5 [f(t) - g(t)]\, dt \qquad \text{(By Property 4, Section 6.5)}$$

This example shows that some practical problems can be solved by finding the area of a region between two curves, which in turn can be found by evaluating an appropriate definite integral.

FINDING THE AREA BETWEEN TWO CURVES

We now turn our attention to the general problem of finding the area of a plane region bounded both above and below by the graphs of functions. First, consider the situation in which the graph of one function lies above that of another. More specifically, let R be the region in the xy-plane (Figure 6.30) that is bounded above by the graph of a continuous function f, below by a continuous function g where $f(x) \geq g(x)$ on $[a, b]$ and to the left and right by the vertical lines $x = a$ and $x = b$, respectively. From the figure, we see that

$$\text{Area of } R = \text{Area under } f(x) - \text{Area under } g(x)$$

$$= \int_a^b f(x)\, dx - \int_a^b g(x)\, dx$$

$$= \int_a^b [f(x) - g(x)]\, dx$$

upon using Property 4 of the definite integral.

FIGURE 6.30
Area of $R = \int_a^b [f(x) - g(x)]\, dx$

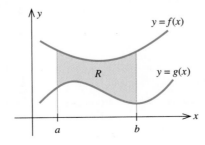

19. $f(x) = -x^2 + 2x + 3$, $g(x) = -x + 3$; $a = 0$, $b = 2$

20. $f(x) = 9 - x^2$, $g(x) = 2x + 3$; $a = -1$, $b = 1$

21. $f(x) = x^2 + 1$, $g(x) = \frac{1}{3}x^3$; $a = -1$, $b = 2$

22. $f(x) = \sqrt{x}$, $g(x) = -\frac{1}{2}x - 1$; $a = 1$, $b = 4$

23. $f(x) = \frac{1}{x}$, $g(x) = 2x - 1$; $a = 1$, $b = 4$

24. $f(x) = x^2$, $g(x) = \frac{1}{x^2}$; $a = 1$, $b = 3$

25. $f(x) = e^x$, $g(x) = \frac{1}{x}$; $a = 1$, $b = 2$

26. $f(x) = x$, $g(x) = e^{2x}$; $a = 1$, $b = 3$

In Exercises 27–34, sketch the graph and find the area of the region bounded by the graph of the function f and the lines y = 0, x = a, and x = b.

27. $f(x) = x$; $a = -1$, $b = 2$

28. $f(x) = x^2 - 2x$; $a = -1$, $b = 1$

29. $f(x) = -x^2 + 4x - 3$; $a = -1$, $b = 2$

30. $f(x) = x^3 - x^2$; $a = -1$, $b = 1$

31. $f(x) = x^3 - 4x^2 + 3x$; $a = 0$, $b = 2$

32. $f(x) = 4x^{1/3} + x^{4/3}$; $a = -1$, $b = 8$

33. $f(x) = e^x - 1$; $a = -1$, $b = 3$

34. $f(x) = xe^{x^2}$; $a = 0$, $b = 2$

In Exercises 35–40, sketch the graph and find the area of the region completely enclosed by the graphs of the given functions f and g.

35. $f(x) = x + 2$ and $g(x) = x^2 - 4$

36. $f(x) = -x^2 + 4x$ and $g(x) = 2x - 3$

37. $f(x) = x^2$ and $g(x) = x^3$

38. $f(x) = x^3 - 6x^2 + 9x$ and $g(x) = x^2 - 3x$

39. $f(x) = \sqrt{x}$ and $g(x) = x^2$

40. $f(x) = 2x$ and $g(x) = x\sqrt{x + 1}$

41. **EFFECT OF ADVERTISING ON REVENUE** In the accompanying figure, the function f gives the rate of change of Odyssey Travel's revenue with respect to the amount x it spends on advertising with their current advertising agency. By engaging the services of a different advertising agency, it is expected that Odyssey's revenue will grow at the rate given by the function g. Give an interpretation of the area A of the region S and find an expression for A in terms of a definite integral involving f and g.

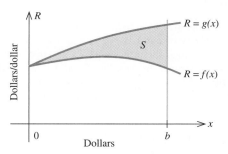

42. **PULSE RATE DURING EXERCISE** In the accompanying figure, the function f gives the rate of increase of an individual's pulse rate when he walked a prescribed course on a treadmill 6 mo ago. The function g gives the rate of increase of his pulse rate when he recently walked the same prescribed course. Give an interpretation of the area A of the region S and find an expression for A in terms of a definite integral involving f and g.

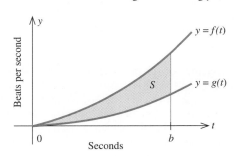

43. **AIR PURIFICATION** To study the effectiveness of air purifiers in removing smoke, engineers run each purifier in a smoke-filled 10 × 20-ft room. In the accompanying figure, the function f gives the rate of change of the smoke level/minute, t min after the start of the test, when a brand A purifier is used. The function g gives the rate of change of the smoke level/minute when a brand B purifier is used.
   a. Give an interpretation of the area of the region S.
   b. Find an expression for the area of S in terms of a definite integral involving f and g.

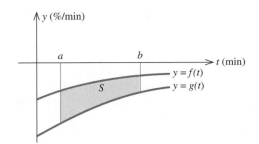

44. Two cars start out side by side and travel along a straight road. The velocity of car 1 is $f(t)$ ft/sec, the velocity of car 2 is $g(t)$ ft/sec over the interval $[0, T]$, and $0 < T_1 < T$. Furthermore, suppose the graphs of f and g are as depicted in the accompanying figure. Denote the area of region I by A_1 and the area of region II by A_2.

a. Write the number

$$\int_{T_1}^{T} [g(t) - f(t)] \, dt - \int_{0}^{T_1} [f(t) - g(t)] \, dt$$

in terms of A_1 and A_2.

b. What does the number obtained in part (a) represent?

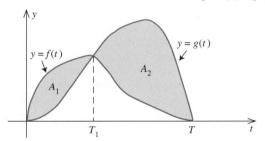

45. The rate of change of the revenue of company A over the (time) interval $[0, T]$ is $f(t)$ dollars/week, whereas the rate of change of the revenue of company B over the same period is $g(t)$ dollars/week. Suppose the graphs of f and g are depicted in the accompanying figure. Find an expression in terms of definite integrals involving f and g giving the additional revenue that company B will have over company A in the period $[0, T]$.

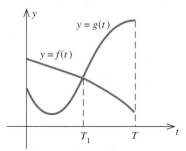

46. TURBO-CHARGED ENGINE VS. STANDARD ENGINE In tests conducted by *Auto Test Magazine* on two identical models of the Phoenix Elite—one equipped with a standard engine and the other with a turbo-charger—it was found that the acceleration of the former is given by

$$a = f(t) = 4 + 0.8t \qquad (0 \le t \le 12)$$

ft/sec/sec, t sec after starting from rest at full throttle, whereas the acceleration of the latter is given by

$$a = g(t) = 4 + 1.2t + 0.03t^2 \qquad (0 \le t \le 12)$$

ft/sec/sec. How much faster is the turbo-charged model moving than the model with the standard engine, at the end of a 10-sec test run at full throttle?

47. ALTERNATIVE ENERGY SOURCES Because of the increasingly important role played by coal as a viable alternative energy source, the production of coal has been growing at the rate of

$$3.5e^{0.05t}$$

billion metric tons/year t yr from 1980 (which corresponds to $t = 0$). Had it not been for the energy crisis, the rate of production of coal since 1980 might have been only

$$3.5e^{0.01t}$$

billion metric tons/year t yr from 1980. Determine how much additional coal was produced between 1980 and the end of the century as an alternate energy source.

48. EFFECT OF TV ADVERTISING ON CAR SALES Carl Williams, the new proprietor of Carl Williams Auto Sales, estimates that with extensive television advertising, car sales over the next several years could be increasing at the rate of

$$5e^{0.3t}$$

thousand cars/year t yr from now, instead of at the current rate of

$$(5 + 0.5t^{3/2})$$

thousand cars/year t yr from now. Find how many more cars Carl expects to sell over the next 5 yr by implementing his advertising plans.

49. POPULATION GROWTH In an endeavor to curb population growth in a Southeast Asian island state, the government has decided to launch an extensive propaganda campaign. Without curbs, the government expects the rate of population growth to have been

$$60e^{0.02t}$$

thousand people/year t yr from now, over the next 5 yr. However, successful implementation of the proposed campaign is expected to result in a population growth rate of

$$-t^2 + 60$$

thousand people/yr t yr from now, over the next 5 yr. Assuming that the campaign is mounted, how many fewer people will there be in that country 5 yr from now than there would have been if no curbs had been imposed?

In Exercises 50 and 51, determine whether the statement is true or false. If it is true, explain why it is true. If it is false, give an example to show why it is false.

50. If f and g are continuous on $[a, b]$ and either $f(x) \geq g(x)$ for all x in $[a, b]$ or $f(x) \leq g(x)$ for all x in $[a, b]$, then the area of the region bounded by the graphs of f and g and the vertical lines $x = a$ and $x = b$ is given by $\int_a^b |f(x) - g(x)| \, dx$.

51. The area of the region bounded by the graphs of $f(x) = 2 - x$ and $g(x) = 4 - x^2$ and the vertical lines $x = 0$ and $x = 2$ is given by $\int_0^2 [f(x) - g(x)] \, dx$.

52. Show that the area of a region R bounded above by the graph of a function f and below by the graph of a function g from $x = a$ to $x = b$ is given by

$$\int_a^b [f(x) - g(x)] \, dx$$

Hint: The validity of the formula was verified earlier for the case when both f and g were nonnegative. Now, let f and g be two functions such that $f(x) \geq g(x)$ for $a \leq x \leq b$. Then, there exists some nonnegative constant c such that the curves $y = f(x) + c$ and $y = g(x) + c$ are translated in the y-direction in such a way that the region R' has the same area as the region R (see the accompanying figures). Show that the area of R' is given by

$$\int_a^b \{[f(x) + c] - [g(x) + c]\} \, dx = \int_a^b [f(x) - g(x)] \, dx$$

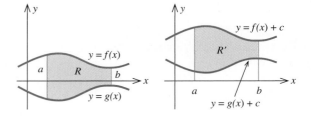

SOLUTIONS TO SELF-CHECK EXERCISES 6.6

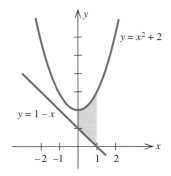

1. The region in question is shown in the accompanying figure. Since the graph of the function f lies above that of the function g for $0 \leq x \leq 1$, we see that the required area is given by

$$\int_0^1 [(x^2 + 2) - (1 - x)] \, dx = \int_0^1 (x^2 + x + 1) \, dx$$

$$= \frac{1}{3}x^3 + \frac{1}{2}x^2 + x \Big|_0^1$$

$$= \frac{1}{3} + \frac{1}{2} + 1$$

$$= \frac{11}{6}$$

or $\frac{11}{6}$ square units.

2. The region in question is shown in the figure on page 518. To find the points of intersection of the two curves, we solve the equation

$$-x^2 + 6x + 5 = x^2 + 5$$

$$2x^2 - 6x = 0$$

$$2x(x - 3) = 0$$

giving $x = 0$ or $x = 3$. Therefore, the points of intersection are $(0, 5)$ and $(3, 14)$.

(continued on p. 518)

FINDING THE AREA BETWEEN TWO CURVES

The numerical integral operation can also be used to find the area between two curves. We do this by using the numerical integral operation to evaluate an appropriate definite integral or the sum (difference) of appropriate definite integrals. In the following example, the intersection operation is also used to advantage to help us find the limits of integration.

EXAMPLE 1

Use a graphing utility to find the area of the region R that is completely enclosed by the graphs of the functions

$$f(x) = 2x^3 - 8x^2 + 4x - 3 \quad \text{and} \quad g(x) = 3x^2 + 10x - 11$$

SOLUTION ✔

The graphs of f and g in the viewing rectangle $[-3, 4] \times [-20, 5]$ are shown in Figure T1.

FIGURE T1

The region R is completely enclosed by the graphs of f and g.

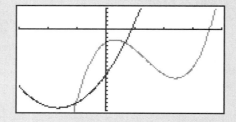

Using the intersection operation of a graphing utility, we find the x-coordinates of the points of intersection of the two graphs to be approximately -1.04 and 0.65, respectively. Since the graph of f lies above that of g on the interval $[-1.04, 0.65]$, we see that the area of R is given by

$$A = \int_{-1.04}^{0.65} [(2x^3 - 8x^2 + 4x - 3) - (3x^2 + 10x - 11)] \, dx$$

$$= \int_{-1.04}^{0.65} (2x^3 - 11x^2 - 6x + 8) \, dx$$

Using the numerical integral function of a graphing utility, we find $A \approx 9.87$, and so the area of R is approximately 9.87 square units. ∎∎∎∎

Exercises

In Exercises 1–6, use a graphing utility to (a) plot the graphs of the functions *f* and *g* and (b) find the area of the region enclosed by these graphs and the vertical lines *x* = *a* and *x* = *b*. Express your answers accurate to four decimal places.

1. $f(x) = x^3(x-5)^4$, $g(x) = 0$; $a = 1$, $b = 3$

2. $f(x) = x - \sqrt{1-x^2}$, $g(x) = 0$; $a = -\dfrac{1}{2}$, $b = \dfrac{1}{2}$

3. $f(x) = x^{1/3}(x+1)^{1/2}$, $g(x) = x^{-1}$; $a = 1.2$, $b = 2$

4. $f(x) = 2$, $g(x) = \ln(1+x^2)$; $a = -1$, $b = 1$

5. $f(x) = \sqrt{x}$, $g(x) = \dfrac{x^2-3}{x^2+1}$; $a = 0$, $b = 3$

6. $f(x) = \dfrac{4}{x^2+1}$, $g(x) = x^4$; $a = -1$, $b = 1$

In Exercises 7–12, use a graphing utility to (a) plot the graphs of the functions *f* and *g* and (b) find the area of the region totally enclosed by the graphs of these functions.

7. $f(x) = 2x^3 - 8x^2 + 4x - 3$ and $g(x) = -3x^2 + 10x - 10$

8. $f(x) = x^4 - 2x^2 + 2$ and $g(x) = 4 - 2x^2$

9. $f(x) = 2x^3 - 3x^2 + x + 5$ and $g(x) = e^{2x} - 3$

10. $f(x) = \dfrac{1}{2}x^2 - 3$ and $g(x) = \ln x$

11. $f(x) = xe^{-x}$ and $g(x) = x - 2\sqrt{x}$

12. $f(x) = e^{-x^2}$ and $g(x) = x^4$

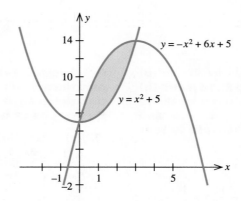

$y = -x^2 + 6x + 5$

$y = x^2 + 5$

Since the graph of f always lies above that of g for $0 \le x \le 3$, we see that the required area is given by

$$\int_0^3 [(-x^2 + 6x + 5) - (x^2 + 5)]\, dx = \int_0^3 (-2x^2 + 6x)\, dx$$

$$= -\frac{2}{3}x^3 + 3x^2 \Big|_0^3$$

$$= -18 + 27$$

$$= 9$$

or 9 square units.

3. The additional profits realizable over the next 10 yr are given by

$$\int_0^{10} [(t - 2\sqrt{t} + 4) - (1 + t^{2/3})]\, dt$$

$$= \int_0^{10} (t - 2t^{1/2} + 3 - t^{2/3})\, dt$$

$$= \frac{1}{2}t^2 - \frac{4}{3}t^{3/2} + 3t - \frac{3}{5}t^{5/3} \Big|_0^{10}$$

$$= \frac{1}{2}(10)^2 - \frac{4}{3}(10)^{3/2} + 3(10) - \frac{3}{5}(10)^{5/3}$$

$$\approx 9.99$$

or approximately \$10 million.

6.7 Applications of the Definite Integral to Business and Economics

In this section we consider several applications of the definite integral in the fields of business and economics.

CONSUMERS' AND PRODUCERS' SURPLUS

FIGURE 6.38
$D(x)$ is a demand function.

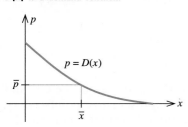

We begin by deriving a formula for computing the consumers' surplus. Suppose $p = D(x)$ is the demand function that relates the unit price p of a commodity to the quantity x demanded of it. Furthermore, suppose a fixed unit market price has been established for the commodity and corresponding to this unit price the quantity demanded is $\bar{x}$ units (Figure 6.38). Then, those consumers who would be willing to pay a unit price higher than $\bar{p}$ for the commodity would in effect experience a savings. This difference between what the consumers *would* be willing to pay for $\bar{x}$ units of the commodity and what they *actually* pay for them is called the **consumers' surplus.**

To derive a formula for computing the consumers' surplus, divide the interval $[0, \bar{x}]$ into n subintervals, each of length $\Delta x = \bar{x}/n$, and denote the right end points of these subintervals by $x_1, x_2, \ldots, x_n = \bar{x}$ (Figure 6.39).

We observe in Figure 6.39 that there are consumers who would pay a unit price of at least $D(x_1)$ dollars for the first Δx units of the commodity instead of the market price of $\bar{p}$ dollars per unit. The savings to these consumers is approximated by

$$D(x_1)\Delta x - \bar{p}\Delta x = [D(x_1) - \bar{p}]\Delta x$$

FIGURE 6.39
Approximating consumers' surplus by the sum of the rectangles $r_1, r_2, \ldots, r_n$

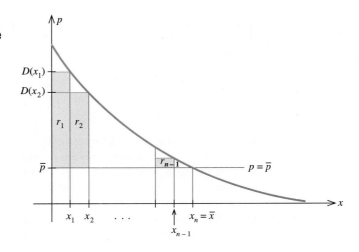

FIGURE 6.47
The closer the Lorentz curve is to the line, the more equitable the income distribution.

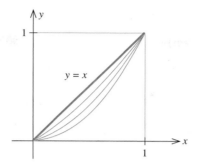

of income recipients, and so on. Now, it is evident that the closer a Lorentz curve is to this line, the more equitable the income distribution is among the income recipients. But the proximity of a Lorentz curve to the line of complete equality is reflected by the area between the Lorentz curve and the line $y = x$ (Figure 6.47). The closer the curve is to the line, the smaller the enclosed area.

This observation suggests that we may define a number, called the coefficient of inequality of a Lorentz curve, as the ratio of the area between the line of complete equality and the Lorentz curve to the area under the line of complete equality. Since the area under the line of complete equality is $\frac{1}{2}$, we see that the coefficient of inequality is given by the following formula,

Coefficient of Inequality of a Lorentz Curve

The **coefficient of inequality**, or **Gini index**, of a Lorentz curve is

$$L = 2 \int_0^1 [x - f(x)] \, dx \qquad (21)$$

The coefficient of inequality is a number between 0 and 1. For example, a coefficient of zero implies that the income distribution is perfectly uniform.

EXAMPLE 7

In a study conducted by a certain country's Economic Development Board with regard to the income distribution of certain segments of the country's workforce, it was found that the Lorentz curves for the distribution of income of medical doctors and of movie actors are described by the functions

$$f(x) = \frac{14}{15}x^2 + \frac{1}{15}x \qquad \text{and} \qquad g(x) = \frac{5}{8}x^4 + \frac{3}{8}x$$

respectively. Compute the coefficient of inequality for each Lorentz curve. Which profession has a more equitable income distribution?

| SOLUTION ✔ | The required coefficients of inequality are, respectively,

$$L_1 = 2 \int_0^1 \left[x - \left(\frac{14}{15} x^2 + \frac{1}{15} x \right) \right] dx = 2 \int_0^1 \left(\frac{14}{15} x - \frac{14}{15} x^2 \right) dx$$

$$= \frac{28}{15} \int_0^1 (x - x^2) \, dx = \frac{28}{15} \left(\frac{1}{2} x^2 - \frac{1}{3} x^3 \right) \Big|_0^1$$

$$= \frac{14}{45} \approx 0.311$$

$$L_2 = 2 \int_0^1 \left[x - \left(\frac{5}{8} x^4 + \frac{3}{8} x \right) \right] dx = 2 \int_0^1 \left(\frac{5}{8} x - \frac{5}{8} x^4 \right) dx$$

$$= \frac{5}{4} \int_0^1 (x - x^4) \, dx = \frac{5}{4} \left(\frac{1}{2} x^2 - \frac{1}{5} x^5 \right) \Big|_0^1$$

$$= \frac{15}{40} \approx 0.375$$

We conclude that in this country the incomes of medical doctors are more evenly distributed than the incomes of movie actors.   ■ ■ ■ ■

SELF-**C**HECK **E**XERCISE **6.7**

The demand function for a certain make of exercise bicycle that is sold exclusively through cable television is

$$p = d(x) = \sqrt{9 - 0.02x}$$

where p is the unit price in hundreds of dollars and x is the quantity demanded/week. The corresponding supply function is given by

$$p = s(x) = \sqrt{1 + 0.02x}$$

where p has the same meaning as before and x is the number of exercise bicycles the supplier will make available at price p. Determine the consumers' surplus and the producers' surplus if the unit price is set at the equilibrium price.

The solution to Self-Check Exercise 6.7 can be found on page 534.

6.7 Exercises

A calculator is recommended for this exercise set.

1. **CONSUMERS' SURPLUS** The demand function for a certain make of replacement cartridges for a water purifier is given by

$$p = -0.01x^2 - 0.1x + 6$$

where p is the unit price in dollars and x is the quantity demanded each week, measured in units of a thousand. Determine the consumers' surplus if the market price is set at \$4/cartridge.

2. **CONSUMERS' SURPLUS** The demand function for a certain brand of compact disc is given by

$$p = -0.01x^2 - 0.2x + 8$$

where p is the wholesale unit price in dollars and x is the quantity demanded each week, measured in units of a thousand. Determine the consumers' surplus if the wholesale market price is set at \$5/disc.

3. **CONSUMERS' SURPLUS** It is known that the quantity demanded of a certain make of portable hair dryer is x hundred units/week and the corresponding wholesale unit price is

$$p = \sqrt{225 - 5x}$$

dollars. Determine the consumers' surplus if the wholesale market price is set at \$10/unit.

4. **PRODUCERS' SURPLUS** The supplier of the portable hair dryers in Exercise 3 will make x hundred units of hair dryers available in the market when the wholesale unit price is

$$p = \sqrt{36 + 1.8x}$$

dollars. Determine the producers' surplus if the wholesale market price is set at \$9/unit.

5. **PRODUCERS' SURPLUS** The supply function for the compact discs of Exercise 2 is given by

$$p = 0.01x^2 + 0.1x + 3$$

where p is the unit wholesale price in dollars and x stands for the quantity that will be made available in the market by the supplier, measured in units of a thousand. Determine the producers' surplus if the wholesale market price is set at the equilibrium price.

6. **CONSUMERS' AND PRODUCERS' SURPLUS** The management of the Titan Tire Company has determined that the quantity demanded x of their Super Titan tires/week is related to the unit price p by the relation

$$p = 144 - x^2$$

where p is measured in dollars and x is measured in units of a thousand. Titan will make x units of the tires available in the market if the unit price is

$$p = 48 + \frac{1}{2}x^2$$

dollars. Determine the consumers' surplus and the producers' surplus when the market unit price is set at the equilibrium price.

7. **CONSUMERS' AND PRODUCERS' SURPLUS** The quantity demanded x (in units of a hundred) of the Mikado miniature cameras/week is related to the unit price p (in dollars) by

$$p = -0.2x^2 + 80$$

and the quantity x (in units of a hundred) that the supplier is willing to make available in the market is related to the unit price p (in dollars) by

$$p = 0.1x^2 + x + 40$$

If the market price is set at the equilibrium price, find the consumers' surplus and the producers' surplus.

8. Refer to Example 3, page 525. Verify that

$$\int_0^3 580{,}000e^{-0.1t}\, dt - 180{,}000 \approx 1{,}323{,}254$$

9. **PRESENT VALUE OF AN INVESTMENT** Suppose an investment is expected to generate income at the rate of

$$R(t) = 200{,}000$$

dollars/year for the next 5 yr. Find the present value of this investment if the prevailing interest rate is 8%/year compounded continuously.

10. **FRANCHISES** Camille purchased a 15-yr franchise for a computer outlet store that is expected to generate income at the rate of

$$R(t) = 400{,}000$$

dollars/year. If the prevailing interest rate is 10%/year compounded continuously, find the present value of the franchise.

11. **THE AMOUNT OF AN ANNUITY** Find the amount of an annuity if $250/month is paid into it for a period of 20 yr earning interest at the rate of 8%/year compounded continuously.

12. **THE AMOUNT OF AN ANNUITY** Find the amount of an annuity if $400/month is paid into it for a period of 20 yr earning interest at the rate of 8%/year compounded continuously.

13. **THE AMOUNT OF AN ANNUITY** Aiso deposits $150/month in a savings account paying 8%/year compounded continuously. Estimate the amount that will be in his account after 15 yr.

14. **CUSTODIAL ACCOUNTS** The Armstrongs wish to establish a custodial account to finance their children's education. If they deposit $200 monthly for 10 yr in a savings account paying 9%/year compounded continuously, how much will their savings account be worth at the end of this period?

15. **IRA ACCOUNTS** Refer to Example 4, page 526. Suppose Marcus makes his IRA payment on April 1, 1990, and annually thereafter. If interest is paid at the same initial rate, approximately how much will Marcus have in his account at the beginning of the year 2006?

16. **PRESENT VALUE OF AN ANNUITY** Estimate the present value of an annuity if payments are $800 monthly for 12 yr and the account earns interest at the rate of 10%/year compounded continuously.

17. **PRESENT VALUE OF AN ANNUITY** Estimate the present value of an annuity if payments are $1200 monthly for 15 yr and the account earns interest at the rate of 10%/year compounded continuously.

18. **LOTTERY PAYMENTS** A state lottery commission pays the winner of the "Million Dollar" lottery 20 annual installments of $50,000 each. If the prevailing interest rate is 8%/year compounded continuously, find the present value of the winning ticket.

19. **REVERSE ANNUITY MORTGAGES** Sinclair wishes to supplement his retirement income by $300/month for the next 10 yr. He plans to obtain a reverse annuity mortgage (RAM) on his home to meet this need. Estimate the amount of the mortgage he will require if the prevailing interest rate is 12%/year compounded continuously.

20. **REVERSE ANNUITY MORTGAGE** Refer to Exercise 19. Leah wishes to supplement her retirement income by $400/month for the next 15 yr by obtaining a RAM. Estimate the amount of the mortgage she will require if the pre-

vailing interest rate is 9%/year compounded continuously.

21. **LORENTZ CURVES** A certain country's income distribution is described by the function

$$f(x) = \frac{15}{16}x^2 + \frac{1}{16}x$$

   a. Sketch the Lorentz curve for this function.
   b. Compute $f(0.4)$ and $f(0.9)$ and interpret your results.

22. **LORENTZ CURVES** In a study conducted by a certain country's Economic Development Board, it was found that the Lorentz curve for the distribution of income of college teachers was described by the function

$$f(x) = \frac{13}{14}x^2 + \frac{1}{14}x$$

and that of lawyers by the function

$$g(x) = \frac{9}{11}x^4 + \frac{2}{11}x$$

   a. Compute the coefficient of inequality for each Lorentz curve.
   b. Which profession has a more equitable income distribution?

23. **LORENTZ CURVES** A certain country's income distribution is described by the function

$$f(x) = \frac{14}{15}x^2 + \frac{1}{15}x$$

   a. Sketch the Lorentz curve for this function.
   b. Compute $f(0.3)$ and $f(0.7)$.

24. **LORENTZ CURVES** In a study conducted by a certain country's Economic Development Board, it was found that the Lorentz curve for the distribution of income of stockbrokers was described by the function

$$f(x) = \frac{11}{12}x^2 + \frac{1}{12}x$$

and that of high school teachers by the function

$$g(x) = \frac{5}{6}x^2 + \frac{1}{6}x$$

   a. Compute the coefficient of inequality for each Lorentz curve.
   b. Which profession has a more equitable income distribution?

CONSUMERS' SURPLUS AND PRODUCERS' SURPLUS

1. Resolve Example 1, Section 6.7, using a graphing utility.

Hint: Use the intersection operation to find the equilibrium quantity and the equilibrium price. Use the numerical integral operation to evaluate the definite integral.

2. Resolve Exercise 7, Section 6.7, using a graphing utility.

Hint: See Exercise 1.

3. The demand function for a certain brand of travel alarm clocks is given by

$$p = -0.01x^2 - 0.3x + 10$$

where p is the wholesale unit price in dollars and x is the quantity demanded each month, measured in units of a thousand. The supply function for this brand of clocks is given by

$$p = -0.01x^2 + 0.2x + 4$$

where p has the same meaning as before and x is the quantity, in thousands, the supplier will make available in the marketplace per month. Determine the consumers' surplus and the producers' surplus when the market unit price is set at the equilibrium price.

4. The quantity demanded of a certain make of compact disc organizer is x thousand units per week, and the corresponding wholesale unit price is

$$p = \sqrt{400 - 8x}$$

dollars. The supplier of the organizers will make x thousand units available in the market when the unit wholesale price is

$$p = 0.02x^2 + 0.04x + 5$$

dollars. Determine the consumers' surplus and the producers' surplus when the market unit price is set at the equilibrium price.

Solution to Self-Check Exercise 6.7

We find the equilibrium price and equilibrium quantity by solving the system of equations

$$p = \sqrt{9 - 0.02x}$$

$$p = \sqrt{1 + 0.02x}$$

simultaneously. Substituting the first equation into the second, we have

$$\sqrt{9 - 0.02x} = \sqrt{1 + 0.02x}$$

Squaring both sides of the equation then leads to

$$9 - 0.02x = 1 + 0.02x$$
$$x = 200$$

Therefore,

$$p = \sqrt{9 - 0.02(200)}$$
$$= \sqrt{5} \approx 2.24$$

The equilibrium price is $224, and the equilibrium quantity is 200. The consumers' surplus is given by

$$CS = \int_0^{200} \sqrt{9 - 0.02x}\, dx - (2.24)(200)$$

$$= \int_0^{200} (9 - 0.02x)^{1/2}\, dx - 448$$

$$= -\frac{1}{0.02}\left(\frac{2}{3}\right)(9 - 0.02x)^{3/2}\Big|_0^{200} - 448 \qquad \text{(Integrating by substitution)}$$

$$= -\frac{1}{0.03}(5^{3/2} - 9^{3/2}) - 448$$

$$\approx 79.32$$

or approximately $7932.

Next, the producers' surplus is given by

$$PS = (2.24)(200) - \int_0^{200} \sqrt{1 + 0.02x}\, dx$$

$$= 448 - \int_0^{200} (1 + 0.02x)^{1/2}\, dx$$

$$= 448 - \frac{1}{0.02}\left(\frac{2}{3}\right)(1 + 0.02x)^{3/2}\Big|_0^{200}$$

$$= 448 - \frac{1}{0.03}(5^{3/2} - 1)$$

$$\approx 108.66$$

or approximately $10,866.

Besides the basic rules of integration developed in Chapter 6, there are more sophisticated techniques for finding the antiderivatives of functions. We begin this chapter by looking at the method of integration by parts. We then look at a technique of integration that involves using tables of integrals that have been compiled for this purpose. We also look at numerical methods of integration, which enable us to obtain approximate solutions to definite integrals, especially those whose exact value cannot be found otherwise. More specifically, we study the trapezoidal rule and Simpson's rule. Numerical integration methods are especially useful when the integrand is known only at discrete points. Finally, we learn how to evaluate integrals in which the intervals of integration are unbounded. Such integrals, called *improper integrals*, play an important role in the study of probability.

What is the area of the oil spill caused by a grounded tanker? In Example 5, page 574, you will see how to determine the area of the oil spill.

7.1 Integration by Parts

THE METHOD OF INTEGRATION BY PARTS

Integration by parts is another technique of integration that, like the method of substitution discussed in Chapter 6, is based on a corresponding rule of differentiation. In this case, the rule of differentiation is the product rule, which asserts that if f and g are differentiable functions, then

$$\frac{d}{dx}[f(x)g(x)] = f(x)g'(x) + g(x)f'(x) \tag{1}$$

If we integrate both sides of Equation (1) with respect to x, we obtain

$$\int \frac{d}{dx}f(x)g(x)\,dx = \int f(x)g'(x)\,dx + \int g(x)f'(x)\,dx$$

$$f(x)g(x) = \int f(x)g'(x)\,dx + \int g(x)f'(x)\,dx$$

This last equation, which may be written in the form

$$\int f(x)g'(x)\,dx = f(x)g(x) - \int g(x)f'(x)\,dx \tag{2}$$

is called the formula for **integration by parts.** This formula is useful since it enables us to express one indefinite integral in terms of another that may be easier to evaluate. Formula (2) may be simplified by letting

$$u = f(x) \qquad dv = g'(x)\,dx$$
$$du = f'(x)\,dx \qquad v = g(x)$$

giving the following version of the formula for integration by parts.

Integration by Parts Formula

$$\int u\,dv = uv - \int v\,du \tag{3}$$

EXAMPLE 1   Evaluate $\int xe^x\,dx$.

SOLUTION ✔   No method of integration developed thus far enables us to evaluate the given indefinite integral in its present form. Therefore, we attempt to write it in terms of an indefinite integral that will be easier to evaluate. Let's use the integration by parts Formula (3) by letting

$$u = x \qquad \text{and} \qquad dv = e^x\,dx$$

so that

$$du = dx \qquad \text{and} \qquad v = e^x$$

Therefore,

$$\int xe^x \, dx = \int u \, dv$$

$$= uv - \int v \, du$$

$$= xe^x - \int e^x \, dx$$

$$= xe^x - e^x + C$$

$$= (x - 1)e^x + C$$ ■ ■ ■ ■

The success of the method of integration by parts depends on the proper choice of u and dv. For example, if we had chosen

$$u = e^x \quad \text{and} \quad dv = x \, dx$$

in the last example, then

$$du = e^x \, dx \quad \text{and} \quad v = \frac{1}{2}x^2$$

Thus, (3) would have yielded

$$\int xe^x \, dx = \int u \, dv$$

$$= uv - \int v \, du$$

$$= \frac{1}{2}x^2 e^x - \int \frac{1}{2}x^2 e^x \, dx$$

Since the indefinite integral on the right-hand side of this equation is not readily evaluated (it is in fact more complicated than the original integral!), choosing u and dv as shown has not helped us evaluate the given indefinite integral.

In general, we can use the following guidelines.

Guidelines for Choosing u and dv

Choose u and dv so that

1. du is simpler than u.

2. dv is easy to integrate.

EXAMPLE **2**   Evaluate $\int x \ln x \, dx$.

SOLUTION ✔   Letting

$$u = \ln x \quad \text{and} \quad dv = x \, dx$$

we have

$$du = \frac{1}{x} \, dx \quad \text{and} \quad v = \frac{1}{2}x^2$$

Therefore,

$$\int x \ln x \, dx = \int u \, dv = uv - \int v \, du$$

$$= \frac{1}{2}x^2 \ln x - \int \frac{1}{2}x^2 \cdot \left(\frac{1}{x}\right) dx$$

$$= \frac{1}{2}x^2 \ln x - \frac{1}{2}\int x \, dx$$

$$= \frac{1}{2}x^2 \ln x - \frac{1}{4}x^2 + C$$

$$= \frac{1}{4}x^2(2 \ln x - 1) + C$$

■■■■

EXAMPLE 3 Evaluate $\int \frac{xe^x}{(x + 1)^2} \, dx$.

SOLUTION ✔ Let

$$u = xe^x \qquad \text{and} \qquad dv = \frac{1}{(x + 1)^2} \, dx$$

Then,

$$du = (xe^x + e^x) \, dx = e^x(x + 1) \, dx \qquad \text{and} \qquad v = -\frac{1}{x + 1}$$

Therefore,

$$\int \frac{xe^x}{(x + 1)^2} \, dx = \int u \, dv = uv - \int v \, du$$

$$= xe^x\left(\frac{-1}{x + 1}\right) - \int \left(-\frac{1}{x + 1}\right) e^x(x + 1) \, dx$$

$$= -\frac{xe^x}{x + 1} + \int e^x \, dx$$

$$= -\frac{xe^x}{x + 1} + e^x + C$$

$$= \frac{e^x}{x + 1} + C$$

■■■■

The next example shows that repeated applications of the technique of integration by parts is sometimes required to evaluate an integral.

EXAMPLE 4 Evaluate $\int x^2 e^x \, dx$.

SOLUTION ✔ Let

$$u = x^2 \qquad \text{and} \qquad dv = e^x \, dx$$

so that

$$du = 2x \, dx \qquad \text{and} \qquad v = e^x$$

Therefore,

$$\int x^2 e^x \, dx = \int u \, dv = uv - \int v \, du$$

$$= x^2 e^x - \int e^x (2x) \, dx = x^2 e^x - 2 \int x e^x \, dx$$

To complete the solution of the problem, we need to evaluate the integral

$$\int x e^x \, dx$$

But this integral may be found using integration by parts. In fact, you will recognize that this integral is precisely that of Example 1. Using the results obtained there, we now find

$$\int x^2 e^x \, dx = x^2 e^x - 2[(x-1)e^x] + C = e^x(x^2 - 2x + 2) + C \qquad ■ ■ ■ ■$$

Group Discussion

1. Use the method of integration by parts to derive the formula

$$\int x^n e^{ax} \, dx = \frac{1}{a} x^n e^{ax} - \frac{n}{a} \int x^{n-1} e^{ax} \, dx$$

where n is a positive integer and a is a real number.

2. Use the formula of part 1 to evaluate

$$\int x^3 e^x \, dx.$$

Hint: You may find the results of Example 4 helpful.

APPLICATION

EXAMPLE 5

The estimated rate at which oil will be produced from a certain oil well t years after production has begun is given by

$$R(t) = 100 t e^{-0.1t}$$

thousand barrels per year. Find an expression that describes the total production of oil at the end of year t.

SOLUTION ✔

Let $T(t)$ denote the total production of oil from the well at the end of year t ($t \geq 0$). Then, the rate of oil production will be given by $T'(t)$ thousand barrels per year. Thus,

$$T'(t) = R(t) = 100 t e^{-0.1t}$$

so

$$T(t) = \int 100 t e^{-0.1t} \, dt$$

$$= 100 \int t e^{-0.1t} \, dt$$

1. Use the table of integrals to evaluate

$$\int_0^2 \frac{dx}{(5 - x^2)^{3/2}}$$

2. During a flu epidemic, the number of children in the Easton Middle School who contracted influenza t days after the outbreak began was given by

$$N(t) = \frac{200}{1 + 9e^{-0.8t}}$$

Determine the average number of children who contracted the flu in the first 10 days of the epidemic.

Solutions to Self-Check Exercises 7.2 can be found on page 565.

7.2 Exercises

In Exercises 1–32, use the table of integrals in this section to evaluate each of the given integrals.

1. $\int \frac{2x}{2 + 3x} \, dx$

2. $\int \frac{x}{(1 + 2x)^2} \, dx$

3. $\int \frac{3x^2}{2 + 4x} \, dx$

4. $\int \frac{x^2}{3 + x} \, dx$

5. $\int x^2 \sqrt{9 + 4x^2} \, dx$

6. $\int x^2 \sqrt{4 + x^2} \, dx$

7. $\int \frac{dx}{x \sqrt{1 + 4x}}$

8. $\int_0^2 \frac{x + 1}{\sqrt{2 + 3x}} \, dx$

9. $\int_0^2 \frac{dx}{\sqrt{9 + 4x^2}}$

10. $\int \frac{dx}{x \sqrt{4 + 8x^2}}$

11. $\int \frac{dx}{(9 - x^2)^{3/2}}$

12. $\int \frac{dx}{(2 - x^2)^{3/2}}$

13. $\int x^2 \sqrt{x^2 - 4} \, dx$

14. $\int_3^5 \frac{dx}{x^2 \sqrt{x^2 - 9}}$

15. $\int \frac{\sqrt{4 - x^2}}{x} \, dx$

16. $\int_0^1 \frac{dx}{(4 - x^2)^{3/2}}$

17. $\int xe^{2x} \, dx$

18. $\int \frac{dx}{1 + e^{-x}}$

19. $\int \frac{dx}{(x + 1)\ln(1 + x)}$

Hint: First use the substitution $u = x + 1$.

20. $\int \frac{x}{(x^2 + 1)\ln(x^2 + 1)} \, dx$

Hint: First use the substitution $u = x^2 + 1$.

21. $\int \frac{e^{2x}}{(1 + 3e^x)^2} \, dx$

22. $\int \frac{e^{2x}}{\sqrt{1 + 3e^x}} \, dx$

23. $\int \frac{3e^x}{1 + e^{(1/2)x}} \, dx$

24. $\int \frac{dx}{1 - 2e^{-x}}$

25. $\int \frac{\ln x}{x(2 + 3 \ln x)} \, dx$

26. $\int_1^e (\ln x)^2 \, dx$

27. $\int_0^1 x^2 e^x \, dx$

28. $\int x^3 e^{2x} \, dx$

29. $\int x^2 \ln x \, dx$

30. $\int x^3 \ln x \, dx$

31. $\int (\ln x)^3 \, dx$

32. $\int (\ln x)^4 \, dx$

33. CONSUMERS' SURPLUS Refer to Section 6.7. The demand function for Apex ladies' boots is

$$p = \frac{250}{\sqrt{16 + x^2}}$$

where p is the wholesale unit price in dollars and x is the quantity demanded daily, in units of a hundred. Find the consumers' surplus if the wholesale price is set at $50/pair.

34. **PRODUCERS' SURPLUS** Refer to Section 6.7. The supplier of Apex ladies' boots will make x hundred pairs of the boots available in the market daily when the wholesale unit price is

$$p = \frac{30x}{5 - x}$$

dollars. Find the producers' surplus if the wholesale price is set at $50/pair.

35. **AMUSEMENT PARK ATTENDANCE** The management of Astro World ("The Amusement Park of the Future") estimates that the number of visitors (in thousands) entering the amusement park t hr after opening time at 9 A.M. is given by

$$R(t) = \frac{60}{(2 + t^2)^{3/2}}$$

per hour. Determine the number of visitors admitted by noon.

36. **VOTER REGISTRATION** The number of voters in a certain district of a city is expected to grow at the rate of

$$R(t) = \frac{3000}{\sqrt{4 + t^2}}$$

people/year t yr from now. If the number of voters at present is 20,000, how many voters will be in the district 5 yr from now?

37. **GROWTH OF FRUIT FLIES** Based on data collected during an experiment, a biologist found that the number of fruit flies (*Drosophila*) with a limited food supply could be approximated by the exponential model

$$N(t) = \frac{1000}{1 + 24e^{-0.02t}}$$

where t denotes the number of days since the beginning of the experiment. Find the average number of fruit flies in the colony in the first 10 days of the experiment and in the first 20 days.

38. **VCR OWNERSHIP** According to estimates by Paul Kroger Associates, the percentage of households that own videocassette recorders (VCRs) is given by

$$P(t) = \frac{68}{1 + 21.67e^{-0.62t}} \qquad (0 \le t \le 12)$$

where t is measured in years, with $t = 0$ corresponding to the beginning of 1981. Find the average percentage of households owning VCRs from the beginning of 1981 to the beginning of 1993.
Source: Paul Kroger Associates

39. **RECYCLING PROGRAMS** The commissioner of the City of Newton Department of Public Works estimates that the number of people in the city who have been recycling their magazines in year t following the introduction of the recycling program at the beginning of 1990 is

$$N(t) = \frac{100,000}{2 + 3e^{-0.2t}}$$

Find the average number of people who will have recycled their magazines during the first 5 yr since the program was introduced.

40. **FRANCHISES** Elaine purchased a 10-yr franchise for a fast-food restaurant that is expected to generate income at the rate of $R(t) = 250,000 + 2000t^2$ dollars/year, t yr from now. If the prevailing interest rate is 10%/year compounded continuously, find the present value of the franchise.
Hint: Use Formula (18), Section 6.7.

41. **ACCUMULATED VALUE OF AN INCOME STREAM** The revenue of Virtual Reality, a video-game arcade, is generated at the rate of $R(t) = 20,000t$ dollars. If the revenue is invested t yr from now in a business earning interest at the rate of 15%/year compounded continuously, find the accumulated value of this stream of income at the end of 5 yr.
Hint: Use Formula (18), Section 6.7.

42. **LORENTZ CURVES** In a study conducted by a certain country's Economic Development Board regarding the income distribution of certain segments of the country's workforce, it was found that the Lorentz curve for the distribution of income of college professors is described by the function

$$g(x) = \frac{1}{3}x\sqrt{1 + 8x}$$

Compute the coefficient of inequality of the Lorentz curve.
Hint: Use Formula (21), Section 6.7.

SOLUTIONS TO SELF-CHECK EXERCISES 7.2

1. Using Formula 22, page 559, with $a^2 = 5$ and $u = x$, we see that

$$\int_0^2 \frac{dx}{(5 - x^2)^{3/2}} = \frac{x}{5\sqrt{5 - x^2}} \bigg|_0^2$$

$$= \frac{2}{5\sqrt{5 - 4}}$$

$$= \frac{2}{5}$$

2. The average number of children who contracted the flu in the first 10 days of the epidemic is given by

$$A = \frac{1}{10} \int_0^{10} \frac{200}{1 + 9e^{-0.8t}} \, dt = 20 \int_0^{10} \frac{dt}{1 + 9e^{-0.8t}}$$

$$= 20 \left[t + \frac{1}{0.8} \ln(1 + 9e^{-0.8t}) \right] \bigg|_0^{10} \qquad \text{(Formula 25, } a = -0.8,\\ \qquad\qquad\qquad\qquad\qquad\qquad\qquad\qquad\qquad b = 9, u = t)$$

$$= 20 \left[10 + \frac{1}{0.8} \ln(1 + 9e^{-8}) \right] - 20 \left(\frac{1}{0.8} \right) \ln 10$$

$$\approx 200.07537 - 57.56463$$

$$\approx 143$$

or 143 students.

7.3 Numerical Integration

APPROXIMATING DEFINITE INTEGRALS

One method of measuring cardiac output is to inject 5 to 10 milligrams (mg) of a dye into a vein leading to the heart. After making its way through the lungs, the dye returns to the heart and is pumped into the aorta, where its concentration is measured at equal time intervals. The graph of the function c in Figure 7.1 shows the concentration of dye in a person's aorta, measured at 2-second intervals after 5 mg of dye have been injected. The person's cardiac output, measured in liters per minute (L/min), is computed using the formula

$$R = \frac{60D}{\int_0^{28} c(t) \, dt} \tag{4}$$

where D is the quantity of dye injected (see Exercise 44).

Now, to use Formula (4), we need to evaluate the definite integral

$$\int_0^{28} c(t) \, dt$$

FIGURE 7.1
The function c gives the concentration of a dye measured at the aorta. The graph is constructed by drawing a smooth curve through a set of discrete points.

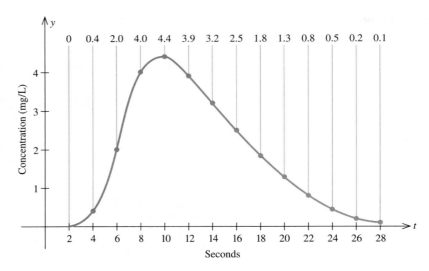

But we do not have the algebraic rule defining the integrand c for all values of t in $[0, 28]$. In fact, we are given its values only at a set of discrete points in that interval. In situations such as this, the fundamental theorem of calculus proves useless because we cannot find an antiderivative of c. (We will complete the solution to this problem in Example 4.)

Other situations also arise in which an integrable function has an antiderivative that cannot be found in terms of elementary functions (functions that can be expressed as a finite combination of algebraic, exponential, logarithmic, and trigonometric functions). Examples of such functions are

$$f(x) = e^{x^2}, \qquad g(x) = x^{-1/2}e^x, \qquad h(x) = \frac{1}{\ln x}$$

Riemann sums provide us with a good approximation of a definite integral, provided the number of subintervals in the partitions is large enough. But there are better techniques and formulas, called *quadrature formulas,* that give a more efficient way of computing approximate values of definite integrals. In this section we look at two rather simple but effective ways of approximating definite integrals.

THE TRAPEZOIDAL RULE

We assume that $f(x) \geq 0$ on $[a, b]$ in order to simplify the derivation of the trapezoidal rule, but the result is valid without this restriction. We begin by subdividing the interval $[a, b]$ into n subintervals of equal length Δx, by means of the $(n + 1)$ points $x_0 = a, x_1, x_2, \ldots, x_n = b$, where n is a positive integer (Figure 7.2).

Then, the length of each subinterval is given by

$$\Delta x = \frac{b - a}{n}$$

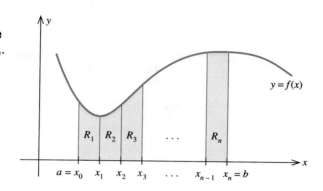

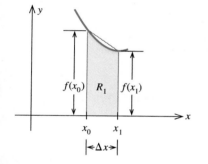

Furthermore, as we saw earlier, we may view the definite integral

$$\int_a^b f(x)\, dx$$

as the area of the region R under the curve $y = f(x)$ between $x = a$ and $x = b$. This area is given by the sum of the areas of the n nonoverlapping subregions $R_1, R_2, \ldots, R_n$, such that R_1 represents the region under the curve $y = f(x)$ from $x = x_0$ to $x = x_1$, and so on.

The basis for the trapezoidal rule lies in the approximation of each of the regions $R_1, R_2, \ldots, R_n$ by a suitable trapezoid. This often leads to a much better approximation than one obtained by means of rectangles (a Riemann sum).

Let's consider the subregion R_1, shown magnified for the sake of clarity in Figure 7.3. Observe that the area of the region R_1 may be approximated by the trapezoid of width Δx whose parallel sides are of lengths $f(x_0)$ and $f(x_1)$. The area of the trapezoid is given by

$$\left[\frac{f(x_0) + f(x_1)}{2}\right] \Delta x \qquad \text{(Average of the lengths of the parallel sides times the width)}$$

square units. Similarly, the area of the region R_2 may be approximated by the trapezoid of width Δx and sides of lengths $f(x_1)$ and $f(x_2)$. The area of the trapezoid is given by

$$\left[\frac{f(x_1) + f(x_2)}{2}\right] \Delta x$$

Similarly, we see that the area of the last (nth) approximating trapezoid is given by

$$\left[\frac{f(x_{n-1}) + f(x_n)}{2}\right] \Delta x$$

Then, the area of the region R is approximated by the sum of the areas of the n trapezoids—that is,

$$\left[\frac{f(x_0) + f(x_1)}{2}\right]\Delta x + \left[\frac{f(x_1) + f(x_2)}{2}\right]\Delta x + \cdots + \left[\frac{f(x_{n-1}) + f(x_n)}{2}\right]\Delta x$$

$$= \frac{\Delta x}{2}[f(x_0) + f(x_1) + f(x_1) + f(x_2) + \cdots + f(x_{n-1}) + f(x_n)]$$

$$= \frac{\Delta x}{2}[f(x_0) + 2f(x_1) + 2f(x_2) + \cdots + 2f(x_{n-1}) + f(x_n)]$$

Since the area of the region R is given by the value of the definite integral we wished to approximate, we are led to the following approximation formula, which is called the **trapezoidal rule**.

Trapezoidal Rule

$$\int_a^b f(x)\, dx \approx \frac{\Delta x}{2}[f(x_0) + 2f(x_1) + 2f(x_2)$$

$$+ \cdots + 2f(x_{n-1}) + f(x_n)]$$

(5)

where $\Delta x = \dfrac{b - a}{n}$.

The approximation generally improves with larger values of n.

EXAMPLE 1   Approximate the value of

$$\int_1^2 \frac{1}{x}\, dx$$

using the trapezoidal rule with $n = 10$. Compare this result with the exact value of the integral.

SOLUTION ✔   Here, $a = 1$, $b = 2$, and $n = 10$, so

$$\Delta x = \frac{b - a}{n} = \frac{1}{10} = 0.1$$

and

$$x_0 = 1, \qquad x_1 = 1.1, \qquad x_2 = 1.2, \qquad x_3 = 1.3, \ldots, x_9 = 1.9, \qquad x_{10} = 2$$

The trapezoidal rule yields

$$\int_1^2 \frac{1}{x}\, dx \approx \frac{0.1}{2}\left[1 + 2\left(\frac{1}{1.1}\right) + 2\left(\frac{1}{1.2}\right) + 2\left(\frac{1}{1.3}\right) + \cdots + 2\left(\frac{1}{1.9}\right) + \frac{1}{2}\right]$$

$$\approx 0.693771$$

In this case we can easily compute the actual value of the definite integral under consideration. In fact,

$$\int_1^2 \frac{1}{x}\, dx = \ln x \Big|_1^2 = \ln 2 - \ln 1 = \ln 2$$

$$\approx 0.693147$$

Thus, the trapezoidal rule with $n = 10$ yields a result with an error of 0.000624 to six decimal places.   ◼ ◼ ◼ ◼

EXAMPLE 2   The demand function for a certain brand of perfume is given by

$$p = D(x) = \sqrt{10{,}000 - 0.01x^2}$$

where p is the unit price in dollars and x is the quantity demanded each week, measured in ounces. Find the consumers' surplus if the market price is set at $60 per ounce.

SOLUTION ✔   When $p = 60$, we have

$$\sqrt{10{,}000 - 0.01x^2} = 60$$
$$10{,}000 - 0.01x^2 = 3{,}600$$
$$x^2 = 640{,}000$$

or $x = 800$ since x must be nonnegative. Next, using the consumers' surplus formula (page 520) with $\bar{p} = 60$ and $\bar{x} = 800$, we see that the consumers' surplus is given by

$$CS = \int_0^{800} \sqrt{10{,}000 - 0.01x^2}\, dx - (60)(800)$$

It is not easy to evaluate this definite integral by finding an antiderivative of the integrand. Instead, let's use the trapezoidal rule with $n = 10$.

With $a = 0$ and $b = 800$, we find that

$$\Delta x = \frac{b - a}{n} = \frac{800}{10} = 80$$

and

$$x_0 = 0, \qquad x_1 = 80, \qquad x_2 = 160, \qquad x_3 = 240, \ldots, x_9 = 720, \qquad x_{10} = 800$$

so

$$\int_0^{800} \sqrt{10{,}000 - 0.01x^2}\, dx$$

$$\approx \frac{80}{2} [100 + 2\sqrt{10{,}000 - (0.01)(80)^2}$$

$$+ 2\sqrt{10{,}000 - (0.01)(160)^2} + \cdots + 2\sqrt{10{,}000 - (0.01)(720)^2}$$

$$+ \sqrt{10{,}000 - (0.01)(800)^2}]$$

$$= 40[100 + 199.3590 + 197.4234 + 194.1546 + 189.4835$$
$$+ 183.3030 + 175.4537 + 165.6985$$
$$+ 153.6750 + 138.7948 + 60]$$
$$\approx 70{,}293.82$$

Therefore, the consumers' surplus is approximately $70{,}294 - 48{,}000$, or $22,294.   ■■■■

Group Discussion

Explain how you would approximate the value of $\int_0^2 f(x)\, dx$ using the trapezoidal rule with $n = 10$, where

$$f(x) = \begin{cases} \sqrt{1 + x^2} & \text{if } 0 \le x \le 1 \\ \dfrac{2}{\sqrt{1 + x^2}} & \text{if } 1 < x \le 2 \end{cases}$$

and find the value.

SIMPSON'S RULE

Before stating Simpson's rule, let's review the two rules we have used in approximating a definite integral. Let f be a continuous nonnegative function defined on the interval $[a, b]$. Suppose the interval $[a, b]$ is partitioned by means of the $n + 1$ equally spaced points $x_0 = a, x_1, x_2, \ldots, x_n = b$, where n is a positive integer, so that the length of each subinterval is $\Delta x = (b - a)/n$ (Figure 7.4).

FIGURE 7.4
The area under the curve is equal to the sum of the n subregions $R_1, R_2, \ldots, R_n$.

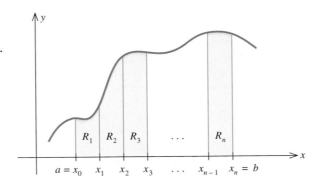

Let's concentrate on the portion of the graph of $y = f(x)$ defined on the interval $[x_0, x_2]$. In using a Riemann sum to approximate the definite integral, we are in effect approximating the function $f(x)$ on $[x_0, x_1]$ by the *constant* function $y = f(p_1)$, where p_1 is chosen to be a point in $[x_0, x_1]$; the function

$f(x)$ on $[x_1, x_2]$ by the constant function $y = f(p_2)$, where p_2 lies in $[x_1, x_2]$; and so on. Using a Riemann sum, we see that the area of the region under the curve $y = f(x)$ between $x = a$ and $x = b$ is approximated by the area under the approximating "step" function (Figure 7.5a).

FIGURE 7.5

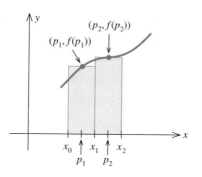

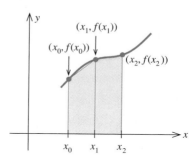

(a) The area under the curve is approximated by the area of the rectangles.

(b) The area under the curve is approximated by the area of the trapezoids.

When we use the trapezoidal rule, we are in effect approximating the function $f(x)$ on the interval $[x_0, x_1]$ by a *linear* function through the two points $(x_0, f(x_0))$ and $(x_1, f(x_1))$; the function $f(x)$ on $[x_1, x_2]$ by a *linear* function through the two points $(x_1, f(x_1))$ and $(x_2, f(x_2))$; and so on. Thus, the trapezoidal rule simply approximates the actual area of the region under the curve $y = f(x)$ from $x = a$ to $x = b$ by the area under the approximating polygonal curve (Figure 7.5b).

A natural extension of the preceding idea is to approximate portions of the graph of $y = f(x)$ by means of portions of the graphs of second-degree polynomials (parts of parabolas). It can be shown that given any three noncollinear points there is a unique parabola that passes through the given points. Choose the points $(x_0, f(x_0))$, $(x_1, f(x_1))$, and $(x_2, f(x_2))$ corresponding to the first three points of the partition. Then, we can approximate the function $f(x)$ on $[x_0, x_2]$ by means of a quadratic function whose graph contains these three points (Figure 7.6).

Although we will not do so here, it can be shown that the area under the parabola between $x = x_0$ and $x = x_2$ is given by

$$\frac{\Delta x}{3} [f(x_0) + 4f(x_1) + f(x_2)]$$

square units. Repeating this argument on the interval $[x_2, x_4]$, we see that the area under the curve between $x = x_2$ and $x = x_4$ is approximated by the area under the parabola between x_2 and x_4—that is, by

$$\frac{\Delta x}{3} [f(x_2) + 4f(x_3) + f(x_4)]$$

square units. Proceeding, we conclude that if n is even (Why?), then the area under the curve $y = f(x)$ from $x = a$ to $x = b$ may be approximated by the

FIGURE 7.6

Simpson's rule approximates the area under the curve by the area under the parabola.

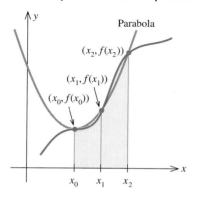

sum of the areas under the $n/2$ approximating parabolas—that is,

$$\frac{\Delta x}{3}\,[\,f(x_0) + 4f(x_1) + f(x_2)\,] + \frac{\Delta x}{3}\,[\,f(x_2) + 4f(x_3) + f(x_4)\,] + \cdots$$

$$+ \frac{\Delta x}{3}\,[\,f(x_{n-2}) + 4f(x_{n-1}) + f(x_n)\,]$$

$$= \frac{\Delta x}{3}\,[\,f(x_0) + 4f(x_1) + f(x_2) + f(x_2) + 4f(x_3) + f(x_4) + \cdots$$

$$+ f(x_{n-2}) + 4f(x_{n-1}) + f(x_n)\,]$$

$$= \frac{\Delta x}{3}\,[\,f(x_0) + 4f(x_1) + 2f(x_2) + 4f(x_3)$$

$$+ 2f(x_4) + \cdots + 4f(x_{n-1}) + f(x_n)\,]$$

The preceding is the derivation of the approximation formula known as **Simpson's rule.**

Simpson's Rule

$$\int_a^b f(x)\,dx \approx \frac{\Delta x}{3}\,[\,f(x_0) + 4f(x_1) + 2f(x_2) + 4f(x_3) + 2f(x_4)$$

$$+ \cdots + 4f(x_{n-1}) + f(x_n)\,] \tag{6}$$

where $\Delta x = \dfrac{b-a}{n}$ and n is even.

In using this rule, remember that n must be even.

 EXAMPLE 3   Find an approximation of

$$\int_1^2 \frac{1}{x}\,dx$$

using Simpson's rule with $n = 10$. Compare this result with that of Example 1 and also with the exact value of the integral.

SOLUTION ✔   We have $a = 1$, $b = 2$, $f(x) = \dfrac{1}{x}$, and $n = 10$, so

$$\Delta x = \frac{b-a}{n} = \frac{1}{10} = 0.1$$

and

$$x_0 = 1, \qquad x_1 = 1.1, \qquad x_2 = 1.2, \qquad x_3 = 1.3, \ldots, x_9 = 1.9, \qquad x_{10} = 2$$

Simpson's rule yields

$$\int_1^2 \frac{1}{x}\,dx \approx \frac{0.1}{3}[f(1) + 4f(1.1) + 2f(1.2) + \cdots + 4f(1.9) + f(2)]$$

$$= \frac{0.1}{3}\left[1 + 4\left(\frac{1}{1.1}\right) + 2\left(\frac{1}{1.2}\right) + 4\left(\frac{1}{1.3}\right) + 2\left(\frac{1}{1.4}\right) + 4\left(\frac{1}{1.5}\right)\right.$$

$$\left. + 2\left(\frac{1}{1.6}\right) + 4\left(\frac{1}{1.7}\right) + 2\left(\frac{1}{1.8}\right) + 4\left(\frac{1}{1.9}\right) + \frac{1}{2}\right]$$

$$\approx 0.693150$$

The trapezoidal rule with $n = 10$ yielded an approximation of 0.693771, which is 0.000624 off the value of ln 2 $\approx$ 0.693147 to six decimal places. Simpson's rule yields an approximation with an error of 0.000003, a definite improvement over the trapezoidal rule. ■ ■ ■ ■

EXAMPLE 4 Solve the problem posed at the beginning of this section. Recall that we wished to find a person's cardiac output by using the formula

$$R = \frac{60D}{\displaystyle\int_0^{28} c(t)\,dt}$$

where D (the quantity of dye injected) is equal to 5 milligrams and the function c has the graph shown in Figure 7.7. Use Simpson's rule with $n = 14$ to estimate the value of the integral.

FIGURE 7.7
The function c gives the concentration of a dye measured at the aorta. The graph is constructed by drawing a smooth curve through a set of discrete points.

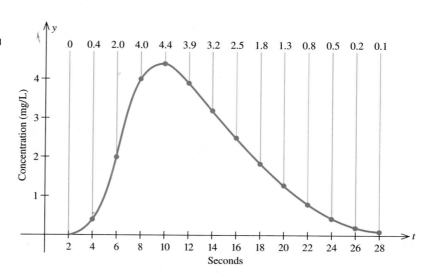

SOLUTION ✔ Using Simpson's rule with $n = 14$ and $\Delta t = 2$ so that

$$t_0 = 0, \qquad t_1 = 2, \qquad t_2 = 4, \qquad t_3 = 6, \ldots, t_{14} = 28$$

we obtain

$$\int_0^{28} c(t)\, dt \approx \frac{2}{3}[c(0) + 4c(2) + 2c(4) + 4c(6) + \cdots$$
$$+ 4c(26) + c(28)]$$
$$= \frac{2}{3}[0 + 4(0) + 2(0.4) + 4(2.0) + 2(4.0)$$
$$+ 4(4.4) + 2(3.9) + 4(3.2) + 2(2.5) + 4(1.8)$$
$$+ 2(1.3) + 4(0.8) + 2(0.5) + 4(0.2) + 0.1]$$
$$\approx 49.9$$

Therefore, the person's cardiac output is

$$R \approx \frac{60(5)}{49.9} \approx 6.0$$

or 6.0 L/min.

EXAMPLE 5   An oil spill off the coastline was caused by a ruptured tank in a grounded oil tanker. Using aerial photographs, the Coast Guard was able to obtain the dimensions of the oil spill (Figure 7.8). Using Simpson's rule with $n = 10$, estimate the area of the oil spill.

FIGURE 7.8
Simpson's rule can be used to calculate the area of the oil spill.

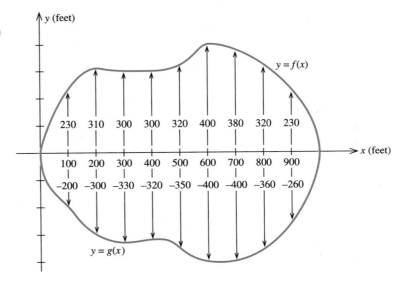

SOLUTION ✔   We may think of the area affected by the oil spill as the area of the plane region bounded above by the graph of the function $f(x)$ and below by the graph of the function $g(x)$ between $x = 0$ and $x = 1000$ (Figure 7.8). Then, the required area is given by

$$A = \int_0^{1000} [f(x) - g(x)]\, dx$$

Using Simpson's rule with $n = 10$ and $\Delta x = 100$ so that

$$x_0 = 0, \qquad x_1 = 100, \qquad x_2 = 200, \ldots, x_{10} = 1000$$

we have

$$
\begin{aligned}
A &= \int_0^{1000} [f(x) - g(x)]\, dx \\
&\approx \frac{\Delta x}{3}\{[f(x_0) - g(x_0)] + 4[f(x_1) - g(x_1)] + 2[f(x_2) - g(x_2)] \\
&\qquad + \cdots + 4[f(x_9) - g(x_9)] + [f(x_{10}) - g(x_{10})]\} \\
&= \frac{100}{3}\{[0 - 0] + 4[230 - (-200)] + 2[310 - (-300)] \\
&\qquad + 4[300 - (-330)] + 2[300 - (-320)] + 4[320 - (-350)] \\
&\qquad + 2[400 - (-400)] + 4[380 - (-400)] + 2[320 - (-360)] \\
&\qquad + 4[230 - (-260)] + [0 - 0]\} \\
&= \frac{100}{3}[0 + 4(430) + 2(610) + 4(630) + 2(620) + 4(670) \\
&\qquad + 2(800) + 4(780) + 2(680) + 4(490) + 0] \\
&= \frac{100}{3}(17{,}420) \\
&\approx 580{,}667
\end{aligned}
$$

or approximately 580,667 square feet.   ■ ■ ■ ■

Group Discussion

Explain how you would approximate the value of $\int_0^2 f(x)\, dx$ using Simpson's rule with $n = 10$, where

$$
f(x) = \begin{cases} \sqrt{1 + x^2} & \text{if } 0 \le x \le 1 \\[2mm] \dfrac{2}{\sqrt{1 + x^2}} & \text{if } 1 < x \le 2 \end{cases}
$$

and find the value.

ERROR ANALYSIS

The following results give the bounds on the errors incurred when the trapezoidal rule and Simpson's rule are used to approximate a definite integral (proof omitted).

Errors in the Trapezoidal and Simpson Approximations

Suppose the definite integral

$$\int_a^b f(x)\, dx$$

is approximated with n subintervals.

1. The *maximum* error incurred in using the trapezoidal rule is

$$\frac{M(b-a)^3}{12n^2} \qquad (7)$$

where M is a number such that $|f''(x)| \le M$ for all x in $[a, b]$.

2. The *maximum* error incurred in using Simpson's rule is

$$\frac{M(b-a)^5}{180n^4} \qquad (8)$$

where M is a number such that $|f^{(4)}(x)| \le M$ for all x in $[a, b]$.

REMARK In many instances, the actual error is less than the upper error bounds given. ■■■

EXAMPLE 6   Find bounds on the errors incurred when

$$\int_1^2 \frac{1}{x}\, dx$$

is approximated using (a) the trapezoidal rule and (b) Simpson's rule with $n = 10$. Compare these with the actual errors found in Examples 1 and 3.

SOLUTION ✔   **a.** Here, $a = 1$, $b = 2$, and $f(x) = 1/x$. Next, to find a value for M, we compute

$$f'(x) = -\frac{1}{x^2} \qquad \text{and} \qquad f''(x) = \frac{2}{x^3}$$

Since $f''(x)$ is positive and decreasing on $(1, 2)$ (Why?), it attains its maximum value of 2 at $x = 1$, the left end point of the interval. Therefore, if we take $M = 2$, then $|f''(x)| \le 2$. Using (7), we see that the maximum error incurred is

$$\frac{2(2-1)^3}{12(10)^2} = \frac{2}{1200} = 0.0016667$$

The actual error found in Example 1, 0.000624, is much less than the upper bound just found.

b. We compute

$$f'''(x) = \frac{-6}{x^4} \qquad \text{and} \qquad f^{(4)}(x) = \frac{24}{x^5}$$

Since $f^{(4)}(x)$ is positive and decreasing on $(1, 2)$ (just look at $f^{(5)}$ to verify this fact), it attains its maximum at the left end point of $[1, 2]$. Now,

$$f^{(4)}(1) = 24$$

and so we may take $M = 24$. Using (8), we obtain the maximum error of

$$\frac{24(2 - 1)^5}{180(10)^4} = 0.0000133$$

The actual error is 0.000003 (see Example 3).   ■ ■ ■ ■

> **Group Discussion**
> Refer to the Group Discussions on pages 570 and 575. Explain how you would find the maximum error incurred in using (1) the trapezoidal rule and (2) Simpson's rule with $n = 10$ to approximate $\int_0^2 f(x)\, dx$.

SELF-CHECK EXERCISES 7.3

1. Use the trapezoidal rule and Simpson's rule with $n = 8$ to approximate the value of the definite integral

$$\int_0^2 \frac{1}{\sqrt{1 + x^2}}\, dx$$

2. The graph in the accompanying figure shows the consumption of petroleum in the United States in quadrillion BTU, from 1976 to 1990. Using Simpson's rule with $n = 14$, estimate the average consumption during the 14-yr period.

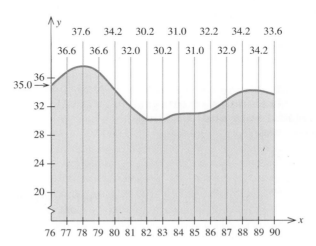

Source: The World Almanac

Solutions to Self-Check Exercises 7.3 can be found on page 581.

7.3 Exercises

A calculator is recommended for this exercise set. In Exercises 1–14, use the trapezoidal rule and Simpson's rule to approximate the value of each definite integral. Compare your result with the exact value of the integral.

1. $\int_0^2 x^2\, dx; n = 6$

2. $\int_1^3 (x^2 - 1)\, dx; n = 4$

3. $\int_0^1 x^3\, dx; n = 4$

4. $\int_1^2 x^3\, dx; n = 6$

5. $\int_1^2 \frac{1}{x}\, dx; n = 4$

6. $\int_1^2 \frac{1}{x}\, dx; n = 8$

7. $\int_1^2 \frac{1}{x^2}\, dx; n = 4$

8. $\int_0^1 \frac{1}{1+x}\, dx; n = 4$

9. $\int_0^4 \sqrt{x}\, dx; n = 8$

10. $\int_0^2 x\sqrt{2x^2 + 1}\, dx; n = 6$

11. $\int_0^1 e^{-x}\, dx; n = 6$

12. $\int_0^1 xe^{-x^2}\, dx; n = 6$

13. $\int_1^2 \ln x\, dx; n = 4$

14. $\int_0^2 x \ln(x^2 + 1)\, dx; n = 8$

In Exercises 15–22, use the trapezoidal rule and Simpson's rule to approximate the value of each definite integral.

15. $\int_0^1 \sqrt{1 + x^3}\, dx; n = 4$

16. $\int_0^2 x\sqrt{1 + x^3}\, dx; n = 4$

17. $\int_0^2 \frac{1}{\sqrt{x^3 + 1}}\, dx; n = 4$

18. $\int_0^1 \sqrt{1 - x^2}\, dx; n = 4$

19. $\int_0^2 e^{-x^2}\, dx; n = 4$

20. $\int_0^1 e^{x^2}\, dx; n = 6$

21. $\int_1^2 x^{-1/2}e^x\, dx; n = 4$

22. $\int_2^4 \frac{dx}{\ln x}; n = 6$

In Exercises 23–28, find a bound on the error in approximating the given definite integral using (a) the trapezoidal rule and (b) Simpson's rule with _n_ intervals.

23. $\int_{-1}^2 x^5\, dx; n = 10$

24. $\int_0^1 e^{-x}\, dx; n = 8$

25. $\int_1^3 \frac{1}{x}\, dx; n = 10$

26. $\int_1^3 \frac{1}{x^2}\, dx; n = 8$

27. $\int_0^2 \frac{1}{\sqrt{1 + x}}\, dx; n = 8$

28. $\int_1^3 \ln x\, dx; n = 10$

29. TRIAL RUN OF AN ATTACK SUBMARINE In a submerged trial run of an attack submarine, a reading of the sub's velocity was made every quarter hour, as shown in the accompa-

nying table. Use the trapezoidal rule to estimate the distance traveled by the submarine during the 2-hr period.

Time, t (hr)	0	$\frac{1}{4}$	$\frac{1}{2}$	$\frac{3}{4}$
Velocity, $V(t)$ (mph)	19.5	24.3	34.2	40.5

Time, t (hr)	1	$\frac{5}{4}$	$\frac{3}{2}$	$\frac{7}{4}$	2
Velocity, $V(t)$ (mph)	38.4	26.2	18	16	8

30. REAL ESTATE Cooper Realty is considering development of a time-sharing condominium resort complex along the oceanfront property illustrated in the accompanying graph. To obtain an estimate of the area of this property, measurements of the distances from the edge of a straight road, which defines one boundary of the property, to the corresponding points on the shoreline are made at 100-ft intervals. Using Simpson's rule with $n = 10$, estimate the area of the oceanfront property.

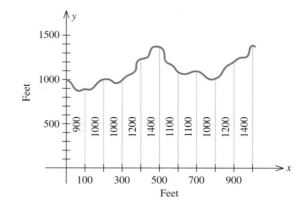

31. FUEL CONSUMPTION OF DOMESTIC CARS Thanks to smaller and more fuel-efficient models, American carmakers have doubled their average fuel economy over a 13-yr period, from 1974 to 1987. The graph depicted in the figure on p. 579 gives the average fuel consumption in miles per gallon (mpg) of domestic-built cars over the period under consideration ($t = 0$ corresponds to the beginning of 1974). Use the trapezoidal rule to estimate the average fuel consumption of the domestic car built during this period.

Hint: Approximate the integral $\frac{1}{13}\int_0^{13} f(t)\,dt$.

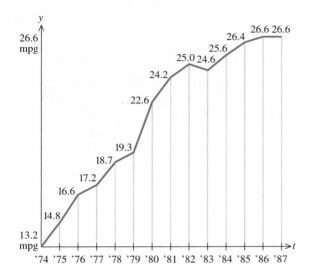

32. **AVERAGE TEMPERATURE**   The graph depicted in the accompanying figure shows the daily mean temperatures recorded during one September in Cameron Highlands. Using (a) the trapezoidal rule and (b) Simpson's rule with $n = 10$, estimate the average temperature during that month.

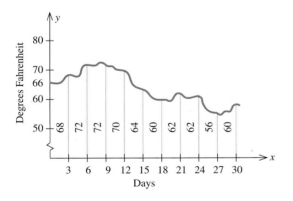

33. **CONSUMERS' SURPLUS**   Refer to Section 6.7. The demand equation for the Sicard wristwatch is given by

$$p = \frac{50}{0.01x^2 + 1} \qquad (0 \le x \le 20)$$

where x (measured in units of a thousand) is the quantity demanded per week and p is the unit price in dollars. Use (a) the trapezoidal rule and (b) Simpson's rule (take $n = 8$) to estimate the consumers' surplus if the market price is $25/watch.

34. **PRODUCERS' SURPLUS**   Refer to Section 6.7. The supply function for the audio compact disc manufactured by the Herald Record Company is given by

$$p = S(x) = \sqrt{0.01x^2 + 0.11x + 38}$$

where p is the unit wholesale price in dollars and x stands for the quantity that will be made available in the market by the supplier, measured in units of a thousand. Use (a) the trapezoidal rule and (b) Simpson's rule (take $n = 8$) to estimate the producers' surplus if the wholesale price is $8/disc.

35. **AIR POLLUTION**   The amount of nitrogen dioxide, a brown gas that impairs breathing, present in the atmosphere on a certain May day in the city of Long Beach has been approximated by

$$A(t) = \frac{136}{1 + 0.25(t - 4.5)^2} + 28 \qquad (0 \le t \le 11)$$

where $A(t)$ is measured in pollutant standard index (PSI), t is measured in hours, and $t = 0$ corresponds to 7 A.M. Use the trapezoidal rule with $n = 10$ to estimate the average PSI between 7 A.M. and noon.

Hint: $\frac{1}{5}\int_0^5 A(t)\,dt$.

36. **GROWTH OF SERVICE INDUSTRIES**   It has been estimated that service industries, which currently make up 30% of the nonfarm work force in a certain country, will continue to grow at the rate of

$$R(t) = 5e^{1/(t+1)}$$

percent/decade t decades from now. Estimate the percentage of the nonfarm workforce in the service industries one decade from now.

Hint: (a) Show that the desired answer is given by $30 + \int_0^1 5e^{1/(t+1)}\,dt$ and (b) use Simpson's rule with $n = 10$ to approximate the definite integral.

37. **LENGTH OF INFANTS AT BIRTH**   Medical records of infants delivered at the Kaiser Memorial Hospital show that the percentage of infants whose length at birth is between 19 and 21 in. is given by

$$P = 100 \int_{19}^{21} \frac{1}{2.6\sqrt{2\pi}} e^{-1/2[(x-20)/2.6]^2}\,dx$$

Use Simpson's rule with $n = 10$ to estimate P.

Portfolio

JAMES H. CHESEBRO, M.D.

TITLE: Professor of Medicine
INSTITUTION: Harvard Medical School, Massachusetts General Hospital

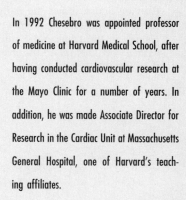

For over 20 years, James Chesebro has worked as an investigative cardiologist, diagnosing and treating patients suffering from heart and blood vessel diseases. He specializes in researching ways to prevent blood clots from narrowing the heart's arteries. Even when patients have coronary-bypass operations to correct problems, stresses Chesebro, "the piece of vein used to detour around blocked arteries is prone to plug up with clots."

Throughout his career, Chesebro has investigated the body's clotting mechanisms, as well as substances that may prevent, or even dissolve, clots. A study conducted in 1982 showed that bypass patients taking dipyridamole and aspirin were nearly three times less likely to develop clots in vein grafts than patients given placebos.

Recently, Chesebro has been working with a substance called hirudin derived from leech saliva. Hirudin has proved quite beneficial in blocking arterial blood-clot formations.

Determining the correct medication and dosage involves intense quantitative research. Colleagues from other branches of science such as nuclear medicine, molecular biology, and biochemistry play key roles in the research process. Chesebro and his colleagues have to understand how the heart's physiology and pharmacological variables such as distribution rates, concentration levels, elimination rates, and biological effects of particular substances dictate the choice and dosage of medication.

In considering whether a particular medication will bind effectively with specific body cells, researchers must determine bond tightness, speed, and length of time to reach optimal concentration. Researchers rely on complicated equations such as "integrating the area under a curve to find the amount of medication in the body at a given time," notes Chesebro. Without calculus, these equations can't be solved.

Whether physicians rely on medication or surgical procedures such as balloon angioplasty to open clogged passages, other factors play a major role in the outcome. Cholesterol and blood sugar levels, patient age, blood pressure, the geometry of the blockage, and the velocity and turbulence of blood flow all warrant consideration. According to Chesebro "linear modeling can be used to predict the contribution of patient variables and local blood vessel variables in the outcome of opening blocked arteries."

The bottom line? Calculus has been a key contributor to Chesebro's success in preventing disabling arterial blood clots.

In 1992 Chesebro was appointed professor of medicine at Harvard Medical School, after having conducted cardiovascular research at the Mayo Clinic for a number of years. In addition, he was made Associate Director for Research in the Cardiac Unit at Massachusetts General Hospital, one of Harvard's teaching affiliates.

38. TREAD LIVES OF TIRES   Under normal driving conditions the percentage of Super Titan radial tires expected to have a useful tread life of between 30,000 and 40,000 mi is given by

$$P = 100 \int_{30,000}^{40,000} \frac{1}{2000\sqrt{2\pi}} e^{-1/2[(x-40,000)/2000]^2} \, dx$$

Use Simpson's rule with $n = 10$ to estimate P.

39. MEASURING CARDIAC OUTPUT   Eight milligrams of a dye are injected into a vein leading to an individual's heart. The concentration of the dye in the aorta measured at 2-sec intervals is shown in the accompanying table. Use Simpson's rule and the formula of Example 4 to estimate the person's cardiac output.

t	0	2	4	6	8	10	12
$C(t)$	0	0	2.8	6.1	9.7	7.6	4.8

t	14	16	18	20	22	24
$C(t)$	3.7	1.9	0.8	0.3	0.1	0

In Exercises 40–43, determine whether the statement is true or false. If it is true, explain why it is true. If it is false, give an example to show why it is false.

40. In using the trapezoidal rule, the number of subintervals n must be even.

41. In using Simpson's rule, the number of subintervals n may be chosen to be odd or even.

42. Simpson's rule is more accurate than the trapezoidal rule.

43. If f is a polynomial function of degree less than or equal to 3, then the approximation $\int_a^b f(x) \, dx$ using Simpson's rule is exact.

44. Derive the formula

$$R = \frac{60D}{\int_0^T c(t)\, dt}$$

for calculating the cardiac output of a person in L/min. Here, $c(t)$ is the concentration of dye in the aorta (in mg/L) at time t (in seconds) for t in $[0, T]$, and D is the amount of dye (in mg) injected into a vein leading to the heart.

Hint: Partition the interval $[0, T]$ into n subintervals of equal length Δt. The amount of dye that flows past the measuring point in the aorta during the time interval $[0, \Delta t]$ is approximately $c(t_i)(R\Delta t)/60$ (concentration times volume). Therefore, the total amount of dye measured at the aorta is

$$\frac{[c(t_1)R\Delta t + c(t_2)R\Delta t + \cdots + c(t_n)R\Delta t]}{60} = D$$

Take the limit of the Riemann sum to obtain

$$R = \frac{60D}{\int_0^T c(t)\, dt}$$

SOLUTIONS TO SELF-CHECK EXERCISES 7.3

1. We have $x = 0$, $b = 2$, and $n = 8$, so

$$\Delta x = \frac{b-a}{n} = \frac{2}{8} = 0.25$$

and $x_0 = 0$, $x_1 = 0.25$, $x_2 = 0.50$, $x_3 = 0.75, \ldots, x_7 = 1.75$, and $x_8 = 2$. The trapezoidal rule gives

$$\int_0^2 \frac{1}{\sqrt{1+x^2}} \, dx \approx \frac{0.25}{2}\left[1 + \frac{2}{\sqrt{1+(0.25)^2}} + \frac{2}{\sqrt{1+(0.5)^2}} + \cdots \right.$$

$$\left. + \frac{2}{\sqrt{1+(1.75)^2}} + \frac{1}{\sqrt{5}} \right]$$

$$\approx 0.125(1 + 1.9403 + 1.7889 + 1.6000 + 1.4142 + 1.2494$$

$$+ 1.1094 + 0.9923 + 0.4472)$$

$$\approx 1.4427$$

28. Using aerial photographs, the Coast Guard was able to determine the dimensions of an oil spill along an embankment on a coastline, as shown in the accompanying figure. Using (a) the trapezoidal rule and (b) Simpson's rule with $n = 10$, estimate the area of the oil spill.

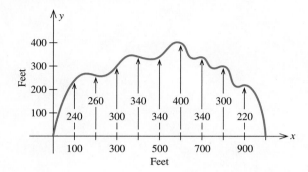

29. Lindsey wishes to establish a memorial fund at the New-town Hospital in the amount of $10,000/year beginning next year. If the fund earns interest at a rate of 9%/year compounded continuously, find the amount of endowment that he is required to make now.

CALCULUS OF SEVERAL VARIABLES

8

Up to now we have dealt with functions involving one variable. In many real-life situations, however, we encounter quantities that depend on two or more quantities. For example, the Consumer Price Index (CPI) compiled every month by the Bureau of Labor Statistics depends on the price of more than 95,000 consumer items from gas to groceries. To study such relationships, we need the notion of a function of several variables, the first topic in this chapter. Next, generalizing the concept of the derivative of a function of one variable, we study the *partial derivatives* of a function of two or more variables. Partial derivatives enable us to study the rate of change of a function with respect to one variable while holding all other variables constant. We then learn how to find the extremum values of a function of several variables. As an application of optimization theory, we learn how to find an equation of the straight line that "best" fits a set of data points scattered about a straight line. Finally, we generalize the notion of the integral to the case involving a function of two variables.

What should the dimensions of the swimming pool be? The operators of the *Viking Princess,* a luxury cruise ship, are thinking about adding another swimming pool to the *Princess.* The chief engineer has suggested that an area in the form of an ellipse, located in the rear of the promenade deck, would be suitable for this purpose. Subject to this constraint, what are the dimensions of the largest pool that can be built? See Example 5, page 657, to see how to solve this problem.

8.1 Functions of Several Variables

Up to now, our study of calculus has been restricted to functions of one variable. In many practical situations, however, the formulation of a problem results in a mathematical model that involves a function of two or more variables. For example, suppose the Ace Novelty Company determines that the profits are $6, $5, and $4 for three types of souvenirs it produces. Let x, y, and z denote the number of type-A, type-B, and type-C souvenirs to be made; then the company's profit is given by

$$P = 6x + 5y + 4z$$

and P is a function of the three variables x, y, and z.

FUNCTIONS OF TWO VARIABLES

Although this chapter deals with real-valued functions of several variables, most of our definitions and results are stated in terms of a function of two variables. One reason for adopting this approach, as you will soon see, is that there is a geometric interpretation for this special case, which serves as an important visual aid. We can then draw upon the experience gained from studying the two-variable case to help us understand the concepts and results connected with the more general case, which, by and large, is just a simple extension of the lower-dimensional case.

A Function of Two Variables

> A **real-valued function of two variables**, f, consists of
>
> 1. A set A of ordered pairs of real numbers (x, y) called the **domain** of the function.
> 2. A rule that associates with each ordered pair in the domain of f one and only one real number, denoted by $z = f(x, y)$.

The variables x and y are called **independent variables,** and the variable z, which is dependent on the values of x and y, is referred to as a **dependent variable.**

As in the case of a real-valued function of one real variable, the number $z = f(x, y)$ is called the **value of f** at the point (x, y). And, unless specified, the domain of the function f will be taken to be the largest possible set for which the rule defining f is meaningful.

EXAMPLE 1   Let f be the function defined by

$$f(x, y) = x + xy + y^2 + 2$$

Compute $f(0, 0)$, $f(1, 2)$, and $f(2, 1)$.

SOLUTION ✔   We have

$$f(0, 0) = 0 + (0)(0) + 0^2 + 2 = 2$$
$$f(1, 2) = 1 + (1)(2) + 2^2 + 2 = 9$$
$$f(2, 1) = 2 + (2)(1) + 1^2 + 2 = 7$$

The domain of a function of two variables $f(x, y)$ is a set of ordered pairs of real numbers and may therefore be viewed as a subset of the xy-plane.

EXAMPLE **2**   Find the domain of each of the following functions.

a. $f(x, y) = x^2 + y^2$ **b.** $g(x, y) = \dfrac{2}{x - y}$

c. $h(x, y) = \sqrt{1 - x^2 - y^2}$

SOLUTION ✔   **a.** $f(x, y)$ is defined for all real values of x and y, so the domain of the function f is the set of all points (x, y) in the xy-plane.

b. $g(x, y)$ is defined for all $x \neq y$, so the domain of the function g is the set of all points in the xy-plane except those lying on the line $y = x$ (Figure 8.1a).

c. We require that $1 - x^2 - y^2 \geq 0$ or $x^2 + y^2 \leq 1$, which is just the set of all points (x, y) lying on and inside the circle of radius 1 with center at the origin (Figure 8.1b).

FIGURE **8.1**

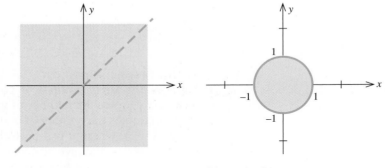

(a) Domain of g **(b)** Domain of h

APPLICATIONS

EXAMPLE **3**   The Acrosonic Company manufactures a bookshelf loudspeaker system that may be bought fully assembled or in a kit. The demand equations that relate the unit prices, p and q, to the quantities demanded weekly, x and y, of the assembled and kit versions of the loudspeaker systems are given by

$$p = 300 - \frac{1}{4}x - \frac{1}{8}y \qquad \text{and} \qquad q = 240 - \frac{1}{8}x - \frac{3}{8}y$$

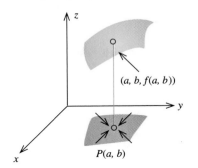

when we study the rate of change of a function of two or more variables. For example, the domain D of a function of two variables $f(x, y)$ is a subset of the plane (Figure 8.12), so if $P(a, b)$ is any point in the domain of f, there are infinitely many directions from which one can approach the point P. We may therefore ask for the rate of change of f at P along any of these directions.

However, we will not deal with this general problem. Instead, we will restrict ourselves to studying the rate of change of the function $f(x, y)$ at a point $P(a, b)$ in each of two *preferred directions*—namely, the direction parallel to the x-axis and the direction parallel to the y-axis. Let $y = b$, where b is a constant, so that $f(x, b)$ is a function of the one variable x. Since the equation $z = f(x, y)$ is the equation of a surface, the equation $z = f(x, b)$ is the equation of the curve C on the surface formed by the intersection of the surface and the plane $y = b$ (Figure 8.13).

FIGURE 8.13
The curve C is formed by the intersection of the plane $y = b$ with the surface $z = f(x, y)$.

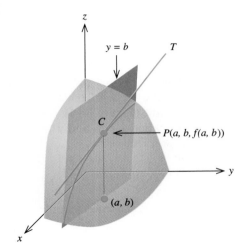

Because $f(x, b)$ is a function of one variable x, we may compute the derivative of f with respect to x at $x = a$. This derivative, obtained by keeping the variable y fixed and differentiating the resulting function $f(x, y)$ with respect to x, is called the **first partial derivative of f with respect to x** at (a, b), written

$$\frac{\partial z}{\partial x}(a, b) \quad \text{or} \quad \frac{\partial f}{\partial x}(a, b) \quad \text{or} \quad f_x(a, b)$$

Thus,

$$\frac{\partial z}{\partial x}(a, b) = \frac{\partial f}{\partial x}(a, b) = f_x(a, b) = \lim_{h \to 0} \frac{f(a + h, b) - f(a, b)}{h}$$

provided that the limit exists. The first partial derivative of f with respect to x at (a, b) measures both the slope of the tangent line T to the curve C and the rate of change of the function f in the x-direction when $x = a$ and $y = b$. We also write

$$\frac{\partial f}{\partial x}\bigg|_{(a,b)} \equiv f_x(a, b)$$

Similarly, we define the **first partial derivative of f with respect to y** at (a, b), written

$$\frac{\partial z}{\partial y}(a, b) \quad \text{or} \quad \frac{\partial f}{\partial y}(a, b) \quad \text{or} \quad f_y(a, b)$$

as the derivative obtained by keeping the variable x fixed and differentiating the resulting function $f(x, y)$ with respect to y. That is,

$$\frac{\partial z}{\partial y}(a, b) = \frac{\partial f}{\partial y}(a, b) = f_y(a, b)$$

$$= \lim_{k \to 0} \frac{f(a, b + k) - f(a, b)}{k}$$

if the limit exists. The first partial derivative of f with respect to y at (a, b) measures both the slope of the tangent line T to the curve C, obtained by holding x constant (Figure 8.14), and the rate of change of the function f in the y-direction when $x = a$ and $y = b$. We write

$$\frac{\partial f}{\partial y}\bigg|_{(a,b)} \equiv f_y(a, b)$$

Before looking at some examples, let's summarize these definitions.

FIGURE 8.14
The first partial derivative of f with respect to y at (a, b) measures the slope of the tangent line T to the curve C with x held constant.

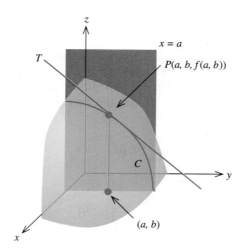

First Partial Derivatives of $f(x, y)$

Suppose $f(x, y)$ is a function of the two variables x and y. Then, the **first partial derivative of f** with respect to x at the point (x, y) is

$$\frac{\partial f}{\partial x} = \lim_{h \to 0} \frac{f(x + h, y) - f(x, y)}{h}$$

provided the limit exists. The first partial derivative of f with respect to y at the point (x, y) is

$$\frac{\partial f}{\partial y} = \lim_{k \to 0} \frac{f(x, y + k) - f(x, y)}{k}$$

provided the limit exists.

EXAMPLE 1

Find the partial derivatives $\partial f / \partial x$ and $\partial f / \partial y$ of the function

$$f(x, y) = x^2 - xy^2 + y^3$$

What is the rate of change of the function f in the x-direction at the point $(1, 2)$? What is the rate of change of the function f in the y-direction at the point $(1, 2)$?

SOLUTION ✔

To compute $\partial f / \partial x$, think of the variable y as a constant and differentiate the resulting function of x with respect to x. Let's write

$$f(x, y) = x^2 - xy^2 + y^3$$

where the variable y to be treated as a constant is shown in color. Then,

$$\frac{\partial f}{\partial x} = 2x - y^2$$

To compute $\partial f / \partial y$, think of the variable x as being fixed—that is, as a constant—and differentiate the resulting function of y with respect to y. In this case,

$$f(x, y) = x^2 - xy^2 + y^3$$

so that

$$\frac{\partial f}{\partial y} = -2xy + 3y^2$$

The rate of change of the function f in the x-direction at the point $(1, 2)$ is given by

$$f_x(1, 2) = \frac{\partial f}{\partial x}\bigg|_{(1, 2)} = 2(1) - 2^2 = -2$$

That is, f decreases 2 units for each unit increase in the x-direction, y being kept constant ($y = 2$). The rate of change of the function f in the y-direction at the point (1, 2) is given by

$$f_y(1, 2) = \frac{\partial f}{\partial y}\bigg|_{(1, 2)} = -2(1)(2) + 3(2)^2 = 8$$

That is, f increases 8 units for each unit increase in the y-direction, x being kept constant ($x = 1$).   ■ ■ ■ ■

Group Discussion

Refer to the Group Discussion on page 600. Suppose the management of the company has decided that the projected sales of the first product is a units. Describe how you might help management decide how many units of the second product the company should produce and sell in order to maximize the company's total profit. Justify your method to management. Suppose, however, management feels that b units of the second product can be manufactured and sold. How would you help management decide how many units of the first product to manufacture in order to maximize the company's total profit?

EXAMPLE 2   Compute the first partial derivatives of each of the following functions.

a. $f(x, y) = \dfrac{xy}{x^2 + y^2}$ **b.** $g(s, t) = (s^2 - st + t^2)^5$

c. $h(u, v) = e^{u^2 - v^2}$ **d.** $f(x, y) = \ln(x^2 + 2y^2)$

SOLUTION ✔   **a.** To compute $\partial f/\partial x$, think of the variable y as a constant. Thus,

$$f(x, y) = \frac{xy}{x^2 + y^2}$$

so that, upon using the quotient rule, we have

$$\frac{\partial f}{\partial x} = \frac{(x^2 + y^2)y - xy(2x)}{(x^2 + y^2)^2}$$

$$= \frac{y(y^2 - x^2)}{(x^2 + y^2)^2}$$

upon simplification and factorization. To compute $\partial f/\partial y$, think of the variable x as a constant. Thus,

$$f(x, y) = \frac{xy}{x^2 + y^2}$$

so that, upon using the quotient rule once again, we obtain

$$\frac{\partial f}{\partial y} = \frac{(x^2 + y^2)x - xy(2y)}{(x^2 + y^2)^2}$$

$$= \frac{x(x^2 - y^2)}{(x^2 + y^2)^2}$$

b. To compute $\partial g/\partial s$, we treat the variable t as if it were a constant. Thus,

$$g(s, t) = (s^2 - st + t^2)^5$$

Using the general power rule, we find

$$\frac{\partial g}{\partial s} = 5(s^2 - st + t^2)^4 \cdot (2s - t)$$
$$= 5(2s - t)(s^2 - st + t^2)^4$$

To compute $\partial g/\partial t$, we treat the variable s as if it were a constant. Thus,

$$g(s, t) = (s^2 - st + t^2)^5$$
$$\frac{\partial g}{\partial t} = 5(s^2 - st + t^2)^4(-s + 2t)$$
$$= 5(2t - s)(s^2 - st + t^2)^4$$

c. To compute $\partial h/\partial u$, think of the variable v as a constant. Thus,

$$h(u, v) = e^{u^2 - v^2}$$

Using the chain rule for exponential functions, we have

$$\frac{\partial h}{\partial u} = e^{u^2 - v^2} \cdot 2u$$
$$= 2ue^{u^2 - v^2}$$

Next, we treat the variable u as if it were a constant,

$$h(u, v) = e^{u^2 - v^2}$$

and we obtain

$$\frac{\partial h}{\partial v} = e^{u^2 - v^2} \cdot (-2v)$$
$$= -2ve^{u^2 - v^2}$$

d. To compute $\partial f/\partial x$, think of the variable y as a constant. Thus,

$$f(x, y) = \ln(x^2 + 2y^2)$$

so that the chain rule for logarithmic functions gives

$$\frac{\partial f}{\partial x} = \frac{2x}{x^2 + 2y^2}$$

Next, treating the variable x as if it were a constant, we find

$$f(x, y) = \ln(x^2 + 2y^2)$$
$$\frac{\partial f}{\partial y} = \frac{4y}{x^2 + 2y^2}$$

■■■■

To compute the partial derivative of a function of several variables with respect to one variable—say, x—we think of the other variables as if they were constants and differentiate the resulting function with respect to x.

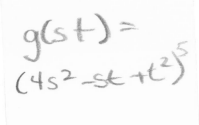

> **Group Discussion**
>
> **1.** Let (a, b) be a point in the domain of $f(x, y)$. Put $g(x) = f(x, b)$ and suppose g is differentiable at $x = a$. Explain why you can find $f_x(a, b)$ by computing $g'(a)$. How would you go about calculating $f_y(a, b)$ using a similar technique? Give a geometric interpretation of these processes.
>
> **2.** Let $f(x, y) = x^2y^3 - 3x^2y + 2$. Use the method of Problem 1 to find $f_x(1, 2)$ and $f_y(1, 2)$.

EXAMPLE 3

Compute the first partial derivatives of the function

$$w = f(x, y, z) = xyz - xe^{yz} + x \ln y$$

SOLUTION ✔

Here we have a function of three variables, x, y, and z, and we are required to compute

$$\frac{\partial f}{\partial x}, \quad \frac{\partial f}{\partial y}, \quad \frac{\partial f}{\partial z}$$

To compute f_x, we think of the other two variables, y and z, as fixed, and we differentiate the resulting function of x with respect to x, thereby obtaining

$$f_x = yz - e^{yz} + \ln y$$

To compute f_y, we think of the other two variables, x and z, as constants, and we differentiate the resulting function of y with respect to y. We then obtain

$$f_y = xz - xze^{yz} + \frac{x}{y}$$

Finally, to compute f_z, we treat the variables x and y as constants and differentiate the function f with respect to z, obtaining

$$f_z = xy - xye^{yz}$$ ■ ■ ■ ■

Exploring with Technology

Refer to the Group Discussion on this page. Let

$$f(x, y) = \frac{e^{\sqrt{xy}}}{(1 + xy^2)^{3/2}}$$

1. Compute $g(x) = f(x, 1)$ and use a graphing utility to plot the graph of g in the viewing rectangle $[0, 2] \times [0, 2]$.
2. Use the differentiation operation of your graphing utility to find $g'(1)$ and hence $f_x(1, 1)$.
3. Compute $h(y) = f(1, y)$ and use a graphing utility to plot the graph of g in the viewing rectangle $[0, 2] \times [0, 2]$.
4. Use the differentiation operation of your graphing utility to find $h'(1)$ and hence $f_y(1, 1)$.

THE COBB–DOUGLAS PRODUCTION FUNCTION

For an economic interpretation of the first partial derivatives of a function of two variables, let's turn our attention to the function

$$f(x, y) = ax^b y^{1-b} \qquad \textbf{(1)}$$

where a and b are positive constants with $0 < b < 1$. This function is called the **Cobb–Douglas production function.** Here, x stands for the amount of money expended for labor, y stands for the cost of capital equipment (buildings, machinery, and other tools of production), and the function f measures the output of the finished product (in suitable units) and is called, accordingly, the production function.

The partial derivative f_x is called the **marginal productivity of labor.** It measures the rate of change of production with respect to the amount of money expended for labor, with the level of capital expenditure held constant. Similarly, the partial derivative f_y, called the **marginal productivity of capital,** measures the rate of change of production with respect to the amount expended on capital, with the level of labor expenditure held fixed.

EXAMPLE 4   A certain country's production in the early years following World War II is described by the function

$$f(x, y) = 30x^{2/3} y^{1/3}$$

units, when x units of labor and y units of capital were used.

a. Compute f_x and f_y.
b. What is the marginal productivity of labor and the marginal productivity of capital when the amounts expended on labor and capital are 125 units and 27 units, respectively?
c. Should the government have encouraged capital investment rather than increasing expenditure on labor to increase the country's productivity?

SOLUTION ✔   **a.** $f_x = 30 \cdot \dfrac{2}{3} x^{-1/3} y^{1/3} = 20 \left(\dfrac{y}{x} \right)^{1/3}$

$f_y = 30x^{2/3} \cdot \dfrac{1}{3} y^{-2/3} = 10 \left(\dfrac{x}{y} \right)^{2/3}$

b. The required marginal productivity of labor is given by

$$f_x(125, 27) = 20 \left(\frac{27}{125} \right)^{1/3} = 20 \left(\frac{3}{5} \right)$$

or 12 units per unit increase in labor expenditure (capital expenditure is held constant at 27 units). The required marginal productivity of capital is given by

$$f_y(125, 27) = 10\left(\frac{125}{27}\right)^{2/3} = 10\left(\frac{25}{9}\right)$$

or $27\frac{7}{9}$ units per unit increase in capital expenditure (labor outlay is held constant at 125 units).

c. From the results of part (b), we see that a unit increase in capital expenditure resulted in a much faster increase in productivity than a unit increase in labor expenditure would have. Therefore, the government should have encouraged increased spending on capital rather than on labor during the early years of reconstruction.   ■■■■

Substitute and Complementary Commodities

For another application of the first partial derivatives of a function of two variables in the field of economics, let's consider the relative demands of two commodities. We say that the two commodities are **substitute** (competitive) **commodities** if a decrease in the demand for one results in an increase in the demand for the other. Examples of competitive commodities are coffee and tea. Conversely, two commodities are referred to as **complementary commodities** if a decrease in the demand for one results in a decrease in the demand for the other as well. Examples of complementary commodities are automobiles and tires.

We now derive a criterion for determining whether two commodities A and B are substitute or complementary. Suppose the demand equations that relate the quantities demanded, x and y, to the unit prices, p and q, of the two commodities are given by

$$x = f(p, q) \quad \text{and} \quad y = g(p, q)$$

Let's consider the partial derivative $\partial f/\partial p$. Since f is the demand function for commodity A, we see that, for fixed q, f is typically a decreasing function of p—that is, $\partial f/\partial p < 0$. Now, if the two commodities were substitute commodities, then the quantity demanded of commodity B would increase with respect to p—that is, $\partial g/\partial p > 0$. A similar argument with p fixed shows that if A and B are substitute commodities, then $\partial f/\partial q > 0$. Thus, the two commodities A and B are substitute commodities if

$$\frac{\partial f}{\partial q} > 0 \quad \text{and} \quad \frac{\partial g}{\partial p} > 0$$

Similarly, A and B are complementary commodities if

$$\frac{\partial f}{\partial q} < 0 \quad \text{and} \quad \frac{\partial g}{\partial p} < 0$$

Substitute and Complementary Commodities

Two commodities A and B are substitute commodities if

$$\frac{\partial f}{\partial q} > 0 \quad \text{and} \quad \frac{\partial g}{\partial p} > 0 \qquad\qquad (2)$$

Two commodities A and B are complementary commodities if

$$\frac{\partial f}{\partial q} < 0 \quad \text{and} \quad \frac{\partial g}{\partial p} < 0 \qquad\qquad (3)$$

EXAMPLE 5   Suppose that the daily demand for butter is given by

$$x = f(p, q) = \frac{3q}{1 + p^2}$$

and the daily demand for margarine is given by

$$y = g(p, q) = \frac{2p}{1 + \sqrt{q}} \qquad (p > 0, q > 0)$$

where p and q denote the prices per pound (in dollars) of butter and margarine, respectively, and x and y are measured in millions of pounds. Determine whether these two commodities are substitute, complementary, or neither.

SOLUTION ✔   We compute

$$\frac{\partial f}{\partial q} = \frac{3}{1 + p^2} \quad \text{and} \quad \frac{\partial g}{\partial p} = \frac{2}{1 + \sqrt{q}}$$

Since

$$\frac{\partial f}{\partial q} > 0 \quad \text{and} \quad \frac{\partial g}{\partial p} > 0$$

for all values of $p > 0$ and $q > 0$, we conclude that butter and margarine are substitute commodities.   ■ ■ ■ ■

SECOND-ORDER PARTIAL DERIVATIVES

The first partial derivatives $f_x(x, y)$ and $f_y(x, y)$ of a function $f(x, y)$ of the two variables x and y are also functions of x and y. As such, we may differentiate each of the functions f_x and f_y to obtain the second-order partial derivatives of f (Figure 8.15). Thus, differentiating the function f_x with respect to x leads to the second partial derivative

$$f_{xx} \equiv \frac{\partial^2 f}{\partial x^2} = \frac{\partial}{\partial x}(f_x)$$

If the inequalities in this last definition hold for *all* points (x, y) in the domain of f, then f has an **absolute maximum** or **(absolute minimum)** at (a, b) with **absolute maximum value** (or **absolute minimum value**) $f(a, b)$. Figure 8.16 shows the graph of a function with relative maxima at (a, b) and (e, f) and a relative minimum at (c, d). The absolute maximum of f occurs at (e, f) and the absolute minimum of f occurs at (g, h).

FIGURE 8.16

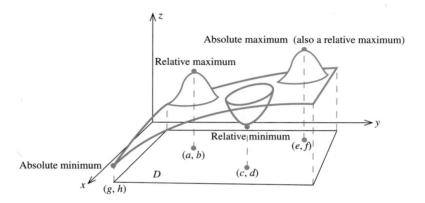

Just as in the case of a function of one variable, a relative extremum (relative maximum or relative minimum) may or may not be an absolute extremum. However, to simplify matters, we will assume that whenever an absolute extremum exists, it will occur at a point where f has a relative extremum.

Just as the first and second derivatives play an important role in determining the relative extrema of a function of one variable, the first and second partial derivatives are powerful tools for locating and classifying the relative extrema of functions of several variables.

Suppose now that a differentiable function $f(x, y)$ of two variables has a relative maximum (relative minimum) at a point (a, b) in the domain of f. From Figure 8.17 it is clear that at the point (a, b) the slope of the "tangent lines" to the surface in any direction must be zero. In particular, this implies that both

$$\frac{\partial f}{\partial x}(a, b) \qquad \text{and} \qquad \frac{\partial f}{\partial y}(a, b)$$

must be zero.

The point (a, b) is called a **critical point** of the function f. In case we are tempted to conclude that a critical point of a function f must automatically be a relative extremum of f, let's consider the graph of the function f in Figure 8.18. Here, we have both

$$\frac{\partial f}{\partial x}(a, b) = 0 \qquad \text{and} \qquad \frac{\partial f}{\partial y}(a, b) = 0$$

The point $(a, b, f(a, b))$ is neither a relative maximum nor a relative minimum of the function f since there are points nearby that are higher and others that

FIGURE 8.17

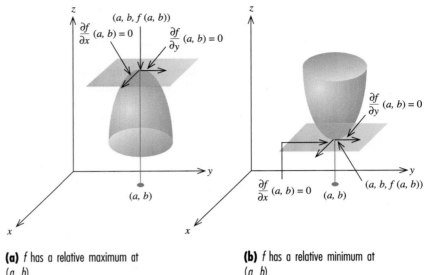

(a) f has a relative maximum at (a, b).

(b) f has a relative minimum at (a, b).

are lower than it. Such a point is called a **saddle point.** Thus, as in the case of a function of one variable, we may conclude that a critical point of a function of two (or more) variables is only a candidate for a relative extremum of f.

FIGURE 8.18
The point $(a, b, f(a, b))$ is called a saddle point.

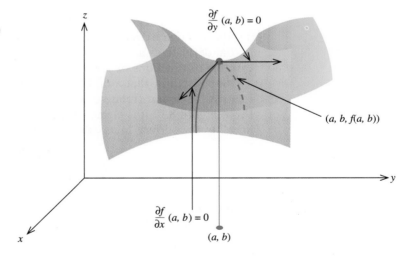

To determine the nature of a critical point of a function $f(x, y)$ of two variables, we use the second partial derivatives of f. The resulting test, which helps us classify these points, is called the **second derivative test** and is incorporated in the following procedure for finding and classifying the relative extrema of f.

Determining Relative Extrema

1. Find the critical points of $f(x, y)$ by solving the system of simultaneous equations

$$f_x = 0$$
$$f_y = 0$$

2. The second derivative test: Let

$$D(x, y) = f_{xx}f_{yy} - f_{xy}^2$$

Then,
a. $D(a, b) > 0$ and $f_{xx}(a, b) < 0$ implies that $f(x, y)$ has a **relative maximum** at the point (a, b).
b. $D(a, b) > 0$ and $f_{xx}(a, b) > 0$ implies that $f(x, y)$ has a **relative minimum** at the point (a, b).
c. $D(a, b) < 0$ implies that $f(x, y)$ has neither a relative maximum nor a relative minimum at the point (a, b).
d. $D(a, b) = 0$ implies that the test is inconclusive, so some other technique must be used to solve the problem.

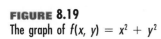

Find the relative extrema of the function

$$f(x, y) = x^2 + y^2$$

SOLUTION ✔ We have

$$f_x = 2x$$
$$f_y = 2y$$

To find the critical point(s) of f, we set $f_x = 0$ and $f_y = 0$ and solve the resulting system of simultaneous equations

$$2x = 0$$
$$2y = 0$$

FIGURE 8.19
The graph of $f(x, y) = x^2 + y^2$

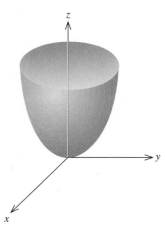

obtaining $x = 0$, $y = 0$, or $(0, 0)$, as the sole critical point of f. Next, we apply the second derivative test to determine the nature of the critical point $(0, 0)$. We compute

$$f_{xx} = 2, \quad f_{xy} = 0, \quad f_{yy} = 2$$

and

$$D(x, y) = f_{xx}f_{yy} - f_{xy}^2 = (2)(2) - 0 = 4$$

In particular, $D(0, 0) = 4$. Since $D(0, 0) > 0$ and $f_{xx}(0, 0) = 2 > 0$, we conclude that $f(x, y)$ has a relative minimum at the point $(0, 0)$. The relative minimum value, 0, also happens to be the absolute minimum of f. The graph of the function f, shown in Figure 8.19, confirms these results.   ■■■■

EXAMPLE 2

Find the relative extrema of the function

$$f(x, y) = 3x^2 - 4xy + 4y^2 - 4x + 8y + 4$$

SOLUTION ✔

We have

$$f_x = 6x - 4y - 4$$
$$f_y = -4x + 8y + 8$$

To find the critical points of f, we set $f_x = 0$ and $f_y = 0$ and solve the resulting system of simultaneous equations

$$6x - 4y = 4$$
$$-4x + 8y = -8$$

Multiplying the first equation by 2 and the second equation by 3, we obtain the equivalent system

$$12x - 8y = 8$$
$$-12x + 24y = -24$$

Adding the two equations gives $16y = -16$, or $y = -1$. We substitute this value for y into either equation in the system to get $x = 0$. Thus, the only critical point of f is the point $(0, -1)$. Next, we apply the second derivative test to determine whether the point $(0, -1)$ gives rise to a relative extremum of f. We compute

$$f_{xx} = 6, \quad f_{xy} = -4, \quad f_{yy} = 8$$

and

$$D(x, y) = f_{xx}f_{yy} - f_{xy}^2 = (6)(8) - (-4)^2 = 32$$

Since $D(0, -1) = 32 > 0$ and $f_{xx}(0, -1) = 6 > 0$, we conclude that $f(x, y)$ has a relative minimum at the point $(0, -1)$. The value of $f(x, y)$ at the point $(0, -1)$ is given by

$$f(0, -1) = 3(0)^2 - 4(0)(-1) + 4(-1)^2 - 4(0) + 8(-1) + 4 = 0$$

■ ■ ■ ■

> **Group Discussion**
>
> Suppose $f(x, y)$ has a relative extremum (relative maximum or relative minimum) at a point (a, b). Let $g(x) = f(x, b)$ and $h(y) = f(a, y)$. Assuming that f and g are differentiable, explain why $g'(a) = 0$ and $h'(b) = 0$. Explain why these results are equivalent to the conditions $f_x(a, b) = 0$ and $f_y(a, b) = 0$.

EXAMPLE 3

Find the relative extrema of the function

$$f(x, y) = 4y^3 + x^2 - 12y^2 - 36y + 2$$

SOLUTION ✔

To find the critical points of f, we set $f_x = 0$ and $f_y = 0$ simultaneously, obtaining

$$f_x = 2x = 0$$
$$f_y = 12y^2 - 24y - 36 = 0$$

The first equation implies that $x = 0$. The second equation implies that

$$y^2 - 2y - 3 = 0$$
$$(y + 1)(y - 3) = 0$$

<table>
<tr><td>

Group Discussion

1. Refer to the second derivative test. Can the condition $f_{xx}(a, b) < 0$ in part 2a be replaced by the condition $f_{yy}(a, b) < 0$? Explain your answer. How about the condition $f_{xx}(a, b) > 0$ in part 2b?

2. Let $f(x, y) = x^4 + y^4$.

a. Show that $(0, 0)$ is a critical point of f and that $D(0, 0) = 0$.

b. Explain why f has a relative (in fact, an absolute) minimum at $(0, 0)$. Does this contradict the second derivative test? Explain your answer.

</td></tr>
</table>

—that is, $y = -1$ or 3. Therefore, there are two critical points of the function f—namely, $(0, -1)$ and $(0, 3)$.

Next, we apply the second derivative test to determine the nature of each of the two critical points. We compute

$$f_{xx} = 2, \quad f_{xy} = 0, \quad f_{yy} = 24y - 24 = 24(y - 1)$$

Therefore,

$$D(x, y) = f_{xx}f_{yy} - f_{xy}^2 = 48(y - 1)$$

For the point $(0, -1)$,

$$D(0, -1) = 48(-1 - 1) = -96 < 0$$

Since $D(0, -1) < 0$, we conclude that the point $(0, -1)$ gives a saddle point of f. For the point $(0, 3)$,

$$D(0, 3) = 48(3 - 1) = 96 > 0$$

Since $D(0, 3) > 0$ and $f_{xx}(0, 3) > 0$, we conclude that the function f has a relative minimum at the point $(0, 3)$. Furthermore, since

$$f(0, 3) = 4(3)^3 + (0)^2 - 12(3)^2 - 36(3) + 2$$
$$= -106$$

we see that the relative minimum value of f is -106.   ▪ ▪ ▪ ▪

APPLICATIONS

As in the case of a practical optimization problem involving a function of one variable, the solution to an optimization problem involving a function of several variables calls for finding the *absolute* extremum of the function. Determining the absolute extremum of a function of several variables is more difficult than merely finding the relative extrema of the function. However, in many situations, the absolute extremum of a function actually coincides with the largest relative extremum of the function that occurs in the interior of its domain. We assume that the problems considered here belong to this category. Furthermore, the existence of the absolute extremum (solution) of a practical problem is often deduced from the geometric or physical nature of the problem.

EXAMPLE 4

The total weekly revenue (in dollars) that the Acrosonic Company realizes in producing and selling its bookshelf loudspeaker systems is given by

$$R(x, y) = -\frac{1}{4}x^2 - \frac{3}{8}y^2 - \frac{1}{4}xy + 300x + 240y$$

where x denotes the number of fully assembled units and y denotes the number of kits produced and sold per week. The total weekly cost attributable to the production of these loudspeakers is

$$C(x, y) = 180x + 140y + 5000$$

dollars, where x and y have the same meaning as before. Determine how many assembled units and how many kits Acrosonic should produce per week to maximize its profit.

SOLUTION ✔   The contribution to Acrosonic's weekly profit stemming from the production and sale of the bookshelf loudspeaker systems is given by

$$P(x, y) = R(x, y) - C(x, y)$$

$$= \left(-\frac{1}{4}x^2 - \frac{3}{8}y^2 - \frac{1}{4}xy + 300x + 240y \right) - (180x + 140y + 5000)$$

$$= -\frac{1}{4}x^2 - \frac{3}{8}y^2 - \frac{1}{4}xy + 120x + 100y - 5000$$

To find the relative maximum of the profit function $P(x, y)$, we first locate the critical point(s) of P. Setting $P_x(x, y)$ and $P_y(x, y)$ equal to zero, we obtain

$$P_x = -\frac{1}{2}x - \frac{1}{4}y + 120 = 0$$

$$P_y = -\frac{3}{4}y - \frac{1}{4}x + 100 = 0$$

Solving the first of these equations for y yields

$$y = -2x + 480$$

which, upon substitution into the second equation, yields

$$-\frac{3}{4}(-2x + 480) - \frac{1}{4}x + 100 = 0$$

$$6x - 1440 - x + 400 = 0$$

$$x = 208$$

We substitute this value of x into the equation $y = -2x + 480$ to get

$$y = 64$$

Therefore, the function P has the sole critical point $(208, 64)$. To show that the point $(208, 64)$ is a solution to our problem, we use the second derivative test. We compute

$$P_{xx} = -\frac{1}{2}, \qquad P_{xy} = -\frac{1}{4}, \qquad P_{yy} = -\frac{3}{4}$$

So,

$$D(x, y) = \left(-\frac{1}{2} \right)\left(-\frac{3}{4} \right) - \left(-\frac{1}{4} \right)^2 = \frac{3}{8} - \frac{1}{16} = \frac{5}{16}$$

In particular, $D(208, 64) = 5/16 > 0$.

Since $D(208, 64) > 0$ and $P_{xx}(208, 64) < 0$, the point $(208, 64)$ yields a relative maximum of P. This relative maximum is also the absolute maximum of P. We conclude that Acrosonic can maximize its weekly profit by manufacturing 208 assembled units and 64 kits of their bookshelf loudspeaker systems. The maximum weekly profit realizable from the production and sale of these loudspeaker systems is given by

$$P(208, 64) = -\frac{1}{4}(208)^2 - \frac{3}{8}(64)^2 - \frac{1}{4}(208)(64)$$

$$+ 120(208) + 100(64) - 5000$$

$$= 10{,}680$$

or \$10,680.   ■ ■ ■ ■

EXAMPLE 5   A television relay station will serve towns A, B, and C, whose relative locations are shown in Figure 8.20. Determine a site for the location of the station if the sum of the squares of the distances from each town to the site is minimized.

SOLUTION ✔   Suppose the required site is located at the point $P(x, y)$. With the aid of the distance formula, we find that the square of the distance from town A to the site is

$$(x - 30)^2 + (y - 20)^2$$

The respective distances from towns B and C to the site are found in a similar manner, so the sum of the squares of the distances from each town to the site is given by

$$f(x, y) = (x - 30)^2 + (y - 20)^2 + (x + 20)^2$$
$$+ (y - 10)^2 + (x - 10)^2 + (y + 10)^2$$

FIGURE 8.20
Locating a site for a television relay station

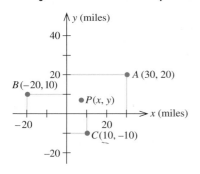

To find the relative minimum of $f(x, y)$, we first find the critical point(s) of f. Using the chain rule to find $f_x(x, y)$ and $f_y(x, y)$ and setting each equal to zero, we obtain

$$f_x = 2(x - 30) + 2(x + 20) + 2(x - 10) = 6x - 40 = 0$$

$$f_y = 2(y - 20) + 2(y - 10) + 2(y + 10) = 6y - 40 = 0$$

from which we deduce that $(20/3, 20/3)$ is the sole critical point of f. Since

$$f_{xx} = 6, \qquad f_{xy} = 0, \qquad f_{yy} = 6$$

we have

$$D(x, y) = f_{xx}f_{yy} - f_{xy}^2 = (6)(6) - 0 = 36$$

Since $D(20/3, 20/3) > 0$ and $f_{xx}(20/3, 20/3) > 0$, we conclude that the point $(20/3, 20/3)$ yields a relative minimum of f. Thus, the required site has coordinates $x = 20/3$ and $y = 20/3$.   ■ ■ ■ ■

1. Let $f(x, y) = 2x^2 + 3y^2 - 4xy + 4x - 2y + 3$.
   a. Find the critical point of f.
   b. Use the second derivative test to classify the nature of the critical point.
   c. Find the relative extremum of f, if it exists.

2. The Robertson Controls Company manufactures two basic models of setback thermostats: a standard mechanical thermostat and a deluxe electronic thermostat. Robertson's monthly revenue (in hundreds of dollars) is

$$R(x, y) = -\frac{1}{8}x^2 - \frac{1}{2}y^2 - \frac{1}{4}xy + 20x + 60y$$

where x (in units of a hundred) denotes the number of mechanical thermostats manufactured and y (in units of a hundred) denotes the number of electronic thermostats manufactured per month. The total monthly cost incurred in producing these thermostats is

$$C(x, y) = 7x + 20y + 280$$

hundred dollars. Find how many thermostats of each model Robertson should manufacture per month in order to maximize its profits. What is the maximum profit?

Solutions to Self-Check Exercises 8.3 can be found on page 636.

Solutions to Self-Check Exercises 8.3 can be found on page 636.

8.3 Exercises

In Exercises 1–20, find the critical point(s) of the function. Then use the second derivative test to classify the nature of each point, if possible. Finally, determine the relative extrema of the function.

1. $f(x, y) = 1 - 2x^2 - 3y^2$

2. $f(x, y) = x^2 - xy + y^2 + 1$

3. $f(x, y) = x^2 - y^2 - 2x + 4y + 1$

4. $f(x, y) = 2x^2 + y^2 - 4x + 6y + 3$

5. $f(x, y) = x^2 + 2xy + 2y^2 - 4x + 8y - 1$

6. $f(x, y) = x^2 - 4xy + 2y^2 + 4x + 8y - 1$

7. $f(x, y) = 2x^3 + y^2 - 9x^2 - 4y + 12x - 2$

8. $f(x, y) = 2x^3 + y^2 - 6x^2 - 4y + 12x - 2$

9. $f(x, y) = x^3 + y^2 - 2xy + 7x - 8y + 4$

10. $f(x, y) = 2y^3 - 3y^2 - 12y + 2x^2 - 6x + 2$

11. $f(x, y) = x^3 - 3xy + y^3 - 2$

12. $f(x, y) = x^3 - 2xy + y^2 + 5$

13. $f(x, y) = xy + \dfrac{4}{x} + \dfrac{2}{y}$

14. $f(x, y) = \dfrac{x}{y^2} + xy$

15. $f(x, y) = x^2 - e^{y^2}$

16. $f(x, y) = e^{x^2 - y^2}$

17. $f(x, y) = e^{x^2 + y^2}$

18. $f(x, y) = e^{xy}$

19. $f(x, y) = \ln(1 + x^2 + y^2)$

20. $f(x, y) = xy + \ln x + 2y^2$

21. MAXIMIZING PROFIT The total weekly revenue (in dollars) of the Country Workshop realized in manufacturing and selling its rolltop desks is given by

$$R(x, y) = -0.2x^2 - 0.25y^2 - 0.2xy + 200x + 160y$$

where x denotes the number of finished units and y denotes the number of unfinished units manufactured and sold per week. The total weekly cost attributable to the manufacture of these desks is given by

$$C(x, y) = 100x + 70y + 4000$$

dollars. Determine how many finished units and how many unfinished units the company should manufacture per week in order to maximize its profit. What is the maximum profit realizable?

22. **MAXIMIZING PROFIT** The total daily revenue (in dollars) that the Weston Publishing Company realizes in publishing and selling its English-language dictionaries is given by

$$R(x, y) = -0.005x^2 - 0.003y^2 - 0.002xy$$
$$+ 20x + 15y$$

where x denotes the number of deluxe copies and y denotes the number of standard copies published and sold daily. The total daily cost of publishing these dictionaries is given by

$$C(x, y) = 6x + 3y + 200$$

dollars. Determine how many deluxe copies and how many standard copies Weston should publish per day to maximize its profits. What is the maximum profit realizable?

23. **MAXIMUM PRICE** The rectangular region R shown in the accompanying figure represents the financial district of a city. The price of land within the district is approximated by the function

$$p(x, y) = 200 - 10\left(x - \frac{1}{2}\right)^2 - 15(y - 1)^2$$

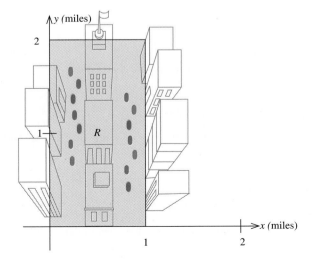

where $p(x, y)$ is the price of land at the point (x, y) in dollars per square foot and x and y are measured in miles. At what point within the financial district is the price of land highest?

24. **MAXIMIZING PROFIT** C&G Imports, Inc., imports two brands of white wine, one from Germany and the other from Italy. The German wine costs $4/bottle, and the Italian wine can be obtained for $3/bottle. It has been estimated that if the German wine retails at p dollars/bottle and the Italian wine is sold for q dollars/bottle, then

$$2000 - 150p + 100q$$

bottles of the German wine and

$$1000 + 80p - 120q$$

bottles of the Italian wine will be sold per week. Determine the unit price for each brand that will allow C&G to realize the largest possible weekly profit.

25. **DETERMINING THE OPTIMAL SITE** An auxiliary electric power station will serve three communities, A, B, and C, whose relative locations are shown in the accompanying figure. Determine where the power station should be located if the sum of the squares of the distances from each community to the site is minimized.

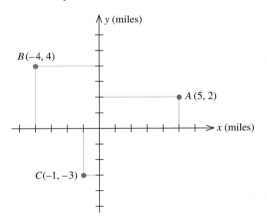

26. **PACKAGING** An open rectangular box having a volume of 108 in.^{3} is to be constructed from a tin sheet. Find the dimensions of such a box if the amount of material used in its construction is to be minimal.

Hint: Let the dimensions of the box be x'' by y'' by z''. Then, $xyz = 108$ and the amount of material used is given by $S = xy + 2yz + 2xz$. Show that

$$S = f(x, y) = xy + \frac{216}{x} + \frac{216}{y}$$

Minimize $f(x, y)$.

27. **PACKAGING** Postal regulations specify that the combined length and girth of a parcel sent by parcel post may not exceed 108 in. Find the dimensions of the rectangular package that would have the greatest possible volume under these regulations.

    Hint: Let the dimensions of the box be x'' by y'' by z'' (see the figure below). Then, $2x + 2z + y = 108$, and the volume $V = xyz$. Show that

    $$V = f(x, z) = 108xz - 2x^2z - 2xz^2$$

    Maximize $f(x, z)$.

    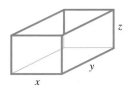

28. **MINIMIZING HEATING AND COOLING COSTS** A building in the shape of a rectangular box is to have a volume of 12,000 ft^3 (see the figure). It is estimated that the annual heating and cooling costs will be $2/square foot for the top, $4/square foot for the front and back, and $3/square foot for the sides. Find the dimensions of the building that will result in a minimal annual heating and cooling cost. What is the minimal annual heating and cooling cost?

    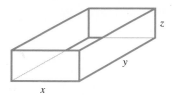

29. **PACKAGING** An open box having a volume of 48 in.^{3} is to be constructed. If the box is to include a partition that is parallel to a side of the box, as shown in the figure, and the amount of material used is to be minimal, what should be the dimensions of the box?

    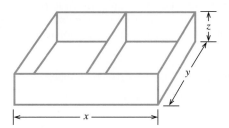

In Exercises 30 and 31, determine whether the statement is true or false. If it is true, explain why it is true. If it is false, give an example to show why it is false.

30. If $f_x(a, b) = 0$ and $f_y(a, b) = 0$, then f must have a relative extremum at (a, b).

31. If (a, b) is a critical point of f and both the conditions $f_{xx}(a, b) < 0$ and $f_{yy}(a, b) < 0$ hold, then f has a relative maximum at (a, b).

SOLUTIONS TO SELF-CHECK EXERCISES 8.3

1. **a.** To find the critical point(s) of f, we solve the system of equations

$$f_x = 4x - 4y + 4 = 0$$
$$f_y = -4x + 6y - 2 = 0$$

obtaining $x = -2$ and $y = -1$. Thus, the only critical point of f is the point $(-2, -1)$.

b. We have $f_{xx} = 4$, $f_{xy} = -4$, and $f_{yy} = 6$, so

$$D(x, y) = f_{xx}f_{yy} - f_{xy}^2$$
$$= (4)(6) - (-4)^2 = 8$$

Since $D(-2, -1) > 0$ and $f_{xx}(-2, -1) > 0$, we conclude that f has a relative minimum at the point $(-2, -1)$.

c. The relative minimum value of $f(x, y)$ at the point $(-2, -1)$ is

$$f(-2, -1) = 2(-2)^2 + 3(-1)^2 - 4(-2)(-1) + 4(-2) - 2(-1) + 3$$
$$= 0$$

2. Robertson's monthly profit is

$$P(x, y) = R(x, y) - C(x, y)$$

$$= \left(-\frac{1}{8}x^2 - \frac{1}{2}y^2 - \frac{1}{4}xy + 20x + 60y\right) - (7x + 20y + 280)$$

$$= -\frac{1}{8}x^2 - \frac{1}{2}y^2 - \frac{1}{4}xy + 13x + 40y - 280$$

The critical point of P is found by solving the system

$$P_x = -\frac{1}{4}x - \frac{1}{4}y + 13 = 0$$

$$P_y = -\frac{1}{4}x - \quad y + 40 = 0$$

giving $x = 16$ and $y = 36$. Thus, $(16, 36)$ is the critical point of P. Next,

$$P_{xx} = -\frac{1}{4}, \qquad P_{xy} = -\frac{1}{4}, \qquad P_{yy} = -1$$

and

$$D(x, y) = f_{xx}f_{yy} - f_{xy}^2$$

$$= \left(-\frac{1}{4}\right)(-1) - \left(-\frac{1}{4}\right)^2 = \frac{3}{16}$$

Since $D(16, 36) > 0$ and $P_{xx}(16, 36) < 0$, the point $(16, 36)$ yields a relative maximum of P. We conclude that the monthly profit is maximized by manufacturing 1600 mechanical and 3600 electronic setback thermostats per month. The maximum monthly profit realizable is

$$P(16, 36) = -\frac{1}{8}(16)^2 - \frac{1}{2}(36)^2 - \frac{1}{4}(16)(36) + 13(16) + 40(36) - 280$$

$$= 544$$

or $54,400.

8.4 The Method of Least Squares

THE METHOD OF LEAST SQUARES

In Section 1.4, Example 10, we saw how a linear equation can be used to approximate the sales trend for a local sporting goods store. As we saw there, one use of a **trend line** is to predict a store's future sales. Recall that we obtained the line by requiring that it pass through two data points, the rationale being that such a line seems to *fit* the data reasonably well.

FIGURE 8.21

A scatter diagram

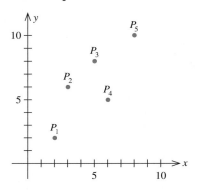

FIGURE 8.21

A scatter diagram

FIGURE 8.22

The approximating line misses each point by the amounts $d_1, d_2, \ldots, d_5$.

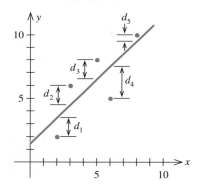

In this section we describe a general method, known as the **method of least squares,** for determining a straight line that, in some sense, *best* fits a set of data points when the points are scattered about a straight line. To illustrate the principle behind the method of least squares, suppose, for simplicity, that we are given five data points,

$$P_1(x_1, y_1), \qquad P_2(x_2, y_2), \qquad P_3(x_3, y_3), \qquad P_4(x_4, y_4), \qquad P_5(x_5, y_5)$$

that describe the relationship between the two variables x and y. By plotting these data points, we obtain a graph called a **scatter diagram** (Figure 8.21).

If we try to fit a straight line to these data points, the line will miss the first, second, third, fourth, and fifth data points by the amounts d_1, d_2, d_3, d_4, and d_5, respectively (Figure 8.22).

The **principle of least squares** states that the straight line L that fits the data points best is the one chosen by requiring that the sum of the squares of $d_1, d_2, \ldots, d_5$—that is,

$$d_1^2 + d_2^2 + d_3^2 + d_4^2 + d_5^2$$

be made as small as possible. If we think of the amount d_1 as the error made when the value y_1 is approximated by the corresponding value of y lying on the straight line L, and d_2 as the error made when the value y_2 is approximated by the corresponding value of y, and so on, then it can be seen that the least-squares criterion calls for minimizing the sum of the squares of the errors. The line L obtained in this manner is called the **least-squares line,** or **regression line.**

To find a method for computing the regression line L, suppose L has representation $y = f(x) = mx + b$, where m and b are to be determined. Observe that

$$
\begin{aligned}
d_1^2 &+ d_2^2 + d_3^2 + d_4^2 + d_5^2 \\
&= [f(x_1) - y_1]^2 + [f(x_2) - y_2]^2 + [f(x_3) - y_3]^2 \\
&\quad + [f(x_4) - y_4]^2 + [f(x_5) - y_5]^2 \\
&= (mx_1 + b - y_1)^2 + (mx_2 + b - y_2)^2 + (mx_3 + b - y_3)^2 \\
&\quad + (mx_4 + b - y_4)^2 + (mx_5 + b - y_5)^2
\end{aligned}
$$

and may be viewed as a function of the two variables m and b. Thus, the least-squares criterion is equivalent to minimizing the function

$$
\begin{aligned}
f(m, b) &= (mx_1 + b - y_1)^2 + (mx_2 + b - y_2)^2 + (mx_3 + b - y_3)^2 \\
&\quad + (mx_4 + b - y_4)^2 + (mx_5 + b - y_5)^2
\end{aligned}
$$

with respect to m and b. Using the chain rule, we compute

$$
\begin{aligned}
\frac{\partial f}{\partial m} &= 2(mx_1 + b - y_1)x_1 + 2(mx_2 + b - y_2)x_2 + 2(mx_3 + b - y_3)x_3 \\
&\quad + 2(mx_4 + b - y_4)x_4 + 2(mx_5 + b - y_5)x_5 \\
&= 2[mx_1^2 + bx_1 - x_1 y_1 + mx_2^2 + bx_2 - x_2 y_2 + mx_3^2 + bx_3 - x_3 y_3 \\
&\quad + mx_4^2 + bx_4 - x_4 y_4 + mx_5^2 + bx_5 - x_5 y_5] \\
&= 2[(x_1^2 + x_2^2 + x_3^2 + x_4^2 + x_5^2)m + (x_1 + x_2 + x_3 + x_4 + x_5)b \\
&\quad - (x_1 y_1 + x_2 y_2 + x_3 y_3 + x_4 y_4 + x_5 y_5)]
\end{aligned}
$$

and

$$\frac{\partial f}{\partial b} = 2(mx_1 + b - y_1) + 2(mx_2 + b - y_2) + 2(mx_3 + b - y_3)$$

$$+ 2(mx_4 + b - y_4) + 2(mx_5 + b - y_5)$$

$$= 2[(x_1 + x_2 + x_3 + x_4 + x_5)m + 5b - (y_1 + y_2 + y_3 + y_4 + y_5)]$$

Setting

$$\frac{\partial f}{\partial m} = 0 \qquad \text{and} \qquad \frac{\partial f}{\partial b} = 0$$

gives

$$(x_1^2 + x_2^2 + x_3^2 + x_4^2 + x_5^2)m + (x_1 + x_2 + x_3 + x_4 + x_5)b$$

$$= x_1 y_1 + x_2 y_2 + x_3 y_3 + x_4 y_4 + x_5 y_5$$

and

$$(x_1 + x_2 + x_3 + x_4 + x_5)m + 5b = y_1 + y_2 + y_3 + y_4 + y_5$$

Solving these two simultaneous equations for m and b then leads to an equation $y = mx + b$ of a straight line.

Before looking at an example, we state a more general result whose derivation is identical to the special case involving the five data points just discussed.

The Method of Least Squares

Suppose we are given n data points:

$$P_1(x_1, y_1), \qquad P_2(x_2, y_2), \qquad P_3(x_3, y_3), \ldots, P_n(x_n, y_n)$$

Then, the least-squares (regression) line for the data is given by the linear equation

$$y = f(x) = mx + b$$

where the constants m and b satisfy the equations

$$(x_1^2 + x_2^2 + x_3^2 + \cdots + x_n^2)m + (x_1 + x_2 + x_3 + \cdots + x_n)b$$

$$= x_1 y_1 + x_2 y_2 + x_3 y_3 + \cdots + x_n y_n \tag{4}$$

and

$$(x_1 + x_2 + x_3 + \cdots + x_n)m + nb$$

$$= y_1 + y_2 + y_3 + \cdots + y_n \tag{5}$$

simultaneously. Equations (4) and (5) are called **normal equations.**

EXAMPLE 1

Find an equation of the least-squares line for the data

$$P_1(1, 1), \qquad P_2(2, 3), \qquad P_3(3, 4), \qquad P_4(4, 3), \qquad P_5(5, 6)$$

SOLUTION ✔

Here, we have $n = 5$ and

$$x_1 = 1, \qquad x_2 = 2, \qquad x_3 = 3, \qquad x_4 = 4, \qquad x_5 = 5$$
$$y_1 = 1, \qquad y_2 = 3, \qquad y_3 = 4, \qquad y_4 = 3, \qquad y_5 = 6$$

so Equation (4) becomes

$$(1 + 4 + 9 + 16 + 25)m + (1 + 2 + 3 + 4 + 5)b = 1 + 6 + 12 + 12 + 30$$

or

$$55m + 15b = 61 \tag{6}$$

and (5) becomes

$$(1 + 2 + 3 + 4 + 5)m + 5b = 1 + 3 + 4 + 3 + 6$$

or

$$15m + 5b = 17 \tag{7}$$

Solving Equation (7) for b gives

$$b = -3m + \frac{17}{5} \tag{8}$$

which, upon substitution into (6), gives

$$15\left(-3m + \frac{17}{5}\right) + 55m = 61$$

$$-45m + 51 + 55m = 61$$

$$10m = 10$$

$$m = 1$$

Substituting this value of m into (8) gives

$$b = -3 + \frac{17}{5} = \frac{2}{5} = 0.4$$

Therefore, the required least-squares line is

$$y = x + 0.4$$

The scatter diagram and the regression line are shown in Figure 8.23.

■ ■ ■ ■

FIGURE 8.23

The scatter diagram and the least-squares line $y = x + 0.4$

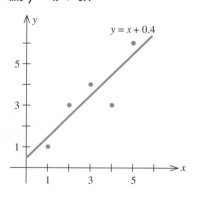

APPLICATIONS

EXAMPLE 2

The proprietor of the Leisure Travel Service compiled the following data relating the firm's annual profit to its annual advertising expenditure (both measured in thousands of dollars).

Annual Advertising Expenditure, x	12	14	17	21	26	30
Annual Profit, y	60	70	90	100	100	120

a. Determine an equation of the least-squares line for these data.
b. Draw a scatter diagram and the least-squares line for these data.
c. Use the result obtained in part (a) to predict Leisure Travel's annual profit if the annual advertising budget is $20,000.

SOLUTION ✔ **a.** The calculations required for obtaining the normal equations may be summarized as follows:

x	y	x^2	xy
12	60	144	720
14	70	196	980
17	90	289	1,530
21	100	441	2,100
26	100	676	2,600
30	120	900	3,600
Sum 120	540	2,646	11,530

The normal equations are

$$6b + 120m = 540 \tag{9}$$
$$120b + 2646m = 11{,}530 \tag{10}$$

Solving Equation (9) for b gives

$$b = -20m + 90 \tag{11}$$

which, upon substitution into Equation (10), gives

$$120(-20m + 90) + 2646m = 11{,}530$$
$$-2400m + 10{,}800 + 2646m = 11{,}530$$
$$246m = 730$$
$$m \approx 2.97$$

Substituting this value of m into Equation (11) gives

$$b = -20(2.97) + 90 = 30.6$$

Therefore, the required least-squares line is given by

$$y = f(x) = 2.97x + 30.6$$

b. The scatter diagram and the least-squares line are shown in Figure 8.24.
c. Leisure Travel's predicted annual profit corresponding to an annual budget of $20,000 is given by

$$f(20) = 2.97(20) + 30.6$$
$$= 90$$

or $90,000.

FIGURE 8.24
The scatter diagram and the least-squares line $y = 2.97x + 30.6$

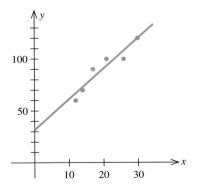

EXAMPLE 3   A market research study conducted for the Century Communications Company provided the following data based on the projected monthly sales x (in thousands) of Century's videocassette version of a box-office hit adventure movie with a proposed wholesale unit price of p dollars.

p	38	36	34.5	30	28.5
x	2.2	5.4	7.0	11.5	14.6

a. Find the demand equation if the demand curve is the least-squares line for these data.

b. Suppose the total monthly cost function associated with producing and distributing the videocassette movies is given by

$$C(x) = 4x + 25$$

where x denotes the number of units (in thousands) produced and sold and $C(x)$ is in thousands of dollars. Determine the unit wholesale price that will maximize Century's monthly profit.

SOLUTION ✔   **a.** The calculations required for obtaining the normal equations may be summarized as follows:

	x	p	x^2	xp
	2.2	38	4.84	83.6
	5.4	36	29.16	194.4
	7.0	34.5	49	241.5
	11.5	30	132.25	345
	14.6	28.5	213.16	416.1
Sum	40.7	167	428.41	1280.6

The normal equations are

$$5b + 40.7m = 167$$
$$40.7b + 428.41m = 1280.6$$

Solving this system of linear equations simultaneously, we find that

$$m \approx -0.81 \quad \text{and} \quad b \approx 39.99$$

Therefore, the required least-squares line is given by

$$p = f(x) = -0.81x + 39.99$$

which is the required demand equation, provided $0 \le x \le 49.37$.

b. The total revenue function in this case is given by

$$R(x) = xp = -0.81x^2 + 39.99x$$

and since the total cost function is

$$C(x) = 4x + 25$$

we see that the profit function is

$$P(x) = -0.81x^2 + 39.99x - (4x + 25)$$
$$= -0.81x^2 + 35.99x - 25$$

To find the absolute minimum of $P(x)$ over the closed interval $[0, 49.37]$, we compute

$$P'(x) = -1.62x + 35.99$$

Since $P'(x) = 0$, we find $x \approx 22.22$ as the only critical point of P. Finally, from the table

x	0	22.22	49.37
$P(x)$	-25	374.78	-222.47

we see that the optimal wholesale price is $22.22 per videocassette.

■ ■ ■ ■

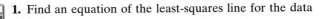

SELF-CHECK EXERCISES 8.4

1. Find an equation of the least-squares line for the data

$$P_1(0, 3), \quad P_2(2, 6.5), \quad P_3(4, 10), \quad P_4(6, 16), \quad P_5(7, 16.5)$$

2. The following data obtained from the U.S. Department of Commerce give the percentage of people over age 65 who have high school diplomas.

Year, x	0	6	11	16	22	26
Percentage with Diplomas, y	19	25	30	35	44	48

Here, $x = 0$ corresponds to the beginning of the year 1959.
a. Find an equation of the least-squares line for the given data.
b. Assuming that this trend continues, what percentage of people over 65 years of age will have high school diplomas at the beginning of the year 2003 ($x = 44$)?
Source: U.S. Department of Commerce

Solutions to Self-Check Exercises 8.4 can be found on page 649.

8.4 Exercises

A calculator is recommended for this exercise set. In Exercises 1–6, (a) find the equation of the least-squares line for the given data and (b) draw a scatter diagram for the given data and graph the least-squares line.

1.

x	1	2	3	4
y	4	6	8	11

2.

x	1	3	5	7	9
y	9	8	6	3	2

3.

x	1	2	3	4	4	6
y	4.5	5	3	2	3.5	1

4.

x	1	1	2	3	4	4	5
y	2	3	3	3.5	3.5	4	5

5. $P_1(1, 3)$, $P_2(2, 5)$, $P_3(3, 5)$, $P_4(4, 7)$, $P_5(5, 8)$

6. $P_1(1, 8)$, $P_2(2, 6)$, $P_3(5, 6)$, $P_4(7, 4)$, $P_5(10, 1)$

7. COLLEGE ADMISSIONS The following data were compiled by the admissions office at Faber College during the past 5 yr. The data relate the number of college brochures and follow-up letters (x) sent to a preselected list of high school juniors who had taken the PSAT and the number of completed applications (y) received from these students (both measured in units of 1000).

x	4	4.5	5	5.5	6
y	0.5	0.6	0.8	0.9	1.2

a. Determine the equation of the least-squares line for these data.
b. Draw a scatter diagram and the least-squares line for these data.

c. Use the result obtained in part (a) to predict the number of completed applications that might be expected if 6400 brochures and follow-up letters are sent out during the next year.

8. NET SALES The management of Kaldor, Inc., a manufacturer of electric motors, submitted the following data in the annual report to its stockholders. The table shows the net sales (in millions of dollars) during the 5 yr that have elapsed since the new management team took over. (The first year the firm operated under the new management corresponds to the time period $x = 1$, and the four subsequent years correspond to $x = 2, 3, 4, 5$.)

Year, x	1	2	3	4	5
Net Sales, y	426	437	460	473	477

a. Determine the equation of the least-squares line for these data.
b. Draw a scatter diagram and the least-squares line for these data.
c. Use the result obtained in part (a) to predict the net sales for the upcoming year.

9. SAT VERBAL SCORES The following data were compiled by the superintendent of schools in a large metropolitan area. The table shows the average SAT verbal scores of high school seniors during the 5 yr since the district implemented the "back-to-basics" program.

Year, x	1	2	3	4	5
Average Score, y	436	438	428	430	426

a. Determine the equation of the least-squares line for these data.
b. Draw a scatter diagram and the least-squares line for these data.
c. Use the result obtained in part (a) to predict the average SAT verbal score of high school seniors 2 yr from now ($x = 7$).

10. AUTO OPERATING COSTS The following figures were compiled by Clarke, Kingsley, and Company, a consulting firm that specializes in auto operating costs, relating the annual mileage (in thousands of miles) that an average new compact car is driven to the cost per mile (in cents) of operating the car.

Annual Mileage, x	5	10	15	20	25	30
Cost per Mile, y	50.3	34.8	30.1	27.4	25.6	23.5

a. Determine an equation of the least-squares line for these data.
b. Draw a scatter diagram and the least-squares line for these data.
c. Use the result obtained in part (a) to estimate the cost per mile of operating a new company car if it is driven 8000 mi during the first year of ownership.
Source: Clarke, Kingsley, and Company

11. **SIZE OF AVERAGE FARM** The size of the average farm in the United States has been growing steadily over the years. The following data, obtained from the U.S. Department of Agriculture, give the size of the average farm y (in acres) from 1940 through 1991. (Here, $x = 0$ corresponds to the beginning of the year 1940).

Year, x	0	10	20	30	40	51
Size of Farm, y	168	213	297	374	427	467

a. Find the equation of the least-squares line for these data.
b. Use the result of part (a) to estimate the size of the average farm in the year 2003.
Source: The World Almanac

12. **WELFARE COSTS** According to the Massachusetts Department of Welfare, the spending (in billions of dollars) by Medicaid, the national health-care plan for the poor, over the 5-yr period from 1988 to 1992 is summarized in the following table. (Here, $x = 0$ represents the beginning of the year 1988.)

Year, x	0	1	2	3	4
Expenditure, y	1.550	1.662	1.786	1.888	2.009

a. Find an equation of the least-squares line for these data.
b. Use the result of part (a) to estimate Medicaid spending for the year 2002, assuming the trend continued.
Source: Massachusetts Department of Welfare

13. **MASS TRANSIT SUBSIDIES** The following table gives the projected state subsidies (in millions of dollars) to the Massachusetts Bay Transit Authority (MBTA) over a 5-yr period.

Year, x	1	2	3	4	5
Subsidy, y	20	24	26	28	32

a. Find an equation of the least-squares line for these data.
b. Use the result of part (a) to estimate the state subsidy to the MBTA for the eighth year ($x = 8$).
Source: Massachusetts Bay Transit Authority

14. **SOCIAL SECURITY WAGE BASE** The Social Security (FICA) wage base (in thousands of dollars) from 1996 to 2001 is given in the following table.

Year	1996	1997	1998	1999	2000	2001
Wage Base, y	62.7	65.4	68.4	72.6	76.2	80.4

a. Find an equation of the least-squares line for these data. (Let $x = 1$ represent the year 1996.)
b. Use your result of part (a) to estimate the FICA wage base in the year 2004.
Source: The World Almanac

15. **PRODUCTION OF ALL-ALUMINUM CANS** Steel has been playing a decreasing role in the manufacture of beverage cans in the United States. According to the Can Manufacture Institute, the use of bimetallic cans has been dwindling, whereas the use of all-aluminum cans has been growing steadily. The accompanying table gives the production of all-aluminum cans (in billions) over the period from 1975 through 1989.

Year	1975	1977	1979	1981
Number of Cans	16.7	26	33.3	48.3

Year	1983	1985	1987	1989
Number of Cans	57	65.8	74.2	83.3

a. Find an equation of the least-squares line for these data. (Let $x = 1$ represent 1975.)
b. Use the result of part (a) to estimate the number of cans produced in 1993, assuming the trend continued.
Source: Can Manufacturing Institute

FINDING AN EQUATION OF A LEAST-SQUARES LINE

A graphing utility is especially useful in calculating an equation of the least-squares line for a set of data. We simply enter the given data in the form of lists into the calculator and then use the linear regression function to obtain the coefficients of the required equation.

EXAMPLE 1

Find an equation of the least-squares line for the following data:

x	1.1	2.3	3.2	4.6	5.8	6.7	8
y	-5.8	-5.1	-4.8	-4.4	-3.7	-3.2	-2.5

SOLUTION ✓

First, we enter the data as

$$x_1 = 1.1, \quad y_1 = -5.8, \quad x_2 = 2.3, \quad y_2 = -5.1, \quad x_3 = 3.2, \quad y_3 = -4.8, \quad x_4 = 4.6$$

$$y_4 = -4.4, \quad x_5 = 5.8, \quad y_5 = -3.7, \quad x_6 = 6.7, \quad y_6 = -3.2, \quad x_7 = 8, \quad y_7 = -2.5$$

Then, using the linear regression function from the statistics menu, we find

$$a = -6.29996900666, \quad b = 0.460560979389, \quad \text{corr.} = .994488871079, \quad n = 7$$

Therefore, an equation of the least-squares line ($y = a + bx$) is

$$y = -6.3333 + 0.46x$$

The correlation coefficient of .99449 attests to the excellent fit of the regression line.   ▪ ▪ ▪ ▪

EXAMPLE 2

According to Pacific Gas and Electric, the nation's largest utility company, the demand for electricity from 1990 through the year 2000 is as follows:

t	0	2	4	6	8	10
y	333	917	1500	2117	2667	3292

Here, $t = 0$ corresponds to 1990, and y gives the amount of electricity demanded in year t, measured in megawatts. Find an equation of the least-squares line for these data.

Source: Pacific Gas and Electric

SOLUTION ✔ First, we enter the data as

$$x_1 = 0, \quad y_1 = 333, \quad x_2 = 2, \quad y_2 = 917, \quad x_3 = 4, \quad y_3 = 1500,$$
$$x_4 = 6, \quad y_4 = 2117, \quad x_5 = 8, \quad y_5 = 2667, \quad x_6 = 10, \quad y_6 = 3292$$

Then, using the linear regression function from the statistics menu, we find

$$a = 328.476190476 \quad \text{and} \quad b = 295.171428571$$

Therefore, an equation of the least-squares line is

$$y = 328 + 295t$$

■ ■ ■ ■

Exercises

In Exercises 1–4, find an equation of the least-squares line for the data.

1.

x	2.1	3.4	4.7	5.6	6.8	7.2
y	8.8	12.1	14.8	16.9	19.8	21.1

2.

x	1.1	2.4	3.2	4.7	5.6	7.2
y	−0.5	1.2	2.4	4.4	5.7	8.1

3.

x	−2.1	−1.1	0.1	1.4	2.5	4.2	5.1
y	6.2	4.7	3.5	1.9	0.4	−1.4	−2.5

4.

x	−1.12	0.1	1.24	2.76	4.21	6.82
y	7.61	4.9	2.74	−0.47	−3.51	−8.94

5. WASTE GENERATION According to data from the Council on Environmental Quality, the amount of waste (in millions of tons per year) generated in the United States from 1960 to 1990 was:

Year	1960	1965	1970	1975
Amount, y	81	100	120	124

Year	1980	1985	1990
Amount, y	140	152	164

a. Find an equation of the least-squares line for these data. (Let x be in units of 5 and let $x = 1$ represent 1960.)
b. Use the result of part (a) to estimate the amount of waste generated in the year 2000, assuming the trend continues.
Source: Council on Environmental Quality

6. MEDIAN PRICE OF HOMES According to data from the Association of Realtors, the median price (in thousands of dollars) of existing homes in a certain metropolitan area from 1990 to 1999 was:

Year	1990	1991	1992	1993	1994
Price, y	92.3	96.0	99.5	101.6	104.9

Year	1995	1996	1997	1998	1999
Price, y	108.1	112.3	117.6	122.3	126.8

a. Find an equation of the least-squares line for these data. (Let $x = 1$ represent 1990.)
b. Use the result of part (a) to estimate the median price of a house in the year 2003, assuming the trend continued.
Source: Association of Realtors

16. ON-LINE BANKING According to industry sources, on-line banking is expected to take off in the near future. The projected number of households (in millions) using this service is given in the following table. (Here, $x = 0$ corresponds to the beginning of 1997.)

Year, x	0	1	2	3	4	5
Number of Households, y	4.5	7.5	10.0	13.0	15.6	18.0

a. Find an equation of the least-squares line for these data.
b. Use the result of part (a) to estimate the number of households using on-line banking at the beginning of 2003, assuming the projection is accurate.
Source: Jupiter Communications, Forrester Research Inc.

17. NET-CONNECTED COMPUTERS IN EUROPE The projected number of computers (in millions) connected to the Internet in Europe from 1998 through 2002 is summarized in the accompanying table. (Here, $x = 0$ corresponds to the beginning of 1998.)

Year, x	0	1	2	3	4
Net-Connected Computers, y	21.7	32.1	45.0	58.3	69.6

a. Find an equation of the least-squares line for these data.
b. Use the result of part (a) to estimate the projected number of computers connected to the Internet in Europe at the beginning of 2003, assuming the trend continues.
Source: Dataquest, Inc.

18. PORTABLE-PHONE SERVICES The projected number of wireless subscribers y (in millions) from the year 2000 through 2006 is summarized in the accompanying table. (Here, $x = 0$ corresponds to the beginning of the year 2000.)

Year, x	0	1	2	3	4	5	6
Number of Subscribers, y	90.4	100.0	110.4	120.4	130.8	140.4	150.0

a. Find an equation of the least-squares line for these data.

b. Use the result of part (a) to estimate the projected number of wireless subscribers at the beginning of 2006. How does this result compare with the given data for that year?
Source: BancAmerica Robertson Stephens

19. HEALTH-CARE SPENDING The following data, compiled by the Organization for Economic Cooperation and Development (OECD) in 1990, gives the per capita Gross Domestic Product (GDP) (in thousands of dollars) and the corresponding per capita spending on health care (in dollars) for selected countries.

Country	Turkey	Spain	Netherlands
GDP	4.25	10	14
Health-Care Spending	178	667	1194

Country	Sweden	Switzerland	Canada
GDP	15.5	17.8	19.5
Health-Care Spending	1500	1388	1640

a. Letting x denote a country's GDP (in thousands of dollars per capita) and y denote the per capita health-care spending (in dollars), find an equation of the least-squares line for these data giving the typical relationship between GDP and health-care spending for the selected countries.
b. The per capita GDP of the United States is \$20,000. If the health-care spending of the United States were in line with that of these sample OECD countries, what would it be? (*Note:* The actual per capita health-care spending of the United States in 1990 was \$2444.)
Source: Organization for Economic Cooperation and Development

In Exercises 20 and 21, determine whether the statement is true or false. If it is true, explain why it is true. If it is false, give an example to show why it is false.

20. The least-squares line must pass through at least one of the data points.

21. The sum of the squares of the errors incurred in approximating n data points using the least-squares linear function is zero if and only if the n data points lie along a straight line.

SOLUTIONS TO SELF-CHECK EXERCISES 8.4

1. We first construct the table:

	x	y	x^2	xy
	0	3	0	0
	2	6.5	4	13
	4	10	16	40
	6	16	36	96
	7	16.5	49	115.5
Sum	19	52	105	264.5

The normal equations are

$$5b + 19m = 52$$
$$19b + 105m = 264.5$$

Solving the first equation for b gives

$$b = -3.8m + 10.4$$

which, upon substitution into the second equation, gives

$$19(-3.8m + 10.4) + 105m = 264.5$$
$$-72.2m + 197.6 + 105m = 264.5$$
$$32.8m = 66.9$$
$$m \approx 2.04$$

Substituting this value of m into the expression for b found earlier gives

$$b = -3.8(2.04) + 10.4 \approx 2.65$$

Therefore, the required least-squares line has the equation given by

$$y = 2.04x + 2.65$$

2. a. The calculations required for obtaining the normal equations may be summarized as follows:

	x	y	x^2	xy
	0	19	0	0
	6	25	36	150
	11	30	121	330
	16	35	256	560
	22	44	484	968
	26	48	676	1248
Sum	81	201	1573	3256

The normal equations are

$$6b + \quad 81m = \quad 201$$
$$81b + 1573m = 3256$$

Solving this system of linear equations simultaneously, we find

$$m \approx 1.13 \quad \text{and} \quad b \approx 18.23$$

Therefore, the required least-squares line has the equation given by

$$y = f(x) = 1.13x + 18.23$$

b. The percentage of people over the age of 65 who will have high school diplomas at the beginning of the year 2003 is given by

$$f(44) = 1.13(44) + 18.23$$
$$= 67.95$$

or approximately 68%.

8.5 Constrained Maxima and Minima and the Method of Lagrange Multipliers

CONSTRAINED RELATIVE EXTREMA

In Section 8.3 we studied the problem of determining the relative extremum of a function $f(x, y)$ without placing any restrictions on the independent variables x and y—except, of course, that the point (x, y) lies in the domain of f. Such a relative extremum of a function f is referred to as an **unconstrained relative extremum** of f. However, in many practical optimization problems, we must maximize or minimize a function in which the independent variables are subjected to certain further constraints.

In this section we discuss a powerful method for determining the relative extrema of a function $f(x, y)$ whose independent variables x and y are required to satisfy one or more constraints of the form $g(x, y) = 0$. Such a relative extremum of a function f is called a **constrained relative extremum of f**. We can see the difference between an unconstrained extremum of a function $f(x, y)$ of two variables and a constrained extremum of f, where the independent variables x and y are subjected to a constraint of the form $g(x, y) = 0$, by considering the geometry of the two cases. Figure 8.25a depicts the graph of a function $f(x, y)$ that has an unconstrained relative minimum at the point $(0, 0)$. However, when the independent variables x and y are subjected to an equality constraint of the form $g(x, y) = 0$, the points (x, y, z) that satisfy both $z = f(x, y)$ and the constraint equation $g(x, y) = 0$ lie on a curve C. Therefore, the constrained relative minimum of f must also lie on C (Figure 8.25b).

Our first example involves an equality constraint $g(x, y) = 0$ in which we solve for the variable y explicitly in terms of x. In this case we may apply the

FIGURE 8.25

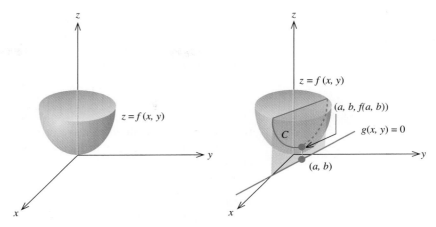

(a) $f(x, y)$ has an unconstrained relative extremum at $(0, 0)$.

(b) $f(x, y)$ has a constrained relative extremum at $(a, b, f(a, b))$.

technique used in Chapter 4 to find the relative extrema of a function of one variable.

EXAMPLE 1   Find the relative minimum of the function

$$f(x, y) = 2x^2 + y^2$$

subject to the constraint $g(x, y) = x + y - 1 = 0$.

SOLUTION ✔   Solving the constraint equation for y explicitly in terms of x, we obtain $y = -x + 1$. Substituting this value of y into the function $f(x, y) = 2x^2 + y^2$ results in a function of x,

$$h(x) = 2x^2 + (-x + 1)^2 = 3x^2 - 2x + 1$$

The function h describes the curve C lying on the graph of f on which the constrained relative minimum of f occurs. To find this point, use the technique developed in Chapter 4 to determine the relative extrema of a function of one variable:

$$h'(x) = 6x - 2 = 2(3x - 1)$$

Setting $h'(x) = 0$ gives $x = \frac{1}{3}$ as the sole critical point of the function h. Next, we find

$$h''(x) = 6$$

and, in particular,

$$h''\left(\frac{1}{3}\right) = 6 > 0$$

Therefore, by the second derivative test, the point $x = \frac{1}{3}$ gives rise to a relative minimum of h. Substitute this value of x into the constraint equation $x + y - 1 = 0$ to get $y = \frac{2}{3}$. Thus, the point $(\frac{1}{3}, \frac{2}{3})$ gives rise to the required constrained relative minimum of f. Since

$$f\left(\frac{1}{3}, \frac{2}{3}\right) = 2\left(\frac{1}{3}\right)^2 + \left(\frac{2}{3}\right)^2 = \frac{2}{3}$$

the required constrained relative minimum value of f is $\frac{2}{3}$ at the point $(\frac{1}{3}, \frac{2}{3})$. It may be shown that $\frac{2}{3}$ is in fact a constrained absolute minimum value of f (Figure 8.26).

FIGURE 8.26
f has a constrained absolute minimum of $\frac{2}{3}$ at $(\frac{1}{3}, \frac{2}{3})$.

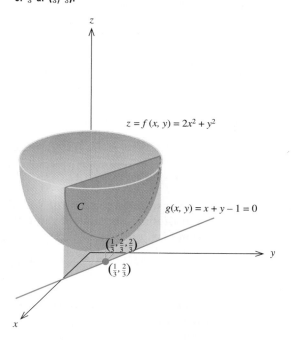

THE METHOD OF LAGRANGE MULTIPLIERS

The major drawback of the technique used in Example 1 is that it relies on our ability to solve the constraint equation $g(x, y) = 0$ for y explicitly in terms of x. This is not always an easy task. Moreover, even when we can solve the constraint equation $g(x, y) = 0$ for y explicitly in terms of x, the resulting function of one variable that is to be optimized may turn out to be unnecessarily complicated. Fortunately, an easier method exists. This method, called the **method of Lagrange multipliers** (Joseph Lagrange, 1736–1813), is as follows:

The Method of Lagrange Multipliers

To find the relative extremum of the function $f(x, y)$ subject to the constraint $g(x, y) = 0$ (assuming that these extreme values exist),

1. Form an auxiliary function

$$F(x, y, \lambda) = f(x, y) + \lambda g(x, y)$$

called the Lagrangian function (the variable λ is called the Lagrange multiplier).

2. Solve the system that consists of the equations

$$F_x = 0, \qquad F_y = 0, \qquad F_\lambda = 0$$

for all values of x, y, and λ.

3. Evaluate f at each of the points (x, y) found in step 2. The largest (smallest) of these values is the maximum (minimum) value of f.

Let's re-solve Example 1 using the method of Lagrange multipliers.

EXAMPLE 2   Using the method of Lagrange multipliers, find the relative minimum of the function

$$f(x, y) = 2x^2 + y^2$$

subject to the constraint $x + y = 1$.

SOLUTION ✔   Write the constraint equation $x + y = 1$ in the form $g(x, y) = x + y - 1 = 0$. Then, form the Lagrangian function

$$\begin{aligned} F(x, y, \lambda) &= f(x, y) + \lambda g(x, y) \\ &= 2x^2 + y^2 + \lambda(x + y - 1) \end{aligned}$$

To find the critical point(s) of the function F, solve the system composed of the equations

$$\begin{aligned} F_x &= 4x + \lambda = 0 \\ F_y &= 2y + \lambda = 0 \\ F_\lambda &= x + y - 1 = 0 \end{aligned}$$

Solving the first and second equations in this system for x and y in terms of λ, we obtain

$$x = -\frac{1}{4}\lambda \qquad \text{and} \qquad y = -\frac{1}{2}\lambda$$

which, upon substitution into the third equation, yields

$$-\frac{1}{4}\lambda - \frac{1}{2}\lambda - 1 = 0 \qquad \text{or} \qquad \lambda = -\frac{4}{3}$$

Therefore, $x = \frac{1}{3}$ and $y = \frac{2}{3}$, and $(\frac{1}{3}, \frac{2}{3})$ affords a constrained minimum of the function f, in agreement with the result obtained earlier.   ■ ■ ■ ■

The method of Lagrange multipliers may be used to solve a problem involving a function of three or more variables, as illustrated in the next example.

EXAMPLE 3

Use the method of Lagrange multipliers to find the minimum of the function

$$f(x, y, z) = 2xy + 6yz + 8xz$$

subject to the constraint

$$xyz = 12,000$$

(*Note:* The existence of the minimum is suggested by the geometry of the problem.)

SOLUTION ✔

Write the constraint equation $xyz = 12,000$ in the form $g(x, y, z) = xyz - 12,000$. Then, the Lagrangian function is

$$F(x, y, z, \lambda) = f(x, y, z) + \lambda g(x, y, z)$$
$$= 2xy + 6yz + 8xz + \lambda(xyz - 12,000)$$

To find the critical point(s) of the function F, we solve the system composed of the equations

$$F_x = 2y + 8z + \lambda yz = 0$$
$$F_y = 2x + 6z + \lambda xz = 0$$
$$F_z = 6y + 8x + \lambda xy = 0$$
$$F_\lambda = xyz - 12,000 = 0$$

Solving the first three equations of the system for λ in terms of x, y, and z, we have

$$\lambda = -\frac{2y + 8z}{yz}$$

$$\lambda = -\frac{2x + 6z}{xz}$$

$$\lambda = -\frac{6y + 8x}{xy}$$

Equating the first two expressions for λ leads to

$$\frac{2y + 8z}{yz} = \frac{2x + 6z}{xz}$$
$$2xy + 8xz = 2xy + 6yz$$
$$x = \frac{3}{4}y$$

Next, equating the second and third expressions for λ in the same system yields

$$\frac{2x + 6z}{xz} = \frac{6y + 8x}{xy}$$

$$2xy + 6yz = 6yz + 8xz$$

$$z = \frac{1}{4}y$$

Finally, substituting these values of x and z into the equation $xyz - 12{,}000 = 0$, the fourth equation of the first system of equations, we have

$$\left(\frac{3}{4}y\right)(y)\left(\frac{1}{4}y\right) - 12{,}000 = 0$$

$$y^3 = \frac{(12{,}000)(4)(4)}{3} = 64{,}000$$

or

$$y = 40$$

The corresponding values of x and z are given by $x = \frac{3}{4}(40) = 30$ and $z = \frac{1}{4}(40) = 10$. Therefore, we see that the point $(30, 40, 10)$ gives the constrained minimum of f. The minimum value is

$$f(30, 40, 10) = 2(30)(40) + 6(40)(10) + 8(30)(10) = 7200$$

■ ■ ■ ■

APPLICATIONS

EXAMPLE **4**   Refer to Example 3, Section 8.1. The total weekly profit (in dollars) that the Acrosonic Company realized in producing and selling its bookshelf loudspeaker systems is given by the profit function

$$P(x, y) = -\frac{1}{4}x^2 - \frac{3}{8}y^2 - \frac{1}{4}xy + 120x + 100y - 5000$$

where x denotes the number of fully assembled units and y denotes the number of kits produced and sold per week. Acrosonic's management decides that production of these loudspeaker systems should be restricted to a total of exactly 230 units per week. Under this condition, how many fully assembled units and how many kits should be produced per week to maximize Acrosonic's weekly profit?

SOLUTION ✔   The problem is equivalent to the problem of maximizing the function

$$P(x, y) = -\frac{1}{4}x^2 - \frac{3}{8}y^2 - \frac{1}{4}xy + 120x + 100y - 5000$$

subject to the constraint

$$g(x, y) = x + y - 230 = 0$$

The Lagrangian function is

$$F(x, y, \lambda) = P(x, y) + \lambda g(x, y)$$

$$= -\frac{1}{4}x^2 - \frac{3}{8}y^2 - \frac{1}{4}xy + 120x + 100y$$

$$-5000 + \lambda(x + y - 230)$$

To find the critical point(s) of F, solve the following system of equations:

$$F_x = -\frac{1}{2}x - \frac{1}{4}y + 120 + \lambda = 0$$

$$F_y = -\frac{3}{4}y - \frac{1}{4}x + 100 + \lambda = 0$$

$$F_\lambda = x + y - 230 = 0$$

Solving the first equation of this system for λ, we obtain

$$\lambda = \frac{1}{2}x + \frac{1}{4}y - 120$$

which, upon substitution into the second equation, yields

$$-\frac{3}{4}y - \frac{1}{4}x + 100 + \frac{1}{2}x + \frac{1}{4}y - 120 = 0$$

$$-\frac{1}{2}y + \frac{1}{4}x - 20 = 0$$

Solving the last equation for y gives

$$y = \frac{1}{2}x - 40$$

When we substitute this value of y into the third equation of the system, we have

$$x + \frac{1}{2}x - 40 - 230 = 0$$

$$x = 180$$

The corresponding value of y is $(\frac{1}{2})(180) - 40$, or 50. Thus, the required constrained relative maximum of P occurs at the point (180, 50). Again, we can show that the point (180, 50) in fact yields a constrained absolute maximum for P. Thus, Acrosonic's profit is maximized by producing 180 assembled and 50 kit versions of their bookshelf loudspeaker systems. The maximum weekly profit realizable is given by

$$P(180, 50) = -\frac{1}{4}(180)^2 - \frac{3}{8}(50)^2 - \frac{1}{4}(180)(50)$$

$$+ 120(180) + 100(50) - 5000$$

$$= 10{,}312.5$$

or $10,312.50.

■ ■ ■ ■

EXAMPLE **5**

The operators of the *Viking Princess,* a luxury cruise liner, are contemplating the addition of another swimming pool to the ship. The chief engineer has suggested that an area in the form of an ellipse located in the rear of the promenade deck would be suitable for this purpose. This location would provide a poolside area with sufficient space for passenger movement and placement of deck chairs (Figure 8.27). It has been determined that the shape of the ellipse may be described by the equation $x^2 + 4y^2 = 3600$, where x and y are measured in feet. *Viking*'s operators would like to know the dimensions of the rectangular pool with the largest possible area that would meet these requirements.

FIGURE 8.27
A rectangular-shaped pool will be built in the elliptical-shaped poolside area.

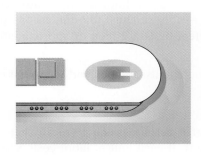

SOLUTION ✔

To solve this problem, we need to find the rectangle inscribed in the ellipse with equation $x^2 + 4y^2 = 3600$ and having the largest area. Letting the sides of the rectangle be $2x$ and $2y$ feet, we see that the area of the rectangle is $A = 4xy$ (Figure 8.28). Furthermore, the point (x, y) must be constrained to move along the ellipse so that it satisfies the equation $x^2 + 4y^2 = 3600$. Thus, the problem is equivalent to the problem of maximizing the function

FIGURE 8.28
We want to find the largest rectangle that can be inscribed in the ellipse described by $x^2 + 4y^2 = 3600$.

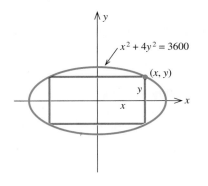

$$f(x, y) = 4xy$$

subject to the constraint $g(x, y) = x^2 + 4y^2 - 3600 = 0$. The Lagrangian function is

$$F(x, y, \lambda) = f(x, y) + \lambda g(x, y)$$
$$= 4xy + \lambda(x^2 + 4y^2 - 3600)$$

To find the critical point(s) of F, we solve the following system of equations:

$$F_x = 4y + 2\lambda x = 0$$
$$F_y = 4x + 8\lambda y = 0$$
$$F_\lambda = x^2 + 4y^2 - 3600 = 0$$

Solving the first equation of this system for λ, we obtain

$$\lambda = -\frac{2y}{x}$$

which, upon substitution into the second equation, yields

$$4x + 8\left(-\frac{2y}{x}\right)(y) = 0 \quad \text{or} \quad x^2 - 4y^2 = 0$$

—that is, $x = \pm 2y$. Substituting these values of x into the third equation of the system, we have

$$4y^2 + 4y^2 - 3600 = 0$$

or, upon solving $y = \pm\sqrt{450} = \pm 15\sqrt{2}$. The corresponding values of x are $\pm 30\sqrt{2}$. Because both x and y must be nonnegative, we have $x = 30\sqrt{2}$ and $y = 15\sqrt{2}$. Thus, the dimensions of the pool with maximum area are $30\sqrt{2}$ feet by $60\sqrt{2}$ feet, or approximately 42 feet × 85 feet. ▪▪▪▪

EXAMPLE 6

Suppose x units of labor and y units of capital are required to produce

$$f(x, y) = 100x^{3/4}y^{1/4}$$

units of a certain product (recall that this is a Cobb–Douglas production function). If each unit of labor costs $200 and each unit of capital costs $300 and a total of $60,000 is available for production, determine how many units of labor and how many units of capital should be used in order to maximize production.

SOLUTION ✔

The total cost of x units of labor at $200 per unit and y units of capital at $300 per unit is equal to $200x + 300y$ dollars. But $60,000 is budgeted for production, so $200x + 300y = 60,000$, which we rewrite as

$$g(x, y) = 200x + 300y - 60,000 = 0$$

To maximize $f(x, y) = 100x^{3/4}y^{1/4}$ subject to the constraint $g(x, y) = 0$, we form the Lagrangian function

$$\begin{aligned} F(x, y, \lambda) &= f(x, y) + \lambda g(x, y) \\ &= 100x^{3/4}y^{1/4} + \lambda(200x + 300y - 60,000) \end{aligned}$$

To find the critical point(s) of F, we solve the following system of equations:

$$\begin{aligned} F_x &= 75x^{-1/4}y^{1/4} + 200\lambda = 0 \\ F_y &= 25x^{3/4}y^{-3/4} + 300\lambda = 0 \\ F_\lambda &= 200x + 300y - 60,000 = 0 \end{aligned}$$

Solving the first equation for λ, we have

$$\lambda = -\frac{75x^{-1/4}y^{1/4}}{200} = -\frac{3}{8}\left(\frac{y}{x}\right)^{1/4}$$

which, when substituted into the second equation, yields

$$25 \left(\frac{x}{y}\right)^{3/4} + 300 \left(-\frac{3}{8}\right)\left(\frac{y}{x}\right)^{1/4} = 0$$

Multiplying the last equation by $(x/y)^{1/4}$ then gives

$$25 \left(\frac{x}{y}\right) - \frac{900}{8} = 0$$

$$x = \left(\frac{900}{8}\right)\left(\frac{1}{25}\right)y = \frac{9}{2}y$$

Substituting this value of x into the third equation of the first system of equations, we have

$$200 \left(\frac{9}{2}y\right) + 300y - 60{,}000 = 0$$

from which we deduce that $y = 50$. Hence, $x = 225$. Thus, maximum production is achieved when 225 units of labor and 50 units of capital are used. ■ ■ ■ ■

When used in the context of Example 6, the negative of the Lagrange multiplier λ is called the **marginal productivity of money.** That is, if one additional dollar is available for production, then approximately $-\lambda$ units of a product can be produced. Here,

$$\lambda = -\frac{3}{8}\left(\frac{y}{x}\right)^{1/4} = -\frac{3}{8}\left(\frac{50}{225}\right)^{1/4} \approx -0.257$$

so, in this case, the marginal productivity of money is 0.257. For example, if \$65,000 is available for production instead of the originally budgeted figure of \$60,000, then the maximum production may be boosted from the original

$$f(225, 50) = 100(225)^{3/4}(50)^{1/4}$$

or 15,448 units, to

$$15{,}448 + 5000(0.257)$$

or 16,733 units.

SELF-CHECK EXERCISES 8.5

 1. Use the method of Lagrange multipliers to find the relative maximum of the function

$$f(x, y) = -2x^2 - y^2$$

subject to the constraint $3x + 4y = 12$.

2. The total monthly profit of the Robertson Controls Company in manufacturing and selling x hundred of its standard mechanical setback thermostats and y hundred of its deluxe electronic setback thermostats per month is given by the total profit function

$$P(x, y) = -\frac{1}{8}x^2 - \frac{1}{2}y^2 - \frac{1}{4}xy + 13x + 40y - 280$$

where P is in hundreds of dollars. If the production of setback thermostats is to be restricted to a total of exactly 4000/month, how many of each model should Robertson manufacture in order to maximize its monthly profits? What is the maximum monthly profit?

Solutions to Self-Check Exercises 8.5 can be found on page 662.

8.5 Exercises

In Exercises 1–16, use the method of Lagrange multipliers to optimize the given function subject to the given constraint.

1. Minimize the function $f(x, y) = x^2 + 3y^2$ subject to the constraint $x + y - 1 = 0$.

2. Minimize the function $f(x, y) = x^2 + y^2 - xy$ subject to the constraint $x + 2y - 14 = 0$.

3. Maximize the function $f(x, y) = 2x + 3y - x^2 - y^2$ subject to the constraint $x + 2y = 9$.

4. Maximize the function $f(x, y) = 16 - x^2 - y^2$ subject to the constraint $x + y - 6 = 0$.

5. Minimize the function $f(x, y) = x^2 + 4y^2$ subject to the constraint $xy = 1$.

6. Minimize the function $f(x, y) = xy$ subject to the constraint $x^2 + 4y^2 = 4$.

7. Maximize the function $f(x, y) = x + 5y - 2xy - x^2 - 2y^2$ subject to the constraint $2x + y = 4$.

8. Maximize the function $f(x, y) = xy$ subject to the constraint $2x + 3y - 6 = 0$.

9. Maximize the function $f(x, y) = xy^2$ subject to the constraint $9x^2 + y^2 = 9$.

10. Minimize the function $f(x, y) = \sqrt{y^2 - x^2}$ subject to the constraint $x + 2y - 5 = 0$.

11. Find the maximum and minimum values of the function $f(x, y) = xy$ subject to the constraint $x^2 + y^2 = 16$.

12. Find the maximum and minimum values of the function $f(x, y) = e^{xy}$ subject to the constraint $x^2 + y^2 = 8$.

13. Find the maximum and minimum values of the function $f(x, y) = xy^2$ subject to the constraint $x^2 + y^2 = 1$.

14. Maximize the function $f(x, y, z) = xyz$ subject to the constraint $2x + 2y + z = 84$.

15. Minimize the function $f(x, y, z) = x^2 + y^2 + z^2$ subject to the constraint $3x + 2y + z = 6$.

16. Find the maximum value of the function $f(x, y, z) = x + 2y - 3z$ subject to the constraint $z = 4x^2 + y^2$.

17. MAXIMIZING PROFIT The total weekly profit (in dollars) realized by the Country Workshop in manufacturing and selling its rolltop desks is given by the profit function

$$P(x, y) = -0.2x^2 - 0.25y^2 - 0.2xy$$
$$+ 100x + 90y - 4000$$

where x stands for the number of finished units and y denotes the number of unfinished units manufactured and sold per week. The company's management decides to restrict the manufacture of these desks to a total of exactly 200 units/week. How many finished and how many unfinished units should be manufactured per week to maximize the company's weekly profit?

18. MAXIMIZING PROFIT The total daily profit (in dollars) realized by the Weston Publishing Company in publishing and selling its dictionaries is given by the profit function

$$P(x, y) = -0.005x^2 - 0.003y^2 - 0.002xy$$
$$+ 14x + 12y - 200$$

where x stands for the number of deluxe editions and y denotes the number of standard editions sold daily. Weston's management decides that publication of these dictionaries should be restricted to a total of exactly 400 copies/day. How many deluxe copies and how many standard copies should be published per day to maximize Weston's daily profit?

19. **MINIMIZING CONSTRUCTION COSTS** The management of UNICO Department Store decides to enclose an 800-ft^2 area outside their building to display potted plants. The enclosed area will be a rectangle, one side of which is provided by the external walls of the store. Two sides of the enclosure will be made of pine board, and the fourth side will be made of galvanized steel fencing material. If the pine board fencing costs $6/running foot and the steel fencing costs $3/running foot, determine the dimensions of the enclosure that will cost the least to erect.

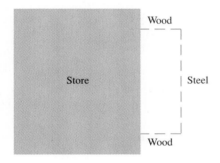

20. **MINIMIZING CONTAINER COSTS** The Betty Moore Company requires that its corned beef hash containers have a capacity of 64 in.^{3}, be right circular cylinders, and be made of a tin alloy. Find the radius and height of the least expensive container that can be made if the metal for the side and bottom costs 4 cents/square inch and the metal for the pull-off lid costs 2 cents/square inch.
Hint: Let the radius and height of the container be r and h inches, respectively. Then, the volume of the container is $\pi r^2 h = 64$, and the cost is given by $C(r, h) = 8\pi rh + 6\pi r^2$.

21. **MINIMIZING CONSTRUCTION COSTS** An open rectangular box is to be constructed from material that costs $3/square foot for the bottom and $1/square foot for its sides. Find the dimensions of the box of greatest volume that can be constructed for $36.

22. **MINIMIZING CONSTRUCTION COSTS** A closed rectangular box having a volume of 4 ft^3 is to be constructed. If the material for the sides costs $1/square foot and the material for the top and bottom costs $1.50/square foot, find the dimensions of the box that can be constructed with minimum cost.

23. **MAXIMIZING SALES** The Ross–Simons Company has a monthly advertising budget of $60,000. Their marketing department estimates that if they spend x dollars on newspaper advertising and y dollars on television advertising, then the monthly sales will be given by

$$z = f(x, y) = 90x^{1/4}y^{3/4}$$

dollars. Determine how much money Ross–Simons should spend on newspaper ads and on television ads per month to maximize its monthly sales.

24. **MAXIMIZING PRODUCTION** John Mills, the proprietor of the Mills Engine Company, a manufacturer of model airplane engines, finds that it takes x units of labor and y units of capital to produce

$$f(x, y) = 100x^{3/4}y^{1/4}$$

units of the product. If a unit of labor costs $100 and a unit of capital costs $200 and $200,000 is budgeted for production, determine how many units should be expended on labor and how many units should be expended on capital in order to maximize production.

25. Use the method of Lagrange multipliers to solve Exercise 28, Exercises 8.3.

In Exercises 26 and 27, determine whether the statement is true or false. If it is true, explain why it is true. If it is false, give an example to show why it is false.

26. If (a, b) gives rise to a (constrained) relative extremum of f subject to the constraint $g(x, y) = 0$, then (a, b) also gives rise to the unconstrained relative extremum of f.

27. If (a, b) gives rise to a (constrained) relative extremum of f subject to the constraint $g(x, y) = 0$, then $f_x(a, b) = 0$ and $f_y(a, b) = 0$, simultaneously.

SOLUTIONS TO SELF-CHECK EXERCISES 8.5

1. Write the constraint equation in the form $g(x, y) = 3x + 4y - 12 = 0$. Then, the Lagrangian function is

$$F(x, y, \lambda) = -2x^2 - y^2 + \lambda(3x + 4y - 12)$$

To find the critical point(s) of F, we solve the system

$$F_x = -4x + 3\lambda = 0$$
$$F_y = -2y + 4\lambda = 0$$
$$F_\lambda = 3x + 4y - 12 = 0$$

Solving the first two equations for x and y in terms of λ, we find $x = \frac{3}{4}\lambda$ and $y = 2\lambda$. Substituting these values of x and y into the third equation of the system yields

$$3\left(\frac{3}{4}\lambda\right) + 4(2\lambda) - 12 = 0$$

or $\lambda = \frac{48}{41}$. Therefore, $x = \left(\frac{3}{4}\right)\left(\frac{48}{41}\right) = \frac{36}{41}$ and $y = 2\left(\frac{48}{41}\right) = \frac{96}{41}$, and we see that the point $\left(\frac{36}{41}, \frac{96}{41}\right)$ gives the constrained maximum of f. The maximum value is

$$f\left(\frac{36}{41}, \frac{96}{41}\right) = -2\left(\frac{36}{41}\right)^2 - \left(\frac{96}{41}\right)^2$$

$$= -\frac{11{,}808}{1681} = -\frac{288}{41}$$

2. We want to maximize

$$P(x, y) = -\frac{1}{8}x^2 - \frac{1}{2}y^2 - \frac{1}{4}xy + 13x + 40y - 280$$

subject to the constraint

$$g(x, y) = x + y - 40 = 0$$

The Lagrangian function is

$$F(x, y, \lambda) = P(x, y) + \lambda g(x, y)$$
$$= -\frac{1}{8}x^2 - \frac{1}{2}y^2 - \frac{1}{4}xy + 13x$$
$$+ 40y - 280 + \lambda(x + y - 40)$$

To find the critical points of F, solve the following system of equations:

$$F_x = -\frac{1}{4}x - \frac{1}{4}y + 13 + \lambda = 0$$

$$F_y = -\frac{1}{4}x - y + 40 + \lambda = 0$$

$$F_\lambda = x + y - 40 = 0$$

Subtracting the first equation from the second gives

$$-\frac{3}{4}y + 27 = 0 \qquad \text{or} \qquad y = 36$$

Substituting this value of y into the third equation yields $x = 4$. Therefore, in order to maximize its monthly profits, Robertson should manufacture 400 standard and 3600 deluxe thermostats. The maximum monthly profit is given by

$$P(4, 36) = -\frac{1}{8}(4)^2 - \frac{1}{2}(36)^2 - \frac{1}{4}(4)(36)$$
$$+ 13(4) + 40(36) - 280$$
$$= 526$$

or $52,600.

8.6 Total Differentials

DIFFERENTIALS OF TWO OR MORE VARIABLES

In Section 3.7 we defined the differential of a function of one variable $y = f(x)$ to be

$$dy = f'(x)\, dx$$

In particular, we showed that if the actual change $dx = \Delta x$ in the independent variable x is small, then the differential dy provides us with an approximation of the actual change Δy in the dependent variable y; that is,

$$\Delta y = f(x + \Delta x) - f(x)$$
$$\approx f'(x)\, dx$$

The concept of the differential extends readily to a function of two or more variables.

Total Differential

Let $z = f(x, y)$ define a differentiable function of x and y.

1. The **differentials** of the independent variables x and y are $dx = \Delta x$ and $dy = \Delta y$.

2. The **differential** of the dependent variable z is

$$dz = \frac{\partial f}{\partial x}\, dx + \frac{\partial f}{\partial y}\, dy \qquad (12)$$

Thus, analogous to the one-variable case, the total differential of z is a linear function of dx and dy. Furthermore, it provides us with an approximation of the exact change in z,

$$\Delta z = f(x + \Delta x, y + \Delta y) - f(x, y)$$

corresponding to a net change Δx in x from x to $x + \Delta x$ and a net change Δy in y from y to $y + \Delta y$; that is,

$$\Delta z \approx dz = \frac{\partial f}{\partial x}(x, y)\, dx + \frac{\partial f}{\partial y}(x, y)\, dy \qquad (13)$$

provided $\Delta x = dx$ and $\Delta y = dy$ are sufficiently small.

EXAMPLE 1   Let $z = 2x^2y + y^3$.

a. Find the differential dz of z.
b. Find the approximate change in z when x changes from $x = 1$ to $x = 1.01$ and y changes from $y = 2$ to $y = 1.98$.
c. Find the actual change in z when x changes from $x = 1$ to $x = 1.01$ and y changes from $y = 2$ to $y = 1.98$. Compare the result with that obtained in (b).

SOLUTION ✔   **a.** Let $f(x, y) = 2x^2y + y^3$. Then the required differential is

$$dz = \frac{\partial f}{\partial x}\, dx + \frac{\partial f}{\partial y}\, dy = 4xy\, dx + (2x^2 + 3y^2)\, dy$$

b. Here $x = 1$, $y = 2$, and $dx = 1.01 - 1 = 0.01$ and $dy = 1.98 - 2 = -0.02$. Therefore,

$$\Delta z \approx dz = 4(1)(2)(0.01) + [2(1) + 3(4)](-0.02) = -0.20$$

c. The actual change in z is given by

$$\begin{aligned}
\Delta z &= f(1.01, 1.98) - f(1, 2) \\
&= [2(1.01)^2(1.98) + (1.98)^3] - [2(1)^2(2) + (2)^3] \\
&\approx 11.801988 - 12 \\
&= -0.1980
\end{aligned}$$

We see that $\Delta z \approx dz$, as expected.   ■ ■ ■ ■

APPLICATIONS

EXAMPLE 2   The weekly total revenue of Acrosonic Company resulting from the production and sales of x fully assembled bookshelf loudspeaker systems and y kit versions of the same loudspeaker system is

$$R(x, y) = -\frac{1}{4}x^2 - \frac{3}{8}y^2 - \frac{1}{4}xy + 300x + 240y$$

dollars. Determine the approximate change in Acrosonic's weekly total revenue when the level of production is increased from 200 assembled units and 60 kits per week to 206 assembled units and 64 kits per week.

SOLUTION ✔ The approximate change in the weekly total revenue is given by the total differential R at $x = 200$ and $y = 60$, $dx = 206 - 200 = 6$ and $dy = 64 - 60 = 4$; that is, by

$$dR = \frac{\partial R}{\partial x}\,dx + \frac{\partial R}{\partial y}\,dy \,\bigg|_{\substack{x=200,\,y=60 \\ dx=6,\,dy=4}}$$

$$= \left(-\frac{1}{2}x - \frac{1}{4}y + 300\right)\bigg|_{(200,\,60)} \cdot (6)$$

$$+ \left(-\frac{3}{4}y - \frac{1}{4}x + 240\right)\bigg|_{(200,\,60)} \cdot (4)$$

$$= (-100 - 15 + 300)6 + (-45 - 50 + 240)4$$

$$= 1690$$

or $1690. ■ ■ ■ ■

EXAMPLE 3 The production for a certain country in the early years following World War II is described by the function

$$f(x, y) = 30x^{2/3}y^{1/3}$$

units, when x units of labor and y units of capital were utilized. Find the approximate change in output if the amount expended on labor had been decreased from 125 units to 123 units and the amount expended on capital had been increased from 27 to 29 units. Is your result as expected given the result of Example 4c, Section 8.2?

SOLUTION ✔ The approximate change in output is given by the total differential of f at $x = 125$, $y = 27$, $dx = 123 - 125 = -2$, and $dy = 29 - 27 = 2$; that is, by

$$df = \frac{\partial f}{\partial x}\,dx + \frac{\partial f}{\partial y}\,dy \,\bigg|_{\substack{x=125,\,y=27 \\ dx=-2,\,dy=2}}$$

$$= 20x^{-1/3}y^{1/3}\bigg|_{(125,\,27)} \cdot (-2) + 10x^{2/3}y^{-2/3}\bigg|_{(125,\,27)} \cdot (2)$$

$$= 20\left(\frac{27}{125}\right)^{1/3}(-2) + 10\left(\frac{125}{27}\right)^{2/3}(2)$$

$$= -20\left(\frac{3}{5}\right)(2) + 10\left(\frac{25}{9}\right)(2) = \frac{284}{9}$$

or $31\frac{5}{9}$ units. This result is fully compatible with the result of Example 4, where the recommendation was to encourage increased spending on capital rather than on labor. ■ ■ ■ ■

If f is a function of the three variables x, y, and z, then the total differential of $w = f(x, y, z)$ is defined to be

$$dw = \frac{\partial f}{\partial x}\,dx + \frac{\partial f}{\partial y}\,dy + \frac{\partial f}{\partial z}\,dz$$

where $dx = \Delta x$, $dy = \Delta y$, and $dz = \Delta z$ are the actual changes in the independent variables x, y, and z as x changes from $x = a$ to $x = a + \Delta x$, y changes from $y = b$ to $y = b + \Delta y$, and z changes from $z = c$ to $z = c + \Delta z$, respectively.

■ ■ ■ ■

EXAMPLE 4   Find the maximum percentage error in calculating the volume of a rectangular box if an error of at most 1% is made in measuring the length, width, and height of the box.

SOLUTION ✔   Let x, y, and z denote the length, width, and height, respectively, of the rectangular box. Then the volume of the box is given by $V = f(x, y, z) = xyz$ cubic units. Now suppose the true dimensions of the rectangular box are a, b, and c units, respectively. Since the error committed in measuring the length, width, and height of the box is at most 1%, we have

$$|\Delta x| = |x - a| \leq 0.01a$$
$$|\Delta y| = |y - b| \leq 0.01b$$
$$|\Delta z| = |z - c| \leq 0.01c$$

Therefore, the maximum error in calculating the volume of the box is

$$|\Delta V| \approx |dV| = \left| \frac{\partial f}{\partial x} dx + \frac{\partial f}{\partial y} dy + \frac{\partial f}{\partial z} dz \right|_{x=a, y=b, z=c}$$

$$= |yz\, dx + xz\, dy + xy\, dz|_{x=a, y=b, z=c}$$

$$= |bc\, dx + ac\, dy + ab\, dz|$$

$$\leq bc|dx| + ac|dy| + ab|dz|$$

$$\leq bc(0.01a) + ac(0.01b) + ab(0.01c)$$

$$= (0.03)abc$$

Since the actual volume of the box is abc cubic units, we see that the maximum percentage error in calculating its volume is

$$\frac{|\Delta V|}{V}\Big|_{(a,b,c)} \approx \frac{(0.03)abc}{abc} = 0.03$$

—that is, approximately 3%.

■ ■ ■ ■

■ **Group Discussion**
    Refer to Example 4, where we found the maximum percentage error in calculating the volume of the rectangular box to be *approximately* 3%. What is the precise maximum percentage error?

SELF-CHECK EXERCISE 8.6

Let f be a function defined by $z = f(x, y) = 3xy^2 - 4y$. Find the total differential of f at $(-1, 3)$. Then find the approximate change in z when x changes from $x = -1$ to $x = -0.98$ and y changes from $y = 3$ to $y = 3.01$.

The solution to Self-Check Exercise 8.6 can be found on page 669.

8.6 EXERCISES

In Exercises 1–18, find the total differential of the function.

1. $f(x, y) = x^2 + 2y$

2. $f(x, y) = 2x^2 + 3y^2$

3. $f(x, y) = 2x^2 - 3xy + 4x$

4. $f(x, y) = xy^3 - x^2y^2$

5. $f(x, y) = \sqrt{x^2 + y^2}$

6. $f(x, y) = (x + 3y^2)^{1/3}$

7. $f(x, y) = \dfrac{5y}{x - y}$

8. $f(x, y) = \dfrac{x + y}{x - y}$

9. $f(x, y) = 2x^5 - ye^{-3x}$

10. $f(x, y) = xye^{x+y}$

11. $f(x, y) = x^2e^y + y \ln x$

12. $f(x, y) = \ln(x^2 + y^2)$

13. $f(x, y, z) = xy^2z^3$

14. $f(x, y, z) = x\sqrt{y} + y\sqrt{z}$

15. $f(x, y, z) = \dfrac{x}{y + z}$

16. $f(x, y, z) = \dfrac{x + y}{y + z}$

17. $f(x, y, z) = xyz + xe^{yz}$

18. $f(x, y, z) = \sqrt{e^x + e^y + ze^{xy}}$

In Exercises 19–30, find the approximate change in z when the point (x, y) changes from (x_0, y_0) to (x_1, y_1).

19. $f(x, y) = 4x^2 - xy$; from $(1, 2)$ to $(1.01, 2.02)$

20. $f(x, y) = 2x^2 - 2x^3y^2 - y^3$; from $(-1, 2)$ to $(-0.98, 2.01)$

21. $f(x, y) = x^{2/3}y^{1/2}$; from $(8, 9)$ to $(7.97, 9.03)$

22. $f(x, y) = \sqrt{x^2 + y^2}$; from $(1, 3)$ to $(1.03, 3.03)$

23. $f(x, y) = \dfrac{x}{x - y}$; from $(-3, -2)$ to $(-3.02, -1.98)$

24. $f(x, y) = \dfrac{x - y}{x + y}$; from $(-3, -2)$ to $(-3.02, -1.98)$

25. $f(x, y) = 2xe^{-y}$; from $(4, 0)$ to $(4.03, 0.03)$

26. $f(x, y) = \sqrt{x}e^y$; from $(1, 1)$ to $(1.01, 0.98)$

27. $f(x, y) = xe^{xy} - y^2$; from $(-1, 0)$ to $(-0.97, 0.03)$

28. $f(x, y) = xe^{-y} + ye^{-x}$; from $(1, 1)$ to $(1.01, 0.90)$

29. $f(x, y) = x \ln x + y \ln x$; from $(2, 3)$ to $(1.98, 2.89)$

30. $f(x, y) = \ln(xy)^{1/2}$; from $(5, 10)$ to $(5.05, 9.95)$

31. EFFECT OF INVENTORY AND FLOOR SPACE ON PROFIT The monthly profit (in dollars) of Bond and Barker Department Store depends on the level of inventory x (in thousands of dollars) and the floor space y (in thousands of square feet) available for display of the merchandise, as given by the equation

$$P(x, y) = -0.02x^2 - 15y^2 + xy + 39x$$
$$+ 25y - 20,000$$

Currently, the level of inventory is \$4,000,000 ($x = 4000$), and the floor space is 150,000 square feet ($y = 150$). Find the anticipated change in monthly profit if management increases the level of inventory by $500,000 and decreases the floor space for display of merchandise by 10,000 square feet.

32. **Effect of Production on Profit**   The Country Workshop's total weekly profit (in dollars) realized in manufacturing and selling its rolltop desks is given by

$$P(x, y) = -0.2x^2 - 0.25y^2 - 0.2xy$$
$$+ 100x + 90y - 4000$$

where x stands for the number of finished units and y denotes the number of unfinished units manufactured and sold per week. Currently, the weekly output is 190 finished and 105 unfinished units. Determine the approximate change in the total weekly profit if the sole proprietor of the Country Workshop decides to increase the number of finished units to 200 per week and decrease the number of unfinished units to 100 per week.

33. **Revenue of a Travel Agency**   The Odyssey Travel Agency's monthly revenue (in thousands of dollars) depends on the amount of money x (in thousands) spent on advertising per month and the number of agents y in its employ in accordance with the rule

$$R(x, y) = -x^2 - 0.5y^2 + xy + 8x + 3y + 20$$

Currently, the amount of money spent on advertising is $10,000 per month, and there are 15 agents in the agency's employ. Estimate the change in revenue resulting from an increase of $1000 per month in advertising expenditure and a decrease of 1 agent.

34. **Effect of Capital and Labor on Productivity**   The productivity of a South American country is given by the function

$$f(x, y) = 20x^{3/4}y^{1/4}$$

when x units of labor and y units of capital are utilized. Find the approximate change in output if the amount expended on labor is decreased from 256 to 254 units and the amount expended on capital is increased from 16 to 18 units.

35. **Price–Earnings Ratio**   The price–earnings ratio (PE ratio) of a stock is given by

$$R(x,y) = \frac{x}{y}$$

where x denotes the price per share of the stock and y denotes the earnings per share. Estimate the change in the PE ratio of a stock if its price increases from $60/share to $62/share while its earnings decrease from $4/share to $3.80/share.

36. **Error in Calculating the Surface Area of a Human**   The formula

$$S = 0.007184W^{0.425}H^{0.725}$$

gives the surface area S of a human body (in square meters) in terms of its weight W in kilograms and its height H in centimeters. If an error of 1% is made in measuring the weight of a person and an error of 2% is made in measuring the height, what is the percentage error in the measurement of the person's surface area?

37. **Error in Calculating the Volume of a Cylinder**   The radius and height of a right circular cylinder are measured with a maximum error of 0.1 cm in each measurement. Approximate the maximum error in calculating the volume of the cylinder if the measured dimensions $r = 8$ cm and $h = 20$ cm are used.

38. **Effect of Capital and Labor on Productivity**   The production of a certain company is given by the function

$$f(x,y) = 50x^{1/3}y^{2/3}$$

when x units of labor and y units of capital are utilized. Find the approximate percentage change in the production of the company if labor is increased by 2% and capital is increased by 1%.

39. **Error in Calculating Total Resistance**   The total resistance R of three resistors with resistance R_1, R_2, and R_3, connected in parallel, is given by the relationship

$$\frac{1}{R} = \frac{1}{R_1} + \frac{1}{R_2} + \frac{1}{R_3}$$

If R_1, R_2, and R_3 are measured at 100, 200, and 300 ohms, respectively, with a maximum error of 1% in each measurement, find the approximate maximum error in the calculated value of R.

40. **Error in Measuring Arterial Blood Flow**   The flow of blood through an arteriole in cubic centimeters per second is given by

$$V = \frac{\pi p r^4}{8kl}$$

where l (in cm) is the length of the arteriole, r (in cm) is its radius, p (in dyne/cm^2) is the difference in pressure between the two ends of the arteriole, and k is the viscosity of blood (in dyne-sec/cm^2). Find the approximate percentage change in the flow of blood if an error of 2% is made in measuring the length of the arteriole and an error of 1% is made in measuring its radius. Assume that p and k are constant.

SOLUTION TO SELF-CHECK EXERCISE 8.6

We find

$$\frac{\partial f}{\partial x} = 3y^2 \quad \text{and} \quad \frac{\partial f}{\partial y} = 6xy - 4$$

so that

$$\frac{\partial f}{\partial x}(-1, 3) = 3(3)^2 = 27$$

and

$$\frac{\partial f}{\partial y}(-1, 3) = 6(-1)(3) - 4 = -22$$

Therefore, the total differential is

$$dz = \frac{\partial f}{\partial x}(-1, 3)\, dx + \frac{\partial f}{\partial y}(-1, 3)\, dy$$

$$= 27\, dx - 22\, dy$$

Now $dx = -0.98 - (-1) = 0.02$ and $dy = 3.01 - 3 = 0.01$, so that the approximate change in z is

$$dz = 27(0.02) - 22(0.01) = 0.32$$

8.7 Double Integrals

A GEOMETRIC INTERPRETATION OF THE DOUBLE INTEGRAL

To introduce the notion of the integral of a function of two variables, let's first recall the definition of the definite integral of a continuous function of one variable $y = f(x)$ over the interval $[a, b]$. We first divide the interval $[a, b]$ into n subintervals, each of equal length, by the points $x_0 = a < x_1 < x_2 < \cdots < x_n = b$ and define the **Riemann sum** by

$$S_n = f(p_1)h + f(p_2)h + \cdots + f(p_n)h$$

where $h = (b - a)/n$ and p_i is an arbitrary point in the interval $[x_{i-1}, x_i]$. The definite integral of f over $[a, b]$ is defined as the limit of the Riemann sum S_n as n tends to infinity, whenever it exists. Furthermore, recall that when f is a nonnegative continuous function on $[a, b]$, then the ith term of the Riemann sum, $f(p_i)h$, is an approximation (by the area of a rectangle) of the area under that part of the graph of $y = f(x)$ between $x = x_{i-1}$ and $x = x_i$, so that the Riemann sum S_n provides us with an approximation of the area under the curve $y = f(x)$ from $x = a$ to $x = b$. The integral

$$\int_a^b f(x)\, dx = \lim_{n \to \infty} S_n$$

gives the *actual* area under the curve from $x = a$ to $x = b$.

FIGURE 8.29
f(x, y) is a function defined over a rectangular region R.

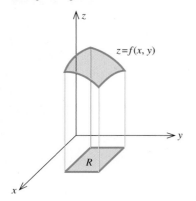

Now suppose $f(x, y)$ is a continuous function of two variables defined over a region R. For simplicity, we assume for the moment that R is a rectangular region in the plane (Figure 8.29). Let's construct a Riemann sum for this function over the rectangle R by following a procedure that parallels the case for a function of one variable over an interval I. We begin by observing that the analogue of a *partition* in the two-dimensional case is a rectangular **grid** composed of mn rectangles, each of length h and width k, as a result of partitioning the side of the rectangle R of length $(b - a)$ into m segments and the side of length $(d - c)$ into n segments. By construction

$$h = \frac{b - a}{m} \quad \text{and} \quad k = \frac{d - c}{n}$$

A sample grid with $m = 5$ and $n = 4$ is shown in Figure 8.30.

Let's label the rectangles $R_1, R_2, R_3, \ldots, R_{mn}$. If (x_i, y_i) is *any* point in R_i ($1 \le i \le mn$), then the **Riemann sum of $f(x, y)$ over the region R** is defined as

$$S(m, n) = f(x_1, y_1)hk + f(x_2, y_2)hk + \cdots + f(x_{mn}, y_{mn})hk$$

If the limit of $S(m, n)$ exists as both m and n tend to infinity, we call this limit the value of the **double integral of $f(x, y)$ over the region R** and denote it by

$$\iint\limits_R f(x, y)\, dA$$

If $f(x, y)$ is a nonnegative function, then the number $f(x_i, y_i)hk$ is the volume of a prism with base R_i and height $f(x_i, y_i)$ that provides us with an approximation of the volume of the solid under the surface $z = f(x, y)$ and bounded below by the rectangular region R_i (Figure 8.31).

FIGURE 8.30
Grid with $m = 5$ and $n = 4$

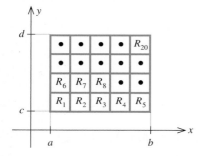

FIGURE 8.31

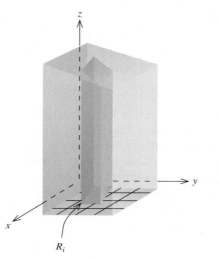

(a) Approximating the volume of a solid with rectangular prisms

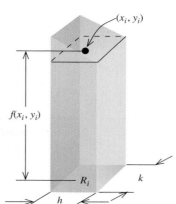

(b) A prism with base R_i and height $f(x_i, y_i)$

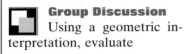

Group Discussion

Using a geometric interpretation, evaluate

$$\iint_R \sqrt{4 - x^2 - y^2}\, dA$$

where $R = \{(x, y)\,|\, x^2 + y^2 \leq 4\}$.

Therefore, the Riemann sum $S(m, n)$ gives us an approximation of the volume of the solid bounded above by the surface $z = f(x, y)$ and below by the plane region R. As both m and n tend to infinity, the Riemann sum $S(m, n)$ approaches the *actual* volume under the solid.

EVALUATING A DOUBLE INTEGRAL OVER A RECTANGULAR REGION

Let's turn our attention to the evaluation of the double integral

$$\iint_R f(x, y)\, dA$$

where R is the rectangular region shown in Figure 8.29. As in the case of the definite integral of a function of one variable, it turns out that the double integral can be evaluated without our having to first find an appropriate Riemann sum and then take the limit of that sum. Instead, as we will now see, the technique calls for evaluating two single integrals—the so-called **iterated integrals**—in succession, using a process that might be called "anti-partial differentiation." The technique is described in the following result, which we state without proof.

Let R be the rectangle defined by the inequalities $a \leq x \leq b$ and $c \leq y \leq d$ (see Figure 8.30). Then,

$$\iint_R f(x, y)\, dA = \int_c^d \left[\int_a^b f(x, y)\, dx \right] dy \tag{14}$$

where the iterated integrals on the right-hand side are evaluated as follows. We first compute the integral

$$\int_a^b f(x, y)\, dx$$

by treating y as if it were a constant and integrating the resulting function of x with respect to x (the "dx" reminds us that we are integrating with respect to x). In this manner we obtain a value for the integral that may contain the variable y. Thus,

$$\int_a^b f(x, y)\, dx = g(y)$$

for some function g. Substituting this value into Equation (14) gives

$$\int_c^d g(y)\,dy$$

which may be integrated in the usual manner.

EXAMPLE 1   Evaluate $\iint_R f(x, y)\,dA$, where $f(x, y) = x + 2y$ and R is the rectangle defined by $1 \le x \le 4$ and $1 \le y \le 2$.

SOLUTION ✔   Using Equation (14), we find

$$\iint_R f(x, y)\,dA = \int_1^2 \left[\int_1^4 (x + 2y)\,dx \right] dy$$

To compute

$$\int_1^4 (x + 2y)\,dx$$

we treat y as if it were a constant (remember that the "dx" reminds us that we are integrating with respect to x). We obtain

$$\int_1^4 (x + 2y)\,dx = \frac{1}{2}x^2 + 2xy \Big|_{x=1}^{x=4}$$

$$= \left[\frac{1}{2}(16) + 2(4)y \right] - \left[\frac{1}{2}(1) + 2(1)y \right]$$

$$= \frac{15}{2} + 6y$$

Thus,

$$\iint_R f(x, y)\,dA = \int_1^2 \left(\frac{15}{2} + 6y \right) dy = \left(\frac{15}{2}y + 3y^2 \right) \Big|_1^2$$

$$= (15 + 12) - \left(\frac{15}{2} + 3 \right) = 16\frac{1}{2} \qquad ■■■■$$

EVALUATING A DOUBLE INTEGRAL OVER A PLANE REGION

Up to now we have assumed that the region over which a double integral is to be evaluated is rectangular. In fact, however, it is possible to compute the double integral of functions over rather arbitrary regions. The next theorem, which we state without proof, expands the number of types of regions over which we may integrate.

THEOREM 1

a. Suppose $g_1(x)$ and $g_2(x)$ are continuous functions on $[a, b]$ and the region R is defined by $R = \{(x, y) \mid g_1(x) \le y \le g_2(x); a \le x \le b\}$. Then,

$$\iint_R f(x, y) \, dA = \int_a^b \left[\int_{g_1(x)}^{g_2(x)} f(x, y) \, dy \right] dx \qquad (15)$$

(Figure 8.32a).

b. Suppose $h_1(y)$ and $h_2(y)$ are continuous functions on $[c, d]$ and the region R is defined by $R = \{(x, y) \mid h_1(y) \le x \le h_2(y); c \le y \le d\}$. Then,

$$\iint_R f(x, y) \, dA = \int_c^d \left[\int_{h_1(y)}^{h_2(y)} f(x, y) \, dx \right] dy \qquad (16)$$

(Figure 8.32b).

FIGURE 8.32

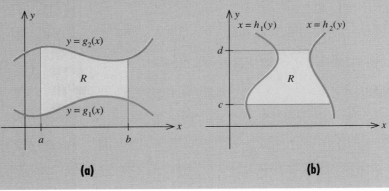

(a)

(b)

REMARK **1.** Observe that in (15) the lower and upper limits of integration with respect to y are given by $y = g_1(x)$ and $y = g_2(x)$. This is to be expected since, for a fixed value of x lying between $x = a$ and $x = b$, y runs between the lower curve defined by $y = g_1(x)$ and the upper curve defined by $y = g_2(x)$ (see Figure 8.32a). Observe, too, that in the special case when $g_1(x) = c$ and $g_2(x) = d$, the region R is rectangular, and (15) reduces to (14).
2. For a fixed value of y, x runs between $x = h_1(y)$ and $x = h_2(y)$, giving the indicated limits of integration with respect to x in (16) (see Figure 8.32b).
3. Note that the two curves in Figure 8.32b are not graphs of functions of x (use the vertical-line test), but they are graphs of functions of y. It is this observation that justifies the approach leading to (16). ■ ■ ■

We now look at several examples.

EXAMPLE **2**

Evaluate $\iint_R f(x, y)\, dA$ given that $f(x, y) = x^2 + y^2$ and R is the region bounded by the graphs of $g_1(x) = x$ and $g_2(x) = 2x$ for $0 \le x \le 2$.

FIGURE 8.33
R is the region bounded by $g_1(x) = x$ and $g_2(x) = 2x$ for $0 \le x \le 2$.

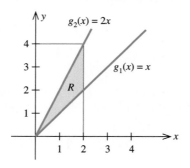

SOLUTION ✔

The region under consideration is shown in Figure 8.33. Using Equation (15), we find

$$\iint_R f(x, y)\, dA = \int_0^2 \left[\int_x^{2x} (x^2 + y^2)\, dy \right] dx$$

$$= \int_0^2 \left[\left(x^2 y + \frac{1}{3} y^3 \right) \Big|_x^{2x} \right] dx$$

$$= \int_0^2 \left[\left(2x^3 + \frac{8}{3} x^3 \right) - \left(x^3 + \frac{1}{3} x^3 \right) \right] dx$$

$$= \int_0^2 \frac{10}{3} x^3\, dx = \frac{5}{6} x^4 \Big|_0^2 = 13\frac{1}{3}$$

■ ■ ■ ■

EXAMPLE **3**

Evaluate $\iint_R f(x, y)\, dA$, where $f(x, y) = xe^y$ and R is the plane region bounded by the graphs of $y = x^2$ and $y = x$.

SOLUTION ✔

FIGURE 8.34
R is the region bounded by $y = x^2$ and $y = x$.

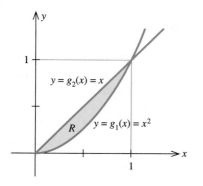

The region in question is shown in Figure 8.34. The point of intersection of the two curves is found by solving the equation $x^2 = x$, giving $x = 0$ and $x = 1$. Using Equation (15), we find

$$\iint_R f(x, y)\, dA = \int_0^1 \left[\int_{x^2}^x xe^y\, dy \right] dx = \int_0^1 \left[xe^y \Big|_{x^2}^x \right] dx$$

$$= \int_0^1 (xe^x - xe^{x^2})\, dx = \int_0^1 xe^x\, dx - \int_0^1 xe^{x^2}\, dx$$

and integrating the first integral on the right-hand side by parts,

$$= \left[(x - 1)e^x - \frac{1}{2} e^{x^2} \right] \Big|_0^1$$

$$= -\frac{1}{2} e - \left(-1 - \frac{1}{2} \right) = \frac{1}{2}(3 - e)$$

■ ■ ■ ■

The next example not only illustrates the use of Equation (16) but also shows that it may be the only viable way to evaluate the given double integral.

EXAMPLE 4   Evaluate

$$\iint_R xe^{y^2}\, dA$$

where R is the plane region bounded by the y-axis, $x = 0$, the horizontal line $y = 4$, and the graph of $y = x^2$.

SOLUTION ✔   The region R is shown in Figure 8.35. The point of intersection of the line $y = 4$ and the graph of $y = x^2$ is found by solving the equation $x^2 = 4$, giving $x = 2$ and the required point $(2, 4)$. Using Equation (15) with $y = g_1(x) = x^2$ and $y = g_2(x) = 4$ leads to

$$\iint_R xe^{y^2}\, dA = \int_0^2 \left[\int_{x^2}^4 xe^{y^2}\, dy \right] dx$$

Now evaluation of the integral

$$\int_{x^2}^4 xe^{y^2}\, dy = x\int_{x^2}^4 e^{y^2}\, dy$$

calls for finding the antiderivative of the integrand e^{y^2} in terms of elementary functions, a task that, as was pointed out in Section 7.3, cannot be done. Let's begin afresh and attempt to make use of Equation (16).

Since the equation $y = x^2$ is equivalent to the equation $x = \sqrt{y}$, which clearly expresses x as a function of y, we may write, with $x = h_1(y) = 0$ and $h_2(y) = \sqrt{y}$,

$$\iint_R xe^{y^2}\, dA = \int_0^4 \left[\int_0^{\sqrt{y}} xe^{y^2}\, dx \right] dy = \int_0^4 \left[\frac{1}{2}x^2 e^{y^2} \Big|_0^{\sqrt{y}} \right] dy$$

$$= \int_0^4 \frac{1}{2} ye^{y^2}\, dy = \frac{1}{4} e^{y^2} \Big|_0^4 = \frac{1}{4}(e^{16} - 1)$$

FIGURE 8.35

R is the region bounded by the y-axis, $x = 0$, $y = 4$, and $y = x^2$.

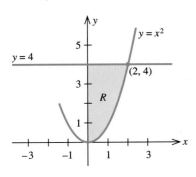

Evaluate $\iint\limits_{R} (x + y)\, dA$, where R is the region bounded by the graphs of $g_1(x) = x$ and $g_2(x) = x^{1/3}$.

The solution to Self-Check Exercise 8.7 can be found on page 677.

8.7 Exercises

In Exercises 1–25, evaluate the double integral $\iint\limits_{R} f(x, y)\, dA$ for the given function $f(x, y)$ and the region R.

1. $f(x, y) = y + 2x$; R is the rectangle defined by $1 \le x \le 2$ and $0 \le y \le 1$.

2. $f(x, y) = x + 2y$; R is the rectangle defined by $-1 \le x \le 2$ and $0 \le y \le 2$.

3. $f(x, y) = xy^2$; R is the rectangle defined by $-1 \le x \le 1$ and $0 \le y \le 1$.

4. $f(x, y) = 12xy^2 + 8y^3$; R is the rectangle defined by $0 \le x \le 1$ and $0 \le y \le 2$.

5. $f(x, y) = \dfrac{x}{y}$; R is the rectangle defined by $-1 \le x \le 2$ and $1 \le y \le e^3$.

6. $f(x, y) = \dfrac{xy}{1 + y^2}$; R is the rectangle defined by $-2 \le x \le 2$ and $0 \le y \le 1$.

7. $f(x, y) = 4xe^{2x^2 + y}$; R is the rectangle defined by $0 \le x \le 1$ and $-2 \le y \le 0$.

8. $f(x, y) = \dfrac{y}{x^2} e^{y/x}$; R is the rectangle defined by $1 \le x \le 2$ and $0 \le y \le 1$.

9. $f(x, y) = \ln y$; R is the rectangle defined by $0 \le x \le 1$ and $1 \le y \le e$.

10. $f(x, y) = \dfrac{\ln y}{x}$; R is the rectangle defined by $1 \le x \le e^2$ and $1 \le y \le e$.

11. $f(x, y) = x + 2y$; R is bounded by $x = 0$, $x = 1$, $y = 0$, and $y = x$.

12. $f(x, y) = xy$; R is bounded by $x = 0$, $x = 1$, $y = 0$, and $y = x$.

13. $f(x, y) = 2x + 4y$; R is bounded by $x = 1$, $x = 3$, $y = 0$, and $y = x + 1$.

14. $f(x, y) = 2 - y$; R is bounded by $x = -1$, $x = 1 - y$, $y = 0$, and $y = 2$.

15. $f(x, y) = x + y$; R is bounded by $x = 0$, $x = \sqrt{y}$, $y = 0$, and $y = 4$.

16. $f(x, y) = x^2 y^2$; R is bounded by $x = 0$, $x = 1$, $y = x^2$, and $y = x^3$.

17. $f(x, y) = y$; R is bounded by $x = 0$, $x = \sqrt{4 - y^2}$, $y = 0$, and $y = 2$.

18. $f(x, y) = \dfrac{y}{x^3 + 2}$; R is bounded by $x = 0$, $x = 1$, $y = 0$, and $y = x$.

19. $f(x, y) = 2xe^y$; R is bounded by $x = 0$, $x = 1$, $y = 0$, and $y = x$.

20. $f(x, y) = 2x$; R is bounded by $x = e^{2y}$, $x = y$, $y = 0$, and $y = 1$.

21. $f(x, y) = ye^x$; R is bounded by $y = \sqrt{x}$ and $y = x$.

22. $f(x, y) = xe^{-y^2}$; R is bounded by $x = 0$, $y = x^2$, and $y = 4$.

23. $f(x, y) = e^{y^2}$; R is bounded by $x = 0$, $x = 1$, $y = 2x$, and $y = 2$.

24. $f(x, y) = y$; R is bounded by $x = 1$, $x = e$, $y = 0$, and $y = \ln x$.

25. $f(x, y) = ye^{x^3}$; R is bounded by $x = y/2$, $x = 1$, $y = 0$, and $y = 2$.

In Exercises 26 and 27, determine whether the statement is true or false. If it is true, explain why it is true. If it is false, give an example to show why it is false.

26. If $h(x, y) = f(x)g(y)$, where f is continuous on $[a, b]$ and g is continuous on $[c, d]$, then $\iint_R h(x, y)\, dA = [\int_a^b f(x)\, dx][\int_c^d g(y)\, dy]$, where $R = \{(x, y)\,|\,a \le x \le b; c \le y \le d\}$.

27. If $\iint_{R_1} f(x, y)\, dA$ exists, where $R_1 = \{(x, y)\,|\,a \le x \le b; c \le y \le d\}$, then $\iint_{R_2} f(x, y)\, dA$ exists, where $R_2 = \{(x, y)\,|\, c \le x \le d; a \le y \le b\}$.

SOLUTIONS TO SELF-CHECK EXERCISES 8.7

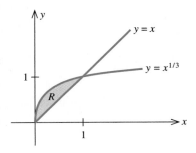

The region R is shown in the figure. The points of intersection of the two curves are found by solving the equation $x = x^{1/3}$, giving $x = 0$ and $x = 1$. Using Equation (15), we find

$$\iint_R (x + y)\, dA = \int_0^1 \left[\int_x^{x^{1/3}} (x + y)\, dy \right] dx$$

$$= \int_0^1 \left[xy + \frac{1}{2}\, y^2 \Big|_x^{x^{1/3}} \right] dx$$

$$= \int_0^1 \left[\left(x^{4/3} + \frac{1}{2}\, x^{2/3} \right) - \left(x^2 + \frac{1}{2}\, x^2 \right) \right] dx$$

$$= \int_0^1 \left(x^{4/3} + \frac{1}{2}\, x^{2/3} - \frac{3}{2}\, x^2 \right) dx$$

$$= \frac{3}{7}\, x^{7/3} + \frac{3}{10}\, x^{5/3} - \frac{1}{2}\, x^3 \Big|_0^1$$

$$= \frac{3}{7} + \frac{3}{10} - \frac{1}{2} = \frac{8}{35}$$

8.8 Applications of Double Integrals

In this section we will give some sample applications involving the double integral.

FINDING THE VOLUME OF A SOLID BY DOUBLE INTEGRALS

As we saw in the last section, the double integral

$$\iint_R f(x, y)\, dA$$

gives the volume of the solid bounded by the graph of $f(x, y)$ over the region R.

The Volume of a Solid Under a Surface

> Let R be a region in the xy-plane and let f be continuous and nonnegative on R. Then, the volume of the solid bounded above by the surface $z = f(x, y)$ and below by R is given by
>
> $$V = \iint_R f(x, y)\, dA$$

EXAMPLE 1

Find the volume of the solid bounded above by the plane $z = f(x, y) = y$ and below by the plane region R defined by $y = \sqrt{1 - x^2}$, $0 \le x \le 1$.

SOLUTION ✓

The region R is sketched in Figure 8.36. Observe that $f(x, y) = y \ge 0$ for $y \in R$. Therefore, the required volume is given by

$$\iint_R y\, dA = \int_0^1 \left[\int_0^{\sqrt{1-x^2}} y\, dy \right] dx = \int_0^1 \left[\frac{1}{2} y^2 \Big|_0^{\sqrt{1-x^2}} \right] dx$$

$$= \int_0^1 \frac{1}{2}(1 - x^2)\, dx = \frac{1}{2}\left(x - \frac{1}{3}x^3 \right) \Big|_0^1 = \frac{1}{3}$$

or $\frac{1}{3}$ cubic unit. The solid is shown in Figure 8.37. Note that it is not necessary to make a sketch of the solid in order to compute its volume.

FIGURE 8.36
The plane region R defined by $y = \sqrt{1 - x^2}$, $0 \le x \le 1$

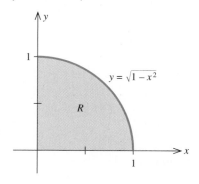

FIGURE 8.37
The solid bounded above by the plane $z = y$ and below by the plane region defined by $y = \sqrt{1 - x^2}$, $0 \le x \le 1$

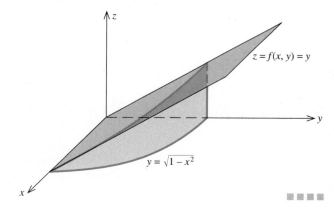

POPULATION OF A CITY

Suppose the plane region R represents a certain district of a city and $f(x, y)$ gives the population density (the number of people per square mile) at any point (x, y) in R. Enclose the set R by a rectangle and construct a grid for it

FIGURE 8.38
The rectangular region R representing a certain district of a city is enclosed by a rectangular grid.

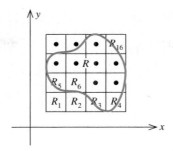

in the usual manner. In any rectangular region of the grid that has no point in common with R, set $f(x_i, y_i)hk = 0$ (Figure 8.38). Then, corresponding to any grid covering the set R, the general term of the Riemann sum $f(x_i, y_i)hk$ (population density times area) gives the number of people living in that part of the city corresponding to the rectangular region R_i. Therefore, the Riemann sum gives an approximation of the number of people living in the district represented by R and, in the limit, the double integral

$$\iint_R f(x, y)\, dA$$

gives the actual number of people living in the district under consideration.

EXAMPLE 2 The population density of a certain city is described by the function

$$f(x, y) = 10{,}000 e^{-0.2|x| - 0.1|y|}$$

where the origin $(0, 0)$ gives the location of the city hall. What is the population inside the rectangular area described by

$$R = \{(x, y) \mid -10 \le x \le 10;\ -5 \le y \le 5\}$$

if x and y are in miles? (See Figure 8.39.)

FIGURE 8.39
The rectangular region R represents a certain district of a city.

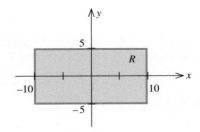

SOLUTION ✔ By symmetry, it suffices to compute the population in the first quadrant. (Why?) Then, upon observing that in this quadrant

$$f(x, y) = 10{,}000 e^{-0.2x - 0.1y} = 10{,}000 e^{-0.2x} e^{-0.1y}$$

we see that the population in R is given by

$$\iint_R f(x, y)\, dA = 4 \int_0^{10} \left[\int_0^5 10{,}000 e^{-0.2x}\, e^{-0.1y}\, dy \right] dx$$

$$= 4 \int_0^{10} \left[-100{,}000 e^{-0.2x} e^{-0.1y} \Big|_0^5 \right] dx$$

$$= 400{,}000 (1 - e^{-0.5}) \int_0^{10} e^{-0.2x}\, dx$$

$$= 2{,}000{,}000 (1 - e^{-0.5})(1 - e^{-2})$$

or approximately 680,438 people.   ■ ■ ■ ■

Group Discussion

1. Consider the improper double integral $\iint_D f(x, y)\, dA$ of the continuous function f of two variables defined over the plane region

$$D = \{(x, y)\mid 0 \le x < \infty;\ 0 \le y < \infty\}$$

Using the definition of improper integrals of functions of one variable (Section 7.4), explain why it makes sense to define

$$\iint_D f(x, y)\, dA = \lim_{N \to \infty} \int_0^N \left[\lim_{M \to \infty} \int_0^M f(x, y)\, dx \right] dy$$

$$= \lim_{M \to \infty} \int_0^M \left[\lim_{N \to \infty} \int_0^N f(x, y)\, dy \right] dx$$

provided the limits exist.

2. Refer to Example 2. Assuming that the population density of the city is described by

$$f(x, y) = 10{,}000 e^{-0.2|x| - 0.1|y|}$$

for $-\infty < x < \infty$ and $-\infty < y < \infty$, show that the population outside the rectangular region

$$R = \{(x, y)\mid -10 < x < 10;\ -5 < y \le 5\}$$

of Example 2 is given by

$$4 \iint_D f(x, y)\, dx\, dy - 680{,}438$$

(recall that 680,438 is the approximate population inside R).

3. Use the results of parts 1 and 2 to determine the population of the city outside the rectangular area R.

Average Value of a Function

In Section 6.5 we showed that the average value of a continuous function $f(x)$ over an interval $[a, b]$ is given by

$$\frac{1}{b-a}\int_a^b f(x)\, dx$$

That is, the average value of a function over $[a, b]$ is the integral of f over $[a, b]$ divided by the length of the interval. An analogous result holds for a function of two variables $f(x, y)$ over a plane region R. To see this, we enclose R by a rectangle and construct a rectangular grid. Let (x_i, y_i) be any point in the rectangle R_i of area hk. Now, the average value of the mn numbers $f(x_1, y_1), f(x_2, y_2), \ldots, f(x_{mn}, y_{mn})$ is given by

$$\frac{f(x_1, y_1) + f(x_2, y_2) + \cdots + f(x_{mn}, y_{mn})}{mn}$$

which can also be written as

$$\frac{hk}{hk}\left[\frac{f(x_1, y_1) + f(x_2, y_2) + \cdots + f(x_{mn}, y_{mn})}{mn}\right]$$

$$= \frac{1}{(mn)hk}[f(x_1, y_1) + f(x_2, y_2) + \cdots + f(x_{mn}, y_{mn})]hk$$

Now the area of R is approximated by the sum of the mn rectangles (*omitting* those having no points in common with R), each of area hk. Note that this is the denominator of the previous expression. Therefore, taking the limit as m and n both tend to infinity, we obtain the following formula for the *average value of $f(x, y)$ over R.*

Average Value of $f(x, y)$ over the Region R

If f is integrable over the plane region R, then its average value over R is given by

$$\frac{\displaystyle\iint_R f(x, y)\, dA}{\text{Area of } R} \quad \text{or} \quad \frac{\displaystyle\iint_R f(x, y)\, dA}{\displaystyle\iint_R dA} \tag{17}$$

REMARK If we let $f(x, y) = 1$ for all (x, y) in R, then

$$\iint_R f(x, y)\, dA = \iint_R dA = \text{Area of } R$$

■ ■ ■

EXAMPLE **3**   Find the average value of the function $f(x, y) = xy$ over the plane region defined by $y = e^x$, $0 \leq x \leq 1$.

SOLUTION ✔   The region R is shown in Figure 8.40. The area of the region R is given by

FIGURE 8.40
The plane region R defined by $y = e^x$, $0 \leq x \leq 1$

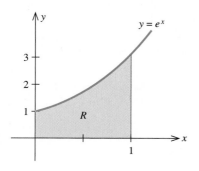

$$\int_0^1 \left[\int_0^{e^x} dy \right] dx = \int_0^1 \left[y \Big|_0^{e^x} \right] dx$$

$$= \int_0^1 e^x \, dx$$

$$= e^x \Big|_0^1$$

$$= e - 1$$

square units. We would obtain the same result had we viewed the area of this region as the area of the region under the curve $y = e^x$ from $x = 0$ to $x = 1$. Next, we compute

$$\iint_R f(x, y) \, dA = \int_0^1 \left[\int_0^{e^x} xy \, dy \right] dx$$

$$= \int_0^1 \left[\frac{1}{2} xy^2 \Big|_0^{e^x} \right] dx$$

$$= \int_0^1 \frac{1}{2} xe^{2x} \, dx$$

$$= \frac{1}{4} xe^{2x} - \frac{1}{8} e^{2x} \Big|_0^1 \qquad \text{(Integrating by parts)}$$

$$= \left(\frac{1}{4} e^2 - \frac{1}{8} e^2 \right) + \frac{1}{8}$$

$$= \frac{1}{8} (e^2 + 1)$$

square units. Therefore, the required average value is given by

$$\frac{\iint_R f(x, y) \, dA}{\iint_R dA} = \frac{\frac{1}{8}(e^2 + 1)}{e - 1} = \frac{e^2 + 1}{8(e - 1)}$$

■ ■ ■ ■

    EXAMPLE **4**   (Refer to Example 2.) The population density of a certain city (number of people per square mile) is described by the function

$$f(x, y) = 10{,}000e^{-0.2|x| - 0.1|y|}$$

where the origin gives the location of the city hall. What is the average population density inside the rectangular area described by

$$R = \{(x, y) \mid -10 \le x \le 10; -5 \le y \le 5\}$$

where x and y are measured in miles?

SOLUTION ✔   From the results of Example 2, we know that

$$\iint\limits_{R} f(x, y)\, dA \approx 680{,}438$$

From Figure 8.39, we see that the area of the plane rectangular region R is $(20)(10)$, or 200, square miles. Therefore, the average population inside R is

$$\frac{\displaystyle\iint\limits_{R} f(x, y)\, dA}{\displaystyle\iint\limits_{R} dA} = \frac{680{,}438}{200} = 3402.19$$

or approximately 3402 people per square mile.   ■ ■ ■ ■

S E L F - C H E C K   E X E R C I S E   8 . 8

The population density of a coastal town located on an island is described by the function

$$f(x, y) = \frac{5000xe^{y}}{1 + 2x^{2}} \qquad (0 \le x \le 4; -2 \le y \le 0)$$

where x and y are measured in miles (see the accompanying figure).

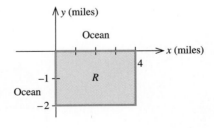

What is the population inside the rectangular area defined by $R = \{(x, y) \mid 0 \le x \le 4; -2 \le y \le 0\}$? What is the average population density in the area?

Solution to Self-Check Exercise 8.8 can be found on page 686.

8.8 Exercises

In Exercises 1–8, use a double integral to find the volume of the solid shown in the figure.

1.

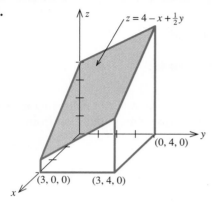

$z = 4 - x + \frac{1}{2}y$

$(0, 4, 0)$
$(3, 0, 0)$ $(3, 4, 0)$

2.

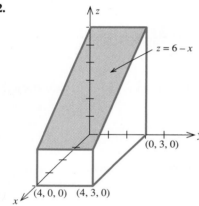

$z = 6 - x$

$(0, 3, 0)$
$(4, 0, 0)$ $(4, 3, 0)$

3.

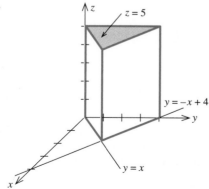

$z = 5$
$y = -x + 4$
$y = x$

4.

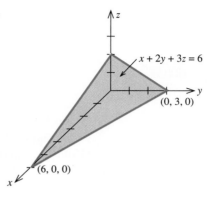

$x + 2y + 3z = 6$
$(0, 3, 0)$
$(6, 0, 0)$

5.

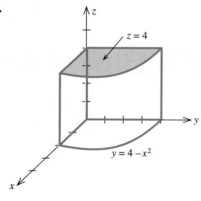

$z = 4$
$y = 4 - x^2$

6.

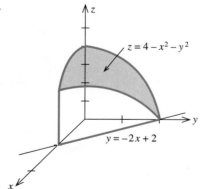

$z = 4 - x^2 - y^2$
$y = -2x + 2$

7.

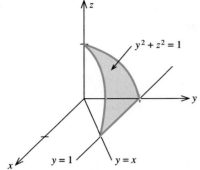

8.

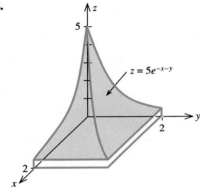

In Exercises 9–16, find the volume of the solid bounded above by the surface $z = f(x, y)$ and below by the plane region R.

9. $f(x, y) = 4 - 2x - y$; $R = \{(x, y) \mid 0 \le x \le 1; 0 \le y \le 2\}$

10. $f(x, y) = 2x + y$; R is the triangle bounded by $y = 2x$, $y = 0$, and $x = 2$.

11. $f(x, y) = x^2 + y^2$; R is the rectangle with vertices $(0, 0)$, $(1, 0)$, $(1, 2)$, and $(0, 2)$.

12. $f(x, y) = e^{x+2y}$; R is the triangle with vertices $(0, 0)$, $(1, 0)$, and $(0, 1)$.

13. $f(x, y) = 2xe^y$; R is the triangle bounded by $y = x$, $y = 2$, and $x = 0$.

14. $f(x, y) = \dfrac{2y}{1 + x^2}$; R is the region bounded by $y = \sqrt{x}$, $y = 0$, and $x = 4$.

15. $f(x, y) = 2x^2y$; R is the region bounded by the graphs of $y = x$ and $y = x^2$.

16. $f(x, y) = x$; R is the region in the first quadrant bounded by the semicircle $y = \sqrt{16 - x^2}$, the x-axis, and the y-axis.

In Exercises 17–22, find the average value of the given function $f(x, y)$ over the plane region R.

17. $f(x, y) = 6x^2y^3$; $R = \{(x, y) \mid 0 \le x \le 2; 0 \le y \le 3\}$

18. $f(x, y) = x + 2y$; R is the triangle with vertices $(0, 0)$, $(1, 0)$, and $(1, 1)$.

19. $f(x, y) = xy$; R is the triangle bounded by $y = x$, $y = 2 - x$, and $y = 0$.

20. $f(x, y) = e^{-x^2}$; R is the triangle with vertices $(0, 0)$, $(1, 0)$, and $(1, 1)$.

21. $f(x, y) = xe^y$; R is the triangle with vertices $(0, 0)$, $(1, 0)$, and $(1, 1)$.

22. $f(x, y) = \ln x$; R is the region bounded by the graphs of $y = 2x$ and $y = 0$ from $x = 1$ to $x = 3$.
Hint: Use integration by parts.

23. POPULATION DENSITY   The population density of a coastal town is described by the function

$$f(x, y) = \frac{10{,}000e^y}{1 + 0.5|x|} \qquad (-10 \le x \le 10; -4 \le y \le 0)$$

where x and y are measured in miles (see the accompanying figure). Find the population inside the rectangular area described by

$$R = \{(x, y) \mid -5 \le x \le 5; -2 \le y \le 0\}$$

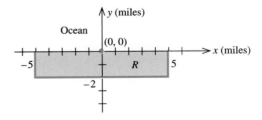

24. AVERAGE POPULATION DENSITY   Refer to Exercise 23. Find the average population density inside the rectangular area R.

25. **AVERAGE PROFIT**   The Country Workshop's total weekly profit (in dollars) realized in manufacturing and selling its rolltop desks is given by the profit function

$$P(x, y) = -0.2x^2 - 0.25y^2 - 0.2xy$$
$$+ 100x + 90y - 4000$$

where x stands for the number of finished units and y stands for the number of unfinished units manufactured and sold per week. Find the average weekly profit if the number of finished units manufactured and sold varies between 180 and 200 and the number of unfinished units varies between 100 and 120/week.

26. **AVERAGE PRICE OF LAND**   The rectangular region R shown in the accompanying figure represents a city's financial district. The price of land in the district is approximated by the function

$$p(x, y) = 200 - 10\left(x - \frac{1}{2}\right)^2 - 15(y - 1)^2$$

where $p(x, y)$ is the price of land at the point (x, y) in dollars per square foot and x and y are measured in miles. What is the average price of land per square foot in the district?

In Exercises 27 and 28, determine whether the statement is true or false. If it is true, explain why it is true. If it is false, give an example to show why it is false.

27. Let R be a region in the xy-plane and let f and g be continuous functions on R that satisfy the condition $f(x, y) \leq g(x, y)$ for all (x, y) in R. Then, $\iint_R g(x, y) - f(x, y)] \, dA$ gives the volume of the solid bounded above by the surface $z = g(x, y)$ and below by the surface $z = f(x, y)$.

28. Suppose f is nonnegative and integrable over the plane region R. Then, the average value of f over R can be thought of as the (constant) height of the cylinder with base R and volume that is exactly equal to the volume of the solid under the graph of $z = f(x, y)$. (*Note:* The cylinder referred to here has sides perpendicular to R.)

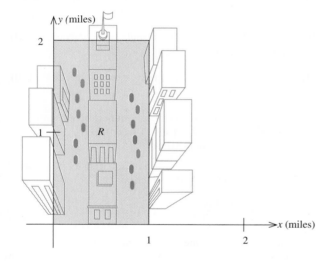

SOLUTION TO SELF-CHECK EXERCISE 8.8

The population in R is given by

$$\iint_R f(x, y) \, dA = \int_0^4 \left[\int_{-2}^0 \frac{5000xe^y}{1 + 2x^2} \, dy \right] dx$$

$$= \int_0^4 \left[\frac{5000xe^y}{1 + 2x^2} \Big|_{-2}^0 \right] dx$$

$$= 5000(1 - e^{-2}) \int_0^4 \frac{x}{1 + 2x^2} \, dx$$

$$= 5000(1 - e^{-2}) \left[\frac{1}{4} \ln(1 + 2x^2) \Big|_0^4 \right]$$

$$= 5000(1 - e^{-2}) \left(\frac{1}{4}\right) \ln 33$$

or approximately 3779 people. The average population density inside R is

$$\frac{\displaystyle\iint_R f(x, y)\, dA}{\displaystyle\iint_R dA} = \frac{3779}{(2)(4)}$$

or approximately 472 people per square mile.

CHAPTER 8 Summary of Principal Terms

Terms

function of two variables	absolute maximum value
domain	absolute minimum
three-dimensional cartesian coordinate system	absolute minimum value
	critical point
level curve	saddle point
first partial derivative	second derivative test
Cobb–Douglas production function	method of least squares
marginal productivity of labor	least-squares line (regression line)
marginal productivity of capital	normal equations
substitute commodities	constrained relative extremum
complementary commodities	method of Lagrange multipliers
second-order partial derivative	total differential
relative maximum	Riemann sum
relative minimum	double integral
absolute maximum	volume of a solid under a surface

CHAPTER 8 REVIEW EXERCISES

1. Let $f(x, y) = \dfrac{xy}{x^2 + y^2}$. Compute $f(0, 1)$, $f(1, 0)$, and $f(1, 1)$. Does $f(0, 0)$ exist?

2. Let $f(x, y) = \dfrac{xe^y}{1 + \ln xy}$. Compute $f(1, 1)$, $f(1, 2)$, and $f(2, 1)$. Does $f(1, 0)$ exist?

3. Let $h(x, y, z) = xye^z + \dfrac{x}{y}$. Compute $h(1, 1, 0)$, $h(-1, 1, 1)$, and $h(1, -1, 1)$.

4. Find the domain of the function $f(u, v) = \dfrac{\sqrt{u}}{u - v}$.

5. Find the domain of the function $f(x, y) = \dfrac{x - y}{x + y}$.

6. Find the domain of the function $f(x, y) = x\sqrt{y} + y\sqrt{1 - x}$.

7. Find the domain of the function $f(x, y, z) = \dfrac{xy\sqrt{z}}{(1 - x)(1 - y)(1 - z)}$.

In Exercises 8–11, sketch the level curves of the function corresponding to the given values of z.

8. $z = f(x, y) = 2x + 3y; z = -2, -1, 0, 1, 2$

9. $z = f(x, y) = y - x^2; z = -2, -1, 0, 1, 2$

10. $z = f(x, y) = \sqrt{x^2 + y^2}; z = 0, 1, 2, 3, 4$

11. $z = f(x, y) = e^{xy}; z = 1, 2, 3$

In Exercises 12–21, compute the first partial derivatives of the function.

12. $f(x, y) = x^2y^3 + 3xy^2 + \dfrac{x}{y}$

13. $f(x, y) = x\sqrt{y} + y\sqrt{x}$

14. $f(u, v) = \sqrt{uv^2 - 2u}$

15. $f(x, y) = \dfrac{x - y}{y + 2x}$　　16. $g(x, y) = \dfrac{xy}{x^2 + y^2}$

17. $h(x, y) = (2xy + 3y^2)^5$　　18. $f(x, y) = (xe^y + 1)^{1/2}$

19. $f(x, y) = (x^2 + y^2)e^{x^2+y^2}$

20. $f(x, y) = \ln(1 + 2x^2 + 4y^4)$

21. $f(x, y) = \ln\left(1 + \dfrac{x^2}{y^2}\right)$

In Exercises 22–27, compute the second-order partial derivatives of the function.

22. $f(x, y) = x^3 - 2x^2y + y^2 + x - 2y$

23. $f(x, y) = x^4 + 2x^2y^2 - y^4$

24. $f(x, y) = (2x^2 + 3y^2)^3$　　25. $g(x, y) = \dfrac{x}{x + y^2}$

26. $g(x, y) = e^{x^2+y^2}$　　27. $h(s, t) = \ln\left(\dfrac{s}{t}\right)$

28. Let $f(x, y, z) = x^3y^2z + xy^2z + 3xy - 4z$. Compute $f_x(1, 1, 0), f_y(1, 1, 0),$ and $f_z(1, 1, 0)$ and interpret your results.

In Exercises 29–34, find the critical point(s) of the functions. Then use the second derivative test to classify the nature of each of these points, if possible. Finally, determine the relative extrema of each function.

29. $f(x, y) = 2x^2 + y^2 - 8x - 6y + 4$

30. $f(x, y) = x^2 + 3xy + y^2 - 10x - 20y + 12$

31. $f(x, y) = x^3 - 3xy + y^2$

32. $f(x, y) = x^3 + y^2 - 4xy + 17x - 10y + 8$

33. $f(x, y) = e^{2x^2+y^2}$

34. $f(x, y) = \ln(x^2 + y^2 - 2x - 2y + 4)$

In Exercises 35–38, use the method of Lagrange multipliers to optimize the function subject to the given constraints.

35. Maximize the function $f(x, y) = -3x^2 - y^2 + 2xy$ subject to the constraint $2x + y = 4$.

36. Minimize the function $f(x, y) = 2x^2 + 3y^2 - 6xy + 4x - 9y + 10$ subject to the constraint $x + y = 1$.

37. Find the maximum and minimum values of the function $f(x, y) = 2x - 3y + 1$ subject to the constraint $2x^2 + 3y^2 - 125 = 0$.

38. Find the maximum and minimum values of the function $f(x, y) = e^{x-y}$ subject to the constraint $x^2 + y^2 = 1$.

In Exercises 39 and 40, find the total differential of each function at the given point.

39. $f(x, y) = (x^2 + y^4)^{3/2}, (3, 2)$

40. $f(x, y) = xe^{x-y} + x \ln y; (1, 1)$

In Exercises 41 and 42, find the approximate change in z when the point (x, y) changes from (x_0, y_0) to (x_1, y_1).

41. $f(x, y) = 2x^2y^3 + 3y^2x^2 - 2xy$; from $(1, -1)$ to $(1.02, -0.98)$

42. $f(x, y) = 4x^{3/4}y^{1/4}$; from $(16, 81)$ to $(17, 80)$

In Exercises 43–46, evaluate the double integrals.

43. $f(x, y) = 3x - 2y$; R is the rectangle defined by $2 \le x \le 4$ and $-1 \le y \le 2$.

44. $f(x, y) = e^{-x-2y}$; R is the rectangle defined by $0 \le x \le 2$ and $0 \le y \le 1$.

45. $f(x, y) = 2x^2y$; R is bounded by $x = 0, x = 1, y = x^2,$ and $y = x^3$.

46. $f(x, y) = \dfrac{y}{x}$; R is bounded by $x = 1, x = 2, y = 1,$ and $y = x$.

FIGURE 9.2

Two possible solutions of a differential equation, each of which describes a logistic function

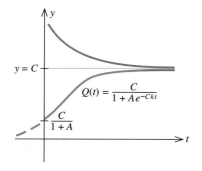

where k is a constant of proportionality. Observe that when $Q(t)$ is small relative to C, the rate of growth of Q is approximately proportional to Q. But as $Q(t)$ approaches C, the growth rate slows down to zero. Next, if $Q > C$, then $dQ/dt < 0$ and the quantity is decreasing with time, with the decay rate slowing down as Q approaches C. We will show later that the solution of the differential Equation (3) is just the logistic function we discussed in Chapter 5. Its graph is shown in Figure 9.2.

STIMULUS RESPONSE In the quantitative theory of psychology, one model that describes the relationship between a stimulus S and the resulting response R is the Weber–Fechner law. This law asserts that the rate of change of a reaction R is inversely proportional to the stimulus S. Mathematically, this law may be expressed as

$$\frac{dR}{dS} = \frac{k}{S} \tag{4}$$

where k is a constant of proportionality. Furthermore, suppose that the threshold level, the lowest level of stimulation at which sensation is detected, is S_0. Then we have the condition $R = 0$ when $S = S_0$; that is, $R(S_0) = 0$. The graph of R versus S is shown in Figure 9.3.

FIGURE 9.3

R, the solution to a differential equation, describes the response to a stimulus.

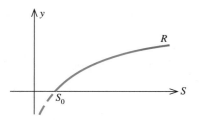

MIXTURE PROBLEMS Our next example is a typical mixture problem. Suppose a tank initially contains 10 gallons of pure water. Brine containing 3 pounds of salt per gallon flows into the tank at a rate of 2 gallons per minute, and the well-stirred mixture flows out of the tank at the same rate. How much salt is in the tank at any given time?

Let's formulate this problem mathematically. Suppose $A(t)$ denotes the amount of salt in the tank at any time t. Then the derivative dA/dt, the rate of change of the amount of salt at any time t, must satisfy the condition

$$\frac{dA}{dt} = (\text{Rate of salt flowing in}) - (\text{Rate of salt flowing out})$$

FIGURE 9.4

The rate of change of the amount of salt at time t = (Rate of salt flowing in) − (Rate of salt flowing out)

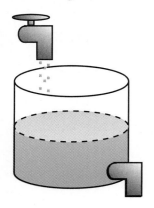

(Figure 9.4). But the rate at which salt flows into the tank is given by

$$(2 \text{ gal/min})(3 \text{ lb/gal}) \qquad [(\text{Rate of flow}) \times (\text{Concentration})]$$

or 6 pounds per minute. Since the rate at which the solution leaves the tank is the same as the rate at which the brine is poured into it, that tank contains 10 gallons of the mixture at any time t. Since the salt content at any time t is A pounds, the concentration of the mixture is $(A/10)$ pounds per gallon. Therefore, the rate at which salt flows out of the tank is given by

$$(2 \text{ gal/min}) \left(\frac{A}{10} \text{ lb/gal} \right)$$

or $(A/5)$ pounds per minute. Therefore, we are led to the differential equation

$$\frac{dA}{dt} = 6 - \frac{A}{5} \tag{5}$$

An additional condition arises from the fact that initially there is no salt in the solution. This condition may be expressed mathematically as $A = 0$ when $t = 0$ or, more concisely, $A(0) = 0$.

We will solve each of the differential equations we have introduced here in Section 9.3.

SOLUTIONS OF DIFFERENTIAL EQUATIONS

Suppose we are given a differential equation involving the derivative(s) of a function y. Recall that a **solution** to a differential equation is any function $f(x)$ that satisfies the differential equation. Thus, $y = f(x)$ is a solution of the differential equation, provided that the replacement of y and its derivative(s) by the function $f(x)$ and its corresponding derivatives reduces the given differential equation to an identity for all values of x.

EXAMPLE 1   Show that the function $f(x) = e^{-x} + x - 1$ is a solution of the differential equation

$$y' + y = x$$

SOLUTION ✔   Let

$$y = f(x) = e^{-x} + x - 1$$

so that

$$y' = f'(x) = -e^{-x} + 1$$

Substituting this last equation into the left side of the given equation yields

$$\overbrace{(-e^{-x} + 1)}^{y'} + \overbrace{(e^{-x} + x - 1)}^{y} = -e^{-x} + 1 + e^{-x} + x - 1 = x$$

which is equal to the right side of the given equation for all values of x. Therefore, $f(x) = e^{-x} + x - 1$ is a solution of the given differential equation.

■ ■ ■ ■

In the preceding example, we verified that $y = e^{-x} + x - 1$ is a solution of the differential equation $y' + y = x$. This is by no means the only solution of the differential equation, as the next example shows.

EXAMPLE 2   Show that any function of the form $f(x) = ce^{-x} + x - 1$, where c is a constant, is a solution of the differential equation

$$y' + y = x$$

SOLUTION ✔   Let

$$y = f(x) = ce^{-x} + x - 1$$

so that

$$y' = f'(x) = -ce^{-x} + 1$$

Substituting the last equation into the left side of the given differential equation yields

$$\overbrace{-ce^{-x} + 1}^{y'} + \overbrace{ce^{-x} + x - 1}^{y} = x$$

and we have verified the assertion.   ■ ■ ■ ■

It can be shown that *every* solution of the differential equation $y' + y = x$ must have the form $y = ce^{-x} + x - 1$, where c is a constant; therefore, this is the **general solution** of the differential equation $y' + y = x$. Figure 9.5 shows a family of solutions of this differential equation for selected values of c.

FIGURE 9.5
Some solutions of $y' + y = x$

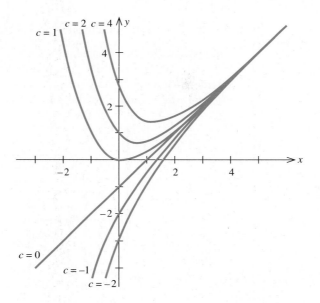

Recall that a solution obtained by assigning a specific value to the constant c is called a **particular solution** of the differential equation. For example, the particular solution $y = e^{-x} + x - 1$ of Example 1 is obtained from the general solution by taking $c = 1$. In practice, a particular solution of a differential equation is obtained from the general solution of the differential equation by requiring that the solution and/or its derivative(s) satisfy certain conditions at one or more values of x.

EXAMPLE 3   Use the results of Example 2 to find the particular solution of the equation $y' + y = x$ that satisfies the condition $y(0) = 0$; that is, $f(0) = 0$, where f denotes the solution.

SOLUTION ✔   From the results of Example 2, we see that the general solution of the given differential equation is given by

$$y = f(x) = ce^{-x} + x - 1$$

Using the given condition, we see that

$$f(0) = ce^0 + 0 - 1 = c - 1 = 0 \qquad \text{or} \qquad c = 1$$

Therefore, the required particular solution is $y = e^{-x} + x - 1$.   ■ ■ ■ ■

Group Discussion

Consider the differential equation $\dfrac{dy}{dx} = F(x, y)$ and suppose $y = f(x)$ is a solution of the differential equation.

1. If (a, b) is a point in the domain of F, explain why $F(a, b)$ gives the slope of f at $x = a$.

2. For the differential equation $\dfrac{dy}{dx} = \dfrac{x}{y}$, compute $F(x, y)$ for selected integral values of x and y. (For example, try $x = \pm 2, \pm 1, 0, \pm 1, \pm 2$ and $y = \pm 2, \pm 1, \pm 2, \pm 3$.) Verify that if you draw a lineal element (a tiny line segment) having slope $F(x, y)$ through each point (x, y), you obtain a *direction field* similar to the one shown in the figure:

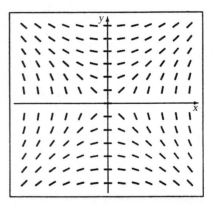

3. The direction field associated with the differential equation hints at the solution curves for the differential equation. Sketch a few solution curves for the differential equation. (You will be asked to verify your answer to part (3) in the next section.)

SELF-CHECK EXERCISES 9.1

1. Consider the differential equation

$$xy' + 2y = 4x^2$$

   a. Show that $y = x^2 + (c/x^2)$ is the general solution of the differential equation.

   b. Find the particular solution of the differential equation that satisfies $y(1) = 4$.

2. The population of a certain species grows at a rate directly proportional to the square root of its size. If the initial population is N_0, find the population at any time t. Formulate but do not solve the problem.

Solutions to Self-Check Exercises 9.1 can be found on page 699.

9.1 Exercises

In Exercises 1–12, verify that y is a solution of the given differential equation.

1. $y = x^2$; $xy' + y = 3x^2$

2. $y = e^x$; $y' - y = 0$

3. $y = \dfrac{1}{2} + ce^{-x^2}$, c any constant; $y' + 2xy = x$

4. $y = Ce^{kx}$, C any constant; $\dfrac{dy}{dx} = ky$

5. $y = e^{-2x}$; $y'' + y' - 2y = 0$

6. $y = C_1e^x + C_2e^{2x}$; $y'' - 3y' + 2y = 0$

7. $y = C_1e^{-2x} + C_2xe^{-2x}$; $y'' + 4y' + 4y = 0$

8. $y = C_1 + C_2x^{1/3}$; $3xy'' + 2y' = 0$

9. $y = \dfrac{C_1}{x} + C_2\dfrac{\ln x}{x}$; $x^2y'' + 3xy' + y = 0$

10. $y = C_1e^x + C_2xe^x + C_3x^2e^x$; $y''' - 3y'' + 3y' - y = 0$

11. $y = C - Ae^{-kt}$, A and C constants; $\dfrac{dy}{dt} = k(C - y)$

12. $y = \dfrac{C}{1 + Ae^{-Ckt}}$, A and C constants; $\dfrac{dy}{dt} = ky(C - y)$

In Exercises 13–18, verify that y is a general solution of the differential equation. Then find a particular solution of the differential equation that satisfies the side condition.

13. $y = Cx^2 - 2x$; $y' - 2\left(\dfrac{y}{x}\right) = 2$; $y(1) = 10$

14. $y = Ce^{-x^2}$; $y' = -2xy$; $y(0) = y_0$

15. $y = \dfrac{C}{x}$; $y' + \left(\dfrac{1}{x}\right)y = 0$; $y(1) = 1$

16. $y = Ce^{2x} - 2x - 1$; $y' - 2y - 4x = 0$; $y(0) = 3$

17. $y = \dfrac{Ce^x}{x} + \dfrac{1}{2}xe^x$; $y' + \left(\dfrac{1-x}{x}\right)y = e^x$; $y(1) = -\dfrac{1}{2}e$

18. $y = C_1x^3 + C_2x^2$; $x^2y'' - 4xy' + 6y = 0$; $y(2) = 0$ and $y'(2) = 4$

19. RADIOACTIVE DECAY A radioactive substance decays at a rate directly proportional to the amount present. If the substance is present in the amount of Q_0 g initially ($t = 0$), find the amount present at some later time. Formulate the problem in terms of a differential equation with a side condition. Do not solve it.

20. SUPPLY AND DEMAND Let $S(t)$ denote the supply of a certain commodity as a function of time t. Suppose that the rate of change of the supply is proportional to the difference between the demand $D(t)$ and the supply. Find a differential equation that describes this situation.

21. NET INVESTMENT The management of a company has decided that the level of investment should not exceed C dollars. Furthermore, management has decided that the rate of net investment (the rate of change of the total capital invested) should be proportional to the difference between C and the total capital invested. Formulate but do not solve the problem in terms of a differential equation.

22. LAMBERT'S LAW OF ABSORPTION Lambert's law of absorption states that the percentage of incident light L, absorbed in passing through a thin layer of material x, is proportional to the thickness of the material. If, for a certain material, x_0 in. of the material reduces the light to half its intensity, how much additional material is needed to reduce the intensity to a quarter of its initial value? Formulate but do not solve the problem in terms of a differential equation with a side condition.

23. CONCENTRATION OF A DRUG IN THE BLOODSTREAM The rate at which the concentration of a drug in the bloodstream decreases is proportional to the concentration at any time t. Initially, the concentration of the drug in the bloodstream is C_0 g/mL. What is the concentration of the drug in the bloodstream at any time t? Formulate but do not solve the problem in terms of a differential equation with a side condition.

24. **AMOUNT OF GLUCOSE IN THE BLOODSTREAM** Suppose glucose is infused into the bloodstream at a constant rate of C g/min and, at the same time, the glucose is converted and removed from the bloodstream at a rate proportional to the amount of glucose present. Show that the amount of glucose $A(t)$ present in the bloodstream at any time t is governed by the differential equation

$$A' = C - kA$$

where k is a constant.

25. **NEWTON'S LAW OF COOLING** Newton's law of cooling states that the temperature of a body drops at a rate that is proportional to the difference between the temperature y of the body and the constant temperature C of the surrounding medium (assume that the temperature of the body is initially greater than C). Show that Newton's law of cooling may be expressed as the differential equation

$$\frac{dy}{dt} = -k(y - C), \quad y(0) = y_0$$

where y_0 denotes the temperature of the body before immersion in the medium.

26. **FISK'S LAW** Suppose a cell of volume V cc is surrounded by a homogeneous chemical solution of concentration C g/cc. Let y denote the concentration of the solute inside the cell at any time t and suppose that, initially, the concentration is y_0. Fisk's law, named after the German physiologist Adolf Fisk (1829–1901), states that the rate of change of the concentration of solute inside the cell at any time t is proportional to the difference between the concentration of the solute outside the cell and the concentration inside the cell and inversely proportional to the volume of the cell. Show that Fisk's law may be expressed as the differential equation

$$\frac{dy}{dt} = \frac{k}{V}(C - y), \quad y(0) = y_0$$

where k is a constant. (*Note:* The constant of proportionality k depends on the area and permeability of the cell membrane.)

27. **ALLOMETRIC LAWS** Suppose $x(t)$ denotes the weight of an animal's organ at time t and $g(t)$ denotes the size of another organ in the same animal at the same time t.

An allometric law (allometry is the study of the relative growth of a part in relation to an entire organism) states that the relative growth rate of one organ, $(dx/dt)/x$, is proportional to the relative growth rate of the other, $(dy/dt)/y$. Show that this allometric law may be stated in terms of the differential equation

$$\frac{1}{x}\frac{dx}{dt} = k\frac{1}{y}\frac{dy}{dt}$$

where k is a constant.

28. **GOMPERTZ GROWTH CURVE** Suppose a quantity $Q(t)$ does not exceed some number C; that is, $Q(t) \leq C$ for all t. Suppose further that the rate of growth of $Q(t)$ is jointly proportional to its current size and the difference between its upper bound and the natural logarithm of its current size. What is the size of the quantity $Q(t)$ at any time t? Show that the mathematical formulation of this problem leads to the differential equation

$$\frac{dQ}{dt} = kQ(C - \ln Q), \quad Q(0) = Q_0$$

where Q_0 denotes the size of the quantity present initially. The graph of $Q(t)$ is called the *Gompertz growth curve*. This model, like the ones leading to the learning curve and the logistic curve, describes restricted growth.

In Exercises 29–34, determine whether the statement is true or false. If it is true, explain why it is true. If it is false, give an example to show why it is false.

29. The function $f(x) = x^2 + 2x + \dfrac{1}{x}$ is a solution of the differential equation $xy' + y = 3x^2 + 4x$.

30. The function $f(x) = \dfrac{1}{4}e^{3x} + ce^{-x}$ is a solution of the differential equation $y' + y = e^{3x}$.

31. The function $f(x) = 2 + ce^{-x^3}$ is a solution of the differential equation $y' + 3x^2y = x^2$.

32. The function $f(x) = 1 + cx^{-2}$ is a solution of the differential equation $xy' + 2y = 3$.

33. If $y = f(x)$ is a solution of a first-order differential equation, then $y = Cf(x)$ is also a solution.

34. If $y = f(x)$ is a solution of a first-order differential equation, then $y = f(x) + C$ is also a solution.

SOLUTIONS TO SELF-CHECK EXERCISES 9.1

1. **a.** We compute

$$y' = 2x - \frac{2c}{x^3}$$

Substituting this into the left side of the given differential equation gives

$$x\left(2x - \frac{2c}{x^3}\right) + 2\left(x^2 + \frac{c}{x^2}\right) = 2x^2 - \frac{2c}{x^2} + 2x^2 + \frac{2c}{x^2} = 4x^2$$

which equals the expression on the right side of the differential equation, and this verifies the assertion.

b. Using the given condition, we have

$$4 = 1^2 + \frac{c}{1^2} \qquad \text{or} \qquad c = 3$$

and the required particular solution is

$$y = x^2 + \frac{3}{x^2}$$

2. Let N denote the size of the population at any time t. Then the required differential equation is

$$\frac{dN}{dt} = kN^{1/2}$$

and the initial condition is $N(0) = N_0$.

9.2 Separation of Variables

THE METHOD OF SEPARATION OF VARIABLES

Differential equations are classified according to their basic form. A compelling reason for this categorization is that different methods are used to solve different types of equations.

A differential equation may be classified by the order of its derivative. A differential equation is of **order n** if the highest derivative of the unknown function appearing in the equation is of order n. For example, the differential equations

$$y' = xe^x \qquad \text{and} \qquad y' + 2y = x^2$$

are **first-order equations,** whereas the differential equation

$$\frac{d^2y}{dt^2} + \left(\frac{dy}{dt}\right)^3 + ty - 8 = 0$$

is a second-order equation. For the remainder of this chapter, we restrict our study to first-order differential equations.

In this section we describe a method for solving an important class of first-order differential equations that can be written in the form

$$\frac{dy}{dx} = f(x)g(y)$$

where $f(x)$ is a function of x only and $g(y)$ is a function of y only. Such differential equations are said to be **separable** because the variables can be separated. Equations (1) through (5) are first-order separable differential equations. As another example, the equation

$$\frac{dQ}{dt} = kQ(C - Q)$$

has the form $dQ/dt = f(t)g(Q)$, where $f(t) = k$ and $g(Q) = Q(C - Q)$, and so is separable. On the other hand, the differential equation

$$\frac{dy}{dx} = xy^2 + 2$$

is *not* separable.

Separable first-order equations can be solved using the *method of separation of variables*.

Method of Separation of Variables

Suppose we are given a first-order separable differential equation in the form

$$\frac{dy}{dx} = f(x)g(y) \tag{6}$$

Step 1 Write Equation (6) in the form

$$\frac{dy}{g(y)} = f(x)\, dx \tag{7}$$

When written in this form, the variables in (7) are said to be *separated*.
Step 2 Integrate each side of Equation (7) with respect to the appropriate variable.

We will justify this method at the end of this section.

SOLVING SEPARABLE DIFFERENTIAL EQUATIONS

EXAMPLE 1   Find the general solution of the first-order differential equation

$$y' = \frac{xy}{x^2 + 1}$$

SOLUTION ✔   **Step 1**   Observe that the given differential equation has the form

$$\frac{dy}{dx} = \left(\frac{x}{x^2 + 1}\right) y = f(x)g(y)$$

where $f(x) = x/(x^2 + 1)$ and $g(y) = y$, and is therefore separable. Separating the variables, we obtain

$$\frac{dy}{y} = \left(\frac{x}{x^2 + 1}\right) dx$$

Step 2   Integrating each side of the last equation with respect to the appropriate variable, we have

$$\int \frac{dy}{y} = \int \frac{x}{x^2 + 1} dx$$

or
$$\ln|y| + C_1 = \frac{1}{2}\ln(x^2 + 1) + C_2$$

$$\ln|y| = \frac{1}{2}\ln(x^2 + 1) + C_2 - C_1$$

where C_1 and C_2 are arbitrary constants of integration. Letting C denote the constant such that $C_2 - C_1 = \ln|C|$, we have

$$\ln|y| = \frac{1}{2}\ln(x^2 + 1) + \ln|C|$$

$$= \ln \sqrt{x^2 + 1} + \ln|C|$$

$$= \ln|C\sqrt{x^2 + 1}| \qquad (\ln A + \ln B = \ln AB)$$

so the general solution is

$$y = C\sqrt{x^2 + 1} \qquad\qquad ■ ■ ■ ■$$

Exploring with Technology

Refer to Example 1, where it was shown that the general solution of the given differential equation is $y = C\sqrt{x^2 + 1}$. Use a graphing utility to plot the graphs of the members of this family of solutions corresponding to $C = -3, -2, -1, 0, 1, 2$, and 3. Use the standard viewing rectangle.

EXAMPLE 2   Find the particular solution of the differential equation

$$ye^x + (y^2 - 1)y' = 0$$

that satisfies the condition $y(0) = 1$.

SOLUTION ✔   **Step 1**   Writing the given differential equation in the form

$$ye^x + (y^2 - 1)\frac{dy}{dx} = 0 \qquad \text{or} \qquad (y^2 - 1)\frac{dy}{dx} = -ye^x$$

and separating the variables, we obtain

$$\frac{y^2 - 1}{y}\,dy = -e^x\,dx$$

Step 2   Integrating each side of this equation with respect to the appropriate variable, we have

$$\int \frac{y^2 - 1}{y}\,dy = -\int e^x\,dx$$

$$\int \left(y - \frac{1}{y}\right) dy = -\int e^x\,dx$$

$$\frac{1}{2}y^2 - \ln|y| = -e^x + C_1$$

$$y^2 - \ln y^2 = -2e^x + C \qquad (C = 2C_1)$$

Using the condition $y(0) = 1$, we have

$$1 - \ln 1 = -2 + C \qquad \text{or} \qquad C = 3$$

Therefore, the required solution is given by

$$y^2 - \ln y^2 = -2e^x + 3 \qquad\qquad ■■■■$$

Example 2 is an initial value problem. In general, an **initial value problem** is a differential equation with enough conditions specified at a point to determine a particular (unique) solution.* Also observe that the solution of Example 2 appeared as an implicit equation involving x and y. This often happens when we solve separable differential equations.

EXAMPLE 3   Find an equation describing f given that (1) the slope of the tangent line to the graph of f at any point $P(x, y)$ is given by the expression $-x/(2y)$ and (2) the graph of f passes through the point $P(1, 2)$.

SOLUTION ✔   The slope of the tangent line to the graph of f at any point $P(x, y)$ is given by the derivative

$$y' = \frac{dy}{dx} = -\frac{x}{2y}$$

which is a separable first-order differential equation. Separating the variables, we obtain

$$2y\,dy = -x\,dx$$

---

* One condition is sufficient for a first-order differential equation.

which, upon integration, yields

$$y^2 = -\frac{1}{2}x^2 + C_1$$

or $$x^2 + 2y^2 = C \qquad (C = 2C_1)$$

where C is an arbitrary constant.

To evaluate C, we use the second condition, which implies that when $x = 1$, $y = 2$. This gives

$$1^2 + 2(2^2) = C \qquad \text{or} \qquad C = 9$$

Hence, the required equation is

$$x^2 + 2y^2 = 9$$

The graph of the equation f appears in Figure 9.6.   ■ ■ ■ ■

FIGURE 9.6
The graph of $x^2 + 2y^2 = 9$

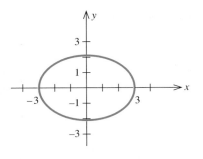

Group Discussion
Refer to the Group Discussion Problem on page 696. Use the method of separation of variables to solve the differential equation $\dfrac{dy}{dx} = \dfrac{x}{y}$ and thus verify that your solution to part 3 in the problem is indeed correct.

JUSTIFICATION OF THE METHOD OF SEPARATION OF VARIABLES

To justify the method of separation of variables, let's consider the separable Equation (6) in its general form:

$$\frac{dy}{dx} = f(x)g(y)$$

If $g(y) \neq 0$, we may rewrite the equation in the form

$$\frac{1}{g(y)}\frac{dy}{dx} - f(x) = 0$$

Now, suppose that G is an antiderivative of $1/g$ and F is an antiderivative of f. Using the chain rule, we see that

$$\frac{d}{dx}[G(y) - F(x)] = G'(y)\frac{dy}{dx} - F'(x) = \frac{1}{g(y)}\frac{dy}{dx} - f(x)$$

Therefore,

$$\frac{d}{dx}[G(y) - F(x)] = 0$$

and so $$G(y) - F(x) = C \qquad (C, \text{a constant})$$

But the last equation is equivalent to

$$G(y) = F(x) + C$$

or $$\int \frac{dy}{g(y)} = \int f(x)\,dx$$

which is precisely the result of step 2 in the method of separation of variables.

SELF-CHECK EXERCISE 9.3

Newton's law of cooling states that the temperature of an object drops at a rate that is proportional to the difference in the temperature between the object and that of the surrounding medium. Suppose that an apple pie is taken out of the oven at a temperature of 200°F and placed on the counter in a room where the temperature is 70°F. If the temperature of the apple pie is 150°F after 5 min, find its temperature $y(t)$ as a function of time t.

The solution to Self-Check Exercise 9.3 can be found on page 715.

9.3 Exercises

A calculator is recommended for these exercises.

1. **CHEMICAL DECOMPOSITION** The rate of decomposition of a certain chemical substance is directly proportional to the amount present at any time t. If y_0 g of the chemical are present at time $t = 0$, find an expression for the amount present at any time t.

2. **GROWTH OF BACTERIA** Under ideal laboratory conditions, the rate of growth of bacteria in a culture is proportional to the size of the culture at any time t. Suppose that 2000 bacteria are present initially in the culture and 5000 are present 1 hr later. How many bacteria will be in the culture at the end of 2 hr?

3. **WORLD POPULATION GROWTH** The world population at the beginning of 1980 was 4.5 billion. Assuming that the population continues to grow at its present rate of approximately 2%/year, find a function $Q(t)$ that expresses the world population (in billions) as a function of time t (in years). What will be the world population at the beginning of 2005?

4. **POPULATION GROWTH** The population of a certain community is increasing at a rate directly proportional to the population at any time t. In the last 3 yr, the population has doubled. How long will it take for the population to triple?

5. **LAMBERT'S LAW OF ABSORPTION** According to Lambert's law of absorption, the percentage of incident light L, absorbed in passing through a thin layer of material x, is proportional to the thickness of the material. For a certain material, if $\frac{1}{2}$ in. of the material reduces the light to half of its intensity, how much additional material is needed to reduce the intensity to one-fourth of its initial value?

6. **SAVINGS ACCOUNTS** An amount of money deposited in a savings account grows at a rate proportional to the amount present. (It can be shown that an amount of money grows in this manner if it earns interest compounded continuously.) Suppose $10,000 is deposited in a fixed account earning interest at the rate of 10%/year compounded continuously.
   a. What is the accumulated amount after 5 yr?
   b. How long does it take for the original deposit to double in value?

7. **CHEMICAL REACTIONS** In a certain chemical reaction, a substance is converted into another substance at a rate proportional to the square of the amount of the first substance present at any time t. Initially ($t = 0$) 50 g of the first substance was present; 1 hr later, only 10 g of it remained. Find an expression that gives the amount of the first substance present at any time t. What is the amount present after 2 hr?

8. **NEWTON'S LAW OF COOLING** Newton's law of cooling states that the rate at which the temperature of an object changes is directly proportional to the difference in temperature between the object and that of the surrounding medium. A horseshoe heated to a temperature of 100°C is immersed in a large tank of water at a (constant) temperature at 30°C at time $t = 0$. Three minutes later the temperature of the horseshoe is reduced to 70°C. Derive an expression that gives the temperature of the horseshoe at any time t. What is the temperature of the horseshoe 5 min after it has been immersed in the water?

9. **NEWTON'S LAW OF COOLING** Newton's law of cooling states that the rate at which the temperature of an object changes is directly proportional to the difference in temperature between the object and that of the surrounding medium. A cup of coffee is prepared with boiling water (212°F) and left to cool on the counter in a room where the temperature is 72°F. If the temperature of the coffee is 140°F after 2 min, determine when the coffee will be cool enough to drink (say, 110°F).

10. **LEARNING CURVES** The American Stenographic Institute finds that the average student taking Elementary Shorthand will progress at a rate given by

$$\frac{dQ}{dt} = k(80 - Q)$$

in a 20-wk course, where $Q(t)$ measures the number of words of dictation a student can take per minute after t wk in the course. If the average student can take 50 words of dictation per minute after 10 wk in the course, how many words per minute can the average student take after completing the course?

11. **TRAINING PERSONNEL** The personnel manager of the Gibraltar Insurance Company estimates that the number of insurance claims an experienced clerk can process in a day is 40. Furthermore, the rate at which a clerk can process insurance claims during the tth wk of training is proportional to the difference between the maximum number possible (40) and the number he or she can process in the tth wk. If the number of claims the average trainee can process after 2 wk on the job is 10/day, determine how many claims the average trainee can process after 6 wk on the job.

12. **EFFECT OF IMMIGRATION ON POPULATION GROWTH** Suppose a country's population at any time t grows in accordance with the rule

$$\frac{dP}{dt} = kP + I$$

where P denotes the population at any time t, k is a positive constant reflecting the natural growth rate of the population, and I is a constant giving the (constant) rate of immigration into the country. If the total population of the country at time $t = 0$ is P_0, find an expression for the population at any time t.

13. **EFFECT OF IMMIGRATION ON POPULATION GROWTH** Refer to Exercise 12. The population of the United States in the year 1980 ($t = 0$) was 226.5 million. Suppose that the natural growth rate is 0.8% annually ($k = 0.008$) and net immigration is allowed at the rate of .5 million people/year ($I = 0.5$) until the end of the century. What will be the U.S. population in 2005?

14. **SINKING FUNDS** The proprietor of the Carson Hardware Store has decided to set up a sinking fund for the purpose of purchasing a computer 2 yr from now. It is expected that the purchase will involve a sum of $30,000. The fund grows at the rate of

$$\frac{dA}{dt} = rA + P$$

where A denotes the size of the fund at any time t, r is the annual interest rate earned by the fund compounded continuously, and P is the amount (in dollars) paid into the fund by the proprietor per year (assume this is done on a frequent basis in small deposits over the year so that it is essentially continuous). If the fund earns 10% interest per year compounded continuously, determine the size of the yearly investment the proprietor should pay into the fund.

15. **SPREAD OF A RUMOR** The rate at which a rumor spreads through an Alpine village of 400 residents is jointly proportional to the number of residents who have heard it and the number who have not. Initially, 10 residents heard the rumor, but 2 days later this number increased to 80. Find the number of people who will have heard the rumor after 1 wk.

16. **GROWTH OF A FRUIT-FLY COLONY** A biologist has determined that the maximum number of fruit flies that can be sustained in a carefully controlled environment (with a limited supply of space and food) is 400. Suppose that the rate at which the population of the colony increases obeys the rule

$$\frac{dQ}{dt} = kQ(C - Q)$$

where C is the carrying capacity (400) and Q denotes the number of fruit flies in the colony at any time t. If the initial population of fruit flies in the experiment is 10 and it grows to 45 after 10 days, determine the population of the colony of fruit flies on the 20th day.

17. **GOMPERTZ GROWTH CURVES** Refer to Exercise 28, Section 9.1. Consider the differential equation

$$\frac{dQ}{dt} = kQ(C - \ln Q)$$

with the side condition $Q(0) = Q_0$. The solution $Q(t)$ describes restricted growth and has a graph known as the Gompertz curve. Using separation of variables, solve this differential equation.

18. **CHEMICAL REACTION RATES** Two chemical solutions, one containing N molecules of chemical A and another containing M molecules of chemical B, are mixed together at time $t = 0$. The molecules from the two chemicals combine to form another chemical solution containing y (AB) molecules. The rate at which the AB molecules are formed, dy/dt, is called the *reaction rate* and is jointly proportional to $(N - y)$ and $(M - y)$. Thus,

$$\frac{dy}{dt} = k(N - y)(M - y)$$

where k is a constant (we assume the temperature of the chemical mixture remains constant during the interaction). Solve this differential equation with the side condition $y(0) = 0$ assuming that $N - y > 0$ and $M - y > 0$.

Hint: Use the identity

$$\frac{1}{(N-y)(M-y)} = \frac{1}{M-N}\left(\frac{1}{N-y} - \frac{1}{M-y}\right)$$

19. MIXTURE PROBLEMS A tank initially contains 20 gal of pure water. Brine containing 2 lb of salt per gallon flows into the tank at a rate of 3 gal/min, and the well-stirred mixture flows out of the tank at the same rate. How much salt is present in the tank at any time? How much salt is present at the end of 20 min? How much salt is present in the long run?

20. MIXTURE PROBLEMS A tank initially contains 50 gal of brine, in which 10 lb of salt is dissolved. Brine containing 2 lb of dissolved salt per gallon flows into the tank at the rate of 2 gal/min, and the well-stirred mixture flows out of the tank at the same rate. How much salt is present in the tank at the end of 10 min?

SOLUTION TO SELF-CHECK EXERCISE 9.3

We are required to solve the initial value problem

$$\frac{dy}{dt} = -k(y - 70) \qquad (k, \text{ a constant of proportionality})$$

$$y(0) = 200$$

To solve the differential equation, we separate variables and integrate, obtaining

$$\int \frac{dy}{y - 70} = \int -k\,dt$$

$$\ln|y - 70| = -kt + d \qquad (d, \text{ an arbitrary constant})$$

$$y - 70 = e^{-kt+d} = Ae^{-kt} \qquad (A = e^d)$$

$$y = 70 + Ae^{-kt}$$

Using the initial condition $y(0) = 200$, we find

$$200 = 70 + A$$
$$A = 130$$

Therefore,

$$y = 70 + 130e^{-kt}$$

To determine the value of k, we use the fact that $y(5) = 150$, obtaining

$$150 = 70 + 130e^{-5k}$$

$$130e^{-5k} = 80$$

$$e^{-5k} = \frac{80}{130}$$

$$-5k = \ln\frac{80}{130}$$

$$k = -\frac{1}{5}\ln\frac{80}{130}$$

$$= 0.097$$

Therefore, the required expression is

$$y(t) = 70 + 130e^{-0.097t}$$

9.4 Approximate Solutions of Differential Equations

EULER'S METHOD

As in the case of definite integrals, there are many differential equations whose exact solutions cannot be found using any of the available methods. In such cases, we must once again resort to approximate solutions to the problems at hand.

Many numerical methods have been developed for efficient computation of approximate solutions to differential equations. In this section, we will look at a method for solving the problem

$$\frac{dy}{dx} = F(x, y), \quad y(x_0) = y_0 \tag{12}$$

Euler's method, named after its discoverer, Leonhard Euler (1707–1783), illustrates quite clearly the idea behind the method for finding the approximate solution of Equation (12). Basically, the technique calls for approximating the actual solution $y = f(x)$ at certain selected values of x. The values of f between two adjacent values of x are then found by linear interpolation. This situation is depicted geometrically in Figure 9.13. Thus, in Euler's method, the actual solution curve of the differential equation is approximated by a suitable polygonal curve. We now describe the method in greater detail.

FIGURE 9.13
Using Euler's method, the actual solution curve of the differential equation is approximated by a polygonal curve.

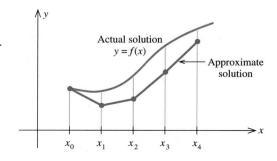

Let h be a small positive number and let $x_n = x_0 + nh$ ($n = 0, 1, 2, 3, \ldots$). Then,

$$x_1 = x_0 + h, \qquad x_2 = x_0 + 2h, \qquad x_3 = x_0 + 3h, \qquad \ldots$$

Thus, the points $x_0, x_1, x_2, x_3, \ldots$ are spaced evenly apart, and the distance between any two adjacent points is h units.

We begin by finding an approximation $y = \tilde{f}_1(x)$ to the actual solution $y = f(x)$ on the subinterval $[x_0, x_1]$. Observe that the *initial* condition $y(x_0) = y_0$ of (12) tells us that the point (x_0, y_0) lies on the solution curve. Now, Euler's method calls for approximating the part of the graph of f on the interval $[x_0, x_1]$ by a straight-line segment that is tangent to the graph of f at the point (x_0, y_0). To find this approximating linear function $y = \tilde{f}_1(x)$, recall that the differential equation $y' = F(x, y)$ gives the slope of the tangent line to the graph of $y = f(x)$ at any point (x, y) lying on the graph. In particular, the slope of the required straight-line segment is equal to $F(x_0, y_0)$. Therefore, using the point-slope form of the equation of a line, we see that the equation of the straight line segment is

$$y - y_0 = F(x_0, y_0)(x - x_0)$$
$$y = y_0 + F(x_0, y_0)(x - x_0)$$

Thus, the required approximation to the actual solution $y = f(x)$ on the interval $[x_0, x_1]$ is given by the linear function

$$\tilde{f}_1(x) = y_0 + F(x_0, y_0)(x - x_0) \tag{13}$$

This situation is depicted graphically in Figure 9.14.

FIGURE 9.14
$\tilde{f}_1(x) = y_0 + F(x_0, y_0)(x - x_0)$ is the equation of the straight line used to approximate $f(x)$ on $[x_0, x_1]$.

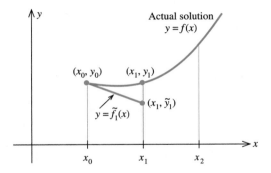

Next, to find an approximation $y = \tilde{f}_2(x)$ to the actual solution $y = f(x)$ on the subinterval $[x_1, x_2]$, observe that at $x = x_1$ the true value $y_1 = f(x)$ is approximated by the number

$$\tilde{y}_1 = \tilde{f}_1(x_1) = y_0 + F(x_0, y_0)(x_1 - x_0)$$
$$= y_0 + F(x_0, y_0)h \quad \text{(Since } x_1 - x_0 = h)$$

According to Euler's method, the approximating function in this interval has a graph that is just a line segment beginning at the point $(x_1, \tilde{y}_1)$ and having an appropriate slope. Ideally, this slope should be the slope of the tangent line to the graph of $y = f(x)$ at the point (x_1, y_1). But since this point is supposedly unknown to us, a practical alternative is to use the point $(x_1, \tilde{y}_1)$ in lieu of the point (x_1, y_1) to compute an approximation of the desired slope—namely, $F(x_1, \tilde{y}_1)$. Once again, using the point-slope form of the equation of a line, we find that the equation of the required tangent line is

$$y - \tilde{y}_1 = F(x_1, \tilde{y}_1)(x - x_1)$$
$$y = \tilde{y}_1 + F(x_1, \tilde{y}_1)(x - x_1)$$

so the required approximating linear function on the interval $[x_1, x_2]$ is given by

$$\tilde{f}_2(x) = \tilde{y}_1 + F(x_1, \tilde{y}_1)(x - x_1) \tag{14}$$

(Figure 9.15).

FIGURE 9.15
$\tilde{f}_2(x) = \tilde{y}_1 + F(x_1, \tilde{y}_1)(x - x_1)$ is the equation of the straight line used to approximate $f(x)$ on $[x_1, x_2]$.

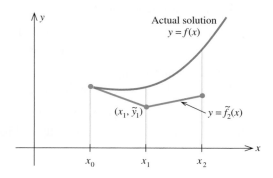

Proceeding, we find that the actual value $y_2 = f(x_2)$ is approximated by the number

$$\tilde{y}_2 = \tilde{f}_2(x_2) = \tilde{y}_1 + F(x_1, \tilde{y}_1)(x_2 - x_1)$$
$$= \tilde{y}_1 + F(x_1, \tilde{y}_1)h$$

and the approximating function on the interval $[x_2, x_3]$ is given by

$$\tilde{f}_3(x) = \tilde{y}_2 + F(x_2, \tilde{y}_2)(x - x_2)$$

and so on.

In many practical applications, we are interested only in finding an approximation of the solution of the differential Equation (12) at some specific value of x—say, $x = b$. In such cases, it is not important to compute the approximating functions $\tilde{f}_1, \tilde{f}_2, \tilde{f}_3, \ldots$. It suffices to compute the numbers $\tilde{y}_1, \tilde{y}_2, \tilde{y}_3, \ldots, \tilde{y}_n$ at the points $x_1, x_2, x_3, \ldots, x_n$ with h chosen to be

$$h = \frac{b - x_0}{n}$$

where n is a sufficiently large positive integer. In this case, we see that

$$x_1 = x_0 + h, \qquad x_2 = x_0 + 2h, \ldots, \qquad x_n = x_0 + nh = x_0 + (b - x_0) = b$$

so that $\tilde{y}_n = \tilde{f}_n(x_n) = \tilde{f}_n(b)$ does in fact yield the required approximation. We may summarize the procedure as follows.

Euler's Method

Suppose we are given the differential equation

$$\frac{dy}{dx} = F(x, y)$$

subject to the initial condition $y(x_0) = y_0$ and we wish to find an approximation of $y(b)$, where b is a number greater than x_0. Compute

$$h = \frac{b - x_0}{n}$$

$$x_1 = x_0 + h, \qquad x_2 = x_0 + 2h, \qquad x_3 = x_0 + 3h, \ldots, \qquad x_n = x_0 + nh = b$$

and

$$\begin{aligned}
\bar{y}_0 &= y_0 \\
\bar{y}_1 &= \bar{y}_0 + hF(x_0, \bar{y}_0) \\
\bar{y}_2 &= \bar{y}_1 + hF(x_1, \bar{y}_1) \\
&\;\;\vdots \\
\bar{y}_n &= \bar{y}_{n-1} + hF(x_{n-1}, \bar{y}_{n-1})
\end{aligned}$$

Then, $\bar{y}_n$ gives an approximation of the true value $y(b)$ of the solution to the initial value problem at $x = b$.

SOLVING DIFFERENTIAL EQUATIONS WITH EULER'S METHOD

EXAMPLE 1   Use Euler's method with $n = 8$ to obtain an approximation of the initial value problem

$$y' = x - y, \quad y(0) = 1$$

when $x = 2$.

SOLUTION ✔   Here, $x_0 = 0$ and $b = 2$ so that, taking $n = 8$, we find

$$h = \frac{2 - 0}{8} = \frac{1}{4}$$

and

$$x_0 = 0, \qquad x_1 = \frac{1}{4}, \qquad x_2 = \frac{1}{2}, \qquad x_3 = \frac{3}{4}, \qquad x_4 = 1,$$

$$x_5 = \frac{5}{4}, \qquad x_6 = \frac{3}{2}, \qquad x_7 = \frac{7}{4}, \qquad x_8 = b = 2$$

Also,

$$F(x, y) = x - y \qquad \text{and} \qquad y_0 = y(0) = 1$$

Therefore, the approximations of the actual solution at the points x_0, x_1, $x_2, \ldots, x_n = b$ are

$$\tilde{y}_0 = y_0 = 1$$

$$\tilde{y}_1 = \tilde{y}_0 + hF(x_0, \tilde{y}_0) = 1 + \frac{1}{4}(0 - 1) = \frac{3}{4}$$

$$\tilde{y}_2 = \tilde{y}_1 + hF(x_1, \tilde{y}_1) = \frac{3}{4} + \frac{1}{4}\left(\frac{1}{4} - \frac{3}{4}\right) = \frac{5}{8}$$

$$\tilde{y}_3 = \tilde{y}_2 + hF(x_2, \tilde{y}_2) = \frac{5}{8} + \frac{1}{4}\left(\frac{1}{2} - \frac{5}{8}\right) = \frac{19}{32}$$

$$\tilde{y}_4 = \tilde{y}_3 + hF(x_3, \tilde{y}_3) = \frac{19}{32} + \frac{1}{4}\left(\frac{3}{4} - \frac{19}{32}\right) = \frac{81}{128}$$

$$\tilde{y}_5 = \tilde{y}_4 + hF(x_4, \tilde{y}_4) = \frac{81}{128} + \frac{1}{4}\left(1 - \frac{81}{128}\right) = \frac{371}{512}$$

$$\tilde{y}_6 = \tilde{y}_5 + hF(x_5, \tilde{y}_5) = \frac{371}{512} + \frac{1}{4}\left(\frac{5}{4} - \frac{371}{512}\right) = \frac{1753}{2048}$$

$$\tilde{y}_7 = \tilde{y}_6 + hF(x_6, \tilde{y}_6) = \frac{1753}{2048} + \frac{1}{4}\left(\frac{3}{2} - \frac{1753}{2048}\right) = \frac{8331}{8192}$$

$$\tilde{y}_8 = \tilde{y}_7 + hF(x_7, \tilde{y}_7) = \frac{8331}{8192} + \frac{1}{4}\left(\frac{7}{4} - \frac{8331}{8192}\right) = \frac{39{,}329}{32{,}768}$$

Thus, the approximate value of $y(2)$ is

$$\frac{39{,}329}{32{,}768} \approx 1.2002$$

■ ■ ■ ■

  EXAMPLE **2**   Use Euler's method with (a) $n = 5$ and (b) $n = 10$ to approximate the solution of the initial value problem

$$y' = -2xy^2, \quad y(0) = 1$$

on the interval $[0, 0.5]$. Find the actual solution of the initial value problem. Finally, sketch the graphs of the approximate solutions and the actual solution for $0 \le x \le 0.5$ on the same set of axes.

SOLUTION ✔   **a.** Here, $x_0 = 0$ and $b = 0.5$. Taking $n = 5$, we find

$$h = \frac{0.5 - 0}{5} = 0.1$$

and $x_0 = 0$, $x_1 = 0.1$, $x_2 = 0.2$, $x_3 = 0.3$, $x_4 = 0.4$, and $x_5 = b = 0.5$. Also,

$$F(x, y) = -2xy^2 \quad \text{and} \quad y_0 = y(0) = 1$$

Therefore,

$$\tilde{y}_0 = y_0 = 1$$
$$\tilde{y}_1 = \tilde{y}_0 + hF(x_0, \tilde{y}_0) = 1 + 0.1(-2)(0)(1)^2 = 1$$
$$\tilde{y}_2 = \tilde{y}_1 + hF(x_1, \tilde{y}_1) = 1 + 0.1(-2)(0.1)(1)^2 = 0.98$$
$$\tilde{y}_3 = \tilde{y}_2 + hF(x_2, \tilde{y}_2) = 0.98 + 0.1(-2)(0.2)(0.98)^2 = 0.9416$$
$$\tilde{y}_4 = \tilde{y}_3 + hF(x_3, \tilde{y}_3) = 0.9416 + 0.1(-2)(0.3)(0.9416)^2 = 0.8884$$
$$\tilde{y}_5 = \tilde{y}_4 + hF(x_4, \tilde{y}_4) = 0.8884 + 0.1(-2)(0.4)(0.8884)^2 = 0.8253$$

b. Here $x_0 = 0$ and $b = 0.5$. Taking $n = 10$, we find

$$h = \frac{0.5 - 0}{10}$$

and $x_0 = 0$, $x_1 = 0.05$, $x_2 = 0.10$, $\ldots$, $x_9 = 0.45$, and $x_{10} = 0.5 = b$. Proceeding as in part (a), we obtain the approximate solutions listed in the following table:

x	0.00	0.05	0.10	0.15	0.20	0.25
$\tilde{y}_n$	1.0000	1.0000	0.9950	0.9851	0.9705	0.9517

x	0.30	0.35	0.40	0.45	0.50
$\tilde{y}_n$	0.9291	0.9032	0.8746	0.8440	0.8119

To obtain the actual solution of the differential equation, we separate variables, obtaining

$$\frac{dy}{y^2} = -2x \, dx$$

Integrating each side of the last equation with respect to the appropriate variables, we have

$$\int \frac{dy}{y^2} = -\int 2x \, dx$$

or
$$-\frac{1}{y} + C_1 = -x^2 + C_2$$

$$\frac{1}{y} = x^2 + C \qquad (C = C_1 - C_2)$$

$$y = \frac{1}{x^2 + C}$$

Using the condition $y(0) = 1$, we have

$$1 = \frac{1}{0 + C} \qquad \text{or} \qquad C = 1$$

Therefore, the required solution is given by

$$y = \frac{1}{x^2 + 1}$$

The graphs of the approximate solutions and the actual solution are sketched in Figure 9.16.

FIGURE 9.16
The approximate solutions and the actual so-lution to an initial value problem

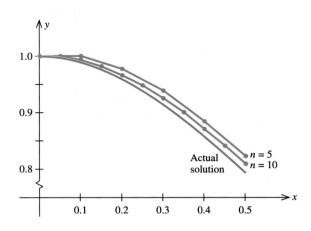

SELF-CHECK EXERCISE 9.4

Use Euler's method with $n = 5$ to obtain an approximation of the initial value problem

$$y' = 2x + y, \quad y(0) = 1$$

when $x = 1$.

The solution to Self-Check Exercise 9.4 can be found on page 723.

9.4 Exercises

A calculator is recommended for this exercise set. In Exercises 1–10, use Euler's method with (a) $n = 4$ and (b) $n = 6$ to obtain an approximation of the initial value problem when $x = b$.

1. $y' = x + y,\ y(0) = 1;\ b = 1$

2. $y' = x - 2y,\ y(0) = 1;\ b = 2$

3. $y' = 2x - y + 1,\ y(0) = 2;\ b = 2$

4. $y' = 2xy,\ y(0) = 1;\ b = 0.5$

5. $y' = -2xy^2,\ y(0) = 1;\ b = 0.5$

6. $y' = x^2 + y^2,\ y(0) = 1;\ b = 1.5$

7. $y' = \sqrt{x + y},\ y(1) = 1;\ b = 1.5$

8. $y' = (1 + x^2)^{-1},\ y(0) = 0;\ b = 1$

9. $y' = \dfrac{x}{y},\ y(0) = 1;\ b = 1$

10. $y' = xy^{1/3},\ y(0) = 1;\ b = 1$

In Exercises 11–15, use Euler's method with $n = 5$ **to obtain an approximate solution to the initial value problem over the indicated interval.**

11. $y' = \dfrac{1}{2}xy$, $y(0) = 1$; $0 \leq x \leq 1$

12. $y' = x^2y$, $y(0) = 2$; $0 \leq x \leq 0.6$

13. $y' = 2x - y + 1$, $y(0) = 2$; $0 \leq x \leq 1$

14. $y' = x + y^2$, $y(0) = 0$; $0 \leq x \leq 0.5$

15. $y' = x^2 + y$, $y(0) = 1$; $0 \leq x \leq 0.5$

16. GROWTH OF SERVICE INDUSTRIES It has been estimated that service industries, which currently make up 30% of the nonfarm workforce in a certain country, will continue to grow at the rate of

$$R(t) = 5e^{1/(t+1)}$$

percent per decade t decades from now. Estimate the percentage of the nonfarm workforce in the service industries one decade from now.

Hint: (a) Show that the desired answer is $P(1)$, where P is the solution of the initial value problem

$$P' = 5e^{1/(t+1)} \qquad [P(0) = 30]$$

(b) Use Euler's method with $n = 10$ to approximate the solution.

SOLUTION TO SELF-CHECK EXERCISE 9.4

Here, $x_0 = 0$ and $b = 1$ so that, taking $n = 5$, we find

$$h = \frac{1-0}{5} = \frac{1}{5}$$

and

$$x_0 = 0, \qquad x_1 = \frac{1}{5}, \qquad x_2 = \frac{2}{5}, \qquad x_3 = \frac{3}{5}, \qquad x_4 = \frac{4}{5}, \qquad x_5 = b = 1$$

Also,

$$F(x, y) = 2x + y \qquad \text{and} \qquad y_0 = y(0) = 1$$

Therefore, the approximations of the actual solution at the points $x_0, x_1, x_2, \ldots, x_5 = 1$ are

$$\tilde{y}_0 = y_0 = 1$$

$$\tilde{y}_1 = \tilde{y}_0 + hF(x_0, \tilde{y}_0) = 1 + \frac{1}{5}(0 + 1) = \frac{6}{5}$$

$$\tilde{y}_2 = \tilde{y}_1 + hF(x_1, \tilde{y}_1) = \frac{6}{5} + \frac{1}{5}\left(\frac{2}{5} + \frac{6}{5}\right) = \frac{38}{25}$$

$$\tilde{y}_3 = \tilde{y}_2 + hF(x_2, \tilde{y}_2) = \frac{38}{25} + \frac{1}{5}\left(\frac{4}{5} + \frac{38}{25}\right) = \frac{248}{125}$$

$$\tilde{y}_4 = \tilde{y}_3 + hF(x_3, \tilde{y}_3) = \frac{248}{125} + \frac{1}{5}\left(\frac{6}{5} + \frac{248}{125}\right) = \frac{1638}{625}$$

$$\tilde{y}_5 = \tilde{y}_4 + hF(x_4, \tilde{y}_4) = \frac{1638}{625} + \frac{1}{5}\left(\frac{8}{5} + \frac{1638}{625}\right) = \frac{10,828}{3125}$$

Thus, the approximate value of $y(1)$ is

$$\frac{10,828}{3125} \approx 3.4650$$

CHAPTER 9 Summary of Principal Terms

Terms

differential equation
general solution of a
   differential equation
particular solution of a
   differential equation
first-order differential
   equation

separable differential equation
method of separation of variables
initial value problem
Euler's method

CHAPTER 9 REVIEW EXERCISES

In Exercises 1–3, verify that *y* is a solution of the differential equation.

1. $y = C_1 e^{2x} + C_2 e^{-3x}$; $y'' + y' - 6y = 0$

2. $y = 2e^{2x} + 3x - 2$; $y'' - y' - 2y = -6x + 1$

3. $y = Cx^{-4/3}$; $4xy^3\,dx + 3x^2y^2\,dy = 0$

In Exercises 4 and 5, verify that *y* is a general solution of the differential equation and find a particular solution of the differential equation satisfying the side condition.

4. $y = \dfrac{1}{x^2 - C}$; $\dfrac{dy}{dx} = -2xy^2$; $y(0) = 1$

5. $y = (9x + C)^{-1/3}$; $\dfrac{dy}{dx} = -3y^4$; $y(0) = \dfrac{1}{2}$

In Exercises 6–11, solve the differential equation.

6. $y' = \dfrac{x^3 + 1}{y^2}$

7. $\dfrac{dy}{dt} = 2(4 - y)$

8. $y' = \dfrac{y \ln x}{x}$

9. $y' = 3x^2y^2 + y^2$; $y(0) = -2$

10. $y' = x^2(1 - y)$; $y(0) = -2$

11. $\dfrac{dy}{dx} = -\dfrac{3}{2}x^2y$; $y(0) = 3$

12. Find a function f given that (1) the slope of the tangent line to the graph of f at any point $P(x, y)$ is given by the expression

$$y' = -\frac{4xy}{x^2 + 1}$$

and (2) the graph of f passes through the point $(1, 1)$.

In Exercises 13–16, use Euler's method with (a) *n* = 4 and (b) *n* = 6 to obtain an approximation of the initial value problem when *x* = *b*.

13. $y' = x + y^2$, $y(0) = 0$; $b = 1$

14. $y' = x^2 + 2y^2$; $y(0) = 0$; $b = 1$

15. $y' = 1 + 2xy^2$, $y(0) = 0$; $b = 1$

16. $y' = e^x + y^2$, $y(0) = 0$; $b = 1$

In Exercises 17 and 18, use Euler's method with n = 5 to obtain an approximate solution to the initial value problem over the indicated interval.

17. $y' = 2xy$, $y(0) = 1$; $0 \le x \le 1$

18. $y' = x^2 + y^2$, $y(0) = 1$; $0 \le x \le 1$

HISTOGRAMS

Table 10.1	
Number of Cars	**Frequency of Occurrence**
0	2
1	9
2	16
3	12
4	8
5	6
6	4
7	2
8	1

A discrete probability function or distribution may be exhibited graphically by means of a **histogram.** To construct a histogram of a particular probability distribution, first locate the values of the random variable on a number line. Then, above each such number, erect a rectangle with width 1 and height equal to the probability associated with that value of the random variable.

For example, consider the data in Table 10.1, the number of cars observed waiting in line at 2-minute intervals between 3 and 5 P.M. on a certain Friday at the drive-in teller of the Westwood Savings Bank and the corresponding frequency of occurrence. If we divide each number on the right of the table by 60 (the sum of these numbers), then we obtain the respective probabilities associated with the random variable X, when X assumes the values 0, 1, 2, . . . , 8. For example,

$$P(X = 0) = \frac{2}{60} \approx 0.03$$

$$P(X = 1) = \frac{9}{60} = 0.15, \ldots$$

The resulting probability distribution is shown in Table 10.2.

Table 10.2	Probability Distribution for the Random Variable X
x	$P(X = x)$
0	0.03
1	0.15
2	0.27
3	0.20
4	0.13
5	0.10
6	0.07
7	0.03
8	0.02

The histogram associated with this probability distribution is shown in Figure 10.1 on the next page.

Observe that the area of a rectangle in a histogram is associated with a value of a random variable X and that this area gives precisely the probability associated with that value of X. This follows since each rectangle, by construction, has width 1 and height corresponding to the probability associated with the value of the random variable.

Another consequence arising from the method of construction of a histogram is that the probability associated with more than one value of the random

FIGURE 10.1
Probability distribution of the number of
cars waiting in line

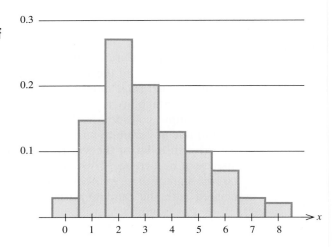

variable X is given by the sum of the areas of the rectangles associated with those values of X. For example, the probability that three or four cars are in line is given by

$$P(X = 3) + P(X = 4)$$

which may be obtained from the histogram by adding the areas of the rectangles associated with the values 3 and 4 of the random variable X. Thus, the required probability is

$$P(X = 3) + P(X = 4) = 0.20 + 0.13 = 0.33$$

CONTINUOUS RANDOM VARIABLES

A random variable x that can assume any value in an interval is called a **continuous random variable.** Examples of continuous random variables are the life span of a light bulb, the length of a telephone call, the length of an infant at birth, the daily amount of rainfall in Boston, and the life span of a certain plant species. For the remainder of this chapter, we will be interested primarily in continuous random variables.

Consider an experiment in which the associated random variable x has the interval $[a, b]$ as its sample space. Then an event of the experiment is any subset of $[a, b]$. For example, if x denotes the life span of a light bulb, then the sample space associated with the experiment is $[0, \infty)$, and the event that a light bulb selected at random has a life span between 500 and 600 hours inclusive is described by the interval $[500, 600]$ or, equivalently, by the inequality $500 \leq x \leq 600$. The probability that the light bulb will have a life span of between 500 and 600 hours is denoted by $P(500 \leq x \leq 600)$.

In general, we will be interested in computing $P(a \leq x \leq b)$, the probability that a random variable x assumes a value in the interval $a \leq x \leq b$. This computation is based on the notion of a probability density function, which we now introduce.

Probability Density Function

A **probability density function** of a random variable x is a nonnegative function f having the following properties:

1. The total area of the region under the graph of f is equal to 1 (see Figure 10.2a).
2. The probability that an observed value of the random variable x lies in the interval $[a, b]$ is given by

$$P(a \leq x \leq b) = \int_a^b f(x)\, dx$$

(see Figure 10.2b).

FIGURE 10.2

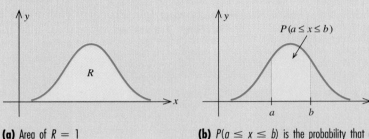

(a) Area of $R = 1$

(b) $P(a \leq x \leq b)$ is the probability that an outcome of an experiment will lie between a and b.

A few comments are in order. First, a probability density function of a random variable x may be constructed using methods that range from theoretical considerations of the problem on the one extreme to an interpretation of data associated with the experiment on the other. Second, Property 1 states that the probability that a continuous random variable takes on a value lying in its range is 1, a certainty, which is expected. Third, Property 2 states that the probability that the random variable x assumes a value in an interval $a \leq x \leq b$ is given by the area of the region between the graph of f and the x-axis from $x = a$ to $x = b$. Because the area under one point of the graph of f is equal to zero, we see immediately that $P(a \leq x \leq b) = P(a < x \leq b) = P(a \leq x < b) = P(a < x < b)$.

EXAMPLE 1 Show that each of the following functions satisfies the nonnegativity condition and Property 1 of probability density functions:

a. $f(x) = \dfrac{2}{27} x(x - 1) \qquad (1 \leq x \leq 4)$

b. $f(x) = \dfrac{1}{3} e^{(-1/3)x} \qquad (0 \leq x < \infty)$

SOLUTION ✔   **a.** Since the factors x and $(x - 1)$ are both nonnegative, we see that $f(x) \geq 0$ on $[1, 4]$. Next, we compute

$$\int_1^4 \frac{2}{27} x(x - 1)\, dx = \frac{2}{27} \int_1^4 (x^2 - x)\, dx$$

$$= \frac{2}{27} \left(\frac{1}{3} x^3 - \frac{1}{2} x^2 \right) \Big|_1^4$$

$$= \frac{2}{27} \left[\left(\frac{64}{3} - 8 \right) - \left(\frac{1}{3} - \frac{1}{2} \right) \right]$$

$$= \frac{2}{27} \left(\frac{27}{2} \right)$$

$$= 1$$

showing that Property 1 of probability density functions holds as well.

b. First, $f(x) = \frac{1}{3} e^{(-1/3)x} \geq 0$ for all values of x in $[0, \infty)$. Next,

$$\int_0^\infty \frac{1}{3} e^{(-1/3)x}\, dx = \lim_{b \to \infty} \int_0^b \frac{1}{3} e^{(-1/3)x}\, dx$$

$$= \lim_{b \to \infty} -e^{(-1/3)x} \Big|_0^b$$

$$= \lim_{b \to \infty} \left(-e^{(-1/3)b} + 1 \right)$$

$$= 1$$

so the area under the graph of $f(x) = \frac{1}{3} e^{(-1/3)x}$ is equal to 1, as we set out to show.   ■ ■ ■ ■

EXAMPLE 2   **a.** Determine the value of the constant k such that the function $f(x) = kx^2$ is a probability density function on the interval $[0, 5]$.

b. If x is a continuous random variable with the probability density function given in part (a), compute the probability that x will assume a value between $x = 1$ and $x = 2$.

c. Find the probability that x will assume a value at $x = 3$.

SOLUTION ✔   **a.** We compute

$$\int_0^5 kx^2\, dx = k \int_0^5 x^2\, dx$$

$$= \frac{k}{3} x^3 \Big|_0^5$$

$$= \frac{125}{3} k$$

Since this value must be equal to 1, we find that $k = \dfrac{3}{125}$.

b. The required probability is given by

$$P(1 \le x \le 2) = \int_1^2 f(x)\, dx$$

$$= \int_1^2 \frac{3}{125} x^2\, dx$$

$$= \frac{1}{125} x^3 \Big|_1^2$$

$$= \frac{1}{125}(8 - 1)$$

$$= \frac{7}{125}$$

FIGURE 10.3
$P(1 \le x \le 2)$ for the probability density function
$$y = \frac{3}{125} x^2$$

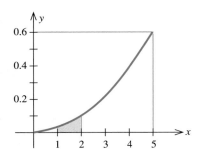

The graph of the probability density function f and the area corresponding to the probability $P(1 \le x \le 2)$ are shown in Figure 10.3.

c. The required probability is given by

$$P(x = 3) = \int_3^3 f(x)\, dx$$

$$= \int_3^3 \frac{3}{125} x^2\, dx = 0$$

REMARK Observe that, in general, the probability that x will assume a value at a point $x = a$ is zero since $P(x = a) = \int_a^a f(x)\, dx = 0$ by Property 1 of the definite integral (page 491). ■ ■ ■

EXAMPLE 3 TKK Products Corporation manufactures a 200-watt electric light bulb. Laboratory tests show that the life spans of these light bulbs have a distribution described by the probability density function

$$f(x) = 0.001 e^{-0.001x}$$

Determine the probability that a light bulb will have each of these life spans:

a. 500 hours or less

b. More than 500 hours

c. More than 1000 hours but less than 1500 hours

SOLUTION ✔ Let x denote the life span of a light bulb.

a. The probability that a light bulb will have a life span of 500 hours or less is given by

$$P(0 \le x \le 500) = \int_0^{500} 0.001 e^{-0.001x}\, dx$$

$$= -e^{-0.001x} \Big|_0^{500}$$

$$= -e^{-0.5} + 1$$

$$\approx 0.3935$$

b. The probability that a light bulb will have a life span of more than 500 hours is given by

$$P(x > 500) = \int_{500}^{\infty} 0.001 e^{-0.001x} \, dx$$

$$= \lim_{b \to \infty} \int_{500}^{b} 0.001 e^{-0.001x} \, dx$$

$$= \lim_{b \to \infty} -e^{-0.001x} \Big|_{500}^{b}$$

$$= \lim_{b \to \infty} \left(-e^{-0.001b} + e^{-0.5} \right)$$

$$= e^{-0.5} \approx 0.6065$$

This result may also be obtained by observing that

$$P(x > 500) = 1 - P(x \le 500)$$
$$= 1 - 0.3935 \qquad \text{[Using the result from part (a)]}$$
$$\approx 0.6065$$

c. The probability that a light bulb will have a life span of more than 1000 hours but less than 1500 hours is given by

$$P(1000 < x < 1500) = \int_{1000}^{1500} 0.001 e^{-0.001x} \, dx$$

$$= -e^{-0.001x} \Big|_{1000}^{1500}$$

$$= -e^{-1.5} + e^{-1}$$

$$\approx -0.2231 + 0.3679$$

$$= 0.1448 \qquad\qquad ▪▪▪▪$$

The probability density function of Example 3 has the form

$$f(x) = ke^{-kx}$$

where $x \ge 0$ and k is a positive constant. Its graph is shown in Figure 10.4. This probability function is called an **exponential density function,** and the random variable associated with it is said to be **exponentially distributed.** Exponential random variables are used to represent the life span of electronic components, the duration of telephone calls, the waiting time in a doctor's office, and the time between successive flight arrivals and departures in an airport, to mention but a few applications.

FIGURE 10.4
The area under the graph of the exponential density function is equal to 1.

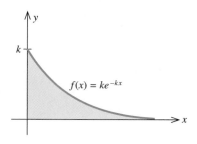

SELF-CHECK EXERCISES 10.1

1. Determine the value of the constant k such that the function $f(x) = k(4x - x^2)$ is a probability density function on the interval $[0, 4]$.

2. Suppose that x is a continuous random variable with the probability density function of Exercise 1. Find the probability that x will assume a value between $x = 1$ and $x = 3$.

Solutions to Self-Check Exercises 10.1 can be found on page 739.

10.1 Exercises

In Exercises 1–12, show that the given function is a probability density function on the given interval.

1. $f(x) = \dfrac{1}{3}$; $(3 \leq x \leq 6)$

2. $f(x) = \dfrac{1}{\pi}$; $(\pi \leq x \leq 2\pi)$

3. $f(x) = \dfrac{2}{32} x$; $(2 \leq x \leq 6)$

4. $f(x) = \dfrac{3}{8} x^2$; $(0 \leq x \leq 2)$

5. $f(x) = \dfrac{2}{9}(3x - x^2)$; $(0 \leq x \leq 3)$

6. $f(x) = \dfrac{3}{32}(x - 1)(5 - x)$; $(1 \leq x \leq 5)$

7. $f(x) = \dfrac{12 - x}{72}$; $(0 \leq x \leq 12)$

8. $f(x) = 20(x^3 - x^4)$; $(0 \leq x \leq 1)$

9. $f(x) = \dfrac{8}{7x^2}$; $(1 \leq x \leq 8)$

10. $f(x) = \dfrac{3}{14}\sqrt{x}$; $(1 \leq x \leq 4)$

11. $f(x) = \dfrac{x}{(x^2 + 1)^{3/2}}$; $(0 \leq x < \infty)$

12. $f(x) = 4xe^{-2x^2}$; $(0 \leq x < \infty)$

In Exercises 13–20, find the value of the constant k so that the given function is a probability density function in the given interval.

13. $f(x) = k$; $[1, 4]$

14. $f(x) = kx$; $[0, 4]$

15. $f(x) = k(4 - x)$; $[0, 4]$

16. $f(x) = kx^3$; $[0, 1]$

17. $f(x) = k\sqrt{x}$; $[0, 4]$

18. $f(x) = \dfrac{k}{x}$; $[1, 5]$

19. $f(x) = \dfrac{k}{x^3}$; $[1, \infty)$

20. $f(x) = ke^{-x/2}$; $[0, \infty)$

In Exercises 21–29, f is a probability density function defined on the given interval. Find the indicated probabilities.

21. $f(x) = \dfrac{1}{12} x$; $[1, 5]$

   a. $P(2 \leq x \leq 4)$ **b.** $P(1 \leq x \leq 4)$

   c. $P(x \geq 2)$ **d.** $P(x = 2)$

22. $f(x) = \dfrac{1}{9} x^2$; $[0, 3]$

   a. $P(1 \leq x \leq 2)$ **b.** $P(1 < x \leq 3)$

   c. $P(x \leq 2)$ **d.** $P(x = 1)$

23. $f(x) = \dfrac{3}{32}(4 - x^2)$; $[-2, 2]$

   a. $P(-1 \leq x \leq 1)$ **b.** $P(x \leq 0)$

   c. $P(x > -1)$ **d.** $P(x = 0)$

24. $f(x) = \dfrac{3}{16}\sqrt{x}$; $[0, 4]$

   a. $P(1 < x < 3)$ **b.** $P(x \leq 2)$

   c. $P(x = 2)$ **d.** $P(x \geq 1)$

25. $f(x) = \dfrac{1}{4\sqrt{x}}$; $[1, 9]$

   a. $P(x \geq 4)$ **b.** $P(1 \leq x < 8)$

   c. $P(x = 3)$ **d.** $P(x \leq 4)$

26. $f(x) = \dfrac{1}{2} e^{-x/2}$; $[0, \infty)$

   a. $P(x \leq 4)$ **b.** $P(1 < x < 2)$

   c. $P(x = 50)$ **d.** $P(x \geq 2)$

27. $f(x) = 4xe^{-2x^2}$; $[0, \infty)$

   a. $P(0 \leq x \leq 4)$ **b.** $P(x \geq 1)$

28. $f(x) = \dfrac{1}{9} xe^{-x/3}$; $[0, \infty)$

   a. $P(0 \leq x \leq 3)$ **b.** $P(x \geq 1)$

(continued on p. 738)

GRAPHING A HISTOGRAM

A graphing calculator can be used to plot the histogram for a given set of data, as illustrated by the example.

EXAMPLE 1

A survey of 90,000 households conducted in 1995 revealed the following percentage of women who wear each given shoe size:

Shoe Size	<5	5–5½	6–6½	7–7½	8–8½	9–9½	10–10½	>10½
Percentage of Women	1	5	15	27	29	14	7	2

Source: Footwear Market Insights survey

a. Plot a histogram for the given data.

b. What percentage of women in the survey wear size 7–7½ or 8–8½ shoes?

SOLUTION ✔

a. Let X denote the random variable taking on the values 1 through 8, where 1 corresponds to a shoe size less than 5, 2 corresponds to a shoe size of 5–5½, and so on. Entering the values of X as $x_1 = 1$, $x_2 = 2$, ... , $x_8 = 8$ and the corresponding values of Y as $y_1 = 1$, $y_2 = 5$, ... , $y_8 = 2$ and then using the DRAW function from the Statistics menu, we obtain the histogram shown in Figure T1.

FIGURE T1
The histogram for the given data, using the viewing rectangle $[0, 9] \times [0, 35]$

b. The probability that a woman participating in the survey wears size 7–7½ or 8–8½ shoes is given by

$$P(X = 4) + P(X = 5) = 0.27 + 0.29 = 0.56$$

which tells us that 56% of the women wear either size 7–7½ or size 8–8½ shoes.

■ ■ ■ ■

1. Graph the histogram associated with the data given in Table 10.2 page 729. Compare your graph with that given in Figure 10.1, page 730.

2. **DISTRIBUTION OF FAMILIES BY SIZE** A survey was conducted by the Public Housing Authority in a certain community among 1000 families to determine the distribution of families by size. The results follow.

Family Size	2	3	4	5
Frequency of Occurrence	350	200	245	125

Family Size	6	7	8
Frequency of Occurrence	66	10	4

a. Find the probability distribution of the random variable X, where X denotes the number of persons in a randomly chosen family.

b. Graph the histogram corresponding to the probability distribution found in part (a).

3. **WAITING LINES** The accompanying data were obtained in a study conducted by the manager of Sav-More Supermarket. In this study the number of customers waiting in line at the express checkout at the beginning of each 3-min interval between 9 A.M. and 12 noon on Saturday was observed.

Number of Customers	0	1	2	3	4
Frequency of Occurrence	1	4	2	7	14

Number of Customers	5	6	7	8	9	10
Frequency of Occurrence	8	10	6	3	4	1

a. Find the probability distribution of the random variable X, where X denotes the number of customers observed waiting in line.

b. Graph the histogram representing this probability distribution.

4. **TELEVISION PILOTS** After the private screening of a new television pilot, audience members were asked to rate the new show on a scale of 1 to 10 (10 being the highest rating). From a group of 140 people, these responses were obtained:

Rating	1	2	3	4	5
Frequency of Occurrence	1	4	3	11	23

Rating	6	7	8	9	10
Frequency of Occurrence	21	28	29	16	4

Let the random variable X denote the rating given to the show by a randomly chosen audience member. Find the probability distribution and graph the histogram associated with these data.

29. $f(x) = \begin{cases} x & \text{if } 0 \leq x \leq 1 \\ 2 - x & \text{if } 1 \leq x \leq 2 \end{cases}; [0,2]$

  a. $P\left(\dfrac{1}{2} \leq x \leq 1\right)$ **b.** $P\left(\dfrac{1}{2} \leq x \leq \dfrac{3}{2}\right)$

  c. $P(x \geq 1)$ **d.** $P\left(x \leq \dfrac{3}{2}\right)$

30. $f(x) = \begin{cases} \dfrac{3}{40}\sqrt{x} & 0 \leq x \leq 4 \\ \dfrac{12}{5x^2} & 4 < x < \infty \end{cases}; [0,\infty)$

  a. $P(1 \leq x \leq 4)$ **b.** $P(0 \leq x \leq 5)$

31. LIFE SPAN OF A PLANT The life span of a certain plant species (in days) is described by the probability density function

$$f(x) = \frac{1}{100}e^{-x/100}$$

  a. Find the probability that a plant of this species will live for 100 days or less.
  b. Find the probability that a plant of this species will live longer than 120 days.
  c. Find the probability that a plant of this species will live longer than 60 days but less than 140 days.

32. WAITING TIME AT A HEALTH CLINIC The average waiting time in minutes for patients arriving at the Newtown Health Clinic between 1 and 4 P.M. on a weekday is an exponentially distributed random variable x with associated probability density function $f(x) = \frac{1}{15}e^{-(1/15)x}$.
  a. What is the probability that a patient arriving at the clinic between 1 and 4 P.M. will have to wait longer than 15 min?
  b. What is the probability that a patient arriving at the clinic between 1 and 4 P.M. will have to wait between 10 and 12 min?

33. RELIABILITY OF ROBOTS The National Welding Company uses industrial robots in some of its assembly-line operations. Management has determined that the lengths of time in hours between breakdowns are exponentially distributed with probability density function $f(t) = 0.001e^{-0.001t}$.
  a. What is the probability that a robot selected at random will break down after between 600 and 800 hr of use?
  b. What is the probability that a robot will break down after 1200 hr of use?

34. WAITING TIME AT AN EXPRESSWAY TOLLBOOTH Suppose the time intervals in seconds between arrivals of successive cars at an expressway tollbooth during rush hour are exponentially distributed with associated probability density function $f(t) = \frac{1}{8}e^{-(1/8)t}$. Find the probability that the average time interval between arrivals of successive cars is more than 8 sec.

35. MAIL-ORDER PHONE CALLS A study conducted by Uni-Mart, a mail-order department store, reveals that the time intervals in minutes between incoming telephone calls on its toll-free 800 line between 10 A.M. and 2 P.M. are exponentially distributed with probability density function $f(t) = \frac{1}{30}e^{-t/30}$. What is the probability that the time interval between successive calls is more than 2 min?

36. RELIABILITY OF A MICROPROCESSOR The microprocessors manufactured by United Motor Works Company, which are used in automobiles to regulate fuel consumption, are guaranteed against defects for 20,000 mi of use. Tests conducted in the laboratory under simulated driving conditions reveal that the distances driven in miles before the microprocessors break down are exponentially distributed with probability density function $f(x) = 0.00001e^{-0.00001x}$. What is the probability that a microprocessor selected at random will fail during the warranty period?

37. PROBABILITY OF SNOWFALL The amount of snowfall (in feet) in a remote region of Alaska in the month of January is a continuous random variable with probability density function $f(x) = \frac{2}{9}x(3 - x)$, $0 \leq x \leq 3$. Find the probability that the amount of snowfall will be between 1 and 2 ft; more than 1 ft.

38. NUMBER OF CHIPS IN CHOCOLATE CHIP COOKIES The number of chocolate chips in each cookie of a certain brand has a distribution described by the probability density function

$$f(x) = \frac{1}{36}(6x - x^2) \qquad (0 \leq x \leq 6)$$

Find the probability that the number of chocolate chips in a randomly chosen cookie is fewer than two.

39. LIFE EXPECTANCY OF COLOR TELEVISION TUBES The life expectancy (in years) of a certain brand of color television tube is a continuous random variable with probability density function

$$f(t) = 9(9 + t^2)^{-3/2} \qquad (0 \leq t < \infty)$$

Find the probability that a randomly chosen television tube will last more than 4 yr.
Hint: Integrate using a table of integrals.

40. PRODUCT RELIABILITY The tread life (in thousands of miles) of a certain make of tire is a continuous random variable with a probability density function

$$f(x) = 0.02e^{-0.02x} \qquad (0 \le x < \infty)$$

a. Find the probability that a randomly selected tire of this make will have a tread life of at most 30,000 mi.
b. Find the probability that a randomly selected tire of this make will have a tread life between 40,000 and 60,000 mi.
c. Find the probability that a randomly selected tire of this make will have a tread life of at least 70,000 mi.

In Exercises 41 and 42, determine whether the statement is true or false. If it is true, explain why it is true. If it is false, give an example to show why it is false.

41. If $\int_a^b f(x)\, dx = 1$, then f is a probability density function on $[a, b]$.

42. If f is a probability function on an interval $[a, b]$, then f is a probability function on $[c, d]$ for any real numbers c and d satisfying $a < c < d < b$.

SOLUTIONS TO SELF-CHECK EXERCISES 10.1

1. We compute

$$\int_0^4 k(4x - x^2)\, dx = k\left(2x^2 - \frac{1}{3}x^3\right)\Big|_0^4$$

$$= k\left(32 - \frac{64}{3}\right)$$

$$= \frac{32}{3}k$$

Since this value must be equal to 1, we have $k = \dfrac{3}{32}$.

2. The required probability is given by

$$P(1 \le x \le 3) = \int_1^3 f(x)\, dx$$

$$= \int_1^3 \frac{3}{32}(4x - x^2)\, dx$$

$$= \frac{3}{32}\left(2x^2 - \frac{1}{3}x^3\right)\Big|_1^3$$

$$= \frac{3}{32}\left[(18 - 9) - \left(2 - \frac{1}{3}\right)\right] = \frac{11}{16}$$

10.2 Expected Value and Standard Deviation

EXPECTED VALUE

The average value of a set of numbers is a familiar notion to most people. For example, to compute the average of the four numbers 12, 16, 23, and 37, we simply add these numbers and divide the resulting sum by 4, giving the

EXAMPLE 8   Find the expected value, variance, and standard deviation of the random variable x associated with the probability density function

$$f(x) = \frac{32}{15x^3}$$

on $[1, 4]$.

SOLUTION ✔   Using (3), we find the mean of x:

$$\mu = \int_a^b xf(x)\, dx = \int_1^4 x \cdot \frac{32}{15x^3}\, dx$$

$$= \frac{32}{15}\int_1^4 x^{-2}\, dx = \frac{32}{15}\left[-\frac{1}{x}\right]_1^4$$

$$= \frac{32}{15}\left(-\frac{1}{4} + 1\right) = \frac{8}{5}$$

Next, using (8), we find

$$\text{Var}\,(x) = \int_a^b (x - \mu)^2 f(x)\, dx = \int_1^4 \left(x - \frac{8}{5}\right)^2 \cdot \frac{32}{15x^3}\, dx$$

$$= \frac{32}{15}\int_1^4 \left(x^2 - \frac{16}{5}x + \frac{64}{25}\right)\frac{1}{x^3}\, dx$$

$$= \frac{32}{15}\int_1^4 \left(\frac{1}{x} - \frac{16}{5}x^{-2} + \frac{64}{25}x^{-3}\right) dx$$

$$= \frac{32}{15}\left[\ln x + \frac{16}{5x} - \frac{32}{25x^2}\right]_1^4$$

$$= \frac{32}{15}\left[\left(\ln 4 + \frac{16}{5 \cdot 4} - \frac{32}{25 \cdot 16}\right) - \left(\ln 1 + \frac{16}{5} - \frac{32}{25}\right)\right]$$

$$= \frac{32}{15}\left(\ln 4 + \frac{4}{5} - \frac{2}{25} - \frac{16}{5} + \frac{32}{25}\right)$$

$$= \frac{32}{15}\left(\ln 4 - \frac{6}{5}\right) \approx 0.40$$

Finally, using (9), we find the required standard deviation to be

$$\sigma = \sqrt{\text{Var}\,(x)} \approx 0.63$$

■ ■ ■ ■

ALTERNATIVE FORMULA FOR VARIANCE Using (8) to calculate the variance of a continuous random variable can be rather tedious. The following formula often makes this task easier:

$$\text{Var}\,(x) = \int_a^b x^2 f(x)\, dx - \mu^2 \tag{10}$$

This equation follows from these computations:

$$\text{Var}(x) = \int_a^b (x - \mu)^2 f(x)\, dx$$

$$= \int_a^b (x^2 - 2\mu x + \mu^2) f(x)\, dx$$

$$= \int_a^b x^2 f(x)\, dx - 2\mu \int_a^b x f(x)\, dx + \mu^2 \int_a^b f(x)\, dx$$

$$= \int_a^b x^2 f(x)\, dx - 2\mu \cdot \mu + \mu^2 \qquad \left[\text{Since } \int_a^b x f(x)\, dx = \mu \quad \text{and} \quad \int_a^b f(x)\, dx = 1 \right]$$

$$= \int_a^b x^2 f(x)\, dx - \mu^2$$

EXAMPLE 9 Use Formula (10) to calculate the variance of the random variable of Example 8.

SOLUTION ✓ Using (10), we have

$$\text{Var}(x) = \int_a^b x^2 f(x)\, dx - \mu^2 = \int_1^4 x^2 \cdot \frac{32}{15x^3}\, dx - \left(\frac{8}{5}\right)^2$$

$$= \frac{32}{15} \int_1^4 \frac{1}{x}\, dx - \frac{64}{25} = \frac{32}{15} \ln x \Big|_1^4 - \frac{64}{25}$$

$$= \frac{32}{15} \ln 4 - \frac{64}{25} = \frac{32}{15} \left(\ln 4 - \frac{6}{5} \right)$$

as obtained earlier.

■ ■ ■ ■

10.2 Exercises

In Exercises 1–14, find the mean, variance, and standard deviation of the random variable x associated with the given probability density function over the given interval.

1. $f(x) = \frac{1}{3}$; [3, 6]

2. $f(x) = \frac{1}{4}$; [2, 6]

3. $f(x) = \frac{3}{125} x^2$; [0, 5]

4. $f(x) = \frac{3}{8} x^2$; [0, 2]

5. $f(x) = \frac{3}{32} (x - 1)(5 - x)$; [1, 5]

6. $f(x) = 20(x^3 - x^4)$; [0, 1]

7. $f(x) = \frac{8}{7x^2}$; [1, 8]

8. $f(x) = \frac{4}{3x^2}$; [1, 4]

9. $f(x) = \frac{3}{14} \sqrt{x}$; [1, 4]

10. $f(x) = \frac{5}{2} x^{3/2}$; [0, 1]

11. $f(x) = \frac{3}{x^4}$; [1, ∞)

12. $f(x) = 3.5 x^{-4.5}$; [1, ∞)

13. $f(x) = \frac{1}{4} e^{-x/4}$; [0, ∞)   [Hint: $\lim_{x \to \infty} x^n e^{kx} = 0, k < 0$]

14. $f(x) = \frac{1}{9} x e^{-x/3}$; [0, ∞)   [Hint: $\lim_{x \to \infty} x^n e^{kx} = 0, k < 0$]

15. LIFE SPAN OF A PLANT The life span of a certain plant species (in days) is described by the probability density function

$$f(x) = \frac{1}{100} e^{-x/100}$$

If a plant of this species is selected at random, how long can the plant be expected to live?

16. **NUMBER OF CHIPS IN CHOCOLATE CHIP COOKIES** The number of chocolate chips in a certain brand of cookies has a distribution described by the probability density function

$$f(x) = \frac{1}{36}(6x - x^2) \qquad (0 \le x \le 6)$$

Find the expected number of chips in a cookie selected at random.

17. **SHOPPING HABITS** The amount of time t (in minutes) a shopper spends browsing in the magazine section of a supermarket is a continuous random variable with probability density function

$$f(t) = \frac{2}{25}t \qquad (0 \le t \le 5)$$

How much time is a shopper chosen at random expected to spend in the magazine section?

18. **REACTION TIME OF A MOTORIST** The amount of time t (in seconds) it takes a motorist to react to a road emergency is a continuous random variable with probability density function

$$f(t) = \frac{9}{4t^3} \qquad (1 \le t \le 3)$$

What is the expected reaction time for a motorist chosen at random?

19. **EXPECTED SNOWFALL** The amount of snowfall in feet in a remote region of Alaska in the month of January is a continuous random variable with probability density function

$$f(x) = \frac{2}{9}x(3 - x) \qquad (0 \le x \le 3)$$

Find the amount of snowfall one can expect in any given month of January in Alaska.

20. **GAS STATION SALES** The amount of gas (in thousands of gallons) Al's Gas Station sells on a typical Monday is a continuous random variable with probability density function

$$f(x) = 4(x - 2)^3 \qquad (2 \le x \le 3)$$

How much gas can the gas station expect to sell each Monday?

21. **DEMAND FOR BUTTER** The quantity demanded x (in thousands of pounds) of a certain brand of butter per week is a continuous random variable with probability density function

$$f(x) = \frac{6}{125}x(5 - x) \qquad (0 \le x \le 5)$$

What is the expected demand for this brand of butter per week?

22. **LIFE EXPECTANCY OF COLOR TELEVISION TUBES** The life expectancy (in years) of a certain brand of color television tube is a continuous random variable with probability density function

$$f(t) = 9(9 + t^2)^{-3/2} \qquad (0 \le t < \infty)$$

How long is one of these color television tubes expected to last?

In Exercises 23–28, find the median of the random variable x with the given probability density function defined on the given interval I. The median of x is defined to be the number m such that $P(x \le m) = \frac{1}{2}$. Observe that half of the x-values lie below m and the other half lie above m.

23. $f(x) = \frac{1}{6}; [2, 8]$ 24. $f(x) = \frac{2}{15}x; [1, 4]$

25. $f(x) = \frac{3}{16}\sqrt{x}; [0, 4]$ 26. $f(x) = \frac{1}{6\sqrt{x}}; [1, 16]$

27. $f(x) = \frac{3}{x^2}; [1, \infty)$ 28. $f(x) = \frac{1}{2}e^{-x/2}; [0, \infty)$

In Exercises 29 and 30, determine whether the statement is true or false. If it is true, explain why it is true. If it is false, give an example to show why it is false

29. If f is a probability density function of a continuous random variable x in the interval $[a, b]$, then the expected value of x is given by $\int_a^b x^2 f(x)\, dx$.

30. If f is a probability density function of a continuous random variable x in the interval $[a, b]$, then

$$\text{Var}(x) = \int_a^b x^2 f(x)\, dx - \left[\int_a^b x f(x)\, dx\right]^2$$

FINDING THE MEAN AND STANDARD DEVIATION

The calculation of the mean and standard deviation of a random variable is facilitated by the use of a graphing utility.

EXAMPLE 1

A survey conducted in 1995 of the Fortune 1000 companies revealed the following age distribution of company directors.

Age	20–25	25–30	30–35	35–40	40–45	45–50	50–55
Number of Directors	1	6	28	104	277	607	1142

Age	55–60	60–65	65–70	70–75	75–80	80–85	85–90
Number of Directors	1413	1424	494	159	62	31	5

Source: Directorship

a. Plot a histogram for the given data.

b. Find the mean age and the standard deviation of the company directors.

SOLUTION ✔

a. Let X denote the random variable taking on the values 1 through 14, where 1 corresponds to the age bracket 20–25, 2 corresponds to the age bracket 25–30, and so on. Entering the values of X as $x_1 = 1$, $x_2 = 2$, ... , $x_{14} = 14$ and the corresponding values of Y as $y_1 = 1$, $y_2 = 6$, ... , $y_{14} = 5$, and then using the **DRAW** function from the Statistics menu, we obtain the histogram shown in Figure T1.

FIGURE T1
The histogram for the given data, using the viewing rectangle $[0, 16] \times [0, 1500]$

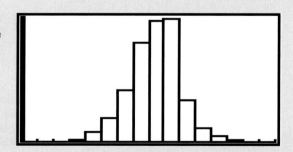

b. Using the appropriate function from the Statistics menu, we find that $\bar{x} = 7.9193$ and $\sigma x = 1.6378$; that is, the mean of X is $\mu \approx 7.9$ and the standard deviation is $\sigma \approx 1.6$. Thus, the average age of the directors is in the 55- to 60-year-old bracket. ■■■■

1. **a.** Graph the histogram associated with the random variable X in Example 5, page 747.
   b. Find the mean and the standard deviation for these data.

2. **a.** Graph the histogram associated with the random variable Y in Example 5, page 747.
   b. Find the mean and the standard deviation for these data.

3. **DRIVING AGE REQUIREMENTS** The minimum age requirement for a regular driver's license differs from state to state. The frequency distribution for this age requirement in the 50 states is given in the following table:

Minimum Age	15	16	17	18	19	21
Frequency of Occurrence	1	15	4	28	1	1

   a. Describe a random variable X that is associated with these data.
   b. Find the probability distribution for the random variable X.
   c. Graph the histogram associated with these data.
   d. Find the mean and the standard deviation for these data.

4. The distribution of the number of chocolate chips in a cookie is shown in the following table:

Number of Chocolate Chips, x	0	1	2
$P(X = x)$	0.01	0.03	0.05

Number of Chocolate Chips, x	3	4	5
$P(X = x)$	0.11	0.13	0.24

Number of Chocolate Chips, x	6	7	8
$P(X = x)$	0.22	0.16	0.05

   a. Describe a random variable X that is associated with these data.
   b. Find the probability distribution for the random variable X.
   c. Graph the histogram associated with these data.
   d. Find the mean and the standard deviation for these data.

5. A sugar refiner uses a machine to pack sugar in 5-lb cartons. In order to check the machine's accuracy, cartons are selected at random and weighed. The results follow:

```
4.98  5.02  4.96  4.97  5.03
4.96  4.98  5.01  5.02  5.06
4.97  5.04  5.04  5.01  4.99
4.98  5.04  5.01  5.03  5.05
4.96  4.97  5.02  5.04  4.97
5.03  5.01  5.00  5.01  4.98
```

   a. Describe a random variable X that is associated with these data.
   b. Find the probability distribution for the random variable X.
   c. Find the mean and the standard deviation for these data.

6. The scores of 25 students in a mathematics examination follow:

```
90  85  74  92  68  94  66
87  85  70  72  68  73  72
69  66  58  70  74  88  90
98  71  75  68
```

   a. Describe a random variable X that is associated with these data.
   b. Find the probability distribution for the random variable X.
   c. Find the mean and the standard deviation for these data.

10.3 Normal Distributions

NORMAL DISTRIBUTIONS

In Section 10.2 we saw the useful role played by exponential density functions in many applications. In this section we look at yet another class of continuous probability distributions known as **normal distributions.** The normal distribution is without doubt the most important of all the probability distributions. Many phenomena, such as the heights of people in a given population, the weights of newborn infants, the IQs of college students, and the actual weights of 16-ounce packages of cereals, have probability distributions that are normal. The normal distribution also provides us with an accurate approximation to the distributions of many random variables associated with random sampling problems.

The general **normal probability density function** with mean μ and standard deviation σ is defined to be

$$f(x) = \frac{e^{-(1/2)[(x-\mu)/\sigma]^2}}{\sigma\sqrt{2\pi}} \qquad (-\infty < x < \infty)$$

The graph of f, which is bell shaped, is called a **normal curve** (Figure 10.8).

FIGURE 10.8
A normal curve

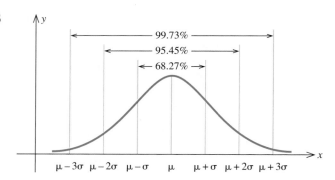

The normal curve (and therefore the corresponding normal distribution) is completely determined by its mean μ and standard deviation σ. In fact, the normal curve has the following characteristics, described in terms of these two parameters:

1. The curve has a peak at $x = \mu$.
2. The curve is symmetrical with respect to the vertical line $x = \mu$.
3. The curve always lies above the x-axis but approaches the x-axis as x extends indefinitely in either direction.

4. The area under the curve is 1.
5. For any normal curve, 68.27% of the area under the curve lies within 1 standard deviation of the mean (that is, between $\mu - \sigma$ and $\mu + \sigma$), 95.45% of the area lies within 2 standard deviations of the mean, and 99.73% of the area lies within 3 standard deviations of the mean.

Figure 10.9 shows two normal curves with different means μ_1 and μ_2 but the same deviation. Figure 10.10 shows two normal curves with the same mean but different standard deviations σ_1 and σ_2. (Which number is smaller?)

FIGURE 10.9
Two normal curves that have the same standard deviation but different means

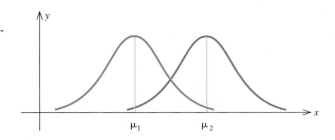

FIGURE 10.10
Two normal curves that have the same mean but different standard deviations

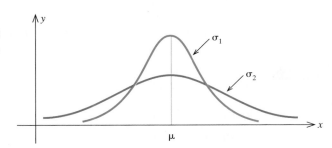

In general, the mean μ of a normal distribution determines where the center of the curve is located, whereas the standard deviation σ of a normal distribution determines the sharpness (or flatness) of the curve.

As this discussion reveals, infinitely many normal curves correspond to different choices of the parameters μ and σ, which characterize such curves. Fortunately, any normal curve may be transformed into any other normal curve (as we see later on), so in the study of normal curves it suffices to single out one such particular curve for special attention. The normal curve with mean $\mu = 0$ and standard deviation $\sigma = 1$ is called the **standard normal curve.** The corresponding distribution is called the **standard normal distribution.** The random variable itself is called the **standard normal variable** and is commonly denoted by Z.

the probability that a student randomly selected from the course could take dictation at the following speeds:
a. More than 120 words/minute
b. Between 80 and 120 words/minute
c. Less than 80 words/minute

25. **IQs** The IQs of students at Wilson Elementary School were measured recently and found to be normally distributed with a mean of 100 and a standard deviation of 15. Find the probability that a student selected at random will have an IQ of the following:
a. 140 or higher   **b.** 120 or higher
c. Between 100 and 120
d. 90 or less

26. **PRODUCT RELIABILITY** The tread lives of Super Titan radial tires under normal driving conditions are normally distributed with a mean of 40,000 mi and a standard deviation of 2000 mi. What is the probability that a tire selected at random will have a tread life of more than 35,000 mi? If four new tires are installed on a car and they experience even wear, determine the probability that all four tires still have useful tread lives after 35,000 mi of driving.

27. **FEMALE FACTORY WORKERS' WAGES** According to data released by a city's Chamber of Commerce, the weekly wages (in dollars) of female factory workers are normally distributed with a mean of 475 and a standard deviation of 50. Find the probability that a female factory worker selected at random from the city has a weekly wage of $450 to $550.

28. **CIVIL SERVICE EXAMS** To be eligible for further consideration, applicants for certain Civil Service positions must first pass a written qualifying examination on which a score of 70 or more must be obtained. In a recent examination, it was found that the scores were normally distributed with a mean of 60 points and a standard deviation of 10 points. Determine the percentage of applicants who passed the written qualifying examination.

29. **WARRANTIES** The general manager of the Service Department of MCA Television Company has estimated that the time that elapses between the dates of purchase and the dates on which the 19-in. sets manufactured by the company first require service is normally distributed with a mean of 22 mo and a standard deviation of 4 mo. If MCA gives a 1-yr warranty on parts and labor for these sets, determine what percentage of sets manufactured and sold may require service before the warranty period runs out.

30. **GRADE DISTRIBUTIONS** The scores on an economics examination are normally distributed with a mean of 72 and a standard deviation of 16. If the instructor assigns a grade of A to 10% of the class, what is the lowest score a student may have and still get an A?

31. **GRADE DISTRIBUTIONS** The scores on a sociology examination are normally distributed with a mean of 70 and a standard deviation of 10. If the instructor assigns A's to 15%, B's to 25%, C's to 40%, D's to 15%, and F's to 5% of the class, find the cutoff points for these grades.

SOLUTIONS TO SELF-CHECK EXERCISES 10.3

1. a. The probability $P(-1.2 < Z < 2.1)$ is given by the shaded area in the accompanying figure. We have

$$P(-1.2 < Z < 2.1) = P(Z < 2.1) - P(Z < -1.2)$$
$$= 0.9821 - 0.1151$$
$$= 0.867$$

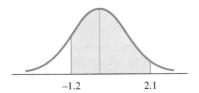

b. The region associated with $P(-z < Z < z)$ is shown in the accompanying figure. Observe that we have the following relationship:

$$P(Z < z) = \frac{1}{2}[1 + P(-z < Z < z)]$$

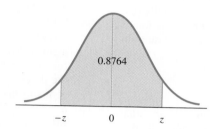

(see Example 2c). With $P(-z < Z < z) = 0.8764$, we find

$$P(Z < z) = \frac{1}{2}(1 + 0.8764)$$

$$= 0.9382$$

Consulting the table, we find $z = 1.54$.

2. Using the transformations (11), (12), and (13) and the table of values of Z, we have

a. $P(X < 100) = P\left(Z < \dfrac{100 - 80}{10}\right)$

$= P(Z < 2)$

$= 0.9772$

b. $P(X > 60) = P\left(Z > \dfrac{60 - 80}{10}\right)$

$= P(Z > -2)$

$= P(Z < 2)$

$= 0.9772$

c. $P(70 < X < 90) = P\left(\dfrac{70 - 80}{10} < Z < \dfrac{90 - 80}{10}\right)$

$= P(-1 < Z < 1)$

$= P(Z < 1) - P(Z < -1)$

$= 0.8413 - 0.1587$

$= 0.6826$

3. Let X be the normal random variable denoting the serum cholesterol levels (in mg/dL) in the current Mediterranean population under consideration. Thus, the percentage of the population having blood cholesterol levels between 160 and 180 mg/dL is given by $P(160 < X < 180)$. To compute $P(160 < X < 180)$, we use

(11), with $\mu = 160$, $\sigma = 50$, $a = 160$, and $b = 180$. We find

$$P(160 < X < 180) = P\left(\frac{160 - 160}{50} < Z < \frac{180 - 160}{50}\right)$$

$$= P(0 < Z < 0.4)$$

$$= P(Z < 0.4) - P(Z < 0)$$

$$= 0.6554 - 0.5000$$

$$= 0.1554$$

so approximately 15.5% of the population has blood cholesterol levels between 160 and 180 mg/dL.

CHAPTER 10 Summary of Principal Formulas and Terms

Formulas

1. Expected value $E(x) = x_1p_1 + x_2p_2 + \cdots + x_np_n$

2. Expected value of a continuous $E(x) = \int_a^b xf(x)\,dx$
   random variable

3. Exponential density function $f(x) = ke^{-kx}$

4. Expected value of an exponen- $E(x) = \dfrac{1}{k}$
   tial density function

5. Variance of a continuous ran- $\text{Var}(x) = \int_a^b (x - \mu)^2 f(x)\,dx$
   dom variable

6. Standard deviation of a continu- $\sigma = \sqrt{\text{Var}(x)}$
   ous random variable

Terms

experiment
outcomes (sample points)
sample space
event
probability of an event
random variable
discrete probability
    function
histogram
continuous random
    variable

probability density
    function
average (mean)
expected value
variance of a random
    variable
standard deviation of a
    random variable
normal distribution
standard normal
    random variable

CHAPTER 10 REVIEW EXERCISES

In Exercises 1–4, show that the function is a probability density function on the given interval.

1. $f(x) = \dfrac{1}{28}(2x + 3); [0, 4]$

2. $f(x) = \dfrac{3}{16}\sqrt{x}; [0, 4]$

3. $f(x) = \dfrac{1}{4}; [7, 11]$

4. $f(x) = \dfrac{4}{x^5}; [1, \infty)$

In Exercises 5–8, find the value of the constant k so that the function is a probability density function in the given interval.

5. $f(x) = kx^2; [0, 9]$

6. $f(x) = \dfrac{k}{\sqrt{x}}; [1, 16]$

7. $f(x) = \dfrac{k}{x^2}; [1, 3]$

8. $f(x) = \dfrac{k}{x^{2.5}}; [1, \infty)$

In Exercises 9–12, f is a probability density function defined on the given interval. Find the indicated probabilities.

9. $f(x) = \dfrac{2}{21}x; [2, 5]$

   a. $P(x \le 4)$ **b.** $P(x = 4)$
   c. $P(3 \le x \le 4)$

10. $f(x) = \dfrac{1}{4}; [1, 5]$

   a. $P(2 \le x \le 4)$ **b.** $P(x \le 3)$
   c. $P(x \ge 2)$

11. $f(x) = \dfrac{3}{16}\sqrt{x}; [0, 4]$

   a. $P(1 \le x \le 3)$ **b.** $P(x \le 3)$
   c. $P(x = 2)$

12. $f(x) = \dfrac{1}{x^2}; [1, \infty)$

   a. $P(x \le 10)$ **b.** $P(2 \le x \le 4)$
   c. $P(x \ge 2)$

In Exercises 13–16, find the mean, variance, and standard deviation of the random variable x associated with the probability density function f over the given interval.

13. $f(x) = \dfrac{1}{5}; [2, 7]$

14. $f(x) = \dfrac{1}{28}(2x + 3); [0, 4]$

15. $f(x) = \dfrac{1}{4}(3x^2 + 1); [-1, 1]$

16. $f(x) = \dfrac{4}{x^5}; [1, \infty)$

In Exercises 17–20, Z is the standard normal variable. Find the given probability.

17. $P(Z < 2.24)$

18. $P(Z > -1.24)$

19. $P(0.24 \le Z \le 1.28)$

20. $P(-1.37 \le Z \le 1.37)$

21. Suppose X is a normal random variable with $\mu = 80$ and $\sigma = 8$. Find these values:

   a. $P(X \le 84)$ **b.** $P(X \ge 70)$
   c. $P(75 \le X \le 85)$

22. Suppose X is a normal random variable with $\mu = 45$ and $\sigma = 3$. Find these values:

   a. $P(X \le 50)$ **b.** $P(X \ge 40)$
   c. $P(40 \le X \le 50)$

23. Records at Centerville Hospital indicate that the length of time in days that a maternity patient stays in the hospital has a probability density function given by

$$P(t) = \frac{1}{4}e^{-(1/4)t}$$

a. What is the probability that a woman entering the maternity wing will be there longer than 6 days?

b. What is the probability that a woman entering the maternity wing will be there less than 2 days?

c. What is the average length of time that a woman entering the maternity wing stays in the hospital?

24. The life span (in years) of a certain make of car battery is an exponentially distributed random variable with an expected value of 5. Find the probability that the life span of a battery is (a) less than 4 yr, (b) more than 6 yr, and (c) between 2 and 4 yr.

TAYLOR POLYNOMIALS AND INFINITE SERIES

11

In this chapter we show how certain functions can be represented by a *power series*. A power series involves infinitely many terms, but when truncated it is just a polynomial. By approximating a function with a *Taylor polynomial*, we are often able to obtain approximate solutions to problems that we cannot otherwise solve.

We also look at a method, called the Newton–Raphson method, for finding the zeros of a function. For example, Newton's method can be used to find the critical points of a function, which, as you may recall, are candidates for the solution of the optimization problems considered in Chapter 4.

What percentage of the nonfarm workforce will be in the service industries one decade from now? In Example 4, page 780, you will see how a Taylor polynomial can be used to help answer this question.

11.1 Taylor Polynomials

As we saw earlier, obtaining an exact solution to a problem is not always possible; in such cases we have to settle for an approximate solution. In this section we show how a function may be approximated near a given point by a polynomial. Polynomials, as we have seen time and again, are easy to work with; for example, they are easy to evaluate, differentiate, and integrate. Thus, by using polynomials rather than working with the original function itself, we can often obtain approximate solutions to a problem that we might otherwise not be able to solve.

TAYLOR POLYNOMIALS

Suppose we are given a differentiable function f and a point $x = a$ in the domain of f. Then the polynomial of degree 0 that best approximates f *near* $x = a$ is the constant polynomial

$$P_0(x) = f(a)$$

which coincides with f at $x = a$ (Figure 11.1a).

FIGURE 11.1

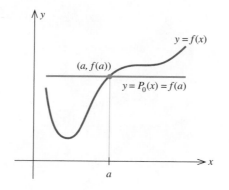

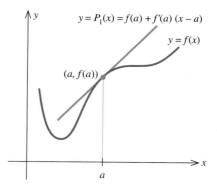

(a) $P_0(x) = f(a)$ is a zero-degree polynomial that approximates f near $x = a$.

(b) $P_1(x) = f(a) + f'(a)(x - a)$ is a first-degree polynomial that approximates f near $x = a$.

Now, unless f itself is a constant function, it is possible in most cases to obtain a better approximation of f *near* $x = a$ by using a polynomial function of degree 1. Recall that the linear function

$$L(x) = f(a) + f'(a)(x - a)$$

is just an equation of the tangent line to the graph of the function f at the point $(a, f(a))$ (Figure 11.1b). As such, the value of the function L coincides

c. Here

$$a_1 = \frac{\pi}{1!} = \pi, \qquad a_2 = \frac{\pi^2}{2!} = \frac{\pi^2}{2}, \qquad a_3 = \frac{\pi^3}{3!} = \frac{\pi^3}{6}, \qquad a_4 = \frac{\pi^4}{4!} = \frac{\pi^4}{24}, \qquad \ldots$$

so the required sequence is

$$\pi, \frac{\pi^2}{2!}, \frac{\pi^3}{3!}, \frac{\pi^4}{4!}, \qquad \ldots, \qquad \frac{\pi^n}{n!}, \qquad \ldots \qquad \blacksquare\blacksquare\blacksquare\blacksquare$$

EXAMPLE 2 Find the general term of each sequence:

a. $1, \dfrac{1}{4}, \dfrac{1}{9}, \dfrac{1}{16}, \dfrac{1}{25}, \ldots$ **b.** $-1, 1, -1, 1, -1, 1, \ldots$

SOLUTION ✔ **a.** Observe that the terms may be written as

$$a_1 = \frac{1}{1^2}, \qquad a_2 = \frac{1}{2^2}, \qquad a_3 = \frac{1}{3^2}, \qquad a_4 = \frac{1}{4^2}, \qquad a_5 = \frac{1}{5^2}, \qquad \ldots$$

and we conclude that the nth term is $a_n = \dfrac{1}{n^2}$.

b. Here
$$a_1 = -1, \qquad a_2 = 1, \qquad a_3 = -1, \qquad a_4 = 1, \qquad a_5 = -1, \qquad \ldots$$

which may also be written as

$$a_1 = (-1)^1, \qquad a_2 = (-1)^2, \qquad a_3 = (-1)^3,$$
$$a_4 = (-1)^4, \qquad a_5 = (-1)^5, \qquad \ldots$$

and so $a_n = (-1)^n$. $\blacksquare\blacksquare\blacksquare\blacksquare$

GRAPHS OF INFINITE SEQUENCES

Since an infinite sequence is a function, we can sketch its graph. Because the domain of the function is the set of positive integers, the graph of a sequence consists of an infinite collection of points in the xy-plane.

EXAMPLE 3 Sketch the graph of each sequence:

a. $\{n + 1\}$ **b.** $\left\{\dfrac{n + 1}{n}\right\}$ **c.** $\{(-1)^n\}$

SOLUTION ✔ **a.** The terms of the sequence $\{n + 1\}$ are 2, 3, 4, 5, ... Recalling that these numbers are precisely the functional values of $f(n) = n + 1$ for $n = 1, 2, 3, \ldots$, we obtain the following table of values of f:

n	1	2	3	4	$\cdots$
$f(n)$	2	3	4	5	$\cdots$

from which we construct the graph of $\{n + 1\}$ shown in Figure 11.3a.

FIGURE 11.3

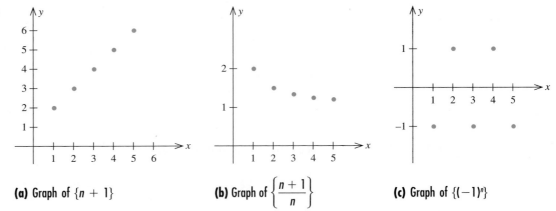

(a) Graph of $\{n + 1\}$ **(b)** Graph of $\left\{\dfrac{n+1}{n}\right\}$ **(c)** Graph of $\{(-1)^n\}$

b. From the following table of values of f,

n	1	2	3	4	$\cdots$
$f(n)$	2	$\frac{3}{2}$	$\frac{4}{3}$	$\frac{5}{4}$	$\cdots$

we sketch the graph shown in Figure 11.3b.

c. We use the following table of values for the sequence

n	1	2	3	4	$\cdots$
$f(n)$	-1	1	-1	1	$\cdots$

to sketch the graph of $\{(-1)^n\}$ shown in Figure 11.3c.   ■ ■ ■ ■

REMARK Notice that the function $f(n) = n + 1$ defining the sequence $\{n + 1\}$ may be viewed as the function $f(x) = x + 1$ $(-\infty < x < \infty)$ with x *restricted* to the set of positive integers. Similarly, the function defining the sequence $\{(n + 1)/n\}$ is just the function $f(x) = (x + 1)/x$ with a similar restriction on its domain. Finally, the function defining the sequence $\{(-1)^n\}$ is the function $f(x) = \cos \pi x$ with x a positive integer.* These observations suggest that it is possible to study the properties of a sequence by analyzing the properties of a corresponding function defined for *all* values of x in some suitable interval.   ■ ■ ■

THE LIMIT OF A SEQUENCE

Given a sequence $\{a_n\}$, we may ask whether the terms a_n of the sequence approach some specific number L as n gets larger and larger. If they do, we say

---

* The trigonometric function $f(x) = \cos \pi x$ will be defined in Chapter 12.

that the sequence a_n *converges* to L. More specifically, we have the following informal definition.

Limit of a Sequence

> Let $\{a_n\}$ be a given sequence. We say that the sequence $\{a_n\}$ **converges** and has the limit L, written
>
> $$\lim_{n \to \infty} a_n = L$$
>
> if the terms of the sequence, a_n, can be made as close to L as we please by taking n sufficiently large. If a sequence is not convergent, it is said to be **divergent**.

EXAMPLE 4   Determine whether each of the following infinite sequences converges or diverges.

a. $\left\{ \dfrac{1}{n} \right\}$ b. $\{\sqrt{n}\}$ c. $\{(-1)^n\}$

SOLUTION ✓

a. The terms of the sequence,

$$1, \frac{1}{2}, \frac{1}{3}, \frac{1}{4}, \frac{1}{5}, \quad \ldots, \quad \frac{1}{n}, \quad \ldots$$

approach the number $L = 0$ as n gets larger and larger. We conclude that

$$\lim_{n \to \infty} a_n = \lim_{n \to \infty} \frac{1}{n} = 0$$

Compare this with

$$\lim_{x \to \infty} \frac{1}{x} = 0$$

b. The terms of the sequence,

$$1, \sqrt{2}, \sqrt{3}, \sqrt{4}, \sqrt{5}, \ldots, \sqrt{n}, \ldots$$

get larger and larger as n gets larger and larger. Consequently, they do not approach any finite number L as n tends to infinity. Therefore, the sequence is divergent. Compare this with

$$\lim_{x \to \infty} \sqrt{x}$$

c. The terms of the sequence are

$$a_1 = -1, \quad a_2 = 1, \quad a_3 = -1, \quad a_4 = 1,$$
$$a_5 = -1, \quad \ldots, \quad a_n = (-1)^n, \quad \ldots$$

Thus, no matter how large n is, there are terms that are equal to -1 (those with odd-numbered subscripts) and also terms that are equal to 1 (those with even-numbered subscripts). This implies that there cannot be a *unique* real number L such that a_n is arbitrarily close to L no matter how large n is. Therefore, the sequence is divergent. ■ ■ ■ ■

Exploring with Technology

Refer to Example 4 and the REMARK on page 790.

1. Plot the graph of $f(x) = 1/x$, using the viewing rectangle $[0, 10] \times [0, 3]$, and thus verify graphically that $\lim\limits_{n\to\infty} (1/n) = 0$.

2. Plot the graph of $f(x) = \sqrt{x}$, using an appropriate viewing rectangle, and thus verify graphically that $\lim\limits_{n\to\infty} \sqrt{n}$ does not exist.

3. Plot the graph of $f(x) = \cos \pi x$, using the viewing rectangle $[0, 10] \times [-2, 2]$, and thus verify graphically that $\lim\limits_{n\to\infty} (-1)^n$ does not exist. [*Note:* Trigonometric functions will be studied in Chapter 12.]

The following properties of sequences, which parallel those of the limit of $f(x)$ at infinity, are helpful in computing limits of sequences.

Limit Properties of Sequences

Suppose that

$$\lim_{n\to\infty} a_n = A \qquad \text{and} \qquad \lim_{n\to\infty} b_n = B$$

Then

1. $\lim\limits_{n\to\infty} ca_n = c \lim\limits_{n\to\infty} a_n = cA \qquad (c, \text{a constant})$

2. $\lim\limits_{n\to\infty} (a_n \pm b_n) = \lim\limits_{n\to\infty} a_n \pm \lim\limits_{n\to\infty} b_n = A \pm B \qquad$ (Sum rule)

3. $\lim\limits_{n\to\infty} a_n b_n = \left(\lim\limits_{n\to\infty} a_n\right)\left(\lim\limits_{n\to\infty} b_n\right) = AB \qquad$ (Product rule)

4. $\lim\limits_{n\to\infty} \dfrac{a_n}{b_n} = \dfrac{\lim\limits_{n\to\infty} a_n}{\lim\limits_{n\to\infty} b_n} = \dfrac{A}{B} \quad$ provided $B \neq 0 \qquad$ (Quotient rule)

EXAMPLE 5   Evaluate:

a. $\lim\limits_{n\to\infty} \left(\dfrac{1}{2}\right)^n$ **b.** $\lim\limits_{n\to\infty} \dfrac{2n^2 + 1}{3n^2 + n + 2}$

SOLUTION ✔

a. $\lim\limits_{n\to\infty} \left(\dfrac{1}{2}\right)^n = \lim\limits_{n\to\infty} \dfrac{1}{2^n}$

$= \dfrac{1}{\lim\limits_{n\to\infty} 2^n} = 0$

b. Dividing the numerator and denominator by n^2, we find

$$\lim_{n \to \infty} \frac{2n^2 + 1}{3n^2 + n + 2} = \lim_{n \to \infty} \frac{2 + \dfrac{1}{n^2}}{3 + \dfrac{1}{n} + \dfrac{2}{n^2}}$$

$$= \frac{\lim\limits_{n \to \infty} \left(2 + \dfrac{1}{n^2} \right)}{\lim\limits_{n \to \infty} \left(3 + \dfrac{1}{n} + \dfrac{2}{n^2} \right)}$$

$$= \frac{2}{3}$$

■ ■ ■ ■

Group Discussion

Consider the sequences $\{a_n\}$ and $\{b_n\}$ defined by $a_n = (-1)^{n+1}$ and $b_n = (-1)^n$.

1. Show that $\lim\limits_{n \to \infty} a_n$ and $\lim\limits_{n \to \infty} b_n$ do not exist.

2. Show that $\lim\limits_{n \to \infty} (a_n + b_n) = 0$.

3. Do the results of parts (1) and (2) contradict the limit properties of sequences listed on page 792? Explain your answer.

EXAMPLE 6

Of the spark plugs manufactured by the Parts Division of United Motors Corporation, 2% are defective. It can be shown that the probability of getting at least one defective plug in a random sample of n spark plugs is $f(n) = 1 - (0.98)^n$. Consider the sequence $\{a_n\}$ defined by $a_n = f(n)$.

a. Write down the terms a_5, a_{10}, a_{25}, a_{100}, and a_{200} of the sequence $\{a_n\}$.
b. Evaluate

$$\lim_{n \to \infty} a_n$$

and interpret your results.

SOLUTION ✔

a. The required terms of the sequence are

$$0.10, \ 0.18, \ 0.40, \ 0.87, \ 0.98$$

For example, the probability of getting at least one defective plug in a random sample of 25 is 0.4—that is, a 40% chance.

b. $\lim\limits_{n \to \infty} a_n = \lim\limits_{n \to \infty} [1 - (0.98)^n]$

$= \lim\limits_{n \to \infty} 1 - \lim\limits_{n \to \infty} (0.98)^n$

$= 1 - 0$

$= 1$

The result tells us that if the sample is large enough, we will almost certainly pick at least one defective plug!

■ ■ ■ ■

1. Determine whether the sequence $\left\{ \dfrac{3n^2 + 2n + 1}{n^2 + 4} \right\}$ converges or diverges. If it converges, find its limit.

2. Consider the sequence $\left\{ \dfrac{n}{n^2 + 1} \right\}$.

   a. Sketch the graph of the sequence.
   b. Show that the sequence is decreasing—that is, $a_1 > a_2 > a_3 > \cdots$.
   Hint: Consider $f(x) = x/(x^2 + 1)$ and show that f is decreasing by computing f'.

   Solutions to Self-Check Exercises 11.2 can be found on page 795.

11.2 Exercises

In Exercises 1–9, write down the first five terms of the sequence.

1. $\{a_n\} = \{2^{n-1}\}$

2. $\{a_n\} = \left\{ \dfrac{2n}{1 + n^2} \right\}$

3. $\{a_n\} = \left\{ \dfrac{n-1}{n+1} \right\}$

4. $\{a_n\} = \left\{ \left(-\dfrac{1}{3}\right)^n \right\}$

5. $\{a_n\} = \left\{ \dfrac{2^{n-1}}{n!} \right\}$

6. $\{a_n\} = \left\{ \dfrac{(-1)^n}{(2n)!} \right\}$

7. $\{a_n\} = \left\{ \dfrac{e^n}{n^3} \right\}$

8. $\{a_n\} = \left\{ \dfrac{\sqrt{n}}{\sqrt{n}+1} \right\}$

9. $\{a_n\} = \left\{ \dfrac{3n^2 - n + 1}{2n^2 + 1} \right\}$

In Exercises 10–21, find the general term of the sequence.

10. $\dfrac{1}{2}, \dfrac{1}{4}, \dfrac{1}{6}, \dfrac{1}{8}, \ldots$

11. $1, 4, 7, 10, \ldots$

12. $\dfrac{1}{3}, \dfrac{1}{7}, \dfrac{1}{11}, \dfrac{1}{15}, \ldots$

13. $1, \dfrac{1}{8}, \dfrac{1}{27}, \dfrac{1}{64}, \ldots$

14. $1, \dfrac{2}{3}, \dfrac{4}{9}, \dfrac{8}{27}, \ldots$

15. $2, \dfrac{8}{5}, \dfrac{32}{25}, \dfrac{128}{125}, \ldots$

16. $1, \dfrac{5}{4}, \dfrac{7}{5}, \dfrac{9}{6}, \ldots$

17. $1, -\dfrac{1}{2}, \dfrac{1}{4}, -\dfrac{1}{8}, \ldots$

18. $1 + \dfrac{1}{2}, 1 + \dfrac{1}{3}, 1 + \dfrac{1}{4}, 1 + \dfrac{1}{5}, \ldots$

19. $\dfrac{1}{2 \cdot 3}, \dfrac{2}{3 \cdot 4}, \dfrac{3}{4 \cdot 5}, \dfrac{4}{5 \cdot 6}, \ldots$

20. $1, \dfrac{2}{1 \cdot 3}, \dfrac{4}{1 \cdot 3 \cdot 5}, \dfrac{8}{1 \cdot 3 \cdot 5 \cdot 7}, \ldots$

21. $1, e, \dfrac{e^2}{2}, \dfrac{e^3}{6}, \dfrac{e^4}{24}, \dfrac{e^5}{120}, \ldots$

In Exercises 22–29, sketch the graph of the sequence.

22. $\{n^2\}$

23. $\left\{ \dfrac{2n}{n+1} \right\}$

24. $\left\{ \dfrac{(-1)^n}{n} \right\}$

25. $\{\sqrt{n}\}$

26. $\{\ln n\}$

27. $\{e^n\}$

28. $\{ne^{-n}\}$

29. $\{n - \sqrt{n}\}$

In Exercises 30–44, determine the convergence or divergence of each given sequence $\{a_n\}$. If the sequence converges, find its limit.

30. $a_n = \dfrac{n}{n^2 + 1}$

31. $a_n = \dfrac{n+1}{2n}$

32. $a_n = \sqrt[3]{n}$

33. $a_n = \dfrac{(-1)^n}{\sqrt{n}}$

34. $a_n = \dfrac{1}{n+1} - \dfrac{1}{n+2}$

35. $a_n = \dfrac{\sqrt{n} - 1}{\sqrt{n} + 1}$

36. $a_n = \dfrac{3n^2 + n - 1}{6n^2 + n + 1}$

37. $a_n = \dfrac{2n^3 - 1}{n^3 + 2n + 1}$

38. $a_n = \dfrac{1 + (-1)^n}{3^n}$

39. $a_n = 2 - \dfrac{1}{2^n}$

40. $a_n = \dfrac{2n}{n!}$

41. $a_n = \dfrac{2^n}{3^n}$

42. $a_n = \dfrac{2^n - 1}{2^n}$

43. $a_n = \dfrac{n}{\sqrt{2n^2 + 3}}$

44. $a_n = \dfrac{2^n}{n!}$

45. AUTOMOBILE MICROPROCESSORS Of the microprocessors manufactured by a microelectronics firm for use in regulating fuel consumption in automobiles, $1\frac{1}{2}\%$ are defective. It can be shown that the probability of getting at least one defective microprocessor in a random sample of n microprocessors is $f(n) = 1 - (0.985)^n$. Consider the sequence $\{a_n\}$ defined by $a_n = f(n)$.
  a. Write down the terms a_1, a_{10}, a_{100}, and a_{1000} of the sequence $\{a_n\}$.
  b. Evaluate $\lim\limits_{n \to \infty} a_n$ and interpret your results.

46. SAVINGS ACCOUNTS An amount of $1000 is deposited in a bank that pays 8% interest/year compounded daily (take the number of days in a year to be 365). Let a_n denote the total amount on deposit after n days, assuming no deposits or withdrawals are made during the period in question.
  a. Find the formula for a_n.
  Hint: See Section 5.3.
  b. Compute a_1, a_{10}, a_{50}, and a_{100}.
  c. What amount is on deposit after 1 yr?

47. ACCUMULATED AMOUNT Suppose $100 is deposited into an account earning interest at 12%/year compounded monthly. Let a_n denote the amount on deposit (called the accumulated amount or the future value) at the end of the nth mo.
  a. Show that $a_1 = 100(1.01)$, $a_2 = 100(1.01)^2$, and $a_3 = 100(1.01)^3$.
  b. Find the accumulated amount a_n.
  c. Find the 24th term of the sequence $\{a_n\}$ and interpret your result.

48. TRANSMISSION OF DISEASE In the early stages of an epidemic, the number of persons who have contracted the disease on the $(n + 1)$st day, a_{n+1}, is related to the number of persons who have the disease on the nth day, a_n, by the equation

$$a_{n+1} = (1 + aN - b)a_n$$

where N denotes the total population and a and b are positive constants that depend on the nature of the disease. Put $r = 1 + aN - b$.
  a. Write down the first n terms of the sequence $\{a_n\}$.
  b. Evaluate $\lim\limits_{n \to \infty} a_n$ for each of the three cases $r < 1$, $r = 1$, and $r > 1$.
  c. Interpret the results of part (b).

In Exercises 49–52, determine whether the statement is true or false. If it is true, explain why it is true. If it is false, give an example to show why it is false.

49. If $\lim\limits_{n \to \infty} a_n = L$ and $\lim\limits_{n \to \infty} b_n = 0$, then $\lim\limits_{n \to \infty} a_n b_n = 0$.

50. If $\{a_n\}$ and $\{b_n\}$ are sequences such that $\lim\limits_{n \to \infty} (a_n + b_n)$ exists, then both $\lim\limits_{n \to \infty} a_n$ and $\lim\limits_{n \to \infty} b_n$ must exist.

51. If $\{a_n\}$ is bounded (that is $|a_n| \le M$ for some positive real numbers M and $n = 1, 2, 3, \ldots$) and $\{b_n\}$ converges, then $\lim\limits_{n \to \infty} a_n b_n$ exists.

52. If $\lim\limits_{n \to \infty} a_n b_n$ exists, then both $\lim\limits_{n \to \infty} a_n$ and $\lim\limits_{n \to \infty} b_n$ must exist.

SOLUTIONS TO SELF-CHECK EXERCISES 11.2

1. We compute

$$\lim_{n \to \infty} \frac{3n^2 + 2n + 1}{n^2 + 4} = \lim_{n \to \infty} \frac{3 + \dfrac{2}{n} + \dfrac{1}{n^2}}{1 + \dfrac{4}{n^2}} \qquad \text{(Divide numerator and denominator by } n^2 \text{.)}$$

$$= \frac{3}{1} = 3$$

2. a. From the table

n	1	2	3	4	5	$\cdots$
$f(n)$	$\frac{1}{2}$	$\frac{2}{5}$	$\frac{3}{10}$	$\frac{4}{17}$	$\frac{5}{26}$	$\cdots$

we obtain the graph shown in the figure.

b. We compute

$$f'(x) = \frac{(x^2+1)(1) - x(2x)}{(x^2+1)^2} = \frac{1-x^2}{(x^2+1)^2} < 0$$

for $x > 1$, and so f is decreasing on $(1, \infty)$. We conclude that $\{a_n\}$ is decreasing.

11.3 Infinite Series

THE SUM OF AN INFINITE SERIES

In Section 11.2 we raised a question concerning the "sum" of a series involving infinitely many terms. In this section we show how we define such a sum. Consider the **infinite series**

$$a_1 + a_2 + \cdots + a_n + \cdots$$

which may be abbreviated through the use of sigma notation as

$$\sum_{n=1}^{\infty} a_n = a_1 + a_2 + a_3 + \cdots + a_n + \cdots \qquad (4)$$

and is read "the sum of the numbers a_n for n running from 1 to infinity."

Now, given an infinite series (4) whose terms are drawn from the sequence $\{a_n\}$, let's define the sums

$$S_1 = \sum_{n=1}^{1} a_n = a_1$$

$$S_2 = \sum_{n=1}^{2} a_n = a_1 + a_2$$

$$S_3 = \sum_{n=1}^{3} a_n = a_1 + a_2 + a_3$$

$$S_N = \sum_{n=1}^{N} a_n = a_1 + a_2 + a_3 + \cdots + a_N$$

Observe that each of these sums exists since each is obtained by adding together finitely many numbers. For each N, S_N is called the **Nth partial sum** of the series (4), and the sequence $\{S_n\}$ is called the **sequence of partial sums** of the series (4). We are now in a position to define the sum of an infinite series.

Sum of an Infinite Series

Let

$$\sum_{n=1}^{\infty} a_n = a_1 + a_2 + a_3 + \cdots$$

be an infinite series and let $\{S_n\}$ be the sequence of partial sums of the infinite series. If

$$\lim_{n \to \infty} S_n = S$$

we say that the infinite series $\sum_{n=1}^{\infty} a_n$ converges to S and write

$$\sum_{n=1}^{\infty} a_n = \lim_{n \to \infty} S_n = S$$

In this case, S is called the sum of the series. If $\{S_n\}$ does not converge, we say that the infinite series diverges and has no sum.

REMARK Simply stated, this definition tells us that the partial sums of a convergent series ultimately form a sequence of increasingly accurate approximations to the sum of the series.*   ■ ■ ■

EXAMPLE 1

a. Show that the following infinite series diverges:

$$\sum_{n=1}^{\infty} (-1)^{n+1} = 1 - 1 + 1 - 1 + 1 - 1 + \cdots$$

b. Show that the following infinite series converges:

$$\sum_{n=1}^{\infty} \frac{1}{n(n+1)} = \frac{1}{1 \cdot 2} + \frac{1}{2 \cdot 3} + \frac{1}{3 \cdot 4} + \cdots$$

SOLUTION ✔

a. The partial sums of the given infinite series are

$$S_1 = 1, \qquad S_2 = 1 - 1 = 0, \qquad S_3 = 1 - 1 + 1 = 1$$
$$S_4 = 1 - 1 + 1 - 1 = 0, \qquad \cdots$$

The sequence of partial sums $\{S_n\}$ evidently diverges, and so the given infinite series diverges.

b. Let's write

$$\frac{1}{n(n+1)} = \frac{1}{n} - \frac{1}{n+1}$$

---

* This is true for most of the convergent series that arise in applications.

an equality that is easily verified. The Nth partial sum of the given series is

$$S_N = \sum_{n=1}^{N} \frac{1}{n(n+1)} = \sum_{n=1}^{N} \left(\frac{1}{n} - \frac{1}{n+1} \right)$$

$$= \left(1 - \frac{1}{2} \right) + \left(\frac{1}{2} - \frac{1}{3} \right) + \left(\frac{1}{3} - \frac{1}{4} \right) + \cdots + \left(\frac{1}{N} - \frac{1}{N+1} \right)$$

$$= 1 + \left(-\frac{1}{2} + \frac{1}{2} \right) + \left(-\frac{1}{3} + \frac{1}{3} \right) + \cdots + \left(-\frac{1}{N} + \frac{1}{N} \right) - \frac{1}{N+1}$$

$$= 1 - \frac{1}{N+1}$$

Since

$$\lim_{N \to \infty} S_N = \lim_{N \to \infty} \left(1 - \frac{1}{N+1} \right) = 1$$

we conclude that the given series converges and has a sum equal to 1; that is,

$$\sum_{n=1}^{\infty} \frac{1}{n(n+1)} = 1$$

▪ ▪ ▪ ▪

The series in Example 1(b) is called a *telescoping series* because all the terms between the first and the last in the expression for S_N "collapse."

Exploring with Technology

Refer to Example 1b.

1. Verify graphically the results of Example 1b by plotting the graphs of the partial sums

$$S_1 = \frac{1}{1 \cdot 2}, \qquad S_2 = \frac{1}{1 \cdot 2} + \frac{1}{2 \cdot 3}, \qquad \cdots, \qquad S_6 = \frac{1}{1 \cdot 2} + \frac{1}{2 \cdot 3} + \cdots + \frac{1}{6 \cdot 7}$$

and $S = 1$ in the viewing rectangle $[0, 3] \times [0, 1]$.

2. Refer to the abbreviated form for S_N—namely, $S_N = \left(1 - \frac{1}{N+1} \right)$. By plotting the graphs of $y_1 = \left(1 - \frac{1}{x+1} \right)$ and $y_2 = 1$ in the viewing rectangle $[0, 50] \times [0, 1.1]$, verify graphically that $\lim_{N \to \infty} S_N = 1$ and thus the result $\sum_{n=1}^{\infty} \frac{1}{n(n+1)} = 1$, as obtained in Example 1b.

Group Discussion

1. An example of a divergent infinite series is

$$\sum_{n=0}^{\infty} 1 = 1 + 1 + 1 + \cdots$$

Show that the series is divergent by establishing the following:

a. The nth partial sum of the infinite series is $S_n = n$.

b. $\lim\limits_{n \to \infty} S_n = \infty$ so that $\{S_n\}$ is divergent and the desired result follows.

2. Since the *terms* of the infinite series in part 1 do not decrease, it is evident that the partial sums of the series must grow without bound. Therefore, it is not difficult to see that the infinite series cannot converge. Now consider the infinite series

$$\sum_{n=1}^{\infty} \frac{1}{n} = 1 + \frac{1}{2} + \frac{1}{3} + \cdots$$

Even though the terms of this series (called the *harmonic series*) approach zero as n goes to infinity, it can be shown that the harmonic series is divergent. Show that this result is intuitively true by establishing the following:

a. Observe that $\quad S_2 = 1 + \dfrac{1}{2} > \dfrac{1}{2} + \dfrac{1}{2} = 1$

and $\qquad S_4 = 1 + \dfrac{1}{2} + \dfrac{1}{3} + \dfrac{1}{4} = S_2 + \dfrac{1}{3} + \dfrac{1}{4} > 1 + \left(\dfrac{1}{4} + \dfrac{1}{4}\right) = \dfrac{3}{2}$

Using a similar argument, show that $S_8 > 4/2$ and $S_{16} > 5/2$. Conclude that in general

$$S_{2^n} > \frac{n+1}{2}$$

b. Explain why $S_1 < S_2 < S_3 < \cdots < S_n < \cdots$.

c. Use the results of parts (a) and (b) to explain why the harmonic series is divergent.

GEOMETRIC SERIES

In general, it is no easy task to determine whether a given infinite series is convergent or divergent. It is an even more difficult problem to determine the sum of an infinite series that is known to be convergent. But there is an important and useful series whose sum, when it exists, is easy to find.

Geometric Series

A geometric series with ratio r is a series of the form

$$\sum_{n=0}^{\infty} ar^n = a + ar + ar^2 + ar^3 + \cdots + ar^n + \cdots \qquad (5)$$

The ratio here refers to the ratio of two consecutive terms.

Note that, for this series, we begin the summation with $n = 0$ instead of $n = 1$. To determine the conditions under which the geometric series (5) converges and find its sum, let's consider the nth partial sum of the infinite series

$$S_n = a + ar + ar^2 + \cdots + ar^n$$

Multiplying both sides of the equation by r gives

$$rS_n = ar + ar^2 + ar^3 + \cdots + ar^{n+1}$$

Subtracting the second equation from the first yields

$$S_n - rS_n = a - ar^{n+1}$$
$$(1 - r)S_n = a(1 - r^{n+1})$$
$$S_n = \frac{a(1 - r^{n+1})}{1 - r}$$

provided $r \neq 1$. You are asked to show that if $|r| \geq 1$, then the series (5) diverges (Exercise 44). On the other hand, observe that if $|r| < 1$ (that is, if $-1 < r < 1$), then

$$\lim_{n \to \infty} r^{n+1} = 0$$

For example, if $r = \dfrac{1}{2}$, then

$$\lim_{n \to \infty} r^{n+1} = \lim_{n \to \infty} \left(\frac{1}{2}\right)^{n+1} = \lim_{n \to \infty} \frac{1}{2^{n+1}} = 0$$

Using this fact, together with the properties of limits stated earlier, we see that

$$\lim_{n \to \infty} S_n = \lim_{n \to \infty} \frac{a(1 - r^{n+1})}{1 - r}$$
$$= \frac{a}{1 - r} \lim_{n \to \infty} (1 - r^{n+1})$$
$$= \frac{a}{1 - r}$$

These results are summarized in Theorem 2.

THEOREM 2

The geometric series

$$\sum_{n=0}^{\infty} ar^n = a + ar + ar^2 + \cdots$$

converges and its sum is $\dfrac{a}{1 - r}$; that is,

$$\sum_{n=0}^{\infty} ar^n = a + ar + ar^2 + \cdots = \frac{a}{1 - r} \tag{6}$$

if $|r| < 1$. The series diverges if $|r| \geq 1$.

EXAMPLE **2**   Show that each of the following infinite series is a geometric series, and find its sum if it is convergent.

a. $\displaystyle\sum_{n=0}^{\infty} \frac{1}{2^n}$ **b.** $\displaystyle\sum_{n=0}^{\infty} 3\left(\frac{5}{2}\right)^n$ **c.** $\displaystyle\sum_{n=1}^{\infty} 5\left(-\frac{3}{4}\right)^n$

SOLUTION ✔   **a.** Observe that

$$\frac{1}{2^n} = \left(\frac{1}{2}\right)^n$$

so that

$$\sum_{n=0}^{\infty} \frac{1}{2^n} = \sum_{n=0}^{\infty} \left(\frac{1}{2}\right)^n$$

is a geometric series with $a = 1$ and $r = \frac{1}{2}$. Since $|r| = \frac{1}{2} < 1$, we conclude that the series is convergent. Finally, using (6), we have

$$\sum_{n=0}^{\infty} \left(\frac{1}{2}\right)^n = 1 + \frac{1}{2} + \frac{1}{4} + \cdots = \frac{1}{1 - \frac{1}{2}} = 2$$

b. This is a geometric series with $a = 3$ and $r = \frac{5}{2}$. Since

$$|r| = \left|\frac{5}{2}\right| = \frac{5}{2} > 1$$

we deduce that the series is divergent.

c. The summation here begins with $n = 1$. However, we may rewrite the series as

$$\sum_{n=1}^{\infty} 5\left(-\frac{3}{4}\right)^n = 5\left(-\frac{3}{4}\right) + 5\left(-\frac{3}{4}\right)^2 + 5\left(-\frac{3}{4}\right)^3 + \cdots$$

$$= 5\left(-\frac{3}{4}\right)\left[1 + \left(-\frac{3}{4}\right) + \left(-\frac{3}{4}\right)^2 + \cdots\right]$$

$$= \sum_{n=0}^{\infty} \left(\frac{-15}{4}\right)\left(-\frac{3}{4}\right)^n$$

which is just a geometric series with $a = -\frac{15}{4}$ and $r = -\frac{3}{4}$. Since $|r| = \left|-\frac{3}{4}\right| = \frac{3}{4} < 1$, the series is convergent. Using (6), we find

$$\sum_{n=1}^{\infty} 5\left(-\frac{3}{4}\right)^n = \sum_{n=0}^{\infty} \left(-\frac{15}{4}\right)\left(-\frac{3}{4}\right)^n = \frac{-\frac{15}{4}}{1 - \left(-\frac{3}{4}\right)} = -\frac{\frac{15}{4}}{\frac{7}{4}} = -\frac{15}{7}$$ ■ ■ ■ ■

PROPERTIES OF INFINITE SERIES

The following properties of infinite series enable us to perform algebraic operations on convergent series.

Properties of Infinite Series

If $\sum\limits_{n=1}^{\infty} a_n$ and $\sum\limits_{n=1}^{\infty} b_n$ are convergent infinite series and c is a constant, then

1. $\sum\limits_{n=1}^{\infty} ca_n = c \sum\limits_{n=1}^{\infty} a_n$

2. $\sum\limits_{n=1}^{\infty} (a_n \pm b_n) = \sum\limits_{n=1}^{\infty} a_n \pm \sum\limits_{n=1}^{\infty} b_n$

Thus, we may multiply each term of a convergent series by a constant c, which results in a convergent series whose sum is c times the sum of the original series. We may also add (subtract) the corresponding terms of two convergent series, which gives a convergent series whose sum is the sum (difference) of the sums of the original series.

EXAMPLE 3   Find the sum of the following series if it exists:

$$\sum_{n=0}^{\infty} \frac{2 \cdot 3^n - 2^n}{5^n} = 1 + \frac{4}{5} + \frac{14}{25} + \cdots$$

SOLUTION ✔   Note that this series starts with $n = 0$. We can write

$$\sum_{n=0}^{\infty} \frac{2 \cdot 3^n - 2^n}{5^n} = \sum_{n=0}^{\infty} \left(\frac{2 \cdot 3^n}{5^n} - \frac{2^n}{5^n} \right)$$

Now observe that

$$\sum_{n=0}^{\infty} \frac{2 \cdot 3^n}{5^n} \quad \text{and} \quad \sum_{n=0}^{\infty} \frac{2^n}{5^n}$$

are convergent geometric series with ratios $r = \frac{3}{5} < 1$ and $r = \frac{2}{5} < 1$, respectively. Therefore, using Property 2 of infinite series, we have

$$\sum_{n=0}^{\infty} \frac{2 \cdot 3^n - 2^n}{5^n} = \sum_{n=0}^{\infty} \frac{2 \cdot 3^n}{5^n} - \sum_{n=0}^{\infty} \frac{2^n}{5^n}$$

$$= 2 \sum_{n=0}^{\infty} \frac{3^n}{5^n} - \sum_{n=0}^{\infty} \frac{2^n}{5^n} \qquad \text{(Using Property 1 on the first sum)}$$

$$= 2 \sum_{n=0}^{\infty} \left(\frac{3}{5} \right)^n - \sum_{n=0}^{\infty} \left(\frac{2}{5} \right)^n$$

$$= 2 \left(\frac{1}{1 - \frac{3}{5}} \right) - \left(\frac{1}{1 - \frac{2}{5}} \right)$$

$$= 2 \left(\frac{5}{2} \right) - \frac{5}{3}$$

$$= \frac{10}{3}$$

APPLICATIONS

We now consider some applications of geometric series.

EXAMPLE 4   Find the rational number that has the repeated decimal representation $0.222\ldots$.

SOLUTION ✔   By definition, the decimal representation

$$0.222\ldots = \frac{2}{10} + \frac{2}{100} + \frac{2}{1000} + \cdots$$

$$= \frac{2}{10}\left(1 + \frac{1}{10} + \frac{1}{100} + \cdots\right)$$

$$= \sum_{n=0}^{\infty} \left(\frac{2}{10}\right)\left(\frac{1}{10}\right)^{n}$$

which is a geometric series with $a = \frac{2}{10}$ and $r = \frac{1}{10}$. Since $|r| = r = \frac{1}{10} < 1$, the series converges. In fact, using Formula (6), we have

$$0.222\ldots = \frac{\left(\frac{2}{10}\right)}{1 - \frac{1}{10}} = \frac{\frac{2}{10}}{\frac{9}{10}} = \frac{2}{9} \qquad \blacksquare\blacksquare\blacksquare\blacksquare$$

The following example illustrates a phenomenon in economics known as the **multiplier effect.**

EXAMPLE 5   Suppose the average wage earner saves 10% of her take-home pay and spends the other 90%. Estimate the impact that a proposed $20 billion tax cut will have on the economy over the long run in terms of the additional spending generated.

SOLUTION ✔   Of the $20 billion received by the original beneficiaries of the proposed tax cut, $(0.9)(20)$ billion dollars will be spent. Of the $(0.9)(20)$ billion dollars reinjected into the economy, 90% of it, or $(0.9)(0.9)(20)$ billion dollars, will find its way into the economy again. This process will go on ad infinitum, so this one-time proposed tax cut will result in additional spending over the years in the amount of

$$(0.9)(20) + (0.9)^{2}(20) + (0.9)^{3}(20) + \cdots = (0.9)(20)[1 + 0.9 + 0.9^{2} + 0.9^{3} + \cdots]$$

$$= 18\left[\frac{1}{1 - 0.9}\right]$$

$$= 180$$

or $180 billion. $\qquad \blacksquare\blacksquare\blacksquare\blacksquare$

A **perpetuity** is a sequence of payments made at regular time intervals and continuing on forever. The **capital value of a perpetuity** is the sum of the present values of all future payments. The following example illustrates these concepts.

EXAMPLE 6   The Robinson family wishes to create a scholarship fund at a college. If a scholarship in the amount of $5000 is to be awarded on an annual basis beginning next year, find the amount of the endowment they are required to make now. Assume that this fund will earn interest at a rate of 10% per year compounded continuously.

SOLUTION ✔   The amount of the endowment, A, is given by the sum of the present values of the amounts awarded annually in perpetuity. Now, the present value of the amount of the first award is equal to

$$5000e^{-0.1(1)}$$

(see Section 6.7). The present value of the amount of the second award is

$$5000e^{-0.1(2)}$$

and so on. Continuing, we see that the present value of the amount of the nth award is

$$5000e^{-0.1(n)}$$

Therefore, the amount of the endowment is

$$A = 5000e^{-0.1(1)} + 5000e^{-0.1(2)} + \cdots + 5000e^{-0.1(n)} + \cdots$$

To find the sum of the infinite series on the right-hand side, let

$$r = e^{-0.1}$$

Then

$$A = 5000r^1 + 5000\,r^2 + \cdots + 5000r^n + \cdots$$
$$= 5000r(1 + r + r^2 + \cdots + r^n + \cdots)$$

The series inside the parentheses is a geometric series with $r = e^{-0.1} \approx 0.905 < 1$, so that, using (6), we find

$$A = 5000r\left(\frac{1}{1-r}\right) = \frac{5000e^{-0.1}}{1 - e^{-0.1}} = 47{,}541.66$$

Thus, the amount of the endowment is $47,541.66.   ■ ■ ■ ■

Our final example is an application of a geometric series in the field of medicine.

EXAMPLE 7   A patient is to be given 5 units of a certain drug daily for an indefinite period of time. For this particular drug, it is known that the fraction of a dose that remains in the patient's body after t days is given by $e^{-0.3t}$. Determine the residual amount of the drug that may be expected to be in the patient's body after an extended treatment.

SOLUTION ✔   The amount of the drug in the patient's body 1 day after the first dose is administered, and prior to administration of the second dose, is $5e^{-0.3}$ units. The amount of the drug in the patient's body 2 days later, and prior to administration of the third dose, consists of the residuals from the first two doses. Of the first dose, $5e^{-(0.3)2}$ units of the drug are left in the patient's body, and of the second dose, $5e^{-0.3}$ units of the drug are left. Thus, the amount of the drug two days later is given by

$$5e^{-0.3} + 5e^{-0.3(2)}$$

units. Continuing, we see that the amount of the drug left in the patient's body in the long run, and prior to administration of a fresh dose, is given by

$$R = 5e^{-0.3} + 5e^{-0.3(2)} + 5e^{-0.3(3)} + \cdots$$

To find the sum of the infinite series, we let

$$r = e^{-0.3}$$

Then

$$A = 5r + 5r^2 + 5r^3 + \cdots = 5r(1 + r + r^2 + \cdots)$$
$$= \frac{5r}{1 - r} = \frac{5e^{-0.3}}{1 - e^{-0.3}} \approx 14.29$$

Therefore, after an extended treatment, the residual amount of drug in the patient's body is approximately 14.29 units.   ■ ■ ■ ■

SELF-CHECK EXERCISES 11.3

1. Determine whether the geometric series

$$\sum_{n=0}^{\infty} 5\left(-\frac{1}{3}\right)^n = 5 - 5\left(\frac{1}{3}\right) + 5\left(\frac{1}{9}\right) - \cdots$$

   is convergent or divergent. If it is convergent, find its sum.

2. Suppose the average wage earner in a certain country saves 12% of his take-home pay and spends the other 88%. Estimate the impact that a proposed $10 billion tax cut will have on the economy over the long run due to the additional spending generated.

Solutions to Self-Check Exercises 11.3 can be found on page 807.

11.3 Exercises

In Exercises 1–4, find the Nth partial sum of the infinite series and evaluate its limit to determine whether the series converges or diverges. If the series is convergent, find its sum.

1. $\displaystyle\sum_{n=1}^{\infty} (-2)^n$

2. $\displaystyle\sum_{n=1}^{\infty} \left(\frac{1}{n+1} - \frac{1}{n+2}\right)$

3. $\displaystyle\sum_{n=1}^{\infty} \frac{1}{n^2 + 3n + 2}$

Hint: $\dfrac{1}{n^2 + 3n + 2} = \dfrac{1}{n+1} - \dfrac{1}{n+2}$

4. $\displaystyle\sum_{n=2}^{\infty} \left(\frac{1}{\ln n} - \frac{1}{\ln(n+1)}\right)$

In Exercises 5–16, determine whether the geometric series converges or diverges. If it converges, find its sum.

5. $\displaystyle\sum_{n=0}^{\infty} \left(\frac{1}{3}\right)^n$

6. $\displaystyle\sum_{n=0}^{\infty} 4\left(-\frac{2}{3}\right)^n$

7. $\displaystyle\sum_{n=0}^{\infty} 2(1.01)^n$

8. $\displaystyle\sum_{n=0}^{\infty} 3(0.9)^n$

9. $\displaystyle\sum_{n=0}^{\infty} \frac{(-2)^n}{3^n}$

10. $\displaystyle\sum_{n=0}^{\infty} \frac{3}{2^n}$

11. $\displaystyle\sum_{n=0}^{\infty} \frac{2^n}{3^{n+2}}$

12. $\displaystyle\sum_{n=0}^{\infty} \frac{3^{n+1}}{4^{n-1}}$

13. $\displaystyle\sum_{n=0}^{\infty} e^{-0.2n}$

14. $\displaystyle\sum_{n=0}^{\infty} 2e^{-0.1n}$

15. $\displaystyle\sum_{n=0}^{\infty} \left(-\frac{3}{\pi}\right)^n$

16. $\displaystyle\sum_{n=1}^{\infty} \frac{e^n}{3^{n+1}}$

In Exercises 17–26, determine whether the series converges or diverges. If it converges, find its sum.

17. $8 + 4 + \dfrac{1}{2} + \dfrac{1}{4} + \dfrac{1}{8} + \cdots$

18. $1 + 0.2 + 0.04 + 0.0016 + \cdots$

19. $3 - \dfrac{1}{3} + \dfrac{1}{9} - \dfrac{1}{27} + \cdots$

20. $5 - 1.01 + (1.01)^2 - (1.01)^3 + \cdots$

21. $\displaystyle\sum_{n=0}^{\infty} \frac{3 + 2^n}{3^n}$

22. $\displaystyle\sum_{n=0}^{\infty} \frac{2^n - 3^n}{4^n}$

23. $\displaystyle\sum_{n=0}^{\infty} \frac{3 \cdot 2^n + 4^n}{3^n}$

24. $\displaystyle\sum_{n=0}^{\infty} \frac{2 \cdot 3^n - 3 \cdot 5^n}{7^n}$

25. $\displaystyle\sum_{n=1}^{\infty} \left[\left(\frac{e}{\pi}\right)^n + \left(\frac{\pi}{e^2}\right)^n\right]$

26. $\displaystyle\sum_{n=1}^{\infty} \left[\frac{1}{2^n} - \frac{1}{n(n+1)}\right]$ Hint: $\dfrac{1}{n(n+1)} = \dfrac{1}{n} - \dfrac{1}{n+1}$

In Exercises 27–30, express the decimal as a rational number.

27. $0.3333\ldots$

28. $0.121212\ldots$

29. $1.213213213\ldots$

30. $6.2314314314\ldots$

In Exercises 31–34, find the values of x for which the given series converges and find the sum of the series.

Hint: First show that the series is a geometric series.

31. $\displaystyle\sum_{n=0}^{\infty} (-x)^n$

32. $\displaystyle\sum_{n=0}^{\infty} (x - 2)^n$

33. $\displaystyle\sum_{n=1}^{\infty} 2^n(x - 1)^n$

34. $\displaystyle\sum_{n=0}^{\infty} \frac{x^{2n}}{3^n}$

35. Effect of a Tax Cut on Spending Suppose the average wage earner saves 9% of her take-home pay and spends the other 91%. Estimate the impact that a proposed $30 billion tax cut will have on the economy over the long run due to the additional spending generated.

36. A Bouncing Ball A ball is dropped from a height of 10 m. After hitting the ground, it rebounds to a height of 5 m and then continues to rebound at one-half of its former height thereafter. Find the total distance traveled by the ball before it comes to rest.

37. Winning a Toss Peter and Paul take turns tossing a pair of dice. The first to throw a 7 wins. If Peter starts the game, then it can be shown that his chances of winning are given by

$$p = \frac{1}{6} + \left(\frac{1}{6}\right)\left(\frac{5}{6}\right)^2 + \left(\frac{1}{6}\right)\left(\frac{5}{6}\right)^4 + \cdots$$

Find p.

38. Endowments Hal Corporation wants to establish a fund to provide the art center of a large metropolitan area with an annual grant of $250,000 beginning next year. If the fund will earn interest at a rate of 10%/year compounded continuously, find the amount of endowment the corporation must make at this time.

39. Transmission of Disease Refer to Exercise 48, page 795. It can be shown that the total number of individuals, S_n, who have contracted the disease some time between the first and nth day is approximated by

$$S_n = a_1 + aNa_1 + aNa_2 + \cdots + aNa_{n-1}$$

Show that if $r < 1$, then the total number of persons who will have contracted the disease at some stage of the epidemic is no larger than $a_1b/(b - aN)$.

40. Capital Value of a Perpetuity Find a formula for the capital value of a perpetuity involving payments of P dollars each, paid at the end of each of m periods/year into a fund that earns interest at the nominal rate of $r\%$/year compounded m times/year by verifying the following:

a. The present value of the nth payment is

$$P\left(1 + \frac{r}{m}\right)^{-n}$$

b. The capital value is

$$A = P\left(1 + \frac{r}{m}\right)^{-1} + P\left(1 + \frac{r}{m}\right)^{-2}$$

$$+ P\left(1 + \frac{r}{m}\right)^{-3} + \cdots$$

c. Using the fact that the series in part (b) is a geometric series, its sum is

$$A = \frac{mP}{r}$$

41. CAPITAL VALUE OF A PERPETUITY Find a formula for the capital value of a perpetuity involving payments of P dollars paid at the end of each investment period into a fund that earns interest at the rate of r%/year compounded continuously.
Hint: Study Exercise 40.

42. RESIDUAL DRUG IN THE BLOODSTREAM Ten units of a certain drug are administered to a patient on a daily basis. The fraction of this drug that remains in the patient's bloodstream after t days is given by

$$f(t) = e^{(-1/4)t}$$

Determine the residual amount of the drug in the patient's bloodstream after a period of extended treatment with the drug.

43. RESIDUAL DRUG IN THE BLOODSTREAM Suppose a dose of C units of a certain drug is administered to a patient and the fraction of the dose remaining in the patient's bloodstream t hr after the dose is administered is given by Ce^{-kt}, where k is a positive constant.
a. Show that the residual concentration of the drug in the bloodstream after extended treatment when a dose of C units is administered at intervals of t hr is

given by

$$R = \frac{Ce^{-kt}}{1 - e^{-kt}}$$

b. If the highest concentration of this particular drug that is considered safe is S units, find the minimal time that must exist between doses.
Hint: $C + R \le S$

44. Let

$$\sum_{n=0}^{\infty} ar^n = a + ar + ar^2 + ar^3 + \cdots + ar^n + \cdots$$

be a geometric series with common ratio r. Show that if $|r| \ge 1$, then the series diverges.

In Exercises 45–48, determine whether the statement is true or false. If it is true, explain why it is true. If it is false, give an example to show why it is false.

45. If $\displaystyle\sum_{n=0}^{\infty} (a_n + b_n)$ converges, then both $\displaystyle\sum_{n=0}^{\infty} a_n$ and $\displaystyle\sum_{n=0}^{\infty} b_n$ must converge.

46. If $\displaystyle\sum_{n=0}^{\infty} a_n$ converges and $\displaystyle\sum_{n=0}^{\infty} b_n$ converges, then $\displaystyle\sum_{n=0}^{\infty} (ca_n + db_n)$ also converges, where c and d are constants.

47. If $|r| < 1$, then $\displaystyle\sum_{n=0}^{\infty} |r^n| = \frac{1}{1 - |r|}$.

48. If $|r| > 1$, then $\displaystyle\sum_{n=1}^{\infty} \frac{1}{r^n} = \frac{1}{r - 1}$.

SOLUTIONS TO SELF-CHECK EXERCISES 11.3

1. This is a geometric series with $r = -\frac{1}{3}$ and $a = 5$. Since $0 < \left|-\frac{1}{3}\right| < 1$, we see that the series is convergent. Its sum is

$$\frac{5}{1 - (-\frac{1}{3})} = \frac{5}{\frac{4}{3}} = \frac{15}{4} = 3\frac{3}{4}$$

2. Of the $10 billion received by the original beneficiaries of the proposed cut, $(0.88)(10)$ billion dollars will be spent. Of the $(0.88)(10)$ billion dollars reinjected into the economy, 88% of it, or $(0.88)(0.88)(10)$ billion dollars, will find its way into the economy again. This process will go on ad infinitum, so this one-time proposed tax cut will result in additional spending over the years in the amount of

$$(0.88)(10) + (0.88)^2(10) + (0.88)^3(10) + \cdots$$
$$= (0.88)(10)[1 + 0.88 + (0.88)^2 + \cdots]$$
$$= 8.8 \left[\frac{1}{1 - 0.88} \right]$$
$$= 73.3$$

or $73.3 billion.

11.4 Series with Positive Terms

The convergence or divergence of a telescoping series or a geometric series is relatively easy to determine because we can find a simple formula for the nth partial sum S_n of such a series. In fact, as we have seen earlier in this chapter, the knowledge of such a formula helps us find the actual sum of a convergent series by simply evaluating $\lim_{n \to \infty} S_n$. More often than not, obtaining a simple formula for the nth partial sum of an infinite series is very difficult or impossible, and we are forced to look for alternative ways to investigate the convergence or divergence of the series.

In what follows, we look at several tests for determining the convergence or divergence of an infinite series by examining the nth term a_n of the series. These tests will confirm the convergence of a series without yielding a value for its sum. From the practical point of view, however, this is all that is required. Once it has been ascertained that a series is convergent, we can approximate its sum to any degree of accuracy desired by adding up the terms of its nth partial sum S_n, provided that n is chosen large enough.

Our first test tells us how to identify a divergent series.

THE TEST FOR DIVERGENCE

The following theorem tells us that the terms of a convergent series must ultimately approach zero.

THEOREM 3

If $\displaystyle\sum_{n=1}^{\infty} a_n$ converges, then $\displaystyle\lim_{n \to \infty} a_n = 0$.

To prove this result, let

$$S_n = a_1 + a_2 + \cdots + a_{n-1} + a_n = S_{n-1} + a_n$$

and so

$$a_n = S_n - S_{n-1}$$

Since $\sum_{n=1}^{\infty} a_n$ is convergent, the sequence $\{S_n\}$ is convergent. Let $\lim_{n \to \infty} S_n = S$. Then

$$\lim_{n \to \infty} a_n = \lim_{n \to \infty} (S_n - S_{n-1}) = \lim_{n \to \infty} S_n - \lim_{n \to \infty} S_{n-1} = S - S = 0$$

An important consequence of Theorem 3 is the following useful test for *divergence*.

THEOREM 4	**The Test for Divergence** If $\lim_{n \to \infty} a_n$ does not exist or $\lim_{n \to \infty} a_n \neq 0$, then $\sum_{n=1}^{\infty} a_n$ diverges.

It is important to realize that the test for divergence does *not* say

$$\text{If } \lim_{n \to \infty} a_n = 0, \text{ then } \sum_{n=1}^{\infty} a_n \text{ must converge.}$$

In other words, the converse of Theorem 4 is not true in general. For example, $\lim_{n \to \infty} \frac{1}{n} = 0$, and yet, as we will see later, the **harmonic series** $\sum_{n=1}^{\infty} \frac{1}{n}$ is divergent. (See Example 3.) In short, the test for divergence rules out convergence for a series whose nth term does not approach zero but yields no information if the nth term of a series does approach zero—that is, the series may or may not converge.

EXAMPLE 1   Show that the following series are divergent:

a. $\displaystyle\sum_{n=1}^{\infty} (-1)^{n-1}$ **b.** $\displaystyle\sum_{n=1}^{\infty} \frac{2n^2 + 1}{3n^2 - 1}$

SOLUTION ✔   **a.** Here $a_n = (-1)^{n-1}$, and since

$$\lim_{n \to \infty} a_n = \lim_{n \to \infty} (-1)^{n-1}$$

does not exist, we conclude by the test for divergence that the series diverges.

b. Here

$$\lim_{n \to \infty} a_n = \lim_{n \to \infty} \frac{2n^2 + 1}{3n^2 - 1} = \lim_{n \to \infty} \frac{2 + \dfrac{1}{n^2}}{3 - \dfrac{1}{n^2}} = \frac{2}{3} \neq 0$$

and so, by the test for divergence, the series diverges.   ■ ■ ■ ■

We now look at several tests that tell us if a series is convergent. These tests apply only to series with positive terms.

THE INTEGRAL TEST

The integral test ties the convergence or divergence of an infinite series $\sum_{n=1}^{\infty} a_n$ to the convergence or divergence of the improper integral $\int_1^{\infty} f(x)\,dx$ where $f(n) = a_n$.

THEOREM 5	**The Integral Test**

The Integral Test

Suppose f is a continuous, positive, and decreasing function on $[1, \infty)$. If $f(n) = a_n$ for $n \geq 1$, then

$$\sum_{n=1}^{\infty} a_n \qquad \text{and} \qquad \int_1^{\infty} f(x)\,dx$$

either both converge or both diverge.

Let's give an intuitive justification for this theorem. If you examine Figure 11.4a, you will see that the height of the first rectangle is $a_2 = f(2)$.

FIGURE **11.4**

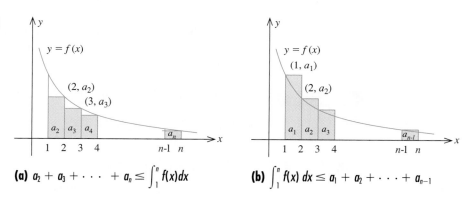

(a) $a_2 + a_3 + \cdots + a_n \leq \int_1^n f(x)\,dx$ 　　　 **(b)** $\int_1^n f(x)\,dx \leq a_1 + a_2 + \cdots + a_{n-1}$

Since this rectangle has width one, the area of the rectangle is also $a_2 = f(2)$. Similarly, the area of the second is a_3, and so on. Comparing the sum of the areas of the first $(n-1)$ inscribed rectangles with the area under the graph of f over the interval $[1, n]$, we see that

$$a_2 + a_3 + \cdots + a_n \leq \int_1^n f(x)\,dx$$

which implies that

$$S_n = a_1 + a_2 + a_3 + \cdots + a_n \leq a_1 + \int_1^n f(x)\,dx$$

If $\int_1^\infty f(x)\,dx$ is convergent and has value L, then

$$S_n \le a_1 + \int_1^n f(x)\,dx \le a_1 + L$$

This shows that $\{S_n\}$ is bounded above. Also,

$$S_{n+1} = S_n + a_{n+1} \ge S_n \qquad [\text{Because } a_{n+1} = f(n+1) \ge 0]$$

shows that $\{S_n\}$ is increasing as well.

Now, intuitively, such a sequence must converge to a number no greater than its upper bound. (Although this result can be demonstrated with mathematical rigor, we will not do so here.) In other words, $\sum_{n=1}^\infty a_n$ is convergent.

Next, by examining Figure 11.4b, you will see that

$$\int_1^n f(x)\,dx \le a_1 + a_2 + \cdots + a_{n-1} = S_{n-1}$$

So, if

$$\int_1^\infty f(x)\,dx$$

diverges to infinity (because $f(x) \ge 0$), then $\lim_{n\to\infty} S_n = \infty$ and so $\sum_{n=1}^\infty a_n$ is divergent.

REMARKS

1. The integral test simply tells us whether a series converges or diverges. If it indicates that a series converges, we may not conclude that the (finite) value of the improper integral used in conjunction with the test is the *sum* of the convergent series.

2. Since the convergence of an infinite series is not affected by the omission or addition of a finite number of terms to the series, we sometimes study the series

$$\sum_{n=N}^\infty a_n = a_N + a_{N+1} + \cdots$$

rather than the series $\sum_{n=1}^\infty a_n$. In this case, the series is compared to the improper integral

$$\int_N^\infty f(x)\,dx$$

as we will see in Example 4.   ■ ■ ■

EXAMPLE 2   Use the integral test to determine whether

$$\sum_{n=1}^{\infty} \frac{1}{n^2}$$

converges or diverges.

SOLUTION ✔   Here

$$a_n = f(n) = \frac{1}{n^2}$$

and so we consider the function $f(x) = 1/x^2$. Since f is continuous, positive, and decreasing on $[1, \infty)$, we may use the integral test. Now

$$\int_1^{\infty} \frac{1}{x^2} dx = \lim_{b \to \infty} \int_1^b x^{-2} dx = \lim_{b \to \infty} \left[-\frac{1}{x} \Big|_1^b \right]$$

$$= \lim_{b \to \infty} \left(-\frac{1}{b} + 1 \right) = 1$$

Since

$$\int_1^{\infty} \frac{1}{x^2} dx$$

converges, we conclude that

$$\sum_{n=1}^{\infty} \frac{1}{n^2}$$

converges as well.

EXAMPLE 3   Use the integral test to determine whether the harmonic series $\sum_{n=1}^{\infty} \frac{1}{n}$ converges or diverges.

SOLUTION ✔   Here $a_n = f(n) = 1/n$ and so we consider the function $f(x) = 1/x$. Since f is continuous, positive, and decreasing on $[1, \infty)$, we may use the integral test. But as you may verify

$$\int_1^{\infty} \frac{1}{x} dx = \infty$$

We conclude that $\sum_{n=1}^{\infty} \frac{1}{n}$ diverges.

EXAMPLE 4   Use the integral test to determine whether $\sum_{n=2}^{\infty} \dfrac{\ln n}{n}$ converges or diverges.

SOLUTION ✔   Here $a_n = (\ln n)/n$ and so we consider the function $f(x) = (\ln x)/x$. Observe that f is continuous and positive on $[2, \infty)$. Next, we compute

$$f'(x) = \frac{x\left(\dfrac{1}{x}\right) - \ln x}{x^2} = \frac{1 - \ln x}{x^2}$$

Note that $f'(x) < 0$ if $\ln x > 1$; that is, if $x > e$. This shows that f is decreasing on $[3, \infty)$. Therefore, we may use the integral test. Now,

$$\int_3^{\infty} \frac{\ln x}{x}\, dx = \lim_{b \to \infty} \int_3^b \frac{\ln x}{x}\, dx = \lim_{b \to \infty} \left[\frac{1}{2}(\ln x)^2 \Big|_3^b \right]$$

$$= \lim_{b \to \infty} \frac{1}{2}[(\ln b)^2 - (\ln 3)^2] = \infty$$

and we conclude that $\sum_{n=2}^{\infty} \dfrac{\ln n}{n}$ diverges.

THE p-SERIES

The following series will play an important role in our work later on.

p-Series

A p-series is a series of the form

$$\sum_{n=1}^{\infty} \frac{1}{n^p} = 1 + \frac{1}{2^p} + \frac{1}{3^p} + \cdots$$

where p is a constant.

Observe that for $p = 1$, the p-series is just the harmonic series $\sum_{n=1}^{\infty} \dfrac{1}{n}$. The conditions for the convergence or divergence of the p-series can be found by applying the integral test to the series. We have the following result.

THEOREM 6

Convergence of p-Series

The p-series $\sum_{n=1}^{\infty} \dfrac{1}{n^p}$ converges if $p > 1$ and diverges if $p \leq 1$.

PROOF If $p < 0$, then

$$\lim_{n \to \infty} \frac{1}{n^p} = \infty$$

If $p = 0$, then

$$\lim_{n \to \infty} \frac{1}{n^p} = 1$$

In either case,

$$\lim_{n \to \infty} \frac{1}{n^p} \neq 0$$

and so the p-series diverges by the test for divergence. If $p > 0$, then the function $f(x) = 1/x^p$ is continuous, positive, and decreasing on $[1, \infty)$. It can be shown that

$$\int_1^\infty \frac{1}{x^p} \, dx \text{ converges if } p > 1 \text{ and diverges if } p \leq 1$$

(see Exercise 54). Using this result and the integral test, we conclude that $\sum_{n=1}^\infty \frac{1}{n^p}$ converges if $p > 1$ and diverges if $0 < p \leq 1$. Therefore, $\sum_{n=1}^\infty \frac{1}{n^p}$ converges if $p > 1$ and diverges if $p \leq 1$.

EXAMPLE 5   Determine whether each series converges or diverges.

a. $\sum_{n=1}^\infty \frac{1}{n^2}$ **b.** $\sum_{n=1}^\infty \frac{1}{\sqrt{n}}$ **c.** $\sum_{n=1}^\infty n^{-1.001}$

SOLUTION ✔   **a.** This is a p-series with $p = 2 > 1$, and so by Theorem 6, the series converges.

b. Rewriting the series in the form $\sum_{n=1}^\infty \frac{1}{n^{1/2}}$, we see that the series is a p-series with $p = \frac{1}{2} < 1$, and so by Theorem 6, it diverges.

c. We rewrite the series in the form $\sum_{n=1}^\infty \frac{1}{n^{1.001}}$, which we recognize to be a p-series with $p = 1.001 > 1$ and conclude accordingly that the series converges.

■ ■ ■ ■

THE COMPARISON TEST

The convergence or divergence of a given series $\sum a_n$ can be determined by comparing its terms with the terms of a *test series* that is known to be convergent or divergent. This is the basis for the comparison test for series that follows.

In the rest of this section, we assume that all series under consideration have positive terms.

Suppose the terms of a series $\Sigma\, a_n$ are smaller than the corresponding terms of a series $\Sigma\, b_n$. This situation is illustrated in Figure 11.5, where the respective terms are represented by rectangles, each of width one and appropriate height.

FIGURE 11.5

Each rectangle representing a_n is contained in the rectangle representing b_n.

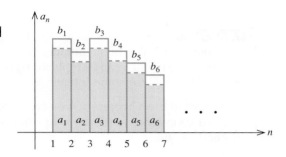

If $\Sigma\, b_n$ is convergent, the total area of the rectangles representing this series is finite. Since each rectangle representing the series $\Sigma\, a_n$ is contained in a corresponding rectangle representing the terms of $\Sigma\, b_n$, the total area of the rectangles representing $\Sigma\, a_n$ must also be finite: that is, the series $\Sigma\, a_n$ must be convergent. A similar argument would seem to suggest that if all the terms of a series $\Sigma\, a_n$ are larger than the corresponding terms of a series $\Sigma\, b_n$ that is known to be divergent, then $\Sigma\, a_n$ must itself be divergent. These observations lead to the following theorem.

THEOREM 7

The Comparison Test

Suppose $\Sigma\, a_n$ and $\Sigma\, b_n$ are series with positive terms.

a. If $\Sigma\, b_n$ is convergent and $a_n \leq b_n$ for all n, then $\Sigma\, a_n$ is also convergent.

b. If $\Sigma\, b_n$ is divergent and $a_n \geq b_n$ for all n, then $\Sigma\, a_n$ is also divergent.

Here is an intuitive justification for Theorem 7. Let

$$S_n = \sum_{k=1}^{n} a_k \qquad \text{and} \qquad T_n = \sum_{k=1}^{n} b_k$$

be the nth terms of the sequence of partial sums of $\Sigma\, a_n$ and $\Sigma\, b_n$, respectively. Since both series have positive terms, $\{S_n\}$ and $\{T_n\}$ are increasing.

1. If $\displaystyle\sum_{n=1}^{\infty} b_n$ is convergent, then there exists a number L such that $\displaystyle\lim_{n\to\infty} T_n = L$ and $T_n \leq L$ for all n. Since $a_n \leq b_n$ for all n, we have $S_n \leq T_n$, and this implies that $S_n \leq L$ for all n. We have shown that $\{S_n\}$ is increasing and bounded above, and so, as before, we can argue intuitively that S_n and therefore $\Sigma \, a_n$ converges.

2. If $\Sigma \, b_n$ is divergent, then $\lim\limits_{n \to \infty} T_n = \infty$ since $\{T_n\}$ is increasing. But $a_n \geq b_n$ for all n, and this implies that $S_n \geq T_n$, which in turn implies that $\lim\limits_{n \to \infty} S_n = \infty$. Therefore, $\Sigma \, a_n$ diverges.

REMARK Since the convergence or divergence of a series is not affected by the omission of a finite number of terms of the series, the condition $a_n \leq b_n \, (a_n \geq b_n)$ for all n can be replaced by the condition that these inequalities hold for all $n \geq N$ for some integer N. ■ ■ ■

In order to use Theorem 7, we need a catalog of test series whose convergence and divergence are known. In what follows, we will use the geometric series and the p-series as test series.

EXAMPLE 6

Determine whether the series

$$\sum_{n=1}^{\infty} \frac{1}{n^2 + 2}$$

converges or diverges.

SOLUTION ✔

Let

$$a_n = \frac{1}{n^2 + 2}$$

Observe that when n is large, $n^2 + 2$ behaves like n^2 and so a_n behaves like

$$b_n = \frac{1}{n^2}$$

This observation suggests that we compare $\Sigma \, a_n$ with the test series $\Sigma \, b_n$, which is a convergent p-series with $p = 2$. Now

$$0 < \frac{1}{n^2 + 2} < \frac{1}{n^2} \qquad (n \geq 1)$$

and the given series is indeed "smaller" than the test series

$$\sum \frac{1}{n^2}$$

Since the test series converges, we conclude by the comparison test that

$$\sum \frac{1}{n^2 + 2}$$

also converges. ■ ■ ■ ■

EXAMPLE 7   Determine whether the series

$$\sum_{n=1}^{\infty} \frac{1}{3 + 2^n}$$

converges or diverges.

SOLUTION ✔   Let

$$a_n = \frac{1}{3 + 2^n}$$

Observe that if n is large, $3 + 2^n$ behaves like 2^n, and so a_n behaves like $b_n = 1/2^n$. This observation suggests that we compare $\Sigma\, a_n$ with $\Sigma\, b_n$. Now the series $\Sigma\, 1/2^n = \Sigma\, (1/2)^n$ is a geometric series with $r = 1/2 < 1$ and so it is convergent. Since

$$a_n = \frac{1}{3 + 2^n} < \frac{1}{2^n} = b_n \qquad (n \geq 1)$$

the comparison test tells us that the given series is convergent.   ■ ■ ■ ■

EXAMPLE 8   Determine whether the series

$$\sum_{n=2}^{\infty} \frac{1}{\sqrt{n} - 1}$$

is convergent or divergent.

SOLUTION ✔   Let

$$a_n = \frac{1}{\sqrt{n} - 1}$$

If n is large, $\sqrt{n} - 1$ behaves like $\sqrt{n}$, and so a_n behaves like

$$b_n = \frac{1}{\sqrt{n}}$$

Now the series

$$\sum b_n = \sum \frac{1}{\sqrt{n}} = \sum \frac{1}{n^{1/2}}$$

is a p-series with $p = 1/2 < 1$ and so it is divergent. Since

$$a_n = \frac{1}{\sqrt{n} - 1} > \frac{1}{\sqrt{n}} = b_n \qquad (\text{for } n \geq 2) \qquad \text{(See the remark following Theorem 7.)}$$

the comparison test implies that the given series is divergent.   ■ ■ ■ ■

11.4 Exercises

In Exercises 1–10, show that the series is divergent.

1. $\frac{1}{2} + \frac{2}{3} + \frac{3}{4} + \cdots$

2. $1 - \frac{3}{2} + \frac{9}{4} - \frac{27}{8} + \cdots$

3. $\sum_{n=1}^{\infty} \frac{2n}{3n+1}$

4. $\sum_{n=1}^{\infty} \frac{n^2}{2n^2+1}$

5. $\sum_{n=1}^{\infty} 2(1.5)^n$

6. $\sum_{n=0}^{\infty} \frac{(-1)^n 3^n}{2^{n-1}}$

7. $\sum_{n=1}^{\infty} \frac{1}{2+3^{-n}}$

8. $\sum_{n=1}^{\infty} \frac{n}{\sqrt{2n^2+1}}$

9. $\sum_{n=0}^{\infty} \left(-\frac{\pi}{3}\right)^n$

10. $\sum_{n=1}^{\infty} \frac{3^{n+1}}{e^n}$

In Exercises 11–20, use the integral test to determine whether the series is convergent or divergent.

11. $\sum_{n=1}^{\infty} \frac{1}{n+1}$

12. $\sum_{n=1}^{\infty} \frac{3}{2n-1}$

13. $\sum_{n=1}^{\infty} \frac{n}{2n^2+1}$

14. $\sum_{n=1}^{\infty} ne^{-n^2}$

15. $\sum_{n=1}^{\infty} ne^{-n}$

16. $\sum_{n=1}^{\infty} \frac{1}{n(2n-1)}$

17. $\sum_{n=1}^{\infty} \frac{n}{(n^2+1)^{3/2}}$

18. $\sum_{n=2}^{\infty} \frac{1}{n\sqrt{\ln n}}$

19. $\sum_{n=9}^{\infty} \frac{1}{n\ln^3 n}$

20. $\sum_{n=0}^{\infty} \frac{1}{e^n+1}$

In Exercises 21–26, determine whether the _p_-series is convergent or divergent.

21. $\sum_{n=1}^{\infty} \frac{1}{n^3}$

22. $\sum_{n=1}^{\infty} \frac{1}{n^{2/3}}$

23. $\sum_{n=1}^{\infty} \frac{1}{n^{1.01}}$

24. $\sum_{n=1}^{\infty} \frac{1}{n^e}$

25. $\sum_{n=1}^{\infty} n^{-\pi}$

26. $\sum_{n=1}^{\infty} n^{-0.98}$

In Exercises 27–36, use the comparison test to determine whether the series is convergent or divergent.

27. $\sum_{n=1}^{\infty} \frac{1}{2n^2+1}$

28. $\sum_{n=1}^{\infty} \frac{1}{n^2+2n}$

29. $\sum_{n=3}^{\infty} \frac{1}{n-2}$

30. $\sum_{n=2}^{\infty} \frac{1}{n^{2/3}-1}$

31. $\sum_{n=2}^{\infty} \frac{1}{\sqrt{n^2-1}}$

32. $\sum_{n=0}^{\infty} \frac{1}{\sqrt{n^3+1}}$

33. $\sum_{n=0}^{\infty} \frac{2^n}{3^n+1}$

34. $\sum_{n=3}^{\infty} \frac{3^n}{2^n-4}$

35. $\sum_{n=2}^{\infty} \frac{\ln n}{n}$

36. $\sum_{n=1}^{\infty} \frac{1}{n^n}$

In Exercises 37–48, determine whether the series is convergent or divergent.

37. $\sum_{n=0}^{\infty} \frac{1}{\sqrt{n+1}}$

38. $\sum_{n=1}^{\infty} \frac{n}{\sqrt{2n^2+1}}$

39. $\sum_{n=2}^{\infty} \frac{1}{n\sqrt{n^2+1}}$

40. $\sum_{n=2}^{\infty} \frac{\sqrt{n^2+1}}{n^2}$

41. $\sum_{n=1}^{\infty} \left(\frac{1}{n\sqrt{n}} + \frac{2}{n^2}\right)$

42. $\sum_{n=1}^{\infty} \left[\left(\frac{2}{3}\right)^n + \frac{1}{n^{3/2}}\right]$

43. $\sum_{n=2}^{\infty} \frac{\ln n}{\sqrt{n}}$

44. $\sum_{n=2}^{\infty} \frac{\ln n}{n^{2.1}}$

45. $\sum_{n=2}^{\infty} \frac{1}{n(\ln n)^2}$

46. $\sum_{n=1}^{\infty} \frac{e^{1/n}}{n^2}$

47. $\sum_{n=1}^{\infty} \frac{1}{\sqrt{n}+4}$

Hint: $\frac{1}{\sqrt{n}+4} > \frac{1}{3\sqrt{n}}$ if $n > 4$

48. $\sum_{n=1}^{\infty} \frac{1}{4n^2-1}$

Hint: $\frac{1}{4n^2-1} < \frac{1}{2n^2}$ if $n \geq 1$

In Exercises 49 and 50, find the value of _p_ for which the series is convergent.

49. $\sum_{n=2}^{\infty} \frac{1}{n(\ln n)^p}$

50. $\sum_{n=1}^{\infty} \frac{\ln n}{n^p}$

51. Find the value(s) of _a_ for which the series

$$\sum_{n=1}^{\infty} \left(\frac{a}{n+1} - \frac{1}{n+2}\right)$$

converges. Justify your answer.

52. Consider the series

$$\sum_{n=0}^{\infty} e^{-n}$$

a. Evaluate

$$\int_0^{\infty} e^{-x}\, dx$$

and deduce from the integral test that the given series is convergent.
b. Show that the given series is a geometric series and find its sum.
c. Conclude that although the convergence of

$$\int_0^{\infty} e^{-x}\, dx$$

implies convergence of the infinite series, its value does not give the sum of the infinite series.

53. Show that $\int_1^{\infty} \dfrac{1}{x^p}\, dx$ converges if $p > 1$ and diverges if $p \le 1$.

54. Suppose

$$\sum_{n=1}^{\infty} a_n$$

is a convergent series with positive terms. Let $f(n) = a_n$, where f is a continuous and decreasing function for $x \ge N$, where N is some positive integer. Show that the error incurred in approximating the sum of the given

series by the Nth partial sum of the series

$$S_N = \sum_{n=1}^{N} a_n$$

is less than $\int_N^{\infty} f(x)\, dx$.

In Exercises 55–60, determine whether the statement is true or false. If it is true, explain why it is true. If it is false, give an example to show why it is false.

55. True or false. The series

$$\sum_{n=1}^{\infty} \frac{x}{n}$$

converges only for $x = 0$.

56. If $\lim\limits_{n\to\infty} a_n = 0$, then $\sum\limits_{n=0}^{\infty} a_n$ converges.

57. If $\sum\limits_{n=0}^{\infty} a_n$ diverges, then $\lim\limits_{n\to\infty} a_n \neq 0$.

58. $\int_1^{\infty} \dfrac{2}{(x^2 + 1)^{1.1}}\, dx$ converges.

59. Suppose $\sum a_n$ and $\sum b_n$ are series with positive terms. If $\sum a_n$ is convergent and $b_n \ge a_n$ for all n, then $\sum b_n$ is divergent.

60. Suppose $\sum a_n$ and $\sum b_n$ are series with positive terms. If $\sum b_n$ is divergent and $a_n \le b_n$, for all n, then $\sum a_n$ may or may not converge.

11.5 Power Series and Taylor Series

POWER SERIES AND INTERVALS OF CONVERGENCE

Recall that one of our goals in this chapter is to see what happens when the number of terms of a Taylor polynomial is allowed to increase without bound. In other words, we wish to study the expression

$$f(a) + f'(a)(x - a) + \frac{f''(a)}{2!}(x - a)^2 + \cdots + \frac{f^{(n)}(a)}{n!}(x - a)^n + \cdots \quad \textbf{(7)}$$

called the **Taylor series** of $f(x)$ at $x = a$. If the series (7) is truncated after $(n + 1)$ terms, the result is a Taylor polynomial of degree n of $f(x)$ at $x = a$

(see Section 11.1). Note that when $a = 0$, (7) reduces to

$$f(0) + f'(0)x + \frac{f''(0)}{2!}x^2 + \cdots + \frac{f^{(n)}(0)}{n!}x^n + \cdots \qquad \textbf{(8)}$$

which is referred to as the **Maclaurin series of** $f(x)$. Thus, the Maclaurin series is a special case of the Taylor series when $a = 0$.

Observe that the Taylor series (7) has the form

$$\sum_{n=0}^{\infty} a_n(x - a)^n = a_0 + a_1(x - a) + a_2(x - a)^2 + \cdots$$
$$+ a_n(x - a)^n + \cdots \qquad \textbf{(9)}$$

In fact, comparing the Taylor series (7) with the infinite series (9), we see that

$$a_0 = f(a), \qquad a_1 = f'(a), \qquad a_2 = \frac{f''(a)}{2!}, \qquad \ldots, \qquad a_n = \frac{f^{(n)}(a)}{n!}, \qquad \ldots$$

The infinite series (9) is called a **power series centered at** $x = a$. Thus, the Taylor series is just a power series with coefficients that involve the values of some function f and its derivatives at the point $x = a$.

Let's examine some important properties of the power series given in (9). Observe that when x is assigned a value, then this power series becomes an infinite series with constant terms. Accordingly, we can determine, at least theoretically, whether this infinite series converges or diverges. Now the totality of all values of x for which this power series *converges* comprises a set of points in the real line called the **interval of convergence** of the power series. This observation suggests that we may view the power series (9) as a function f whose domain coincides with the interval of convergence of the series and whose functional values are the sums of the infinite series obtained by allowing x to take on all values in the interval of convergence. In this case, we also say the function f is **represented by the power series** (9).

Given such a power series, how does one determine its interval of convergence? The following theorem, whose proof we will omit, provides the answer to this question.

THEOREM 8

Suppose that we are given the power series

$$\sum_{n=0}^{\infty} a_n(x - a)^n$$

Let

$$R = \lim_{n \to \infty} \left| \frac{a_n}{a_{n+1}} \right|$$

a. If $R = 0$, the series converges only for $x = a$.

b. If $0 < R < \infty$, the series converges for x in the interval $(a - R, a + R)$ and diverges for x outside this interval (Figure 11.6).

FIGURE 11.6

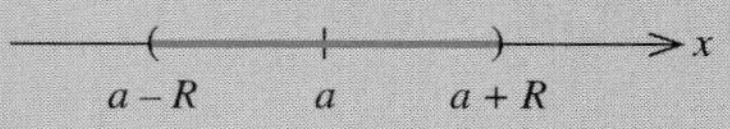

c. If $R = \infty$, the series converges for all x.

REMARKS

1. The domain of convergence of a power series is an interval called the interval of convergence. This interval of convergence is determined by R, the **radius of convergence.** Depending on whether $R = 0$, $0 < R < \infty$, or $R = \infty$, the "interval" of convergence may just be the degenerate interval consisting of the point a, a bona fide interval $(a - R, a + R)$, or the entire real line.

2. As mentioned earlier, a power series, inside its interval of convergence, represents a function f whose domain of definition coincides with the interval of convergence; that is,

$$f(x) = \sum_{n=0}^{\infty} a_n(x - a)^n \qquad x \in (a - R, a + R)$$

3. The power series may or may not converge at an endpoint. In general, it is difficult to determine whether a power series is convergent at an end point. For this reason, we restrict our attention to points inside the interval of convergence of a power series. ■ ■ ■

EXAMPLE 1 Find the radius of convergence and the interval of convergence of the power series

$$\sum_{n=0}^{\infty} \frac{(x - 1)^n}{2^n}$$

Show that $f(2)$ exists and find its value, where

$$f(x) = \sum_{n=0}^{\infty} \frac{(x - 1)^n}{2^n}$$

SOLUTION ✔ Since $a_n = 1/2^n$, we have

$$R = \lim_{n \to \infty} \left| \frac{a_n}{a_{n+1}} \right| = \lim_{n \to \infty} \left| \frac{\dfrac{1}{2^n}}{\dfrac{1}{2^{n+1}}} \right|$$

$$= \lim_{n \to \infty} \frac{2^{n+1}}{2^n} = \lim_{n \to \infty} 2 = 2$$

Therefore, bearing in mind that $a = 1$, we see that the interval of convergence of the given series is $(-1, 3)$. Thus, in the interval $(-1, 3)$, the given power series defines a function

$$f(x) = \sum_{n=0}^{\infty} \frac{(x - 1)^n}{2^n} \qquad x \in (-1, 3)$$

Since $2 \in (-1, 3)$, $f(2)$ exists. In fact,

$$f(2) = \sum_{n=0}^{\infty} \frac{(2 - 1)^n}{2^n} = \sum_{n=0}^{\infty} \frac{1}{2^n}$$

which is a geometric series with $a = 1$ and $r = \frac{1}{2}$, so that

$$f(2) = \frac{1}{1 - \frac{1}{2}} = 2$$

■ ■ ■ ■

EXAMPLE 2   Find the interval of convergence of each of the following power series:

a. $\displaystyle\sum_{n=0}^{\infty} n^3(x + 2)^n$ **b.** $\displaystyle\sum_{n=0}^{\infty} n!(x - 1)^n$ **c.** $\displaystyle\sum_{n=0}^{\infty} \frac{x^n}{n!}$

SOLUTION ✔   **a.** Here $a_n = n^3$, so

$$R = \lim_{n\to\infty}\left|\frac{a_n}{a_{n+1}}\right| = \lim_{n\to\infty}\frac{n^3}{(n + 1)^3}$$

$$= \lim_{n\to\infty}\frac{1}{\left(1 + \dfrac{1}{n}\right)^3} = 1 \qquad \text{(Dividing numerator and denominator by } n^3)$$

Since $a = -2$, we find that the interval of convergence of the series is $(-3, -1)$.

b. Here $a_n = n!$, so

$$R = \lim_{n\to\infty}\left|\frac{a_n}{a_{n+1}}\right| = \lim_{n\to\infty}\left|\frac{n!}{(n + 1)!}\right|$$

$$= \lim_{n\to\infty}\frac{1}{n + 1} = 0$$

Therefore, the series converges only at the point $a = 1$.

c. Here $a_n = 1/n!$ and

$$R = \lim_{n\to\infty}\left|\frac{a_n}{a_{n+1}}\right| = \lim_{n\to\infty}\left|\frac{\dfrac{1}{n!}}{\dfrac{1}{(n + 1)!}}\right| = \lim_{n\to\infty}(n + 1) = \infty$$

Therefore, the interval of convergence of the series is $(-\infty, \infty)$; that is, it converges for any value of x.   ■ ■ ■ ■

Group Discussion
Suppose the power series $\displaystyle\sum_{n=0}^{\infty} a_n(x - a)^n$ has a radius of convergence R. What can you deduce about the radius of convergence of the series $\displaystyle\sum_{n=0}^{\infty} a_n(x - a)^{2n}$?

FINDING A TAYLOR SERIES

Theorem 8 guarantees that a power series represents a function whose domain is precisely the interval of convergence of the series. We now show that if a function is defined in this manner, then the power series must be a Taylor series. To see this, we need the following theorem, which we state without proof.

THEOREM 9

Suppose the function f is defined by

$$f(x) = \sum_{n=0}^{\infty} a_n(x - a)^n = a_0 + a_1(x - a) + a_2(x - a)^2 + \cdots$$

with radius of convergence $R > 0$. Then

$$f'(x) = \sum_{n=1}^{\infty} na_n(x - a)^{n-1} = a_1 + 2a_2(x - a) + 3a_3(x - a)^2 + \cdots$$

on the interval $(a - R, a + R)$.

Thus, the derivative of f may be found by differentiating the power series term by term.

We now prove the statement asserted earlier. Suppose that f is represented by a power series centered about $x = a$; that is,

$$f(x) = a_0 + a_1(x - a) + a_2(x - a)^2 + a_3(x - a)^3 + \cdots$$
$$+ a_n(x - a)^n + \cdots$$

Then, applying Theorem 9 repeatedly, we find

$$f'(x) = a_1 + 2a_2(x - a) + 3a_3(x - a)^2 + \cdots + na_n(x - a)^{n-1} + \cdots$$
$$f''(x) = 2a_2 + 3 \cdot 2a_3(x - a) + 4 \cdot 3a_4(x - a)^2 + \cdots$$
$$+ n(n - 1)a_n(x - a)^{n-2} + \cdots$$
$$f'''(x) = 3 \cdot 2a_3 + 4 \cdot 3 \cdot 2a_4(x - a) + \cdots$$
$$+ n(n - 1)(n - 2)a_n(x - a)^{n-3} + \cdots$$
$$= 3!a_3 + 4!a_4(x - a) + \cdots + n(n - 1)(n - 2)a_n(x - a)^{n-3} + \cdots$$
$$\vdots$$
$$f^{(n)}(x) = n!a_n + (n + 1)!a_{n+1}(x - a) + \cdots$$

Evaluating $f(x)$ and each of these derivatives at $x = a$ yields

$$f(a) = a_0$$
$$f'(a) = a_1$$
$$f''(a) = 2a_2 = 2!a_2$$
$$f'''(a) = 3 \cdot 2a_3 = 3!a_3$$
$$\vdots$$
$$f^{(n)}(a) = n!a_n$$

Thus,

$$a_0 = f(a), \qquad a_1 = f'(a), \qquad a_2 = \frac{f''(a)}{2!}$$

$$a_3 = \frac{f'''(a)}{3!}, \qquad \ldots, \qquad a_n = \frac{f^{(n)}(a)}{n!}, \qquad \ldots$$

as we set out to show.

Next we turn our attention to the converse problem. More precisely, suppose we are given a function f that has derivatives of *all* orders in an open interval I. Can we find a power series representation of f in that interval? The answer to this question is contained in the following theorem.

THEOREM 10	**Taylor Series Representation of a Function**

If a function f has derivatives of all orders in an open interval $I = (a - R, a + R)$ $(R > 0)$ centered at $x = a$, then

$$f(x) = \sum_{n=1}^{\infty} \frac{f^{(n)}(a)}{n!} (x - a)^n$$

if and only if

$$\lim_{n \to \infty} R_n(x) = 0$$

for all x in I, where $R_n(x) = f(x) - P_n(x)$ is as defined in Theorem 1.

Thus, f has a power series representation in the form of a Taylor series provided the error term associated with the Taylor polynomial tends to zero as the number of terms of the Taylor polynomial increases without bound. (A proof of this theorem is sketched in Exercise 33, page 828.) In what follows, we assume that the functions under consideration have Taylor series representations.

EXAMPLE 3   Find the Taylor series of the function $f(x) = \dfrac{1}{x - 1}$ at $x = 2$.

SOLUTION ✔   Here $a = 2$ and

$$f(x) = \frac{1}{x - 1}$$

$$f'(x) = -\frac{1}{(x - 1)^2}$$

$$f''(x) = \frac{2}{(x - 1)^3}$$

$$f'''(x) = -\frac{3 \cdot 2}{(x - 1)^4}$$

$$\vdots$$

$$f^{(n)}(x) = (-1)^n \frac{n!}{(x - 1)^{n+1}}$$

$$\vdots$$

so that

$$f(2) = 1, \quad f'(2) = -1, \quad f''(2) = 2$$
$$f'''(2) = -6 = -3!, \quad \ldots, \quad f^{(n)}(2) = (-1)^n n!, \quad \ldots$$

Therefore, the required Taylor series is

$$\frac{1}{x-1} = f(2) + f'(2)(x-2) + \frac{f''(2)}{2!}(x-2)^2 + \frac{f'''(2)}{3!}(x-2)^3 + \cdots$$

$$+ \frac{f^{(n)}(2)}{n!}(x-2)^n + \cdots$$

$$= 1 - (x-2) + \frac{2!}{2!}(x-2)^2 - \frac{3!}{3!}(x-2)^3 + \cdots$$

$$+ (-1)^n \frac{n!}{n!}(x-2)^n + \cdots$$

$$= 1 - (x-2) + (x-2)^2 - (x-2)^3 + \cdots + (-1)^n(x-2)^n + \cdots$$

$$= \sum_{n=0}^{\infty} (-1)^n(x-2)^n$$

To find the radius of convergence of the series, we compute

$$R = \lim_{n \to \infty} \left| \frac{a_n}{a_{n+1}} \right| = \lim_{n \to \infty} \left| \frac{(-1)^n}{(-1)^{n+1}} \right|$$

$$= \lim_{n \to \infty} 1 = 1$$

Thus, the series converges in the interval $(1, 3)$. By Theorem 10, we see that the function $f(x) = 1/(x-1)$ is represented by the Taylor series in the interval $(1, 3)$.   ■ ■ ■ ■

 The representation of the function $f(x) = 1/(x-1)$ by its Taylor series at $x = 2$ is valid only in the interval $(1, 3)$, despite the fact that f itself has a domain that is the set of all real numbers except $x = 1$. This serves to remind us of the local nature of this representation.

Exploring with Technology

Refer to Example 3, where it was shown that the Taylor series at $x = 2$ representing $f(x) = 1/(x-1)$ is

$$P(x) = \sum_{n=0}^{\infty} (-1)^n(x-2)^n = 1 - (x-2) + (x-2)^2 - (x-2)^3 + \cdots$$

for $1 < x < 3$. This means that if c is any number satisfying $1 < c < 3$, then $f(c) = 1/(c-1) = P(c)$. In particular, this means that $\lim_{n \to \infty} |P_n(c) - f(c)| = 0$, where $\{P_n(x)\}$ is the sequence of partial sums of $P(x)$.

1. Plot the graphs of f, P_0, P_1, P_2, P_3, P_4, P_5, and P_6 on the same set of axes, using the viewing rectangle $[2.5, 3.1] \times [-0.1, 1.1]$.

2. Do the results of part 1 give a visual confirmation of the statement $\lim_{n \to \infty} P_n(c) = f(c)$ for the special case where $c = 2.8$? Explain what happens when $c = 3$.

EXAMPLE **4**   Find the Taylor series of the function $f(x) = \ln x$ at $x = 1$.

SOLUTION ✔   Here $a = 1$ and

$$f(x) = \ln x$$

$$f'(x) = \frac{1}{x}$$

$$f''(x) = -\frac{1}{x^2}$$

$$f'''(x) = \frac{2}{x^3} = \frac{2!}{x^3}$$

$$f^{(4)}(x) = -\frac{3 \cdot 2}{x^4} = -\frac{3!}{x^4}$$

$$\vdots$$

$$f^{(n)}(x) = (-1)^{n+1}\frac{(n-1)!}{x^n}$$

so that

$$f(1) = 0, \quad f'(1) = 1, \quad f''(1) = -1, \quad f'''(1) = 2!$$
$$f^{(4)}(1) = -3!, \quad \ldots, \quad f^{(n)}(1) = (-1)^{n+1}(n-1)!, \quad \ldots$$

Therefore, the required Taylor series is

$$f(x) = \ln x = f(1) + f'(1)(x-1) + \frac{f''(1)}{2!}(x-1)^2 + \frac{f'''(1)}{3!}(x-1)^3$$

$$+ \cdots + \frac{f^{(n)}(1)}{n!}(x-1)^n + \cdots$$

$$= (x-1) - \frac{1}{2!}(x-1)^2 + \frac{2!}{3!}(x-1)^3 - \frac{3!}{4!}(x-1)^4 + \cdots$$

$$+ (-1)^{n+1}\frac{(n-1)!}{n!}(x-1)^n + \cdots$$

$$= (x-1) - \frac{1}{2}(x-1)^2 + \frac{1}{3}(x-1)^3 - \frac{1}{4}(x-1)^4 + \cdots$$

$$+ \frac{(-1)^{n+1}}{n}(x-1)^n + \cdots$$

$$= \sum_{n=1}^{\infty} \frac{(-1)^{n+1}}{n}(x-1)^n$$

Since

$$R = \lim_{n \to \infty}\left|\frac{a_n}{a_{n+1}}\right| = \lim_{n \to \infty}\left|\frac{\frac{(-1)^{n+1}}{n}}{\frac{(-1)^{n+2}}{n+1}}\right|$$

$$= \lim_{n \to \infty}\frac{n+1}{n} = \lim_{n \to \infty}\left(1 + \frac{1}{n}\right) = 1$$

we see that the power series representation of $f(x) = \ln x$ is valid in the interval $(0, 2)$. It can be shown that the representation is valid at $x = 2$ as well. ■ ■ ■ ■

EXAMPLE 5   Find the power series representation about $x = 0$ for the function $f(x) = e^x$.

SOLUTION ✔   Here $a = 0$ and, since

$$f(x) = f'(x) = f''(x) = \cdots = e^x$$

we see that

$$f(0) = f'(0) = f''(0) = \cdots = e^0 = 1$$

Therefore, the required power series (Taylor series) is

$$f(x) = e^x = f(0) + f'(0)(x - 0) + \frac{f''(0)}{2!}(x - 0)^2 + \cdots + \frac{f^{(n)}}{n!}(x - 0)^n + \cdots$$

$$= 1 + x + \frac{x^2}{2!} + \frac{x^3}{3!} + \cdots + \frac{x^n}{n!} + \cdots$$

$$= \sum_{n=0}^{\infty} \frac{x^n}{n!}$$

Since

$$R = \lim_{n \to \infty} \left| \frac{a_n}{a_{n+1}} \right| = \lim_{n \to \infty} \left| \frac{\dfrac{1}{n!}}{\dfrac{1}{(n+1)!}} \right|$$

$$= \lim_{n \to \infty} (n + 1) = \infty$$

we see that the series representation is valid for all x. ■ ■ ■ ■

SELF-CHECK EXERCISES 11.5

1. Find the interval of convergence of the power series $\sum_{n=0}^{\infty} \dfrac{x^n}{n^2 + 1}$ (disregard the end points).

2. Find the Taylor series of the function $f(x) = e^{-x}$ at $x = 1$ and determine its interval of convergence.

Solutions to Self-Check Exercises 11.5 can be found on page 828.

11.5 Exercises

In Exercises 1–20, find the radius of convergence and the interval of convergence of the power series.

1. $\sum_{n=0}^{\infty} (x - 1)^n$

2. $\sum_{n=0}^{\infty} \left(\frac{x}{2}\right)^n$

3. $\sum_{n=1}^{\infty} n^2 x^n$

4. $\sum_{n=0}^{\infty} \frac{(n + 1)(x + 2)^n}{2^n}$

5. $\sum_{n=0}^{\infty} \frac{(-1)^n x^n}{4^n}$

6. $\sum_{n=0}^{\infty} \frac{(2x)^n}{3^n}$

7. $\sum_{n=0}^{\infty} \dfrac{(x-1)^n}{n!2^n}$

8. $\sum_{n=0}^{\infty} (2n)!x^n$

9. $\sum_{n=0}^{\infty} \dfrac{(-1)^n n!(x+2)^n}{2^n}$

10. $\sum_{n=2}^{\infty} \dfrac{x^n}{n(n+1)}$

11. $\sum_{n=2}^{\infty} \dfrac{(x+3)^n}{(n+1)^2}$

12. $\sum_{n=0}^{\infty} \dfrac{n!(x+1)^n}{(3n)!}$

13. $\sum_{n=1}^{\infty} \dfrac{2n(x-3)^n}{(n+1)!}$

14. $\sum_{n=0}^{\infty} \dfrac{(-2x)^{2n}}{4^n(n+1)}$

15. $\sum_{n=1}^{\infty} \dfrac{n(-2x)^n}{n+1}$

16. $\sum_{n=0}^{\infty} \dfrac{x^{2n+1}}{(2n+1)!}$

17. $\sum_{n=0}^{\infty} \dfrac{n!(x+1)^n}{3^n}$

18. $\sum_{n=0}^{\infty} (n+1)(x+2)^n$

19. $\sum_{n=0}^{\infty} \dfrac{n^3(x-3)^n}{3^n}$

20. $\sum_{n=0}^{\infty} \dfrac{(-1)^{n+1}(x-2)^n}{n2^n}$

In Exercises 21–32, find the Taylor series of the function at the indicated point and give its radius and interval of convergence (disregard the end points).

21. $f(x) = \dfrac{1}{x}; x = 1$

22. $f(x) = \dfrac{1}{x+1}; x = 0$

23. $f(x) = \dfrac{1}{x+1}; x = 2$

24. $f(x) = \dfrac{1}{1-x}; x = 0$

25. $f(x) = \dfrac{1}{1-x}; x = 2$

26. $f(x) = \ln(x+1); x = 0$

27. $f(x) = \sqrt{x}; x = 1$

28. $f(x) = \sqrt{1-x}; x = 0$

29. $f(x) = e^{2x}; x = 0$

30. $f(x) = e^{2x}; x = 1$

31. $f(x) = \dfrac{1}{\sqrt{x+1}}; x = 0$

32. $f(x) = \sqrt{x+1}; x = 1$

33. Prove Theorem 10. Hint: The nth partial sum of the Taylor series is $S_n(x) = P_n(x)$ where $P_n(x)$ is the nth Taylor polynomial. Use this fact to show that $\lim_{n\to\infty} S_n(x) = f(x)$ if and only if $\lim_{n\to\infty} R_n(x) = 0$.

In Exercises 34–36, determine whether the statement is true or false. If it is true, explain why it is true. If it is false, give an example to show why it is false.

34. If $\sum_{n=0}^{\infty} a_n(x-2)^n$ converges for $x = 4$, then it converges for $x = 1$.

35. If $\sum_{n=0}^{\infty} a_n(x-a)^n$ has radius of convergence R, then $\sum_{n=0}^{\infty} na_n(x-a)^n$ has radius of convergence R.

36. If $\sum_{n=0}^{\infty} a_n(x-a)^n$ has radius of convergence R, then $\sum_{n=0}^{\infty} a_n^2(x-a)^n$ has radius of convergence $\sqrt{R}$.

SOLUTIONS TO SELF-CHECK EXERCISES 11.5

1. We first find the radius of convergence of the power series. Since $a_n = \dfrac{1}{n^2+1}$, we have

$$R = \lim_{n\to\infty} \left|\frac{a_n}{a_{n+1}}\right| = \lim_{n\to\infty} \frac{\dfrac{1}{n^2+1}}{\dfrac{1}{(n+1)^2+1}}$$

$$= \lim_{n\to\infty} \frac{(n+1)^2+1}{n^2+1}$$

$$= \lim_{n\to\infty} \frac{\left(1+\dfrac{1}{n}\right)^2 + \dfrac{1}{n^2}}{1 + \dfrac{1}{n^2}} \quad \text{(Dividing numerator and denominator by } n^2\text{)}$$

$$= 1$$

Therefore, the interval of convergence of the series is $(-1, 1)$.

2. Here $a = 1$ and

$$f(x) = e^{-x}$$
$$f'(x) = -e^{-x}$$
$$f''(x) = e^{-x}$$
$$f'''(x) = -e^{-x}$$
$$\vdots$$
$$f^{(n)}(x) = (-1)^n e^{-x}$$

so that

$$f(1) = e^{-1}, \qquad f'(1) = -e^{-1}, \qquad f''(1) = e^{-1}$$
$$f'''(1) = -e^{-1}, \qquad \ldots, \qquad f^{(n)}(1) = (-1)^n e^{-1}$$

Therefore, the required Taylor series is

$$e^{-x} = f(1) + f'(1)(x - 1) + \frac{f''(1)}{2!}(x - 1)^2$$
$$+ \frac{f'''(1)}{3!}(x - 1)^3 + \cdots + \frac{f^{(n)}(1)}{n!}(x - 1)^n + \cdots$$
$$= e^{-1} - \frac{e^{-1}}{1!}(x - 1) + \frac{e^{-1}}{2!}(x - 1)^2$$
$$- \frac{e^{-1}}{3!}(x - 1)^3 + \cdots + \frac{(-1)^n e^{-1}}{n!}(x - 1)^n + \cdots$$
$$= \frac{1}{e} - \frac{1}{e}(x - 1) + \frac{1}{2!e}(x - 1)^2 - \frac{1}{3!e}(x - 1)^3 + \cdots$$
$$+ \frac{(-1)^n}{n!e}(x - 1)^n + \cdots$$
$$= \sum_{n=0}^{\infty} \frac{(-1)^n}{n!e}(x - 1)^n$$

11.6 More on Taylor Series

A USEFUL TECHNIQUE FOR FINDING TAYLOR SERIES

In the last section, we showed how to find the power series representation of certain functions. This representation turned out to be the Taylor series of f at $x = a$. The method we used to compute the series relies solely on our ability to find the higher-order derivatives of the function f. This method, however, is rather tedious.

In this section, we show how the Taylor series of a function can often be found by manipulating some well-known power series. For this purpose, we first catalog the power series of some of the most commonly used functions (Table 11.1). Recall that each of these representations was derived in Sections 11.1 and 11.5.

Table 11.1   Power Series Representations for Some Common Functions

1. $\dfrac{1}{1-x} = 1 + x + x^2 + \cdots + x^n + \cdots \qquad (-1 < x < 1)$

2. $e^x = 1 + x + \dfrac{1}{2!}x^2 + \dfrac{1}{3!}x^3 + \cdots + \dfrac{1}{n!}x^n + \cdots \qquad (-\infty < x < \infty)$

3. $\ln x = (x-1) - \dfrac{1}{2}(x-1)^2 + \dfrac{1}{3}(x-1)^3 - \cdots$

   $\qquad + \dfrac{(-1)^{n+1}}{n}(x-1)^n + \cdots \qquad (0 < x \le 2)$

EXAMPLE 1   Find the Taylor series of each function at the indicated point.

a. $f(x) = \dfrac{1}{1+x}; x = 0$ $\qquad$ **b.** $f(x) = \dfrac{1}{1+x}; x = 2$

c. $f(x) = \dfrac{1}{1-3x}; x = 0$ $\qquad$ **d.** $f(x) = \dfrac{x}{1+x^2}; x = 0$

SOLUTION ✔ **a.** We write

$$f(x) = \frac{1}{1-(-x)}$$

and use the power series representation of $1/(1-x)$ with x replaced by $-x$ to obtain

$$f(x) = \frac{1}{1+x}$$
$$= 1 + (-x) + (-x)^2 + \cdots + (-x)^n + \cdots$$
$$= 1 - x + x^2 - x^3 + \cdots + (-1)^n x^n + \cdots \qquad (-1 < x < 1)$$

b. We write

$$f(x) = \frac{1}{1+x} = \frac{1}{3 + (x-2)} = \frac{1}{3\left[1 + \left(\frac{x-2}{3}\right)\right]} = \frac{1}{3}\left[\frac{1}{1 + \left(\frac{x-2}{3}\right)}\right]$$

From the result of part (a), we have

$$\frac{1}{1+u} = 1 - u + u^2 - u^3 + \cdots + (-1)^n u^n + \cdots \qquad (-1 < u < 1)$$

Thus, with $u = \dfrac{x-2}{3}$, we find

$$f(x) = \frac{1}{3}\left[\frac{1}{1 + \left(\frac{x-2}{3}\right)}\right]$$

$$= \frac{1}{3}\left[1 - \left(\frac{x-2}{3}\right) + \left(\frac{x-2}{3}\right)^2 - \left(\frac{x-2}{3}\right)^3 + \cdots\right.$$

$$\left. + (-1)^n \left(\frac{x-2}{3}\right)^n + \cdots\right]$$

$$= \frac{1}{3} - \frac{1}{3^2}(x-2) + \frac{1}{3^3}(x-2)^2 - \frac{1}{3^4}(x-2)^3 + \cdots$$

$$+ \frac{(-1)^n}{3^{n+1}}(x-2)^n + \cdots$$

which converges when

$$-1 < \frac{x-2}{3} < 1 \qquad \text{or} \qquad -1 < x < 5$$

c. Here we simply replace x in the power series representation of $1/(1-x)$ by $3x$ to get

$$f(x) = \frac{1}{1 - 3x} = 1 + 3x + (3x)^2 + (3x)^3 + \cdots + (3x)^n + \cdots$$

$$= 1 + 3x + 3^2 x^2 + 3^3 x^3 + \cdots + 3^n x^n + \cdots$$

The series converges for $-1 < 3x < 1$, or $-\dfrac{1}{3} < x < \dfrac{1}{3}$.

d. We write

$$f(x) = \frac{x}{1 + x^2} = x\left[\frac{1}{1 - (-x^2)}\right]$$

$$= x[1 + (-x^2) + (-x^2)^2 + (-x^2)^3 + \cdots + (-x^2)^n + \cdots]$$

$$= x - x^3 + x^5 - x^7 + \cdots + (-1)^n x^{2n+1} + \cdots$$

The series converges for $-1 < x < 1$. ■ ■ ■

EXAMPLE 2 Find the Taylor series of each of the following functions at the indicated point.

a. $f(x) = xe^{-x}; x = 0$　　　　**b.** $f(x) = \ln(1 + x); x = 0$

SOLUTION ✔ **a.** First, we replace x with $-x$ in the expression

$$e^x = 1 + x + \frac{x^2}{2!} + \frac{x^3}{3!} + \cdots + \frac{x^n}{n!} + \cdots \qquad \text{(See Table 11.1)}$$

to obtain

$$e^{-x} = 1 + (-x) + \frac{(-x)^2}{2!} + \frac{(-x)^3}{3!} + \cdots + \frac{(-x)^n}{n!} + \cdots$$

$$= 1 - x + \frac{x^2}{2!} - \frac{x^3}{3!} + \cdots + \frac{(-1)^n x^n}{n!} + \cdots$$

Then, multiplying both sides of this expression by x gives the required expression

$$f(x) = xe^{-x} = x - x^2 + \frac{x^3}{2!} - \frac{x^4}{3!} + \cdots + \frac{(-1)^n x^{n+1}}{n!} + \cdots$$

b. From Table 11.1 we have

$$\ln x = (x - 1) - \frac{1}{2}(x - 1)^2 + \frac{1}{3}(x - 1)^3 - \cdots$$

$$+ \frac{(-1)^{n+1}}{n}(x - 1)^n \cdots \qquad (0 < x \le 2)$$

Replacing x with $1 + x$ in this expression, we obtain the required expression

$$f(x) = \ln(1 + x) = x - \frac{1}{2}x^2 + \frac{1}{3}x^3 - \cdots + \frac{(-1)^{n+1}}{n}x^n + \cdots$$

valid for $-1 < x \le 1$. (Why?) ■ ■ ■

■ **Group Discussion**
The formula

$$\ln x = (x - 1) - \frac{1}{2}(x - 1)^2 + \frac{1}{3}(x - 1)^3 - \cdots \qquad (0 < x \le 2) \quad \textbf{(A)}$$

from Table 11.1 can be used to compute the value of $\ln x$ for $0 < x \le 2$. However, the restriction on x and the slow convergence of the series limit its effectiveness from the computational point of view. A more effective formula, first obtained by the Scottish mathematician James Gregory (1638–1675), follows:

1. Using Formula (A), derive the formulas

$$\ln(1 + x) = x - \frac{x^2}{2} + \frac{x^3}{3} - \frac{x^4}{4} - \cdots \qquad (-1 < x \le 1)$$

and $\quad \ln(1 - x) = -x - \frac{x^2}{2} - \frac{x^3}{3} - \frac{x^4}{4} - \cdots \qquad (-1 \le x < 1)$

2. Use part 1 to show that

$$\ln\left(\frac{1 + x}{1 - x}\right) = 2\left(x + \frac{x^3}{3} + \frac{x^5}{5} + \frac{x^7}{7} + \cdots\right) \qquad (-1 < x < 1)$$

3. To compute the natural logarithm of a positive number p, let

$$p = \frac{1 + x}{1 - x}$$

and show that

$$x = \frac{p - 1}{p + 1} \qquad (-1 < x < 1)$$

4. Use parts 2 and 3 to show that

$$\ln 2 = 2\left[\left(\frac{1}{3}\right) + \frac{(\frac{1}{3})^3}{3} + \frac{(\frac{1}{3})^5}{5} + \frac{(\frac{1}{3})^7}{7} + \cdots\right]$$

and this yields $\ln 2 \approx 0.6931$ when we add the first four terms of the series. This approximation of $\ln 2$ is accurate to four decimal places.
5. Compare this method of computing $\ln 2$ with that of using Formula (A) directly.

In Theorem 10 in the last section, we saw that a power series may be differentiated term by term to yield another power series whose interval of convergence coincides with that of the original series. The latter is of course the power series representation of the derivative f' of the function f represented by the first series. The following theorem tells us that we may integrate a power series term by term.

THEOREM 11

Suppose

$$f(x) = \sum_{n=0}^{\infty} a_n(x-a)^n$$

$$= a_0 + a_1(x-a) + a_2(x-a)^2 + a_3(x-a)^3 + \cdots$$
$$+ a_n(x-a)^n + \cdots \qquad x \in (a-R, a+R)$$

Then

$$\int f(x)\, dx = \sum_{n=0}^{\infty} \frac{a_n}{n+1}(x-a)^{n+1}$$

$$= a_0(x-a) + \frac{a_1}{2}(x-a)^2 + \frac{a_2}{3}(x-a)^3 + \cdots$$

$$+ \frac{a_n}{n+1}(x-a)^{n+1} + \cdots \qquad x \in (a-R, a+R)$$

Theorems 10 and 11 can also be used to help us find the power-series representation of a function starting from the series representation of some appropriate function, as the next two examples show.

EXAMPLE 3   Differentiate the power series for the function $1/(1-x)$ at $x = 0$ (see Table 11.1) to obtain a Taylor series representation of the function $f(x) = 1/(1-x)^2$ at $x = 0$.

SOLUTION ✔   From Table 11.1, we have

$$\frac{1}{1-x} = 1 + x + x^2 + x^3 + \cdots + x^n + \cdots \qquad (-1 < x < 1)$$

Differentiating both sides of the equation with respect to x, and using Theorem 10, we obtain

$$\frac{d}{dx}\left(\frac{1}{1-x}\right) = \frac{d}{dx}(1 + x + x^2 + x^3 + \cdots + x^n + \cdots)$$

or

$$f(x) = \frac{1}{(1-x)^2}$$
$$= 1 + 2x + 3x^2 + \cdots + nx^{n-1} + \cdots \qquad (-1 < x < 1) \quad \blacksquare\blacksquare\blacksquare\blacksquare$$

EXAMPLE 4   Integrate the Taylor series for the function $1/(1+x)$ at $x = 0$ (see Table 11.1) to obtain a power series representation for the function $f(x) = \ln(1+x)$ centered at $x = 0$. Compare your result with that of Example 2b.

SOLUTION ✔ From Table 11.1 we see that

$$\frac{1}{1-x} = 1 + x + x^2 + x^3 + \cdots + x^n + \cdots \qquad (-1 < x < 1)$$

Replacing x by $-x$ gives

$$\frac{1}{1-(-x)} = 1 + (-x) + (-x)^2 + (-x)^3 + \cdots + (-x)^n + \cdots$$

or

$$\frac{1}{1+x} = 1 - x + x^2 - x^3 + \cdots + (-1)^n x^n + \cdots \qquad (-1 < x < 1)$$

Finally, integrating both sides of this equation with respect to x and using Theorem 11, we obtain

$$\int \frac{1}{1+x}\, dx = \int [1 - x + x^2 - x^3 + \cdots + (-1)^n x^n + \cdots]\, dx$$

or

$$f(x) = \ln(1 + x) = x - \frac{1}{2}x^2 + \frac{1}{3}x^3 - \cdots + \frac{(-1)^n}{n+1}x^{n+1} + \cdots$$

$$= x - \frac{1}{2}x^2 + \frac{1}{3}x^3 - \cdots + \frac{(-1)^{n+1}}{n}x^n + \cdots \qquad (-1 < x < 1)$$

This result is the same as that of Example 2b, as expected. ■ ■ ■ ■

APPLICATION

In many practical applications, all that is required is an approximation of the actual solution to a problem. Thus, rather than working with the Taylor series of a function at a point that is *equal* to $f(x)$ inside its interval of convergence, one often works with the truncated Taylor series. The truncated Taylor series is a Taylor polynomial that gives an acceptable approximation to the values of $f(x)$ in the neighborhood of the point about which the function is expanded, provided the degree of the polynomial or, equivalently, the number of terms of the Taylor series retained is large enough.

Before looking at an example, we want to point out the comparative ease with which one can obtain the Taylor polynomial approximation of a function using the method of this section, rather than obtaining it directly as was done in Section 11.1.

EXAMPLE 5 The serum cholesterol levels (in mg/dL) in a current Mediterranean population are found to be normally distributed with a probability density function given by

$$f(x) = \frac{1}{50\sqrt{2\pi}}\, e^{-1/2[(x-160)/50]^2}$$

Scientists at the National Heart, Lung, and Blood Institute consider this pattern ideal for a minimal risk of heart attacks. Find the percentage of the population who have blood cholesterol levels between 160 and 180 mg/dL.

SOLUTION ✔   The required probability is given by

$$P(160 \le x \le 180) = \frac{1}{50\sqrt{2\pi}} \int_{160}^{180} e^{-1/2[(x-160)/50]^2} \, dx$$

(see Section 10.3). Let's approximate the integrand by a sixth-degree Taylor polynomial about $x = 160$. From Table 11.1, we have

$$e^x = 1 + x + \frac{1}{2!}x^2 + \frac{1}{3!}x^3 + \cdots$$

Replacing x with $-\frac{1}{2}[(x - 160)/50]^2$, we obtain

$$e^{-1/2[(x-160)/50]^2} \approx 1 - \frac{1}{2}\left(\frac{x-160}{50}\right)^2 + \frac{1}{2!}\left[-\frac{1}{2}\left(\frac{x-160}{50}\right)^2\right]^2$$
$$+ \frac{1}{3!}\left[-\frac{1}{2}\left(\frac{x-160}{50}\right)^2\right]^3$$
$$= 1 - \frac{(x-160)^2}{5000} + \frac{(x-160)^4}{5 \cdot 10^7} - \frac{(x-160)^6}{7.5 \cdot 10^{11}}$$

Therefore,

$$P(160 \le x \le 180)$$

$$\approx \frac{1}{50\sqrt{2\pi}} \int_{160}^{180} \left[1 - \frac{(x-160)^2}{5000} + \frac{(x-160)^4}{5 \cdot 10^7} - \frac{(x-160)^6}{7.5 \cdot 10^{11}}\right] dx$$

$$\approx \frac{1}{50\sqrt{2\pi}} \left[x - \frac{(x-160)^3}{15,000} + \frac{(x-160)^5}{2.5 \cdot 10^8} - \frac{(x-160)^7}{5.25 \cdot 10^{12}}\right]\Bigg|_{160}^{180}$$

$$\approx \frac{1}{50\sqrt{2\pi}} [(180 - 0.53333 + 0.0128 - 0.00024) - 160]$$

$$\approx 0.1554$$

and so approximately 15.5% of the population has blood cholesterol levels between 160 and 180 mg/dL.   ■ ■ ■ ■

SELF-CHECK EXERCISES **11.6**

1. Find the Taylor series of the function $f(x) = xe^{-x^2}$ at $x = 0$.

2. Use the result from Exercise 1 to write the seventh Taylor polynomial of f about $x = 0$, and use this polynomial to approximate

$$\int_0^{0.5} xe^{-x^2} \, dx$$

Compare this result with the exact value of the integral.

Solutions to Self-Check Exercises 11.6 can be found on page 838.

11.6 Exercises

In Exercises 1–20, find the Taylor series of each function at the indicated point. Give the interval of convergence for each series.

1. $f(x) = \dfrac{1}{1-x}; x = 2$

2. $f(x) = \dfrac{1}{1+x}; x = 1$

3. $f(x) = \dfrac{1}{1+3x}; x = 0$

4. $f(x) = \dfrac{x}{1-2x}; x = 0$

5. $f(x) = \dfrac{1}{4-3x}; x = 0$

6. $f(x) = \dfrac{1}{4-3x}; x = 1$

7. $f(x) = \dfrac{1}{1-x^2}; x = 0$

8. $f(x) = \dfrac{x^2}{1+x^3}; x = 0$

9. $f(x) = e^{-x}; x = 0$

10. $f(x) = e^x; x = 1$

11. $f(x) = xe^{-x^2}; x = 0$

12. $f(x) = xe^{x/2}; x = 0$

13. $f(x) = \dfrac{1}{2}(e^x + e^{-x}); x = 0$

14. $f(x) = \dfrac{1}{2}(e^x - e^{-x}); x = 0$

15. $f(x) = \ln(1 + 2x); x = 0$

16. $f(x) = \ln\left(1 + \dfrac{x}{2}\right); x = 0$

17. $f(x) = \ln(1 + x^2); x = 0$

18. $f(x) = \ln(1 + 2x); x = 2$

19. $f(x) = (x - 2)\ln x; x = 2$

20. $f(x) = x^2 \ln\left(1 + \dfrac{x}{2}\right); x = 0$

21. Differentiate the power series for $\ln(1 + x)$ at $x = 0$ to obtain a series representation for the function $f(x) = 1/(1 + x)$.

22. Differentiate the power series for $1/(1 + x)$ at $x = 0$ to obtain a series representation for the function $f(x) = 1/(1 + x)^2$.

23. Integrate the power series for $1/(1 + x)$ to obtain a power series representation for the function $f(x) = \ln(1 + x)$.

24. Integrate the power series for $2x/(1 + x^2)$ to obtain a power series representation for the function $f(x) = \ln(1 + x^2)$.

25. Use the sixth-degree Taylor polynomial to approximate

$$\int_0^{0.5} \frac{1}{\sqrt{1 + x^2}}\, dx$$

Hint: $P_6(x) = 1 - \frac{1}{2}x^2 + \frac{3}{8}x^4 - \frac{5}{16}x^6$

26. Use the sixth-degree Taylor polynomial to approximate

$$\int_0^{0.4} \ln(1 + x^2)\, dx$$

27. Use the eighth-degree Taylor polynomial to approximate

$$\int_0^1 e^{-x^2}\, dx$$

28. Use the sixth-degree Taylor polynomial to approximate

$$\int_0^{0.5} \frac{\ln(1 + x)}{x}\, dx$$

29. Use the eighth-degree Taylor polynomial of $f(x) = 1/(1 + x^2)$ at $x = 0$ and the relationship

$$\pi = 4\int_0^1 \frac{dx}{1 + x^2}$$

to obtain an approximation of π.

30. FACTORY WORKER WAGES According to data released by a city's Chamber of Commerce, the weekly wages of factory workers are normally distributed according to the probability density function

$$f(x) = \frac{1}{50\sqrt{2\pi}}\, e^{-1/2[(x-500)/50]^2}$$

Find the probability that a worker selected at random from the city has a weekly wage of $450–$550.

31. TV SET RELIABILITY The General Manager of the Service Department of MCA Television Company has estimated that the time that elapses between the dates of purchase and the dates on which the 19-in. sets manufactured by the company first require service is normally distributed according to the probability density function

$$f(x) = \frac{1}{10\sqrt{2\pi}}\, e^{-1/2[(x-30)/10]^2}$$

where x is measured in months. Determine the percentage of sets manufactured and sold by MCA that may require service 28–32 mo after purchase.

32. DEMAND FOR WRISTWATCHES The demand equation for the "Tempus" quartz wristwatch is given by

$$p = 50e^{-0.1(x+1)^2}$$

where x (measured in units of a thousand) is the quantity demanded per week and p is the unit wholesale price in dollars. National Importers, the supplier of the watches, will make x units (in thousands) available in the market if the unit wholesale price is

$$p = 10 + 5x^2$$

dollars. Use an appropriate Taylor polynomial approximation of the function

$$p = f(x) = 50e^{-0.1(x+1)^2}$$

to find the consumers' surplus.
Hint: The equilibrium quantity is 1689 units.

SOLUTIONS TO SELF-CHECK EXERCISES 11.6

1. First, we replace x in the expression

$$e^x = 1 + x + \frac{x^2}{2!} + \frac{x^3}{3!} + \cdots + \frac{x^n}{n!} + \cdots$$

(see Table 11.1) with $-x^2$ to obtain

$$e^{-x^2} = 1 + (-x^2) + \frac{(-x^2)^2}{2!}$$

$$+ \frac{(-x^2)^3}{3!} + \cdots + \frac{(-x^2)^n}{n!} + \cdots$$

$$= 1 - x^2 + \frac{x^4}{2!} - \frac{x^6}{3!} + \cdots + \frac{(-1)^n x^{2n}}{n!} + \cdots$$

Then multiplying both sides of this expression by x gives the required expression

$$f(x) = xe^{-x^2} = x - x^3 + \frac{x^5}{2!}$$

$$- \frac{x^7}{3!} + \cdots + \frac{(-1)^n x^{2n+1}}{n!} + \cdots$$

$$= \sum_{n=0}^{\infty} \frac{(-1)^n x^{2n+1}}{n!}$$

2. Using the result from Exercise 1, we see that the seventh Taylor polynomial of f about $x = 0$ is

$$P_7(x) = x - x^3 + \frac{x^5}{2!} - \frac{x^7}{3!}$$

$$= x - x^3 + \frac{1}{2}x^5 - \frac{1}{6}x^7$$

Therefore,

$$\int_0^{0.5} xe^{-x^2}\,dx \approx \int_0^{0.5} \left(x - x^3 + \frac{1}{2}x^5 - \frac{1}{6}x^7\right) dx$$

$$= \frac{1}{2}x^2 - \frac{1}{4}x^4 + \frac{1}{12}x^6 - \frac{1}{48}x^8 \Big|_0^{0.5}$$

$$\approx 0.1105957$$

The exact value of the integral is

$$\int_0^{0.5} xe^{-x^2}\,dx = -\frac{1}{2}e^{-x^2}\Big|_0^{0.5} \qquad \text{(Use the substitution } u = x^2.)$$

$$= -\frac{1}{2}(e^{-0.25} - 1)$$

which is approximately 0.1105996. Thus, the error is 0.0000039.

11.7 The Newton–Raphson Method

THE NEWTON–RAPHSON METHOD

Recall that in much of our previous work we had occasion to find the zeros of a function f or, equivalently, the **roots** of the equation $f(x) = 0$. For example, the x-intercepts of a function f are precisely the values of x when $f(x) = 0$; the critical points of f include the roots of the equation $f'(x) = 0$; and the candidates for the inflection points of f include the roots of the equation $f''(x) = 0$.

If $f(x)$ is a linear or quadratic function or $f(x)$ is a polynomial that is easily factored, the roots of f are readily found. In practice, however, we often encounter functions with zeros that cannot be found as readily. For example, the function

$$f(t) = -t^3 + 96t^2 + 195t + 5$$

of Example 8, page 189, which gives the altitude (in feet) of a rocket t seconds into flight, is not easily factored, so its zeros cannot be found by elementary algebraic methods. (We will find the zeros of this function in Example 2.) Another example of a function with roots that are not easily found is

$$g(x) = e^{2x} - 3x - 2$$

In this section we develop an algorithm, based on the approximation of a function $f(x)$ by a first-degree Taylor polynomial, to approximate the value of a zero of $f(x)$ to any desired degree of accuracy.

Suppose the function f has a zero at $x = c$ (Figure 11.7). Let x_0 be an initial estimate of the actual zero c of $f(x)$. Now, the first-degree Taylor polynomial of $f(x)$ at $x = x_0$ is

$$P_1(x) = f(x_0) + f'(x_0)(x - x_0)$$

and, as observed earlier, is just the equation of the tangent line to the graph of $y = f(x)$ at the point $(x_0, f(x_0))$. Since the tangent line T is an approximation of the graph of $y = f(x)$ near $x = x_0$ (and x_0 is assumed to be close to c!), it may be expected that the zero of $P_1(x)$ is close to the zero $x = c$ of $f(x)$. But

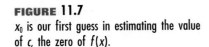

FIGURE 11.7

x_0 is our first guess in estimating the value of c, the zero of $f(x)$.

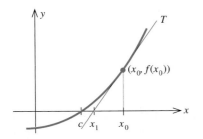

the zero of $P_1(x)$, a linear function, is found by setting $P_1(x) = 0$. Thus,

$$f(x_0) + f'(x_0)(x - x_0) = 0$$
$$f'(x_0)(x - x_0) = -f(x_0)$$
$$x - x_0 = -\frac{f(x_0)}{f'(x_0)}$$
$$x = x_0 - \frac{f(x_0)}{f'(x_0)}$$

This number provides us with another estimate of the zero of $f(x)$:

$$x_1 = x_0 - \frac{f(x_0)}{f'(x_0)}$$

In general, the estimate x_1 is better than the initial estimate x_0.

This process may be repeated with the initial estimate x_0 replaced by the recent estimate x_1. This leads to yet another estimate,

$$x_2 = x_1 - \frac{f(x_1)}{f'(x_1)}$$

which is usually better than x_1. In this manner, we generate a sequence of approximations $x_0, x_1, x_2, \ldots, x_n, x_{n+1}, \ldots$, with

$$x_{n+1} = x_n - \frac{f(x_n)}{f'(x_n)}$$

which, in most instances, approaches the zero $x = c$ of $f(x)$.

A summary of this algorithm follows.

Newton–Raphson Algorithm

1. Pick an initial estimate x_0 of the root c.
2. Find a new estimate using the iterative formula

$$x_{n+1} = x_n - \frac{f(x_n)}{f'(x_n)} \qquad (n = 0, 1, 2, \ldots) \qquad \textbf{(11)}$$

3. Compute $|x_n - x_{n+1}|$. If this number is less than a prescribed positive number, stop. The required approximation to the root $x = c$ is $x = x_{n+1}$.

SOLVING EQUATIONS USING THE NEWTON–RAPHSON METHOD

    EXAMPLE 1   Use the Newton–Raphson algorithm to approximate the zero of $f(x) = x^2 - 2$. Start the iteration with initial guess $x_0 = 1$ and terminate the process when two successive approximations differ by less than 0.00001.

SOLUTION ✔

We have

$$f(x) = x^2 - 2$$
$$f'(x) = 2x$$

so that, by (11), the required iterative formula is

$$x_{n+1} = x_n - \frac{x_n^2 - 2}{2x_n} = \frac{x_n^2 + 2}{2x_n}$$

With $x_0 = 1$, we find

$$x_1 = \frac{1^2 + 2}{2(1)} = 1.5$$

$$x_2 = \frac{(1.5)^2 + 2}{2(1.5)} \approx 1.416667$$

$$x_3 = \frac{(1.416667)^2 + 2}{2(1.416667)} \approx 1.414216$$

$$x_4 = \frac{(1.414216)^2 + 2}{2(1.414216)} \approx 1.414214$$

Since $x_3 - x_4 = 0.000002 < 0.00001$, we terminate the process. The sequence generated converges to $\sqrt{2}$, which is one of the two roots of the equation $x^2 - 2 = 0$. Note that, to six places, $\sqrt{2} = 1.414214$!   ■ ■ ■ ■

EXAMPLE 2

Refer to Example 8, page 189. The altitude in feet of a rocket t seconds into flight is given by

$$s = f(t) = -t^3 + 96t^2 + 195t + 5 \qquad (t \geq 0)$$

Find the time T when the rocket hits Earth.

SOLUTION ✔

The rocket hits Earth when the altitude is equal to zero. So we are required to solve the equation

$$s = f(t) = -t^3 + 96t^2 + 195t + 5 = 0$$

Let's use the Newton–Raphson algorithm with initial guess $t_0 = 100$ (Figure 11.8). Here

$$f'(t) = -3t^2 + 192t + 195$$

so the required iterative formula takes the form

$$t_{n+1} = t_n - \frac{-t_n^3 + 96t_n^2 + 195t_n + 5}{-3t_n^2 + 192t_n + 195}$$

$$= t_n - \frac{t_n^3 - 96t_n^2 - 195t_n - 5}{3t_n^2 - 192t_n - 195}$$

$$= \frac{3t_n^3 - 192t_n^2 - 195t_n - t_n^3 + 96t_n^2 + 195t_n + 5}{3t_n^2 - 192t_n - 195}$$

$$= \frac{2t_n^3 - 96t_n^2 + 5}{3t_n^2 - 192t_n - 195}$$

FIGURE **11.8**

The rocket's altitude t seconds into flight is given by $f(t)$.

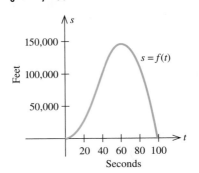

We have

$$t_1 = \frac{2(100)^3 - 96(100)^2 + 5}{3(100)^2 - 192(100) - 195} \approx 98.0674$$

$$t_2 = \frac{2(98.0674)^3 - 96(98.0674)^2 + 5}{3(98.0674)^2 - 192(98.0674) - 195} \approx 97.9906$$

$$t_3 = \frac{2(97.9906)^3 - 96(97.9906)^2 + 5}{3(97.9906)^2 - 192(97.9906) - 195} \approx 97.9905$$

Thus, we see that $T \approx 97.99$, so the rocket hits Earth approximately 98 seconds after liftoff.   ■ ■ ■ ■

The next example shows how the Newton-Raphson Method may be used to help determine a certain commodity's equilibrium quantity and price.

EXAMPLE 3   The demand equation for the "Tempus" quartz wristwatch is given by

$$p = 50e^{-0.1(x+1)^2}$$

where x (measured in units of a thousand) is the quantity demanded per week and p is the unit wholesale price in dollars. National Importers, the supplier of the watches, will make x units (in thousands) available in the market if the unit wholesale price is

$$p = 10 + 5x^2$$

dollars. Find the equilibrium quantity and price.

SOLUTION ✔   We determine the equilibrium point by finding the point of intersection of the demand curve and the supply curve (Figure 11.9). To solve the system of equations

$$p = 50e^{-0.1(x+1)^2}$$
$$p = 10 + 5x^2$$

we substitute the second equation into the first, obtaining

$$10 + 5x^2 = 50e^{-0.1(x+1)^2}$$
$$10 + 5x^2 - 50e^{-0.1(x+1)^2} = 0$$

To solve the last equation, we use the Newton–Raphson method with

$$f(x) = 10 + 5x^2 - 50e^{-0.1(x+1)^2}$$
$$= 5[2 + x^2 - 10e^{-0.1(x+1)^2}]$$
$$f'(x) = 10x - 50e^{-0.1(x+1)^2} \cdot [-0.2(x+1)]$$
$$= 10[x + (x+1)e^{-0.1(x+1)^2}]$$

leading to the iterative formula

$$x_{n+1} = x_n - \frac{2 + x_n^2 - 10e^{-0.1(x_n+1)^2}}{2[x_n + (x_n + 1)e^{-0.1(x_n+1)^2}]}$$

FIGURE 11.9
The equilibrium point is the point of intersection of the demand curve and the supply curve.

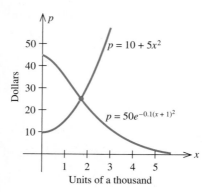

Dollars

Units of a thousand

Referring to Figure 11.9, we see that a reasonable initial estimate is $x_0 = 2$. We have

$$x_1 = 2 - \frac{2 + 2^2 - 10e^{-0.1(3)^2}}{2[2 + 3e^{-0.1(3)^2}]} \approx 1.69962$$

$$x_2 = 1.69962 - \frac{2 + (1.69962)^2 - 10e^{-0.1(2.69962)^2}}{2[1.69962 + 2.69962e^{-0.1(2.69962)^2}]} \approx 1.68899$$

$$x_3 = 1.68899 - \frac{2 + (1.68899)^2 - 10e^{-0.1(2.68899)^2}}{2[1.68899 + 2.68899e^{-0.1(2.68899)^2}]} \approx 1.68898$$

Thus, the equilibrium quantity is approximately 1.689 units, and the equilibrium wholesale price is given by

$$p = 10 + 5(1.68898)^2 \approx 24.263$$

or approximately $24.26 per watch.

■ ■ ■ ■

THE INTERNAL RATE OF RETURN ON AN INVESTMENT

Yet another use of the Newton–Raphson algorithm is in finding a quantity called the *internal rate of return* on an investment. Suppose a company has an initial outlay of C dollars in an investment that yields returns of R_1, $R_2, \ldots, R_n$ dollars at the end of the first, second, $\ldots$, nth periods, respectively. Then the *net present value* of the investment is

$$\frac{R_1}{1 + r} + \frac{R_2}{(1 + r)^2} + \frac{R_3}{(1 + r)^3} + \cdots + \frac{R_n}{(1 + r)^n} - C$$

where r denotes the interest rate per period earned on the investment. Now the internal rate of return on the investment is defined as the rate of return for which the net present value of the investment is equal to zero; that is, it is the value of r that satisfies the equation

$$\frac{R_1}{1 + r} + \frac{R_2}{(1 + r)^2} + \frac{R_3}{(1 + r)^3} + \cdots + \frac{R_n}{(1 + r)^n} - C = 0$$

or, equivalently, the equation

$$C(1 + r)^n - R_1(1 + r)^{n-1} - R_2(1 + r)^{n-2} - R_3(1 + r)^{n-3} - \cdots - R_n = 0$$

obtained by multiplying both sides of the former by $(1 + r)^n$. The internal rate of return is used by management to decide on the worthiness or profitability of an investment.

EXAMPLE 4

The management of A-1 Rental—a tool and equipment rental service for industry, contractors, and homeowners—is contemplating purchasing new equipment. The initial outlay for the equipment, which has a useful life of 4 years, is $45,000, and it is expected that the investment will yield returns of $15,000 at the end of the first year, $18,000 at the end of the second year, $14,000 at the end of the third year, and $10,000 at the end of the fourth year. Find the internal rate of return on this investment.

SOLUTION ✓   Here $n = 4$, $C = 45,000$, $R_1 = 15,000$, $R_2 = 18,000$, $R_3 = 14,000$, and $R_4 = 10,000$. So we are required to solve the equation

$$45,000(1 + r)^4 - 15,000(1 + r)^3 - 18,000(1 + r)^2 - 14,000(1 + r) - 10,000 = 0$$

for r. The equation may be written more simply by letting $x = 1 + r$. Thus,

$$f(x) = 45,000x^4 - 15,000x^3 - 18,000x^2 - 14,000x - 10,000 = 0$$

To solve the equation $f(x) = 0$, using the Newton–Raphson algorithm, we first compute

$$f'(x) = 180,000x^3 - 45,000x^2 - 36,000x - 14,000$$

The required iterative formula is

$$x_{n+1} = x_n - \frac{45,000x_n^4 - 15,000x_n^3 - 18,000x_n^2 - 14,000x_n - 10,000}{180,000x_n^3 - 45,000x_n^2 - 36,000x_n - 14,000}$$

Starting with the initial estimate $x_0 = 1.1$, we find

$$x_1 = 1.1 - \frac{45,000(1.1)^4 - 15,000(1.1)^3 - 18,000(1.1)^2 - 14,000(1.1) - 10,000}{180,000(1.1)^3 - 45,000(1.1)^2 - 36,000(1.1) - 14,000}$$

$$\approx 1.10958$$

$$x_2 \approx 1.10941$$

$$x_3 \approx 1.10941$$

Therefore, we may take $x \approx 1.1094$, in which case we see that $r \approx 0.1094$. Thus, the rate of return on the investment is approximately 10.94% per year.

■ ■ ■ ■

Having seen how effective the Newton–Raphson method can be in finding the zeros of a function, we want to close this section by reminding you that there are situations in which the method fails and that care must be exercised in applying it. Figure 11.10a illustrates a situation where $f'(x_n) = 0$ for some n (in this case, $n = 2$). Since the iterative Formula (11) involves division by $f'(x_n)$, it should be clear why the method fails to work in this case. However, if you choose a different initial estimate x_0, the situation may yet be salvaged (Figure 11.10b).

FIGURE 11.10
In (a), the Newton–Raphson method fails to work because $f'(x_2) = 0$, but this situation is remedied in (b) by selecting a different initial estimate x_0.

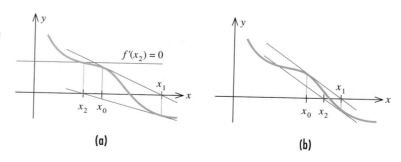

(a)   (b)

> **Group Discussion**
>
> For a concrete example of a situation similar to that depicted in Figure 11.10, consider the function
>
> $$f(x) = x^3 - 1.5x^2 - 6x + 2$$
>
> **1.** Show that the Newton–Raphson method fails to work if we choose $x_0 = -1$ or $x_0 = 2$ for an initial estimate.
>
> **2.** Using the initial estimates $x_0 = -2.5$, $x_0 = 1$, and $x_0 = 2.5$, show that the three roots of $f(x) = 0$ are -2, 0.313859, and 3.186141, respectively.
>
> **3.** Using a graphing utility, plot the graph of f in the viewing rectangle $[-3, 4] \times [-10, 7]$. Verify the results of part 2, using TRACE and ZOOM or the root-finding function of your calculator.

The next situation, shown in Figure 11.11, is more serious, and the method will not work for any choice of the initial estimate x_0 other than the actual zero of the function $f(x)$ because the sequence $x_1, x_2, \ldots x_n$ diverges.

FIGURE 11.11

The Newton–Raphson method fails here because the sequence of estimates diverges.

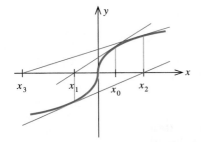

> **Group Discussion**
>
> For a concrete example of a situation similar to that depicted in Figure 11.11, consider the function $f(x) = x^{1/3}$.
>
> **1.** Show that the Newton–Raphson iteration for solving the equation $f(x) = 0$ is $x_{n+1} = -2x_n$, with x_0 being an initial guess.
>
> **2.** Show that the sequence $x_0, x_1, x_2, \ldots$ diverges for any choice of x_0 other than zero and thus the Newton–Raphson method does not lead to the unique solution $x = 0$ for the problem under consideration.
>
> **c.** Illustrate this situation geometrically.

SELF-CHECK EXERCISES 11.7

1. Use three iterations of the Newton–Raphson algorithm on an appropriate function $f(x)$ and initial guess x_0 to obtain an estimate of $\sqrt[3]{10}$.

2. A study prepared for a certain Sunbelt town's Chamber of Commerce projected that the town's population in the next 3 yr will grow according to the rule

$$P(t) = 50,000 + 30t^{3/2} + 20t$$

where $P(t)$ denotes the population t mo from now. Using this model, estimate the time when the population will reach 55,000.

Solutions to Self-Check Exercises 11.7 can be found on page 848.

11.7 Exercises

A calculator is recommended for this exercise set. In Exercises 1–6, estimate the value of each radical by using three iterations of the Newton–Raphson method with the indicated initial guess for each function.

1. $\sqrt{3}$; $f(x) = x^2 - 3$; $x_0 = 1.5$

2. $\sqrt{5}$; $f(x) = x^2 - 5$; $x_0 = 2$

3. $\sqrt{7}$; $f(x) = x^2 - 7$; $x_0 = 2.5$

4. $\sqrt[3]{6}$; $f(x) = x^3 - 6$; $x_0 = 2$

5. $\sqrt[3]{14}$; $f(x) = x^3 - 14$; $x_0 = 2.5$

6. $\sqrt[4]{50}$; $f(x) = x^4 - 50$; $x_0 = 2.5$

In Exercises 7–14, use the Newton–Raphson method to approximate the indicated zero of each function. Continue with the iteration until two successive approximations differ by less than 0.0001.

7. The zero of $f(x) = x^2 - x - 3$ between $x = 2$ and $x = 3$

8. The zero of $f(x) = x^3 + x - 3$ between $x = 1$ and $x = 2$

9. The zero of $f(x) = x^3 + 2x^2 + x - 5$ between $x = 1$ and $x = 2$

10. The zero of $f(x) = x^5 + x - 1$ between $x = 0$ and $x = 1$

11. The zero of $f(x) = \sqrt{x+1} - x$ between $x = 1$ and $x = 2$

12. The zero of $f(x) = e^{-x} - x$ between $x = 0$ and $x = 1$

13. The zero of $f(x) = e^x - (1/x)$ between $x = 0$ and $x = 1$

14. The zero of $f(x) = \ln x^2 - 0.7x + 1$ between $x = 6$ and $x = 7$

15. Let $f(x) = 2x^3 - 9x^2 + 12x - 2$.
  a. Show that $f(x) = 0$ has a root between $x = 0$ and $x = 1$.
  Hint: Compute $f(0)$ and $f(1)$ and use the fact that f is continuous.
  b. Use the Newton–Raphson method to find the zero of f in the interval $(0, 1)$.

16. Let $f(x) = x^3 - x - 1$.
  a. Show that $f(x) = 0$ has a root between $x = 1$ and $x = 2$.
  Hint: See Exercise 15.
  b. Use the Newton–Raphson method to find the zero of f in the interval $(1, 2)$.

17. Let $f(x) = x^3 - 3x - 1$.
  a. Show that f has a zero between $x = 1$ and $x = 2$.
  Hint: See Exercise 15.
  b. Use the Newton–Raphson method to find the zero of f.

18. Let $f(x) = x^4 - 4x^3 + 10$.
  a. Show that $f(x) = 0$ has a root between $x = 1$ and $x = 2$.
  b. Use the Newton–Raphson method to find the zero of f.

In Exercises 19–24, make a rough sketch of the graphs of each of the given pairs of functions. Use your sketch to approximate the point(s) of intersection of the two graphs and then apply the Newton–Raphson method to refine the approximation of the x-coordinate of the point of intersection.

19. $f(x) = 2\sqrt{x+3}$; $g(x) = 2x - 1$

20. $f(x) = e^{-x^2}$; $g(x) = x^2$

21. $f(x) = e^{-x}$; $g(x) = x - 1$

22. $f(x) = \ln x$; $g(x) = 2 - x$

23. $f(x) = \sqrt{x}$; $g(x) = e^{-x}$

24. $f(x) = 2 - x^2$; $g(x) = \ln x$

25. MINIMIZING AVERAGE COST A division of Ditton Industries manufactures the "Futura" model microwave oven. Given that the daily cost of producing these microwave ovens (in dollars) obeys the rule

$$C(x) = 0.0002x^3 - 0.06x^2 + 120x + 5000$$

where x stands for the number of units produced, find the level of production that minimizes the daily average cost per unit.

26. ALTITUDE OF A ROCKET The altitude (in feet) of a rocket t sec into flight is given by

$$s = f(t) = -2t^3 + 114t^2 + 480t + 1 \qquad (t \geq 0)$$

Find the time T when the rocket hits Earth.

27. TEMPERATURE The temperature at 6 A.M. on a certain December day was measured at 15.6°F. In the next t hr, the temperature was given by the function

$$T = -0.05t^3 + 0.4t^2 + 3.8t + 15.6 \qquad (0 \leq t \leq 15)$$

where T is measured in degrees Fahrenheit. At what time was the temperature 0°F?

28. INTERNAL RATE OF RETURN ON AN INVESTMENT The proprietor of Qwik Film Lab recently purchased $12,000 of new film-processing equipment. She expects that this investment, which has a useful life of 4 yr, will yield returns of $4000 at the end of the first year, $5000 at the end of the second year, $4000 at the end of the third year, and $3000 at the end of the fourth year. Find the internal rate of return on the investment.

29. INTERNAL RATE OF RETURN ON AN INVESTMENT Executive Limousine Service recently acquired limousines worth $120,000. The projected returns over the next 3 yr, the time period the limousines will be in service, are $80,000 at the end of the first year, $60,000 at the end of the second year, and $40,000 at the end of the third year. Find the internal rate of return on the investment.

30. INTERNAL RATE OF RETURN ON AN INVESTMENT Suppose an initial outlay of $\$C$ in an investment yields returns of $\$R$ at the end of each period over N periods.
a. Show that the internal rate of return on the investment, r, may be obtained by solving the equation

$$Cr + R[(1 + r)^{-N} - 1] = 0$$

Hint: $1 + x + x^2 + \cdots + x^{N-1} = \dfrac{1 - x^{N+1}}{1 - x}$

b. Show that r can be found by performing the iteration

$$r_{n+1} = r_n - \frac{Cr_n + R[(1 + r_n)^{-N} - 1]}{C - NR(1 + r_n)^{-N-1}}$$

[r_0 (positive), an initial guess]

Hint: Apply the Newton–Raphson method to the function $f(r) = Cr + R[(1 + r)^{-N} - 1]$.

31. HOME MORTGAGES Refer to Exercise 30. The Flemings secured a loan of $100,000 from a bank to finance the purchase of a house. They have agreed to repay the loan in equal monthly installments of $1053 over 25 yr. The bank charges interest at the rate of 12r/year on the unpaid balance, and interest computations are made at the end of each month. Find r.

32. HOME MORTGAGES The Blakelys borrowed a sum of $80,000 from a bank to help finance the purchase of a house. The bank charges interest at the rate of 12r/year on the unpaid balance, with interest being computed at the end of each month. The Blakelys have agreed to repay the loan in equal monthly installments of $643.70 over 30 yr. What is the true rate of interest charged by the bank?
Hint: See Exercise 30.

33. CAR LOANS The price of a certain new car is $8000. Suppose an individual makes a down payment of 25% toward the purchase of the car and secures financing for the balance over 4 yr. If the monthly payment is $152.18, what is the true rate of interest charged by the finance company?
Hint: See Exercise 30.

34. REAL ESTATE INVESTMENT GROUPS Refer to Exercise 30. A group of private investors purchased a condominium complex for $2 million. They made an initial down payment of 10% and have obtained financing for the balance. If the loan is amortized over 15 yr with quarterly repayments of $65,039, determine the interest rate charged by the bank. Assume that interest is calculated at the end of each quarter and is based on the unpaid balance.

35. DEMAND FOR WRISTWATCHES The quantity of "Sicard" wristwatches demanded per month is related to the unit price by the equation

$$p = d(x) = \frac{50}{0.01x^2 + 1} \qquad (1 \leq x \leq 20)$$

where p is measured in dollars and x is measured in units of a thousand. The supplier is willing to make x thousand wristwatches available per month when the price per watch is given by $p = s(x) = 0.1x + 20$ dollars. Find the equilibrium quantity and price.

5. Identify the quadrant in which each of the angles lies.
   a. 220°   **b.** −110°
   c. 460°   **d.** −310°

6. Identify the quadrant in which each of the angles lies.
   a. $\frac{13}{6}\pi$ radians   **b.** $-\frac{11}{4}\pi$ radians

   c. $\frac{17}{3}\pi$ radians   **d.** $-\frac{25}{12}\pi$ radians

In Exercises 7–12, convert each of the given angles to radian measure.

   7. 75°   **8.** 330°   **9.** 160°

   10. −210°   **11.** 630°   **12.** −420°

In Exercises 13–18, convert each of the given angles to degree measure.

   13. $\frac{2}{3}\pi$ radians   **14.** $\frac{7}{6}\pi$ radians   **15.** $-\frac{3}{2}\pi$ radians

   16. $-\frac{13}{12}\pi$ radians   **17.** $\frac{22}{18}\pi$ radians   **18.** $-\frac{21}{6}\pi$ radians

In Exercises 19–22, make a sketch of each of the given angles on a unit circle centered at the origin.

   19. 225°   **20.** −120°

   21. $\frac{7}{3}\pi$ radians   **22.** $-\frac{13}{6}\pi$ radians

In Exercises 23–26, determine a positive angle and a negative angle that are coterminal angles of the angle θ. Use degree measure.

23.

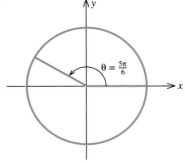

24.

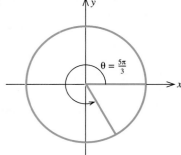

25.

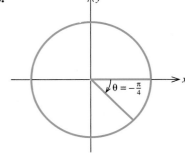

26.

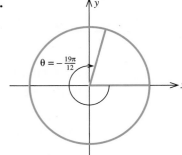

In Exercises 27–30, determine whether the statement is true or false. If it is true, explain why it is true. If it is false, give an example to show why it is false.

27. The angle 3630° lies in the first quadrant.

28. The angle $\frac{103\pi}{6}$ radians lies in the second quadrant.

29. If θ is any angle, then $\theta + n(360)$, where n is a nonzero integer, is coterminal with θ (all angles measured in degrees).

30. If x, y, and z are measured in degrees, then $(x + y + z)$ degrees is equal to $\frac{\pi}{180}(x + y + z)$ radians.

SOLUTIONS TO SELF-CHECK EXERCISES 12.1

1. a. Using Formula (3), we find that the required radian measure is

$$f(315) = \frac{\pi}{180}(315) = \frac{7\pi}{4}, \quad \text{or} \quad \frac{7\pi}{4} \text{ radians}$$

b. Using Formula (4), we find that the required degree measure is

$$g\left(-\frac{5\pi}{4}\right) = \frac{180}{\pi}\left(-\frac{5\pi}{4}\right) = -225, \quad \text{or} \quad -225°$$

2.

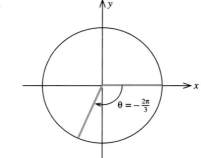

$$\theta = -\frac{2\pi}{3}$$

12.2 The Trigonometric Functions

THE TRIGONOMETRIC FUNCTIONS AND THEIR GRAPHS

Let $P(x, y)$ be a point on the unit circle so that the radius OP forms an angle of θ radians ($0 \le \theta < 2\pi$) with respect to the positive x-axis (see Figure 12.8).

FIGURE 12.8
P is a point on the unit circle with coordinates $x = \cos \theta$ and $y = \sin \theta$.

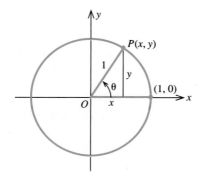

We define the **sine** of the angle θ, written sin θ, to be the y-coordinate of P. Similarly, the **cosine** of the angle θ, written cos θ, is defined to be the x-coordinate of P. The other trigonometric functions, tangent, cosecant, secant, and cotangent of θ—written tan θ, csc θ, sec θ, and cot θ, respectively—are defined in terms of the sine and cosine functions.

Trigonometric Functions

If P is a point on the unit circle and the coordinates of P are (x, y), then

$$\cos \theta = x \quad \text{and} \quad \sin \theta = y$$

$$\tan \theta = \frac{y}{x} = \frac{\sin \theta}{\cos \theta} \qquad (x \neq 0)$$

$$\csc \theta = \frac{1}{y} = \frac{1}{\sin \theta} \qquad (y \neq 0)$$

$$\sec \theta = \frac{1}{x} = \frac{1}{\cos \theta} \qquad (x \neq 0)$$

$$\cot \theta = \frac{x}{y} = \frac{\cos \theta}{\sin \theta} \qquad (y \neq 0)$$

As you work with trigonometric functions, it is useful to remember the values of the sine, cosine, and tangent of some important angles, such as $\theta = 0$, $\pi/6$, $\pi/4$, $\pi/3$, $\pi/2$, and so on. These values may be found using elementary trigonometry. For example, if $\theta = 0$, then the point P has coordinates $(1, 0)$ (see Figure 12.9a), and we see that

$$\sin 0 = y = 0, \quad \cos 0 = x = 1, \quad \tan 0 = \frac{y}{x} = 0$$

As another example, the diagram for $\theta = \pi/4$ (Figure 12.9b) suggests that $x = y$, so that, by the Pythagorean theorem, we have

$$x^2 + y^2 = 2x^2 = 1$$

FIGURE 12.9

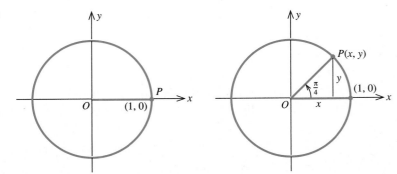

(a) $P = (1, 0)$, $\theta = 0$, and
$\cos \theta = 1$, $\sin \theta = 0$

(b) $\theta = \pi/4$, $x = y$, and $\cos \theta = \sin \theta = \sqrt{2}/2$

and $x = y = \dfrac{\sqrt{2}}{2}$. Therefore,

$$\sin \frac{\pi}{4} = y = \frac{\sqrt{2}}{2}, \qquad \cos \frac{\pi}{4} = x = \frac{\sqrt{2}}{2}, \qquad \tan \frac{\pi}{4} = \frac{y}{x} = 1$$

The values of the sine, cosine, and tangent of some common angles are given in Table 12.2.

Table 12.2

θ in radians	0	$\dfrac{\pi}{6}$	$\dfrac{\pi}{4}$	$\dfrac{\pi}{3}$	$\dfrac{\pi}{2}$	$\dfrac{2}{3}\pi$	$\dfrac{3}{4}\pi$	$\dfrac{5}{6}\pi$	π	$\dfrac{7}{6}\pi$	$\dfrac{5}{4}\pi$	$\dfrac{4}{3}\pi$	$\dfrac{3}{2}\pi$	$\dfrac{5}{3}\pi$	$\dfrac{7}{4}\pi$	$\dfrac{11}{6}\pi$	2π
$\sin \theta$	0	$\dfrac{1}{2}$	$\dfrac{\sqrt{2}}{2}$	$\dfrac{\sqrt{3}}{2}$	1	$\dfrac{\sqrt{3}}{2}$	$\dfrac{\sqrt{2}}{2}$	$\dfrac{1}{2}$	0	$-\dfrac{1}{2}$	$-\dfrac{\sqrt{2}}{2}$	$-\dfrac{\sqrt{3}}{2}$	-1	$-\dfrac{\sqrt{3}}{2}$	$-\dfrac{\sqrt{2}}{2}$	$-\dfrac{1}{2}$	0
$\cos \theta$	1	$\dfrac{\sqrt{3}}{2}$	$\dfrac{\sqrt{2}}{2}$	$\dfrac{1}{2}$	0	$-\dfrac{1}{2}$	$-\dfrac{\sqrt{2}}{2}$	$-\dfrac{\sqrt{3}}{2}$	-1	$-\dfrac{\sqrt{3}}{2}$	$-\dfrac{\sqrt{2}}{2}$	$-\dfrac{1}{2}$	0	$\dfrac{1}{2}$	$\dfrac{\sqrt{2}}{2}$	$\dfrac{\sqrt{3}}{2}$	1
$\tan \theta$	0	$\dfrac{1}{\sqrt{3}}$	1	$\sqrt{3}$	∞	$-\sqrt{3}$	-1	$-\dfrac{1}{\sqrt{3}}$	0	$\dfrac{1}{\sqrt{3}}$	1	$\sqrt{3}$	$-\infty$	$-\sqrt{3}$	-1	$-\dfrac{1}{\sqrt{3}}$	0

Since the rotation of the radius OP by 2π radians leaves it in its original configuration (see Figure 12.10), we see that

$$\sin(\theta + 2\pi) = \sin \theta \qquad \text{and} \qquad \cos(\theta + 2\pi) = \cos \theta$$

FIGURE 12.10

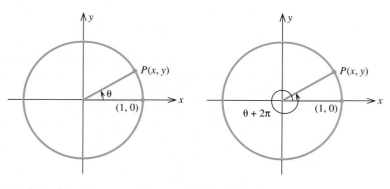

(a) The angle θ **(b)** The angle $(\theta + 2\pi)$

It can be shown that the number 2π is the smallest positive number such that the preceding equation holds true. That is, the sine and cosine functions are **periodic** with **period 2π.** It follows that

$$\sin(\theta + 2n\pi) = \sin \theta \qquad \text{and} \qquad \cos(\theta + 2n\pi) = \cos \theta \qquad (5)$$

whenever n is an integer. Also, from Figure 12.11, we see that

$$\sin(-\theta) = -\sin \theta \qquad \text{and} \qquad \cos(-\theta) = \cos \theta \qquad (6)$$

FIGURE 12.11
$\sin(-\theta) = -\sin\theta$ and
$\cos(-\theta) = \cos\theta$.

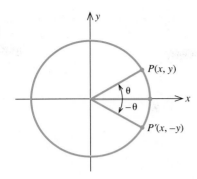

EXAMPLE 1

Evaluate:

a. $\sin\dfrac{7\pi}{2}$ **b.** $\cos 5\pi$ **c.** $\sin\left(-\dfrac{5\pi}{2}\right)$ **d.** $\cos\left(-\dfrac{11\pi}{4}\right)$

SOLUTION ✔

a. Using (5) and Table 12.2, we find

$$\sin\left(\frac{7\pi}{2}\right) = \sin\left(2\pi + \frac{3\pi}{2}\right) = \sin\frac{3\pi}{2} = -1$$

b. Using (5) and Table 12.2, we have

$$\cos 5\pi = \cos(4\pi + \pi) = \cos\pi = -1$$

c. Using (5), (6), and Table 12.2, we have

$$\sin\left(-\frac{5\pi}{2}\right) = -\sin\left(\frac{5\pi}{2}\right) = -\sin\left(2\pi + \frac{\pi}{2}\right) = -\sin\frac{\pi}{2} = -1$$

d. Using (5), (6), and Table 12.2, we have

$$\cos\left(-\frac{11\pi}{4}\right) = \cos\frac{11\pi}{4} = \cos\left(2\pi + \frac{3\pi}{4}\right) = \cos\frac{3\pi}{4} = -\frac{\sqrt{2}}{2}$$ ■ ■ ■ ■

To draw the graph of the function $y = f(x) = \sin x$, we first note that $\sin x$ is defined for every real number x so that the domain of the sine function is $(-\infty, \infty)$. Next, since the sine function is periodic with period 2π, it suffices to concentrate on sketching that part of the graph of $y = \sin x$ on the interval $[0, 2\pi]$ and repeating it as necessary. With the help of Table 12.2, we sketch Figure 12.12, the graph of $y = \sin x$.

FIGURE 12.12
The graph of $y = \sin x$

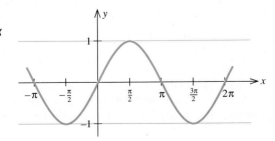

In a similar manner, we can sketch the graph of $y = \cos x$ (Figure 12.13).

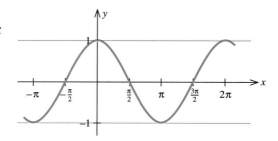

Plot the graphs of $f(x) = \sin x$ and $g(x) = \cos x$ in the viewing rectangle $[-10, 10] \times [-1.1, 1.1]$.

1. What do the graphs suggest about the relationship between the two trigonometric functions?

2. Confirm your observation by plotting the graphs of $h(x) = \sin\left(x + \dfrac{\pi}{2}\right)$ and $g(x) = \cos x$ in the viewing rectangle $[-10, 10] \times [-1.1, 1.1]$.

To sketch the graph of $y = \tan x$, note that $\tan x = \sin x / \cos x$, and so $\tan x$ is not defined when $\cos x = 0$—that is, when $x = (\pi/2) \pm n\pi$ ($n = 0, 1, 2, 3, \ldots$). The function is defined at all other points so that the domain of the tangent function is the set of all real numbers with the exception of the points just noted. Next, we can show that the vertical lines with equation $x = (\pi/2) \pm n\pi (n = 0, 1, 2, 3, \ldots)$ are vertical asymptotes of the function $f(x) = \tan x$. For example, since

$$\lim_{x \to \frac{\pi}{2}^-} \tan x = \infty \qquad \text{and} \qquad \lim_{x \to \frac{\pi}{2}^+} \tan x = -\infty$$

which we readily verify with the help of a calculator, we conclude that $x = \pi/2$ is a vertical asymptote of $y = \tan x$. Finally, using Table 12.2, we can sketch the graph of $y = \tan x$ (Figure 12.14).

FIGURE 12.14
The graph of $y = \tan x$

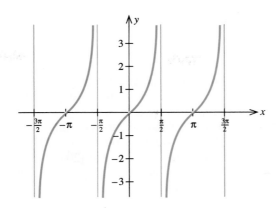

Observe that the tangent function is periodic with period π. It follows that

$$\tan(x + n\pi) = \tan x \qquad\qquad \textbf{(7)}$$

whenever n is an integer.

The graphs of $y = \sec x$, $y = \csc x$, and $y = \cot x$ may be sketched in a similar manner (see Exercises 23–25).

THE PREDATOR–PREY POPULATION MODEL

We will now look at a specific mathematical model of a phenomenon exhibiting cyclical behavior—the so-called **predator-prey population model.**

EXAMPLE 2   The population of owls (predators) in a certain region over a 2-year period is estimated to be

$$P(t) = 1000 + 100 \sin\left(\frac{\pi t}{12}\right)$$

in month t, and the population of mice (prey) in the same area at time t is given by

$$p(t) = 20{,}000 + 4000 \cos\left(\frac{\pi t}{12}\right)$$

Sketch the graphs of these two functions and explain the relationship between the sizes of the two populations.

SOLUTION ✔   We first observe that both of the given functions are periodic with period 24 (months). To see this, recall that both the sine and cosine functions are periodic with period 2π. Now the smallest value of $t > 0$ such that

$\sin(\pi t/12) = 0$ is obtained by solving the equation

$$\frac{\pi t}{12} = 2\pi$$

giving $t = 24$ as the period of $\sin(\pi t/12)$. Since $P(t + 24) = P(t)$, we see that the function P is periodic with period 24. Similarly, one verifies that the function p is also periodic with period 24, as asserted. Next, recall that both the sine and cosine functions oscillate between -1 and $+1$ so that $P(t)$ is seen to oscillate between $[1000 + 100(-1)]$, or 900, and $[1000 + 100(1)]$, or 1100, while $p(t)$ oscillates between $[20,000 + 4000(-1)]$, or 16,000, and $[20,000 + 4000(1)]$, or 24,000. Finally, plotting a few points on each graph for—say, $t = 0, 2, 3$, and so on—we obtain the graphs of the functions P and p as shown in Figure 12.15.

FIGURE 12.15

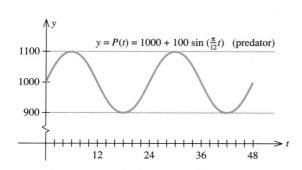

(a) The graph of the predator function $P(t)$

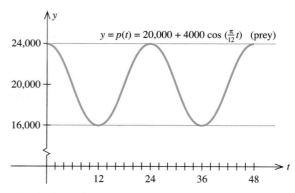

(b) The graph of the prey function $p(t)$

From the graphs, we see that at time $t = 0$ the predator population stands at 1000 owls. As it increases, the prey population decreases from 24,000 mice at that instant of time. Eventually, this decrease in the food supply causes the predator population to decrease, which in turn causes an increase in the prey population. But as the prey population increases, resulting in an increase in food supply, the predator population once again increases. The cycle is complete and starts all over again. ■ ■ ■ ■

TRIGONOMETRIC IDENTITIES

Equations expressing the relationships between trigonometric functions, such as

$$\sin(-\theta) = -\sin \theta \quad \text{and} \quad \cos(-\theta) = \cos \theta$$

are called **trigonometric identities.** Some other important trigonometric identities are listed in Table 12.3. Each identity holds true for every value of θ in the domain of the specified function. The proofs of these identities may be found in any elementary trigonometry book.

Table 12.3 **Trigonometric Identities**		
Pythagorean Identities	**Half-Angle Formulas**	**Sum and Difference Formulas**
$\sin^2 \theta + \cos^2 \theta = 1$	$\cos^2 \theta = \dfrac{1}{2}(1 + \cos 2\theta)$	$\sin(A \pm B) = \sin A \cos B \pm \cos A \sin B$
$\tan^2 \theta + 1 = \sec^2 \theta$	$\sin^2 \theta = \dfrac{1}{2}(1 - \cos 2\theta)$	$\cos(A \pm B) = \cos A \cos B \mp \sin A \sin B$
$\cot^2 \theta + 1 = \csc^2 \theta$		
Double-Angle Formulas	**Cofunctions of Complementary Angles**	
$\sin 2A = 2 \sin A \cos A$	$\sin \theta = \cos\left(\dfrac{\pi}{2} - \theta\right)$	
$\cos 2A = \cos^2 A - \sin^2 A$	$\cos \theta = \sin\left(\dfrac{\pi}{2} - \theta\right)$	

As we will see later, these identities are useful in simplifying trigonometric expressions and equations and in deriving other trigonometric relationships. They are also used to verify other trigonometric identities, as illustrated in the next example.

EXAMPLE 3   Verify the identity

$$\sin \theta(\csc \theta - \sin \theta) = \cos^2 \theta$$

SOLUTION ✔   We verify this identity by showing that the expression on the left side of the equation can be transformed into the expression on the right side. Thus,

$$\sin \theta(\csc \theta - \sin \theta) = \sin \theta \csc \theta - \sin^2 \theta$$
$$= \sin \theta \frac{1}{\sin \theta} - \sin^2 \theta$$
$$= 1 - \sin^2 \theta$$
$$= \cos^2 \theta$$

Exploring with Technology

1. You can confirm some of the trigonometric identities graphically in many ways. For example, explain why an identity given in the form $f(\theta) = g(\theta)$ over some domain D is equivalent to the following statements:

(a) The graphs of f and g coincide when viewed in any appropriate viewing rectangle.

(b) The graphs of f and $-g$ are reflections of each other with respect to the θ-axis when viewed in any appropriate viewing rectangle.

(c) The graph of $h = f - g$ is the graph of the zero function $h(\theta) = 0$ for all θ in the domain of f (and g).

2. Use the observations made in part 1 to verify graphically the identity $\sin^2 \theta + \cos^2 \theta = 1$ by taking $f(x) = \sin^2 x + \cos^2 x$ and $g(x) = 1$.

Hint: Enter $f(x)$ as $y_1 = (\sin x)^{\wedge}2 + (\cos x)^{\wedge}2$ and use the viewing rectangle $[-10, 10] \times [-1.1, 1.1]$.

3. Verify graphically the half-angle formula $\cos^2 \theta = \dfrac{1}{2}(1 + \cos 2\theta)$.

4. Verify graphically the formula $\sin \theta = \cos\left(\dfrac{\pi}{2} - \theta\right)$.

SELF-CHECK EXERCISES 12.2

1. Evaluate $\cos\left(-\dfrac{13\pi}{3}\right)$.

2. Solve the equation $\cos \theta = -\dfrac{\sqrt{2}}{2}$ for $0 \leq \theta \leq 2\pi$.

3. Sketch the graph of $y = 2 \cos x$.

Solutions to Self-Check Exercises 12.2 can be found on page 870.

12.2 Exercises

In Exercises 1–10, evaluate the trigonometric function.

1. $\sin 3\pi$

2. $\cos\left(-\dfrac{3}{2}\pi\right)$

3. $\sin \dfrac{9}{2}\pi$

4. $\cos \dfrac{13}{6}\pi$

5. $\sin\left(-\dfrac{4}{3}\pi\right)$

6. $\cos\left(-\dfrac{5}{4}\pi\right)$

7. $\tan \dfrac{\pi}{6}$

8. $\cot\left(-\dfrac{\pi}{3}\right)$

9. $\sec\left(-\dfrac{5}{8}\pi\right)$

10. $\csc \dfrac{9}{4}\pi$

In Exercises 11–14, find the six trigonometric functions of the angle.

11. $\dfrac{\pi}{2}$

12. $-\dfrac{\pi}{6}$

13. $\dfrac{5}{3}\pi$

14. $-\dfrac{3}{4}\pi$

In Exercises 15–22, find all values of θ that satisfy the equation over the interval $[0, 2\pi]$.

15. $\sin\theta = -\dfrac{1}{2}$

16. $\tan\theta = 1$

17. $\cot\theta = -\sqrt{3}$

18. $\csc\theta = \sqrt{2}$

19. $\sec\theta = -1$

20. $\cos\theta = \sin\theta$

21. $\sin\theta = \sin\left(-\dfrac{4}{3}\pi\right)$

22. $\cos\theta = \cos\left(-\dfrac{\pi}{6}\right)$

In Exercises 23–26, sketch the graph of the function over the interval $[0, 2\pi]$.

23. $y = \csc x$

24. $y = \sec x$

25. $y = \cot x$

26. $y = \tan 2x$

In Exercises 27–30, sketch the graph of the functions over the interval $[0, 2\pi]$ by comparing it to the graph of $y = \sin x$.

27. $y = \sin 2x$

28. $y = 2\sin x$

29. $y = -\sin x$

30. $y = -2\sin 2x$

In Exercises 31–38, verify each identity.

31. $\cos^2\theta - \sin^2\theta = 2\cos^2\theta - 1$

32. $1 - 2\sin^2\theta = 2\cos^2\theta - 1$

33. $(\sec\theta + \tan\theta)(1 - \sin\theta) = \cos\theta$

34. $\dfrac{\sin^2\theta}{1 + \cos^2\theta} = \dfrac{1 - \cos^2\theta}{2 - \sin^2\theta}$

35. $(1 + \cot^2\theta)\tan^2\theta = \sec^2\theta$

36. $\dfrac{\sec\theta - \cos\theta}{\tan\theta} = \sin\theta$

37. $\dfrac{\csc\theta}{\tan\theta + \cot\theta} = \cos\theta$

38. $\tan(A + B) = \dfrac{\tan A + \tan B}{1 - \tan A \tan B}$

39. The accompanying figure shows a right triangle ABC superimposed over a unit circle in the xy-coordinate system. By considering similar triangles, show that

$$\sin\theta = \frac{BC}{AC} = \frac{\text{Opposite side}}{\text{Hypotenuse}}$$

$$\cos\theta = \frac{AB}{AC} = \frac{\text{Adjacent side}}{\text{Hypotenuse}}$$

$$\tan\theta = \frac{BC}{AB} = \frac{\text{Opposite side}}{\text{Adjacent side}}$$

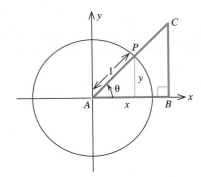

40. Refer to Exercise 39. Find the values of the six trigonometric functions of θ, where θ is the angle shown in the right triangle in the accompanying figure.

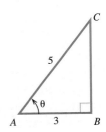

41. Refer to Exercise 39. Find the values of the six trigonometric functions of θ, where θ is the angle shown in the right triangle in the accompanying figure.

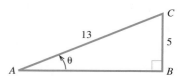

42. PREDATOR–PREY POPULATION The population of foxes (predators) in a certain region over a 2-yr period is estimated to be

$$P(t) = 400 + 50 \sin\left(\frac{\pi t}{12}\right)$$

in month t, and the population of rabbits (prey) in the same region at time t is given by

$$p(t) = 3000 + 500 \cos\left(\frac{\pi t}{12}\right)$$

Sketch the graphs of each of these two functions and explain the relationship between the sizes of the two populations.

43. BLOOD PRESSURE The arterial blood pressure of an individual in a state of relaxation is given by

$$P(t) = 100 + 20 \sin 6t$$

where $P(t)$ is measured in mm of mercury (Hg) and t is the time in seconds.
  a. Show that the individual's systolic pressure (maximum blood pressure) is 120 and his diastolic pressure (minimum blood pressure) is 80.
  b. Find the values of t when the individual's blood pressure is highest and lowest.

In Exercises 44–48, determine whether the statement is true or false. If it is true, explain why it is true. If it is false, give an example to show why it is false.

44. $\sin(a - b) = -\sin(b - a)$

45. If $\sin\theta = -\dfrac{\sqrt{3}}{2}$ and $0 \le \theta \le 2\pi$, then $\theta = \dfrac{5\pi}{3}$.

46. If $\tan\theta = \dfrac{1}{3}$, then $\sin\theta = \dfrac{\sqrt{10}}{10}$.

47. $\cos 2\theta = 2\cos^2\theta - 1$

48. $\sin 3\theta = 3\sin\theta + 4\sin^3\theta$

SOLUTIONS TO SELF-CHECK EXERCISES 12.2

1.
$$\cos\left(-\frac{13\pi}{3}\right) = \cos\frac{13\pi}{3}$$
$$= \cos\left(4\pi + \frac{\pi}{3}\right)$$
$$= \cos\frac{\pi}{3} = \frac{1}{2}$$

2. Cos θ is negative for θ in Quadrants II and III. From Table 12.2, we see that the required values of θ are $3\pi/4$ and $5\pi/4$.

3. By comparison with the graph of $y = \cos x$, we obtain the accompanying graph of $y = 2\cos x$.

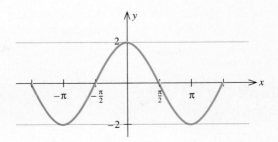

12.3 Differentiation of Trigonometric Functions

In this section we develop rules for differentiating trigonometric functions. Knowledge of the derivatives of the trigonometric functions helps us to study the properties of functions involving the trigonometric functions in much the same way we analyzed the algebraic, exponential, and logarithmic functions in our earlier work.

DERIVATIVES OF THE SINE AND COSINE FUNCTIONS

First we develop the rule for differentiating the sine and cosine functions. Then, using these rules and the rules of differentiation, we derive the rules for differentiating the other trigonometric functions.

To derive the rule for differentiating the sine function, we need the following two results:

$$\lim_{h \to 0} \frac{\sin h}{h} = 1 \quad \text{and} \quad \lim_{h \to 0} \frac{\cos h - 1}{h} = 0 \tag{8}$$

The plausibility of the first limit can be seen by examining Table 12.4, which is constructed with the aid of a calculator.

Table 12.4 Values of $\dfrac{\sin h}{h}$ for Selected Values of h (in Radians) Approaching Zero			
h	± 0.5	± 0.1	± 0.01	± 0.001
$\dfrac{\sin h}{h}$	0.9588511	0.9983342	0.9999833	0.9999998

12.3 Exercises

In Exercises 1–30, find the derivative of the function.

1. $f(x) = \cos 3x$

2. $f(x) = \sin 5x$

3. $f(x) = 2 \cos \pi x$

4. $f(x) = \pi \sin 2x$

5. $f(x) = \sin(x^2 + 1)$

6. $f(x) = \cos \pi x^2$

7. $f(x) = \tan 2x^2$

8. $f(x) = \cot \sqrt{x}$

9. $f(x) = x \sin x$

10. $f(x) = x^2 \cos x$

11. $f(x) = 2 \sin 3x + 3 \cos 2x$

12. $f(x) = 2 \cot 2x + \sec 3x$

13. $f(x) = x^2 \cos 2x$

14. $f(x) = \sqrt{x} \sin \pi x$

15. $f(x) = \sin\sqrt{x^2 - 1}$

16. $f(x) = \csc(x^2 + 1)$

17. $f(x) = e^x \sec x$

18. $f(x) = e^{-x} \csc x$

19. $f(x) = x \cos \dfrac{1}{x}$

20. $f(x) = x^2 \sin \dfrac{1}{x}$

21. $f(x) = \dfrac{x - \sin x}{1 + \cos x}$

22. $f(x) = \dfrac{\sin 2x}{1 + \cos 3x}$

23. $f(x) = \sqrt{\tan x}$

24. $f(x) = \sqrt{\cos x + \sin x}$

25. $f(x) = \dfrac{\sin x}{x}$

26. $f(x) = \dfrac{\cos x}{x^2 + 1}$

27. $f(x) = \tan^2 x$

28. $f(x) = \cot \sqrt{x}$

29. $f(x) = e^{\cot x}$

30. $f(x) = e^{\tan x + \sec x}$

31. Find the equation of the tangent line to the graph of the function $f(x) = \cot 2x$ at the point $(\pi/4, 0)$.

32. Find an equation of the tangent line to the graph of the function $f(x) = e^{\sec x}$ at the point $(\pi/4, e^{\sqrt{2}})$.

33. Determine the intervals where the function

$$f(x) = e^x \cos x \qquad (0 \le x \le 2\pi)$$

is increasing and where it is decreasing.

34. Determine the point(s) of inflection of the function

$$f(x) = x + \sin x \qquad (-2\pi \le x \le 2\pi)$$

In Exercises 35–38, sketch the graph of the function over the specified interval by obtaining the following information:

a. The intervals where f is increasing and where it is decreasing

b. The relative extrema of f

c. The concavity of f

d. The inflection points of f

35. $f(x) = \sin x + \cos x \qquad (0 \le x \le 2\pi)$

36. $f(x) = x - \sin x \qquad (0 \le x \le 2\pi)$

37. $f(x) = 2 \sin x + \sin 2x \qquad (0 \le x \le 2\pi)$

38. $f(x) = e^x \cos x \qquad (0 \le x \le 2\pi)$

39. a. Find the Taylor series about $x = 0$ for the function $f(x) = \sin x$.

b. Use the result of part (a) to show that

$$\lim_{x \to 0} \frac{\sin x}{x} = 1$$

40. a. Find the Taylor series about $x = 0$ for the function $f(x) = \cos x$.

b. Use the result of part (a) to show that

$$\lim_{x \to 0} \frac{\cos x - 1}{x} = 0$$

41. Predator–Prey Population The wolf population in a certain northern region is estimated to be

$$P(t) = 8000 + 1000 \sin\left(\frac{\pi t}{24}\right)$$

in month t, and the caribou population in the same region is given by

$$p(t) = 40{,}000 + 12{,}000 \cos\left(\frac{\pi t}{24}\right)$$

Find the rate of change of each population when $t = 12$.

42. Stock Prices The closing price (in dollars) per share of stock of Tempco Electronics on the tth day it was traded is approximated by

$$P(t) = 20 + 12 \sin \frac{\pi t}{30} - 6 \sin \frac{\pi t}{15} + 4 \sin \frac{\pi t}{10}$$

$$- 3 \sin \frac{2\pi t}{15} \qquad (0 \le t \le 20)$$

where $t = 0$ corresponds to the time the stock was first listed in a major stock exchange. What was the rate of change of the stock's price at the close of the 15th day of trading? What was the closing price on that day?

43. NUMBER OF HOURS OF DAYLIGHT The number of hours of daylight on a particular day of the year in Boston is approximated by the function

$$f(t) = 3 \sin \frac{2\pi}{365} (t - 79) + 12$$

where $t = 0$ corresponds to January 1. Compute $f'(79)$ and interpret your result.

44. WATER LEVEL IN A HARBOR The water level (in feet) in Boston Harbor during a certain 24-hr period is approximated by the formula

$$H = 4.8 \sin \frac{\pi}{6} (t - 10) + 7.6 \qquad (0 \le t \le 24)$$

where $t = 0$ corresponds to 12 A.M. When is the water level rising and when is it falling? Find the relative extrema of H and interpret your results.

45. AVERAGE DAILY TEMPERATURE The average daily temperature (in degrees Fahrenheit) at a tourist resort in Cameron Highlands is approximated by

$$T = 62 - 18 \cos \frac{2\pi(t - 23)}{365}$$

on the tth day ($t = 1$ corresponds to January 1). Which day was the warmest day of the year? The coldest day?

46. VOLUME OF AIR INHALED DURING RESPIRATION Suppose the volume of air inhaled by a person during respiration is given by

$$V(t) = \frac{6}{5\pi} \left(1 - \cos \frac{\pi t}{2}\right)$$

liters at time t (in seconds). When is the volume of inhaled air at a maximum? What is the maximum volume?

47. RESTAURANT REVENUE The revenue of McMenamy's Fish Shanty located at a popular summer resort is approximately

$$R(t) = 2\left(5 - 4 \cos \frac{\pi}{6} t\right) \qquad (0 \le t \le 12)$$

during the tth week ($t = 1$ corresponds to the first week of June), where R is measured in thousands of dollars. On what week does the restaurant realize the greatest revenue?

48. TELEVISING A ROCKET LAUNCH A major network is televising the launching of a rocket. A camera tracking the liftoff of the rocket is located at point A, as shown in accompanying figure, where ϕ denotes the angle of elevation of the camera at A. How fast is ϕ changing at the instant when the rocket is at a distance of 13,000 ft from the camera and this distance is increasing at the rate of 480 ft/sec?

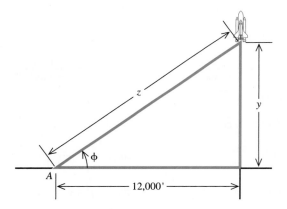

49. MAXIMUM AIR FLOW DURING RESPIRATION Refer to Exercise 46. Find the rate of flow of air in and out of the lungs. At what time is the rate of flow of air at a maximum? A minimum?

50. Refer to Exercise 60, page 259. Use the result of that exercise to find the rate at which the angle θ between the ladder and the ground is changing at the instant the bottom of the ladder is 12 ft from the wall.

51. TRACKING A SUSPECT A police cruiser hunting for a suspect pulls over and stops at a point 20 ft from a straight wall. The flasher on top of the cruiser revolves at a constant rate of 90°/second, and the light beam casts a spot of light as it strikes the wall. How fast is the spot of light moving along the wall at a point 30 ft from the point on the wall closest to the cruiser?

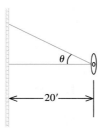

52. PERCENTAGE ERROR IN MEASURING HEIGHT From a point on level ground 150 ft from the base of a derrick, José measures the angle of elevation to the top of the derrick as 60°. If José's measurement is subject to an error of ±1%, find the percentage error in the measured height of the derrick. (See the accompanying figure.)

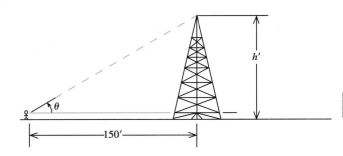

53. MAXIMIZING DRAINAGE CAPACITY The cross section of a drain is a trapezoid, as shown in the figure. The sides and the bottom of the trapezoid are each 5 ft long. Determine the angle θ so that the drain will have a maximal cross-sectional area.

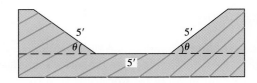

54. SURFACE AREA OF A HONEYCOMB The accompanying figure depicts a prism-shaped single cell in a honeycomb. The front end of the prism is a regular hexagon, and the back is formed by the sides of the cell coming together at a

point. It can be shown that the surface area of a cell is given by

$$S(\theta) = 6ab + \frac{3}{2}b^2\left(\frac{\sqrt{3} - \cos\theta}{\sin\theta}\right) \qquad \left(0 < \theta < \frac{\pi}{2}\right)$$

where θ is the angle between one of the (three) upper surfaces and the altitude. Show that the surface area is minimized if $\cos\theta = 1/\sqrt{3}$, or $\theta \approx 54.7°$. Measurements of actual honeycombs have confirmed that this is, in fact, the angle found in beehives.

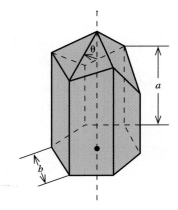

55. FINDING THE POSITION OF A PLANET As shown in the accompanying figure, the position of a planet that revolves about the Sun with an elliptical orbit can be located by calculating the central angle θ. Suppose the central angle sustained by a planet on a certain day satisfies the equation

$$\theta - 0.5\sin\theta = 1$$

Use the Newton–Raphson method to find θ.

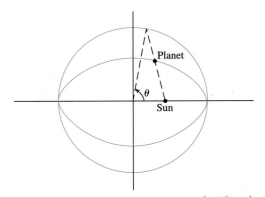

(continued on p. 886)

ANALYZING TRIGONOMETRIC FUNCTIONS

Graphing utilities can be used to analyze complicated trigonometric functions, as this example shows.

EXAMPLE 1

Let $f(x) = \dfrac{\sqrt{x}\cos 2x}{\sqrt{x^2 + \sin 3x}}$, where $0 < x \le 2$.

a. Use a graphing utility to plot the graph of f in the viewing rectangle $[0, 2] \times [-2, 1]$.

b. Use the numerical derivative operation of a graphing utility to find the rate of change of $f(x)$ at $x = 1$.

c. Find the inflection points of f.

d. Evaluate $\lim\limits_{x \to 0^+} f(x)$. Explain why $f(0)$ is not defined. What happens if you use the evaluation function of the graphing utility to find $f(0)$?

e. Find the absolute maximum and absolute minimum values of f on the interval $[0.5, 2]$.

SOLUTION ✓   **a.** The graph of f in the viewing rectangle $[0, 2] \times [-2, 1]$ is shown in Figure T1.

FIGURE T1

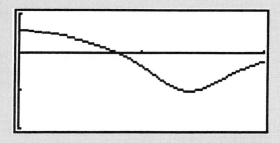

b. We find $f'(1) \approx -2.0628$.

c. The inflection points are $(1.0868, -0.5734)$ and $(1.9244, -0.5881)$. (*Note:* Move the cursor near $x = 1$ to locate the first inflection point and near $x = 1.5$ to locate the other.)

d. Using ZOOM-IN, we find $\lim\limits_{x \to 0^+} f(x) \approx 0.5773$.

e. Using the operation to find the absolute minimum and the absolute maximum in the viewing window $[0.5, 2] \times [-2, 1]$, we find the absolute minimum value to be -1.0823 and the absolute maximum value to be 0.3421.

■ ■ ■ ■

In Exercises 1–6, use the numerical derivative operation of a graphing utility to find the rate of change of $f(x)$ at the value of x. Give your answer accurate to four decimal places.

1. $f(x) = x^2 \sqrt{1 + \sin^2 x}; x = 0.5$

2. $f(x) = \dfrac{1 + \cos x}{\sqrt{1 + \sin 3x}}; x = \dfrac{\pi}{4}$

3. $f(x) = \dfrac{x \tan x}{\sqrt{1 + x^2}}; x = 0.4$

4. $f(x) = e^{\cot x}(1 + x^2)^{3/2}; x = 1$

5. $f(x) = \cos[\csc(\sqrt{x} + 1)]; x = \dfrac{\pi}{3}$

6. $f(x) = \dfrac{\ln(\cos x)}{(1 + e^{-x})^{3/2}}; x = 0.8$

In Exercises 7–10, use a graphing utility to find the absolute maximum and the absolute minimum values of f in the given interval. Express your answers accurate to four decimal places.

7. $f(x) = \dfrac{\sin x}{x}; [1, 6]$

8. $f(x) = \dfrac{\cos x^2}{\sqrt{x}}; [1, 3]$

9. $f(x) = \dfrac{x \sec x}{1 + \cot x}; [0.5, 1]$

10. $f(x) = \dfrac{x + \cos x}{1 + 0.5 \sin x}; [0, 2]$

11. Refer to Exercise 42, page 881.
   a. Plot the graph of P, using the viewing rectangle $[0, 20] \times [15, 35]$.
   b. What was the rate of change of the stock's price at the close of the 23rd day of trading?
   c. What was the closing price on that day?

12. Refer to Exercise 45, page 882.
   a. Plot the graph of T, using the viewing rectangle $[0, 365] \times [20, 80]$.
   b. Using TRACE and ZOOM or using the graphing utility's operation for finding the absolute extrema of a function, verify the solution to the problem.

13. Refer to Exercise 46, page 882.
   a. Plot the graph of V, using the viewing rectangle $[0, 12] \times [0, 2]$.
   b. Verify the solution to the problem using a graphing utility.

14. NUMBER OF HOURS OF DAYLIGHT The number of hours of daylight on a particular day of the year in Boston is approximated by the function

$$f(t) = 3 \sin \dfrac{2\pi}{365}(t - 79) + 12$$

where $t = 0$ corresponds to January 1.
   a. Sketch the graph of f on the interval $[0, 365]$.
   b. When does the longest day occur? When does the shortest day occur?

15. EFFECT OF AN EARTHQUAKE ON A STRUCTURE To study the effect an earthquake has on a structure, engineers look at the way a beam bends when subjected to an Earth tremor. The equation

$$D = a - a \cos\left(\dfrac{\pi h}{2L}\right) \qquad (0 \le h \le L)$$

where L is the length of a beam and a is the maximum deflection from the vertical, has been used by engineers to calculate the deflection D at a point on the beam h feet from the ground (see the figure). Suppose a 10-ft vertical beam has a maximum deflection of $\frac{1}{2}$ ft when subjected to an external force. Using differentials, estimate the difference in the deflection between the point midway on the beam and the point $\frac{1}{10}$ ft above it.

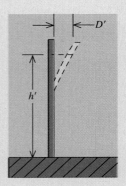

56. FLOW OF BLOOD Suppose some of the fluid flowing along a pipe of radius R is diverted to a pipe of smaller radius r attached to the former at an angle θ (see the figure). Such is the case when blood flowing along an artery is pumped into an arteriole. What should be the angle θ so that the energy loss due to friction in moving the fluid is minimal? Assume that the minimum exists and solve the problem using the following steps:

a. Use Poiseuille's Law, which states that the loss of energy due to friction in nonturbulent flow is proportional to the length of the path and inversely proportional to the fourth power of the radius, to show that the energy loss in moving the fluid from P to S via Q is

$$L = \frac{kd_1}{R^4} + \frac{kd_2}{r^4}$$

where k is a constant.

b. Suppose a and b are fixed. Find d_1 and d_2 in terms of a and b. Then use this result together with the result from part (a) to show that

$$E = k\left(\frac{a - b\cot\theta}{R^4} + \frac{b\csc\theta}{r^4}\right)$$

c. Using the technique of this section, show that E is minimized when

$$\theta = \cos^{-1}\frac{r^4}{R^4}$$

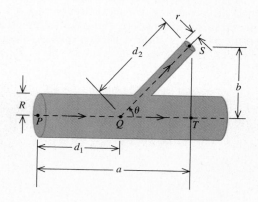

In Exercises 57–60, determine whether the statement is true or false. If it is true, explain why it is true. If it is false, give an example to show why it is false.

57. $\displaystyle\lim_{x\to0}\frac{\sin x}{x} = 1$.

Hint Use the definition of the derivative to study $f'(0)$ where $f(x) = \sin x$.

58. The graph of $f(x) = x - \sin x$ is nondecreasing on $(-\infty, \infty)$.

59. The function $f(x) = \cos x$ has a relative minimum at a point $x = a$ where $g(x) = \sin x$ has a relative maximum.

60. The graph of $f(x) = \sin x + \cos x$ is concave downward on the interval $\left(0, \dfrac{\pi}{2}\right)$.

61. Prove Rule 4:

If $h(x) = \csc f(x)$

then $h'(x) = -[\csc f(x)][\cot f(x)]f'(x)$

Hint: $\csc x = 1/\sin x$

62. Prove Rule 5:

If $h(x) = \sec f(x)$

then $h'(x) = [\sec f(x)][\tan f(x)]f'(x)$

Hint: $\sec x = 1/\cos x$

63. Prove Rule 6:

If $h(x) = \cot f(x)$

then $h'(x) = -[\csc^2 f(x)]f'(x)$

Hint: $\cot x = 1/\tan x = (\cos x)/(\sin x)$

64. Prove $\displaystyle\lim_{h\to0}\frac{\cos h - 1}{h} = 0$

Hint: Multiply by $\dfrac{\cos h + 1}{\cos h + 1}$.

SOLUTIONS TO SELF-CHECK EXERCISES 12.3

1. $f'(x) = x\,\dfrac{d}{dx}\cos 2x + (\cos 2x)\dfrac{d}{dx}(x)$

   $= x(-\sin 2x)(2) + (\cos 2x)(1)$

   $= \cos 2x - 2x\sin 2x$

2. $f'(x) = 1 + \sec^2 x$. Therefore,

$$f'\left(\frac{\pi}{4}\right) = 1 + \sec^2\left(\frac{\pi}{4}\right) = 1 + (\sqrt{2})^2 = 3$$

3. We compute

$$V'(t) = \frac{6}{5\pi}\left(-\cos\frac{\pi t}{2}\right)\left(\frac{\pi}{2}\right) = -\frac{3}{5}\cos\frac{\pi t}{2}$$

Setting $V'(t) = 0$ and solving the resulting equation give $t = 1 + 2n$ ($n = 1, 2, \ldots$) as critical points of V. The points $t = 3, 5, 7, \ldots$ give rise to the relative maxima of V, which coincide with the absolute maximum of V with value $12/(5\pi)$. Thus, the volume of inhaled air is at a maximum when $t = 3, 5, 7, \ldots$, and its value is $12/(5\pi)$ liters.

12.4 Integration of Trigonometric Functions

INTEGRATING TRIGONOMETRIC FUNCTIONS

Each of the six equations pertaining to a rule of differentiation given in the last section may be integrated to yield a corresponding integration formula. For example, integrating the equation $f'(x) = \cos x$ of Rule 1 with respect to x yields

$$\int \cos x \, dx = \int f'(x) \, dx = f(x) + C = \sin x + C$$

where C is a constant of integration. The six rules of integration obtainable in this manner follow.

Trigonometric Integration Formulas

1. $\int \sin x \, dx = -\cos x + C$

2. $\int \cos x \, dx = \sin x + C$

3. $\int \sec^2 x \, dx = \tan x + C$

4. $\int \csc^2 x \, dx = -\cot x + C$

5. $\int \sec x \tan x \, dx = \sec x + C$

6. $\int \csc x \cot x \, dx = -\csc x + C$

EXAMPLE 1   Evaluate $\int \cos 3x \, dx$.

SOLUTION ✓   Let's put $u = 3x$, so that $du = 3 \, dx$ and $dx = \frac{1}{3} \, du$. Then

$$\int \cos 3x \, dx = \frac{1}{3}\int \cos u \, du$$

$$= \frac{1}{3}\sin u + C \qquad \text{(Using Rule 2)}$$

$$= \frac{1}{3}\sin 3x + C$$

■ ■ ■ ■

TABLE 901

The Standard Normal Distribution (continued)

$$F_z(z) = P[Z \le z]$$

z	0.00	0.01	0.02	0.03	0.04	0.05	0.06	0.07	0.08	0.09
-0.4	0.3446	0.3409	0.3372	0.3336	0.3300	0.3264	0.3228	0.3192	0.3156	0.3121
-0.3	0.3821	0.3783	0.3745	0.3707	0.3669	0.3632	0.3594	0.3557	0.3520	0.3483
-0.2	0.4207	0.4168	0.4129	0.4090	0.4052	0.4013	0.3974	0.3936	0.3897	0.3859
-0.1	0.4602	0.4562	0.4522	0.4483	0.4443	0.4404	0.4364	0.4325	0.4286	0.4247
-0.0	0.5000	0.4960	0.4920	0.4880	0.4840	0.4801	0.4761	0.4721	0.4681	0.4641
0.0	0.5000	0.5040	0.5080	0.5120	0.5160	0.5199	0.5239	0.5279	0.5319	0.5359
0.1	0.5398	0.5438	0.5478	0.5517	0.5557	0.5596	0.5636	0.5675	0.5714	0.5753
0.2	0.5793	0.5832	0.5871	0.5910	0.5948	0.5987	0.6026	0.6064	0.6103	0.6141
0.3	0.6179	0.6217	0.6255	0.6293	0.6331	0.6368	0.6406	0.6443	0.6480	0.6517
0.4	0.6554	0.6591	0.6628	0.6664	0.6700	0.6736	0.6772	0.6808	0.6844	0.6879
0.5	0.6915	0.6950	0.6985	0.7019	0.7054	0.7088	0.7123	0.7157	0.7190	0.7224
0.6	0.7257	0.7291	0.7324	0.7357	0.7389	0.7422	0.7454	0.7486	0.7517	0.7549
0.7	0.7580	0.7611	0.7642	0.7673	0.7704	0.7734	0.7764	0.7794	0.7823	0.7852
0.8	0.7881	0.7910	0.7939	0.7967	0.7995	0.8023	0.8051	0.8078	0.8106	0.8133
0.9	0.8159	0.8186	0.8212	0.8238	0.8264	0.8289	0.8315	0.8340	0.8365	0.8389
1.0	0.8413	0.8438	0.8461	0.8485	0.8508	0.8531	0.8554	0.8577	0.8599	0.8621
1.1	0.8643	0.8665	0.8686	0.8708	0.8729	0.8749	0.8770	0.8790	0.8810	0.8830
1.2	0.8849	0.8869	0.8888	0.8907	0.8925	0.8944	0.8962	0.8980	0.8997	0.9015
1.3	0.9032	0.9049	0.9066	0.9082	0.9099	0.9115	0.9131	0.9147	0.9162	0.9177
1.4	0.9192	0.9207	0.9222	0.9236	0.9251	0.9265	0.9278	0.9292	0.9306	0.9319
1.5	0.9332	0.9345	0.9357	0.9370	0.9382	0.9394	0.9406	0.9418	0.9429	0.9441
1.6	0.9452	0.9463	0.9474	0.9484	0.9495	0.9505	0.9608	0.9616	0.9625	0.9633
1.7	0.9554	0.9564	0.9573	0.9582	0.9591	0.9599	0.9608	0.9616	0.9625	0.9633
1.8	0.9641	0.9649	0.9656	0.9664	0.9671	0.9678	0.9686	0.9693	0.9699	0.9706
1.9	0.9713	0.9719	0.9726	0.9732	0.9738	0.9744	0.9750	0.9756	0.9761	0.9767
2.0	0.9772	0.9778	0.9783	0.9788	0.9793	0.9798	0.9803	0.9808	0.9812	0.9817
2.1	0.9821	0.9826	0.9830	0.9834	0.9838	0.9842	0.9846	0.9850	0.9854	0.9857
2.2	0.9861	0.9864	0.9868	0.9871	0.9875	0.9878	0.9881	0.9884	0.9887	0.9890
2.3	0.9893	0.9896	0.9898	0.9901	0.9904	0.9906	0.9909	0.9911	0.9913	0.9916
2.4	0.9918	0.9920	0.9922	0.9925	0.9927	0.9929	0.9931	0.9932	0.9934	0.9936
2.5	0.9938	0.9940	0.9951	0.9943	0.9945	0.9946	0.9948	0.9949	0.9951	0.9952
2.6	0.9953	0.9955	0.9956	0.9957	0.9959	0.9960	0.9961	0.9962	0.9963	0.9964
2.7	0.9965	0.9966	0.9967	0.9968	0.9969	0.9970	0.9971	0.9972	0.9973	0.9974
2.8	0.9974	0.9975	0.9976	0.9977	0.9977	0.9978	0.9979	0.9979	0.9980	0.9981
2.9	0.9981	0.9982	0.9982	0.9983	0.9984	0.9984	0.9985	0.9985	0.9986	0.9986
3.0	0.9987	0.9987	0.9987	0.9988	0.9988	0.9989	0.9989	0.9989	0.9990	0.9990
3.1	0.9990	0.9991	0.9991	0.9991	0.9992	0.9992	0.9992	0.9992	0.9993	0.9993
3.2	0.9993	0.9993	0.9994	0.9994	0.9994	0.9994	0.9994	0.9995	0.9995	0.9995
3.3	0.9995	0.9995	0.9995	0.9996	0.9996	0.9996	0.9996	0.9996	0.9996	0.9997
3.4	0.9997	0.9997	0.9997	0.9997	0.9997	0.9997	0.9997	0.9997	0.9997	0.9998

Answers to Odd-Numbered Exercises

Chapter 1

Exercises 1.1, page 12

1. False   3. False

5.
$$0 \quad 3 \quad 6$$

7.
$$-1 \quad 0 \quad 4$$

9.
$$0$$

11. $(-\infty, 2)$ 13. $(-\infty, -5]$ 15. $(-4, 6)$

17. $(-\infty, -3) \cup (3, \infty)$ 19. $(-2, 3)$ 21. 4

23. 2   25. $5\sqrt{3}$ 27. $\pi + 1$ 29. 2

31. False   33. False   35. True   37. False

39. True   41. False   43. 9   45. 1   47. 4

49. 7   51. $\frac{1}{5}$ 53. 2   55. 2   57. 1

59. True   61. False   63. False   65. False

67. False   69. $\dfrac{1}{(xy)^2}$ 71. $\dfrac{1}{x^{5/6}}$ 73. $\dfrac{1}{(s+t)^3}$

75. $x^{13/3}$ 77. $\dfrac{1}{x^3}$ 79. x 81. $\dfrac{9}{x^2 y^4}$ 83. $\dfrac{y^8}{x^{10}}$

85. $2x^{11/6}$ 87. $-2xy^2$ 89. $2x^{4/3}y^{1/2}$ 91. 2.828

93. 5.196   95. 31.62   97. 316.2   99. $\dfrac{3\sqrt{x}}{2x}$

101. $\dfrac{2\sqrt{3y}}{3}$ 103. $\dfrac{\sqrt[3]{x^2}}{x}$ 105. $\dfrac{2x}{3\sqrt{x}}$ 107. $\dfrac{2y}{\sqrt{2xy}}$

109. $\dfrac{xz}{y\sqrt[3]{xz^2}}$ 111. [362, 488.7]   113. $12,300

115. $22,000   117. $|x - 0.5| < 0.01$

119. Between 1000 and 4000 units

121. False   123. True

Exercises 1.2, page 26

1. $9x^2 + 3x + 1$ 3. $4y^2 + y + 8$ 5. $-x - 1$

7. $\frac{2}{3} + e - e^{-1}$ 9. $6\sqrt{2} + 8 + \frac{1}{2}\sqrt{x} - \frac{11}{4}\sqrt{y}$

11. $x^2 + 6x - 16$ 13. $a^2 + 10a + 25$

15. $x^2 + 4xy + 4y^2$ 17. $4x^2 - y^2$ 19. $-2x$

21. $2t(2\sqrt{t} + 1)$ 23. $2x^3(2x^2 - 6x - 3)$

25. $7a^2(a^2 + 7ab - 6b^2)$ 27. $e^{-x}(1 - x)$

29. $\frac{1}{2}x^{-5/2}(4 - 3x)$ 31. $(2a + b)(3c - 2d)$

33. $(2a + b)(2a - b)$ 35. $-2(3x + 5)(2x - 1)$

37. $3(x - 4)(x + 2)$ 39. $2(3x - 5)(2x + 3)$

41. $(3x - 4y)(3x + 4y)$ 43. $(x^2 + 5)(x^4 - 5x^2 + 25)$

45. $x^3 - xy^2$ 47. $4(x - 1)(3x - 1)(2x + 2)^3$

49. $4(x - 1)(3x - 1)(2x + 2)^3$

51. $2x(x^2 + 2)^2(5x^4 + 20x^2 + 17)$

53. -4 and 3   55. -1 and $\frac{1}{2}$ 57. 2 and 2

59. -2 and $\frac{3}{4}$ 61. $\frac{1}{2} + \frac{1}{4}\sqrt{10}$ and $\frac{1}{2} - \frac{1}{4}\sqrt{10}$

63. $-1 + \frac{1}{2}\sqrt{10}$ and $-1 - \frac{1}{2}\sqrt{10}$ 65. $\dfrac{x - 1}{x - 2}$

67. $\dfrac{3(2t + 1)}{2t - 1}$ 69. $-\dfrac{7}{(4x - 1)^2}$ 71. -8

73. $\dfrac{3x - 1}{2}$ 75. $\dfrac{t + 20}{3t + 2}$ 77. $-\dfrac{x(2x - 13)}{(2x - 1)(2x + 5)}$

79. $-\dfrac{x + 27}{(x - 3)^2(x + 3)}$ 81. $\dfrac{x + 1}{x - 1}$ 83. $\dfrac{4x^2 + 7}{\sqrt{2x^2 + 7}}$

85. $\dfrac{x - 1}{x^2\sqrt{x + 1}}$ 87. $\dfrac{x - 1}{(2x + 1)^{3/2}}$ 89. $\dfrac{\sqrt{3} + 1}{2}$

91. $\dfrac{\sqrt{x} + \sqrt{y}}{x - y}$ 93. $\dfrac{(\sqrt{a} + \sqrt{b})^2}{a - b}$ 95. $\dfrac{x}{3\sqrt{x}}$

97. $-\dfrac{2}{3(1 + \sqrt{3})}$ 99. $-\dfrac{x + 1}{\sqrt{x + 2}(1 - \sqrt{x + 2})}$

101. True   103. False

Exercises 1.3, page 34

1. (3, 3); Quadrant I   **3.** (2, −2); Quadrant IV

5. (−4, −6); Quadrant III   **7.** A **9.** E, F, and G

11. F **13.–19.** See accompanying figure.

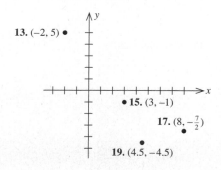

13. (−2, 5) ●

● **15.** (3, −1)

17. $(8, -\frac{7}{2})$

19. (4.5, −4.5)

21. 5   **23.** $\sqrt{61}$ **25.** (−8, −6) and (8, −6)

29. $(x - 2)^2 + (y + 3)^2 = 25$

31. $x^2 + y^2 = 25$

33. $(x - 2)^2 + (y + 3)^2 = 34$

35. 3400 miles   **37.** Route 1   **39.** Model C

41. $10\sqrt{13}t$; 72.1 mi   **43.** True   **45.** True

Exercises 1.4, page 48

1. e   **3.** a   **5.** f   **7.** $\frac{1}{2}$ **9.** Not defined

11. 5   **13.** $\frac{5}{6}$ **15.** $\dfrac{d - b}{c - a}$ $(a \neq c)$

17. a. 4   **b.** −8   **19.** Parallel   **21.** Perpendicular

23. −5   **25.** $y = -3$ **27.** $y = 2x - 10$

29. $y = 2$ **31.** $y = 3x - 2$ **33.** $y = x + 1$

35. $y = 3x + 4$ **37.** $y = 5$

39. $y = \frac{1}{2}x$; $m = \frac{1}{2}$; $b = 0$

41. $y = \frac{2}{3}x - 3$; $m = \frac{2}{3}$; $b = -3$

43. $y = -\frac{1}{2}x + \frac{7}{2}$; $m = -\frac{1}{2}$; $b = \frac{7}{2}$

45. $y = \frac{1}{2}x + 3$ **47.** $y = -6$ **49.** $y = b$

51. $y = \frac{2}{3}x - \frac{2}{3}$ **53.** $k = 8$

55.

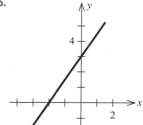

57.

59.

63. $y = -2x - 4$ **65.** $y = \frac{1}{8}x - \frac{1}{2}$ **67.** Yes

69. a. $y = 0.55x$ **b.** 2000   **71.** 82.4% of men's wages

73. a. and **b.**

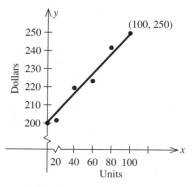

(100, 250)

c. $y = \frac{1}{2}x + 200$ **d.** $227

75. a. and b.

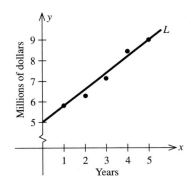

c. $y = 0.8x + 5$ **d.** $12.2 million

77. True   **79.** True   **81.** Yes

Chapter 1 Review Exercises, page 56

1. $[-2, \infty)$ **3.** $(-\infty, -4) \cup (5, \infty)$ **5.** 4

7. $\pi - 6$ **9.** $\frac{27}{8}$ **11.** $\frac{1}{144}$ **13.** $\frac{1}{4}$

15. $4(x^2 + y)^2$ **17.** $\frac{2x}{3z}$ **19.** $6xy^7$

21. $2vw(v^2 + w^2 + u^2)$ **23.** $6t(t + 1)(2t - 3)$

25. -2 and $\frac{1}{3}$ **27.** $\frac{\sqrt{2}}{2}$ and $-\frac{\sqrt{2}}{2}$

29. $-2 + \frac{\sqrt{2}}{2}$ and $-2 - \frac{\sqrt{2}}{2}$ **31.** $\frac{15x^2 + 24x + 2}{4(x + 2)(3x^2 + 2)}$

33. $\frac{2(x + 2)}{\sqrt{x + 1}}$ **35.** $\frac{x - \sqrt{x}}{2x}$ **37.** 2   **39.** $y = 4$

41. $y = -\frac{4}{5}x + \frac{12}{5}$ **43.** $y = \frac{3}{4}x + \frac{11}{2}$

45. $y = -\frac{3}{5}x + \frac{12}{5}$

47.

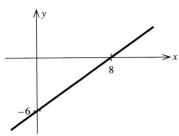

49. $100

Chapter 2

Exercises 2.1, page 70

1. $21, -9, 5a + 6, -5a + 6, 5a + 21$

3. $-3, 6, 3a^2 - 6a - 3, 3a^2 + 6a - 3, 3x^2 - 6$

5. $\frac{8}{15}, 0, \frac{2a}{a^2 - 1}, \frac{2(2 + a)}{a^2 + 4a + 3}, \frac{2(t + 1)}{t(t + 2)}$

7. $8, \frac{2a^2}{\sqrt{a - 1}}, \frac{2(x + 1)^2}{\sqrt{x}}, \frac{2(x - 1)^2}{\sqrt{x - 2}}$

9. $5, 1, 1$ **11.** $\frac{5}{2}, 3, 3, 9$

13. a. -2 **b.** (i) $x = 2$; (ii) $x = 1$ **c.** $[0, 6]$ **d.** $[-2, 6]$

15. Yes   **17.** Yes

19. $(-\infty, \infty)$ **21.** $(-\infty, 0) \cup (0, \infty)$ **23.** $(-\infty, \infty)$

25. $(-\infty, 5]$ **27.** $(-\infty, -1) \cup (-1, 1) \cup (1, \infty)$

29. $[-3, \infty)$ **31.** $(-\infty, -2) \cup (-2, 1]$

33. a. $(-\infty, \infty)$
b. $6, 0, -4, -6, -\frac{25}{4}, -6, -4, 0$
c.

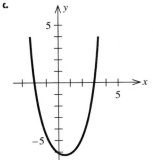

35.

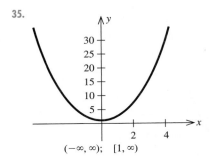

$(-\infty, \infty); \quad [1, \infty)$

37.

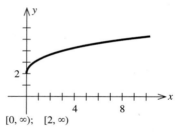

$[0, \infty)$; $[2, \infty)$

39.

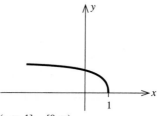

$(-\infty, 1]$; $[0, \infty)$

41.

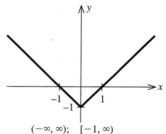

$(-\infty, \infty)$; $[-1, \infty)$

43.

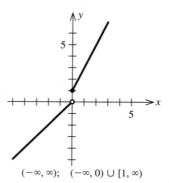

$(-\infty, \infty)$; $(-\infty, 0) \cup [1, \infty)$

45.

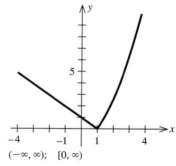

$(-\infty, \infty)$; $[0, \infty)$

47. Yes   **49.** No   **51.** Yes   **53.** Yes

55. 10π in.   **57.** 8   **59. a.** From 1985 to 1990
  b. From 1990 on
  c. 1990; $3.5 billion

61. a.
$$f(t) = \begin{cases} 0.0185t + 0.58 & \text{if } 0 \le t \le 20 \\ 0.015t + 0.65 & \text{if } 20 < t \le 30 \end{cases}$$
  b. 0.0185/yr from 1960 through 1980; 0.015/yr from 1980 through 1990
  c. 1983

63. a. $0.06x$ **b.** $12.00; $0.34   **65.** 160 mg

67. a. $30 + 0.15x$; $25 + 0.20x$
  b.

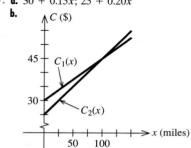

  c. Acme

69. $700,000; $620,000; $540,000

71. a. $0 \le r \le 0.2$
  b. 40; 30; 0; as the distance r increases, the velocity of the blood decreases.

73. 20; 26

75. 0.77. When the proportion of popular votes won by the Democratic presidential candidate is 0.60, the proportion of seats in the House of Representatives won by Democratic candidates is 0.77.

77. a. $(0, 12]$
  b.

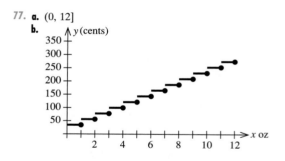

79. True   **81.** False

Using Technology Exercises 2.1, page 79

1.
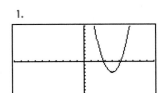

3.
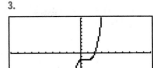

5.
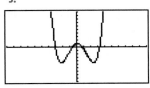

7.
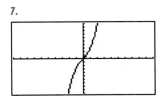

9. a.
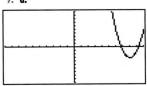

b.
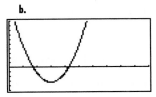

11. a.
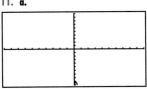

b.
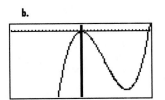

13. a.
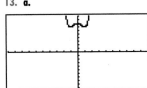

b.
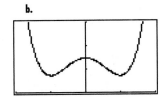

15. a.
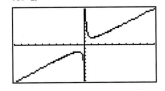

b.
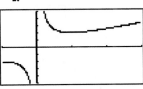

17. a.
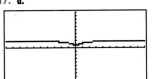

b.
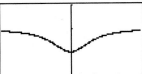

19. a.
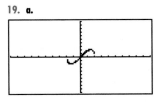

b.
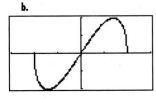

21.
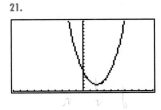

23.
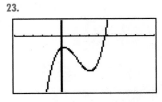

25.
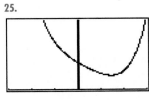

27.
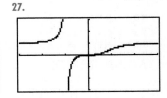

29.
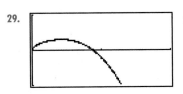

31. 18   **33.** 2   **35.** 18.5505   **37.** 17.3850

39. 4.1616   **41.** 1.7214

43. a.
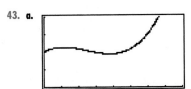

b. 2.1762%; 1.9095%

45. a.
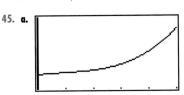

b. 44.7; 52.7; 129.2

Exercises 2.2, page 87

1. $f(x) + g(x) = x^3 + x^2 + 3$

3. $f(x)g(x) = x^5 - 2x^3 + 5x^2 - 10$

5. $\dfrac{f(x)}{g(x)} = \dfrac{x^3 + 5}{x^2 - 2}$

7. $\dfrac{f(x)g(x)}{h(x)} = \dfrac{x^5 - 2x^3 + 5x^2 - 10}{2x + 4}$

9. $f(x) + g(x) = x - 1 + \sqrt{x + 1}$

11. $f(x)g(x) = (x - 1)\sqrt{x + 1}$

13. $\dfrac{g(x)}{h(x)} = \dfrac{\sqrt{x + 1}}{2x^3 - 1}$

15. $\dfrac{f(x)g(x)}{h(x)} = \dfrac{(x - 1)\sqrt{x + 1}}{2x^3 - 1}$

17. $\dfrac{f(x) - h(x)}{g(x)} = \dfrac{x - 2x^3}{\sqrt{x + 1}}$

19. $f(x) + g(x) = x^2 + \sqrt{x} + 3$;
$f(x) - g(x) = x^2 - \sqrt{x} + 7$;
$f(x)g(x) = (x^2 + 5)(\sqrt{x} - 2)$; $\dfrac{f(x)}{g(x)} = \dfrac{x^2 + 5}{\sqrt{x} - 2}$

21. $f(x) + g(x) = \dfrac{(x - 1)\sqrt{x + 3} + 1}{x - 1}$;
$f(x) - g(x) = \dfrac{(x - 1)\sqrt{x + 3} - 1}{x - 1}$;
$f(x)g(x) = \dfrac{\sqrt{x + 3}}{x - 1}$; $\dfrac{f(x)}{g(x)} = (x - 1)\sqrt{x + 3}$

23. $f(x) + g(x) = \dfrac{2(x^2 - 2)}{(x - 1)(x - 2)}$;
$f(x) - g(x) = \dfrac{-2x}{(x - 1)(x - 2)}$;
$f(x)g(x) = \dfrac{(x + 1)(x + 2)}{(x - 1)(x - 2)}$; $\dfrac{f(x)}{g(x)} = \dfrac{(x + 1)(x - 2)}{(x - 1)(x + 2)}$

25. $f(g(x)) = x^4 + x^2 + 1$; $g(f(x)) = (x^2 + x + 1)^2$

27. $f(g(x)) = \sqrt{x^2 - 1} + 1$; $g(f(x)) = x + 2\sqrt{x}$

29. $f(g(x)) = \dfrac{x}{x^2 + 1}$; $g(f(x)) = \dfrac{x^2 + 1}{x}$

31. 49   **33.** $\dfrac{\sqrt{5}}{5}$

35. $f(x) = 2x^3 + x^2 + 1$ and $g(x) = x^5$

37. $f(x) = x^2 - 1$ and $g(x) = \sqrt{x}$

39. $f(x) = x^2 - 1$ and $g(x) = \dfrac{1}{x}$

41. $f(x) = 3x^2 + 2$ and $g(x) = \dfrac{1}{x^{3/2}}$

43. $3h$ **45.** $-h(2a + h)$ **47.** $2a + h$

49. $C(x) = 0.6x + 12{,}100$

51. a. $P(x) = -0.000003x^3 - 0.07x^2 + 300x - 100{,}000$
b. \$182,375

53. a. $N(t) = \dfrac{7}{1 + 0.02\left[\dfrac{10t + 150}{t + 10}\right]^2}$
b. 1.27 million units; 1.74 million units; 1.85 million units

55. $N(x(t)) = 9.94\left[\dfrac{(t + 10)^2}{(t + 10)^2 + 2(t + 15)^2}\right]$; 2.24 million jobs; 2.48 million jobs

57. False   **59.** False

Exercises 2.3, page 99

1. Yes; $y = -\frac{2}{3}x + 2$ **3.** Yes, $y = \frac{1}{2}x + 2$

5. Yes; $y = \frac{1}{2}x + \frac{9}{4}$ **7.** No

9. Polynomial function; degree 6

11. Polynomial function; degree 6

13. Some other function   **15.** $m = -1$; $b = 2$

17. a. $C(x) = 8x + 40{,}000$
b. $R(x) = 12x$
c. $P(x) = 4x - 40{,}000$
d. A loss of \$8000; a profit of \$8000

19. \$28,800   **21.** 104 mg   **23.** \$400,000

25. a.

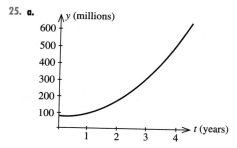

b. 184.84 million

27. a. $R(x) = \dfrac{100x}{40 + x}$ **b.** 60%

29. a. $T = \frac{1}{4}N + 40$ **b.** $N = 4T - 160$; 248 times/min

31. \$72,000   **33. a.** 714,300   **b.** 8,327,800

35. $\dfrac{110}{\frac{1}{2}t + 1} - 26\left(\frac{1}{4}t^2 - 1\right)^2 - 52$; \$32, \$6.71, \$3; the gap was closing.

37. a.

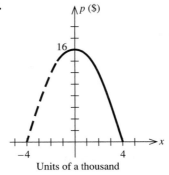

Units of a thousand

b. 3000 units

39. a.

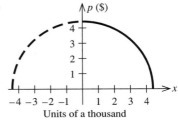

Units of a thousand

b. 3000

41. a.

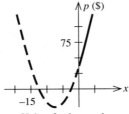

Units of a thousand

b. $76

43. a.

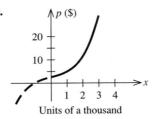

Units of a thousand

b. $15

45. L_2; for each dollar increase in the price of a clock radio, more model A clock radios than model B clock radios will be made available in the marketplace.

47. $p = \sqrt{-x^2 + 100}$; 6614 units

49. $p = \frac{1}{10}\sqrt{x} + 10$; $30   **51.** 2500; $67.50

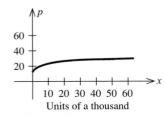

Units of a thousand

53. 11,000; $3   **55.** 500 tents; $32.50

57. a.

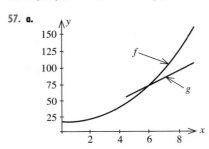

b. 5.6 mph; 71.6 mL/lb/min
c. The oxygen consumption of the walker is greater than that of the runner.

59. $f(x) = 2x + \dfrac{500}{x}$; $x > 0$

61. $f(x) = 0.5x^2 + \dfrac{8}{x}$

63. $f(x) = (22 + x)(36 - 2x)$ bushels/acre

65. True   **67.** False

Using Technology Exercises 2.3, page 106

1. $(-3.0414, 0.1503)$; $(3.0414, 7.4497)$

3. $(-2.3371, 2.4117)$; $(6.0514, -2.5015)$

5. $(-1.0219, -6.3461)$; $(1.2414, -1.5931)$; $(5.7805, 7.9391)$

7. a.

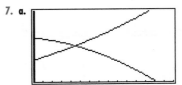

b. 438 wall clocks; $40.92

9. **a.** $y = 0.1375t^2 + 0.675t + 3.1$
b.

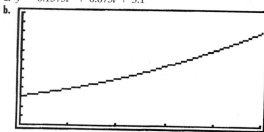

c. 3.1; 3.9; 5; 6.4; 8; 9.9

11. **a.** $y = -0.02028t^3 + 0.31393t^2 + 0.40873t + 0.66024$
b.

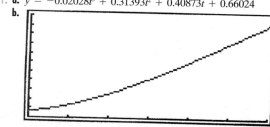

c. 0.66; 1.36; 2.57; 4.16; 6.02; 8.02; 10.03

13. **a.** $y = 0.05833t^3 - 0.325t^2 + 1.8881t + 5.07143$
b.

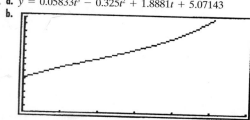

c. 6.7; 8.0; 9.4; 11.2; 13.7

15. **a.** $y = 0.0125t^4 - 0.01389t^3 + 0.55417t^2 + 0.53294t + 4.95238$ $(0 \le t \le 5)$
b.

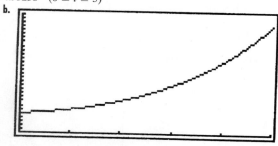

c. 5.0; 6.0; 8.3; 12.2; 18.3; 27.5

Exercises 2.4, page 127

1. $\lim_{x \to -2} f(x) = 3$ 3. $\lim_{x \to 3} f(x) = 3$ 5. $\lim_{x \to -2} f(x) = 3$

7. The limit does not exist.

9.

x	1.9	1.99	1.999
$f(x)$	4.61	4.9601	4.9960

x	2.001	2.01	2.1
$f(x)$	5.004	5.0401	5.41

$\lim_{x \to 2} (x^2 + 1) = 5$

11.

x	−0.1	−0.01	−0.001
$f(x)$	−1	−1	−1

x	0.001	0.01	0.1
$f(x)$	1	1	1

The limit does not exist.

13.

x	0.9	0.99	0.999
$f(x)$	100	10,000	1,000,000

x	1.001	1.01	1.1
$f(x)$	1,000,000	10,000	100

The limit does not exist.

15.

x	0.9	0.99	0.999	1.001	1.01	1.1
$f(x)$	2.9	2.99	2.999	3.001	3.01	3.1

$\lim_{x \to 1} \dfrac{x^2 + x - 2}{x - 1} = 3$

17.

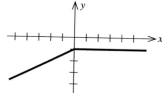

$\lim_{x \to 0} f(x) = -1$

19.

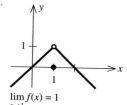

$\lim_{x \to 1} f(x) = 1$

21.

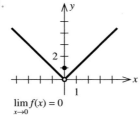

$\lim\limits_{x \to 0} f(x) = 0$

23. 3

25. 3   **27.** −1   **29.** 2   **31.** −4   **33.** $\frac{5}{4}$

35. 2   **37.** $\sqrt{171} = 3\sqrt{19}$ **39.** $\frac{3}{2}$ **41.** −1

43. −6   **45.** 2   **47.** $\frac{1}{6}$ **49.** 2   **51.** −1

53. −10   **55.** The limit does not exist.   **57.** $\frac{5}{3}$

59. $\frac{1}{2}$ **61.** $\frac{1}{3}$ **63.** $\lim\limits_{x \to \infty} f(x) = \infty$; $\lim\limits_{x \to -\infty} f(x) = \infty$

65. 0; 0   **67.** $\lim\limits_{x \to \infty} f(x) = -\infty$; $\lim\limits_{x \to -\infty} f(x) = -\infty$

69.

x	1	10	100	1000
$f(x)$	0.5	0.009901	0.0001	0.000001

x	−1	−10	−100	−1000
$f(x)$	0.5	0.009901	0.0001	0.000001

$\lim\limits_{x \to \infty} f(x) = 0$ and $\lim\limits_{x \to -\infty} f(x) = 0$

71.

x	1	5	10	100
$f(x)$	12	360	2910	2.99×10^6

x	1000	−1	−5
$f(x)$	2.999×10^9	6	−390

x	−10	−100	−1000
$f(x)$	−3090	-3.01×10^6	-3.0×10^9

$\lim\limits_{x \to \infty} f(x) = \infty$ and $\lim\limits_{x \to -\infty} f(x) = -\infty$

73. 3   **75.** 3   **77.** $\lim\limits_{x \to -\infty} f(x) = -\infty$ **79.** 0

81. a. $0.5 million; $0.75 million; $1,166,667; $2 million; $4.5 million; $9.5 million
b. The limit does not exist; as the percentage of pollutant to be removed approaches 100, the cost becomes astronomical.

83. $2.20; the average cost of producing x video discs will approach $2.20/disc in the long run.

85. a. $24 million; $60 million; $83.1 million
b. $120 million

87. a. 76.1 cents/mile; 30.5 cents/mile; 23 cents/mile; 20.6 cents/mile; 19.5 cents/mile
b.

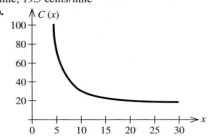

c. It approaches 17.8 cents/mile

89. False   **91.** True   **93.** True

95. a moles/liter/second   **97.** No

Using Technology Exercises 2.4, page 132

1. 5   **3.** 3   **5.** $\frac{2}{3}$ **7.** $\frac{1}{2}$ **9.** e^2

13. a.

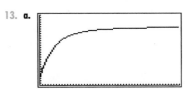

b. 25,000

Exercises 2.5, page 145

1. 3; 2; the limit does not exist.

3. The limit does not exist; 2; the limit does not exist.

5. 0; 2; the limit does not exist.

7. −2; 2; the limit does not exist.

9. True   **11.** True   **13.** False   **15.** True

17. False   **19.** True   **21.** 6   **23.** $-\frac{1}{4}$

25. The limit does not exist.   **27.** −1   **29.** 0

31. −4   **33.** The limit does not exist.   **35.** 4

37. 0   **39.** 0; 0   **41.** 2; 3

43. $x = 0$; conditions 2 and 3

45. Continuous everywhere   **47.** $x = 0$; condition 3

49. $x = 0$; condition 3   **51.** $(-\infty, \infty)$ **53.** $(-\infty, \infty)$

55. $(-\infty, \frac{1}{2}) \cup (\frac{1}{2}, \infty)$ **57.** $(-\infty, -2) \cup (-2, 1) \cup (1, \infty)$

59. $(-\infty, \infty)$ **61.** $(-\infty, \infty)$ **63.** $(-\infty, \infty)$

65. $(-\infty, \infty)$ **67.** -1 and 1 **69.** 1 and 2

71. f is discontinuous at $x = 1, 2, \ldots, 11$.

73. Michael makes progress toward solving the problem until $x = x_1$. Between $x = x_1$ and $x = x_2$, he makes no further progress. But at $x = x_2$ he suddenly achieves a breakthrough, and at $x = x_3$ he proceeds to complete the problem.

75. Conditions 2 and 3 are not satisfied at each of these points.

77.

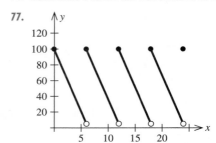

79.

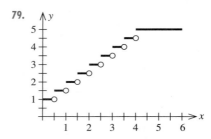

f is discontinuous at $x = \frac{1}{2}, 1, 1\frac{1}{2}, \ldots, 4$.

81. a. ∞; if the speed of the fish is very close to that of the current, the energy expended by the fish is enormous.
b. ∞; if the speed of the fish increases greatly, so does the amount of energy required to swim L feet.

83. $k = -4$ **85. a.** No   **b.** No

87. a. f is a polynomial of degree 3.
b. $f(-1) = -4$ and $f(1) = 4$

89. f is continuous on $[14, 16]$, $f(14) \approx -6.06$, and $f(16) \approx 1.60$.

91. $x = 2$ **93.** ≈ 1.34 **95.** False   **97.** False

99. True   **101. c.** No

Using Technology Exercises 2.5, page 152

1. $x = 0, 1$ **3.** $x = 2$ **5.** $x = 0, \frac{1}{2}$

7. $x = -\frac{1}{2}, 2$ **9.** $x = -2, 1$

11.

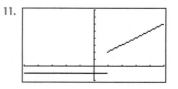

13.

15.

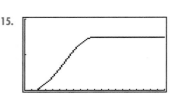

Exercises 2.6, page 169

1. 1.5 lb/month; 0.5833 lb/month; 1.25 lb/month

3. 3.075%/hr; -21.15%/hr

5. a. Car A
b. They are traveling at the same speed.
c. Car B
d. Both cars covered the same distance.

7. a. P_2 **b.** P_1 **c.** Bactericide B; bactericide A

9. 0   **11.** 2   **13.** $6x$ **15.** $-2x + 3$

17. 2; $y = 2x + 7$ **19.** 6; $y = 6x - 3$

21. $\frac{1}{9}$; $y = \frac{1}{9}x - \frac{2}{3}$

23. a. $4x$
b. $y = 4x - 1$
c.

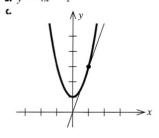

25. a. $2x - 2$
b. $(1, 0)$
c.

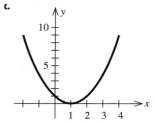

d. 0

27. a. 6; 5.5; 5.1
  b. 5
  c. The computations in part (a) show that as h approaches zero, the average velocity approaches the instantaneous velocity.

29. a. 130 ft/sec; 128.2 ft/sec; 128.02 ft/sec
  b. 128 ft/sec
  c. The computations in part (a) show that as the time intervals over which the average velocity are computed become smaller and smaller, the average velocity approaches the instantaneous velocity of the car at $t = 20$.

31. a. 5 sec   **b.** 80 ft/sec   **c.** 160 ft/sec

33. a. $-\frac{1}{6}$ liter/atmosphere   **b.** $-\frac{1}{4}$ liter/atmosphere

35. a. $-\frac{2}{3}x + 7$ **b.** \$333/quarter; $-$\$13,000/quarter

37. \$6 billion/yr; \$10 billion/yr

39. Average rate of change of the seal population over $[a, a + h]$; instantaneous rate of change of the seal population at $x = a$

41. Average rate of change of the country's industrial production over $[a, a + h]$; instantaneous rate of change of the country's industrial production at $x = a$

43. Average rate of change of atmospheric pressure over $[a, a + h]$; instantaneous rate of change of atmospheric pressure at $x = a$

45. a. Yes   **b.** No   **c.** No

47. a. Yes   **b.** Yes   **c.** No

49. a. No   **b.** No   **c.** No

51. 5.06060; 5.06006; 5.060006; 5.0600006; 5.06000006; \$5.06/case

53. True

55.
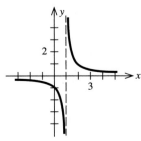

57.
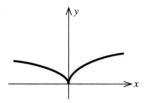

Yes; no; the graph of f has a kink at $x = 0$.

Using Technology Exercises 2.6, page 178

1. a. $y = 4x - 3$
  b.

3. a. $y = -7x - 8$
  b.
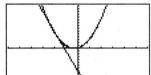

5. a. $y = 9x - 11$
  b.
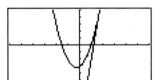

7. a. $y = 2$
  b.
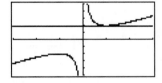

9. a. $y = \frac{1}{4}x + 1$
  b.
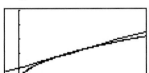

11. a. 4
  b. $y = 4x - 1$
  c.
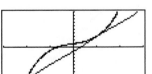

13. a. 20
  b. $y = 20x - 35$
  c.
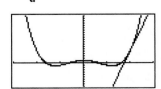

15. a. 0.75
  b. $y = 0.75x - 1$
  c.
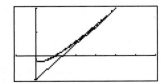

17. a. -0.25
b. $y = -0.25x + 0.75$
c.

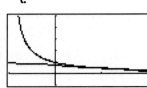

19. a. 4.02
b. $y = 4.02x - 3.57$
c.

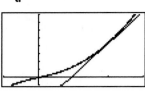

Chapter 2 Review Exercises, page 179

1. a. $(-\infty, 9]$
b. $(-\infty, -1) \cup (-1, \frac{3}{2}) \cup (\frac{3}{2}, \infty)$

3. a.

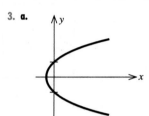

b. No
c. Yes

5. a. $\dfrac{2x+3}{x}$ **b.** $\dfrac{1}{x(2x+3)}$ **c.** $\dfrac{1}{2x+3}$ **d.** $\dfrac{2}{x}+3$

7. 2   **9.** 0   **11.** The limit does not exist.

13. $\frac{9}{2}$ **15.** $\frac{1}{2}$ **17.** 1   **19.** The limit does not exist.

21.

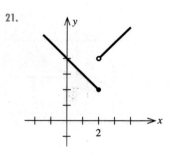

4; 2; the limit does not exist.

23. $x = -\frac{1}{2}, 1$ **25.** $x = 0$ **27.** 3

29. $\frac{3}{2}; y = \frac{3}{2}x + 5$ **31. a.** Yes   **b.** No

33. a. $S(t) = t + 2.4$ **b.** $5.4 million

35. $(6, \frac{21}{2})$ **37.** 6000; $22   **39.** $45,000

41.

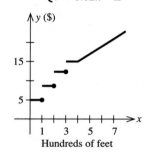

As the length of the list increases, the time taken to learn the list increases by a very large amount.

43. $C(x) = \begin{cases} 5 & \text{if } 1 \le x \le 100 \\ 9 & \text{if } 100 < x \le 200 \\ 12.50 & \text{if } 200 < x \le 300 \\ 15.00 & \text{if } 300 < x \le 400 \\ 7 + 0.02x & \text{if } \quad x > 400 \end{cases}$

The function is discontinuous at $x = 100$, 200, and 300.

CHAPTER 3

Exercises 3.1, page 191

1. 0   **3.** $5x^4$ **5.** $2.1x^{1.1}$ **7.** $6x$ **9.** $2\pi r$

11. $\dfrac{3}{x^{2/3}}$ **13.** $\dfrac{3}{2\sqrt{x}}$ **15.** $-84x^{-13}$ **17.** $10x - 3$

19. $-3x^2 + 4x$ **21.** $0.06x - 0.4$ **23.** $2x - 4 - \dfrac{3}{x^2}$

25. $16x^3 - 7.5x^{3/2}$ **27.** $-\dfrac{3}{x^2} - \dfrac{8}{x^3}$ **29.** $-\dfrac{16}{t^5} + \dfrac{9}{t^4} - \dfrac{2}{t^2}$

31. $2 - \dfrac{5}{2\sqrt{x}}$ **33.** $-\dfrac{4}{x^3} + \dfrac{1}{x^{4/3}}$

35. a. 20   **b.** -4 **c.** 20   **37.** 3   **39.** 11

41. $m = 5; y = 5x - 4$ **43.** $m = -2; y = -2x + 2$

45. a. $(0, 0)$ **b.**

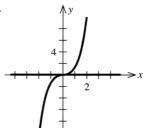

47. a. $(-2, -7), (2, 9)$
  b. $y = 12x + 17$ and $y = 12x - 15$
  c.

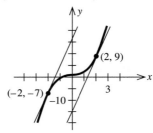

49. a. $(0, 0); \left(1, -\frac{13}{12}\right)$
  b. $(0, 0); \left(2, -\frac{8}{3}\right); \left(-1, -\frac{5}{12}\right)$
  c. $(0, 0); \left(4, \frac{80}{3}\right); \left(-3, \frac{81}{4}\right)$

51. a. $\dfrac{16\pi}{9}$ cm³/cm   **b.** $\dfrac{25\pi}{4}$ cm³/cm

53. $-89.01; -4.36$; if you make 0.25 stop/mile, your average speed will decrease at the rate of approximately 89.01 mph/stop/mile. If you make 2 stops/mile, your average speed will decrease at the rate of approximately 4.36 mph/stop/mile.

55. a. 15 pts/yr; 12.6 pts/yr; 0 pts/yr   **b.** 10 pts/yr

57. 155/mo; 200/mo

59. 32 turtles/yr; 428 turtles/yr; 3260 turtles

61. a. $120 - 30t$ **b.** 120 ft/sec   **c.** 240 ft

63. a. 2.495 million   **b.** 405,000/yr

65. a. $(0.0001)\left(\frac{5}{4}\right) x^{1/4}$ **b.** $0.00125/radio

67. a. $20\left(1 - \dfrac{1}{\sqrt{t}}\right)$
  b. 50 mph; 30 mph; 33.43 mph
  c. $-8.28; 0; 5.86$; at 6:30 A.M., the average velocity is decreasing at the rate of 8.28 mph/hr; at 7 A.M., it is unchanged; and at 8 A.M., it is increasing at the rate of 5.86 mph.

69. a. $f'(t) = 0.225t^2 + 0.05t + 2.45$
  b. $4.625 billion/yr
  c. $12 billion/yr

71. False

Using Technology Exercises 3.1, page 193

1. 1   **3.** 0.4226   **5.** 0.1613

7. a.

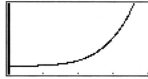

  b. 3.4295 ppm; 105.4332 ppm

9. a.

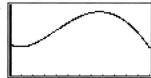

  b. Decreasing at the rate of 9 days/yr; increasing at the rate of 13 days/yr

11. a.

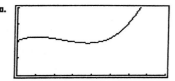

  b. Decreasing at the rate of 0.188887%/yr; increasing at the rate of 0.0777812%/yr

Exercises 3.2, page 205

1. $6x^2 + 2$ **3.** $4t - 1$ **5.** $9x^2 + 2x - 6$

7. $4x^3 + 3x^2 - 1$ **9.** $5w^4 - 4w^3 + 9w^2 - 6w + 2$

11. $\dfrac{25x^2 - 10x^{3/2} + 1}{x^{1/2}}$ **13.** $\dfrac{3x^4 - 10x^3 + 4}{x^2}$

15. $\dfrac{-1}{(x - 2)^2}$ **17.** $\dfrac{3}{(2x + 1)^2}$ **19.** $-\dfrac{2x}{(x^2 + 1)^2}$

21. $\dfrac{s^2 + 2s + 4}{(s + 1)^2}$ **23.** $\dfrac{1 - 3x^2}{2\sqrt{x}(x^2 + 1)^2}$ **25.** $\dfrac{x^2 - 2x - 2}{(x^2 + x + 1)^2}$

27. $\dfrac{2x^3 - 5x^2 - 4x - 3}{(x - 2)^2}$

29. $\dfrac{-2(x^5 + 12x^4 - 8x^3 + 16x + 64)}{(x^4 - 16)^2}$ **31.** 8   **33.** -9

35. $2(3x^2 - x + 3); 10$ **37.** $\dfrac{-3x^4 + 2x^2 - 1}{(x^4 - 2x^2 - 1)^2}; -\dfrac{1}{2}$

39. 60; $y = 60x - 102$ **41.** $-\frac{1}{2}$; $y = -\frac{1}{2}x + \frac{3}{2}$

43. $y = 7x - 5$ **45.** $(\frac{1}{3}, \frac{50}{27})$; $(1, 2)$

47. $(\frac{4}{3}, -\frac{770}{27})$; $(2, -30)$ **49.** $y = -\frac{1}{2}x + 1$; $y = 2x - \frac{3}{2}$

51. 0.125; 0.5; 2; 50

53. -5000/min; -1600/min; 7000; 4000

55. a. $\dfrac{180}{(t + 6)^2}$ **b.** 3.7; 2.2; 1.8; 1.1

c.
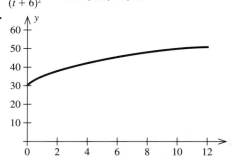

d. 50 words/min

57. Dropping at the rate of 0.0375 ppm/yr; dropping at the rate of 0.006 ppm/yr

59. False   **61.** False

Using Technology Exercises 3.2, page 209

1. 0.8750 **3.** 0.0774 **5.** -0.5000

7. $87,322$/yr

Exercises 3.3, page 220

1. $8(2x - 1)^3$ **3.** $10x(x^2 + 2)^4$

5. $6x^2(1 - x)(2 - x)^2$ **7.** $\dfrac{-4}{(2x + 1)^3}$ **9.** $3x\sqrt{x^2 - 4}$

11. $\dfrac{3}{2\sqrt{3x - 2}}$ **13.** $\dfrac{-2x}{3(1 - x^2)^{2/3}}$ **15.** $-\dfrac{6}{(2x + 3)^4}$

17. $\dfrac{-1}{(2t - 3)^{3/2}}$ **19.** $-\dfrac{3(16x^3 + 1)}{2(4x^4 + x)^{5/2}}$

21. $\dfrac{-4(3x + 1)}{(3x^2 + 2x + 1)^3}$ **23.** $6x(2x^2 - x + 1)$

25. $3(t^{-1} - t^{-2})^2(-t^{-2} + 2t^{-3})$ **27.** $\dfrac{1}{2\sqrt{x - 1}} + \dfrac{1}{2\sqrt{x + 1}}$

29. $-12x(4x - 1)(3 - 4x)^3$

31. $6(x - 1)(2x - 1)(2x + 1)^3$ **33.** $-\dfrac{15(x + 3)^2}{(x - 2)^4}$

35. $\dfrac{3\sqrt{t}}{2(2t + 1)^{5/2}}$ **37.** $\dfrac{-1}{2\sqrt{u + 1}\,(3u + 2)^{3/2}}$

39. $-\dfrac{2x(3x^2 + 1)}{(x^2 - 1)^5}$ **41.** $-\dfrac{2x(3x^2 + 1)^2(3x^2 + 13)}{(x^2 - 1)^5}$

43. $-\dfrac{3x^2 + 2x + 1}{\sqrt{2x + 1}\,(x^2 - 1)^2}$ **45.** $-\dfrac{t^2 + 2t - 1}{2\sqrt{t + 1}\,(t^2 + 1)^{3/2}}$

47. $\frac{4}{3}u^{1/3}$; $6x$; $8x(3x^2 - 1)^{1/3}$

49. $-\dfrac{2}{3u^{5/3}}$; $6x^2 - 1$; $-\dfrac{2(6x^2 - 1)}{3(2x^3 - x + 1)^{5/3}}$

51. $\frac{1}{2}u^{-1/2} - \frac{1}{2}u^{-3/2}$; $3x^2 - 1$; $\dfrac{(3x^2 - 1)(x^3 - x - 1)}{2(x^3 - x)^{3/2}}$

53. -12 **55.** 6 **57.** No

59. $y = -33x + 57$ **61.** $y = \frac{43}{5}x - \frac{54}{5}$

63. 0.333 million/wk; 0.305 million/wk; 16 million; 22.7 million

65. a. $0.027(0.2t^2 + 4t + 64)^{-1/3}(0.1t + 1)$
b. 0.0091 ppm

67. a. $0.21t^2(t - 3)(t - 7)^3$
b. 90.72; 0; -90.72; at 8 A.M., the level of nitrogen dioxide is increasing; at 10 A.M., the level stops increasing; and at 11 A.M., the level is decreasing.

69. $\dfrac{3450t}{(t + 25)^2\sqrt{\frac{1}{2}t^2 + 2t + 25}}$; 2.9 beats/min²; 0.7 beats/min²; 0.2 beats/min²; 179 beats/min

71. 160π ft²/sec   **73.** -27 mph/decade; 19 mph

75. $\dfrac{1.42(140t^2 + 3500t + 21,000)}{(3t^2 + 80t + 550)^2}$; $87,322$ jobs/yr

77. Decreasing at the rate of $\$5.86$/passenger/yr

79. True   **81.** False

Using Technology Exercises 3.3, page 219

1. 0.5774 **3.** 0.9390 **5.** -4.9498

7. $10,146,200$/decade; $7,810,520$/decade

Exercises 3.4, page 236

1. a. $C(x)$ is always increasing because as the number of units x produced increases, the greater the amount of money that must be spent on production.
b. 4000

3. a. $\$114$; $\$120.16$; $\$138.12$ **b.** $\$114$; $\$120$; $\$138$

5. a. $\dfrac{5000}{x} + 2$ **b.** $-\dfrac{5000}{x^2}$

7. $0.0002x^2 - 0.06x + 120 + \dfrac{5000}{x}$; $0.0004x - 0.06 - \dfrac{5000}{x^2}$

9. a. $-0.04x^2 + 800x$ **b.** $-0.08x + 800$
c. 400; when the level of production is 5000 units, the production of the next speaker system will bring in additional revenue of $400.

11. a. $-0.04x^2 + 600x - 300,000$ **b.** $-0.08x + 600$ **c.** 200; -40
d. The profit increases as production increases, peaking at 7500 units; beyond this level, profit falls.

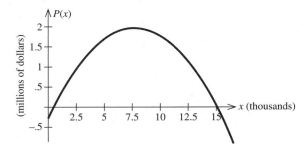

13. a. $-0.006x^2 + 180x$; $-0.000002x^3 + 0.014x^2 + 60x - 60,000$
b. $0.000006x^2 - 0.04x + 120$; $-0.012x + 180$; $-0.000006x^2 + 0.028x + 60$
c. 64; 156; 92   **d.**

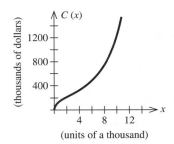

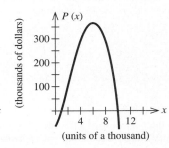

15. a. $0.000004x - 0.02 - \dfrac{60,000}{x^2}$

b. -0.0024; 0.0194; the average cost is decreasing when 5000 TV sets are produced and increasing when 10,000 units are produced.

17. 0.712 billion/billion dollars   **21.** $-\$0.209$ billion/billion dollars

23. $\frac{5}{3}$; elastic   **25.** 1; unitary   **27.** 0.104; inelastic

29. a. Inelastic; elastic   **b.** When $p = 8.66$
c. Increase   **d.** Increase

31. a. Inelastic   **b.** Increase

33. $\dfrac{25}{50 - p}$; for $p > 25$, demand is elastic; for $p = 25$, demand is unitary; and for $p < 25$, demand is inelastic.

35. False

Exercises 3.5, page 245

1. $8x - 2$; 8   **3.** $6x^2 - 6x$; $6(2x - 1)$

5. $4t^3 - 6t^2 + 12t - 3$; $12(t^2 - t + 1)$

7. $10x(x^2 + 2)^4$; $10(x^2 + 2)^3(9x^2 + 2)$

9. $6t(2t^2 - 1)(6t^2 - 1)$; $6(60t^4 - 24t^2 + 1)$

11. $14x(2x^2 + 2)^{5/2}$; $28(2x^2 + 2)^{3/2}(6x^2 + 1)$

13. $(x^2 + 1)(5x^2 + 1)$; $4x(5x^2 + 3)$

15. $\dfrac{1}{(2x + 1)^2}$; $-\dfrac{4}{(2x + 1)^3}$

17. $\dfrac{2}{(s + 1)^2}$; $-\dfrac{4}{(s + 1)^3}$

19. $-\dfrac{3}{2(4-3u)^{1/2}}; \ -\dfrac{9}{4(4-3u)^{3/2}}$

21. $72x - 24$ 23. $-\dfrac{6}{x^4}$

25. $\frac{81}{8}(3s-2)^{-5/2}$ 27. $192(2x-3)$

29. 128 ft/sec; 32 ft/sec^2

31. **a.** and **b.**

t	0	1	2	3	4	5	6	7
$N'(t)$	0	2.7	4.8	6.3	7.2	7.5	7.2	6.3
$N''(t)$				0.6	0	−0.6	−1.2	

33. **a.** $\frac{1}{4}t(t^2-12t+32)$ **b.** 0 ft/sec; 0 ft/sec; 0 ft/sec
   c. $\frac{3}{4}t^2 - 6t + 8$ **d.** 8 ft/sec^2; -4 ft/sec^2; 8 ft/sec^2
   e. 0 ft; 16 ft; 0 ft

35. -3.09; 0.35; 10 min after the test began, the smoke was being removed at the rate of 3%/min and the rate of the rate of change of the amount of smoke remaining was .35%/min^2.

37. True   39. True

41. $f(x) = x^{n+1/2}$

Using Technology Exercises 3.5, page 247

1. -18 3. 15.2762 5. -0.6255 7. 0.1973

9. -68.46214; at the beginning of 1988, the rate of the rate of the rate at which banks were failing was 68 banks/yr/yr/yr.

Exercises 3.6, page 257

1. **a.** $-\frac{1}{2}$ **b.** $-\frac{1}{2}$ 3. **a.** $-\dfrac{1}{x^2}$ **b.** $-\dfrac{y}{x}$

5. **a.** $2x - 1 + \dfrac{4}{x^2}$ **b.** $3x - 2 - \dfrac{y}{x}$

7. **a.** $\dfrac{1-x^2}{(1+x^2)^2}$ **b.** $-2y^2 + \dfrac{y}{x}$

9. $-\dfrac{x}{y}$ 11. $\dfrac{x}{2y}$ 13. $1 - \dfrac{y}{x}$ 15. $-\dfrac{y}{x}$

17. $-\dfrac{\sqrt{y}}{\sqrt{x}}$ 19. $2\sqrt{x+y} - 1$ 21. $-\dfrac{y^3}{x^3}$

23. $\dfrac{2\sqrt{xy}-y}{x-2\sqrt{xy}}$ 25. $\dfrac{6x-3y-1}{3x+1}$ 27. $\dfrac{2(2x-y^{3/2})}{3x\sqrt{y}-4y}$

29. $-\dfrac{2x^2+2xy+y^2}{x^2+2xy+2y^2}$ 31. $y = 2$ 33. $y = -\frac{3}{2}x + \frac{5}{2}$

35. $\dfrac{2y}{x^2}$ 37. $\dfrac{2y(y-x)}{(2y-x)^3}$

39. **a.** $\dfrac{dV}{dt} = \pi r \left(r\dfrac{dh}{dt} + 2h\dfrac{dr}{dt}\right)$ **b.** 3.6π cu in./sec

41. Dropping at the rate of 111 tires/wk

43. Increasing at the rate of 44 ten packs/wk

45. Dropping at the rate of 3.7 cents/carton/wk

47. 0.37; inelastic   49. 160π ft^2/sec   51. 17 ft/sec

53. 7.69 ft/sec

57. 196.8 ft/sec   59. 9 ft/sec   61. 19.2 ft/sec   63. True

Exercises 3.7, page 268

1. $4x\,dx$ 3. $(3x^2 - 1)\,dx$ 5. $\dfrac{dx}{2\sqrt{x+1}}$

7. $\dfrac{6x+1}{2\sqrt{x}}dx$ 9. $\dfrac{x^2-2}{x^2}dx$

11. $\dfrac{-x^2+2x+1}{(x^2+1)^2}dx$ 13. $\dfrac{6x-1}{2\sqrt{3x^2-x}}dx$

15. **a.** $2x\,dx$ **b.** 0.04 **c.** 0.0404

17. **a.** $-\dfrac{dx}{x^2}$ **b.** -0.05 **c.** -0.05263

19. 3.167 21. 7.0358 23. 1.983 25. 0.298

27. 2.50146 29. ± 8.64 cm^3 31. 18.85 ft^3

33. It will drop by 40%.   35. 274 sec   37. $111{,}595$

39. Decrease of $1.33   41. $\pm 64{,}800$

43. Decrease of 11 crimes/yr   45. True

Using Technology Exercises 3.7, page 271

1. 7.5787 3. 0.031220185778 5. -0.0198761598

7. $26.60/month; $35.47/month; $44.34/month

9. 625

Chapter 3 Review Exercises, page 274

1. $15x^4 - 8x^3 + 6x - 2$ 3. $\dfrac{6}{x^4} - \dfrac{3}{x^2}$

5. $-\dfrac{1}{t^{3/2}} - \dfrac{6}{t^{5/2}}$ 7. $1 - \dfrac{2}{t^2} - \dfrac{6}{t^3}$

9. $2x + \dfrac{3}{x^{5/2}}$ 11. $\dfrac{2t}{(2t^2+1)^2}$

13. $\dfrac{1}{\sqrt{x}(\sqrt{x}+1)^2}$ **15.** $\dfrac{2x(x^4-2x^2-1)}{(x^2-1)^2}$

17. $72x^2(3x^3-2)^7$ **19.** $\dfrac{2t}{\sqrt{2t^2+1}}$

21. $\dfrac{-4(3t-1)}{(3t^2-2t+5)^3}$ **23.** $\dfrac{2(x^2+1)(x^2-1)}{x^3}$

25. $4t^2(5t+3)(t^2+t)^3$, or $4t^5(5t+3)(t+1)^3$

27. $\dfrac{1}{2\sqrt{x}}(x^2-1)^2(13x^2-1)$

29. $-\dfrac{12x+25}{2\sqrt{3x+2}\,(4x-3)^2}$

31. $2(12x^2-9x+2)$ **33.** $\dfrac{2t(t^2-12)}{(t^2+4)^3}$

35. $\dfrac{2}{(2x^2+1)^{3/2}}$ **37.** $\dfrac{2x}{y}$

39. $-\dfrac{2x}{y^2-1}$ **41.** $\dfrac{x-2y}{2x+y}$

43. a. $(2,-25)$ and $(-1,14)$
b. $y=-4x-17$; $y=-4x+10$

45. $y=-\dfrac{\sqrt{3}}{3}x+\dfrac{4}{3}\sqrt{3}$

47. $-\dfrac{48}{(2x-1)^4}$; $(-\infty,\frac{1}{2})\cup(\frac{1}{2},\infty)$

49. 200/wk

51. a. $-0.02x^2+600x$
b. $-0.04x+600$
c. 200; the sale of the 10,001st phone will bring in a revenue of $200.

CHAPTER 4

Exercises 4.1, page 292

1. Decreasing on $(-\infty,0)$ and increasing on $(0,\infty)$

3. Increasing on $(-\infty,-1)\cup(1,\infty)$ and decreasing on $(-1,1)$

5. Decreasing on $(-\infty,0)\cup(2,\infty)$ and increasing on $(0,2)$

7. Decreasing on $(-\infty,-1)\cup(1,\infty)$ and increasing on $(-1,1)$

9. Increasing on $(20.2,20.6)\cup(21.7,21.8)$, constant on $(19.6,20.2)\cup(20.6,21.1)$, and decreasing on $(21.1,21.7)\cup(21.8,22.7)$

11. Increasing on $(-\infty,\infty)$

13. Decreasing on $(-\infty,\frac{3}{2})$ and increasing on $(\frac{3}{2},\infty)$

15. Decreasing on $(-\infty,-\sqrt{3}/3)\cup(\sqrt{3}/3,\infty)$ and increasing on $(-\sqrt{3}/3,\sqrt{3}/3)$

17. Increasing on $(-\infty,-2)\cup(0,\infty)$ and decreasing on $(-2,0)$

19. Increasing on $(-\infty,3)\cup(3,\infty)$

21. Decreasing on $(-\infty,0)\cup(0,3)$ and increasing on $(3,\infty)$

23. Decreasing on $(-\infty,2)\cup(2,\infty)$

25. Decreasing on $(-\infty,1)\cup(1,\infty)$

27. Increasing on $(-\infty,0)\cup(0,\infty)$

29. Increasing on $(-1,\infty)$

31. Increasing on $(-4,0)$; decreasing on $(0,4)$

33. Increasing on $(-\infty,0)\cup(0,\infty)$

35. Increasing on $(-\infty,1)$; decreasing on $(1,\infty)$

37. Relative maximum: $f(0)=1$; relative minima: $f(-1)=0$ and $f(1)=0$

39. Relative maximum: $f(-1)=2$; relative minimum: $f(1)=-2$

41. Relative maximum: $f(1)=3$; relative minimum: $f(2)=2$

43. Relative minimum: $f(0)=2$

45. a   **47.** d

49. Relative minimum: $f(2)=-4$

51. Relative minimum: $f(2)=2$

53. Relative minimum: $f(0)=2$

55. Relative maximum: $g(0)=4$; relative minimum: $g(2)=0$

57. Relative minimum: $F(3)=-5$; relative maximum: $F(-1)=\frac{17}{3}$

59. Relative minimum: $g(3)=-19$

61. Relative minimum: $f(\frac{1}{2})=\frac{63}{16}$

63. None

65. Relative maximum: $f(-3)=-4$; relative minimum: $f(3)=8$

67. Relative maximum: $f(1)=\frac{1}{2}$; relative minimum: $f(-1)=-\frac{1}{2}$

69. Relative maximum: $f(0)=0$

71. Relative minimum: $f(1)=0$

73. Increasing on $(0,4000)$ and decreasing on $(4000,\infty)$

75. Decreasing on (0, 5) and increasing on (5, 10); after declining from 1984 through 1989, the index began to increase after 1989.

79. Increasing on (0, 4) and decreasing on (4, 5); the cash in the Central Provident Fund will be increasing from 1995 to 2035 and decreasing from 2035 to 2045.

81. Increasing on (0, 1) and decreasing on (1, 4)

83. Increasing on (0, 4.5) and decreasing on (4.5, 11); the pollution is increasing from 7 A.M. to 11:30 A.M. and decreasing from 11:30 A.M. to 6 P.M.

87. False   **89.** False   **91.** False

93. a. $f'(x) = \begin{cases} -3 & \text{if } x < 0 \\ 2 & \text{if } x > 0 \end{cases}$ **b.** No

Using Technology Exercises 4.1, page 298

1. a. f is decreasing on $(-\infty, -0.2934)$ and increasing on $(-0.2934, \infty)$
b. Relative minimum: $f(-0.2934) = -2.5435$

3. a. f is increasing on $(-\infty, -1.6144) \cup (0.2390, \infty)$ and decreasing on $(-1.6144, 0.2390)$
b. Relative maximum: $f(-1.6144) = 26.7991$; relative minimum: $f(0.2390) = 1.6733$

5. a. f is decreasing on $(-\infty, -1) \cup (0.33, \infty)$ and increasing on $(-1, 0.33)$
b. Relative maximum: $f(0.33) = 1.11$; relative minimum: $f(-1) = -0.63$

7. a. f is decreasing on $(-1, -0.71)$ and increasing on $(-0.71, 1)$
b. Relative minimum: $f(-0.71) = -1.41$

9. a.

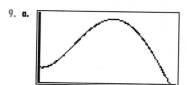

b. f is decreasing on $(0, 0.2398) \cup (6.8758, 12)$ and increasing on $(0.2398, 6.8758)$
c. (6.8758, 200.14)

11. a.

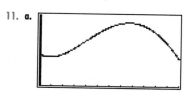

b. f is decreasing on $(0, 0.8343) \cup (7.6726, 12)$ and increasing on $(0.8343, 7.6726)$

13. Increasing from 7 A.M. to 11:30 A.M. and decreasing from 11:30 A.M. to 6 P.M.; highest at 11:30 A.M. when it reaches 164 PSI.

Exercises 4.2, page 313

1. Concave downward on $(-\infty, 0)$ and concave upward on $(0, \infty)$; inflection point: (0, 0)

3. Concave downward on $(-\infty, 0) \cup (0, \infty)$

5. Concave upward on $(-\infty, 0) \cup (1, \infty)$ and concave downward on (0, 1); inflection points: (0, 0) and (1, -1)

7. Concave downward on $(-\infty, -2) \cup (-2, 2) \cup (2, \infty)$

9. a   **11.** b

13. a. $D_1'(t) > 0$, $D_2'(t) > 0$, $D_1''(t) > 0$, and $D_2''(t) < 0$ on (0, 12)
b. With or without the proposed promotional campaign, the deposits will increase; with the promotion, the deposits will increase at an increasing rate; without the promotion, the deposits will increase at a decreasing rate.

15. At the time t_0, corresponding to its t-coordinate, the restoration process is working at its peak.

23. Concave upward on $(-\infty, \infty)$

25. Concave downward on $(-\infty, 0)$; concave upward on $(0, \infty)$

27. Concave upward on $(-\infty, 0) \cup (3, \infty)$; concave downward on (0, 3)

29. Concave downward on $(-\infty, 0) \cup (0, \infty)$

31. Concave downward on $(-\infty, 4)$

33. Concave downward on $(-\infty, 2)$; concave upward on $(2, \infty)$

35. Concave upward on $(-\infty, -\sqrt{6}/3) \cup (\sqrt{6}/3, \infty)$; concave downward on $(-\sqrt{6}/3, \sqrt{6}/3)$

37. Concave downward on $(-\infty, 1)$; concave upward on $(1, \infty)$

39. Concave upward on $(-\infty, 0) \cup (0, \infty)$

41. Concave upward on $(-\infty, 2)$; concave downward on $(2, \infty)$

43. Concave upward on $(-\infty, -1) \cup (1, \infty)$; concave downward on $(-1, 1)$

45. (0, -2)   **47.** (1, -15)   **49.** (0, 1) and $(\frac{2}{3}, \frac{11}{27})$

51. (0, 0)   **53.** (1, 2)   **55.** $(-\sqrt{3}/3, 3/2)$ and $(\sqrt{3}/3, 3/2)$

57. Relative maximum: $f(1) = 5$

59. None

61. Relative maximum: $f(-1) = -\frac{22}{3}$; relative minimum: $f(5) = -\frac{130}{3}$

63. Relative maximum: $f(-3) = -6$; relative minimum: $f(3) = 6$

65. None

67. Relative minimum: $f(-2) = 12$

69. Relative maximum: $g(1) = \frac{1}{2}$; relative minimum: $g(-1) = -\frac{1}{2}$

71. Relative maximum: $f(0) = 0$; relative minimum: $f(\frac{4}{3}) = \frac{256}{27}$

73. Relative minimum: $g(4) = -\frac{1}{108}$

75. a. N is increasing on $(0, 12)$.
b. $N''(t) < 0$ on $(0, 6)$ and $N''(t) > 0$ on $(6, 12)$
c. The rate of growth of the number of help-wanted advertisements was decreasing over the first 6 mo of the year and increasing over the last 6 mo.

77.

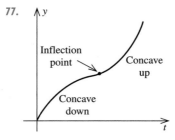

79. $(100, 4600)$; the sales increase rapidly until $100,000 is spent on advertising; after that, any additional expenditure results in increased sales but at a slower rate of increase.

81. 10 A.M.   **83.** $(18, 20{,}310)$; 1452 ft/sec   **85.** True

87. True   **89.** Yes

Using Technology Exercises 4.2, page 319

1. a. f is concave upward on $(-\infty, 0) \cup (1.1667, \infty)$ and concave downward on $(0, 1.1667)$.
b. $(1.1667, 1.1153)$; $(0, 2)$

3. a. f is concave downward on $(-\infty, 0)$ and concave upward on $(0, \infty)$.
b. $(0, 2)$

5. a. f is concave downward on $(-\infty, 0)$ and concave upward on $(0, \infty)$.
b. $(0, 0)$

7. a. f is concave downward on $(-\infty, -2.4495) \cup (0, 2.4495)$ and concave upward on $(-2.4495, 0) \cup (2.4495, \infty)$.
b. $(2.4495, 0.3402)$; $(-2.4495, -0.3402)$

9. a.

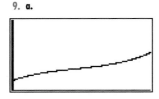

b. $(5.5318, 35.9483)$
c. $t = 5.5318$

11. a.

b. $(3.9024, 77.0919)$

Exercises 4.3, page 331

1. Horizontal asymptote: $y = 0$

3. Horizontal asymptote: $y = 0$; vertical asymptote: $x = 0$

5. Horizontal asymptote: $y = 0$; vertical asymptotes: $x = -1$ and $x = 1$

7. Horizontal asymptote: $y = 3$; vertical asymptote: $x = 0$

9. Horizontal asymptotes: $y = 1$ and $y = -1$

11. Horizontal asymptote: $y = 0$; vertical asymptote: $x = 0$

13. Horizontal asymptote: $y = 0$; vertical asymptote: $x = 0$

15. Horizontal asymptote: $y = 1$; vertical asymptote: $x = -1$

17. None

19. Horizontal asymptote: $y = 1$: vertical asymptotes: $t = -3$ and $t = 3$

21. Horizontal asymptote: $y = 0$; vertical asymptotes: $x = -2$ and $x = 3$

23. Horizontal asymptote: $y = 2$; vertical asymptote: $t = 2$

25. Horizontal asymptote: $y = 1$; vertical asymptotes: $x = -2$ and $x = 2$

27. None

29. f is the derivative function of the function g.

31.

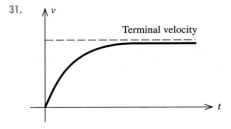

33.

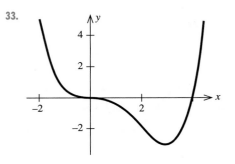

35.

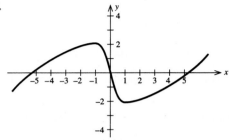

37.

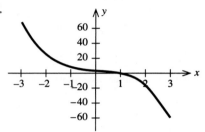

39.

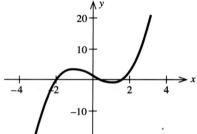

41.

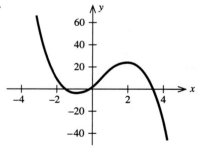

43.

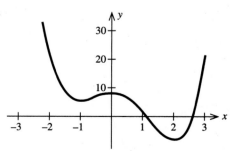

45.

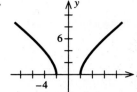

47.

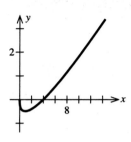

49.

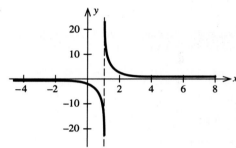

51.

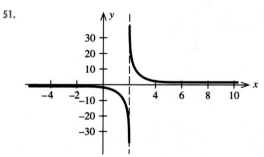

53.

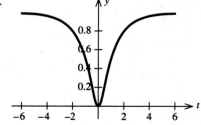

55.

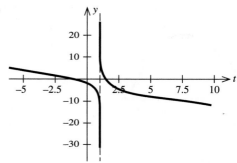

57.

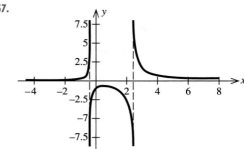

59.

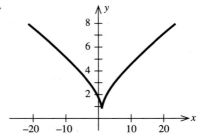

61. a. $y = 2.2$ **b.** \$2.20/disc

63. a. a
  b. The initial speed of the reaction approaches a moles/liter/
  sec as the amount of substrate becomes arbitrarily large.

65.

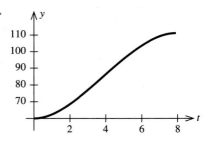

67.

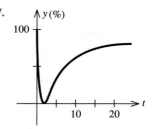

69.

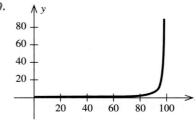

Using Technology Exercises 4.3, page 337

1.

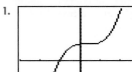

3.

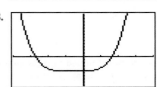

5. -0.9733; 2.3165, 4.6569 **7.** -1.1301; 2.9267

9. 1.5142

Exercises 4.4, page 350

1. None

3. Absolute minimum value: 0

5. Absolute maximum value: 3; absolute minimum value: -2

7. Absolute maximum value: 3; absolute minimum value: $-\frac{27}{16}$

9. Absolute minimum value: $-\frac{41}{8}$

11. No absolute extrema

13. Absolute maximum value: 1

15. Absolute maximum value: 5; absolute minimum value: -4

17. Absolute maximum value: 10; absolute minimum value: 1

19. Absolute maximum value: 19; absolute minimum value: -1

21. Absolute maximum value: 16; absolute minimum value: -1

23. Absolute maximum value: 3; absolute minimum value: $\frac{5}{3}$

25. Absolute maximum value: $\frac{37}{3}$; absolute minimum value: 5

27. Absolute maximum value ≈ 1.04; absolute minimum value: -1.5

29. No absolute extrema

31. Absolute maximum value: 1; absolute minimum value: 0

33. Absolute maximum value: 0; absolute minimum value: -3

35. Absolute maximum value: $\sqrt{2}/4 \approx 0.35$; absolute minimum value: $-\frac{1}{3}$

37. Absolute maximum value: $\sqrt{2}/2$; absolute minimum value: $-\sqrt{2}/2$

39. 144 ft   41. 3000

43. $f(6) = 3.60$, $f(0.5) = 1.13$; the number of nonfarm, full-time, self-employed women over the time interval from 1963 to 1993 reached its highest level, 3.6 million, in 1993.

45. 5000   47. 3333

49. **a.** $0.0025x + 80 + \dfrac{10,000}{x}$ **b.** 2000
   c. 2000   **d.** Same

51. 533

53. **a.** 2 days after the organic waste was dumped into the pond
   b. 3.5 days after the organic waste was dumped into the pond

63. False   65. False

69. **c.**

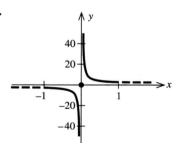

Using Technology Exercises 4.4, page 355

1. Absolute maximum value: 145.8985; absolute minimum value: -4.3834

3. Absolute maximum value: 16; absolute minimum value: -0.1257

5. Absolute maximum value: 2.8889; absolute minimum value: 0

7. **a.**

   b. 200.1410 banks/yr

9. **a.**

   b. Absolute maximum value: 108.8756; absolute minimum value: 49.7773

Exercises 4.5, page 365

1. 750 yd $\times$ 1500 yd; 1,125,000 yd^2

3. $10\sqrt{2}$ ft $\times$ $40\sqrt{2}$ ft

5. $\frac{16}{3}$ in. $\times$ $\frac{16}{3}$ in. $\times$ $\frac{4}{3}$ in.

7. 5.04 in. $\times$ 5.04 in. $\times$ 5.04 in.

9. 18 in. $\times$ 18 in. $\times$ 36 in.; 11,664 in.^{3}

11. $r = \dfrac{36}{\pi}$ in.; $l = 36$ in.; $\dfrac{46,656}{\pi}$ in.^{3}

13. $\frac{2}{3}\sqrt[3]{9}$ ft $\times$ $\sqrt[3]{9}$ ft $\times$ $\frac{2}{3}\sqrt[3]{9}$ ft   15. 250; $62,500; $250

17. $w \approx 13.86$ in.; $h \approx 19.60$ in.

19. $x = 750\sqrt{2}$ ft   21. 56.57 mph   23. 45; 44,445

Chapter 4 Review Exercises, page 369

1. **a.** f is increasing on $(-\infty, 1) \cup (1, \infty)$.
   b. No relative extrema
   c. Concave down on $(-\infty, 1)$; concave up on $(1, \infty)$
   d. $(1, -\frac{17}{3})$

3. **a.** f is increasing on $(-1, 0) \cup (1, \infty)$; decreasing on $(-\infty, -1) \cup (0, 1)$
   b. Relative maximum: 0; relative minimum: -1
   c. Concave up on $(-\infty, -\sqrt{3}/3) \cup (\sqrt{3}/3, \infty)$; concave down on $(-\sqrt{3}/3, \sqrt{3}/3)$
   d. $(-\sqrt{3}/3, -5/9)$; $(\sqrt{3}/3, -5/9)$

5. **a.** f is increasing on $(-\infty, 0) \cup (2, \infty)$ and decreasing on $(0, 1) \cup (1, 2)$.
   b. Relative maximum: 0; relative minimum: 4
   c. Concave up on $(1, \infty)$; concave down on $(-\infty, 1)$
   d. None

7. a. f is decreasing on $(-\infty, 1) \cup (1, \infty)$.
  b. No relative extrema
  c. Concave down on $(-\infty, 1)$; concave up on $(1, \infty)$
  d. $(1, 0)$

9. a. f is increasing on $(-\infty, -1) \cup (-1, \infty)$.
  b. No relative extrema
  c. Concave down on $(-1, \infty)$; concave up on $(-\infty, -1)$
  d. No inflection point

11.

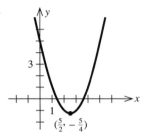

$\left(\frac{5}{2}, -\frac{5}{4}\right)$

13.

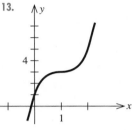

15.

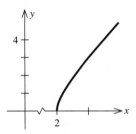

17.

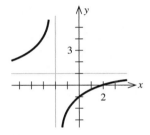

19. Vertical asymptote: $x = -\frac{3}{2}$; horizontal asymptote: $y = 0$

21. Vertical asymptotes: $x = -2$ and $x = 4$; horizontal asymptote: $y = 0$

23. Absolute minimum value: $-\frac{25}{8}$

25. Absolute maximum value: 5; absolute minimum value: 0

27. Absolute maximum value: -16; absolute minimum value: -32

29. Absolute maximum value: $\frac{8}{3}$; absolute minimum value: 0

31. Absolute maximum value: $\frac{1}{2}$; absolute minimum value: $-\frac{1}{2}$

33. $4000   **35.** 168   **37.** 10 A.M.

39. 20,000 cases

CHAPTER 5

Exercises 5.1, page 379

1. a. 16   **b.** 27   **3. a.** 3   **b.** $\sqrt{5}$

5. a. -3 **b.** 8   **7. a.** 25   **b.** $4^{1.8}$

9. a. $4x^3$ **b.** $5xy^2\sqrt{x}$ **11. a.** $\dfrac{2}{a^2}$ **b.** $\frac{1}{3}b^2$

13. a. $8x^9y^6$ **b.** $16x^4y^4z^6$

15. a. $\dfrac{64x^6}{y^4}$ **b.** $(x - y)(x + y)$

17. 2   **19.** 3   **21.** 3   **23.** $\frac{5}{4}$ **25.** 1 or 2

27.

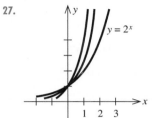

$y = 2^x$

29.

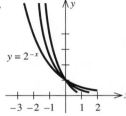

$y = 2^{-x}$

31.

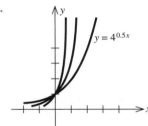

$y = 4^{0.5x}$

33.

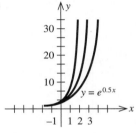

$y = e^{0.5x}$

35.

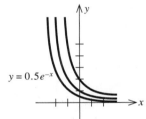

$y = 0.5e^{-x}$

37. False   **39.** True

Using Technology Exercises 5.1, page 381

1.

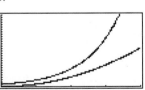

3.

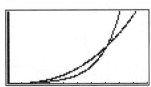

5.

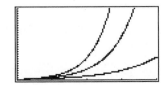

7.

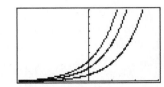

9.

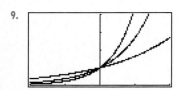

Exercises 5.2, page 389

1. $\log_2 64 = 6$ **3.** $\log_3 \frac{1}{9} = -2$ **5.** $\log_{1/3} \frac{1}{3} = 1$

7. $\log_{32} 8 = \frac{3}{5}$ **9.** $\log_{10} 0.001 = -3$ **11.** 1.0792

13. 1.2042   **15.** 1.6813   **17.** $\log x + 4 \log(x + 1)$

19. $\frac{1}{2} \log(x + 1) - \log(x^2 + 1)$ **21.** $\ln x - x^2$

23. $-\frac{3}{2} \ln x - \frac{1}{2} \ln(1 + x^2)$ **25.** $x \ln x$

27.

29.

31.

33. 5.1986   **35.** −0.0912   **37.** −8.0472

39. −4.9041   **41.** $-2 \ln\left(\dfrac{A}{B}\right)$ **43.** 105.7 mm

45. a. $10^3 I_0$
 b. 100,000 times greater
 c. 10,000,000 times greater

47. $34\frac{1}{2}$ hr earlier, at 1:30 P.M.

49. False   **51.** True

Exercises 5.3, page 401

1. $4974.47   **3.** $223,403.11

5. a. 10.25%/yr   **b.** 9.31%/yr

7. a. $29,227.61   **b.** $29,137.83   **9.** $6885.64

11. $112,926.52   **13.** $3.795 million   **15.** $23,329.48

17. 9.58 yr   **19.** 12.75%   **21.** $40,000

23. a. $33,885.14   **b.** $33,565.38

25. a. $16,262.79   **b.** $12,047.77   **c.** $6611.96

29. Bank B   **31.** 9.531%

Exercises 5.4, page 411

1. $3e^{3x}$ **3.** $-e^{-t}$ **5.** $e^x + 1$ **7.** $x^2 e^x(x + 3)$

9. $\dfrac{2e^x(x - 1)}{x^2}$ **11.** $3(e^x - e^{-x})$ **13.** $-\dfrac{1}{e^w}$

15. $6e^{3x-1}$ **17.** $-2xe^{-x^2}$ **19.** $\dfrac{3e^{-1/x}}{x^2}$

21. $25e^x(e^x + 1)^{24}$ **23.** $\dfrac{e^{\sqrt{x}}}{2\sqrt{x}}$ **25.** $e^{3x+2}(3x - 2)$

27. $\dfrac{2e^x}{(e^x + 1)^2}$ **29.** $2(8e^{-4x} + 9e^{3x})$ **31.** $6e^{3x}(3x + 2)$

33. $y = 2x - 2$

35. f is increasing on $(-\infty, 0)$ and decreasing on $(0, \infty)$.

37. Concave downward on $(-\infty, 0)$; concave upward on $(0, \infty)$

39. $(1, e^{-2})$

41. Absolute maximum value: 1; absolute minimum value: e^{-1}

43. Absolute minimum value: −1; absolute maximum value: $2e^{-3/2}$

45.   **47.**

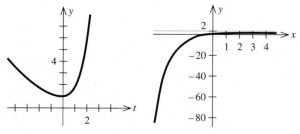

49. −$6065/day; −$3679/day; −$2231/day; −$1353/day

51. b. 4505/yr; 273 cases/yr

53. a. A decrease of 1.63 cents/bottle; a decrease of 1.34 cents/bottle
b. $231.87; $217.03

55. a. 30   **b.** $N'(x) = \dfrac{297{,}000e^{-x}}{(1 + 99e^{-x})^2}$

c.

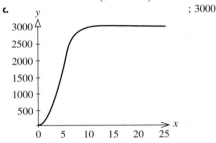

; 3000

57. 7.72 yr; $160,207.69

59. −0.1694, −0.1549, −0.1415; the percentage of the total population relocating was decreasing at the rate of 0.17%/yr in 1970, 0.15%/yr in 1980, and 0.14%/yr in 1990.

61. a. $12/unit
b. Decreasing at the rate of $7/wk
c. $8/unit

65. False   **67.** True

Using Technology Exercises 5.4, page 414

1. 5.4366   **3.** 12.3929   **5.** 0.1861

7. a. 50   **c.**

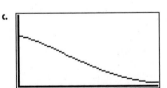

9. a.

b. 4.2720 billion/half century

11. a. 153.024; 235.181
b. −0.634; 18.401

13. a. 69.63%   **b.** 5.094%/decade

Exercises 5.5, page 423

1. $\dfrac{5}{x}$ **3.** $\dfrac{1}{x + 1}$ **5.** $\dfrac{8}{x}$ **7.** $\dfrac{1}{2x}$ **9.** $\dfrac{-2}{x}$

11. $\dfrac{2(4x - 3)}{4x^2 - 6x + 3}$ **13.** $\dfrac{1}{x(x + 1)}$ **15.** $x(1 + 2\ln x)$

17. $\dfrac{2(1 - \ln x)}{x^2}$ **19.** $\dfrac{3}{u - 2}$ **21.** $\dfrac{1}{2x\sqrt{\ln x}}$

23. $\dfrac{3(\ln x)^2}{x}$ **25.** $\dfrac{3x^2}{x^3 + 1}$ **27.** $\dfrac{(x\ln x + 1)e^x}{x}$

29. $\dfrac{e^{2t}[2(t + 1)\ln(t + 1) + 1]}{t + 1}$ **31.** $\dfrac{1 - \ln x}{x^2}$ **33.** $-\dfrac{1}{x^2}$

35. $\dfrac{2(2 - x^2)}{(x^2 + 2)^2}$ **37.** $(x + 1)(5x + 7)(x + 2)^2$

39. $(x - 1)(x + 1)^2(x + 3)^3(9x^2 + 14x - 7)$

41. $\dfrac{(2x^2 - 1)^4(38x^2 + 40x + 1)}{2(x + 1)^{3/2}}$ **43.** $3^x \ln 3$

45. $(x^2 + 1)^{x-1}[2x^2 + (x^2 + 1)\ln(x^2 + 1)]$

47. $y = x - 1$

49. f is decreasing on $(-\infty, 0)$ and increasing on $(0, \infty)$.

51. Concave up: $(-\infty, -1) \cup (1, \infty)$; concave down: $(-1, 0) \cup (0, 1)$

53. $(-1, \ln 2)$ and $(1, \ln 2)$

55. Absolute minimum value: 1; absolute maximum value: $3 - \ln 3$

57.

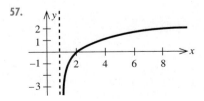

59. False

Exercises 5.6, page 433

1. **a.** 0.05
   b. 400
   c.

t	0	10	20	100	1000
Q	400	659	1087	59,365	2.07×10^{24}

3. **a.** $Q(t) = 100e^{0.035t}$
   b. 266 min
   c. $Q(t) = 1000e^{0.035t}$

5. **a.** 55.5 yr   **b.** 14.25 billion   7. 8.7 lb/in.²; 0.0004 lb/in.²/ft

9. $Q(t) = 100e^{-0.049t}$; 70.6 g; 3.451 g/day   11. 13,412 yr ago

13.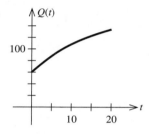

   a. 60 words/min
   b. 107 words/min
   c. 136 words/min

15.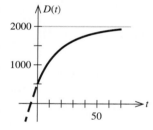

   a. 573 computers; 1177 computers; 1548 computers; 1925 computers
   b. 2000 computers
   c. 46 computers/mo

17. **a.** 11   **b.** 937   **c.** 1000   19. 3%; 65%

21. 1080; 280

23. **a.** r/k
   b.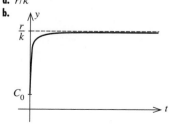

Chapter 5 Review Exercises, page 437

1. **a.–b.**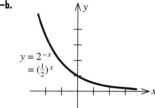

3. $\log_{16} 0.125 = -\frac{3}{4}$ 5. 2   7. $x + 2y - z$

9.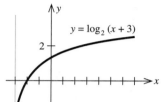

11. $(2x + 1)e^{2x}$ 13. $\dfrac{1 - 4t}{2\sqrt{t}e^{2t}}$ 15. $\dfrac{2(e^{2x} + 2)}{(1 + e^{-2x})^2}$

17. $(1 - 2x^2)e^{-x^2}$ 19. $(x + 1)^2e^x$ 21. $\dfrac{2xe^{x^2}}{e^{x^2} + 1}$

23. $\dfrac{x + 1 - x \ln x}{x(x + 1)^2}$ 25. $\dfrac{4e^{4x}}{e^{4x} + 3}$ 27. $\dfrac{1 + e^x(1 - x \ln x)}{x(1 + e^x)^2}$

29. $-\dfrac{9}{(3x + 1)^2}$ 31. 0

33. $6x(x^2 + 2)^2(3x^3 + 2x + 1)$ 35. $y = -(2x - 3)e^{-2}$

37.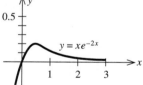

39. Absolute maximum value: $1/e$ 41. 12%/yr

43. **a.** $Q(t) = 2000e^{0.01831t}$ **b.** 161,992

45.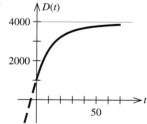

   a. 1175; 2540; 3289   **b.** 4000

CHAPTER 6

Exercises 6.1, page 450

5. b. $y = 2x + C$

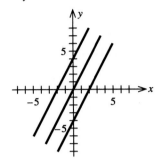

7. b. $y = \frac{1}{3}x^3 + C$

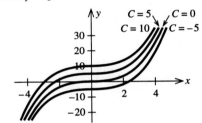

9. $6x + C$ **11.** $\frac{1}{4}x^4 + C$ **13.** $-\dfrac{1}{3x^3} + C$

15. $\frac{3}{5}x^{5/3} + C$ **17.** $-\dfrac{4}{x^{1/4}} + C$ **19.** $-\dfrac{2}{x} + C$

21. $\frac{2}{3}\pi t^{3/2} + C$ **23.** $3x - x^2 + C$

25. $\dfrac{1}{3}x^3 + \dfrac{1}{2}x^2 - \dfrac{1}{2x^2} + C$ **27.** $4e^x + C$

29. $x + \frac{1}{2}x^2 + e^x + C$ **31.** $x^4 + \dfrac{2}{x} - x + C$

33. $\frac{2}{7}x^{7/2} + \frac{4}{5}x^{5/2} - \frac{1}{2}x^2 + C$ **35.** $\frac{2}{3}x^{3/2} + 6\sqrt{x} + C$

37. $\frac{1}{9}u^3 + \frac{1}{3}u^2 - \frac{1}{3}u + C$ **39.** $\frac{2}{3}t^3 - \frac{3}{2}t^2 - 2t + C$

41. $\dfrac{1}{3}x^3 - 2x - \dfrac{1}{x} + C$ **43.** $\frac{1}{3}s^3 + s^2 + s + C$

45. $e^t + \dfrac{t^{e+1}}{e+1} + C$ **47.** $\dfrac{1}{2}x^2 + x - \ln|x| - \dfrac{1}{x} + C$

49. $\ln|x| + \dfrac{4}{\sqrt{x}} - \dfrac{1}{x} + C$ **51.** $x^2 + x + 1$

53. $x^3 + 2x^2 - x - 5$ **55.** $x - \dfrac{1}{x} + 2$ **57.** $x + \ln|x|$

59. $\sqrt{x}$ **61.** $e^x + \frac{1}{2}x^2 + 2$ **63.** $s(t) = \frac{4}{3}t^{3/2}$

65. \$3370   **67.** 5000 units; \$34,000

69. a. $-t^3 + 6t^2 + 45t$ **b.** 212

71. 21,960   **73.** 120,000

75. 1.9424 m²   **77.** $\frac{1}{2}k(R^2 - r^2)$

79. $9\frac{7}{9}$ ft/sec²; 396 ft   **81.** 36 ft/sec²   **83.** True   **85.** True

Exercises 6.2, page 463

1. $\frac{1}{6}(4x + 3)^5 + C$ **3.** $\frac{1}{3}(x^3 - 2x)^3 + C$

5. $-\dfrac{1}{2(2x^2 + 3)^2} + C$ **7.** $\frac{2}{3}(t^3 + 2)^{3/2} + C$

9. $\frac{1}{20}(x^2 - 1)^{10} + C$ **11.** $-\frac{1}{5}\ln|1 - x^5| + C$

13. $\ln(x - 2)^2 + C$ **15.** $\frac{1}{2}\ln(0.3x^2 - 0.4x + 2) + C$

17. $\frac{1}{6}\ln|3x^2 - 1| + C$ **19.** $-\frac{1}{2}e^{-2x} + C$

21. $-e^{2-x} + C$ **23.** $-\frac{1}{2}e^{-x^2} + C$ **25.** $e^x + e^{-x} + C$

27. $\ln(1 + e^x) + C$ **29.** $2e^{\sqrt{x}} + C$

31. $-\dfrac{1}{6(e^{3x} + x^3)^2} + C$ **33.** $\frac{1}{8}(e^{2x} + 1)^4 + C$

35. $\frac{1}{2}(\ln 5x)^2 + C$ **37.** $\ln|\ln x| + C$

39. $\frac{2}{3}(\ln x)^{3/2} + C$ **41.** $\frac{1}{2}e^{x^2} - \frac{1}{2}\ln(x^2 + 2) + C$

43. $\frac{2}{3}(\sqrt{x} - 1)^3 + 3(\sqrt{x} - 1)^2 + 8(\sqrt{x} - 1)$
     $+ 4\ln|\sqrt{x} - 1| + C$

45. $\dfrac{(6x + 1)(x - 1)^6}{42} + C$

47. $5 + 4\sqrt{x} - x - 4\ln(1 + \sqrt{x}) + C$

49. $-\frac{1}{252}(1 - v)^7(28v^2 + 7v + 1) + C$

51. $\frac{1}{2}[(2x - 1)^5 + 5]$ **53.** $e^{-x^2+1} - 1$

55. $21{,}000 - \dfrac{20{,}000}{\sqrt{1 + 0.2t}}$; 6858   **57.** $p(x) = \dfrac{250}{\sqrt{16 + x^2}}$

59. $30(\sqrt{2t + 4} - 2)$; 14,400 π ft²

61. $\dfrac{71.86887}{1 + 2.449e^{-0.3277t}} - 1.43760$; 59.6 in.

63. 24,555 pairs

Exercises 6.3, page 475

1. 4.27 sq units

3. a. 6 sq units

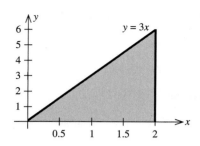

b. 4.5 sq units   **c.** 5.25 sq units   **d.** Yes

5. a. 4 sq units

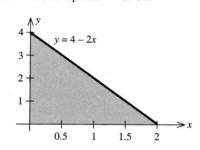

b. 4.8 sq units   **c.** 4.4 sq units   **d.** Yes

7. a. 18.5 sq units   **b.** 18.64 sq units
c. 18.66 sq units   **d.** $\approx$18.7 sq units

9. a. 25 sq units   **b.** 21.12 sq units
c. 19.88 sq units   **d.** $\approx$19.9 sq units

11. a. 0.0625 sq unit   **b.** 0.16 sq unit
c. 0.2025 sq unit   **d.** $\approx$0.2 sq unit

13. 4.64 sq units   **15.** 0.95 sq unit   **17.** 9400 sq ft

Exercises 6.4, page 486

1. 6 sq units   **3.** 8 sq units   **5.** 12 sq units

7. 9 sq units   **9.** ln 2 sq units   **11.** $17\frac{1}{3}$ sq units

13. $18\frac{1}{4}$ sq units   **15.** $(e^2 - 1)$ sq units   **17.** 6

19. 14   **21.** $18\frac{2}{3}$ **23.** $\frac{4}{3}$ **25.** 45   **27.** $\frac{7}{12}$

29. ln 2   **31.** 56   **33.** $\frac{256}{15}$ **35.** $\frac{2}{3}$ **37.** $2\frac{2}{3}$

39. $19\frac{1}{2}$ **41. a.** $4100   **b.** $900

43. a. $2800   **b.** $219.20   **45.** $10,133\frac{1}{3}$ ft

47. 37.7 million   **49.** False   **51.** False

Using Technology Exercises 6.4, page 489

1. 6.1787   **3.** 0.7873   **5.** -0.5888 **7.** 2.7044

9. 3.9973   **11.** 37.7 million   **13.** 333,209   **15.** 903,213

Exercises 6.5, page 498

1. 10   **3.** $\frac{19}{15}$ **5.** $32\frac{4}{15}$ **7.** $\sqrt{3} - 1$ **9.** $24\frac{1}{5}$

11. $\frac{32}{15}$ **13.** $18\frac{2}{15}$ **15.** $\frac{1}{2}(e^4 - 1)$ **17.** $\frac{1}{2}e^2 + \frac{5}{6}$

19. 0   **21.** 2 ln 4   **23.** $\frac{1}{3}(\ln 19 - \ln 3)$

25. $2e^4 - 2e^2 - \ln 2$ **27.** $\frac{1}{2}(e^{-4} - e^{-8} - 1)$ **29.** 5

31. $\frac{17}{3}$ **33.** -1 **35.** $\frac{13}{6}$ **37.** $\frac{1}{4}(e^4 - 1)$

39. 120.3 billion metric tons   **41.** $\approx$$2.24 million

43. $40,339.50   **45.** 26°F   **47.** 16,863

49. 0.071 mg/cm³   **51.** $\dfrac{2k}{3} R^2$ cm/sec

61. -4 **63. a.** -1 **b.** 5   **c.** -13

65. True   **67.** False   **69.** True

Using Technology Exercises 6.5, page 501

1. 7.71667   **3.** 17.56487   **5.** 10,140

Exercises 6.6, page 511

1. 108 sq units   **3.** $\frac{2}{3}$ sq unit   **5.** $2\frac{2}{3}$ sq units

7. $1\frac{1}{2}$ sq units   **9.** 3   **11.** $3\frac{1}{3}$ **13.** 27

15. $2(e^2 - e^{-1})$ **17.** $12\frac{2}{3}$ **19.** $3\frac{1}{3}$ **21.** $4\frac{3}{4}$

23. $12 - \ln 4$ **25.** $e^2 - e - \ln 2$ **27.** $2\frac{1}{2}$ **29.** $7\frac{1}{3}$

31. $\frac{3}{2}$ **33.** $e^3 - 4 + \dfrac{1}{e}$ **35.** $20\frac{5}{6}$ **37.** $\frac{1}{12}$ **39.** $\frac{1}{3}$

41. S is the additional revenue that Odyssey Travel could realize by switching to the new agency;

$$S = \int_0^b [g(x) - f(x)]\, dx$$

43. S is the additional amount of smoke that brand B will remove over brand A in the time interval $[a, b]$;

$$S = \int_a^b [f(t) - g(t)]\, dt$$

45. $\displaystyle\int_{T_1}^{T} [g(t) - f(t)]\, dt - \int_0^{T_1} [f(t) - g(t)]\, dt$

47. 42.79 million metric tons   **49.** 57,179 people

51. False

Using Technology Exercises 6.6, page 517

1. a.

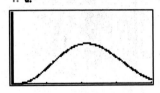

b. 1074.2857

3. a.

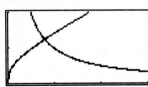

b. 0.9961

5. a.

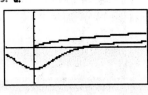

b. 5.4603

7. a.

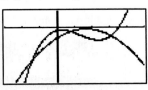

b. 25.8549

9. a.

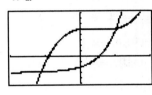

b. 10.5144

11. a.

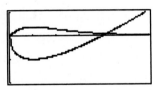

b. 3.5799

Exercises 6.7, page 531

1. $11,667   **3.** $6667   **5.** $11,667

7. Consumers' surplus: $13,333; producers' surplus: $11,667

9. $824,200   **11.** $148,239   **13.** $52,203

15. $76,615   **17.** $111,869   **19.** $20,964

21. a.

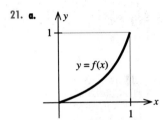

b. 0.175; 0.816

23. a.

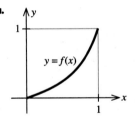

b. 0.104; 0.504

Using Technology Exercises 6.7, page 533

1. Consumers' surplus: $18,000,000; producers' surplus: $11,700,000

3. Consumers' surplus: $33,120; producers' surplus: $2,880

Exercises 6.8, page 541

1. 3π cu units

3. $\dfrac{15\pi}{2}$ cu units

5. $\dfrac{4\pi}{3}$ cu units

7. $\dfrac{16\pi}{15}$ cu units

9. $\dfrac{\pi}{2}(e^2 - 1)$ cu units

11. $\dfrac{2\pi}{15}$ cu units

13. $\dfrac{136\pi}{15}$ cu units

15. $\dfrac{64\sqrt{2}\pi}{3}$ cu units

17. $\dfrac{\pi}{2}(e^2 - 2 + e^{-2})$ cu units

19. $\dfrac{\pi}{6}$ cu units

21. $\dfrac{6517\pi}{240}$ cu units

23. $\dfrac{64\sqrt{2}\pi}{3}$ cu units

25. $\dfrac{8(\sqrt{2} - 1)\pi}{3}$ cu units

27. $\dfrac{4\pi}{3}\, r^3$ cu units

29. $50,000\pi$ cu ft

Chapter 6 Review Exercises, page 536

1. $\frac{1}{4}x^4 + \frac{2}{3}x^3 - \frac{1}{2}x^2 + C$ **3.** $\frac{1}{5}x^5 - \frac{1}{2}x^4 - \frac{1}{x} + C$

5. $\frac{1}{2}x^4 + \frac{2}{5}x^{5/2} + C$ **7.** $\frac{1}{3}x^3 - \frac{1}{2}x^2 + 2\ln|x| + 5x + C$

9. $\frac{3}{8}(3x^2 - 2x + 1)^{4/3} + C$

11. $\frac{1}{2}\ln(x^2 - 2x + 5) + C$ **13.** $\frac{1}{2}e^{x^2+x+1} + C$

15. $\frac{1}{6}(\ln x)^6 + C$ **17.** $\dfrac{(11x^2 - 1)(x^2 + 1)^{11}}{264} + C$

19. $\frac{2}{3}(x + 4)\sqrt{x - 2} + C$ **21.** $\frac{1}{2}$ **23.** $5\frac{2}{3}$

25. -80 **27.** $\frac{1}{2}\ln 5$ **29.** 4   **31.** $\dfrac{e - 1}{2(1 + e)}$

33. $f(x) = x^3 - 2x^2 + x + 1$ **35.** $f(x) = x + e^{-x} + 1$

37. -4.28

39. a. $-0.015x^2 + 60x$ **b.** $p = -0.015x + 60$

41. $3000t - 50,000(1 - e^{-0.04t})$; 16,939

43. \$3100   **45.** 15 sq units   **47.** $\frac{2}{3}$ sq unit

49. $e^2 - 3$ sq units   **51.** $\frac{1}{2}$ sq unit   **53.** $\frac{1}{3}$ sq unit

55. \$2083; \$3333   **57.** \$98,973

59. a.

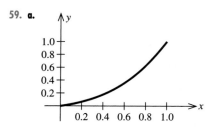

b. 0.1017; 0.3733   **c.** 0.315

61. $\frac{2\pi}{3}$ cu units

CHAPTER 7

Exercises 7.1, page 555

1. $\frac{1}{4}e^{2x}(2x - 1) + C$ **3.** $4(x - 4)e^{x/4} + C$

5. $\frac{1}{2}e^{2x} - 2(x - 1)e^x + \frac{1}{3}x^3 + C$ **7.** $xe^x + C$

9. $\frac{2(x + 2)}{\sqrt{x + 1}} + C$ **11.** $\frac{2}{3}x(x - 5)^{3/2} - \frac{4}{15}(x - 5)^{5/2} + C$

13. $\frac{x^2}{4}(2 \ln 2x - 1) + C$ **15.** $\frac{x^4}{16}(4 \ln x - 1) + C$

17. $\frac{2}{9}x^{3/2}(3 \ln \sqrt{x} - 1) + C$ **19.** $-\frac{1}{x}(\ln x + 1) + C$

21. $x(\ln x - 1) + C$ **23.** $-(x^2 + 2x + 2)e^{-x} + C$

25. $\frac{1}{4}x^2[2(\ln x)^2 - 2 \ln x + 1] + C$ **27.** $2 \ln 2 - 1$

29. $4 \ln 4 - 3$ **31.** $\frac{1}{4}(3e^4 + 1)$

33. $-\frac{1}{2}xe^{-2x} - \frac{1}{4}e^{-2x} + \frac{13}{4}$ **35.** $5 \ln 5 - 4$

37. 1485 ft   **39.** 2.04 mg/mL

41. $-20e^{-0.1t}(t + 10) + 200$ **43.** \$521,087   **45.** True

Exercises 7.2, page 563

1. $\frac{2}{9}[2 + 3x - 2 \ln|2 + 3x|] + C$

3. $\frac{3}{32}[(1 + 2x)^2 - 4(1 + 2x) + 2 \ln|1 + 2x|] + C$

5. $2\left[\frac{x}{8}\left(\frac{9}{4} + 2x^2\right)\sqrt{\frac{9}{4} + x^2} - \frac{81}{128} \ln\left(x + \sqrt{\frac{9}{4} + x^2}\right)\right] + C$

7. $\ln\left(\frac{\sqrt{1 + 4x} - 1}{\sqrt{1 + 4x} + 1}\right) + C$ **9.** $\frac{1}{2}\ln 3$

11. $\frac{x}{9\sqrt{9 - x^2}} + C$

13. $\frac{x}{8}(2x^2 - 4)\sqrt{x^2 - 4} - 2 \ln|x + \sqrt{x^2 - 4}| + C$

15. $\sqrt{4 - x^2} - 2 \ln\left|\frac{2 + \sqrt{4 - x^2}}{x}\right| + C$

17. $\frac{1}{4}(2x - 1)e^{2x} + C$ **19.** $\ln|\ln(1 + x)| + C$

21. $\frac{1}{9}\left[\frac{1}{1 + 3e^x} + \ln(1 + 3e^x)\right] + C$

23. $6[e^{(1/2)x} - \ln(1 + e^{(1/2)x})] + C$

25. $\frac{1}{9}(2 + 3 \ln x - 2 \ln|2 + 3 \ln x|) + C$

27. $e - 2$ **29.** $\frac{x^3}{9}(3 \ln x - 1) + C$

31. $x[(\ln x)^3 - 3(\ln x)^2 + 6 \ln x - 6] + C$

33. $\approx$\$2329   **35.** 27,136   **37.** 44; 49   **39.** 26,157

41. \$418,444

Exercises 7.3, page 578

1. 2.7037; 2.6667; $2\frac{2}{3}$ **3.** 0.2656; 0.2500; $\frac{1}{4}$

5. 0.6970; 0.6933; $\approx$0.6931   **7.** 0.5090; 0.5004; $\frac{1}{2}$

9. 5.2650; 5.3046; $\frac{16}{3}$ **11.** 0.6336; 0.6321; $\approx$0.6321

13. 0.3837; 0.3863; $\approx$0.3863   **15.** 1.1170; 1.1114

17. 1.3973; 1.4052   **19.** 0.8806; 0.8818

21. 3.7757; 3.7625   **23. a.** 3.6   **b.** 0.0324

25. a. 0.013   **b.** 0.00043

27. a. 0.0078125   **b.** 0.0002848

29. 52.84 mi   **31.** 21.65 mpg

33. a. \$142,374   **b.** \$142,698

35. 103.9 PSI   **37.** $\approx$30%   **39.** $\approx$6.42 L/min

41. False   **43.** True

Exercises 7.4, page 590

1. $\frac{2}{3}$ sq unit   **3.** 1 sq unit   **5.** 2 sq units

7. $\frac{2}{3}$ sq unit   **9.** $\frac{1}{2}e^4$ sq units   **11.** 1 sq unit

13. a. $\frac{2}{3}b^{3/2}$ **15.** 1   **17.** 2   **19.** Divergent

21. $-\frac{1}{8}$ **23.** 1   **25.** 1   **27.** $\frac{1}{2}$ **29.** Divergent

31. -1 **33.** Divergent   **35.** 0   **37.** 0

39. Divergent   **41.** Convergent   **43.** \$18,750

45. $\frac{10,000r + 4000}{r^2}$ dollars   **47.** True   **49.** False

Chapter 7 Review Exercises, page 595

1. $-2(1 + x)e^{-x} + C$ **3.** $x(\ln 5x - 1) + C$

5. $\frac{1}{4}(1 - 3e^{-2})$ **7.** $2\sqrt{x}(\ln x - 2) + 2$

9. $\frac{1}{8}\left[3 + 2x - \dfrac{9}{3 + 2x} - 6\ln|3 + 2x|\right] + C$

11. $\frac{1}{32}e^{4x}(8x^2 - 4x + 1) + C$

13. $\dfrac{1}{4}\dfrac{\sqrt{x^2 - 4}}{x} + C$ **15.** $\frac{1}{2}$

17. Divergent   **19.** $\frac{1}{10}$

21. 0.8421; 0.8404   **23.** 2.2379; 2.1791

25. $1,157,641   **27.** $41,100

29. $111,111

CHAPTER 8

Exercises 8.1, page 605

1. $f(0, 0) = -4; f(1, 0) = -2; f(0, 1) = -1; f(1, 2) = 4;$
$f(2, -1) = -3$

3. $f(1, 2) = 7; f(2, 1) = 9; f(-1, 2) = 1; f(2, -1) = 1$

5. $g(1, 2) = 4 + 3\sqrt{2}; g(2, 1) = 8 + \sqrt{2}; g(0, 4) = 2;$
$g(4, 9) = 56$

7. $h(1, e) = 1; h(e, 1) = -1; h(e, e) = 0$

9. $g(1, 1, 1) = e; g(1, 0, 1) = 1; g(-1, -1, -1) = -e$

11. All real values of x and y

13. All real values of u and v except those satisfying the equation $u = v$

15. All real values of r and s satisfying $rs \geq 0$

17. All real values of x and y satisfying $x + y > 5$

19. 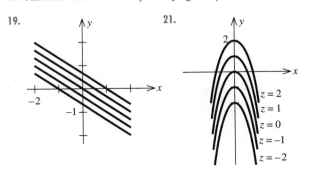   **21.**

23.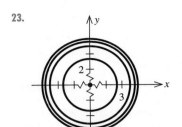

25. 9π ft³   **27.** 40,000k dynes

29. $25,500; $23,580   **31.** $8310; $7910

33. a. The set of all nonnegative values of W and H
b. 1.871 m²

35. $13,498.59   **37.** $76,704.11; $50,814.62   **39.** 18

41. False   **43.** True

Exercises 8.2, page 620

1. 2; 3   **3.** $4x$; 4   **5.** $-\dfrac{4y}{x^3}; \dfrac{2}{x^2}$

7. $\dfrac{2v}{(u + v)^2}; -\dfrac{2u}{(u + v)^2}$

9. $3(2s - t)(s^2 - st + t^2)^2; 3(2t - s)(s^2 - st + t^2)^2$

11. $\dfrac{4x}{3(x^2 + y^2)^{1/3}}; \dfrac{4y}{3(x^2 + y^2)^{1/3}}$ **13.** $ye^{xy+1}; xe^{xy+1}$

15. $\ln y + \dfrac{y}{x}; \dfrac{x}{y} + \ln x$ **17.** $e^u \ln v; \dfrac{e^u}{v}$

19. $yz + y^2 + 2xz; xz + 2xy + z^2; xy + 2yz + x^2$

21. $ste^{rst}; rte^{rst}; rse^{rst}$ **23.** $f_x(1, 2) = 8; f_y(1, 2) = 5$

25. $f_x(2, 1) = 1; f_y(2, 1) = 3$

27. $f_x(1, 2) = \frac{1}{2}; f_y(1, 2) = -\frac{1}{4}$

29. $f_x(1, 1) = e; f_y(1, 1) = e$

31. $f_x(1, 0, 2) = 0; f_y(1, 0, 2) = 8; f_z(1, 0, 2) = 0$

33. $f_{xx} = 2y; f_{xy} = 2x + 3y^2 = f_{yx}; f_{yy} = 6xy$

35. $f_{xx} = 2; f_{xy} = f_{yx} = -2; f_{yy} = 4$

37. $f_{xx} = \dfrac{y^2}{(x^2 + y^2)^{3/2}}; f_{xy} = f_{yx} = -\dfrac{xy}{(x^2 + y^2)^{3/2}};$
$f_{yy} = \dfrac{x^2}{(x^2 + y^2)^{3/2}}$

39. $f_{xx} = \dfrac{1}{y^2}e^{-x/y}; f_{xy} = \dfrac{y - x}{y^3}e^{-x/y} = f_{yx};$
$f_{yy} = \dfrac{x}{y^3}\left(\dfrac{x}{y} - 2\right)e^{-x/y}$

41. a. $f_x = 7.5; f_y = 40$ **b.** Yes

43. $p_x = 10$—at $(0, 1)$, the price of land is changing at the rate of $10/ft^2/mile change to the right; $p_y = 0$—at $(0, 1)$, the price of land is constant/mile change upward.

45. Complementary commodities

47. $30/unit change in finished desks; $-$25/unit change in unfinished desks. The weekly revenue increases by $30/unit for each additional finished desk produced (beyond 300) when the level of production of unfinished desks remains fixed at 250; the revenue decreases by $25/unit when each additional unfinished desk (beyond 250) is produced and the level of production of finished desks remains fixed at 300.

49. 0.039 L/degree; $-$0.015 L/mm of mercury. The volume increases by 0.039 L when the temperature increases by 1 degree (beyond 300 K) and the pressure is fixed at 800 mm of mercury. The volume decreases by 0.015 L when the pressure increases by 1 mm of mercury (beyond 800 mm) and the temperature is fixed at 300 K.

53. True   **55.** False

Using Technology Exercises 8.2, page 623

1. 1.3124; 0.4038   **3.** $-$1.8889; 0.7778

5. $-$0.3863; $-$0.8497

Exercises 8.3, page 634

1. $(0, 0)$; relative maximum value: $f(0, 0) = 1$

3. $(1, 2)$; saddle point: $f(1, 2) = 4$

5. $(8, -6)$; relative minimum value: $f(8, -6) = -41$

7. $(1, 2)$ and $(2, 2)$; saddle point: $f(1, 2) = -1$; relative minimum value: $f(2, 2) = -2$

9. $\left(-\frac{1}{3}, \frac{11}{3}\right)$ and $(1, 5)$; saddle point: $f\left(-\frac{1}{3}, \frac{11}{3}\right) = -\frac{319}{27}$; relative minimum value: $f(1, 5) = -13$

11. $(0, 0)$ and $(1, 1)$; saddle point: $f(0, 0) = -2$; relative minimum value: $f(1, 1) = -3$

13. $(2, 1)$; relative minimum value: $f(2, 1) = 6$

15. $(0, 0)$; saddle point: $f(0, 0) = -1$

17. $(0, 0)$; relative minimum value: $f(0, 0) = 1$

19. $(0, 0)$; relative minimum value: $f(0, 0) = 0$

21. 200 finished units and 100 unfinished units; $10,500

23. Price of land ($200/ft^2$) is highest at $\left(\frac{1}{2}, 1\right)$

25. $(0, 1)$ gives desired location.

27. 18 in. $\times$ 36 in. $\times$ 18 in.

29. 6 in. $\times$ 4 in. $\times$ 2 in.   **31.** False

Exercises 8.4, page 644

1. a. $y = 2.3x + 1.5$
   b.
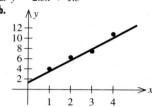

3. a. $y = -0.77x + 5.74$
   b.
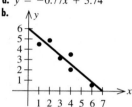

5. a. $y = 1.2x + 2$
   b.
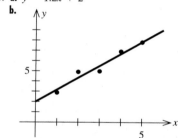

7. a. $y = 0.34x - 0.9$
   b.
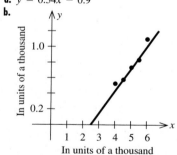

   c. 1276 applications

9. a. $y = -2.8x + 440$
   b.   **c.** 420
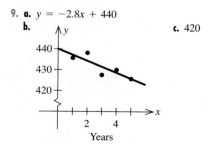

11. a. $y = 6.23x + 167.6$ **b.** 560 acres

13. a. $y = 2.8x + 17.6$ **b.** \$40,000,000

15. a. $y = 4.842x + 11.842$ **b.** 103.8 billion cans

17. a. $y = 12.2x + 20.9$ **b.** 81.9 million

19. a. $y = 98.176x - 231.7$ **b.** \$1732   **21.** True

Using Technology Exercises 8.4, page 647

1. $y = 2.3596x + 3.8639$ **3.** $y = -1.1948x + 3.5525$

5. a. $y = 13.321x + 72.571$ **b.** 192 million tons

Exercises 8.5, page 660

1. Min. of $\frac{3}{4}$ at $\left(\frac{3}{4}, \frac{1}{4}\right)$ **3.** Max. of $-\frac{7}{4}$ at $\left(2, \frac{7}{2}\right)$

5. Min. of 4 at $(\sqrt{2}, \sqrt{2}/2)$ and $(-\sqrt{2}, -\sqrt{2}/2)$

7. Max. of $-\frac{3}{4}$ at $\left(\frac{3}{2}, 1\right)$

9. Max. of $2\sqrt{3}$ at $(\sqrt{3}/3, -\sqrt{6})$ and $(\sqrt{3}/3, \sqrt{6})$

11. Max. of 8 at $(2\sqrt{2}, 2\sqrt{2})$ and $(-2\sqrt{2}, -2\sqrt{2})$;
min. of -8 at $(2\sqrt{2}, -2\sqrt{2})$ and $(-2\sqrt{2}, 2\sqrt{2})$

13. Max.: $\dfrac{2\sqrt{3}}{9}$; min.: $-\dfrac{2\sqrt{3}}{9}$

15. Min. of $\frac{18}{7}$ at $\left(\frac{9}{7}, \frac{6}{7}, \frac{3}{7}\right)$

17. 140 finished and 60 unfinished units

19. $10\sqrt{2}\,\text{ft} \times 40\sqrt{2}\,\text{ft}$ **21.** $2\,\text{ft} \times 2\,\text{ft} \times 3\,\text{ft}$

23. \$15,000 on newspaper and \$45,000 on TV

25. $30\,\text{ft} \times 40\,\text{ft} \times 10\,\text{ft}$; \$7200   **27.** False

Exercises 8.6, page 667

1. $2x\,dx + 2\,dy$

3. $(4x - 3y + 4)\,dx - 3x\,dy$

5. $\dfrac{x}{\sqrt{x^2 + y^2}}\,dx + \dfrac{y}{\sqrt{x^2 + y^2}}\,dy$

7. $-\dfrac{5y}{(x - y)^2}\,dx + \dfrac{5x}{(x - y)^2}\,dy$

9. $(10x^4 + 3ye^{-3x})\,dx - e^{-3x}\,dy$

11. $\left(2xe^y + \dfrac{y}{x}\right)dx + (x^2e^y + \ln x)\,dy$

13. $y^2z^3\,dx + 2xyz^3\,dy + 3xy^2z^2\,dz$

15. $\dfrac{1}{y + z}\,dx - \dfrac{x}{(y + z)^2}\,dy - \dfrac{x}{(y + z)^2}\,dz$

17. $(yz + e^{yz})\,dx + xz(1 + e^{yz})\,dy + xy(1 + e^{yz})\,dz$

19. 0.04   **21.** -0.01 **23.** -0.10

25. -0.18 **27.** 0.06   **29.** -0.1401

31. An increase of \$19,250/month

33. An increase of \$5000/month   **35.** \$1.25

37. 38.4π cu cm   **39.** 0.55 ohm

Exercises 8.7, page 676

1. $\frac{7}{2}$ **3.** 0   **5.** $4\frac{1}{2}$

7. $(e^2 - 1)(1 - e^{-2})$ **9.** 1

11. $\frac{2}{3}$ **13.** $\frac{188}{3}$ **15.** $\frac{84}{5}$

17. $2\frac{2}{3}$ **19.** 1   **21.** $\frac{1}{2}(3 - e)$

23. $\frac{1}{4}(e^4 - 1)$ **25.** $\frac{2}{3}(e - 1)$ **27.** False

Exercises 8.8, page 684

1. 42 cu units   **3.** 20 cu units   **5.** $\frac{64}{3}$ cu units

7. $\frac{1}{3}$ cu unit   **9.** 4 cu units   **11.** $3\frac{1}{3}$ cu units

13. $2(e^2 - 1)$ cu units

15. $\frac{2}{35}$ cu unit   **17.** 54   **19.** $\frac{1}{3}$

21. 1   **23.** 43,329   **25.** \$10,460

27. True

Chapter 8 Review Exercises, page 687

1. $0, 0, \frac{1}{2}$; no   **3.** $2, -(e + 1), -(e + 1)$

5. The set of all ordered pairs (x, y) such that $y \neq -x$

7. The set of all (x, y, z) such that $z \geq 0$ and $x \neq 1$,
$y \neq 1$, and $z \neq 1$

9. 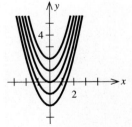   **11.**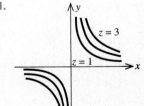

13. $f_x = \sqrt{y} + \dfrac{y}{2\sqrt{x}}; f_y = \dfrac{x}{2\sqrt{y}} + \sqrt{x}$

15. $f_x = \dfrac{3y}{(y + 2x)^2}; f_y = -\dfrac{3x}{(y + 2x)^2}$

17. $10y(2xy + 3y^2)^4; 10(x + 3y)(2xy + 3y^2)^4$

19. $2x(1 + x^2 + y^2)e^{x^2+y^2}$; $2y(1 + x^2 + y^2)e^{x^2+y^2}$

21. $\dfrac{2x}{x^2 + y^2}$; $-\dfrac{2x^2}{y(x^2 + y^2)}$

23. $f_{xx} = 12x^2 + 4y^2$; $f_{xy} = 8xy = f_{yx}$; $f_{yy} = 4x^2 - 12y^2$

25. $g_{xx} = \dfrac{-2y^2}{(x + y^2)^3}$; $g_{yy} = \dfrac{2x(3y^2 - x)}{(x + y^2)^3}$;

$g_{xy} = \dfrac{2y(x - y^2)}{(x + y^2)^3} = g_{yx}$

27. $h_{ss} = -\dfrac{1}{s^2}$; $h_{st} = h_{ts} = 0$; $h_{tt} = \dfrac{1}{t^2}$

29. $(2, 3)$; relative minimum value: $f(2, 3) = -13$

31. $(0, 0)$ and $(\frac{3}{2}, \frac{9}{4})$; saddle point at $f(0, 0) = 0$; relative minimum value: $f(\frac{3}{2}, \frac{9}{4}) = -\frac{27}{16}$

33. $(0, 0)$; relative minimum value: $f(0, 0) = 1$

35. $f(\frac{12}{11}, \frac{20}{11}) = -\frac{32}{11}$

37. Relative maximum value: $f(5, -5) = 26$; relative minimum value: $f(-5, 5) = -24$

39. $45\, dx + 240\, dy$ **41.** 0.04

43. 48 **45.** $\frac{2}{63}$ **47.** $11\frac{1}{3}$ cu units

49. 3 **51.** Complementary

53. $11,000; 14 agents

55. 337.5 yd $\times$ 900 yd

CHAPTER 9

Exercises 9.1, page 697

13. $y = 12x^2 - 2x$ **15.** $y = \dfrac{1}{x}$

17. $y = -\dfrac{e^x}{x} + \dfrac{1}{2} xe^x$ **19.** $\dfrac{dQ}{dt} = -kQ$; $Q(0) = Q_0$

21. $\dfrac{dA}{dt} = k(C - A)$ **23.** $\dfrac{dC}{dt} = -kC$; $C(0) = C_0$

29. True   **31.** False   **33.** False

Exercises 9.2, page 704

1. $\frac{1}{3}y^3 = \frac{1}{2}x^2 + x + C$ **3.** $\frac{1}{3}y^3 = e^x + C$

5. $y = ce^{2x}$ **7.** $-\dfrac{1}{y} = \dfrac{1}{2}x^2 + C$

9. $y = -\frac{4}{3} + ce^{-6x}$ **11.** $y^3 = \frac{1}{3}x^3 + x + C$

13. $y^{1/2} - x^{1/2} = C$ **15.** $\ln|y| = \frac{1}{2}(\ln x)^2 + C$

17. $y^2 = 2x^2 + 2$ **19.** $y = 2 - e^{-x}$

21. $y = \frac{2}{3} + \frac{1}{3}e^{(3/2)x^2}$ **23.** $y = \sqrt{x^2 + 1}$

25. $\ln|y| = (x - 1)e^x$ **27.** $y = \ln(x^3 + e)$

29. $y^2 = x^3 + 8$ **31.** $Q(t) = Q_0 e^{-kt}$

33. $A = \dfrac{C}{k} - d_2 e^{-kt}$

35. $S(t) = D - (D - S_0)e^{-kt}$ **37.** True

39. True   **41.** False

Exercises 9.3, page 713

1. $y = y_0 e^{-kt}$ **3.** $Q(t) = 4.5e^{0.02t}$; 7.4 billion

5. $\frac{1}{2}$ in.   **7.** $Q(t) = \dfrac{50}{4t + 1}$; 5.56 g

9. 3.6 min   **11.** 23

13. 290.5 million   **15.** 395

17. $Q(t) = e^{C - (C - \ln Q_0)e^{-kt}}$

19. $x(t) = 40(1 - e^{-(3/20)t})$; 38 lb; 40 lb

Exercises 9.4, page 722

1. a. $y(1) = \frac{369}{128} \approx 2.8828$ **b.** $y(1) = \frac{70{,}993}{23{,}328} \approx 3.043$

3. a. $y(2) = \frac{51}{16} \approx 3.1875$ **b.** $y(2) = \frac{793}{243} \approx 3.2634$

5. a. $y(0.5) \approx 0.8324$ **b.** $y(0.5) \approx 0.8207$

7. a. $y(1.5) \approx 1.7831$ **b.** $y(1.5) \approx 1.7920$

9. a. $y(1) \approx 1.3390$ **b.** $y(1) \approx 1.3654$

11.

x	0.0	0.2	0.4	0.6	0.8	1
$\tilde{y}_n$	1	1	1.02	1.0608	1.1245	1.2144

13.

x	0.0	0.2	0.4	0.6	0.8	1
$\tilde{y}_n$	2	1.8	1.72	1.736	1.8288	1.9830

15.

x	0.0	0.1	0.2	0.3	0.4	0.5
$\tilde{y}_n$	1	1.1	1.211	1.3361	1.4787	1.6426

Chapter 9 Review Exercises, page 724

5. $y = (9x + 8)^{-1/3}$ **7.** $y = 4 - Ce^{-2t}$

9. $y = -\dfrac{2}{2x^3 + 2x + 1}$ **11.** $y = 3e^{-x^3/2}$

13. a. $y(1) \approx 0.3849$ **b.** $y(1) \approx 0.4361$

15. a. $y(1) \approx 1.3258$ **b.** $y(1) \approx 1.4570$

17.

x	0.0	0.2	0.4	0.6	0.8	1
$\tilde{y}_n$	1	1	1.08	1.2528	1.5535	2.0506

19. a. $S = 50,000(0.8)^t$ **b.** \$16,384

21. $A = \dfrac{P}{r}(e^{rt} - 1)$; \$342,549.50

23. a. $\approx$7:30 P.M.   **25. a.** 183

CHAPTER 10

Exercises 10.1, page 735

13. $k = \frac{1}{3}$ **15.** $k = \frac{1}{8}$ **17.** $k = \frac{3}{16}$ **19.** $k = 2$

21. a. $\frac{1}{2}$ **b.** $\frac{5}{8}$ **c.** $\frac{7}{8}$ **d.** 0

23. a. $\frac{11}{16}$ **b.** $\frac{1}{2}$ **c.** $\frac{27}{32}$ **d.** 0

25. a. $\frac{1}{2}$ **b.** $\frac{1}{2}(2\sqrt{2} - 1)$ **c.** 0   **d.** $\frac{1}{2}$

27. a. 1   **b.** 0.14

29. a. 0.375   **b.** 0.75   **c.** 0.5   **d.** 0.875

31. a. 0.63   **b.** 0.30   **c.** 0.30

33. a. 0.10   **b.** 0.30   **35.** 0.9355

37. 0.4815; 0.7407   **39.** $\frac{1}{5}$ **41.** False

Using Technology Exercises 10.1, page 737

1.

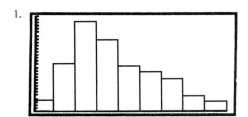

3. a.

x	0	1	2	3	4
$P(X = x)$	.017	.067	.033	.117	.233

x	5	6	7	8	9	10
$P(X = x)$	.133	.167	.1	.05	.067	.017

b.

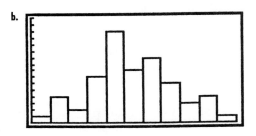

Exercises 10.2, page 752

1. $\mu = \frac{9}{2}$; Var $(x) = \frac{3}{4}$; $\sigma \approx 0.866$

3. $\mu = \frac{15}{4}$; Var $(x) = \frac{15}{16}$; $\sigma \approx 0.9682$

5. $\mu = 3$; Var $(x) = 0.8$; $\sigma \approx 0.8944$

7. $\mu \approx 2.3765$; Var $(x) \approx 2.3522$; $\sigma \approx 1.534$

9. $\mu = \frac{93}{36}$; Var $(x) \approx 0.7151$; $\sigma \approx 0.846$

11. $\mu = \frac{3}{2}$; Var $(x) = \frac{3}{4}$; $\sigma \approx \frac{1}{2}\sqrt{3}$

13. $\mu = 4$; Var $(x) = 16$; $\sigma = 4$

15. 100 days   **17.** $3\frac{1}{3}$ min   **19.** 1.5 ft

21. 2500 lb/wk   **23.** $m = 5$

25. $m \approx 2.52$ **27.** $m = \frac{8}{5}$

29. False

Using Technology Exercises 10.2, page 755

1. a.

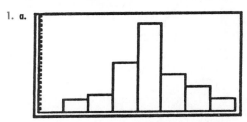

b. $\mu = 4$ and $\sigma = 1.40$

3. a. X gives the minimum age requirement for a regular driver's license.

b.

x	15	16	17	18	19	21
$P(X = x)$	.02	.30	.08	.56	.02	.02

c.

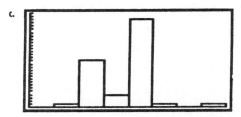

d. $\mu = 17.34$ and $\sigma = 1.11$

5. a. Let X denote the random variable that gives the weight of a carton of sugar.

b.

x	4.96	4.97	4.98	4.99	5.00	5.01
$P(X = x)$	$\frac{3}{30}$	$\frac{4}{30}$	$\frac{4}{30}$	$\frac{1}{30}$	$\frac{1}{30}$	$\frac{5}{30}$

x	5.02	5.03	5.04	5.05	5.06
$P(X = x)$	$\frac{3}{30}$	$\frac{3}{30}$	$\frac{4}{30}$	$\frac{1}{30}$	$\frac{1}{30}$

c. $\mu \approx 5.00$; Var $(X) \approx 0.0009$; $\sigma \approx 0.03$

Exercises 10.3, page 765

1. 0.9265   **3.** 0.0401   **5.** 0.8657

7. a.

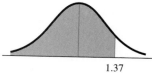

1.37

b. 0.9147

9. a.

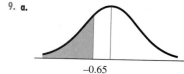

−0.65

b. 0.2578

11. a.

−1.25

b. 0.8944

13. a.

0.68   2.02

b. 0.2266

15. a. 1.23   **b.** −0.81   **17. a.** 1.9   **b.** −1.9

19. a. 0.9772   **b.** 0.9192   **c.** 0.7333

21. a. 0.2206   **b.** 0.2206   **c.** 0.3034

23. a. 0.0228   **b.** 0.0228   **c.** 0.4772   **d.** 0.7258

25. a. 0.0038   **b.** 0.0918   **c.** 0.4082   **d.** 0.2514

27. 0.6247   **29.** 0.62%

31. A: 80; B: 77; C: 73; D: 62; F: 54

Chapter 10 Review Exercises, page 770

5. $\frac{1}{243}$ **7.** $\frac{3}{2}$ **9. a.** $\frac{4}{7}$ **b.** 0   **c.** $\frac{1}{3}$

11. a. ≈ 0.52 **b.** ≈ 0.65 **c.** 0

13. $\mu = \frac{9}{2}$; Var $(x) \approx 2.083$; $\sigma \approx 1.44$

15. $\mu = 0$; Var $(x) \approx \frac{7}{15}$; $\sigma \approx 0.6831$

17. 0.9875   **19.** 0.3049

21. a. 0.6915   **b.** 0.8944   **c.** 0.4681

23. a. 0.22   **b.** 0.39   **c.** 4 days

Chapter 11

Exercises 11.1, page 784

1. $P_1(x) = 1 - x$; $P_2(x) = 1 - x + \frac{1}{2}x^2$;
$P_3(x) = 1 - x + \frac{1}{2}x^2 - \frac{1}{6}x^3$

3. $P_1(x) = 1 - x$; $P_2(x) = 1 - x + x^2$;
$P_3(x) = 1 - x + x^2 - x^3$

5. $P_1(x) = 1 - (x - 1)$; $P_2(x) = 1 - (x - 1) + (x - 1)^2$;
$P_3(x) = 1 - (x - 1) + (x - 1)^2 - (x - 1)^3$

7. $P_1(x) = 1 - \frac{1}{2}x$; $P_2(x) = 1 - \frac{1}{2}x - \frac{1}{8}x^2$;
$P_3(x) = 1 - \frac{1}{2}x - \frac{1}{8}x^2 - \frac{1}{16}x^3$

9. $P_1(x) = -x$; $P_2(x) = -x - \frac{1}{2}x^2$;
$P_3(x) = -x - \frac{1}{2}x^2 - \frac{1}{3}x^3$

11. $P_2(x) = 16 + 32(x - 2) + 24(x - 2)^2$

13. $P_4(x) = (x - 1) - \frac{1}{2}(x - 1)^2 + \frac{1}{3}(x - 1)^3 - \frac{1}{4}(x - 1)^4$

15. $P_4(x) = e + e(x - 1) + \frac{1}{2}e(x - 1)^2 + \frac{1}{6}e(x - 1)^3 + \frac{1}{24}e(x - 1)^4$

17. $P_3(x) = 1 - \frac{1}{2}x - \frac{1}{8}x^2 - \frac{1}{16}x^3$

19. $P_3(x) = \frac{1}{3} - \frac{2}{9}x + \frac{4}{27}x^2 - \frac{8}{81}x^3$

21. $P_n(x) = 1 - x + x^2 - x^3 + \cdots + (-1)^n x^n$; 0.9091; 0.909090...

23. $P_4(x) = 1 - \frac{1}{2}x + \frac{1}{8}x^2 - \frac{1}{48}x^3 + \frac{1}{384}x^4$; 0.90484

25. $P_2(x) = 4 + \frac{1}{8}(x - 16) - \frac{1}{512}(x - 16)^2$; ≈ 3.94969

27. $P_3(x) = x - \frac{1}{2}x^2 + \frac{1}{3}x^3$; 0.109; 0.108

29. 2.01494375; 0.00000042

31. 1.248; 0.00493; 1.25   **33.** 0.095; 0.00033

35. a. 0.04167   **b.** 0.625   **c.** 0.04167
d. 0.007121

37. 0.47995   **39.** 48%   **41.** 2600

43. False   **45.** True

Exercises 11.2, page 794

1. 1, 2, 4, 8, 16

3. $0, \frac{1}{3}, \frac{2}{4}, \frac{3}{5}, \frac{4}{6}$

5. $1, 1, \frac{4}{6}, \frac{8}{24}, \frac{16}{120}$

7. $e, \frac{e^2}{8}, \frac{e^3}{27}, \frac{e^4}{64}, \frac{e^5}{125}$

9. $1, \frac{11}{9}, \frac{25}{19}, \frac{45}{33}, \frac{71}{51}$

11. $a_n = 3n - 2$

13. $a_n = \dfrac{1}{n^3}$

15. $a_n = \dfrac{2^{2n-1}}{5^{n-1}}$

17. $a_n = \dfrac{(-1)^{n+1}}{2^{n-1}}$

19. $a_n = \dfrac{n}{(n+1)(n+2)}$

21. $a_n = \dfrac{e^{n-1}}{(n-1)!}$

23.

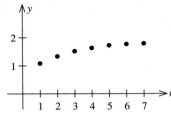

25.

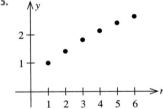

27.

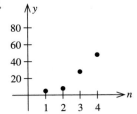

29.

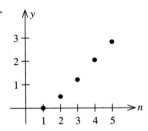

31. Converges; $\frac{1}{2}$

33. Converges; 0

35. Converges; 1

37. Converges; 2

39. Converges; 2

41. Converges; 0

43. Converges; $\dfrac{\sqrt{2}}{2}$

45. a. $a_1 = 0.015$, $a_{10} = 0.140$, $a_{100} = 0.77939$,
$a_{1000} = 0.999999727$
b. 1

47. b. $a_n = 100(1.01)^n$
c. $a_{24} = 126.97$. The accumulated amount at the end of 2 yr is $126.97.

49. True

51. False

Exercises 11.3, page 805

1. $\lim\limits_{N \to \infty} S_N$ does not exist; divergent

3. $S_N = \dfrac{1}{2} - \dfrac{1}{N+2}; \dfrac{1}{2}$

5. Converges; $\frac{3}{2}$

7. Diverges

9. Converges; $\frac{3}{5}$

11. Converges; $\frac{1}{3}$

13. Converges; 5.52

15. $\dfrac{\pi}{\pi + 3}$

17. Converges; 13

19. Converges; $2\frac{3}{4}$

21. Converges; $7\frac{1}{2}$

23. Diverges

25. $\dfrac{e^3 - 2\pi e + \pi^2}{(\pi - e)(e^2 - \pi)}$

27. $\frac{1}{3}$

29. $\frac{404}{333}$

31. $-1 < x < 1; \dfrac{1}{1+x}$

33. $\dfrac{1}{2} < x < \dfrac{3}{2}; \dfrac{2(x-1)}{3-2x}$

35. $303 billion

37. $\frac{6}{11}$

41. $\dfrac{P}{e^r - 1}$

43. b. $\dfrac{1}{k} \ln \dfrac{S}{S - C}$ hr

45. False

47. True

Exercises 11.4, page 818

11. Diverges

13. Diverges

15. Converges

17. Converges

19. Converges

21. Converges

23. Converges

25. Converges

27. Converges

29. Diverges

31. Diverges

33. Converges

35. Diverges

37. Diverges

39. Converges

41. Converges

43. Diverges

45. Converges

47. Diverges

49. $p > 1$

51. $a = 1$

55. True   **57.** False

59. False

Exercises 11.5, page 827

1. $R = 1$; $(0, 2)$ **3.** $R = 1$; $(-1, 1)$

5. $R = 4$; $(-4, 4)$ **7.** $R = \infty$; $(-\infty, \infty)$

9. $R = 0$; $x = -2$ **11.** $R = 1$; $(-4, -2)$

13. $R = \infty$; $(-\infty, \infty)$ **15.** $R = 1$; $(-\frac{1}{2}, \frac{1}{2})$

17. $R = 0$; $x = -1$ **19.** $R = 3$; $(0, 6)$

21. $\sum\limits_{n=0}^{\infty} (-1)^n (x - 1)^n$; $R = 1$; $(0, 2)$

23. $\sum\limits_{n=0}^{\infty} (-1)^n \dfrac{(x - 2)^n}{3^{n+1}}$; $R = 3$; $(-1, 5)$

25. $\sum\limits_{n=0}^{\infty} (-1)^{n+1}(x - 2)^n$; $R = 1$; $(1, 3)$

27. $1 + \dfrac{1}{2}(x - 1) + \sum\limits_{n=2}^{\infty} (-1)^{n+1}\dfrac{1 \cdot 3 \cdot 5 \cdots (2n - 3)}{n!2^n}(x - 1)^n$;
$R = 1$; $(0, 2)$

29. $\sum\limits_{n=0}^{\infty} \dfrac{2^n}{n!} x^n$; $R = \infty$; $(-\infty, \infty)$

31. $\sum\limits_{n=0}^{\infty} (-1)^n \dfrac{1 \cdot 3 \cdot 5 \cdots (2n - 1)}{n!2^n} x^n$; $R = 1$; $(-1, 1)$

35. True

Exercises 11.6, page 837

1. $\sum\limits_{n=0}^{\infty} (-1)^{n+1}(x - 2)^n$; $(1, 3)$

3. $\sum\limits_{n=0}^{\infty} (-1)^n 3^n x^n$; $(-\frac{1}{3}, \frac{1}{3})$

5. $\sum\limits_{n=0}^{\infty} \dfrac{3^n}{4^{n+1}} x^n$; $(-\frac{4}{3}, \frac{4}{3})$ **7.** $\sum\limits_{n=0}^{\infty} x^{2n}$; $(-1, 1)$

9. $\sum\limits_{n=0}^{\infty} (-1)^n \dfrac{x^n}{n!}$; $(-\infty, \infty)$

11. $\sum\limits_{n=0}^{\infty} (-1)^n \dfrac{x^{2n+1}}{n!}$; $(-\infty, \infty)$

13. $f(x) = 1 + \dfrac{x^2}{2!} + \dfrac{x^4}{4!} + \dfrac{x^6}{6!} + \cdots + \dfrac{x^{2n}}{2n!} + \cdots$; $(-\infty, \infty)$

15. $\sum\limits_{n=1}^{\infty} (-1)^{n-1} \dfrac{2^n x^n}{n}$; $(-\frac{1}{2}, \frac{1}{2}]$

17. $\sum\limits_{n=1}^{\infty} (-1)^{n+1} \dfrac{x^{2n}}{n}$; $(-1, 1)$

19. $(\ln 2)(x - 2) + \sum\limits_{n=1}^{\infty} (-1)^{n-1}\left(\dfrac{1}{n2^n}\right)(x - 2)^n$; $(0, 4]$

21. $f'(x) = 1 - x + x^2 + \cdots + (-1)^n x^n + \cdots$

23. $f(x) = x - \dfrac{1}{2}x^2 + \dfrac{1}{3}x^3 + \cdots + \dfrac{(-1)^{n+1}}{n}x^n + \cdots$

25. 0.4812   **27.** 0.7475

29. 3.34   **31.** 15.85%

Exercises 11.7, page 846

1. 1.732051   **3.** 2.645751   **5.** 2.410142

7. 2.30278   **9.** 1.11634   **11.** 1.61803

13. 0.5671   **15. b.** 0.19356   **17. b.** 1.87939

19. 2.9365   **21.** 1.2785   **23.** 0.4263

25. 294 units/day   **27.** 8:39 P.M.

29. 26.82%/yr   **31.** 12%/yr   **33.** 10%/yr

35. 11,671 units; $21.17/unit   **37.** 2.546

Chapter 11 Review Exercises, page 850

1. $f(x) = 1 - (x + 1) + (x + 1)^2 - (x + 1)^3 + (x + 1)^4$

3. $f(x) = x^2 - \frac{1}{2}x^4$

5. $f(x) = 2 + \frac{1}{12}(x - 8) - \frac{1}{288}(x - 8)^2$; 1.983

7. 2.9992591; 8×10^{-11} **9.** 0.37

11. Converges; $\frac{2}{3}$ **13.** Converges; 1

15. 2   **17.** $\dfrac{1}{\sqrt{2} + 1}$ **19.** $\frac{141}{99}$

21. Diverges   **23.** Converges

25. $R = 1$; $(-1, 1)$ **27.** $R = 1$; $(0, 2)$

29. $f(x) = -1 - 2x - 4x^2 - 8x^3 - \cdots - 2^n x^n - \cdots$;
$(-\frac{1}{2}, \frac{1}{2})$

31. $f(x) = 2x - 2x^2 + \frac{8}{3}x^3 - \cdots + \dfrac{(-1)^{n+1}2^n}{n}x^n + \cdots$; $(-\frac{1}{2}, \frac{1}{2}]$

33. 2.28943   **35.** (0.35173, 0.70346)

37. $106,186.10   **39.** 31.08%

CHAPTER 12

Exercises 12.1, page 858

1. $\dfrac{5\pi}{2}$ radians   **3.** $-\dfrac{3\pi}{2}$ radians

5. a. III   **b.** III   **c.** II   **d.** I

7. $\dfrac{5\pi}{12}$ radians   **9.** $\dfrac{8\pi}{9}$ radians   **11.** $\dfrac{7\pi}{2}$ radians

13. 120°   **15.** −270°   **17.** 220°

19.
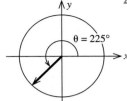

21.
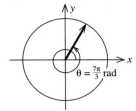

23. $150°, -210°$

25. $315°, -45°$

27. True

29. True

Exercises 12.2, page 868

1. 0

3. 1

5. $\dfrac{\sqrt{3}}{2}$

7. $\dfrac{\sqrt{3}}{3}$

9. -2.6131

11. $\sin \dfrac{\pi}{2} = 1$, $\cos \dfrac{\pi}{2} = 0$, $\tan \dfrac{\pi}{2}$ is undefined, $\csc \dfrac{\pi}{2} = 1$,

$\sec \dfrac{\pi}{2}$ is undefined, $\cot \dfrac{\pi}{2} = 0$

13. $\sin \dfrac{5\pi}{3} = -\dfrac{\sqrt{3}}{2}$, $\cos \dfrac{5\pi}{3} = \dfrac{1}{2}$, $\tan \dfrac{5\pi}{3} = -\sqrt{3}$,

$\csc \dfrac{5\pi}{3} = -\dfrac{2\sqrt{3}}{3}$, $\sec \dfrac{5\pi}{3} = 2$, $\cot \dfrac{5\pi}{3} = -\dfrac{\sqrt{3}}{3}$

15. $\dfrac{7\pi}{6}, \dfrac{11\pi}{6}$

17. $\dfrac{5\pi}{6}, \dfrac{11\pi}{6}$

19. π

21. $\dfrac{\pi}{3}, \dfrac{2\pi}{3}$

23.
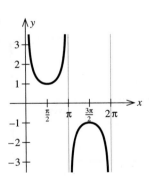

25.
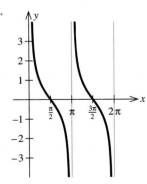

27.
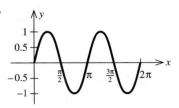

29.
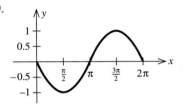

41. $\sin \theta = \dfrac{5}{13}$, $\cos \theta = \dfrac{12}{13}$, $\tan \theta = \dfrac{5}{12}$, $\csc \theta = \dfrac{13}{5}$, $\sec \theta = \dfrac{13}{12}$, $\cot \theta = \dfrac{12}{5}$

43. b. $\dfrac{\pi(4n+1)}{12}$ $(n = 0, 1, 2, \ldots)$;

$\dfrac{\pi(4n+3)}{12}$ $(n = 0, 1, 2, \ldots)$

45. False

47. True

Exercises 12.3, page 881

1. $-3 \sin 3x$

3. $-2\pi \sin \pi x$

5. $2x \cos(x^2 + 1)$

7. $4x \sec^2 2x^2$

9. $x \cos x + \sin x$

11. $6(\cos 3x - \sin 2x)$

13. $2x(\cos 2x - x \sin 2x)$

15. $\dfrac{x \cos \sqrt{x^2 - 1}}{\sqrt{x^2 - 1}}$

17. $e^x \sec x(1 + \tan x)$

19. $\cos \dfrac{1}{x} + \dfrac{1}{x} \sin \dfrac{1}{x}$

21. $\dfrac{x \sin x}{(1 + \cos x)^2}$

23. $\dfrac{\sec^2 x}{2\sqrt{\tan x}}$

25. $\dfrac{x \cos x - \sin x}{x^2}$

27. $2 \tan x \sec^2 x$

29. $-\csc^2 x \cdot e^{\cot x}$

31. $y = -2x + \dfrac{\pi}{2}$

33. Increasing on $\left(0, \dfrac{\pi}{4}\right) \cup \left(\dfrac{5\pi}{4}, 2\pi\right)$; decreasing on $\left(\dfrac{\pi}{4}, \dfrac{5\pi}{4}\right)$

35. $f(x) = \sin x + \cos x$

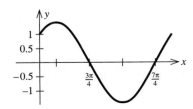

37. $f(x) = 2 \sin x + \sin 2x$

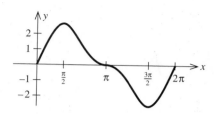

39. a. $f(x) = \sin x = x - \dfrac{x^3}{3!} + \dfrac{x^5}{5!} - \dfrac{x^7}{7!} - \cdots$

$\qquad + (-1)^{n+1} \dfrac{x^{2n+1}}{(2n+1)!} + \cdots$

41. Zero wolves/mo; -1571 caribou/mo

43. 0.05

45. Warmest day is July 25; coldest day is January 23.

47. Sixth week

49. Maximum when $t = 1, 5, 9, 13, \ldots$;
minimum when $t = 3, 7, 11, 15, \ldots$

51. 70.7 ft/sec

53. 60°   **55.** 1.4987 radians

57. True   **59.** False

Using Technology Exercises 12.3, page 885

1. 1.2038

3. 0.7762

5. -0.2368

7. 0.8415; -0.2172

9. 1.1271; 0.2013

11. a.

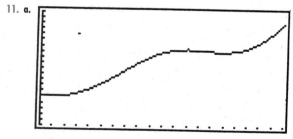

b. $\approx \$0.63$ **c.** $\approx \$27.79$

13. a.

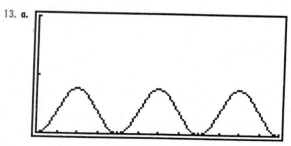

15. ≈ 0.006 ft

Exercises 12.4, page 891

1. $-\frac{1}{3} \cos 3x + C$ **3.** $-3 \cos x + 4 \sin x + C$

5. $\frac{1}{2} \tan 2x + C$ **7.** $\frac{1}{2} \sin x^2 + C$

9. $-\dfrac{1}{\pi} \csc \pi x + C$ **11.** 2

13. $-\frac{1}{2} \ln \frac{1}{2}$ **15.** $\frac{1}{4} \sin^4 x + C$

17. $\dfrac{1}{\pi} \ln|\sec \pi x + \tan \pi x| + C$

19. $\frac{1}{3} \ln(1 + \sqrt{2})$ **21.** $-\frac{2}{3}(\cos x)^{3/2} + C$

23. $-\frac{1}{9}(1 - 2 \sin 3x)^{3/2} + C$ **25.** $\frac{1}{4} \tan^4 x + C$

27. $-\frac{1}{4}(\cot x - 1)^4 + C$ **29.** 2

31. $\frac{1}{2}x[\sin(\ln x) - \cos(\ln x)] + C$

33. $2\sqrt{2}$ sq units   **35.** $\frac{1}{2} \ln 2$ sq units

37. $\frac{1}{2}(\pi^2 - 4)$ sq units   **39.** \$85

41. \$120,000   **43.** $\dfrac{1.2}{\pi}\left(1 - \cos \dfrac{\pi t}{2}\right)$

45. 162 fruit flies   **47.** ≈ 0.9

49. True   **51.** True

Using Technology Exercises 12.4, page 893

1. 0.5419   **3.** 0.7544   **5.** 0.2231

7. 0.6587   **9.** -0.2032 **11.** 0.9045

13. a.

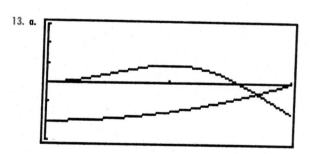

b. 2.2687 sq units

15. a.

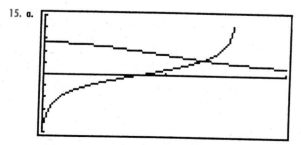

b. 1.8239 sq units

17. a.

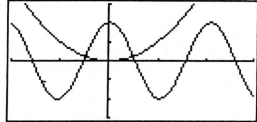

b. 1.2484 sq units

19. a.

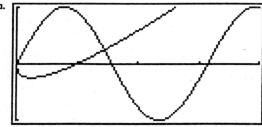

b. 1.0983 sq units

21. 7.6 ft

Chapter 12 Review Exercises, page 897

1. $\dfrac{2\pi}{3}$ radians

3. $-\dfrac{5\pi}{4}$ radians

5. $-450°$

7. $\dfrac{\pi}{3}$ or $\dfrac{5\pi}{3}$

9. $3\cos 3x$

11. $2\cos x + 6\sin 2x$

13. $e^{-x}(3\sec^2 3x - \tan 3x)$

15. $4\cos 2x$ or $4(\cos^2 x - \sin^2 x)$

17. $\dfrac{(\cot x - 1)\sec^2 x - (1 - \tan x)\csc^2 x}{(1 - \cot x)^2}$

19. $\cos(\sin x)\cos x$

21. $y = 4x + 1 - \pi$

23. $\frac{3}{2}\sin\frac{2}{3}x + C$

25. $-\frac{1}{2}\csc x^2 + C$

27. $\frac{1}{3}\sin^3 x + C$

29. $-\csc x + C$

31. 1

33. $2\sqrt{2}$ sq units

35. December 1; September 1

INTRODUCTION

This book is a revised and expanded edition of a similar publication which made its appearance in 2002. The purpose of this book is to show Russia's role and achievements in the development of ekranoplans, or wing-in-ground-effect (WIG) craft – a new means of transportation which remains a fairly exotic domain to this day. These achievements, quite justifiably, attract much interest in many countries. After all, it was in Russia (represented in the shape of the now-defunct Soviet Union) that pioneering work in this field of technology brought about highly impressive results that have yet to be surpassed. Since the 1960s Russia has been occupying a leading position in the world as far as construction of large WIG craft is concerned. The authors intend to present a general review of the work accomplished by Russian design bureaux and production plants in this field of technology (see below in this chapter) and to give a description of specific designs (they will be covered in subsequent chapters). Naturally, relevant events in the period after 2002 are duly covered.

To begin with, a few words about the subject of this book. Wing-in-ground-effect craft making use of a dynamic air cushion are vehicles operating in close proximity to a supporting surface. This is usually water, but basically it makes no difference whether a WIG craft is operating over water or over land – provided that the ground surface is sufficiently even and flat.

A feature common to an aircraft and a WIG craft is wings generating lift due to aerodynamic forces. However, in the case of the WIG craft this lift is augmented owing to the ground effect created by compression of the ram air stream between the wings and the supporting surface. A higher lift/drag ratio enables a WIG craft to obtain the same lift at lower speeds and lower engine power compared to aircraft. As a result, the WIG craft are, in principle, more fuel-efficient compared to aircraft.

Since large flat areas on land are not a common occurrence, WIG craft are in most cases intended for use over water. Operation from the surface of lakes, rivers or seas of necessity introduces some features of water-borne vessels into the design of WIG vehicles. Historically, a number of WIGs emerged as a kind of attempt to lift water-borne craft out of the water for the purpose of achieving higher speeds, and in many cases WIG craft were built at shipyards. Small wonder that the question is posed sometimes whether one must regard these new craft as very low-flying aircraft or as ships that have lifted themselves out of the water.

It would appear that both definitions might be appropriate, since the concept of WIG vehicles embraces a wide variety of craft featuring quite substantial differences. They may tend to be closer to one or the other of the two extremes, but, generally speaking, they are always something of a hybrid. On the one hand, a WIG vehicle in cruise flight is subjected to aerodynamic forces, much in common with conventional aircraft, while the hydrodynamic forces act on it only during take-off and landing – or rather alighting. On the other hand, WIG craft operating in close proximity of the water surface in a marine environment have to be subjected to the same rules and requirements as usual marine vessels in terms of traffic safety.

The latter consideration has played an important role when it came to establishing a formal classification of WIG vehicles with a view to adopting rules concerning their certification and safety regulations. Three basic categories have been formally adopted for this purpose.

The first of them (Type A) encompasses vehicles that can be operated only within the height of the surface effect. They usually feature wings of low aspect ratio (up to 1) and are fitted only with a rudder, there being no elevator; the 'driver' (or should we say helmsman?) does not have to possess piloting skills and steers the vehicle in much the same way as an ordinary speedboat. In Russian parlance, such vehicles are termed Dynamic Air Cushion Vessels, or WIG vessels (ekranoplan boats). Among Russian designs, such examples as the Volga-2, Amphistar and Raketa-2 (described in subsequent chapters) may be cited.

The second category (Type B) includes vehicles which are capable of leaving the surface effect zone for a short while and making brief 'hops'. The altitude of such a 'hop' must not exceed the minimum safe altitude of flight for aircraft, as prescribed by International Civil Aviation Organisation (ICAO) regulations (150 m/500 ft). In Russian parlance such vehicles are regarded as WIG craft (*ekranoplans*) proper; they feature wings with an aspect ratio of up to 3 and are provided with elevators. They are controlled by pilots. Among Russian designs this category is represented by the Orlyonok, KM, Strizh, ESKA-1 and others.

The third category (Type C) covers WIG vehicles capable of flying outside the surface effect zone for a considerable time and of climbing to altitudes in excess of the minimum safe flight altitude for aircraft, as prescribed by ICAO regulations.

This classification subdividing the WIG vehicles into types A, B and C was formulated by Russian organisations and submitted by Russia to the International Maritime Organisation (IMO) and ICAO for their consideration. Thanks in no small degree to determined efforts of the Russian side it has proved possible to reach within the framework of IMO an agreement on a number of basic issues pertaining to legal, technical and operational aspects of WIG craft.

At its 76th session held on 2nd-13th December 2002 the IMO Maritime Safety Committee endorsed a provisional code of WIG craft safety. It contains the classification presented above, along with detailed stipulations concerning technical features of ekranoplans, their modes of operation, maintenance, passenger safety etc.

For the first time international documents were evolved that provide rules for commercial operation of WIG craft and for their safety. These documents represent an important milestone. They have given an expression at a high level for an international recognition of WIG craft as a new and promising means of maritime transport and provided a legal basis for its further development and commercial operation on international sea routes.

Against this background, let us take a look at the story of WIG craft design and production in the Soviet Union and modern Russia.

Early research on ground effect and of efforts aimed at creating practicable WIG vehicles dates back to the 1920s and 1930s when work in this field was started in several countries. As is well known, the first-ever self-propelled WIG vehicle was built by Toivo Kaario, a Finnish engineer, in 1935. His experiments were followed up by research and practical design work in the USA, Germany, Sweden, China, Australia and other countries (a brief review of the work on *ekranoplans* outside Russia can be found in Chapter 7).

The Soviet Union also joined the 'club' of countries working on the WIG craft. Theoretical and experimental work in this direction was started in the USSR in the 1920s (experimental work by Boris N. Yur'yev, 1923). Further work followed in the 1930s, when a whole set of theoretical studies and experiments in the field of ground effect research was performed by Yakov M. Serebriyskiy and Sh. A. Biyachuyev. The results of this work were published in specialised literature.

In the early 1930s the first steps in practical design of WIG craft in the USSR were made by Pavel Ignat'yevich Grokhovsky, an aviation engineer and inventor renowned for his energy and innovative ideas. In 1932 Grokhovskiy, aided by a group of his associates, designed a full-scale mock-up of a sea-going flying vehicle featuring a twin-hull layout. The two hulls served as a sort of endplates for a wide-chord wing centre section; the forward section of each of the hulls housed an M-25 radial (a copy of the Wright R-1820-F3 Cyclone) delivering some 700 hp and driving a tractor propeller. The wing was provided with a flap which was intended to increase the lift on take-off and landing. The flying prototype was never built, but this layout was used in many designs of WIG craft in later decades.

However, it is Rostislav Yevgen'yevich Alekseyev (1916-1980), an outstanding scientist and designer, who must be credited with having played a paramount, decisive role in shaping the course of research, design and construction of WIG vehicles in Russia. His was the conceptual approach and design philosophy; he may truly be regarded as the founder of the Russian wingship construction. Alexeyev started his activities as a builder of hydrofoil ships in his capacity of the chief of the TsKB-19 shipbuilding design bureau based at Gor'kiy (at present Nizhniy Novgorod). In 1958 it was reorganised into the Central Hydrofoil Design Bureau (CHDB, or TsKB po SPK – *Tsen**trahl**'noye kon**strook**torskoye byuro po soo**dahm** na podvodnykh kryl'yakh*). An impressive range of highly successful hydrofoil vessels designed under Alekseyev's guidance was developed and put into operational service. As recognition of R. Alekseyev's outstanding contribution to the creation of hydrofoil ships, he was awarded the degree of Dr. Sc. in technical sciences *honoris causa* in April 1962. Yet, it was his work on WIG craft – work veiled in utmost secrecy for many years – that was destined to become the most prominent and significant part of his creative activities.

The Central Hydrofoil Design Bureau became actively engaged in WIG craft design in the early 1960s. But as early as 1957 Rostislav Alekseyev had demonstrated a small model of a vehicle making use of ground effect. In 1958, at his initiative, work began on the construction of a research and experimenting facility near the town of Chkalovsk on the Gor'kiy Reservoir (later it was known as IS-2 – *ispy**tahtel**'naya **stahn**tsiya*). It incorporated an experimental production facility and unique installations, many of which were created with the express purpose of studying the surface effect. The IS-2 became one of the main subdivisions of the CHDB, supplying it with the scientific information required for the practical design of WIG craft.

This design work was based on the concept of autostabilisation of the wings of a WIG vehicle relative to the interface between the supporting water surface and the air. This concept proved basically sound and was subsequently incorporated in all

An artist's impression of Pavel I. Grokhovskiy's *ekranoplan* project of 1932.

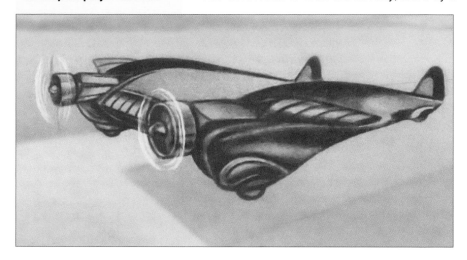

WIG projects issued by the design bureau. On its basis a search was initiated for suitable aero-hydro-dynamic layouts; initially, one of these featured two sets of wings arranged in tandem. The first 3-ton (6,600-lb) SM-1 *ekranoplan* built in 1961 at IS-2 was fitted with two sets of wings. Research revealed that the tandem-wing layout is practicable only in very close proximity to the surface and is unable to ensure the necessary margin of stability and safety, once the craft leaves this close proximity. Experiments with this tandem-wing machine ended in a crash. Alekseyev arrived at the decision to make use of a classic aircraft layout (one set of wings and a tail unit) which was to be subjected to modifications designed to ensure stability and controllability during cruise flight in ground effect. In particular, low-set or mid-set wings of much lower aspect ratio (around 3) were adopted. An important feature was the use of an outsize horizontal tail; it was to be placed sufficiently far aft and high up relative to the main wings so as to minimise the influence of the downwash induced by the wings depending on the flight altitude and pitch angle.

Ten experimental WIG vehicles featuring this layout were built in the Central Hydrofoil Design Bureau, their weight and dimensions growing with every successive machine. These were the machines in the SM series (SM stands for *samokhodnaya model'* – self-propelled model), with an all-up weight of up to 5 tonnes (11,000 lb); a detailed description is given in Chapter 1.

Design experience gained by Rostislav Ye. Alekseyev in developing these machines enabled him to take a bold decision to initiate the design of gigantic WIG vehicles with an all-up weight of more than 400 t (880,000 lb). In 1962 the Central Design Bureau was engaged in project work on a combat WIG craft intended for anti-submarine warfare (ASW) weighing 450 t (990,000 lb); two years later the design team in Nizhniy Novgorod started designing the T-1 troop transport/assault WIG craft.

It should be noted that the very considerable scope attained by the activities of the Central Design Bureau of Hydrofoils was due to the fact that the new means of transport had attracted much interest on the part of the military. The Soviet Navy took much interest in the military potential of *ekranoplans*; it became the main customer for Alekseyev's CHDB. As a consequence, for many years this work was highly classified. Thus, construction of WIG vehicles in the Soviet Union received a boost from military programmes. In the opinion of military specialists both in the Soviet Union (and nowadays in Russia) and in the West, large WIG vehicles can be employed for a wide range of missions in the armed forces, notably in the Navy. These include troop transportation, ASW, anti-shipping strikes

Rostislav Ye. Alekseyev, Chief Designer of the Central Hydrofoil Design Bureau.

One of Alekseyev's sketches from 1947 showing an early concept of WIG vehicles using the tail-first layout. The upper one is particularly unusual, since each wingtip rests on a small auxiliary wing flanked by floats. Note that the foreplanes are attached to the fuselage by a pylon and feature three floats.

with guided missiles, patrol missions etc. The most ambitious projects envisaged the use of WIG craft as flying aircraft carriers! An inherent advantage of WIG vehicles when used in warfare is their ability to remain undetected by enemy radar thanks to the low altitude of their flight; the lack of contact with the sea surface makes them undetectable by acoustic means (sonar devices). WIG vehicles are capable of operating not only over water expanses but also over snow-covered stretches of land and over ice fields. This makes them eminently suitable for use

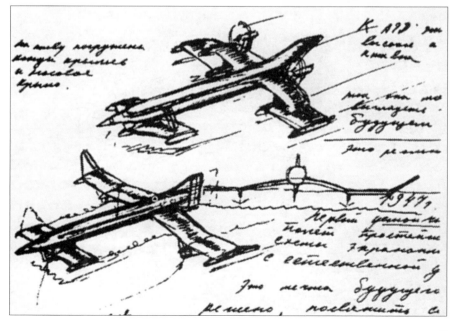

A working meeting at the Central Hydrofoil Design Bureau; Rostislav Alekseyev (standing) heads the meeting. A model of a hydrofoil vessel can be seen on the desk.

Large-scale models of WIG craft built by the CHDB were tested in free flight on open water, using a hydrofoil boat as a tug, as shown here.

in Polar areas. Their high speed ensures their quick response to the changing battlefield situation, and their high load-carrying capacity enhances their capability for accomplishing various missions and carrying a wide range of weapons.

It must be noted that Admiral Sergey G. Gorshkov, the Navy's Commander-in-Chief, was at first somewhat sceptical of WIG craft. Once he asked the CHDB designers: what was a WIG craft – a flying ship or a floating aircraft? If it was an aircraft, the Navy would task Tupolev with preparing a project, if it was a ship, the task would go to CHDB. Naturally, the designers answered in unison that it was *'a ship, but with many distinctive features of a flying vehicle'*!

In assessing the suitability of the WIG craft for ASW, one should bear in mind that, owing to their low flight altitude, WIG vehicles cannot be equipped with sonobuoys. However, they possess a wider range of capabilities for making use of a dunking sonar when afloat. Moreover, thanks to their large dimensions they can, in principle, be fitted with ASW weapons normally carried by surface ships, to be used without becoming airborne.

WIG vehicles are superior to amphibious aircraft in sea-going capabilities and endurance; they can be armed with more potent missiles possessing longer range. However, they have their limitations associated with the need for target designation from an external source (amphibious seaplanes can provide target designation for their weapons when flying at high altitude).

The projects of an ASW WIG vehicle and the T-1 troop-carrying WIG vehicle developed at CHDB never left the drawing board. On the other hand, in 1966 the Design Bureau built, in response to an order from the Navy, a WIG craft designated KM (*korahbl'-maket* – a 'mock-up', i.e., prototype ship). With its fuselage length of 97 m (320 ft), wing span of 37 m (123 ft) and all-up weight of 430 t (948,000 lb), this gigantic machine was a unique piece of engineering. In a record-setting flight its weight reached 540 t (1,190,000 lb), which was an unofficial world record for flying machines at that time. Rostislav Alekseyev argued that the construction and testing of such a WIG craft would provide invaluable experience for the designing of military ekranoplans and civil transport ekranoplans with an AUW of about 2,000 tonnes (4,410,000 lb). He was convinced that large WIG vehicles offered important aerodynamic advantages and represented a promising line of development. Of course, a decision to proceed from small experimental vehicles directly to a giant machine was a technical risk, but it acquitted itself in the end. In 1964, when the activities of CHDB became a subject of critical scrutiny, the design work on the KM was endorsed, among others, by TsAGI, although this authoritative scientific institute had its own ideas, differing from those of Alekseyev, on the optimum layout for ekranoplans.

Creation of large sea-going *ekranoplans* required a new test facility located on a sea shore. Such a test facility was set up on the Caspian Sea, near the town of Kaspiysk in Daghestan. This facility served as base for the testing of the KM and a further two types of large *ekranoplans*.

The abovementioned KM, dubbed 'Caspian Sea Monster' in the West, underwent comprehensive testing in the course of 15 years of operation. It marked the completion of a whole range of research and practical design tasks associated with approbation of the WIG concept as a whole and evolving the scientific basis for their design, construction and testing. The results of this work made it possible to create a theoretical and methodological basis for designing and building practicable examples of WIG vehicles of different classes.

One of these was the *Orlyonok* (Eaglet) troop transport/assault ekranoplan with a take-off weight of 140 t (309,000lb). It was capable of transporting

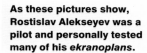
As these pictures show, Rostislav Alekseyev was a pilot and personally tested many of his *ekranoplans*.

a 20-tonne (44,000-lb) cargo at a speed of 400 km/h (248 mph) to a distance of up to 1,500 km (930 miles). Construction of the Orlyonok was conducted at the Volga Shipyard which was set up on the basis of CHDB's experimental production workshops in 1970. Three examples of the Orlyonok (Project 904) were delivered to the Navy for evaluation. Their service career proved to be far from an unqualified success. Normal operation was hampered, above all, by circumstances of bureaucratic nature. The WIG machines were operated by the Navy, yet their crews had to include pilots because in certain operational modes they had to be piloted

like aircraft. However, neither the Air Force nor the Naval Aviation showed any enthusiasm for these machines and sought to 'prove' in every possible way that they could not be regarded as flying machines – unabashed by the fact that provision was made for operating them also out of surface effect and there were plans for long-range ferrying flights at high altitude. Yielding to this pressure, the Navy top brass then decided that WIG craft should be classed as 'ships with aircraft-like properties'.

In turn, the Central Hydrofoil Design Bureau clearly underestimated the 'aviation' aspect of the matter and had failed to consult the Air Force on the methods of testing, which gave rise to justifiable complaints. Arrangements required to facilitate operational use of the machines delivered to the Navy suffered setbacks and delays. Series production of WIG craft for the Navy was expected to amount to several dozens of examples, but these plans failed to materialise. Introduction of new types of weaponry in the USSR, following a pattern common to many countries, depended heavily on lobbying on the part of this or that person in the top echelon. The Soviet Union's Minister of Defence,

The missile-armed Loon' demonstrated the practicality of using WIG vehicles for combat applications.

CHAPTER ONE

ALEKSEYEV, THE PIONEER
STANDARD-SETTING DESIGNS

The history of the establishment and development of the Central Hydrofoil Design Bureau has been presented in the introductory chapter. Here follows a description of its designs.

Vehicles in the SM series

At the initial stage of the work on WIG vehicles in the Central Hydrofoil Design Bureau several light machines were built under the common designation SM (*samokhodnaya model'*,

self-propelled model), individual designs bearing a succession of numbers from 1 to 8. They were proof-of-concept vehicles used to check the surface effect and evolve technical features which came into use later when development of WIG vehicles intended for practical operation was started. Their testing was conducted at a special test facility built on the bank of the Gor'kiy Reservoir (the IS-2 base) which was well equipped with various test facilities.

SM-1

This machine was designed in 1960 and built in 1961at the Chkalovsk test facility, making its first flight on 22nd July with the designer Rostislav Ye. Alekseyev at the controls. It had two sets of wings arranged in tandem and fitted with endplates. The front wings of rectangular shape with an aspect ratio of 1.26 had

Rostislav Yevgen'yevich Alekseyev, head of the Central Hydrofoil Design Bureau.

ailerons/flaps, the aft wings with an aspect ratio of 1.35 were of trapezoidal planform and were mounted at virtually the same height as the front wings. The aft wings were fitted with

The SM-1 during trials. Note the three open cockpits, the uncowled engine and the auxiliary fin immediately ahead of it.

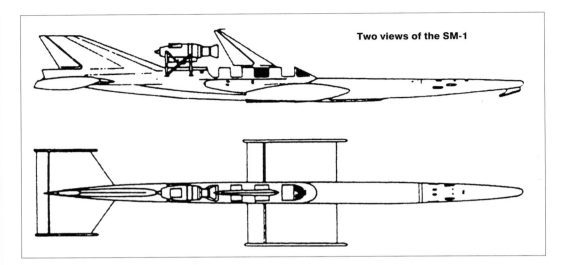

Two views of the SM-1

elevators on their trailing edges. A turbojet engine with no cowling but with a funnel-shaped air intake was mounted on top of the fuselage behind a three-seat open cockpit. The three seats were located in tandem on the vehicle's axis; provision was made for a fourth crew member. An interesting feature of the design was the provision of an extra vertical control surface: in addition to the conventional aft-mounted fin-and-rudder assembly, there was a smaller fin-and-rudder placed atop the fuselage behind the first cockpit.

The tests revealed satisfactory stability and controllability in the ground effect flight mode. As the tests progressed, Alekseyev grew more and more confident of his machine – to such an extent that in the late autumn of 1961 he invited high-ranking guests to take a look at it. Those invited were Marshal Dmitriy F. Ustinov, secretary of the Communist Party's central committee; Boris Ye. Butoma, the Minister of Shipbuilding Industry; and Admiral

Specifications of the SM-1	
Wing span	5.0 m (16 ft 4¾ in)
Length	20.0 m (65 ft 7¹³⁄₃₂ in)
Height	1.53 m (5 ft 0 in)
All-up weight, kg (lb)	2,830 (6,240)
Cruising speed, km/h (mph)	170 (106)
Maximum speed, km/h (mph)	270 (168)
Powerplant	One turbojet of unspecified type
Engine thrust , kgp (lbst)	n.a.

Sergey G. Gorshkov, Commander-in-Chief of the Navy. The guests were so impressed by what they saw that they accepted the invitation to take a ride in the experimental vehicle. Dmitriy Ustinov was very pleased with the ride and instructed Boris Butoma to give all the necessary assistance to Alekseyev.

The SM-2 skims along the water surface before becoming airborne. Note the pitot intake of the forward engine. The vertical tail appears disproportionately large in this view.

Successful tests of the SM-1 corroborated the basic concept of flight in close proximity of a supporting surface. A speed of 200 km/h (124 mph) was reached; the vehicle's stability and handling characteristics during flight close to the supporting surface (water, snow-covered stretches of land) proved to be satisfactory. Yet the tests revealed insufficient seaworthiness and excessively high take-off and landing speeds. The latter may be regarded as one of the main drawbacks of the tandem layout, another being the vehicle's excessive sensitivity to unevenness of the supporting surface.

Furthermore, the SM-1's range of stable flight modes proved to be very limited, as regards the height above the supporting surface and the pitch angle.

In January 1962, during one of the test flights, the SM-1 soared into the air of its own accord, leaving the ground effect zone, and then 'fell through', impacting the ice, when the pilot shut down the engine. The machine was damaged and the crew sustained minor injuries. That brought the SM-1's testing to an end – the machine was not repaired.

The SM-2 in wing-borne flight, showing the air intake and nozzle of the rear-mounted cruise engine. The curved strakes on the aft fuselage underside were probably meant to improve the hydrodynamic properties.

Two stills from a cine film showing the SM-2 in action (in cruise mode).

Starboard side view of the SM-2 in cruise mode. The black spots on the sides of the nose are the forward engine's nozzles.

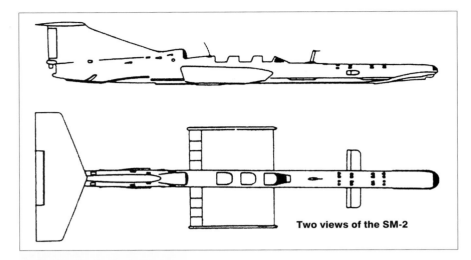

Two views of the SM-2

Specifications of the SM-2

Wing span	5.0 m (16 ft 4⁵⁵⁄₆₄ in)
Length	20.0 m (65 ft 7½ in)
Height	1.5 m (4 ft 11 in)
All-up weight, kg (lb)	3,200 (7,056)
Cruising speed, km/h (mph)	160 (99)
Maximum speed, km/h (mph)	270 (168)
Powerplant	2 x RU-19-300 modified
Engine thrust , kgp (lbst)	2 x 900 (1,984)
Crew	3

SM-2

This machine built in 1962 was initially fitted with two wing sets arranged in tandem. It was damaged by a hangar fire before the trials could begin and reworked in the process of restoration to feature an aircraft layout. This aero-hydrodynamic layout with

one set of wings, T-tail and a device for power-augmented take-off was adopted for all subsequent vehicles designed in the Central Hydrofoil Design Bureau. The SM-2 served for studying the influence of take-off booster blowing devices on the take-off performance of WIG vehicles (that was the first time Alekseyev used such a device). The booster engine located in the front fuselage was fitted with bifurcated flattened nozzle extensions placed wide apart on both sides of the fuselage; they directed the jet efflux under the wings. Initially it also served as a cruise engine, but the thrust of the bifurcated nozzles proved insufficient, and a separate cruise engine was installed in the aft fuselage under the fin. Both engines were 900-kgp (1,984-lb st) Tumanskiy RU19-300 turbojets in a 'marine' version featuring enhanced corrosion protection.

On completion of the testing the vehicle was upgraded: the swept-back wing leading edges were replaced by straight leading edges to improve the effect from blowing, resulting in a rectangular wing planform. In this version the vehicle was redesignated SM-2P (P presumably stands for *pri-amo'ugol'noye* [*krylo*] – rectangular wings).

Thanks to the lift augmentation (blowing) the take-off performance of the SM-2 was considerably improved as compared to the SM-1. The tests corroborated the soundness of the basic design layout and enabled the designers to use it in their further work.

In early May 1962 the SM-2 was shown to the country's government and personally to Nikita S. Khrushchov (this was done at the initiative of Dmitriy Ustinov). The vehicle was urgently transported to Moscow on a sling by a Mi-10 helicopter. This was

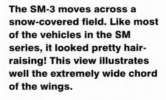

The SM-3 moves across a snow-covered field. Like most of the vehicles in the SM series, it looked pretty hair-raising! This view illustrates well the extremely wide chord of the wings.

Here the SM-3 is seen in wing-borne flight, travelling higher above the ground.

A still from a cine film showing the SM-3 during trials.

vehicle had a T-tail. An unorthodox device for blowing under the wing was used: the air bled from the compressor of the RU19-300 engine placed in the front fuselage was ejected through nozzles located along the wing leading edge on its lower surface. Thereby an efflux curtain was created along the entire leading edge. The tests revealed that the use of the wing with a very low aspect ration had resulted in an oscillatory lateral instability during flights at a height of more than 1.5 m (16 ft) above the supporting surface.

SM-4

This three-seat machine built in 1963 as a derivative of the SM-3, was conceived by Alekseyev as quarter-scale prototype of a fully fledged WIG craft. It differed in having two engines – a booster engine and a cruise engine. The 2,180-kgp (4,800-lbst) Tumanskiy KR7-300 booster turbojet located in the forward

done at short notice, and the test team had no time to fit the rear engine – the vehicle had to be sent for demonstration with the front engine only. Being underpowered, it could not show to advantage all its capabilities, yet the general impression was favourable. Later, with both engines in place, the vehicle displayed good performance and could even fly above sand-bars.

SM-3

This single-seat experimental machine built in 1962 was used for studying the aerodynamic layout with low aspect ratio wings. The wings of this WIG vehicle had a chord twice as big as on the previous machines, the aspect ratio being a mere 0.48. The

Specifications of the SM-3	
Wing span	8.9 m (29 ft 2½ in)
Length	14.5 m (47 ft 7 in)
Height	1.3 m (4 ft 3¼ in)
All-up weight, kg (lb)	3,400 (7,500)
Cruising speed, km/h (mph)	140 (87)
Maximum speed, km/h (mph)	180 (112)
Powerplant	1 x RU19-300
Engine thrust , kgp (lbst)	900 (1,984)

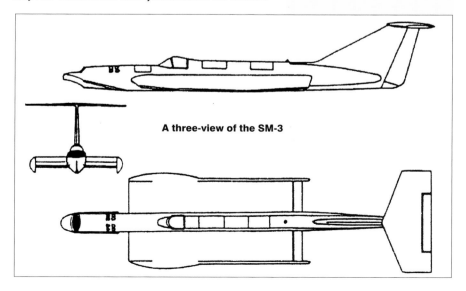

A three-view of the SM-3

Specifications of the SM-2P7	
Wing span	9.5 m (31 ft 8⅜ in)
Length	19.4 m (63 ft 7¾ in)
Height	1.54 m (5 ft 0½ in)
All-up weight, kg (lb)	6,300 (13,890)
Lift-off speed , km/h (mph)	150 (93)
Cruising speed, km/h (mph)	130 (81)
Maximum speed, km/h (mph)	270 (168)
Landing speed, km/h (mph)	140 (67)
Powerplant	1 x KR7-300
Engine thrust , kgp (lbst)	2,000 (4,410)

SM-8

The SM-8 WIG vehicle built in 1967 became the second quarter-scale analogue of the KM *ekranoplan*; it reflected the changes introduced into the layout of the KM in the course of its design. The SM-8 became the last in the family of SM experimental flying vehicles whose testing furnished results essential for the creation of *ekranoplan* theory and for evolving the methods of designing and developing new models of heavy WIG vehicles for military and civil applications. The testing of the SM-8 proceeded in parallel with the testing of the KM; the

Here the SM-2P7 is seen on the point of transitioning to wing-borne flight.

Above and below: The SM-8 was a subscale analogue of the KM vehicle. Except for the dorsal excrescences, from a distance it looked extremely similar to its 'big brother'. The 'smokestack' above the wing leading edge is the protected intake of the forward engine. The full-length rudder acted as a water rudder when afloat.

analogue served for checking the methods of testing its bigger stablemate.

The SM-8 was propelled by a turbojet located in the upper part of the fuselage ahead of the fin. Its air intake was protected from spray by a U-shaped guard and the nozzle was bifurcated, as on the SM-5. To emulate the blowing (booster) engines of the KM, the SM-8 was provided with a special nozzle device in the front fuselage intended to direct part of the gases bled from the engine under the wings. According to some reports, the vehicle was powered by *two* KR7-300 engines, one of them serving as the cruise engine and the other as a booster engine located in the nose. This would explain the purpose of the funnel-shaped structure above the fuselage which must have been the air intake of the booster engine located immediately aft of the cockpit.

Both the SM-8 and the KM were test flown by V. F. Loginov, commander of the flight test detachment of the CHDB, and by leading test pilots D. T. Garbuzov, V. T. Troshin, M. A. Semyonov and V. S. Kudrin, In 1968 the SM-8 was tested in still water conditions and in different sea states with the primary purpose of assessing the craft's seaworthiness, controllability and manoeuvrability, take-off and alighting performance and the machines behaviour during flight at different angles to the waves' movement. Particular attention was paid to mastering the techniques of handling the craft above the water for the purpose of evolving recommendations for piloting the KM *ekranoplan*.

Further seaworthiness tests of the SM-8 were conducted in May-June 1969. Its amphibious qualities were checked on 10th June 1969 when the machine made passes above an unprepared grass-

Above and below: The SM-8 in cruise flight. These photos show the massive intake spray guard and bifurcated nozzle of the main engine, the 'garden rake' immediately aft of the cockpit with nozzles emulating the KM's booster engines, the flap actuator fairings and the strong dihedral of the tailplane.

Another view of the SM-8 over
the Caspian Sea.

Like the SM-5, the SM-8 had a
strut-mounted pitot ahead of
the windscreen. The wing and
tailplane trailing-edge
portions were striped.

covered bank, stretches of marshland, sand-bars
and the like. The SM-8 served as a testbed for a
damping and stabilisation system known as
izdeliye (article) 127, developed by the Leningrad-
based Elektropribor research institute. In 1969 it
was fitted with special devices for spreading evenly
in the horizontal plane the efflux from the nozzles of
the nose-mounted engine.

In addition to research flights, the SM-8 was
used for pilot training.

The construction and testing of the SM series
(SM-1 through SM-8) were directly connected with
the creation of designs that marked the apex of
achievements of the Central Hydrofoil Design
Bureau – the vehicles known as KM, Loon' and Orlyonok. Therefore it is logical to give the description

Kaspiysk to be tested in sea conditions. It was the first WIG vehicle intended for speedy transportation of troops and materiel. Its cargo hold measuring 21 m (68 ft 11 in) in length, 3.2 m (10ft 6 in) in height and 3.0 m (9 ft 10 in) in width made it possible to transport self-propelled vehicles that were on the strength of the Soviet Marines.

The Orlyonok featured an aircraft layout. It was an all-metal cantilever monoplane with a fuselage provided with hydrodynamic elements in its lower portion (planing steps, hydroskis etc.); it had low-set wings and a T-tail with a horizontal tail of considerable dimensions. Its powerplant comprised two Kuznetsov NK-8-4K booster turbofans rated at 10,500 kgp (23,148 lbst) for take-off (provision was made for their eventual replacement with 13,000-kgp/28,660-lbst NK-87 turbofans) and one 15,000-ehp NK-12MK cruise turboprop (a version of the NK-12M used on the Tu-95 bomber) driving AV-90 eight-blade contra-rotating propellers. All the engines were maritime versions of the respective aircraft engines. The booster engines were fitted with special pivoting nozzles and used not only for creating an air cushion on take-off by directing their efflux under the wings (blowing mode) but also for

A model of the Orlyonok assault/transport WIG vehicle undergoes testing in the towing basin.

acceleration to cruising speed. The dorsal air intakes of the turbofan engines were blended into the contours of the forward fuselage, which reduced drag and helped protect the engines from corrosive sea spray. The cruise engine was located at the junction of the fin and horizontal tail; being placed so high, it was less vulnerable to spray ingestion at take-off and landing and to salt contamination from aerosols whose density depends on the height over the sea surface.

The fuselage of the Orlyonok was of beam-and-stringer construction; it was divided into three sec-

The Orlyonok puts to sea in a flurry of spray. Only the prototype wore this civil-style colour scheme. Note the elevator actuator fairings and the four-section elevators.

Above: A full frontal of the Orlyonok prototype on its hardstand at Kaspiysk. From this angle the vehicle has a squat and rather ungainly look. The photo shows well the eight-bladed AV-90 contraprop of the cruise engine and the wing leading edge camber.

This is how the Orlyonok's forward fuselage swung open through 92° for loading and unloading. Note the snap-action vehicle loading ramps, the twin-gun dorsal turret immediately forward of the fuselage break point, the collision avoidance radar in a thimble radome and the pylon-mounted 'saucer' of the navigation radar.

Above: A Soviet BTR-60PB eight-wheel armoured personnel carrier is about to roll off the Orlyonok. This view shows the design of the double-hinged loading ramps, the overhead actuating cylinder of the forward fuselage, the bulkhead with a flight deck access doorway and the numerous securing clamps around the hatch perimeter. The latter is understandable, considering the high stresses in the area.

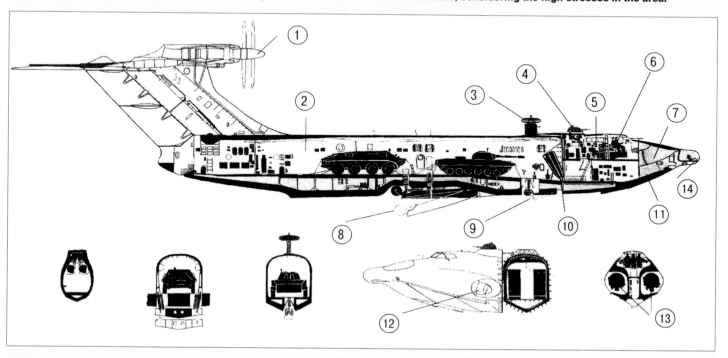

A cutaway drawing of the Orlyonok, with a PT-76 amphibious tank and a turret-less BTR-60P in the cabin.
1. NK-12MK cruise engine. 2. Cargo cabin. 3. Ekran 360° navigation radar. 4. Powered turret with twin 12.7-mm NSV machine-guns.
5. Navigator's workstation. 6. Captain's, co-pilot's and flight engineer's seats. 7. Booster engine inlet ducts. 8. Rear hydroski with shock absorbers and four-wheel main gear unit. 9. Forward hydroski with twin-wheel nose gear unit. 10. Vehicle loading ramps. 11. Forward fuselage section hinged to starboard. 12. Movable booster engine nozzle. 13. NK-8-4K booster engine. 14. Collision avoidance radar.

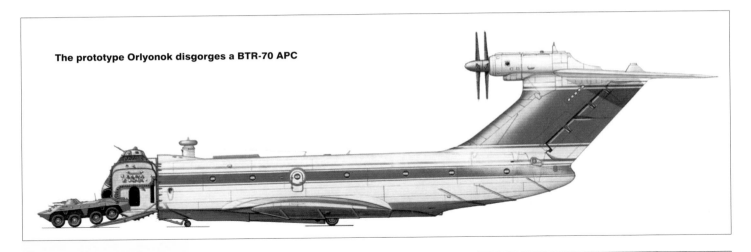

The prototype Orlyonok disgorges a BTR-70 APC

Stills from a cine film showing a BTR-60PB (with the turret and KPVT machine-gun under canvas covers) being unloaded from the Orlyonok prototype.

tions – forward, centre and aft. The centre fuselage accommodated the cargo hold accessed by swinging the hinged forward fuselage 92° to starboard. The hinged part of the fuselage housed the flight deck, the booster engines and a radar in a thimble radome. The aft fuselage housed a compartment for auxiliary power units and accessories required for starting the main engines and operating the vehicle's electrical and hydraulic systems. Placed dorsally on the hull were a turret with twin 12.7-mm (.50 calibre) Nikitin/Sokolov/Volkov NSV-12.7 Utyos heavy machine-guns, the antenna of a 360° naviga-

The Orlyonok's cabin was only just wide enough for the BTR-60 – there were only a few inches to spare.

A production Orlyonok coded '26 White' (the S-26) in Russian Navy markings

The captain of an Orlyonok in the left-hand seat. Interestingly, here he is wearing a simple headset but no helmet.

As this view shows, the captain's instrument panel of the Orlyonok was rather sparsely 'populated' but featured some sort of rectangular display.

Left row: Three interesting aspects of a production Orlyonok, showing the protective grilles on the air intakes (meant to prevent bird ingestion) and open engine cowlings.

The co-pilot of an Orlyonok – in this case wearing a ZSh-7 'bone dome' helmet.

tion radar, direction finder aerials, communication and navigation equipment aerials. To reduce the shock loads in the take-off and landing mode the designers introduced hydroskis shaped as simple deflectable panels. The main hydroski was placed under the centre fuselage. A smaller hydroski was installed in the forward part of the hull. The craft was equipped with a wheeled undercarriage intended for beaching the machine and moving along paved taxiways on the shore. The wheel undercarriage comprised a steerable twin-wheel nose unit and a

The navigator of the Orlyonok at his workstation. Note the flight deck escape tatch.

Above: The Orlyonok prototype before touchdown with the rear hydroski fully deployed. Interestingly, the entry doors amidships are not yet marked externally.

Below: The prototype caught by the camera in a banking turn. Note that the Soviet Navy flag has been added on the tail, revealing the craft's true owner.

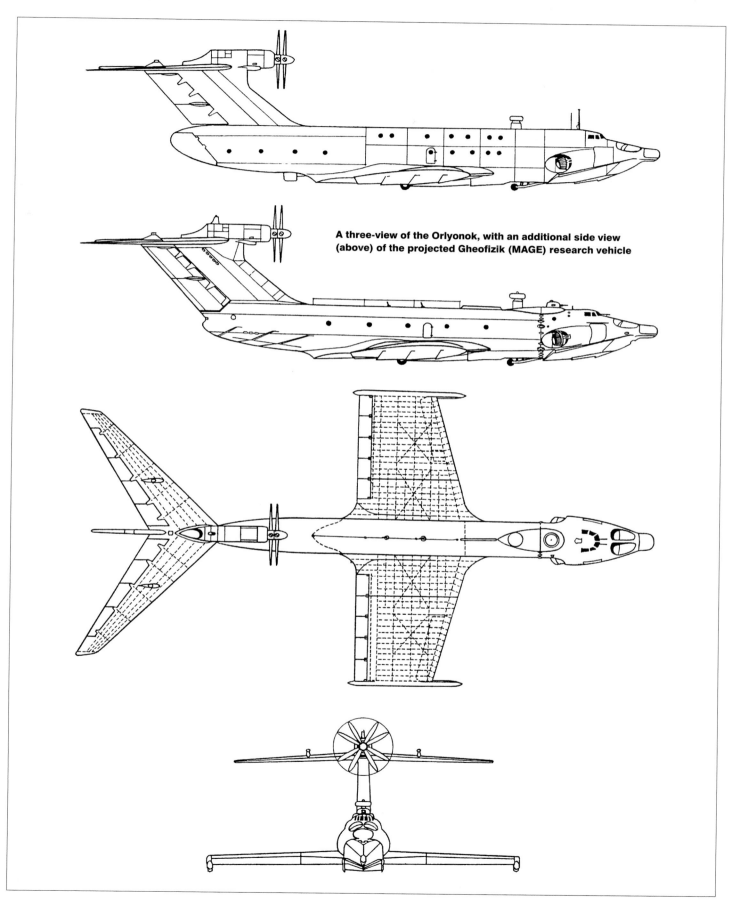

A three-view of the Orlyonok, with an additional side view (above) of the projected Gheofizik (MAGE) research vehicle

on 3rd November 1979, 27th October 1981 and 30th December 1981 respectively. The Naval Command presumed that the WIG vehicles would demonstrate high effectiveness (considerable speed and ensuing capability for surprise actions, capability for overcoming anti-assault obstacles and minefields) and would ensure the seizure of bridgeheads at a coastline defended by the enemy. There were plans in hand for manufacturing 11 Orlyonok (Project 904) machines during the 12th and 13th five-year economic development plan periods (1981-1990), to be followed by the construction of transport and assault WIG craft of a new type (with a new project number) possessing greater cargo carrying capacity. Preparations were made for establishing a WIG vehicle unit in the Red Banner Baltic Fleet. However, for several reasons these plans did not come to fruition. The Orlyonok WIG craft were doomed never to leave the Caspian Sea.

Initially they were operated by the specially established 236th Squadron of WIG vehicles within the brigade of transport and assault ships of the Red Banner Caspian Flotilla. Later an idea cropped up of transferring the WIG vehicles under the authority of headquarters of the Naval Aviation, but these plans met with much opposition on the part of the latter. An end to these disputes was formally put by Order No. 0256 issued by the Minister of Defence on 12th November 1986 under the terms of which ekranoplans became part of the aviation element of the Navy's Fleets. The document prescribed that WIG vehicles, as well as aircraft and helicopters, must be regarded as a class of the Naval Aviation's weaponry. In accordance with directive No. DF-035 dated 21st April 1987 the WIG craft unit, renamed 11th Air Group, was formally placed under the command of the Black Sea Fleet, albeit it retained its former base on the Caspian Sea – the town of Kaspiysk.

The incorporation of the WIG vehicles into the normal activities of the armed forces was not trouble-free and was not pursued all too vigorously. Much time was spent on repairs and modernisation (albeit the machines were almost brand-new!). There were difficulties with crew training. By 1983 four crew captains had received sufficient training; all of them had previously flown the Beriyev Be-12 Chaika (*Mail*) ASW amphibian. Up to 1984 crew training was undertaken in accordance with the 'Provisional course for training the crews of *ekranoplan* ships' prepared by the Combat Training Department of the Navy. Later the manual was reworked with participation of the Combat Training Section of the Naval Aviation.

In 1983 GNII-8 VVS (State Research Institute No. 8 of the Air Force) joined in the testing of one of the Orlyonoks. This was done in response to a request from the Naval Aviation's headquarters, albeit without much enthusiasm. The institute justifiably reproached the developers of the WIG vehicle for having prepared an operation manual for the crew without the participation of test pilots. Its form and contents were quite out of keeping with the standards to which operation manuals for crews of aircraft and helicopters were prepared. There were also other complaints and remarks.

Some interesting details concerning the operational service of the Orlyonoks were disclosed in an internet forum discussion by a person who had served with the *ekranoplan* unit based on the Caspian Sea. Major (in the reserve) Ravil' Dolotkazin served with the 236th *Ekranoplan* Squadron at the Kaspiysk base in 1980-1991. He noted, for example, that the *ekranoplans* were flown, according to instructions, at a height of 1 to 10 m (3 to 30 ft) and at speeds not exceeding 385 km/h (239 mph). Rising to higher altitudes was problematic – not only because of the aerodynamic qualities inherent in the design but also because the necessary equipment for flights in the aircraft mode was lacking. There were no aneroid altimeters, no parachutes and no emergency escape systems.

Yet, the crew of the S-21 craft undertook, at its own initiative and in utmost secrecy, several flights out of ground effect. The crew captain informally obtained personal consent of every member of the crew to such a dangerous experiment – everybody had the right to refuse; yet, nobody backed out. Such flights, performed in excellent weather conditions, lasted 20-30 minutes. The altitude attained was in excess of 100 m (330 ft), maybe even up to 300 m (about 1,000 ft). The captain, a very experienced pilot, later told his colleagues that piloting the *ekranoplan* at such altitudes was difficult. The control surfaces, especially the ailerons, lost their effectiveness. The flight became unstable, involving the risk of stalling and entering a spin, and placed a heavy strain on the pilot who constantly had to parry deviations from the flight path. On the other hand, the Orlyonok was very stable when flying in ground effect.

A different source describing the same situation states that Rostislav Alekseyev was on board the Orlyonok during such flights; they were performed at an altitude of 80-100 m (260-330 ft) and out of sight of the base for concealment. The pilots considered the vehicle to be quite controllable, albeit somewhat sluggish in a bank.

Routinely, the Orlyonoks were flown with a reduced fuel load of 22 tonnes (48,510 lb), the full fuel load being 28 tonnes. A typical mission with the reduced fuel load was a flight on the route from Kaspiysk to Astrakhan' (350 km/217 miles), then from Astrakhan' to Baku (800 km/497 miles) and

The Russian Navy flag (known as the St. Andrew's Flag) as applied to the *ekranoplans*

Two views of the Orlyonok designated S-26 ('26 White') in post-Soviet guise with the Russian Navy flag on the fin replacing the Soviet flag. Note that the planing bottom, wing leading edges and wingtip floats are repainted black.

finally from Baku to Kaspiysk (450 km/280 miles), which adds up to 1,600 km (994 miles).

Unfortunately, mastering the new hardware was not free from incidents and crashes. In 1975 the Orlyonok prototype undergoing tests beached on a rocky sand-bar. Blowing from the booster engines made it possible to lift it off the rocks and safely return to base, but the episode had its consequences. The hull of the prototype machine was made of K282T1 alloy which, while possessing sufficient strength, was on the fragile side. Obviously the contact with the rocks left its traces – cracks in the aft fuselage which went unnoticed at that time. In one of the subsequent test flights, when the craft made a touch-down in rough seas, the aft fuselage/tail unit broke off and sank. Chief Designer Rostislav Ye. Alekseyev, who was on board, reacted instantly by taking over the controls. He ordered the mechanics to reset the thrust deflectors from blowing (vectored thrust) mode into the cruise (horizontal thrust) position and give the engines full throttle. The machine changed its trim and adopted a negative pitch angle, raising the gaping hole in the aft fuselage above the water surface. In this way, exercising a superb mastery of the vehicle, Alekseyev managed to bring the crippled machine safely back to base in planing mode (the conse-

quences of this episode for the Chief Designer are described in the introductory chapter).

The severely damaged machine was returned to Gor'kiy for restoration which, in fact amounted to building a new machine. As a consequence of the accident, Alekseyev reworked the design, utilising the AMG-61 aluminium/magnesium alloy instead of the previously used K282T1 alloy. It was at this stage that the Orlyonok's design came to incorporate the shock-absorbing hydroski coupled with a set of wheels. Some time later the rebuilt machine was ready for flight testing again. This marked the beginning of a period when every year between June and December the factory testing and inter-departmental acceptance trials were conducted – they involved consecutively the three Orlyonok examples and later the Loon' WIG craft (described later in this chapter). By 1989 the three Orlyonok *ekranoplans* performed 438 take-offs and landings between them and performed 118 beachings.

On 12th September 1992 another crash occurred, this time accompanied by a loss of life. The first production Orlyonok with the factory designation S-21 left its base in Kaspiysk to take part in preparations for a demonstration of Russian WIG craft to foreign guests who were to include representatives from the US company Aerocon. While

Right and below: This model of a projected commercial passenger version of the Orlyonok was displayed at the MAKS-93 airshow, complete with water basin with artificial waves. Note the fat dorsal fairing ahead of the fin.

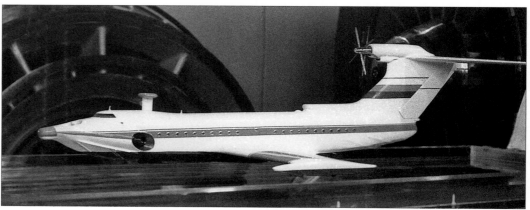

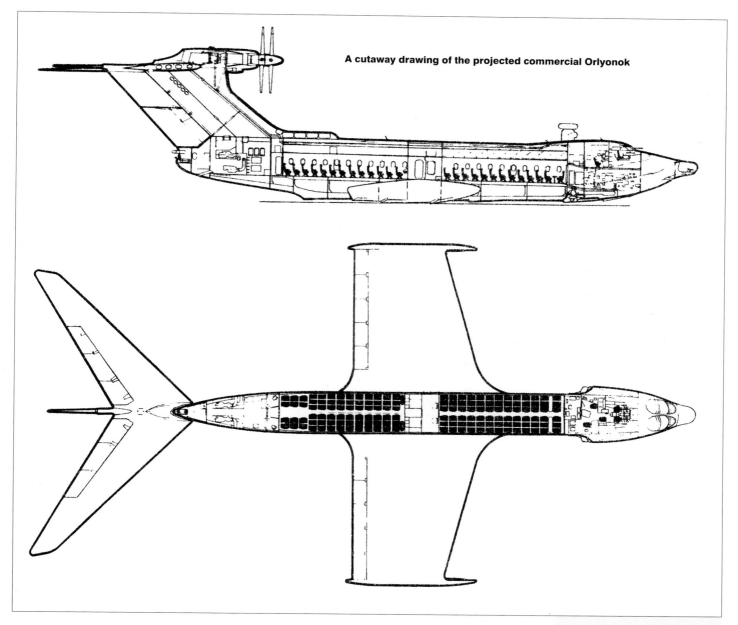

A cutaway drawing of the projected commercial Orlyonok

cruising at 350 km/h (217 mph), the Orlyonok suddenly began to pitch down. To parry this, the crew captain gave full throttle to the cruise engine and hauled back on the control column. As a result, the machine soared in a steep climb. Having reached a height of 40 m (130 ft), it stalled and crashed into the sea, bounced, pulled up a second time and impacted again, sustaining severe damage. Nine of the ten crew survived, albeit with injuries, and were eventually rescued. The tenth crew member – the flight engineer – was killed. The crippled Orlyonok drifted 110 km (60 nm) and was eventually blown up – the Russian Navy could not afford the price asked for its retrieval by salvage companies. It was presumed that the crash had been caused by a failure of the automatic stability system, albeit pilot error is also cited.

After this the remaining WIG complement of the Navy came to include two Project 904 machines (Orlyonok) and one Project 903 machine (Loon'). Quite clearly, for many they were a thorn in their flesh. Gradually, the *ekranoplans* began to sink into oblivion – there were many other things to think of. The vehicles gradually fell into disrepair to the point of no longer being airworthy. Finally, in 1998 the command of the Navy issued an order requiring the Orlyonok WIG vehicles to be written off on account of their alleged unsuitability for repairs and refurbishment.

Fortunately, at least one example of the Orlyonok has been saved for posterity. A few years ago Moscow's Mayor Yuriy M. Luzhkov conceived the idea of setting up a Naval Museum in Moscow. A site for the museum was chosen at the Khimki

A model of another projected passenger version of the Orlyonok designated Orlyonok-P. It features compound-type wings with swept-back dihedral outer panels andan unswept horizontal tail. Again, there is a fat dorsal fairing ahead of the fin.

Two more views of the same model. showing the wing shape. The model bears Ekranoflot titles (a supposed *ekranoplan* operator named by analogy with Aeroflot Russian Airlines, the Russian flag carrier.

Reservoir at the northern outskirts of Moscow. The idea found support from the Government of the Russian Federation. In 2006 the museum was inaugurated, a diesel-powered submarine becoming its first exhibit. But Luzhkov's dream was to obtain one of the Orlyonok *ekranoplans* for the museum, and he succeeded in making it come true. On 7th September 2006 the government adopted a resolution 'On the transfer to the state property of Moscow City of the Project 904 ekranoplan S-26 Orlyonok and the Project 1205 hovercraft D-357 Skat, decommissioned by the Navy and demilitarised, to be placed in Moscow as museum exhibits'.

In 2007 the *Skorosnoy flot* (High-speed Fleet) Joint-Stock Co., which by that time incorporated the Alekseyev Central Hydrofoil Ship Design Bureau, performed all the work associated with the demilitarisation and declassification of the S-26 and effected its transportation by river waterways from its base at Kaspiysk to Moscow. To pass through locks, the machine had one of its wings detached and propellers removed. Upon delivery the ekranoplan was reassembled and prepared for installation as a museum exhibit. The same was done with regard to the mentioned hovercraft. In 2008 the JSC 'Skorosnoy flot' refurbished the Orlyonok and the hovercraft, and they took up their places at Khimki, sharing the company of the Museum's first exhibit – a submarine. The Orlyonok was raised on a pedestal.

ASW version of the Orlyonok

The Project 904 WIG craft served as a basis for a project of an *ekranoplan* intended for ASW duties. The machine was to be equipped with a highly sensitive dipping sonar system which could be used with the machine afloat. No prototype was built. The project is represented by a model in the museum of the CHDB in Nizhniy Novgorod.

The Orlyonok served as a basis for several versions intended for civil applications. These include the following projects:

Sea-going passenger WIG vehicle

Several passenger versions of the Orlyonok were studied by the Design Bureau. In its initial form this variant fully retained the basic configuration of the baseline Orlyonok, differing only in being demilitarised and fitted out in accordance with the new tasks. Importantly, the original wing design of the Orlyonok was retained. Here are its basic characteristics: normal take-off weight, 125 t (275,000 lb); maximum TOW, 140 t (308,000 lb), payload, up to 20 t (44,100 lb) in cargo configuration, 150 passengers in passenger configuration or 30 passengers and 17 t (37,480 lb) of cargo in combi configuration; cruising speed, 350 km/h (217 mph); range, up to 1,500 km (930 miles); wave height during take-off and landing, 1.5 m (5 ft) at normal TOW or 0.5 m (1 ft 8 in) at maximum TOW.

A.90.150

This designation has been applied in Western sources to a somewhat revised projected passenger version of the Orlyonok. According to an advertising booklet from the Design Bureau (which does not mention the A-90.150 designation), it is intended to carry 100 to 150 passengers in a single-deck variant and 300 passengers in a double-deck variant. It was presumed that it could be used on regular passenger routes for transporting 150 passengers or be used as a passenger/cargo transport for speedy delivery of goods and shift crews to off-shore oil rigs, fishing vessels and Polar research stations (the latter case involved landing on ice). With a powerplant identical to that of the baseline version, it had a take-off weight of 110-125 t (242,500-275,000 lb), a cruising speed of 400 km/h (249 mph) and a range of 2,000 km (1,240 miles). Further particulars included a crew of five, a de-luxe seating arrangement for 65-75 passengers, a single-deck seating arrangement for 100-150 passengers and a double-deck configuration seating up to 300 passengers. The single-deck passenger compartment had a length of 25 m (82 ft 0 in), a width of 3.3 m (10 ft 10 in) and a height of 3 m (9 ft 10 in).

Orlyonok-P

This passenger version of Orlyonok (P presumably stands for *passazheerskiy* – passenger, used attributively) features new composite layout wings. A model of the craft shows swept dihedral outer wing panels of smaller chord and higher aspect ratio added to the main wings outboard of the end-plates/floats. This was expected to give the Orlyonok-P a lift/drag ratio that is one-third higher than

that of the basic model; its fuel consumption would be on a level with that of advanced aircraft under development at that time. The horizontal tail was also new, having no sweepback.

The basic specifications and performance of the Orlyonok-P were initially advertised as follows: displacement (take-off weight), up to 140 t (308,640 lb); payload, 40 t (88,180 lb); range, 2,000 km (1,240 miles); speed, 375 km/h (233 mph), seaworthiness at take-off and landing, sea state 4; crew, 6. Later, revised figures were published (see table).

Orlyonok-Gr sea-going cargo WIG vehicle

This was a 'demilitarised' version of the baseline troop transport and assault Orlyonok, retaining its loading feature – the forward fuselage swinging to starboard. It is capable of transporting 30 t (66,000 lb) of cargo to a distance of 1,000 km (620 miles), the maximum range being 2,000 km (1,240 miles). Cruising speed- 400 km/h (249 mph). It is presumably this particular version that has been advertised as the Orlyonok-Gr (*groozovoy* – cargo, used attributively). It was intended for speedy delivery of wheeled and tracked vehicles, including deliveries to sites on a sea shore without any unloading facilities. The dimensions of the cargo hold were as follows: length, 21 m (68 ft 10⁴⁹⁄₆₄ in); width, 3.2 m (10 ft 6 in); height, 3 m (9 ft 10⅞⁄₆₄ in).

More recent official specifications for the Orlyonok-P and Orlyonok-Gr are given in the table below. Note that they differ somewhat from the figures quoted in the text, but this presumably reflects the changes introduced at various stages of the design work.

Provisional specifications of the Orlyonok's projected commercial versions		
	Orlyonok-P	**Orlyonok-Gr**
Register class according to the IMO classification	B	B
Displacement, tonnes (lb)	125 (275,625)	Up to 140 (308,700)
Draught	1.5 m (4 ft 10 in)	1.6 m (5 ft 3 in)
Speed, km/h (mph)	350-400 (217-249)	375-425 (233-264)
Passenger capacity, persons	150	–
Cargo capacity, tonnes (lb)	–	up to 40 (88,200)
Powerplant:		
Cruise engines, number and output, kW	1 x 11,030	1 x 11,030
Booster engines, number and thrust, kgp (lbst)	2 x 10,500 (2 x 23,150)	2 x 10,500 (2 x 23,150)

MAGE

This sea-going Arctic prospecting WIG vehicle (MAGE = *morskoy arkticheskiy gheologorazvedochnyy ekranoplahn*) is intended for prospecting on the shallow-water shelf of the Arctic seas and for transportation tasks related to this work.

It differs from the baseline Orlyonok not only in having the armament and troop-carrying equip-

Commander-in-Chief of the Navy took a decision providing for its preservation at the territory of the 11th Air Group and for transforming it into the air base (for storage of the ekranoplan*), with one crew complement to be retained at the base'.*

In 2002 the Russian Navy command came to the conclusion that the missile-carrying *ekranoplan* could well provide reinforcement to the depleted naval forces of the Russian Federation in the Caspian Sea. This was the result of the then President Vladimir V. Putin's visit to the Caspian Sea region where he, among other things, studied the situation in Russia's regional defences. The President tasked the Navy command with strengthening the Russian military presence on the Caspian Sea to a level commensurate with Russia's status as the strongest of the states sharing the coastline of that sea. As a result, Chief of the General Staff Anatoliy Kvashnin signed a directive calling for bringing the missile-carrying WIG craft back into service.

In May 2002 the Russian MoD daily newspaper *Krasnaya Zvezda* (Red Star) reported that a special commission set up by the Chief of General Staff was busy with refurbishing the Loon' and getting it ready for renewed service. The newspaper asserted that

The port engine package of the Loon' with the lower cowling panels open. Minor damage to the No. 1 engine intake is evident.

the necessary funding had been provided and a new crew had been selected. However, all this fuss came to naught and the Loon' was doomed to remain in a pitiful state to this day.

Spasatel' SAR vehicle (Project 9037)

The second example of the Loon' attack *ekranoplan* (factory designation S-33) ordered by the military was 80% complete when the work was halted and the machine was preserved with an imminent

The Loon' sits in storage at the Kaspiysk base, with maintenance platforms under the engine nacelles. Note the upper section of the rudder deflected differentially from the lower section. The tail in the foreground belongs to another *ekranoplan*.

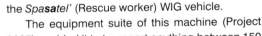

Two towing basin models of the Loon' at the CHDB. The third model in the background represents a projected but unbuilt commercial WIG with twin vertical tails topped by an unswept stabiliser.

Two wind tunnel models of the Spasatel' suspended belly to belly in the wind tunnel.

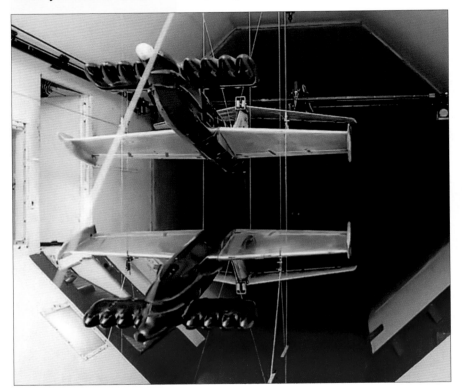

example of the Loon', then under construction, into the *Spasatel'* (Rescue worker) WIG vehicle.

The equipment suite of this machine (Project 9037) enabled it to transport anything between 150 and 500 persons. In the floating hospital version it would provide accommodation for 80 patients on beds. The internal volume was such that the Spasatel' could take 700 to 800 persons on board and remain afloat in a stormy sea waiting for aid to come. This WIG vehicle can take off and alight in sea state 5 and perform cruise flight regardless of the sea state; the rescue means can be deployed directly on its wings. The project provides for an onboard hospital with a surgery room and an intensive therapy ward, as well as a separate place for special treatment needed by those affected by radiation during nuclear powerplant accidents, or by fire or chemical burns. The idea was that improved WIG craft of the Spasatel' type could become the basis of a worldwide SAR system that could be set up under the auspices of the United Nations.

Seven crew members (apart from the navigator and the medical personnel) were accommodated in a flight deck in the forward fuselage, while the navigator and a team of rescue workers sat in an aft compartment at the top of the fin affording excellent, nearly all-round view. The crew was to be provided with inflatable boats, ladders and other rescue means suited for actions under the conditions of a stormy sea.

In addition to aiding those afflicted by 'usual' disasters at sea (a sinking ship etc.), the Spasatel' was to be provided with the means of handling special situations, such as a fire on a ship or a burning oil slick and an emergency on a submerged submarine. For extinguishing fires, the Spasatel' would be equipped with a fire fighting system, including a special tank with 20 tonnes (44,100 lb) of foaming agent. In addition to traditional methods of extinguishing fires, the WIG craft can make use of the powerful jet efflux from its eight engines which, when operated in the boost (blowing) mode, raise a mighty cloud of spray.

In the case of a submarine in distress, the craft would carry piggy-back a deep-sea diving vehicle capable of accommodating 24 persons; this vehicle would be able to carry out rescue operations under the water. It would also carry a decompression chamber. To combat pollution from oil spillages at sea caused by tanker disasters, the Spasatel' could carry a container with floating barriers to an overall length of 3,000 m (9,840 ft). All these sets of equipment for different missions could be used on the same machine that would be properly equipped according to the situation. Furthermore, the Spasatel' was capable of delivering rescue parties and their equipment to offshore oil rigs and of evacuat-

prospect of being scrapped. However, it was saved from that fate through a conversion for civil use. Research performed on the Loon' airframe in 1990-91 confirmed that this machine possessed a considerable potential for conducting search and rescue (SAR) work. In 1992 a plan cropped up in the Ministry of Defence, providing for the conversion of this machine into a SAR vehicle for the benefit of the Navy. A technical project was prepared under the direction of Chief Designer Vladimir N. Kirillovykh, and work was started at the Volga Shipyard in Nizhniy Novgorod with a view to converting the second

Top left: The Spasatel' sits in the assembly hangar of the Volga Shipyard amidst a jumble of work platforms. The flight deck glazing is seen here.

Top: The flight deck roof (with dorsal escape hatch) and the engine pylon attachment area. The open maintenance access hatch is seen at the foot of the picture.

Above left: A look inside the cabin of the unfinished Spasatel'.

Above: The flight deck interior as it is now, showing the escape hatch and the glazing frame.

Two of the air intakes, with the foreign object damage prevention grilles already in place. These were fitted to all eight intakes.

The substantially complete Spasatel' in the assembly shop. Note the glazed observation cabin at the top of the fin.

ing the personnel of such rigs; it could be used for rendering assistance to the population in distress in the event of natural disasters in littoral areas. The Spasatel' was also capable of setting afloat ships that had run aground.

Performance figures for the Spasatel' include a speed of 400-550 km/h (249-342 mph) during cruise flight in WIG mode and 750 km/h (466 mph) during flight at an altitude out of ground effect. In the latter case the machine's service ceiling is 7,500 m

The outer wing panels of the Spasatel' are temporarily attached to the top of the fuselage and temporary floats mounted on the wing centre section stubs, allowing the craft to be towed through narrow channels.

Overall view of the Spasatel' SAR craft nearing completion.

(24,600 ft). In the search mode the Spasatel' is to fly at an altitude of 500 m (1,640 ft). A different source cites a range of 4,000 km (2,486 miles) and a 3,000-km (1,864-mile) tactical radius of action on condition of refuelling at sea from a ship that would arrive at the site within five days (the limit of self-contained operation for the Spasatel'). The cruise flight in the WIG mode is performed at a height of 1 to 4 m (3 to 13 ft), the search is best performed at an altitude between 100 and 300 m (330-980 ft), the ferry flight

The tail unit of the Spasatel' (minus the outer portions of the horizontal tail), showing the fin-top navigator's station/'crow's nest'. Note the upward-hinged tailcone allowing rescue rafts to be launched.

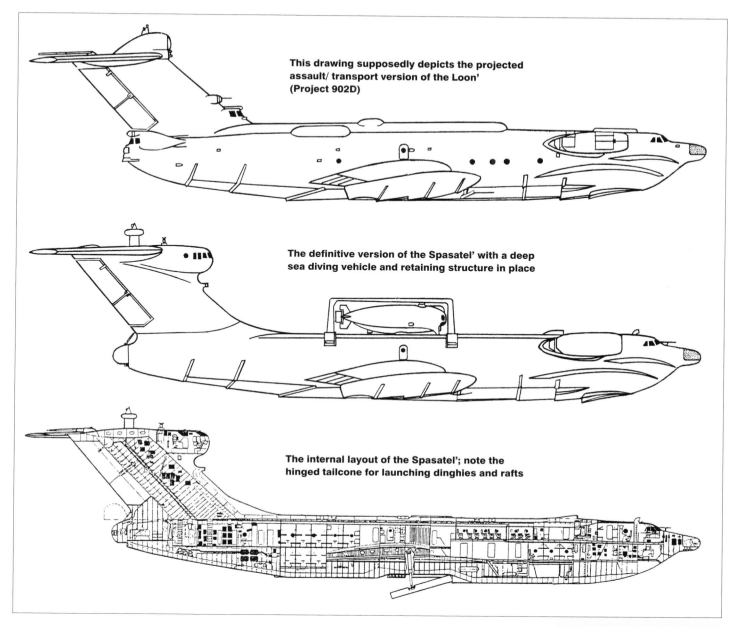

This drawing supposedly depicts the projected assault/ transport version of the Loon' (Project 902D)

The definitive version of the Spasatel' with a deep sea diving vehicle and retaining structure in place

The internal layout of the Spasatel'; note the hinged tailcone for launching dinghies and rafts

approaching its completion – under the influence of a number of tragic accidents, such as the demise of the nuclear missile submarines SNS *Komsomolets* (K-178) and RNS *Kursk* (K-141) and of the car ferry M/V *Estonia* a decision was taken to speed up the construction of this vehicle capable of providing urgent relief in the event of a marine disaster. According to reports, the testing of the Spasatel' was expected to begin at Lake Ladoga as early as 2002, given a favourable course of events.

Unfortunately, these hopes proved ill-founded. To this day the Spasatel' has not been completed and its prospects are bleak.

Until recently, the Spasatel' programme had been 'kept afloat' by the holding company High-Speed Ships which had the Central Hydrofoil Design Bureau as one of its subsidiaries.

Project 902D troop-carrier ekranoplan

Some sources claim that the Project 903 Loon' missile-carrying WIG craft was preceded by a project of a troop-carrier *ekranoplan* designated Project 902 (902D). It is credited with a capacity for carrying 900 troops. This project, the existence of which remains unconfirmed, was allegedly reworked into the project of the missile-carrying Loon', initially designated Project 902R.

Further designs in the SM series

SM-9 Ootka

This was an experimental WIG vehicle featuring a canard (tail-first) layout with the horizontal tail mounted on the front part of the hull, hence the name *Ootka* (Duck, or Canard). Other noteworthy features of the SM-9 were the nose-mounted tractor propeller and the so-called composite layout wings combining a low-aspect-ratio centre section with high-aspect-ratio outer wing panels. Rostislav Alekseyev undertook the design of this craft for the purpose of studying the peculiarities of behaviour of vehicles with this wing layout during flight in ground effect. Furthermore, the nose-mounted propeller served the purpose of studying a novel feature – the use of propellers as a means of blowing in the power-augmented take-off mode (instead of using jet efflux). Finally, this vehicle was intended for studying the use of an inflatable cushioning device in the landing mode.

This three-seat vehicle had a fuselage placed on top of the short-span, wide-chord wing flanked by pylons which rested upon three inflatable multi-section pontoons (two just inboard of the tips and one on the centreline). The fuselage nose housed a 130-hp piston engine which drove a tractor propeller mounted at the end of the elongated nose section. The forward fuselage also carried a steerable foreplane. Underneath the fuselage there was a pylon to which a circular-section inflatable balloon was attached. Attached to the main wing in its rear part were outer wing panels which had moderate dihedral, allowing them to remain within the ground effect zone.

The SM-9 was the first Soviet *ekranoplan* to use a propeller for the blowing-assisted take-off. The use of blowing in the cruise mode made it possible to reduce by half the speed of the craft above the supporting surface as compared with the high-speed *ekranoplans*. Instead of the 200-500-km/h

Above: Poachers beware! ...or what? The SM-9 experimental WIG vehicle bore *Rybnadzor* (Fisheries Control Agency) titles. Here it is shown in late configuration with gull wings and canted wingtips. Only the pilot's cockpit is visible here, the other two seats being closed by a fairing.

This view illustrates the gull wings, the vertical tails and auxiliary engine/propeller packs (used for enhancing directional control) at the wing crank, the engine cooling intake and the movable vanes aft of the main propeller shroud. The outer floats are located inboard of the wing centre section tips, and a third float is barely discernible on the centreline.

(124-311-mph) range of speeds the vehicle could be operated in the speed range between 100 and 250 km/h (62-155 mph). The landing speeds were proportionately reduced by half, too. This considerably reduced the loads and stresses affecting the vehicle. The use of the inflatable cushioning device also helped reduce the stresses and enhanced the vehicle's amphibious qualities.

Rostislav Alekseyev started the testing of the SM-9 in the winter of 1977-78. The first flight, with Alekseyev at the controls, took place on 9th December. The testing revealed that the propeller-effected blowing was much more efficient that the blowing with the use of turbojet efflux. The SM-9 was thoroughly tested both in winter and summer conditions. The tests showed that the layout used in this design made the flight in the WIG mode more secure and considerably reduced the stresses. However, it also became apparent that the vehicle had insufficient longitudinal stability and directional controllability. To remedy this, Alekseyev introduced some modifications. In 1978 the span of the main wing was increased, the outer wing panels were raised to a greater height over the supporting surface (this was done by using a gull-wing configuration). To increase directional controllability, special control devices were installed on the outer wing panels. These were 20-hp piston engines fitted with tractor propellers creating differential thrust.

An available photo of the SM-9 shows the vehicle in this modified configuration. It features an open pilot's cockpit (two more cocpits located further aft are hidden by a fairing) and a shrouded tractor propeller with airflow control vanes at the tip of the elongated fuselage nose. The foreplanes are not readily discernible, but must be there. The fuselage rests on a low aspect ratio wing of rectangular planform, flanked by floats, with a third float on the centreline. The main wing is supplemented by narrow-chord outer panels attached at the aft ends of the floats. The outer wing panels have a marked gull wing configuration and are provided with endplates resembling turned-down tips. Placed atop the outer wing panels at the kinks are two fins with rudders. Also discernible are the auxiliary engines with small propellers. The vehicle carries 'Rybnad**zor**' (Fisheries control) titles. According to the IMO classification, the SM-9 is a Type B vehicle capable of operating out of surface effect for brief periods for the purpose of avoiding obstacles.

The main specifications of the modified SM-9 are as follows: crew, 1-3; length, 11.14 m (36 ft 7 in); span, 9.85 m (32 ft 4 in); height, 2.57 m (8 ft 5 in); all-up weight, 1,750 kg (3,860 lb); cruising speed, 120 km/h (75 mph). In its initial version the SM-9 had a length of 11.69 m (38 ft 4¼ in) and a span of 12.61 m (41 ft 4½ in).

A WIG vehicle based on the SM-9 (project)

Numerous project 'variations on the SM-9 theme' were studied by Alekseyev. One of them, being essentially similar to the initial configuration of the SM-9, differed in having the open forward cockpit and rear fairing replaced by an common canopy enclosing the three seats located one after another. Incidentally, it was also intended for the fisheries control agency, as witnessed by the inscription on the hull in the artist's impression. This machine featured an unshrouded tractor propeller mounted in the similarly elongated fuselage nose. The nose boom carried fairly large-span foreplanes. The wing outer panels featured marked dihedral and sweepback, with no endplates. The fins were placed on top of the outer wing panels at mid-span.

This picture gives some idea of how many project configurations were explored at the CHDB. The models are obviously not to the same scale.

An artist's impression of a vehicle similar to the SM-9 and again wearing Rybnadzor titles.

SM-11

Little is known about this machine, designed and built in 1985. According to some sources this was a Type B WIG vehicle, in the IMO classification, capable of leaving the surface effect zone for brief periods. Some sources claim that in its layout this was a 'flying wing' with compound sweepback on the leading edges. However, a scientific publication (Proceedings of the First International Conference on *Ekranoplans*, St. Petersburg, 1993) carried an artist's impression of a vehicle captioned as *'an SK WIG craft with a small-size horizontal tail'* (SK may stand for *sostavnoye krylo* – composite layout wing). This small single-seat vehicle had large-span wings comprising a wide-chord centre section and swept-back outer panels which were attached to the floats flanking the wing centre section. The outer wing panels had a gull-wing configuration and were fitted with endplates. The vehicle had a T-tail with a small-span horizontal tail. A cruise engine of unknown type, driving a tractor propeller, was mounted at the junction of the fin and the stabilisers. It was supplemented by two propellers located on the sides of the forward fuselage; these were apparently driven through extension shafts by a single engine housed in the fuselage. There are reasons to believe that this drawing depicted the SM-11. The wing layout of this machine was similar to that of the LK project (LK presumably means *letayushcheye krylo*, flying wing) which was also illustrated in the publication. In this project Alekseyev took the 'flying wing' layout with a wing with an aspect ratio that was rather high for a WIG vehicle (of the order of 5). By selecting an optimum airfoil section and making use of automatic oscillation damping and stabilisation systems he tried to solve the problems of stability of flight in ground effect for this new layout. However, a positive solution for 'flying-wing' WIG vehicles was not reached at that time.

The SM-11 was tested in late 1980s, shortly before the break-up of the Soviet Union. It was allegedly destroyed after the completion of testing because its creators *'did not wish it to fall into alien hands'*, whatever this may mean.

Here are some characteristics: crew, 1; length, 6.96 m (22 ft 10 in); span, 9.94 m (32 ft 7 in); all-up weight, 600 kg (1,320 lb); cruising speed, 110 km/h (68 mph).

WIG Vehicles for Military Purposes – Projects that Failed to Materialise

Project 1133 ASW WIG vehicle

Design work on this machine weighing 450 t (990,000 lb) started in 1962, but the design did not reach the hardware stage. No other information is available, apart from a drawing published in 2006 and presented here. The picture shows a machine described as 'a predecessor of the KM'; it is very similar in layout and dimensions to the KM and differs primarily in the arrangement of the cruise engines. Instead of two turbojets flanking the fin, it comprises four turboprops (presumably NK-12Ms) placed on the leading edge of the stabiliser. The project remained on the drawing board, but its basic elements were incorporated in the KM design.

T-1 troop transport WIG vehicle

In 1964 Alekseyev came to the conclusion that ekranoplans could be used by Airborne Troops as a means of delivery of troops and military hardware to sites where they were to be put into action. A WIG vehicle was capable of delivering a landing party to a strip on the coast or in the enemy's rear – something that could not be performed by transport aircraft. In co-operation with representatives of the Airborne Troops the Design Bureau evolved technical requirements for a military transport WIG vehicle which were agreed with the appropriate scientific institutions of the Ministry of Aircraft Industry and of the Air Force and finally approved by the Air Force on 31st July 1964. In October 1964 the Communist Party Central Committee and the Council of Ministers adopted a resolution calling for the development of a preliminary project of a military transport WIG vehicle, which was to be effected in 1965. The project received the designation T-1, with V. V. Sokolov as the project designer. On 28th December 1965 the advanced development project of the T-1 was sent for review to the customer (the Soviet Air Force), to the Ministry of Shipbuilding and to the research institutions of this ministry and of the Navy.

Rostislav Alekseyev officially presented this project in TsAGI on 12th February 1966. The project was praised by specialists, but there was also some criticism. Later in the same month the project passed a critical review at a high-level meeting in the Chkalovsk branch of the CHDB with the participation of the Minister of Defence, Commander of the Airborne Troops, Commander-in-Chief of the Navy, the Minister of Shipbuilding and other high-ranking officials. The machine presented to the in the shape of general arrangement drawings was a flying vehicle with low-set wings, measuring 70 m (229 ft 8 in) in length and 38 m (124 ft 8 in) in wing span. Its empty weight was 105 tonnes (231,525 lb), the normal payload was 20 tonnes (44,100 lb), the cargo-carrying capacity in overload configuration being 40 tonnes (88,160 lb). This enabled the T-1 to carry a medium tank and an infantry platoon with all its weapons and equipment or 150 troops over a distance of 4,000 km (2,486 miles) in the WIG

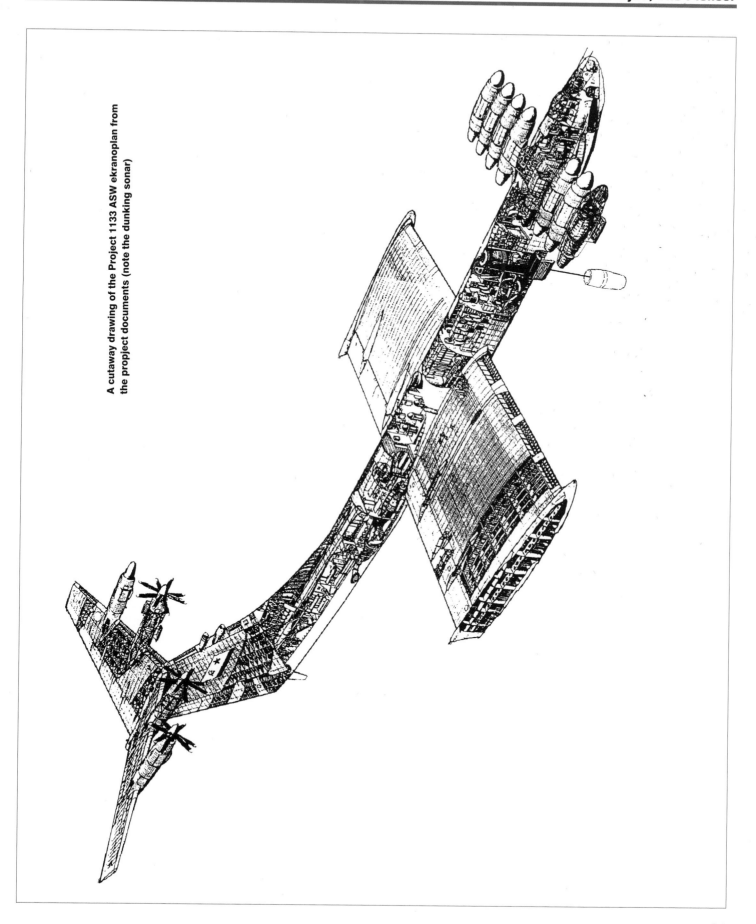

A cutaway drawing of the Project 1133 ASW ekranoplan from the propject documents (note the dunking sonar)

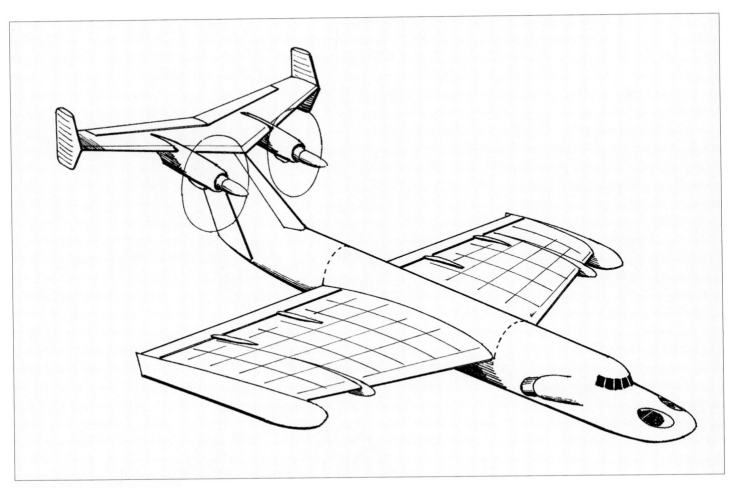

An artist' impression of te T-1 transport/assault WIG vehicle. The drawing does not reveal the location of the loading door, though the same solution as on the Orlyonok was probably envisaged.

mode, or over a distance of 2,000 km (1,243 miles) at an altitude of 4,000 m (13,120 ft). Design performance of the T-1 included a service ceiling of 7,500 m (24,600 ft) which places it into the C category according to the modern IMO classification.)

A published drawing of the T-1 shows a machine featuring an aircraft layout. The wings fitted with endplate floats had a fairly high (for a WIG vehicle) aspect ratio; the craft had a T-tail. The powerplant comprised two booster engines in the forward fuselage arranged in the same fashion as those of Orlyonok, and two cruise turboprop engines. The latter were located in an unorthodox manner – they were mounted at mid-span on the leading edges of the swept stabilisers which was fitted with small endplates.

In May 1966 the Air Force formulated their conclusion on the T-1 project. It took into account the critical remarks made by TsAGI and stated that the project on the whole met the requirements, but a number of drawbacks had to be eliminated in accordance with an annexed list. But the Design Bureau did not get a chance to improve the project. In June 1966 the Ministry of Shipbuilding shut down the funding for the T-1 project on the pretext that it 'had not been accepted by the customer'.

Patrol WIG vehicle

This machine of moderate size featured an aircraft layout and was powered by two turbojets mounted in the forward fuselage. They performed the dual role of booster engines during take-off and cruise engines in forward flight.

Project 905 troop-carrier *ekranoplan*

The Russian reference book Russia's Armament (Vol. III, the Navy, 1996-97), says the Russian state armament programme envisaged at one time the development of strike, ASW and troop-carrier WIG vehicles (Projects 903, 1133 and 905 respectively). Of these, only the Project 903 (Loon') was built. No additional information on the Project 905 is available.

Project 09031

This 'project of a patrol ekranoplan ship' is mentioned in an internet source without any further particulars. The 903 digits in the number may point to a 'family relationship' with the Loon' (project 903).

Ballistic missile transporter WIG vehicle

A project bearing this description (designation not cited) was tailored specially for the transportation of ballistic missiles in an open cargo hold atop the fuselage. The vehicle, comparable in size to an IL-76 transport, basically shared the traditional layout of the CHDB designs in having low-set wings with endplate pontoons (floats) and the use of power-augmented take-off. Its powerplant comprised two turboprop engines mounted on horizontal pylons on each side of the forward fuselage shaped like a boat hull. The engines could be swivelled to direct the air stream under the wings during take-off and were then placed in a horizontal position for cruise flight. The tail unit comprised two swept splayed fins and rudders topped by the horizontal tail.

The open cargo hold occupied the whole length of the top fuselage behind the forward-located flight deck, so that the missile could be loaded and unloaded from the rear between the vertical tails. It appears that the vehicle was expected to be suitable for operation also from flat stretches of land.

Joint Project of the Central Hydrofoil Design Bureau and the Aerocon Company (USA)

In early 1992 negotiations took place between the Alekseyev Central Design Bureau of Hydrofoils and the Aerocon company of the USA on joint development of a new WIG vehicle. The Aerocon company had been set up at the initiative of DARPA (Defense Advanced Research Project Agency) – an agency

working for the US military. Aerocon was tasked with establishing co-operation with the Russian organisation holding leading positions in WIG vehicle construction, with the intention of borrowing advanced Russian technology. Aerocon had received from DARPA a contract worth US$ 546,000 for investigating the concept of a WIG vehicle.

At that time the experts of the US Central Command (CENTCOM) seriously considered the possibility of employing a WIG vehicle in combat. To bring down the cost of creating the WIG machines, US military specialists made a provision for the development of a civil version.

The project envisaged the construction of a giant machine possessing an all-up weight of 5,000 t (11,000,000 lb) and a cargo-carrying capacity of 1,500 t (3,310,000 lb), which would be able to transport 3,000 passengers or 2,000 troops with full military equipment and materiel. The vehicle's dimensions included a span of 170 m (557 ft). a length of 100 m (328 ft) and a height of 35 m (115 ft). The powerplant comprising 20 turbojets with a static thrust of 395 kN each was expected to give the machine a cruising speed of 800 km/h (497 mph) and a range of 16,000 km (9,900 miles) in cruise flight at a height of 2-10 metres (6-33 ft) above the water surface.

This cooperation did not progress far and was terminated without resulting in the emergence of a detailed project.

An artist's impression of a projected fast patrol craft using WIG technology – one of several projects developed by the Central Hydrofoil Design Bureau which never reached the hardware stage.

Chaika-2 (Seagull II), four-engined version

This is a project of a sea-going multipurpose WIG vehicle. Its baseline model was to be developed into a family of shore- or water-based WIG vehicles intended for various duties. Their possible missions include border control, patrolling the 200-mile maritime economic zone, SAR work, passenger and cargo transportation, ecological and legal maritime monitoring.

The machine shares the 'aircraft' layout characteristic of the progeny of the Alekseyev Design Bureau, a novel feature being the composite layout wings with swept outer wing panels of narrower chord outboard of the stabilising floats. The powerplant comprises two 4,500-kgp (9,900-lbst) Gavrilov (Tumanskiy) R-195 booster turbojets in the nose and two 2,500-ehp Klimov (Izotov) TV7-117S turboprop engines driving SV-34 six-blade propellers for cruise. The booster engines used for blowing under the wings during take-off are arranged in a fashion similar to that of the Orlyonok. The cruise engines feature an unusual arrangement: instead of being placed on the fin, they are mounted on small foreplanes with flaps mounted on top of the forward fuselage. Specifications include an all-up weight of 40 to 50 tonnes (88,200-110,250 lb), a payload of 4 tonnes (8,800 lb), a seating capacity of 100 passengers, a speed of 350 km/h (218 mph) and a range of 3,500 km (2,175 miles).

Chaika-2 (Seagull II), twin-engined version

This project bearing the same designation as the preceding one was obviously developed to the same specification, but merits being treated as a separate project due to substantial differences. It shares with the preceding Chaika-2 a basically similar hull (fuselage) and tail unit, but features a totally different powerplant and an unusual wing layout. Instead of a combination of two turbojets and two turboprops, it is powered solely by two turbofans mounted on the forward fuselage sides. They act both as cruise engines and as booster engines during take off; for the latter purpose the engines are fitted with deflector vanes above the nozzles to direct the jet efflux under the wings, thus creating a static air cushion. The nose of the hull, being free from the buried booster engines, is suitably redesigned.

As for the wings, the 'normal' low-set wings similar to those of the four-engined Chaika-2 are supplemented by a pair of wings mounted high on the fuselage above the trailing edge of the main wings. The wingtips of the two sets of wings are joined by vertical endplates with auxiliary floats, creating a biplane wing box with backward stagger. This project exists in several versions intended for both military and civil duties. A model representing a military version of the twin-engined Chaika-2 shows it to be fitted with a peculiar superstructure above the fin in a fashion similar to the Loon' missile-carrying *ekranoplan*. This superstructure is absent in a drawing which apparently shows a civil version. This project apparently failed to reach the hardware stage.

In 1996 mention was made in the Russian press of a baseline project of a sea-going WIG vehicle weighing some 50 t (110,250 lb) which was being developed in the Central Hydrofoil Design Bureau with a view to subsequent construction on its basis

An artist's impression of a twin-turboprop ballistic missile transporter *ekranoplan* for which no designation is known.

of various versions intended both for state agencies and for commercial operation (this may well be a reference to the Chaika-2 design).

Blue Shark

This was a project of an ekranoplan intended for maritime patrol duties related to the 200-mile economic exclusion zone. An available three-view drawing shows this craft to resemble a scaled-down Orlyonok. Like its bigger brother, it has one cruise turboprop engine mounted at the junction of the fin and the high-set horizontal tail. The drawing gives no definite clue as to whether nose-mounted buried booster engines are present (no nozzles are shown, albeit there is something resembling air intakes). The low-set wings, basically similar in proportions to those of the Orlyonok, are supplemented by outer wing panels of higher aspect ratio having a marked sweepback and noticeable dihedral.

The military role of the craft is clearly evidenced by six weapons pylons attached to a kind of a small wing mounted atop the fuselage amidships. However, there was a possibility of commercial applications, too. In 1997 this project was mentioned in a Russian publication as *'a WIG boat intended for export'*.

Projects for Passenger and Transport WIG Vehicles

In addition to passenger versions of the baseline Orlyonok design described above the Central Hydrofoil Design Bureau prepared several more projects of sea-going passenger *ekranoplans*. Details follow of some of these projects.

Chaika (Seagull)

This project of a river-going passenger *ekranoplan* is described as 'the first passenger WIG vehicle design' of the Alekseyev Design Bureau. It was developed in 1970 (incidentally, with the participation of Rostislav Alekseyev's daughter Tatyana, a marine engineer) and was intended to carry 70 passengers at a speed of 250-300 km/h (155-186 mph) over a distance of more than 700 km (435 miles).

The machine is generally similar in its layout to the SM-6 and the Orlyonok with its low-set wings and a T-tail. A notable difference is the absence of a cruise engine on top of the fin. It appears that two engines buried in Chaika's forward fuselage are intended to serve both as booster engines at take-off and as cruise engines, their nozzles rotating to suit the flight mode.

Construction of the prototype started in 1974, but the machine was never completed. The reason was simple: everything pertaining to the design of *ekranoplans* fell into the 'Top Secret' category, which gave those responsible for counter-intelligence a pretext for objecting to the transfer of this technology to the civilian sector. Looking back at this project nowadays, some experts express the opinion that it was unlikely to gain acceptance anyway because its safety level was insufficient and the speed of 250 km/h was clearly too high for the river traffic.

MPE

The Central Hydrofoil Design Bureau worked on a number of projects bearing a common designation MPE (*morskoy passazheerskiy ekranoplahn* – sea-

An artist's impression of the four-engined Chaika-2 in military form with a gun barbette at the base of the fin.

The Volzhanka-1 in cruise flight. Note the increased-area fins, the aft-mounted extra engine and the pipes behind the Nos. 1 and 3 propellers – presumably bleeding off part of the prop wash for inflating the tip floats.

Another view of the modified Volzhanka-1; note the small extra fin aft of the centre engine.

UT-1 Trainer and Strizh Family of Light WIG Vehicles

The Central Hydrofoil Design Bureau has developed and, in part, brought to the hardware stage a number of projects of light WIG vehicles intended for various duties which carry the common designation *Strizh* (Swift, the bird) with varying sequence numbers. Their forerunner was the UT-1, a two-seat training ekranoplan designed and built in 1968.

UT-1

The UT-1 (*oochebno-trenirovochnyy*– trainer, used attributively) was intended for familiarising the pilots with the special features of WIG vehicle control and for providing instruction and training in all opera-

tional modes: during the take-off run and unstick, cruise flight in surface effect and alighting on water or a snow-covered even surface.

The prototype UT-1 was built in 1968. The first phase of the testing was conducted in February and May of that year. The vehicle displayed satisfactory pitch and roll stability, as well as good controllability. In the opinion of the test pilots, the machine's performance fully met the technical requirements, making the UT-1 suitable for use in its intended role. The testing was continued in the subsequent years.

This machine with a wing span of 9.8 m (32 ft 2 in) used the aircraft layout with a T-tail and was powered by a single Czechoslovak-built Walter M-332 six-cylinder in-line engine delivering 120-140 hp. It

A prototype of the Volga-2 WIG craft on the bank of its namesake, the Volga River. Note the exposed engines.

A Volga-2 – possibly the prototype – in cruise flight. Note the cowled engines and the boomerang-shaped aerial above the cockpit.

Another prototype of the Volga-2 (or the same craft following modifications), featuring recontoured engine cowlings, though this is not yet the definitive shape. The tail unit is still unmodified.

A fine shot of a production
Volga-2 in cruise mode. Note
the auxiliary fins on the
stabilisers; the fin bears a
Russian flag.

was Alekseyev's first machine to be provided with a water-ski impact-absorbing device. Testing of the UT-1 fitted with this device was conducted in 1973. It showed a considerable improvement of the craft's seaworthiness – the UT-1 could perform a take-off with a wave height up to 0.4 m (1 ft 4 in).

The UT-1 was used not only for training flights but also for investigating the out-of-ground-effect flight mode ('aircraft mode'). The results of these experiments were incorporated in the design of the Strizh (swift, the bird) described below.

Strizh

This small WIG vehicle was conceived as early as 1981 (the work was headed by chief designer V. Boolanov) and was intended to meet the needs

As these photos show, the Volga-2 can operate both from water and from ice or snow-covered ground.

Right and below: The driver's seat and instrument panel of the Volga-2. Note the side-stick, the rubber-bladed cooling fan, the overhead handle for training the roof-mounted searchlight and the triple windscreen wipers. The latter are a must because a lot of spray is flung up on the windscreen when the craft enters the water.

of the Navy, training the crews for the full-size WIG vehicles taken on strength by the Navy. (According to some sources, it also bore the designation Project 19500).

The technical project was completed in 1985, and the prototype made its first flight in 1991. Flown by project test pilot Yu. A. Chirkin, it successfully passed manufacturer's tests on water and then, during the winter season, in flights above snow-covered fields.

The Strizh has two cockpits in a stepped-tandem arrangement with dual controls and identical sets of instruments. Thanks to the vehicle's special aerohydrodynamic layout it has proved possible to ensure stable movement close to the supporting surface without making use of automation. The Strizh possesses good manoeuvrability and controllability.

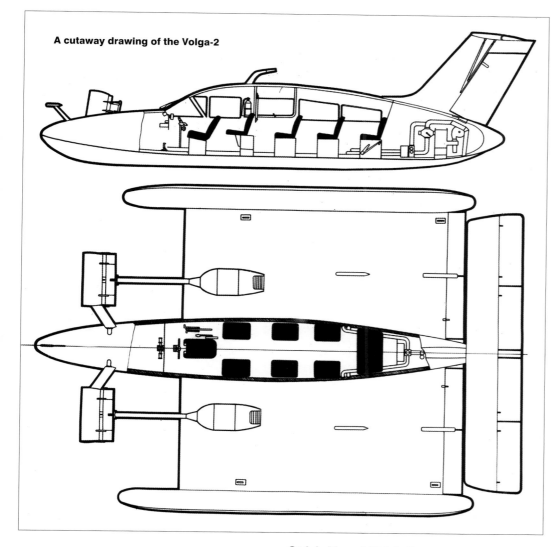

A cutaway drawing of the Volga-2

Pages 94-95: 07 Red, a production Volga-2, 'parked' on the bank of the Volga River in 2001 (note the Russian flag on the fin). The production version has additional vertical tail surfaces and reshaped engine cowlings.

Here are some specifications. The machine has a length of 11.4 m (37 ft 4¾ in), a wing span of 6.6 m (21 ft 7¾ in) and a height of 3.6 m (11 ft 9¾ in). It is powered by two 160-hp VAZ-4133 water-cooled Wankel engines mounted on the wings. The five-bladed fixed-pitch propellers have a diameter of 1.1 m (3 ft 7 in); they are driven by shafts with universal joints via a single-stage reduction gearbox and an automatic coupling.

With an all-up weight of 1,630 kg (3,600 lb) the Strizh has a maximum speed of 200 km/h (124 mph) and a cruising speed of 175 km/h (109 mph) in the WIG mode. Range with a passenger is 500 km (310 miles), the ferrying range is 800 km (500 miles). Flight altitude in ground effect is 0.3-1.0 m (1-3 ft). When flying out of ground effect, the vehicle can rise to an altitude of 800 m (2,620 ft) and reach a speed of 200 km/h (124 mph).

In addition to training, the Strizh can be used for patrol duties, liaison and business flying. The vehicle is placed in the Experimental Class R of the Russian River Fleet Register.

Strizh-M and Strizh-S

A production version designated Strizh-M is to be powered by Voyager 300 engines delivering 220 hp apiece or by 175-200-hp Bakanov M-17 flat-four engines of the Voronezh Engine Design Bureau

Manufacturer's specifications of the Volga-2 and the Volga-2MT		
	Volga-2	**Volga-2MT**
Register class	Experimental R class of the Russian River Fleet Register	Type A in the IMO classification
Displacement, kg (lb)	2,900 (6,395)	3,000 (6,615)
Draught, m (ft)	0.3 m (1 ft 0 in)	0.3 (1 ft 0 in)
Speed, km/h (mph)	120 (75)	120 (75)
Range, km (miles)	300 (186)	300 (186)
Crew	1	1
Number of passengers	8	8
Seaworthiness (wave height), m (ft)		
at take-off and in cruise	0.3 (1 ft 0 in)	0.5 (1 ft 8 in)
when afloat	0.75 (2 ft 5 in)	0.75 (2 ft 5 in)
Engines – number x kW	2 x 114	2 x 147

A production Volga-2 in cruise mode, with propeller vane deflection clearly visible.

In defiance of superstition this vehicle, a subscale analogue of the projected Raketa-2 passenger WIG, was designated SM-13 and marked 013 accordingly. The three engines are well visible here.

(VOKBM). The latter engine, though, has failed to reach production status so far.

Judging by the available three-view drawing, the Strizh-M differs from the prototype Strizh in having two seats in tandem under a common canopy instead of two staggered separate cockpits. However, CHDB advertising materials also mention the Strizh-M with two staggered cockpits which are presumably the standard arrangement for training purposes, while the common two-seat cockpit may be intended for liaison and similar duties. In WIG mode cruise flight the Strizh-M would fly at a height of 0.5-1.2 m (20 in – 3 ft 11 in) above the water surface, with a capacity for making 'hops' to a maximum of 50 m (164 ft) if required.

Mention has been made in the press of a Strizh-S version offered for export. No details are available.

Despite good test results, the Strizh did not go into production. The reason was the curtailment of the Navy's WIG craft procurement programme. The sole prototype was turned over to the CHDB which used it for training test pilots and for demonstration flights at various exhibitions in Russia and abroad. On 5th October 1994 the pilots had the audacity to demonstrate the flying qualities of the Strizh on the Moskva River right in front of the Russian Government building on Krasnopresnenskaya Embankment (popularly known as the 'White House').

The basic layout and main design features of the Strizh were used in developing a family of small WIG vehicles for civil purposes with an all-up weight ranging from 1.6 to 5 t (3,530-11,000 lb). It includes the following machines:

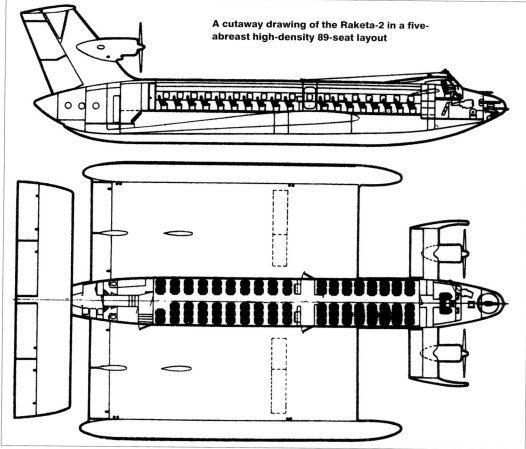

A cutaway drawing of the Raketa-2 in a five-abreast high-density 89-seat layout

An artist's impression of the 90-seat Raketa-2 passenger dynamic air cushion vehicle travelling along one of the inland waterways.

The CHDB in the post-Soviet period

The demise of the Soviet Union and the radical transformation of Russia's political scene and economic structure has had a profound effect on the activities of enterprises engaged in defence-associated industries. The CHDB was affected by these changes – first and foremost due to the virtual disappearance of state funding and defence orders, and, secondly, as a result of the tendency towards privatisation of state enterprises and turning them into joint stock companies.

In the early 1990s the CHDB underwent two major transformations. Firstly, it was privatised, that is, turned into a joint-stock company in which a certain amount of shares was owned by the personnel, while the rest was acquired by other business enterprises. Secondly, the loss of government funding and orders led to a drastic reduction of personnel. Large groups of engineers and designers chose to part company with the CHDB and formed their own new design bureaux with a status of JSCs, taking much of the design work on the *ekranoplans* previously handled by the CHDB with them (more about this in Chapter 6).

The CHDB in its new shape became a part of a group of companies, known as the 'High-Speed Ships' Financial & Industrial Group. This group was formed in 1994 with the declared purpose of uniting Russia's oldest shipbuilding enterprises with a view to ensuring the full cycle of designing and manu-

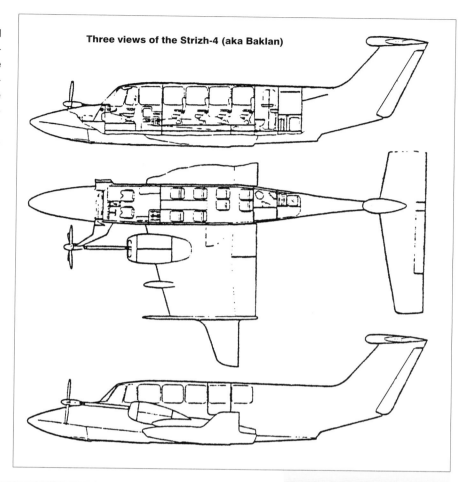

Three views of the Strizh-4 (aka Baklan)

Specifications of ekranoplans of the Strizh family					
	Strizh	**Strizh-M**	**Strizh-3**	**Strizh-4**	**Strizh-5**
Crew	1	1	1	1	1
Passengers	1	1	5-7	7-9	10-12
Length	11.4 m	11.4 m	13.4 m	15.4 m	17.0 m
	(37 ft 4¾ in)	(37 ft 4¾ in)	(43 ft 11⁹⁄₁₆ in)	(50 ft 6¹⁹⁄₆₄ in)	(55 ft 9¹⁹⁄₆₄ in)
Wing span	6.7 m	6.7 m	11.2 m	13.0 m	14.0 m
	(21 ft 11⁴⁹⁄₆₄ in)	(21 ft 11⁴⁹⁄₆₄ in)	(36 ft 8¹⁵⁄₁₆ in)	(42 ft 7¹³⁄₁₆ in)	(45 ft 11³⁄₁₆ in)
Height	3.6 m	3.6 m	4.0 m	4.7 m	5.2 m
	(11 ft 9⁴⁷⁄₆₄ in)	(11 ft 9⁴⁷⁄₆₄ in)	(13 ft 1³¹⁄₆₄ in)	(15 ft 5½ in)	(17 ft 0⁴⁷⁄₆₄ in)
All-up weight, kg (lb)	1,650 (3,640)	1,800 (3,970)	2,700 (5,950)	4,300 (9,480)	5,200 (11,470)
Cruising speed, km/h (mph)	180 (107)	180 (107)	n.a.	210 (131)	220 (137)
Max speed, km/h (mph)	200 (124)	200 (124)	n.a.	240 (149)	270 (168)
Lift-off speed, km/h (mph)	85-90 (53-56)	n.a.	n.a.	n.a.	n.a.
Landing speed, km/h (mph)	115 (71)	n.a.	n.a.	n.a.	n.a.
Engine power, hp	2 x 155	2 x 220	2 x 300	2 x 450	2 x 650
Range	200 (124)	400 (249)	800 (497)	800 (497)	800 (497)
Max wave height, m (ft):					
on take-off	0.4 m	0.4 m	0.5 m	0.6 m	0.7 m
	(1 ft 0⁵⁄₆ in)	(1 ft 0⁵⁄₆ in)	(1 ft 11 in)	(2 ft 3 in)	(2 ft 8 in)
on landing	0.7 m	0.7 m	0.8 m	1.0 m	1.2m
	(2 ft 3in)	(2 ft 3 in)	(2 ft 7 in)	(3 ft 3 in)	(3 ft 11 in)
In cruise over sea surface	0.8 m	0.8 m	1.0 m	1.2 m	1.4 m
	(2 ft 7 in)	(2 ft 7 in)	(3 ft 3 in)	(3 ft 11 in)	(4 ft 7 in)

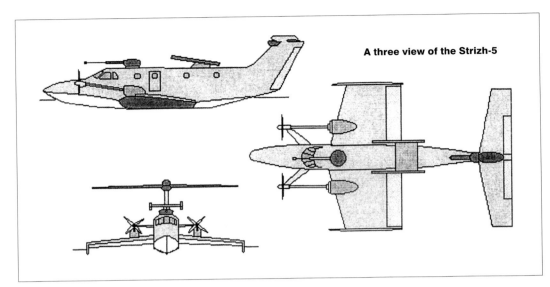

A three view of the Strizh-5

An artist's impression of the TAP-30 craft carrying a load of containers.

A towing basin model of the TAP-700. The blower at the front creating an air cushion is not representative of the projected full-size craft.

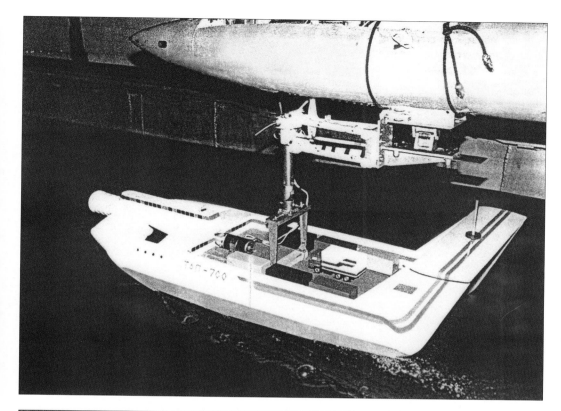

A model of the TAP-700 on test in the towing basin.

A larger radio-controlled model of the TAP-700 used for tests on open water.

facturing high-speed ships. In addition to the CHDB, it comprised several other enterprises (shipyards, an engine plant and a design bureau), one of which was the Volga Shipyard closely associated with the CHDB. An official list of the products of the High-Speed Ships group included ships of various types and, nominally, 'multi-purpose *ekranoplans*'.

In practice, the latter were not regarded a matter of priority and eventually were struck off the list.

The latest development in these matters occurred in June 2009 when the High-Speed Ships group sold its share holding in the CHDB to the St. Petersburg-based company Radar-MMS engaged in the design and manufacture of avionics. The con-

An artist's impression of a civil TAP-700 coming ashore of offload outsize cargo. Note the foreplanes carrying the blower engines.

This artist's imression depicts the TAP-700 as a car ferry. Note the lateral cranes for autonomous cargo handling and the 'danger' markings in the area of the latter.

This model depicts a pure jet craft in the TAP series armed with 16 anti-ship missiles.

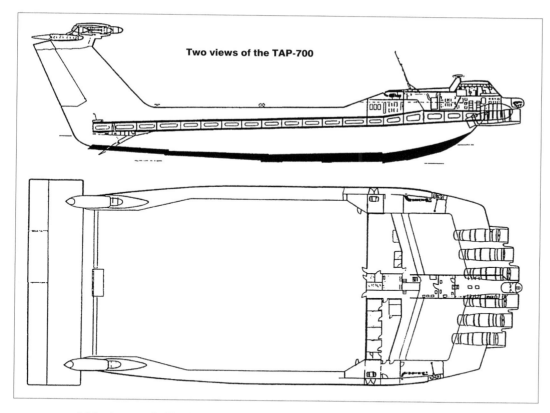

Two views of the TAP-700

sequences of this change of affiliation remain to be seen. The management of Radar-MMS has already declared its willingness to facilitate a revival of the CHDB's activities in the field of ekranoplan design.

In the period since 1992 these activities have been drastically curtailed, no new projects having progressed to the hardware stage. No new designs of military *ekranoplans* from the CHDB (or from the

Design performance characteristics of the Transport Amphibious Platform family					
	TAP-30	**TAP-150**	**TAP-300**	**TAP-500**	**TAP-700**
Displacement, tonnes (lb)	30 (66,150)	150 (330,750)	330 (727,650)	525 (1,157,615)	700 - 750 (1,543,540 - 1,653,750)
Cargo-carrying capacity, tonnes (lb)	10 (22,050)	60 (132,300)	120 (264,600)	200 (441,000)	250 - 300 (551,250 –661,500)
Speed, km/h (mph):					
on still water	100-110 (62-68)	170-180 (106-112)	210-220 (131-137)	240 (149)	250 (155)
on waves	60-80 (37-50)	90-140 (56-87)	110-160 (68-99)	120-200 (75-124)	140-200 (87-124)
on land	90 (56)	100 (62)	100 (62)	100 (62)	100 (62)
on ice or snow	90 (56)	180 (112)	200 (124)	200 (124)	200 (124)
Range, km (miles)					
on still water	125 (78)	480 (298)	750 (466)	1,400 (max) (870)	1,500 (max) (932)
on waves	80 (50)	350 (218)	340 (211)	400 - 1,100 (249 – 683)	700 – 1,100 (435 – 683)
on land	180 (112)	320 (200)	420 (261)	450 – 800 (280 – 497)	500 - 700 (311 – 435)
on ice or snow	180 (112)	580 (360)	850 (528)	900 – 1,100 (559 – 684)	600 – 1,500 (550 – 932)
Crew	3	3	4	4	4

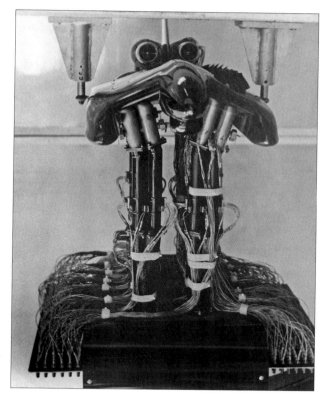

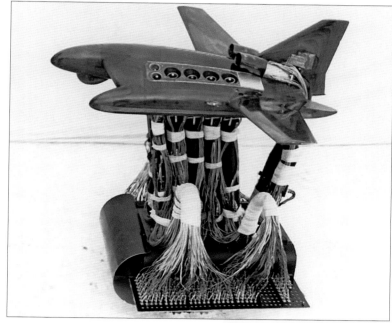

Top left, top and above: A test rig representing the original tailless version of the VVA-14; it features electric fans representing the cruise and lift engines.

Left and above left: A display model representing the final configuration with aft-mouonted thwin vertical tails and a dorsal package of two cruise engines.

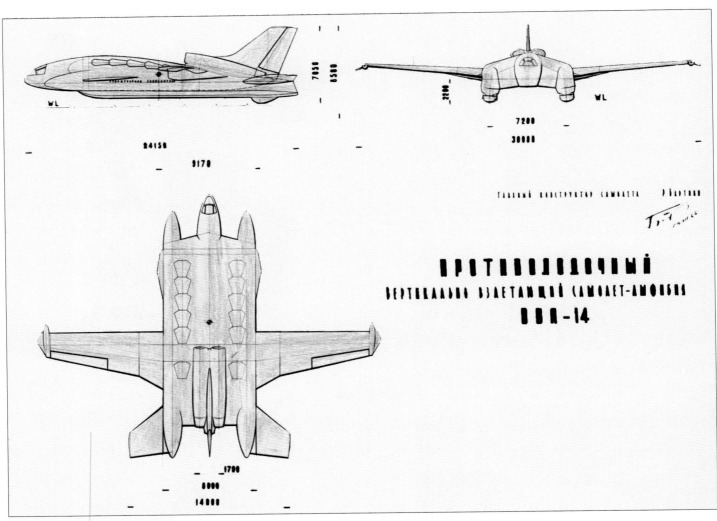

ПРОТИВОЛОДОЧНЫЙ
ВЕРТИКАЛЬНО ВЗЛЕТАЮЩИЙ САМОЛЕТ-АМФИБИЯ
ВВА-14

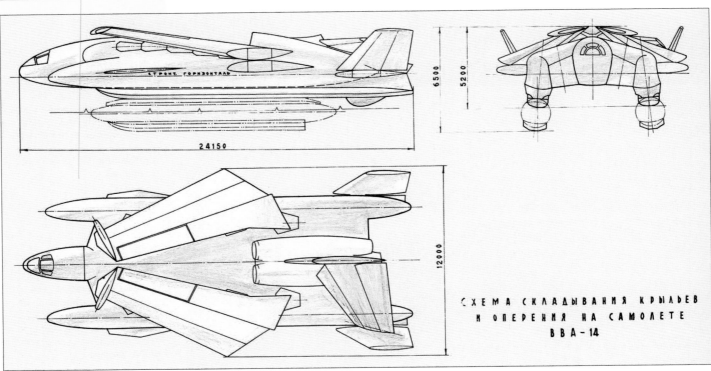

СХЕМА СКЛАДЫВАНИЯ КРЫЛЬЕВ
И ОПЕРЕНИЯ НА САМОЛЕТЕ
ВВА-14

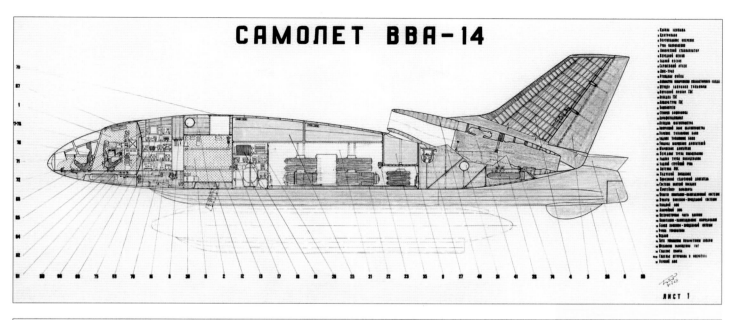

thrust required for vertical take-off and landing to a value less than 1 in relation to the take-off weight.

In 1967 the design team led by Bartini completed an advanced development project of the VVA-14 – a VTOL amphibian with two cruise engines and 12 lift engines. As distinct from later configurations, actually built or projected, this initial project version of the VVA-14 featured a single vertical tail. Dimensions and design performance figures were as follows: length 24.15 m (79 ft 2¾ in), wing span 29.98 m (98 ft 4⅜ in), maximum AUW 50,000 kg (110,250 lb), maximum useful load 2,000 kg (4,410 lb), maximum speed 790 km/h (491 mph), cruising speed 670 km/h (416 mph), loitering speed 350-375 km/h (218-233 mph), service ceiling 9,500 m (31,160 ft), range with maximum fuel load 5,200 km (3,232 miles), time on ASW mission at a distance of 100 km (62 mph) from base – 4.3 hours, crew – 3 persons.

The baseline version was presumably intended to be operated from ground airfields or from sea surface and had non-folding wings. In parallel, a shipboard version of the VVA-14 was projected in 1967. An advanced development project of this version envisioned folding wings and fin that would enable the VVA-14 to be based aboard Type 1123 ASW cruisers (SNS *Moskva* and SNS *Leningrad*) carrying Kamov Ka-25 *Hormone* helicopters, or specially modified large cargo vessels and tankers, or ASW cruisers built specially as VVA-14 carriers.

It was envisaged that the VVA-14 would be operated in three versions differing in equipment

Opposite page, top: A three-view of the VVA-14 in single-fin configuration from the project documents.

Opposite: This is how the VVA-14's wings and tail unit were to fold for deck storage.

Above: Two cutaway drawings of the VVA-14, showing the lift engines with fan attachments and the weapons bay.

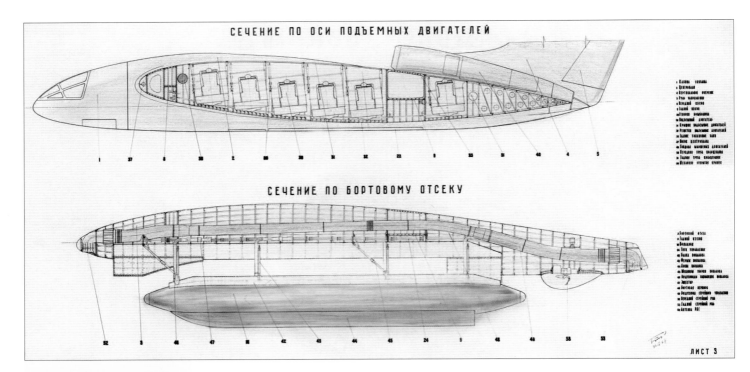

СЕЧЕНИЕ ПО ОСИ ПОДЪЕМНЫХ ДВИГАТЕЛЕЙ

СЕЧЕНИЕ ПО БОРТОВОМУ ОТСЕКУ

ЛИСТ 3

Two more drawings showing sections along the lift engine axes and along the sponson.

One of the ground test rigs used in the VVA-14 programme.

and armament: a search version, a search-and-strike version and a dedicated strike version. In one of the configurations of the latter, studied in 1967, the VVA-14 was to be fitted with the prospective *Polyus* (Pole) search-and-strike system enabling it to deal strikes against missile-carrying submarines

at a distance of not less than 200 km (124 miles) from the aircraft. In this version the VVA-14 was to carry under the fuselage a large air-to-surface missile weighing 3,000-4,000 kg (6,615-8,820 lb) and measuring 9.5 m (31 ft 2 in) in length. Additional equipment comprised a radar ranging device, an

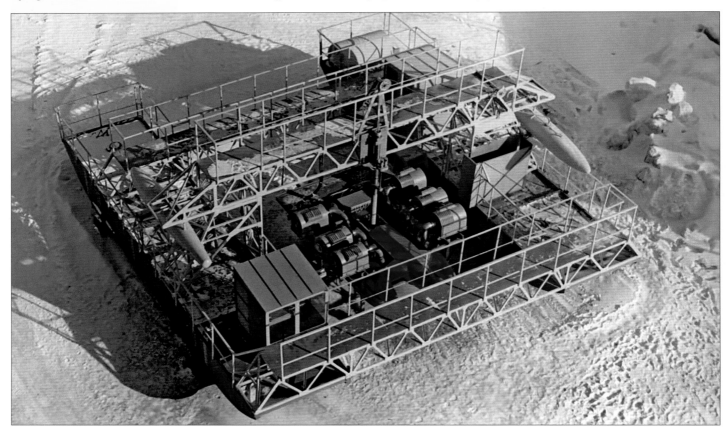

One more view of the VVA-14 waterborne at Taganrog. The starboard entry hatch is open.

Here CCCP-19172 is seen in flight with the floats deployed. Interestingly, the ventral strakes are in place.

Above: One more view of the VVA-14 flying with the floats deployed. Note the characteristic nose-up attityde in level flight.

Page 152: A cutaway drawing of the VVA-14PS in medevac/rescue configuration.

Page 153: A cutaway drawing of the VVA-14PS in space capsule retrieval configuration.

Below: Here CCCP-19172 is seen with the floats stowed for high-speed flight.

VVA-14 – shipboard version (project)

The initial project of the VVA-14 ASW aircraft from 1967 existed in the basic (water- or ground-based) version and the shipborne version. The latter differed in being provided with folding wing panels and tail surfaces; it was intended for being based aboard Type 1123 ASW cruisers, specially adapted large-capacity bulk cargo ships and tankers, or ASW cruisers designed with the VVA-14 in mind.

VVA-14 – transport version (project)

A transport version of the VVA-14 would be able to carry 32 passengers or 5 tons (11,025 lb) of cargo over a distance of up to 3,300 km (2,050 miles).

VVA-14PS – SAR version (project)

The VVA-14 ASW amphibian provided the basis for a project of search-and-rescue version, the VVA-14PS (*poiskovo-spasatel'nyy* – SAR, used attributively) which was studied in the Beriyev Design Bureau in 1973. The aircraft, having all the operational qualities of the baseline version with regard to take-off and landing, was intended for spotting and recovery of spacecraft and reentry modules of manned spacecraft after splashdown in the world ocean. It was also expected to conduct the search and rescue of aircraft, surface ship and submarine crews in distress at sea. In the space programme support version the fuselage of the VVA-14PS housed a compartment with medical

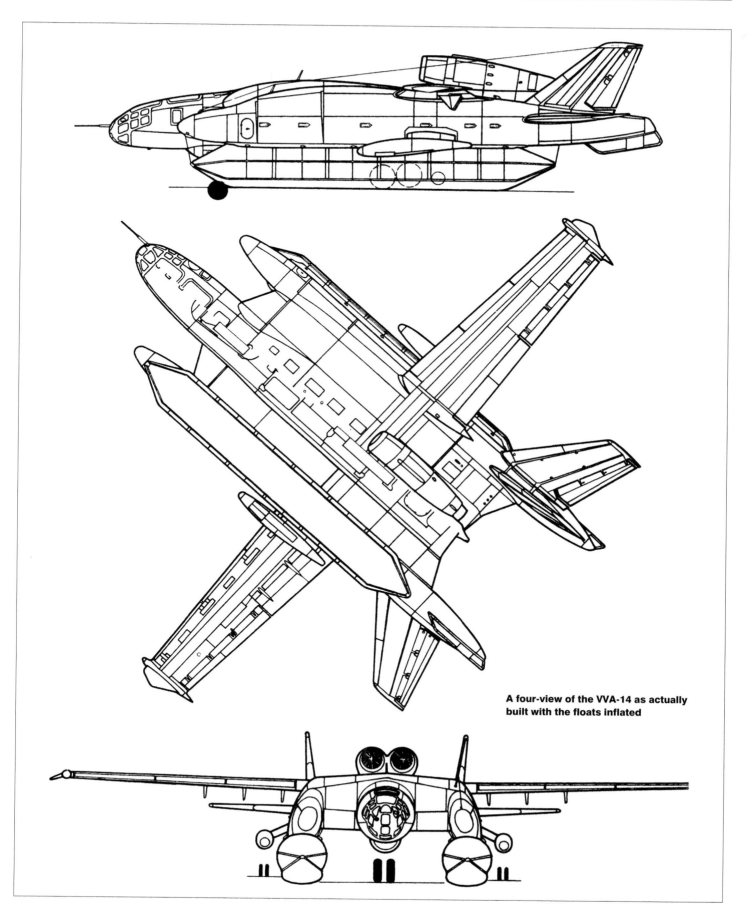

A four-view of the VVA-14 as actually built with the floats inflated

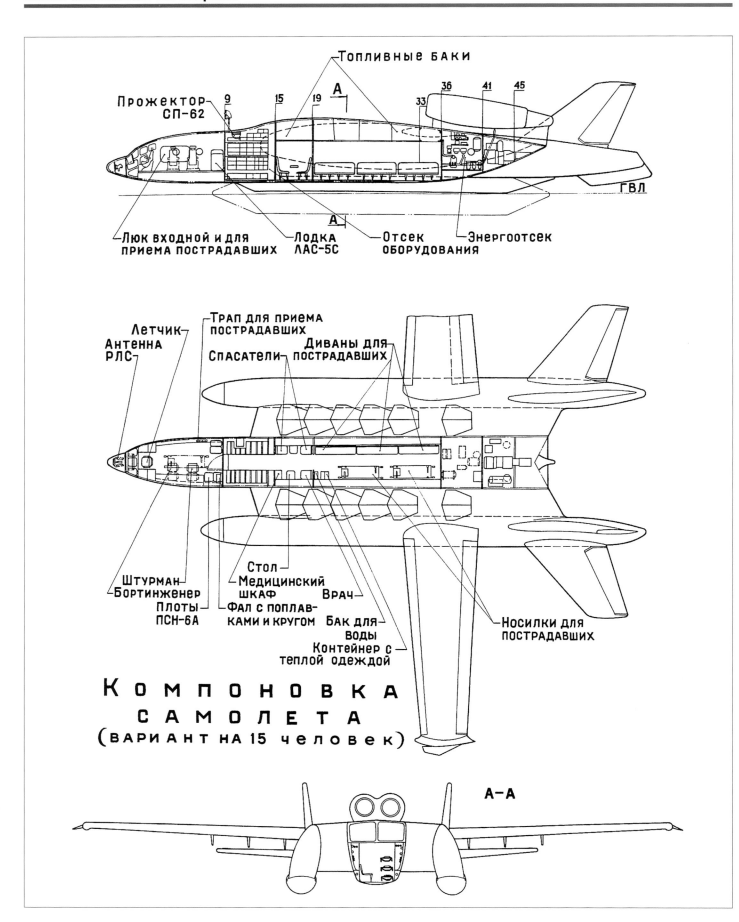

Топливные БАКИ

Прожектор
СП-62

9   15   19   А

33   36   41   45

А

ГВЛ

Люк входной и для
приема пострадавших

Лодка
ЛАС-5С

Отсек
оборудования

Энергоотсек

Летчик
Антенна
РЛС

Трап для приема
пострадавших

Спасатели

Диваны для
пострадавших

Штурман
Бортинженер

Плоты
ПСН-6А

Стол
Медицинский
шкаф

Фал с поплав-
ками и кругом

Врач

Бак для
воды

Контейнер с
теплой одеждой

Носилки для
ПОСТРАДАВШИХ

КОМПОНОВКА
САМОЛЕТА
(вариант на 15 человек)

А-А

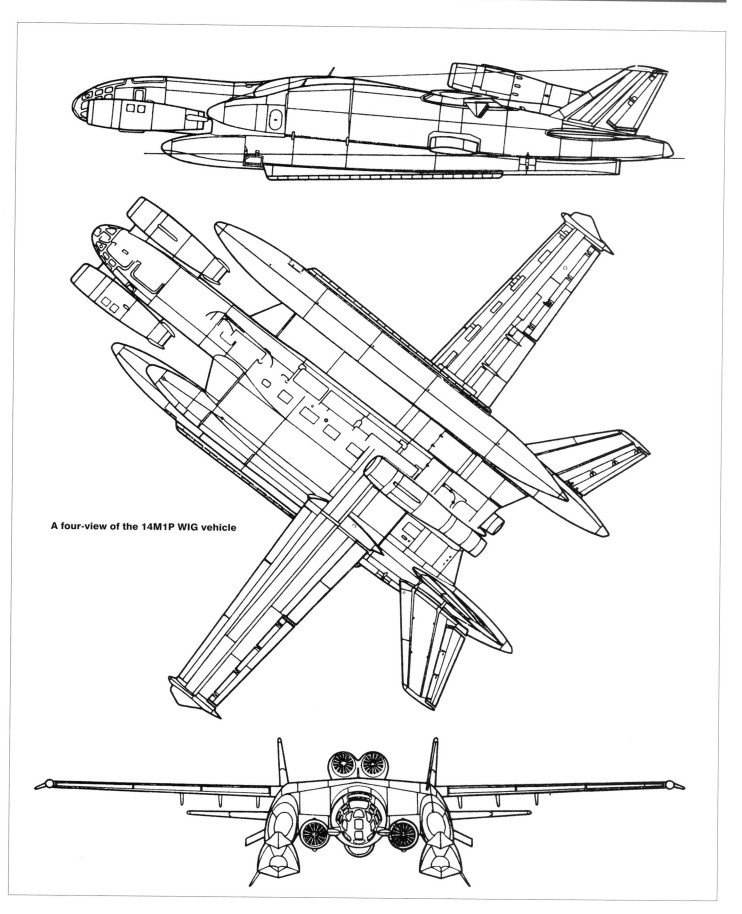

A four-view of the 14M1P WIG vehicle

The disassembled 14M1P at the Monino museum with the fake registration CCCP-10169.

This view shows the ejection hatches. Parts of the airframe are lying beside the vehicle.

'scoop' in which the exhaust gases of the booster turbojets created an air cushion. To enhance the effectiveness of the take-off and landing device, half-submerged skegs were mounted on the floats. The undercarriage was also modified. The original bicycle undercarriage was retained, but it could not be used because of the damage that would be inflicted by the jet efflux of the nose-mounted engines. Therefore the nose and main units were retracted and their wheel well doors were sealed; the outrigger struts were removed. Instead, four non-retractable units of a beaching gear were mounted on the floats, the front units being of a castoring type with a locking mechanism to fix them in flight. Available photos show the 14M1P prior to this modification, with the bicycle undercarriage extended and the original outrigger struts attached to anhedral pylons, but without the streamlined fairings into which they previously retracted.

All the modifications were completed in 1976. The modified aircraft was no longer a VTOL machine, and the letters VVA were deleted from its designation; now it was called 14M1P. It was, in effect, a WIG vehicle in its own right. In Russian sources it is styled as an *ekranolyot*, that is, a vehicle capable of flight both in the WIG mode and in an aircraft mode. The modified machine was subjected to testing which was conducted under the direction of project test engineer I. K. Vinokurov, with the original test crew comprising pilot Yu. M. Kupriyanov and navigator L. F. Kuznetsov (later replaced by pilot V. P. Dem'yanovskiy and navigator E. V. Vedel'). The 14M1P was not intended for fully fledged flight testing; it was, in fact, a full-scale test rig for verifying the viability of the concept.

Testing of the new machine started as pure frustration. The aircraft stubbornly refused to leave the

Close-up of the forward fuselage, showing the booster engine pylon attachments.

Many years after arriving at Monino, the 14M1P still has not been reassembled and is in a pitiful state.

ground during attempts to take off from a runway. This was attributed to dust ingestion by the front engines (the runs were performed on a grass airstrip).

Trials on water proved troublesome as well. The machine was unwilling to make turns, because the turning moments from the differential thrust of the engines were too small. Seagoing trials showed that movement in the dynamic air cushion mode was stable and the use of blowing from the front-mounted engines was a viable solution. The exper-

imental material thus obtained was sufficient for the creation of a flyable example of the machine, but this required a thorough reworking of the design. This was clearly unrealistic, given the lukewarm attitude of the customer.

The work on the 14M1P was abandoned, not least because the Beriyev Design Bureau was over-burdened with other current projects, such as the Tupolev Tu-142MR *Bear-J* communications relay aircraft and the A-50 *Mainstay-A* – the AWACS version of the IL-76MD *Candid-B* transport. The 14M1P

Two artist's impressions of the 'aircraft T' in a steep climb.

The 'T' had several versions; this one had a single fuselage and compound-type wings.

A desktop model of the 'aircraft T' showing the two pairs of booster engines under the nose and the aft-mounted pair of cruise engines located between the twin tails.

Another model of the 'aircraft T' with slightly different booster engine nacelles and forward sweep on the centre section leading edges.

was turned into a floating laboratory, and in 1987 it was sent to the Soviet Air Force Museum in Monino. The machine was delivered by barge to Lytkarino, a small settlement near Moscow. There it was unloaded from a ship to the bank of the Moskva River from where it was to be transported by a helicopter. While awaiting the arrival of the helicopter, the machine remained unattended, with lamentable consequences – it was badly damaged and partly plundered by unknown persons. The damaged aircraft was ferried by a Mi-26 helicopter to the Monino Museum where it remains to this day in a pitiful condition. Incidentally, the vehicle, as shown in Monino, wears the registration CCCP-10687, as distinct from the registration CCCP-19172 which was worn by the VVA-14 (1M) and the 14M1P during tests. One might suppose that this is the second example of the VVA-14, the 2M. However, this sup-

The floats of the 'aircraft T' and their pylons formes a cavity in which the forward engines were to create an air cushion.

The EA-1200 *ekranoplan* project by Bogatyryov, showing the disc-shaped fuselage, the three booms carrying the foreplanes mounting the booster engines, the aft-mouted cruise engines, the wingtip-mounted T-tails and six dorsal loading hatches

edge on the centreline (it also accommodated the flight deck). Being thus placed ahead of the wings, the engines could direct their jet efflux under the wings, augmenting the ram air cushion during take-off. This project also featured additional surfaces attached to the wingtip floats at an angle – again, in keeping with the traditional Lippisch layout. The vehicle apparently had no fuselage, the wings providing the space to accommodate the useful load. The other of the two projects, on the other hand, had a sort of fuselage protruding ahead of the wing leading edge. The forward fuselage was flanked by two packs of turbojet engines, each of them comprising three engines in a common nacelle with separate air intakes. The engines apparently were intended to act in the double role of cruise engines in cruise flight and booster engines during take-off. The engine packs, in turn, were flanked by swept-back foreplanes whose tips were joined to the wing leading edge by special booms. In this case there were no additional aerodynamic surfaces on the wingtip floats.

Another pie-in-the-sky *ekranoplan* project by Bogatyryov with forward-swept anhedral tailplanes tipped by floats. The nose-mounted booster engines appear to be way too low above the water.

Front view of the Be-1 on its beaching gear at Taganrog, showing the float design and the angled forward hydrofoils.

Opposite: This three-quarter view shows well the V-shaped rear hydrofoils. The hydrofoils necessitated a tall beaching gear; all three units were equipped with floats to facilitate installation and removal.

The other two projects by Bogatyryov were altogether different. One on them was a kind of a flying wing of an unorthodox shape. The main part of the wing was nearly circular in planform, and was supplemented by sweptback outer wing panels at the rear. Placed at the tips of these were two T-tails featuring sweepback on all surfaces. A third vertical tail was mounted on the centreline at the wing trailing edge. The powerplant included four booster turbo-

jets placed on the upper surfaces of the foreplanes which were carried by three booms protruding from the wing leading edge (the centre one incorporated the flight deck). A further four turbojets were placed in pairs at the wing trailing edge near the roots of the outer wing panels.

Finally, the fourth project featured, in fact, a tail-first layout with the wings sharply swept back, while the canard foreplanes were swept forward and pro-

Side view of the Be-1, showing the wing endplates, the wing/float junction and the water rudders.

Rear view of the Be-1, showing the three trailing edge flaps and the engine pylon.

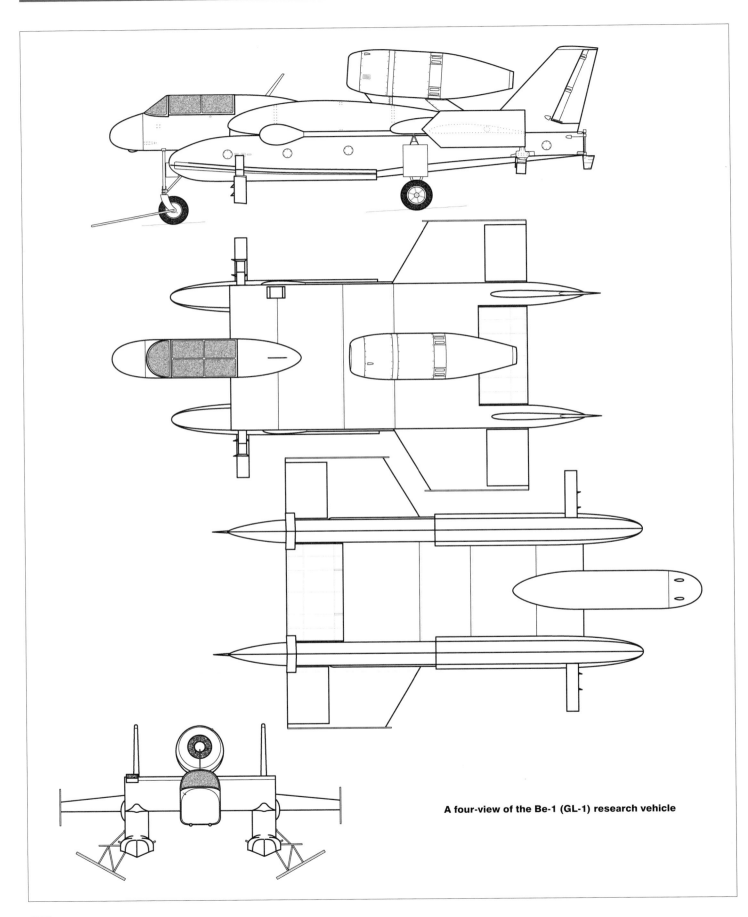

A four-view of the Be-1 (GL-1) research vehicle

ion with the help of blowing from the nose-mounted engines.

A-70 ('D')

This project, sometimes referred to as the Be-70, was presumably similar to the P-2 and served the same purpose. Precise configuration is unknown. Interestingly, two designations – A-70 and A-2500 – accompanied one photo of a towing-basin model of an 'ultra-heavy amphibious aircraft' published in the same source as above. The configuration of the model corresponds to that of the Be-2500 project with aft-mounted engines (see below). Apparently, the A-2500 and the Be-2500 are two designations of the same project. The relation between the A-70 (Be-70) and the A-2500 (Be-2500) is not clear (these may be projects sharing a common layout but differing in dimensions).

Be-750 seaplane/WIG craft (project)

This project was revealed for the first time (without designation) as a drawing published in the materials of a Ghelendzhik symposium on hydro aviation. The designation became known in 2002 when this

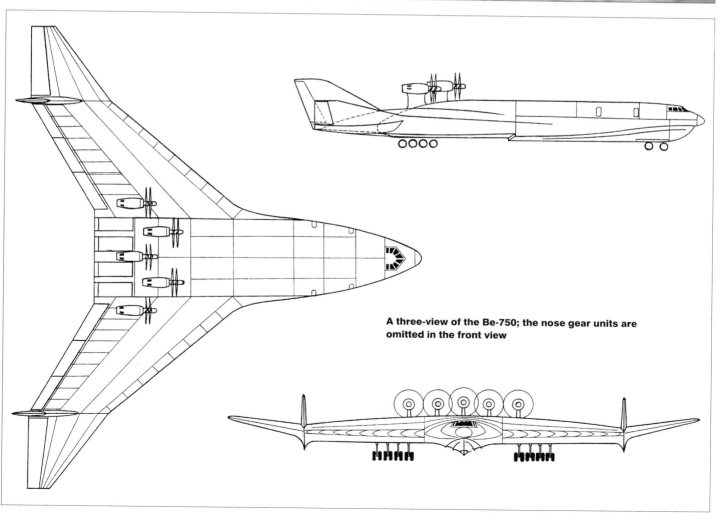

A three-view of the Be-750; the nose gear units are omitted in the front view

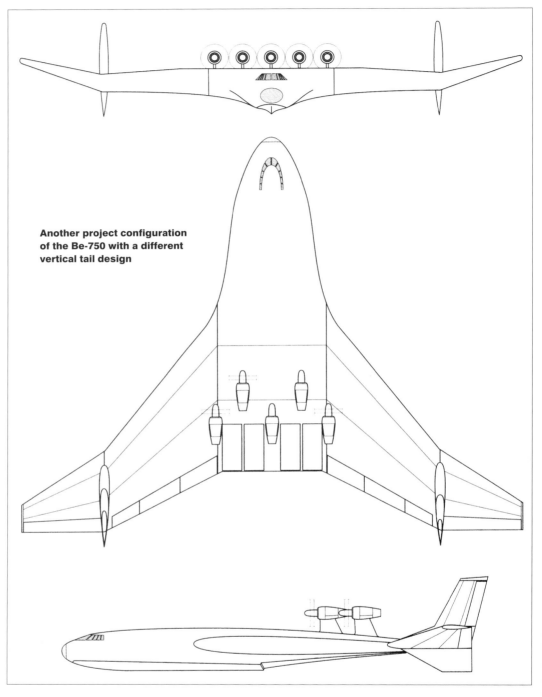

Another project configuration of the Be-750 with a different vertical tail design

drawing appeared in a book on the activities of the Beriyev Design Bureau. Described as 'a super-heavy seaplane', this is a vehicle of the flying wing type with wings featuring double sweepback on the leading edges and blending smoothly into a boat-type fuselage (integral layout). It is powered by five turboprops mounted on pylons above the aft fuselage to drive contra-rotating propellers. Directional stability and control is ensured by two vertical tails mounted on the wings close to wing tips. The wings are provided with ailerons outboard of the fins and elevons (flaps) inboard of them. Other particulars

are unknown. Presumably this design makes use of ground effect in the take-off and landing mode.

Be-750 version powered by three turboprops (project)

A Beriyev TANTK exhibition stand at one of the Ghelendzhik hydro aviation shows carried a drawing of a hydroplane (no designation was given) which was identical in plan view to the Be-750 as described above, the only visible difference being the power-plant. In this case it comprised only three turboprop engines, also driving contra-rotating propellers.

The drawing was accompanied by the following particulars: take-off weight 750 tonnes (1,653 lb); payload 300 tonnes (661,500 lb); cruising speed 800 km/h (497 mph); range 7,800 km (4,848 miles). The take-off weight of 750 tonnes (the same as the designation digits in the previous case) makes it reasonable to surmise that it is actually the same project, probably with the same designation, in which the number of engines was reduced to three due to their greater power (no engine type was indicated in either case). This pattern of providing powerplant options with a different number of engines is also in evidence in other Beriyev designs.

Be-800 seaplane/WIG craft (project)

This *ekranolyot* features a highly unorthodox layout. In addition to the main boat-hull fuselage, it has two smaller boat hulls making the vehicle a sort of trimaran. The vehicle features a joined wing of high aspect ratio. It comprises the high-set swept-back forward wing resting on the central part of the fuselage and the swept-forward aft wing mounted on the twin-fin tail unit and joined to the forward wings at 70% of their span.

The powerplant comprises six Kuznetsov NK-62M turbofans with an aggregate take-off thrust of 35 tonnes (77,175 lbst). Four of these engines are mounted on the aft fuselage and the other two on the upper surface of the forward wing. The flight deck is located in the front part of the main fuselage; the central pressurised part of the fuselage houses six cargo compartments which can be converted into passenger compartments. The fuel is stored in the centre section of the forward wing and in the hulls. Unfortunately, no illustration is available.

Be-1000 amphibian (project)

Broadly similar in layout to the Be-750, this amphibious aircraft an integral type flying wing layout with the wings blending smoothly into the fuselage. The latter is shaped as a boat hull with transverse and longitudinal steps. The thick wing roots are partly submerged when the aircraft is afloat. The wings have a leading-edge sweepback of 50° and are fitted with large winglets serving as twin vertical tail surfaces. Seven Kuznetsov NK-62M turbofans are mounted above the central part of the wing. The aircraft's undercarriage comprises a nose unit, main units and supporting units with six-wheel bogies. The forward fuselage houses the flight deck, while passengers are accommodated on two decks in the pressurised front section of the hull. Volumes in the wing and fuselage aft of the passenger compartments are used for cargo holds. The Be-1000 has a design cargo carrying capacity of up to 600 tonnes (1,323,000 lb) and a maximum cruising speed of 600 km/h (373 mph).

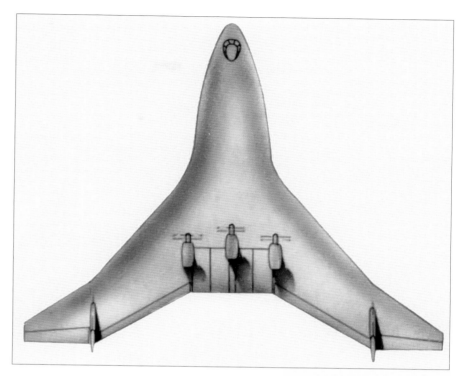

Be-? – WIG craft in the 1,000-tonne AUW class (project)

The Beriyev TANTK exhibition stand mentioned above also carried a plan view of another prospective hydroplane for which the following particulars were cited: take-off weight 1,000 tonnes (2,205,000

A drawing of a three-engined version of the Be-750.

A drawing of the Be-1000 project with eight forward-mounted engines.

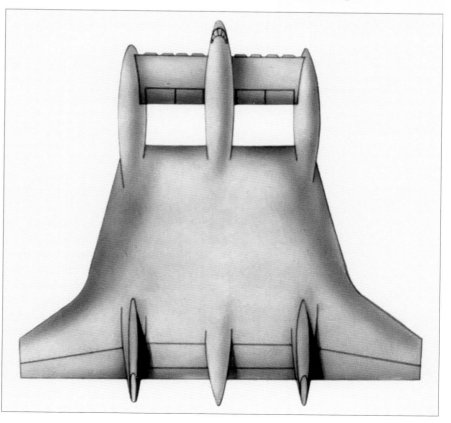

An artist's impression of the Be-1200 at its landing stage with the loading doors in the rear fuselage and wing trailing edge open. Passengers embark through landing gantries at the front.

lb); payload 400 tonnes (881,830 lb); cruising speed 660 km/h (410 mph); range 8,000 km (4,970 miles). The layout of this huge machine was unorthodox indeed. It was a 'flying wing' (tailless aircraft) with three fuselages blended into a rectangular wing centre section of very low aspect ratio and wide chord. The wing centre section was supplemented by outer wing panels with a compound sweepback on the leading edges. The trailing edge of the outer wing panels was occupied by ailerons, while the trailing edge sections located between the fuselages carried control surfaces that might perform the role of elevators or flaps. Vertical tails were mounted on the aft ends of the two outboard fuselages. The front parts of the three fuselages were linked by foreplanes. The drawing showed four turbofans mounted under each foreplane (eight in all). Quite obviously, their efflux could be directed under the wing for creating an air cushion during take-off. This feature, reminiscent of Alekseyev's and Bartini's designs, makes it possible to class this vehicle as having much in common with WIG vehicles proper, even though it may have been intended to be operated mainly in the aircraft mode.

Be-1200 (Be-2000)

This is a project of an amphibious flying boat which was presented in model form by the TANTK at the Ghelendzhik hydro aviation show in 1996 as Be-1200; at Ghelendzhik-98 the same model bore the designation Be-2000. The layout of this gigantic machine powered by six powerful jet engines clearly places it in the category of Type C WIG vehicles, or ekranolyots, in Russian parlance. In its general configuration it comes close to a flying wing, though it features an empennage in the shape of two wing-mounted vertical tail surfaces topped by separate horizontal stabilizers. The wings have considerable sweepback; a thick, wide-chord centre section with a straight trailing edge blends into the fuselage. Outer wing panels of smaller thickness are attached to the centre section, producing what is known as composite layout wings. Large pylons acting as stabilising floats are attached to the underside of the wings at the junction of the centre section and the outer panels; they also act as walls forming a closed space under the wing into which the exhaust of the four nose-mounted engines is directed at take-off,

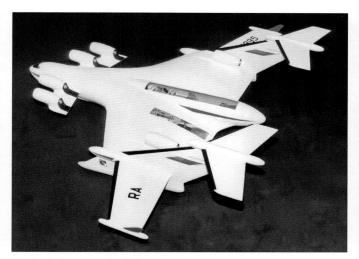

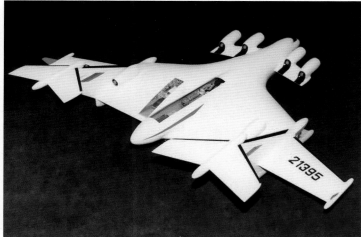

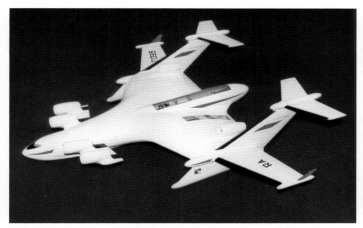

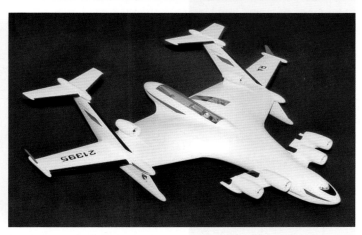

A display model of the giant Be-1200 ekranoplan amphibian, showing the main cargo bay in the fuselage and one of the port wing cargo bays.

Four boost/cruise turbofans are mounted on the foreplanes, augmented by two cruise engines on the wing centre section.

The main cargo cabin of the Be-1200 is large enought to take a Kamov helicopter with its tall rotor mast, as shown here. The wing bays are intended for vehicles and the like.

An artist's impression of the Be-2500P which is similar to the model on the previous pages except for the placement of the cruise engines on the vertical tails.

in the same fashion as on Alekseyev's KM and Loon' vehicles. In cruising flight they presumably supplement the thrust of two cruise engines placed on pylons above the aft part of the wing centre section inboard of the fins.

According to the information subsequently made available by Beriyev, the Be-2000 has a take-off weight of 2,000 tonnes (4,410,000 lb). It has been projected in several versions differing in the wing planform and the number and arrangement of the engines. These versions fall into two basic groups: those featuring the assisted take-off engines and those lacking them. The Be-2000 version described above belongs to the former of the two categories and is designated the Be-2000P (P stands for *pod**doov*** – 'blowing', or power-augmented ram air cushion). In one of the configurations of the Be-2000P six NK-62M turbofans are mounted on the swivelling horizontal pylons installed on the forward fuselage. A further four engines are installed in the aft part of the wing centre section or on pylons attached to vertical tails. The swivelling pylon doubles as a foreplane; the

engines installed on it act as cruise engines in cruise flight.

The Be-2000 *sans suffixe* lacking the booster engines features in one of its configurations the powerplant comprising 12 NK-12M turboprops mounted on pylons above the upper surface of the wing.

The boat hull of the aircraft has a double-deck layout, the upper deck providing accommodation for 730 passengers and the lower deck being configured as a cargo hold measuring 88 x 8 x 4 m (288 ft 8¾ in x 26 ft 3 in x 13 ft 1½ in). In a SAR version the space on both decks can be used for accommodating the survivors (544 heavily wounded and 1,214 rescuees).

Be-2500

Information on this project was presented by the Beriyev TANTK at the Hydro Aviation Show 2000 in Ghelendzhik in September 2000. This huge machine weighing 2,500 tons (5,500,000 lb) at take-off is a further development of the Be-1200 (Be-2000) layout. It combines the properties of a

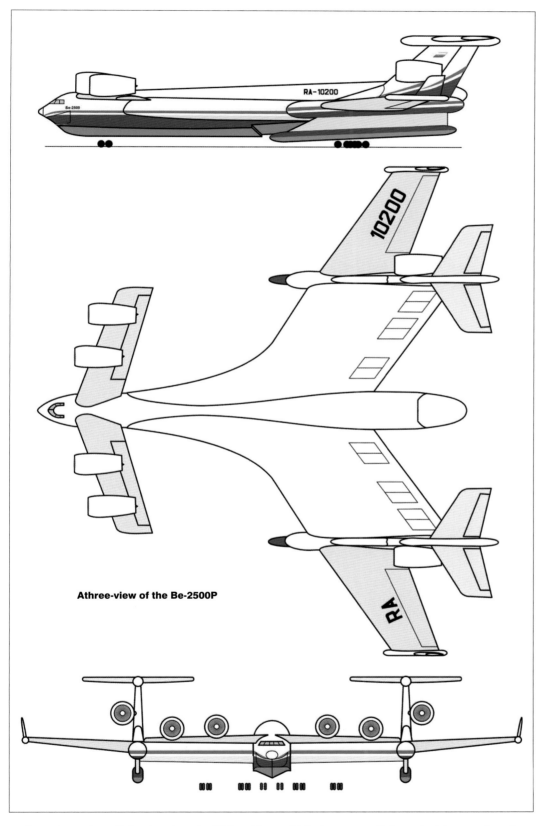

Athree-view of the Be-2500P

traditional seaplane (flying boat) and a WIG vehicle, being capable of flying both in ground effect
at a height of 4-10 m (13-33 ft) and at a high altitude of 8,000 to 12.000 m (26,240 to 39,360 ft). The

vehicle features a 'flying wing' layout, the wings
blending smoothly into the fuselage (integral layout). During take-off the mutual position of the
boat-type fuselage and the wings with a large-area

S-90-200 specifications (provisional)	
Powerplant	2 x NK-12MK
Length	40.0 m (131 ft 2⁵¹⁄₆₄ in)
Wingspan	60.92 m (199 ft 10²⁷⁄₆₄ in)
Height	11.85 m (38 ft 10³⁵⁄₆₄ in)
Wing area, m2 (ft2):	
gross	757.2 (8,150)
centre section	502.0 (5,403)
outer wing panels	254.0 (2,691)
Tailplane area, m² (sq ft)	69.9 (752)
Take-off gross weight, kg (lb)	132,000 (290,930)
Fuel load, kg (lb)	58,000 (127,830)
Max payload, kg (lb)	20,000 (44,080)
Wing loading at normal	
take-off weight, kg/m² (lb/sq ft)	172 (35.08)
Max speed, km/h (mph)	470 (292)
Cruising speed, km/h (mph)	380 (236)
Max range with 220	
passengers, km (miles)	8,000 (4,968)
Max flight altitude, m (ft)	1,500 (4,920)
Ground effect altitude, m (ft)	2.0-5.5 (6.5-18)

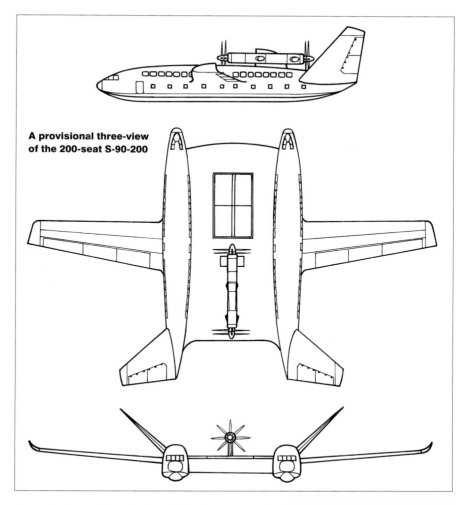

A provisional three-view of the 200-seat S-90-200

The powerplant comprises two 15,000-ehp Kuznetsov NK-12MK turboprops driving eight-blade contra-rotating propellers. The engines are installed in tandem in a pylon-mounted nacelle on the rear part of the wing centre section. During take-off and landing the propeller wash is ingested by a special device and directed through a system of air passages to be discharged under the bottom surface of the wing centre section. Thereby a static air cushion is created, supplementing the dynamic air cushion produced by ram air during cruise flight. There is a provision for a flexible skirt which is retracted during the transition from the static to the dynamic air cushion; the longitudinal segmented components of the skirt act as shock absorbers during landing.

Here are some basic specifications and performance figures: length, 40 m (131 ft 2⁵¹⁄₆₄ in); wing

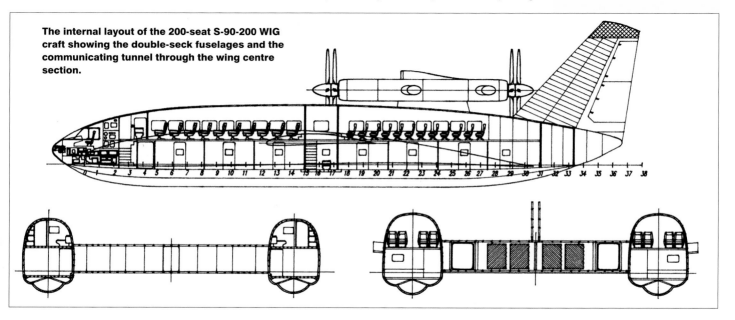

The internal layout of the 200-seat S-90-200 WIG craft showing the double-seck fuselages and the communicating tunnel through the wing centre section.

This model depicts one of the projected WIG craft known as the S-90-40. The three-turbofan craft bears a strong resemblance to the A-90 Orlyonok.

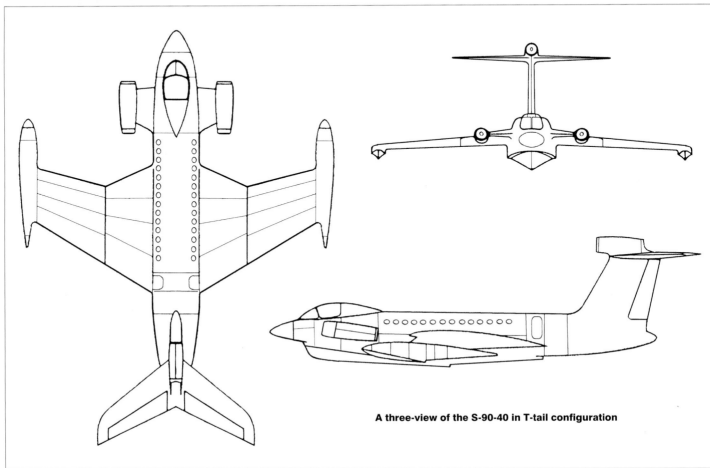

A three-view of the S-90-40 in T-tail configuration

span, 60.92 m (199 ft 10²⁷⁄₆₄ in); take-off gross weight, 132,000 kg (291,000 lb); payload, 20 tons (44,080 lb); maximum speed, 470 km/h (292 mph); cruising speed, 380 km/h (236 mph);* maximum range with 220 passengers, 8,000 km (5,000 miles); maximum flight altitude, 1,500 m (4,920 ft);* ground effect altitude, 2.0-5.5 m (6.6-18 ft).

*Note: An advertising leaflet cites the cruising speed as 470 km/h (292 mph) and the service ceiling as 2,500 m (8,200 ft).

S-90 (twin-fuselage configuration)

This S-90 project was a twin-hulled vehicle, each hull having its own crew cockpit and a passenger cabin, with an entry door amidships. The hulls were connected by a shoulder-mounted rectangular wing centre section occupying about half the length; the high-aspect-ratio outer wing panels were unswept, with strong dihedral, and small winglets. As on the S-90-8 and the S-90-200, the tail unit consisted of trapezoidal fin and rudder assemblies canted strongly outward. The craft had a mixed powerplant: an airfoil-section cross-member

connecting the forward hull sections carried two turboprop engines side by side for creating a static air cushion, while the cruise engine (a turbofan) was mounted on a tall pylon at the trailing edge of the wing centre section.

S-90 (another twin-fuselage configuration)

This configuration was similar to the one described above but had an all-jet powerplant; the pylon carried two cruise turbofans side by side, and two more turbofans of a larger type were installed on top of the wing centre section near the leading edge.

S-90-40 (single-fuselage aircraft-type configuration)

This 40-seat passenger WIG vehicle powered by three turbofans was intended for medium-haul routes over sea and ocean areas. It differed from those described above in being based on a single-fuselage 'aircraft-type' layout reminiscent of Alekseyev's designs, featuring a single cruise engine at the top of the fin and two booster engines in nacelles flanking the forward fuselage. They were

A very different model of a passenger WIG craft bearing the same designation S-90-40, with twin skegs and twin tails carrying the engines.

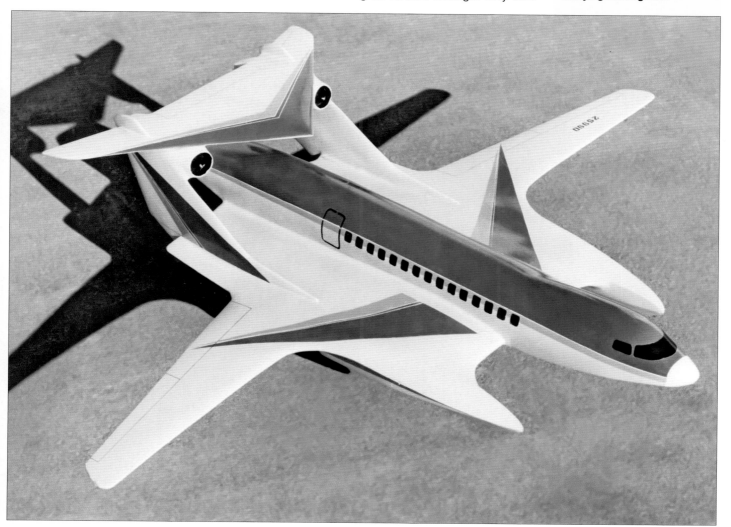

A model of the Su-2000 light passenger WIG craft.

set at an angle permitting their exhaust to be directed under the wings for creating a static air-cushion. A characteristic feature of the design was the planform of the wings which were slightly for-ward-swept, with a leading-edge kink at half-span. The machine had a design cruising speed of 400 km/h (248 mph) and a range of 2,000 km (1,240 miles).

S-90-40 (a different single-fuselage configuration)

A vehicle of approximately the same size, but of a different configuration had an airliner-type fuselage sitting on top of a rectangular wing centre section flanked by long floats (skegs) to which high-aspect-ratio forward-swept outer wing panels equipped with flaps and ailerons were attached. The wing centre section carried twin swept-back vertical tails

connected at the top by an arrowhead-shaped sta-biliser with a three-section elevator. The powerplant consisted of two turbofan engines located at the fin/tailplane junctions on the inboard side.

S-90 (Su-2000)

In 1997 the OKB started work on the S-90 project which featured a layout considerably differing from the two projects described above. Some reports mention Aleksandr Polyakov as the chief designer for this project. In several publications and adver-tising materials this vehicle was referred to simply as the S-90, but there was also a display model of this machine bearing the designation Su-2000.

This time the vehicle makes use of the 'Bartini layout', featuring an integral fuselage (lifting body) and an air cushion undercarriage provided with a special fan and a separate power unit. It is intended

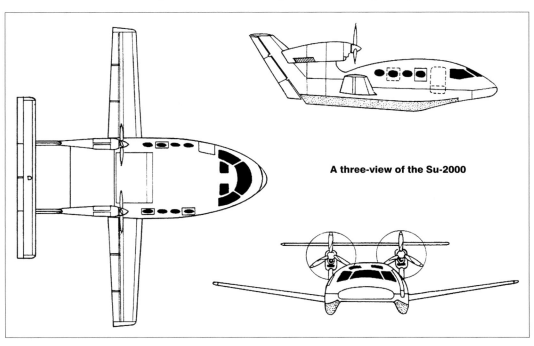

A three-view of the Su-2000

Basic design specifications of the S-90:	
Length	13.83 m (45 ft 4³¹⁄₆₄ in)
Wing span	16.91 m (55 ft 5⁴⁷⁄₆₄ in)
Height	4.85 m (15 ft 10¹⁵⁄₁₆ in)
Crew	2
Number of passengers	19, 26 or 31

Weights and performance figures are indicated for three operation modes: aircraft/WIG craft/ACV

Maximum take-off weight, kg (lb)	7,900/9,500/10,500 (17,420/20,940/23,150)
Empty equipped weight, kg (lb)	4,500/4,500/4,500 (9,920/9,920/9,920)
Maximum payload, kg (lb)	2,500/3,000/3,000 (5,510/6,615/6,615)
Maximum fuel load, kg (lb)	3,000/3,000/3,000 (6,615/6,615/6,615)
Cruising speed, km/h (mph)	400/400/80 (248/248/50)
Flight altitude, m (ft)	0.5 – 4,000/0.05 (1 ft 7 in – 13,120 ft/0 ft 2 in)
Take-off/landing speed, km/h (mph)	120/120/n.a. (74.5/74.5/n.a.)
Range, km (miles):	
with a maximum fuel load	3,100 (1,925)
with 19 passengers	1,260/2,400 (782/1,490)
with 25 passengers	760/2,450 (472/1,521)
with 31 passengers	1,870/1,970 (1,161/1,223)
Take-off/landing run, m (ft)	230/240 (755/790)

for passenger and cargo transportation both in aircraft mode at high altitude and in ground effect mode. In the latter case it can fly over water surface in sea states up to 2 and over flat areas on the ground with hummocks up to 0.5 m (1 ft 7¹¹⁄₁₆ in) high. Initially the designers studied a layout featuring a hydrodynamically shaped displacing hull with planing steps on the bottom and endplates at the wingtips; it was to have booster engines placed ahead of the centre section leading edge and injecting an air stream under the wing (Alekseyev's layout).

A model of the S-90-4 light amphibious WIG craft powered by a four-cylinder piston engine.

the ES-2M was powered by a 32-hp M-63 engine driving a two-blade wooden tractor propeller of 1.6 m (5 ft 3 in) diameter.

For operations from water, the *ekranolyot* was provided with a detachable boat hull which was attached to the fuselage in four points. In that case the undercarriage monowheel was enclosed in a vertical niche placed along the axis of the boat. When extended, the wheel enabled the vehicle to go out onto a gently sloping bank, taxi and perform a take-off from land while being fitted with the flotation gear. Thanks to the thin airfoil section of insignificant curvature, the clean wings and slender fuselage this *ekranolyot* had better aerodynamic characteristics, improved stability and easier handling compared to the original ES-2.

After testing conducted in the summer of 1976 the ES-2M was shown at the Exhibition of Scientific & Technical Achievements of Young Enthusiasts (NTTM-76 – *Na**ooch**no-tekh**nich**eskoye **tvorch**estvo molo**dyo**zhi*) in Moscow where it won a gold medal.

An-2E (first use of designation)

This WIG vehicle project based on the An-2V floatplane version of the well-known Antonov An-2 *Colt* utility biplane was prepared in the TsLST in 1973 by a group of young specialists led by Yevgeniy P. Groonin. The new machine inherited from the An-2 the fuselage and the Shvetsov ASh-62IR radial

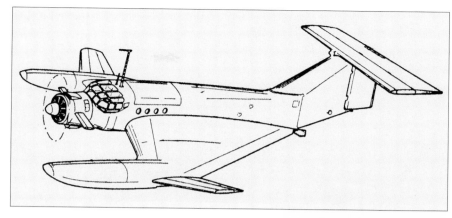

engine with an AV-2 four-blade variable-pitch propeller. These components were mated to new low-set wings of the reversed delta type having marked anhedral and fitted with large wingtip floats; the latter carried small dihedral outer wing panels with ailerons. The vehicle was also given a new T-tail of large span. Since the new wings obstructed the normal portside cargo door, a dorsal loading hatch was provided where the upper wing centre section used to be. Two vanes were installed on the sides of the engine cowling to direct part of the prop wash under the wings, assisting take-off.

The An-2E (*ekranoplan*) was to be provided with a retractable wheel undercarriage which would enable it to operate from land, as well as from the surface of lakes, rivers and littoral areas.

A drawing of the first WIG vehicle to be designated An-2E. Note the dorsal loading hatch and the deflector vanes on the nose.

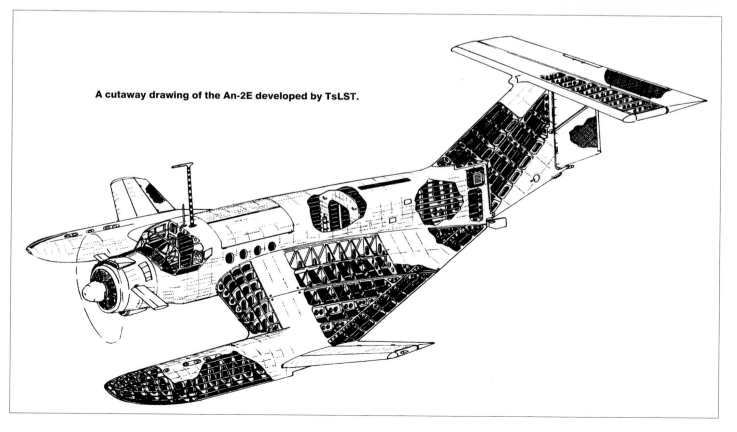

A cutaway drawing of the An-2E developed by TsLST.

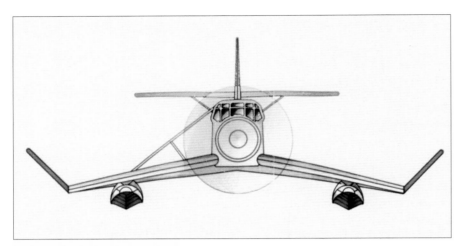

Front view of the An-2E developed by the Moscow Aviation Institute.

One more ekranoplan derivative of the An-2V proposed by the Groonin team.

An artist's impression of the Groonin E-0974 Parawing *ekranoplan*. The large elevator trim tab is noteworthy.

An artist's impression of the TsLST R-1001 Manta. Note the front and rear loading doors.

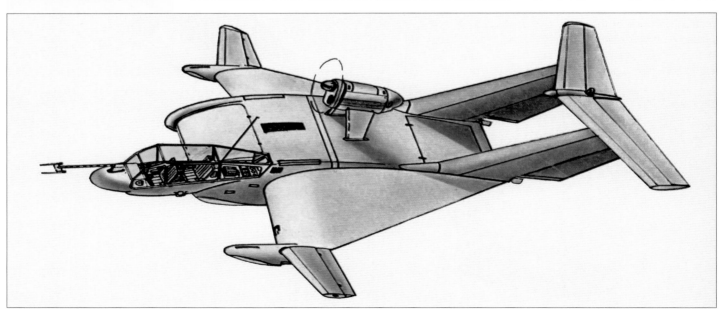

An-2V converted into a WIG vehicle with flexible wings

Another project of an *ekranoplan* based on the An-2V was proposed by a group of young designers in the TsLST. As distinct from the An-2E described above, this project retained practically the whole of the floatplane's airframe – except the wing box, which was replaced by a flexible fabric lifting surface of reversed-delta planform. The rigid leading edges were formed by special outrigger structures attached to both sides of the fuselage and also carrying small surfaces with ailerons, like those used in vehicles of the Lippisch layout.

E-0974 Parawing

This WIG vehicle was designed at the initiative of Yevgeniy P. Groonin. The machine featured an unusual configuration with wings of reversed-delta planform, shaped like a semi-circle when viewed from the front. The wing planform, coupled with the wingtip floats and the aileron-carrying surfaces attached to them, is reminiscent of the Lippisch machines. Originality lies both in the semi-circular front-view shape of the leading edges and in the fact that each of the wings is attached to the fuselage only at the front, while the aft parts serve as booms carrying the twin fins topped by horizontal tail. The vehicle is powered by a 42-hp Czechoslovak Java M-150 motorcycle engine with a pusher propeller of 1.1 m (3 ft 7⁵⁄₁₆ in) diameter.

R-1001 Manta

In 1974 a group of young specialists in the TsLST developed a two-seat *ekranoplan* intended for liaison with the Soviet fishing fleet. The project was designated R-1001 Manta. The machine's configuration was reminiscent of some Lippisch designs. That was seen, among other things, in the asymmetrical location of the cockpit pod which was placed on the port side of the airfoil-shaped body housing a cargo compartment. The outer wing panels followed the Lippisch layout, the tail unit had

twin fins linked by an arrow-shaped horizontal tail. The 210-hp Walter Minor VI engine with a tractor propeller was mounted above the body. The vehicle had an AUW of 1,460 kg (3,220 lb). The Manta did not progress further than the drawing board.

E-120

Built in 1971, that was one of the experimental WIG vehicles designed in the TsLST. This single-seat machine was noteworthy for its wing of circular planform. Such a wing possesses some unique qualities, including the ability to preserve the lift without stalling at very high angles of attack right up to 45°. Such a wing is also capable of using efficiently the ground effect for lift increase and may prove promising for WIG vehicles, provided the controllability issues are solved successfully.

EMA-4

The EMA-4 vehicle, developed and built at TsLST, underwent tests in 1972. It was classed by its

The circular E-120 WIG vehicle, marked '02'. Confusingly, some sources refer to it as the ES-1, though the inscription 'Э-120' can be seen on the starboard wing. The legend on the tail reads *Ispytaniya* (Tests).

Another example of the E-120 marked '03' with the tail unit and powerplant yet to be fitted. Note the lateral strakes. Test pilot Baluyev is shown in the cockpit in the right-hand photo.

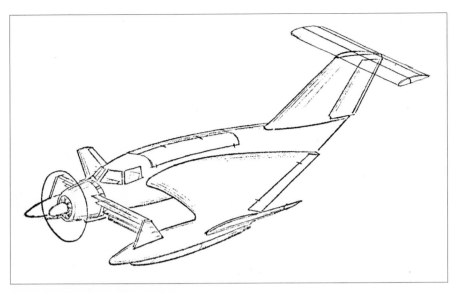

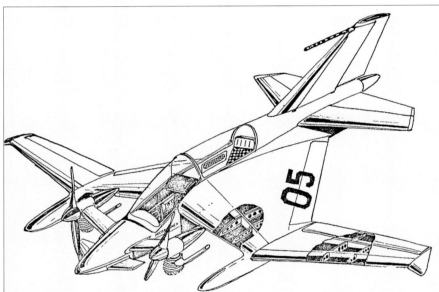

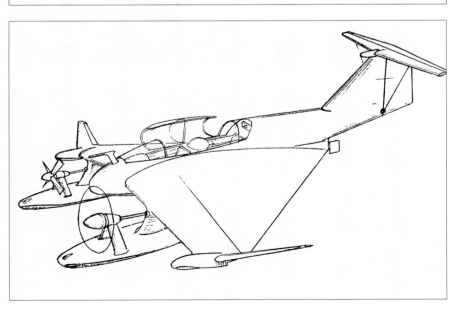

designers as an 'amphibian with aerodynamic off-loading'. The vehicle was a tiny boat of boxy contours whose hull was flanked by aerodynamic lift surfaces forming a disc-shaped wing. A piston engine installed behind an enclosed cockpit drove a ducted two-blade pusher propeller. An all-movable fin located behind the propeller presumably could serve as a rudder. The vehicle apparently could be operated on water in the hydroplaning mode, making use of the WIG effect, and on snow.

E-400

According to an article in the Polish magazine *Skrzydlata Polska*, an experimental WIG vehicle with that designation was designed in the TsLST in 1972. No further details are available.

ESKA EA-6A

This is an improved four-seat version of the ESKA-1. In its contours it is nearly identical to its forerunner, featuring a wider cabin with a curved panoramic windshield and a slightly raised aft fuselage. A scaled-strength model of the EA-6A to 1/4th scale was tested in September 1973. It had a span of 1.75 m (5 ft 8^{57}⁄₆₄ in) and was powered by a 1.8-hp

One of Groonin's WIG vehicle projects resembling a scaled-down version of the An-2E.

A dimilutive WIG vehicle developed by Groonin with two motorcycle engines flanking the nose.

Belowleft: An artist's impression of the Groonin E-0874 with its unusual propeller placement.

Groonin's projects included even an ekranoplan in the form of a human-powered aircraft (HPA).

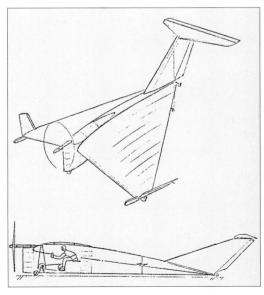

two-cylinder engine driving a propeller of 300 mm (11.8 in) diameter.

TsLST projects for converting aircraft into WIG vehicles

In 1974, at the initiative of Yu. V. Makarov, employees of TsLST in co-operation with specialists from RIIGA (Riga Civil Aviation Engineers Institute) started design work aimed at converting series-built aircraft types into WIG vehicles. They enlisted support from the Section of Ground Use of Aircraft Hardware at the Riga branch of GosNII GA (State Research Institute of Civil Aviation) which dealt, among other things, with converting time-expired aircraft into other types of transport – such as ACVs, aero-sleighs etc.

The TsLST studied the possibility of converting some aircraft types about to be phased out from Aeroflot service (An-2, Lisunov Li-2) and the Be-12 *Mail* ASW amphibian into WIG vehicles. It was presumed that the Be-12 powered by less powerful and, consequently, less fuel-consuming piston engines (as compared to the turboprop-powered baseline aircraft) could find widespread use in the national economy. A project designated Be-12E is described elsewhere in this chapter. Studies involving the An-2, Yak-12 and Yak-40 were conducted by RIIGA (RKIIGA, see below).

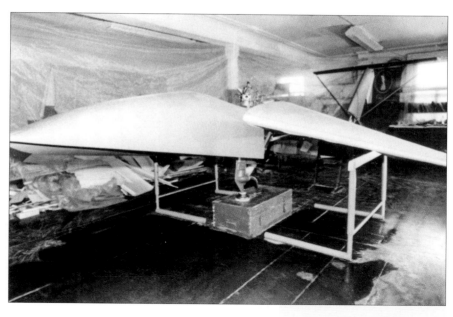

In addition, projects of WIG conversions based on the Il'yushin IL-14 *Crate* airlinerand the An-12 *Cub* transport were developed at TsLST on the level of student graduation projects.

Groonin's projects from 1974-75

During his work in the TsLST Yevgeniy Groonin prepare a big number of interesting projects of WIG vehi-

A speedboat (hydroplane) with aerodynamic offloading developed by Groonin.

Three-view/structural drawings of Groonin's hydroplanes with aerodynamic offloading – the K-550 (left) and MS-298.

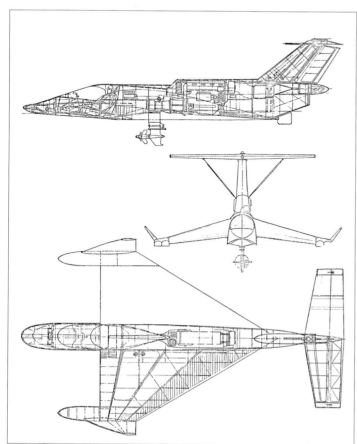

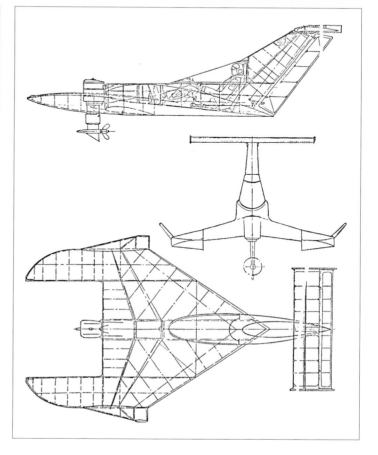

*instit**oot***; currently KnAGTU – Komsomol'sk-on-Amur State Technical University)

ELA-7 Albatross

In 1968-1971 the participants of the student design bureau of the Polytechnical Institute in Komsomolsk-on-Amur created an experimental flying vehicle dubbed ELA-7 Albatross. This was a 'seaplane with a lifting fuselage' - actually a WIG vehicle. It had a wing centre section of rectangular planform and substantial area, supplemented in its rear part by small outer wing panels fitted with ailerons. The wing centre section was flanked with two floats which had a keel surface at the centreline of their bottom. A 32-hp MT-8 flat-twin engine from a Dnepr motorcycle driving a tractor propeller was mounted on the leading edge of the wing centre section. The engine cowling blended into a fairing of the open cockpit. A flap attached to the trailing edge of the centre section, together with the floats, formed a cavity for an air cushion under the wing. The ELA-7 was provided with Vee tail surfaces attached to tail booms linked by a horizontal tail.

The propeller wash served in part for injecting the air under the wings. A small steerable vane mounted behind the propeller on the leading edge of the wing centre section deflected the airflow, increasing the portion of the airstream directed under the wings.

The ELA-7 was tested in the summer of 1971 at a lake near Komsomol'sk-on-Amur. Due to insufficient propeller thrust the vehicle developed a speed of a mere 36 km/h (22 mph). The testing was concentrated mainly on investigating the effect of off-loading the wing by means of injecting (blowing) the air under its surface. The off-loading amounted to 100 kg (220 lb) at the propeller's maximum static thrust. The designers of the vehicle came to the conclusion that the chosen layout was quite satisfactory as far as the aerodynamic off-loading was concerned.

ELA-8

This was a further development of the ELA-7 featuring changes in aerodynamic layout. These included installation of a single-fin empennage and refinement of the hydrodynamic shape of the floats. The horizontal empennage of the ELA-8 was installed in the front part of the floats and had no dihedral. The outer wing panels mounted on the aft sections of the floats had 5° dihedral. Thus, the seaplane featured a tandem-wing aerodynamic layout with a lifting fuselage. The forward set of wings was fitted with elevators, while the aft set of wings, as distinct from the ELA-7, had no control surfaces and was fitted with small endplates. To increase the air blowing effect the propeller axis was moved downward,

closer to the leading edge of the wing centre section; the prop was driven via an extension shaft. The foreplane directing the airstream downward was shaped as a multi-slot deflector. The flap installed on the trailing edge of wing centre section received spring loading and was deflected automatically, as the pressure under the wing decreased.

The all-up weight of the vehicle powered by the same 32-hp MT-8 engine was brought down to 380 kg (840 lb).

The ELA-8 was tested at a lake near Komsomol'sk-on-Amur in the summer of 1974. Several runs were made, the speed reaching 50 km/h (31 mph). The blowing produced an aerodynamic off-loading of up to 280 kg (620 lb).

In 1975 the ELA-8 was modified and redesignated ELA-8M; it appears that the foreplane mentioned above was installed in the process of this modification.

ELA-13

This WIG vehicle built in the SKB of the KnAPI in 1977-78 was, in effect, a research vehicle for investigating the influence of the proximity of the water surface on the take-off and landing performance of

The KnAPI ELA-7 Albatross WIG vehicle on a lake near Komsomol'sk-on-Amur during tests. The tail unit design is well visible.

The logo of the Komsomol'sk-on-Amur State Technical University (KnAGTU, ex KnAPI)

The ELA-8, showing the repositioned engine driving the propeller via an extension shaft, the redesigned fuselage and tail surfaces.

The ELA-13 research vehicle. The apparent boxy shape is due to the sidewalls flanking a 'lifting body'.

seaplanes. It had a lifting fuselage, an open cockpit, a T-tail and was powered by a 32-hp MT-8 flat-twin engine from a Dnepr motorcycle driving a pusher propeller of 1.6 m (5 ft 3 in) in diameter. Two floats were attached to the sides of the fuselage. Attached to the centre part of the fuselage were two outer wing panels with ailerons. The ailerons doubled as flaps, their deflection being geared to the movement of the elevator. This ensured the functioning of a direct wing lift control system which performed well in the course of the ELA-13's flight tests. Endplates installed at the tips of the outer wing panels helped cut induced drag.

The vehicle's all-up weight was 250 kg (551 lb), markedly less than that of the ELA-7 and ELA-8.

During the tests, which commenced on the Amur River in July 1978, the vehicle suffered a crash after pitching-up on take-off. Repairs made in the winter of 1978 were accompanied by modifications, including an increase of the floats' displacement in their nose parts, shifting the centre of gravity forward and selecting a propeller with a static thrust increased to 100 kg (220 lb). The modified vehicle was designated ELA-13M. During the testing conducted on the Amur River in July 1979 the ekranolyot made 32 successful flights, displaying good stability and controllability in different flight modes both in and out of ground effect. The maximum flight speed was in excess of 90 km/h (56 mph). The flight altitude out of ground effect reached 10-12 m (33-40 ft). The vehicle won the first prize at the All-Union contest of scientific work performed by students, held in Leningrad in 1981.

Combat WIG vehicles (projects)

Approximately in the late 1960s or early 1970s the Student Design Bureau of KnAPI conducted work on projects of WIG vehicles intended for attack missions (presumably in response to tactical requirements submitted by the military). One of the project configurations was very similar in layout to Bartini's VVA-14. It featured a wide-chord, low-aspect-ratio wing centre section and narrow-chord, high-aspect-ratio outer wing panels. The wing centre section was flanked by slab-sided sponsons apparently fitted with inflatable pontoons on the underside. The crew cockpit was mounted on the leading edge of the wing centre section, blending into its contours. The tail unit was, in fact, a V-tail with the two surfaces set wide apart and mounted outboard of the aft end of the wing centre section. Experiments were conducted with models, but no actual prototypes were built.

Seen during towing basin tests, this model of a strike WIG craft developed by KnAPI bears a striking resemblance to Bartini's VVA-14.

Komsomol'sk-on-Amur State Technical University (KnAGTU, formerly KnAPI)

In 1996 design work was conducted there aimed at creating a double-purpose transport means. It was envisaged as combining the features of a hydrocycle and a light WIG vehicle, the one being easily transformed into the other. It was to be a WIG vehicle of Type A, requiring no aircraft-type certification.

RKIIGA (formerly RIIGA)

(Riga Red Banner Institute of Civil Aviation Engineers; now Riga Aviation University)

From 1975 the Student Design Bureau of this institute conducted research and development work on transport means intended for use in areas with difficult terrain and no roads. This work encompassed ACVs and vehicles based on dynamic air cushion, that is, vehicles using ground effect. This work was conducted under the direction of V. Z. Shestakov and R. V. Shchavinskiy.

Apart from the ELA-01 (a joint project with MAI, see above), the SKB of this educational institution developed a project of the ELA-3 *ekranolyot*. Its scale model was presented at the Exhibition of scientific and technical work of young specialists in Moscow in 1984 (NTTM-84); further details of the vehicle are not available.

It may be surmised that the following 'anonymous' description contained in the institute's official history referred to the ELA-3. The vehicle was a flying machine with an air cushion undercarriage. Its powerplant drove a propeller creating horizontal thrust in cruise flight, and a fan which was engaged on take-off for creating a static air cushion under the fuselage. The low pressure exerted by the undercarriage on the supporting surface enabled the vehicle to operate from flat stretches of land, as well as from ice, snow and water. In cruise flight it could make use of a dynamic air cushion, that is, ground effect.

Students of the RKIIGA prepared a number of graduation designs of WIG vehicles of various categories. These included some projects prepared at the initiative of the TsLST (as mentioned above) on the basis of the An-2, Yakovlev Yak-12 light cabin monoplane and Yak-40 *Codling* short-haul airliner. In all cases the reversed-delta wing layout was used.

Interestingly, a project based on the An-2 envisaged converting the machine into a WIG vehicle by fitting it with rigid monoplane low-aspect-ratio wings while retaining the floats of the An-2V hydroplane. The 1,000-hp ASh-62IR radial was to be replaced by a 420-hp YaMZ-7511 Vee-8 automotive turbo diesel engine.

The conversion of the Yak-40 was based on using Lippisch-type wings with pronounced anhedral and wingtip floats lifting the fuselage above the water. A similar layout was chosen for the Yak-12 conversion.

Irkutsk Polytechnical Institute (IPI)

EVP-1

In 1982-83 a group of students and engineers under the direction of S. V. Trunov working in the SKB of the aviation faculty of this institute developed a design of the EVP-1 WIG vehicle. The machine featured aircraft layout with a T-shaped tail unit and slab stabiliser. The wings of trapezoidal planform had anhedral on the leading edges. The trailing edge section of the wing behind the rearmost spar was made elastic. The 18-hp Izh-P3 cruise engine (from an Izh Planeta motorcycle) with a pusher propeller was installed on a pylon ahead of the fin. The EVP-1 was also fitted with a 2-hp Sh-52 lift engine which drove a fan producing a static air cushion. The wings with a total area of 5 m² (53 sq. ft) and a span of 4.5 m (14 ft 9 in) ensured aerodynamic offloading of the vehicle amounting to 90% of its all-up weight at speeds of 65-70 km/h (40-43 mph). The ekranoplan had an empty weight of 100 kg (220 lb) and an all-up weight of 180 kg (397 lb). In the course

This unorthodox vehicle is one of the projected configurations of the M-6 passenger *ekranoplan* created in Irkutsk in the mid-1990s. The model shown here was displayed at the MAKS-95 airshow.

An early artist's impression and three-view of the RT-6 (alias Ikar) light WIG vehicle developed by the Kazan' State Technical University. The actual vehicle was somewhat different.

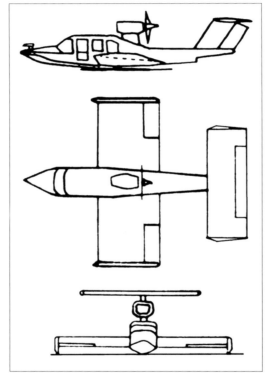

The logo of the Kazan' Aviation Institute (KAI)

of testing conducted at the Irkutsk water reservoir in 1983 the EVP-1 displayed excellent manoeuvrability and stability in all flight modes in ground effect.

M-6

This light ekranoplan was built under the auspices of the IPI in the NPS Scientific Development and Production in the mid-1990s. The machine featured a canard layout with a foreplane that was diamond-shaped in planform. The foreplane was split into two halves, their edges forming a 'reverse slot'; the designers claimed that this layout helped ensure stability in ground-effect flight mode through auto-stabilisation without resorting to the automatic control system. The vehicle was powered by a Rotax engine driving a three-blade tractor propeller, the air flow being directed under the wings during take-off via a large chin intake.

Kazan' State Technical University named after Andrey N. Tupolev (KGTU)

(formerly KAI – the Kazan' Aviation Institute)

RT (RT-6, RT-X, River Taxi, Ikar)

In the early 1990s several KAI graduates who had worked on WIG vehicles at the Alekseyev CHDB suggested to their alma mater that it should take up the summing up of the research data on this subject that had been accumulated by that time. In 1993 a programme, aptly named 'Ekranoplan' and supported by state funding, was initiated; KAI became the leading organisation co-ordinating the efforts of 15 higher learning and research institutions. The work on the systematisation of research data was successfully conducted; in particular, all aerodynamic and structural layouts of WIG vehicles that had existed in the world were carefully studied. KAI researchers came to the conclusion that it was

The actual prototype of the RT-6 (Ikar) in the assembly shop at KAI. Note the original two-bladed propeller. The shape of the glazing differs from the drawing above.

The same procedure is repeated by the shuttle orbiter upon its return, after which it can be re-used.

The WIG vehicle, as presented in the published illustration, is a gigantic machine with a wingspan of some 80 m (260 ft), featuring a twin-fuselage layout with a wide-chord thick wing centre section and outer wing panels of trapezoidal platform. The two boat hulls each carry their own set of tail surfaces located on the outboard side of the hulls, thus affording free access to the wing centre section from the rear and the transfer of a space vehicle to and from the *ekranoplan*. Placed on a crosspiece connecting the front parts of the boats are six turbofans performing the double function of boost engines (during take-off) and cruise engines. The shuttle orbiter is placed above the wing centre section. The WIG vehicle is provided with retractable diesel-driven water propellers at the stern and transverse thrusters at the bows for low-speed manoeuvring when afloat. The front parts of the boat hulls incorporate raised flight decks and compartments for the transported personnel.

A model of the WIG Sea Launch space system proposed by TsNII showing the bizarre WIG launch/retrieval vehicle and the space shuttle mounted piggyback. The Nos. 2 to 5 engines have thrust reversers and, apparently, deflectors directin the jet under the wings for take-off.

Front view of the same model. The *ekranoplan*'s tails are widely spaced to prevent damage by the blast of the orbiter's engines.

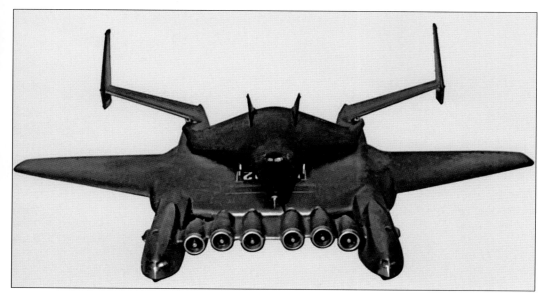

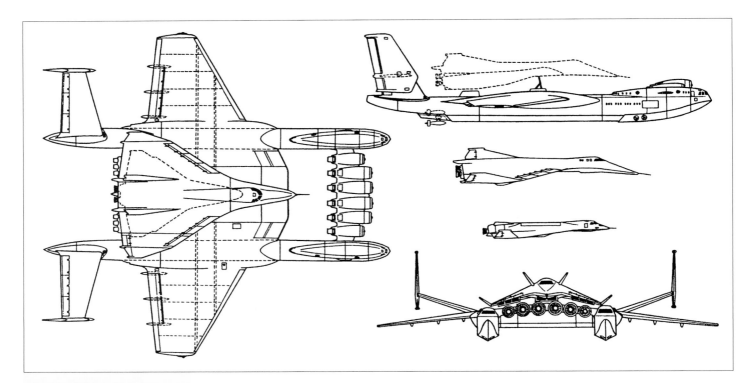

A three-view drawing of the Krylov TsNII's space launch system with two alternative shuttle orbiters. Note the retractable water propulsors and transverse thrusters for manoeuvring when afloat.

Here are some basic specifications. The full take-off weight with the space shuttle on board is 750 t (1,650,000 lb), the main powerplant comprises six turbofans rated at 30,000-35,000 kgp (66,000 to 77,000 lbst) apiece, the powerplant for slow movement on water comprises two high-rpm diesel engines. Cruise flight speed is 550 to 600 km/h (342 to 373 mph), low speed for manoeuvring when afloat is 25 km/h (16 mph).

Burevesnik

The *Burevesnik* (Storm Petrel) appears to be yet another of the Krylov TsNII's projects of aerospace systems based on the use of a WIG vehicle as a carrier for a spaceplane, on the same lines as in the project described above. A photo of a model shows this to be a giant *ekranoplan* of a similar twin-hull configuration, with the wide-chord wing centre section serving as a platform for a space vehicle mounted on top of it. A transverse structural mem-

A model of the broadly similar Burevesnik WIG vehicle (marked as such on the hulls). The shuttle orbiter is marked *Aviatsionno-kosmicheskaya sistema* (Aerospace system).

A model of the SSS twin-hull WIG vehicle. The engines are carried on cross-members ahead of the wing centre section.

ber joining together the front sections of the hulls carries six high bypass ratio turbofans whose efflux presumably can be partly directed into the space under the wing centre section during take-off. No further information is available.

SAR *ekranoplan* for the SSS (Sea Safety System)

In the course of the last two decades the Krylov Shipbuilding Institute has devoted much attention to investigating the possibilities of creating a global international sea safety system, the last three words forming the SSS acronym. This system is called upon to ensure speedy and effective assistance to crews and passengers of ships, aircraft, submarines, oil rigs and other technical means who happen to be in distress at sea. Large *ekranoplans* with an AUW of about 750 tonnes (1,653,400 lb) and a radius of action of 3,000-4,000 km (1,860-2,490 miles) from their bases constitute the main element of this system, which can comprise 12-13 ports located in the Far East, South-East Asia, Africa, Europe. North and South America, and Australia and 50 heavy *ekranoplans* stationed at these ports.

The Krylov Shipbuilding Institute has also prepared an advanced design project of a heavy WIG vehicle weighing some 750 t (1,653,400 lb) and intended to be a part of this global safety system. It is referred to in some sources as the 'SSS' (a display model of this vehicle wears the SSS inscription). The vehicle is capable of taking off and alighting in sea states up to 5 inclusive, staying afloat and drifting for a long time and reaching a port of refuge at low speed if it is prevented from normal flight for some reason. The WIG vehicle can reach speeds of 400-500 km/h (249-311 mph) and operate within a radius of 3,000-4,000 km (1,860-2,490 miles) from a port. The *ekranoplan*-ship can

deliver to the scene of an accident a large number of rescue means ranging from life rafts and motorboats to special boats for setting up floating barriers (used for containing oil spills), a deep sea diving vehicle and a helicopter. On board the WIG vehicle the survivors are provided with the necessary medical aid without waiting for arrival at a seaport. The safety of operation of such a WIG vehicle in sea conditions is far superior to various types of aircraft-based means.

To substantiate the project, the Krylov Institute (TsNII) has conducted a vast amount of design studies and research in wind tunnels, towing basins, on test rigs and on the open water surface. The design of the heavy WIG vehicle was developed on the basis of using available engines, structural materials, onboard equipment and with due regard to the state-of-the-art level of technology and production methods in the construction of high-speed ships and aircraft. It was presumed that further development of the project, construction and operation of the rescue WIG vehicles and of the system as a whole could be put into effect by an international consortium of shipbuilding and aircraft companies with financial support from nations having an interest in maritime rescue operations.

The SSS *ekranoplan* developed by the Krylov TsNII is a catamaran vehicle with two boat hulls and a composite wing layout. Placed between the hulls is a wing centre section of large chord and relative thickness; outboard of the hulls it is supplemented by outer wing panels of higher aspect ratio and lesser relative thickness. The twin-fin tail unit includes a horizontal tail resting on top of the fins. The crew stations are housed in a nacelle attached to the leading edge of the wing centre section and two foreplanes joining the nacelle and the hulls. Suspended under the foreplanes are four turbojets

A computer-generated image of the RT1-360 with tandem wings, hydrofoils and tracrot propellers driven by a single engine via reduction gear.

The basic project version of the RT1-360 with single wings, hydroskis and an M-14PT engine driving a single pusher propeller. Note the yacht-type handrails.

which act both as booster engines at take-off and as cruise engines. For slow-speed movement when afloat the vehicle is fitted with diesel engines driving water propellers in the aft parts of the hulls. The vehicle can carry a Kamov Ka-27PS *Helix-C* SAR helicopter on top of its wing centre section.

Krylov TsNII branch in Nizhniy Novgorod

A branch of the Krylov Central Shipbuilding Research Institute was set up at Gor'kiy (now Nizhniy Novgorod) on the basis of the CHDB test facil-

ity. In the subsequent years this organisation developed a series of projects of small WIG vehicles. The designer behind these projects was E. V. Vasil'yev. Some of them are described below.

KEP-6 (KEP-6A)

This was a project of a passenger WIG vehicle (KEP = *kahter-ekranoplahn* – WIG speedboat) designed to carry four to six persons. Its wings had a composite configuration in planform and consisted of a centre section and outer wing panels with endplates. According to some sources, the KEP-6A had a twin-fin tail unit and a powerplant with two propellers that could change their plane of rotation. The project was presented at the 'Ekranoplan-96' international conference held in Kazan' in 1996.

WIG vehicles in the RT series

The TsNII's Nizhniy Novgorod branch also prepared a technical proposal for a series of business-class light amphibious WIG vehicles powered by indigenous automobile engines from the VAZ (Lada) and ZMZ enterprises and by the production 360-hp Vedeneyev M-14P aircraft radial engine. They were known under the common designation RT (*rechnoye taksi* – river taxi) with sequence numbers. They featured an 'aircraft' layout (like that of the SM-6). The programme included such vehicles as the RT-2, RT-2M, RT-4, RT-5, RT-6MA, also presented at the 'Ekranoplan-96' conference.

But the series started with the RT-1 project, which served as a basis for the further projects evolved by modifying and scaling up the basic 'aircraft' layout. In one source, four projects developed at the branch are singled out. These were the RT2-760 and RT-360 cargo/passenger craft and the RT1-360 and RT1-150 passenger WIG vehicles. They all shared the same basic layout and featured a modular construction. The boat-type hull had a superstructure which could be built of composite materials. The powerplant consisted of series-built automobile and aircraft-type piston engines. The craft were provided with a wheel/hydroski undercarriage enabling them to be operated from water, snow-covered ground and ice.

It remains to be added that the basic RT-1 project was recommended by the TsNII branch to the student design bureau of the Kazan' Aviation Insti-

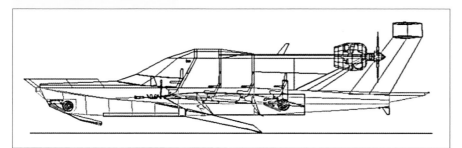

A cutaway drawing of the RT1-360 in M-14PT-powered pusher configuration. Note the retractable beaching gear.

Performance figures on some of the RT series projects			
	RT2-760	RT2-360	RT1-360
Number of passengers	30	21	12
Engine power, hp	2x540	2x195	1x195
Operational speed, km/h (mph)	193 (120)	161 (100)	152 (94)
Range with max fuel load, km (miles)	1,500 (932)	1.500 (932)	1,000 (622)

by floats at their tips. The 140-hp Walter Minor engine with a pusher propeller was placed behind a two-seat enclosed cockpit. The tail unit comprised a fin-and-rudder assembly and a stabiliser with an elevator. The vehicle was tested in 1985. When flown from the water it failed to reach the design speed due to the unsatisfactory hydrodynamic shape of the floats. In tests on snow with a ski undercarriage it showed a top speed of 150 km/h (93 mph). The all-up weight of the rather heavy machine was 1,100 kg (2,425 lb).

P. G. Tkachenko

In 2002 P. G. Tkachenko and Yuriy Makarov, aviation specialists known for their work in student design bureaux, were working in Moscow on a project of a small two-seat WIG vehicle basically repeating the Lippisch layout. The distinctive feature of the design was the use of the powerplant (a small piston engine driving a shrouded pusher propeller) for creating a static air cushion under the boat hull. For this purpose the engine was mounted behind the cockpit in such a way that the lower part of the propeller rotation plane was 'sunk' into the fuselage and the air flow from the propeller was driven through a slot into the space under the hull.

P. Tsymbalyuk

A group of enthusiasts in Arkhangelsk headed by P. Tsymbalyuk designed and built a vehicle called 'Ekranolyot' (again!). In this case, too, the Lippisch layout was used, with a reversed-delta wing having anhedral on the leading edges and supplemented with aileron-carrying winglets. The tail unit included all-movable strut-braced stabilisers placed atop the fin. Floats were attached under the wingtips. A 105-hp M-332 engine was mounted on a truss cabane above the wing centre section; it drove a standard propeller but with cropped blades, which resulted in a substantial loss of thrust. The boat hull was strengthened with glassfibre. The vehicle was tested in 1977, reaching a speed of more than 60 km/h (37 mph) on water and up to 80 km/h (50 mph) on ice.

I. Vorontsov

In 1977-1978 in the city of Perm' a group led by I. I. Vorontsov designed and built an ekranolyot named

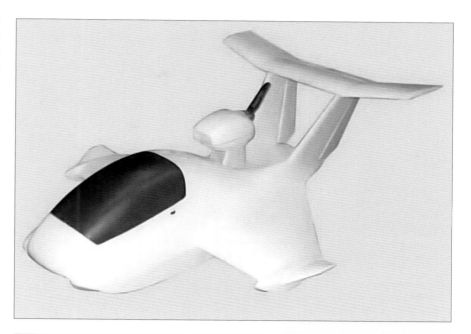

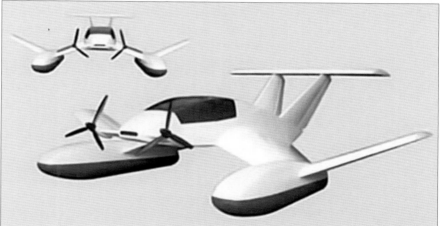

A model of the slightly larger Sigma Proxima.

A compiter-generated image of the Sigma-E. The propellers are driven by a single engine via extension shafts.

An artist' impression of the Sigma-J. Note the nose-mounted prop wash deflector vanes.

A model of an ekranoplan designed by P. G. Tkachenko. Note how part of the prop wash is directed under the vehicle to form a static air cushion.

Elektron. It was of wooden construction and had a configuration similar to that of the TsLST ESKA-1. The vehicle was powered by a 32-hp engine from a Ural motorcycle driving a wooden propeller of 1.7 m (5 ft 7 in) in diameter through reduction gear. Initially it was powered by an engine from a Czech-built CZ motorcycle (pronounced 'chezet'), but its output proved insufficient for a vehicle weighing 400 kg (880 lb).

In March 1979 the ekranolyot was flight-tested, making take-offs and landings on snow. Tests revealed good stability.

P. P. Yablonskiy

In 1985 P. P. Yablonskiy, then a student of the Zhukovskiy Military Engineering Air Academy, submitted a graduation project of a transport WIG vehicle. It had two fuselages joined together by a low-aspect-ratio wing centre section. To reduce the outward air flow from beneath the wing centre section, strakes were fitted to the underside of the fuselages in the area of the wings. A high-set horizontal tail was used. Wind-tunnel tests simulating the ground effect were conducted with models. Test results corroborated the design characteristics of the WIG vehicle.

A computer-generated image of the Manta WIG designed by A. Yukayev.

Nikolay V. Yakubovich

A design belonging to Nikolay V. Yakubovich is mentioned in his article Between two elements published in the July 1998 issue of the *Kryl'ya Rodiny* magazine: *'In 1975 this author submitted a project of a transport WIG vehicle intended for transporting cargoes of up to 365 t to a distance of up to 9,000 km [5,590 miles] at a cruising speed of 410 km/h [254 mph]. It featured a wing centre section with an aspect ratio of 0.5, flanked by boats of a planoconvex shape ./.../ When the vehicle was loaded, special elastic floats were inflated on the sides. /.../ To create an air cushion, a stream of air from turbofan installations was pumped into the cavity formed by the underside of the wing centre section, and the front and rear flaps. As the cruise engines accelerated the vehicle, the front flap was retracted and a transition from the static to the dynamic air cushion took place. The powerplant consisted of two high-output NK-12 turboprops'.*

Al. Yukayev

This enthusiast designer has put forward a project of a 10-seat passenger WIG vehicle dubbed Manta (that is, the manta ray). The vehicle weighing up to 19 tonnes (41,890 lb) in flight configuration has a length of 21 m (68 ft 11 in) and a wing span (including the endplate floats) of 18 m (59 ft). It is a vehicle in the C class, capable of free flight at altitude in addition to the flight in ground effect. Its performance includes an economical cruising speed of 180 km/h (112 mph) in ground effect. The designer claims that fuel consumption in this mode will be 7-8 times smaller in comparison with a yacht with similar dimensions and payload. In free flight at altitudes up to 5,000 m (16,400 ft) the vehicle is expected to attain a cruising speed of nearly 500 km/h (310 mph) at the cost of much greater fuel consumption.

CHAPTER SIX

FREE ENTERPRISE
ENTER NEW COMPANIES

The beginning of the 1990s was characterised by the break-up of the Soviet Union. It gave place to the Russian Federation and new independent states, former Soviet Republics which, together with Russia, formed the Commonwealth of Independent States (CIS). The construction of WIG vehicles as a branch remained virtually 100% within the boundaries of the Russian Federation (apart from Russia, only the Ukraine could boast some scientific work in this field during the Soviet period).

Sweeping market reforms giving full reign to private enterprise in the Russian economy had their effect on the situation also in the aircraft industry and associated branches. Unfortunately, the consequences were mostly on the negative side. The drastic reduction of state funding and of state orders placed many enterprises on the brink of bankruptcy and severely curtailed the scope of their activities. In these circumstances the WIG craft design and construction could hardly be a matter of priority.

The early 1990s saw the emergence of a large number of private companies (design bureaux) which declared their intention to engage in aircraft design. Among these were several firms whose plans encompassed the design and construction of WIG vehicles. These were small enterprises which, as a rule, had no production facilities of their own and possessed very limited financial resources. Their activities only rarely resulted in producing real hardware even in prototype form, to say nothing of series production. However, they produced a number of 'paper designs' which merit some attention because of their innovative spirit.

In addition to new private enterprises, this chapter will touch upon the activities of two design bureaux, whose activities in the field of ekranoplans date back to the Soviet period – simply because they do not merit a separate chapter.

As mentioned in the Introduction, the late 1980s and early 1990s saw a gradual disintegration of the Alekseyev CHDB. Several groups of engineers, having abandoned this design bureau, set up nearly a dozen small companies which declared their intention to continue design work on transport means based on dynamic principles of sustentation. In actual fact, only a few of them remained true to *ekranoplans* – many of the new-born enterprises switched over to work on air cushion vehicles, a line of business that seemed more lucrative, or reverted to the construction of small hydrofoil boats (this was the case with Transall-AKS, for example).

Among the new companies engaged in the design of WIG vehicles only two can boast construction of prototypes intended for series production. These are the Amphibious Transport Technologies Joint-Stock Society (JSC) – initially known as Technology and Transport – and the KOMETEL JSC. Both of them subsequently underwent transformations and changes of name and affiliation. It stands to reason that the following review of designs produced by new firms should start with these two enterprises.

Amphibious Transport Technologies (ATT, Joint Stock Company)

(now a subsidiary of the ATTK – Arctic Trade and Transport Company)

This company was founded in 1992 in Nizhniy Novgorod under the name Technology and Transport (TET) JSC. It was formed by about 130 engineers who, until then, had worked in the Central Design Bureau of Hydrofoils named after Rostislav Ye. Alekseyev. Gradually the staff rose to about 200 employees. The new company was headed by D. N. Sinitsyn, with Aleksandr I. Maskalik as his deputy. The company declared its intention to pursue the design of WIG vehicles for civil duties, making use of some projects which were taken over from the CHDB (this caused a dispute over intellectual property rights which was eventually settled).

The basic problem of finding an investor was solved by establishing co-operation with a businessman from Taiwan who actually became the owner of the bulk of the company's stock. He financed the work on a WIG vehicle, but became the owner of the rights to the design and of all the machines produced, as well as of the right to sell them. In response to his order the newly formed design team embarked on the design of a small WIG boat named Passat (Tradewind); in 1996 it was renamed Amphistar

Amphistar

The Amphistar belongs to the category of small Type A WIG vehicles – high-speed craft intended for operation only within the height of ground effect, powered by an engine of more than 55 kW, carrying not more than 12 passengers, operated only in daytime within 20 miles from the coast and within 100 miles from a place of refuge, and having an all-up weight of not more than 10 t (22,045 lb). The five-seat Amphistar is intended for maritime

An all-white Amphistar moves along a snow-covered field, showing the two large side windows.

A more colourful example of the Amphistar. Note the curved 'bumper' (propeller guard), another distinctive feature of the early model, and the inflatable centreline pontoon.

Probably the same Amphistar is seen here travelling over water.

4⅜₂ in); overall height, 3.35 m (10 ft 11⁵⁷⁄₆₄ in); maximum all-up weight, 2,720 kg (6,000 lb); operating empty weight, 2,200 kg (4,850 lb); crew and equipment weight, 117 kg (258 lb); payload, 300 kg (660 lb); fuel load, 100 kg (220 lb). Range with 100 kg of fuel is 350-400 km (217-248 miles) depending on the weather (wind force, wave height). The vehicle is amphibious – it can negotiate shallow places, come out onto an unprepared bank with a gradient of up to 5°, move on the ground at speeds of 10-15 km/h (6.2-9.3 mph) and go back to the water. The airframe of the Amphistar is made of corrosion-proof composite materials.

The Amphistar's wing set of rectangular planform and low aspect ratio is fitted with flaps and provided with endplates in the shape of inflatable pontoons which ensure the vehicle's stability when afloat and form the sidewalls of the 'scoop' in which the air cushion is created (there is also a retractable inflatable support pontoon on the centreline). A compartment placed amidships in the hull accommodates four passengers and a driver. The aft bay

pleasure rides and tourism. It is a dynamic air cushion speedboat with an enclosed cabin, featuring a forward location of the engine.

Here are the main specifications and performance figures of the machine: overall length, 10.44 m (34 ft 3 in), overall width (wingspan), 5.9 m (19 ft

of the hull houses equipment, while the forward part of the hull is occupied by a 300-hp Mercedes-Benz engine which transmits its torque through a central gearbox and two side gearboxes to two AV-110 tractor propellers with ground-adjustable pitch. The propellers can swivel their plane of rotation and direct the air stream under the wings for take-off.

The tail unit comprises twin fins and rudders mounted on the rear fuselage in a V arrangement, with the horizontal tail placed atop the fins. Directional control is effected by the rudders during movement in ground effect and by a retractable water rudder when afloat.

Confronted with the lack of interest for the EKIP programme at home, the management of the Saratov aircraft plant began to look for potential investors abroad. In 2000 Aleksandr Yermishin, the then director of the plant, went to the USA where he had meetings with the military and with representatives of the aircraft industry. The US side displayed much interest for starting the manufacture of EKIP vehicles in a pilotless version in the USA, but wished to obtain exclusive rights to the EKIP, which was unacceptable for Yermishin. From 2003 the work on EKIP at the Saratov plant was stopped because the plant was in dire straits financially. Eventually a compromise was reached. It was reported that the Saratov plant, the EKIP concern and the US Naval Aviation Research Center would undertake joint production of EKIP vehicles. US agencies were interested in using the EKIP as pilotless vehicles for combating forest fires and eliminating the consequences of natural disasters.

It was expected that in 2007 the testing of a Russian-US vehicle based on the EKIP design should start in the state of Maryland, USA, with a prospect of starting series production five years later. However, these plans obviously failed to be put into practice. At present the future of the programme is uncertain.

Eurasia JSC in co-operation with Latvian institutions

MF-17

In the mid-1990s co-operation between Russian and Latvian institutions resulted in the emergence of this project of a light WIG vehicle designated MF-17. It was developed as a joint venture with the participation of the Russia-based Eurasia JSC (headed by F. Mukhamedov), the design group of the Riga Aviation University (formerly RKIIGA) and the 'Latvijas Laivas' ('Free Latvia') Experimental Sports Shipbuilding Centre based in Jurmala, Latvia. This was a two-seat vehicle featuring a typical Lippisch layout with the reverse-delta wings. Upturned winglets with ailerons were mounted on the wingtip floats; the T-tail incorporated a rudder and elevators. The cockpit accommodating two persons in a side-by-side arrangement had a large streamlined transparency; it could be fitted with single or dual sets of controls. Mounted on a pylon behind the cockpit was an Austrian (presumably Rotax) engine driving a shrouded pusher propeller of Russian manufacture. The avionics included US-manufactured satellite navigation and communication equipment.

The vehicle's principal mode of operation was flight in ground effect, but it was capable of performing short flights at low altitude to avoid obsta-

A model of the MF-17 light WIG vehicle co-developed by Russia and Latvia.

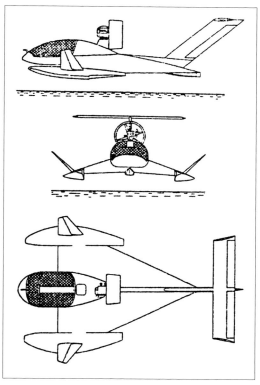

A provisional three-view of the MF-17, showing the 'taildragger' float arrangement.

cles. With an empty weight of 250 kg (551 lb) and an AUW of 450 kg (990 lb), the vehicle had a design speed of 100 km/h (62 mph) and a range of 250 km (155 miles). The admissible wave height was 0.5-0.6 m (1 ft 8 in – 2 ft).

The participants of the project announced their plans for building ad certifying the prototype of the MF-17 in the course of 1996-97, whereupon it was to be put into production in Jurmala and sold to customers at a price of approximately US$ 30,000. All this hinged upon finding investors to back up this venture, but apparently the necessary funding could not be procured and no production ensued.

LAT

LAT (*Lyohkaya Aviahtsiya Taganroga* – Light Aviation of Taganrog). This joint stock company which styles itself as Science and Production Enterprise (*naoochno-proizvodstvennoye predpriyahtiye* – NPP) came into existence as a result of activities of a group of young designers who, to begin with, worked within the framework of the *Krasnyye*

The LAT R-02 Robert on a ground handling dolly at the Hydro Aviation Show-2000 in Ghelendzhik. The aircraft is registered by the Aviation Enthusiasts' Federarion of Russia (FLA RF) as FLA-RF 02558.

A three-view of the R-02, with an additional side view (top) of the forward-swept wing R-01M Robert.

Kryl'ya (Red Wings) Centre of aviation enthusiast work. Later this centre transformed itself into a small private enterprise under the same name; inside this enterprise the LAT company was organised. In 1992 it became an independent structure, headed by Yuriy Usol'tsev.

Under Usol'tsev's guidance several light aircraft designs were developed first by the abovementioned Centre and then by the LAT company. These were the R-01, R-01M, R-02 and R-50 light seaplanes which bore a common name, Robert (in honour of Robert L. Bartini). Owing to some special features of their design they possessed pronounced WIG vehicle properties. All of these machines have low-set wings with a wide-chord centre section which rests on the water surface when the machines are afloat, thus ensuring their stability and buoyancy. During take-off and landing the ram air creates a dynamic air cushion under the wing centre section.

R-01, R-02

The R-01 aircraft weighing 350 kg (771 lb) for take-off was a single-seater with an open cockpit; its outer wing panels had pronounced forward sweep. It was successfully test-flown in 1989. The improved R-01M version powered by a 44-hp Robin-140EC piston engine had a take-off weight of 440 kg (970

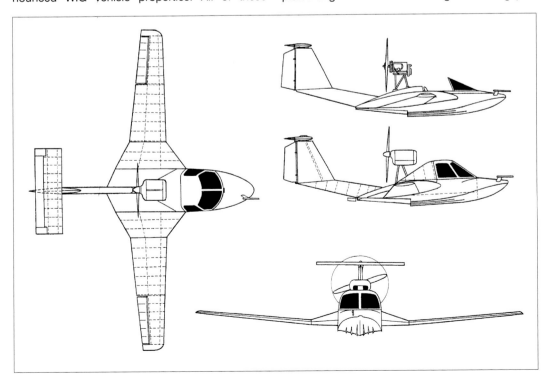

lb). The R-02 differed in having an enclosed two-seat cockpit and straight wings with an extended-chord centre section. It was intended for use as an executive, patrol and trainer aircraft.

The aircraft was powered by an 80-hp Rotax 912A3 engine installed behind the cockpit and driving a tractor propeller. The R-02 was provided with a quick-action parachute rescue system. Here are some basic figures for the R-02: wingspan, 10.0 m (32 ft 9.7 in); length, 6.3 m (20 ft 8 in); wing area, 12.4m² (133.3 sq ft); all-up weight, 560 kg (1,234 lb); speed, 165 km/h (103 mph).

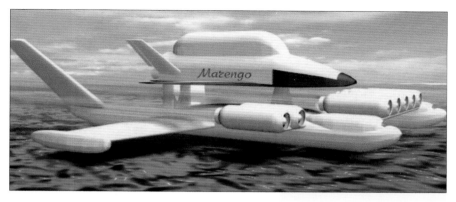

R-50

In 1992-94 the LAT designers prepared an advanced development project of a multi-purpose amphibian designated R-50 Robert. In its baseline version it was a cargo and passenger aircraft with a cabin accommodating a pilot and five passengers; the passenger seats could be easily removed for the carriage of cargo. The basic version could also be used for patrol and liaison duties. Other versions under development included ambulance, fire-fighting, agricultural, SAR, ecological survey and other versions. The baseline version was to be powered by two Czech 140-hp Motorlet (Walter) M-332AR engines which could eventually be replaced by the indigenous MKB Granit TVD-400 turboprops (the definitive version was expected to deliver 500-560 ehp) or 250-hp VOKBM M-17 piston engines. Provision was made for the installation of the MIKBO-43 multi-function integrated avionics suite permitting the aircraft to fly in adverse weather conditions and at night.

In common with its predecessor, the R-50 has a wide-chord wing centre section creating a WIG effect during take-of and landing. In ground effect the lift/drag ratio of the wings rises to 19-20 as against 11-13 in normal cruise flight. One can also use the ground effect mode for cruising flight, but a prolonged flight close to the supporting surface is possible only on condition of prior installation of an automatic pitch stability augmentation system which can be done at the customer's request.

In 1995 there were plans for creating the R-50M version with a cabin lengthened by 0.75 m (2 ft 5.5 in) and twin engines. Design specifications for it included a take-off weight of 2,700 kg (5,950 lb), a payload of 560 kg (1,234 lb) for a range of 600 km (370 miles) with two 210-hp engines or 650 kg (1,433 lb) for a range of 750 km (465 miles) with two 240-hp engines. Its dimensions were: wingspan, 15.40 m (50 ft 6.3 in); wing aspect ratio, 8.227; length, 11.40 m (37 ft 4⁵⁄₆₄ in); height, 4.30 m (14 ft 1¹⁹⁄₆₄ in). Series production of the R-50 was to be launched at an aircraft plant in Doobna, Moscow Region, but financial difficulties prevented these plans from being implemented.

Marengo JSC

This little-known joint stock company is engaged in developing a project of an *ekranoplan* intended as a carrier for an air-launched aerospace vehicle. The configuration of this huge twin-fuselage machine is apparent from an artist's impression published here. It is in many respects similar to the twin-hull *ekranoplan* proposed by the Krylov TsNII for the same purposes (see the preceding chapter). However, the Marengo project has some special features. In addition to the five cruise/boost turbofans mounted on the crossbar between the hulls forward of the wing centre section, there area two pairs of turbofans mounted on horizontal pylons outboard of the forward sections of the hulls. Their efflux presumably can be deflected into the space below the outer wing panels to create a static air cushion for take-off. It is apparently for this purpose that the wing outer panels carry endplates at their tips – together with the hull and the flap they create a

An artist's impression of the huge twin-hull WIG vehicle designed by Marengo JSC as a space shuttle transporter.

'Doctored' images of a model of the Mirel light WIG vehicle showing the machine on a beach and in flight.

A three-view drawing of the NVA-1SM light WIG vehicle showing the horizontal fan creating a static air cushion.

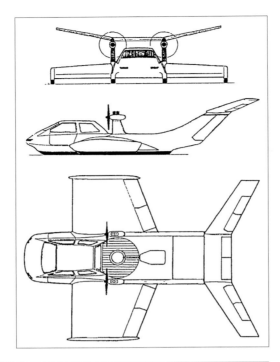

A model of a heavier cargo WIG vehicle in the NVA series, showing the cargo bays inside the wings.

chamber into which the efflux is injected. Unfortunately, no further information is available.

Mirel

This is presumably the name of a company about which no information has been published. Two artist's impressions of a WIG vehicle wearing this name on the fuselage are practically the only materials available at present on this design. The pictures show a vehicle whose layout is basically similar to that of the EK-12 Ivolga. A fuselage resembling that of a 6-8-seat lightplane and featuring a T-tail is flanked by wide-chord, short-span wing sections the tips of which rest on floats. The wing centre section has anhedral. High-aspect-ratio outer wing panels are attached to the wing centre section. Two horizontally opposed piston engines flanking the front fuselage drive three-blade propellers whose wash can be directed under the wing centre section.

NVA projects by N. N. Nazarov (Ekolen, Amphicon and other companies)

A series of highly original WIG vehicle projects was developed on the basis of a concept evolved by N. N. Nazarov, an engineer who had worked at one time in a branch of Alekseyev's CHDB which from 1976 became a Gorkiy-based branch of the Krylov TsNII. He started his work on NVA projects already during his work at the TsNII branch. Later he left this organisation and founded his own company – Science and Production Association (NPO) Ekolen JSC in St. Petersburg. (Ekolen is an abbreviation for *Ekologiya Lenskovo reghiona* – Ecology of the Lena River Region, indicating that the projects were to be tailored to the needs of that region in East Siberia.) It was there that he started formulating his comprehensive programme of *ekranoplan* con-

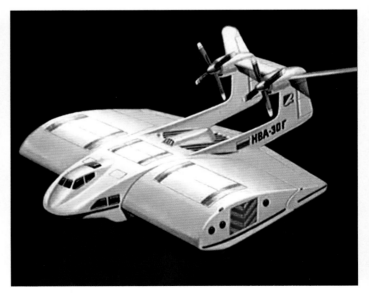

These artist's impression of the NVA-30G show the unusual wing design with detachable full-span cargo modules. Note the loading doors in the tip ribs and upper surface of the centre section and the outer wing modules.

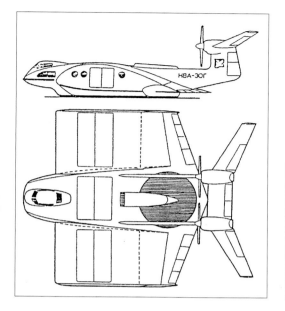

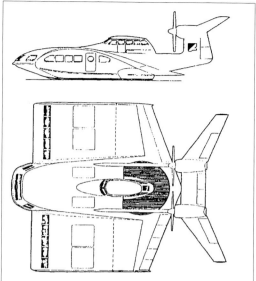

Far left: Two views of the NVA-30G with the wing cargo modules attached.

The NVA-30P passenger verion is rather different, having a cockpit located well aft (on top of the fan creating the static air cushion).

struction based on the NVA concept. This work was carried on by Nazarov in a successor company, named Amphicon, which was set up in Nizhniy Novgorod. Amphicon is an abbreviation for Amfibiynye konstrooktsiï (amphibious designs). The full name of the company is NPF Amphicon, NPF is deciphered as naoochno-proizvodstvennaya firma – Science and Production firm. N. N. Nazarov was its General Director and Chief Designer.

Owing to various vicissitudes Nazarov had to change his place of work and move to Siberia where he worked in the Baikalenergotrans (BET) company based in Ulan-Ude, Buryatia, and in the Amphibia JSC company based in Abakan, Khakassia, in co-operation with a number of other companies and institutions. During a short period some work on Nazarov's designs was conducted at the Krasnoyarsk Machine-building Plant – one of Russia's major enterprises in the field of missile and space technology. N. N. Nazarov passed away in Abakan after a serious illness on 6th July 2008. His work is being continued by his associates. Some of Nazarov's projects are described below.

In the early 1990s the Amphicon company presented a national development programme of means of transportation in Russia based on a new generation of WIG vehicles – the so-called Ground-and-Air Amphibious Vehicles (GAAV), the Russian abbreviation being NVA (nazemno-vozdooshnyye amfibii). The programme covered a 15-year period up to 2008. The range of projects envisaged in the programme represented a gradual transition from a 3-tonne (6,610-lb) vehicle of 'aircraft' layout to a 'flying wing' vehicle weighing 5,000 t (11,022,930 lb). In all, by 1993 Amphicon had developed ten baseline projects, each in several versions. Information about them was presented in the catalogue of the

An artist's imp[ression of the NVA-60P – a scaled-up version of the NVA-30P with turbofan cruise engines.

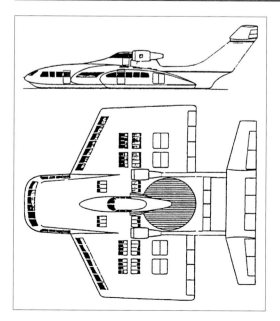

A three-view drawing of the NVA-60P. Note the 'skylights' in the wing upper surface; without them the cabins would be too dark.

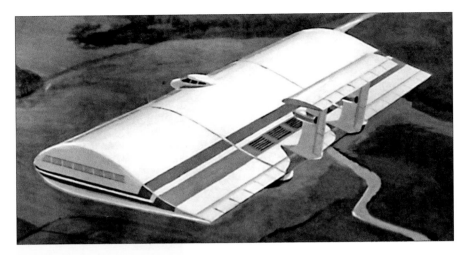

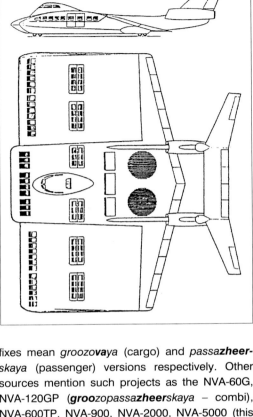

An artist's impression of the huge NVA-120 *ekranoplan* using the 'flying wing' layout.

Right: The passenger-configured NVA-120P featuring a revised tail with swept dihedral outer portions. Again, it has 'skylights' in the wing upper surface.

MAKS-93 air show (September 1993); every project was accompanied by the planned year of construction (between 1996 and 2008). The machine to be built in 2008 was to have the all-up weight of 5,000 t. Drawings of the following six projects were published in the catalogue: NVA-1SM (1996), NVA-30P (1997), NVA-30G (1997), NVA-120P (1997), NVA-60P (1997), NVA-600 (2003). The G and P suf-

An artist's impression of the NVA-120P.

This artist's impression shows a huge WIG vehicle in the NVA series used as a submarine tender.

fixes mean *groozovaya* (cargo) and *passazheerskaya* (passenger) versions respectively. Other sources mention such projects as the NVA-60G, NVA-120GP (*groozopassazheerskaya* – combi), NVA-600TP, NVA-900, NVA-2000, NVA-5000 (this list is not exhaustive). In English-language advertising materials the abbreviation NVA is replaced by GAAV. As witnessed by the drawings, many of the projects envisage the use of powerful fans for creating a static air cushion during take-off, making these craft a combination of ACV and WIG vehicle. The company advertised such an 'important advantage' of the NVA vehicles as 'their load-carrying capacity amounting to 50% of the all-up weight'. A characteristic feature of almost all designs was the use of airfoil-shaped fuselages as lifting bodies. Conversely, wings of very high relative thickness provided the space for cargoes and passengers.

An attempt to bring one of Nazarov's designs to the hardware stage was connected with the Krasnoyarsk Machine-building Plant (Krasmash). The project selected for practical implementation was the NVA-30, a relatively small vehicle. Design work progressed to the preliminary design stage. A specially established private company named Sibiam provided the funding needed for detail design and prototype manufacture which would be followed by operational trials in the Krasnoyarsk region. In 1992 Krasmash began receiving documentation on the NVA-30 from Ekolen. Construction of the prototype was started. Yet, soon the allotted funds 'vanished into thin air'; the machine was not completed.

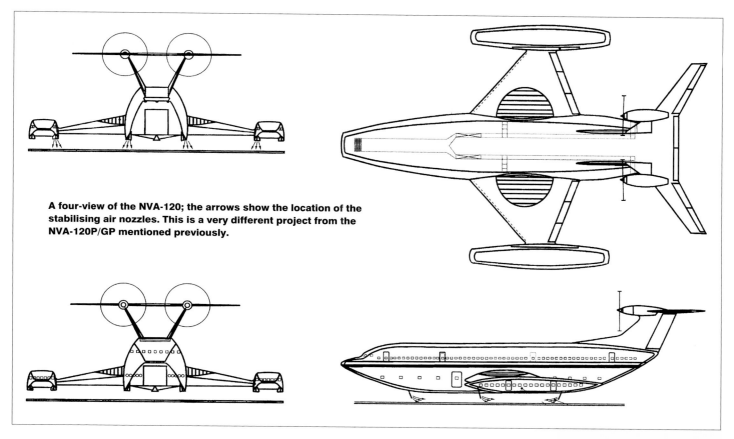

A four-view of the NVA-120; the arrows show the location of the stabilising air nozzles. This is a very different project from the NVA-120P/GP mentioned previously.

The very ambitious programme described above could be implemented only on condition of massive financial support from the Government or other sources – which was far from realistic to expect under the circumstances prevailing in Russia. To date, not a single project from the Amphicon period progressed further than the drawing board.

Nazarov's work in Baikalenergotrans resulted in the emergence of new projects differing in their layout from those described above. Some of them are illustrated by a drawing which presents three types of different size. They differ from the previous generation of Nazarov's projects in making use of the traditional low-aspect-ratio wings rather than airfoil-shaped lifting bodies. These are:

NVA-06-10 GP

This is a small river-going WIG-vehicle, presumably with an AUW of some 10 tonnes (32,800 lb). In its layout it closely resembles Rostislav Alekseyev's machines, although it is powered by a single aft-mounted turboprop, without the use of front-mounted engines for power-augmented take-off. It is intended for the transportation of cargoes and passengers, as indicated by the GP suffix.

NVA-220-300 GP

Described as a sea-going *ekranoplan*, this vehicle in the 300-tonne (661,400-lb) class embodies a tri-

maran layout, the side fuselages flanking a very wide-chord wing centre section. On the outboard sides they are fitted with smaller-chord outer wing panels having endplate floats. The centre fuselage is reminiscent of an airliner. Two turboprop engines

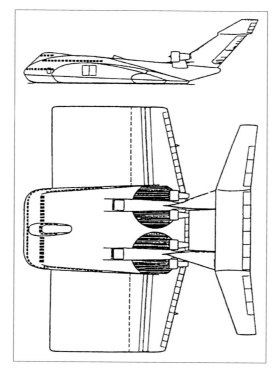

Two views of the NVA-600. Due to its large size the vehicle requires two fans to create a static air cushion.

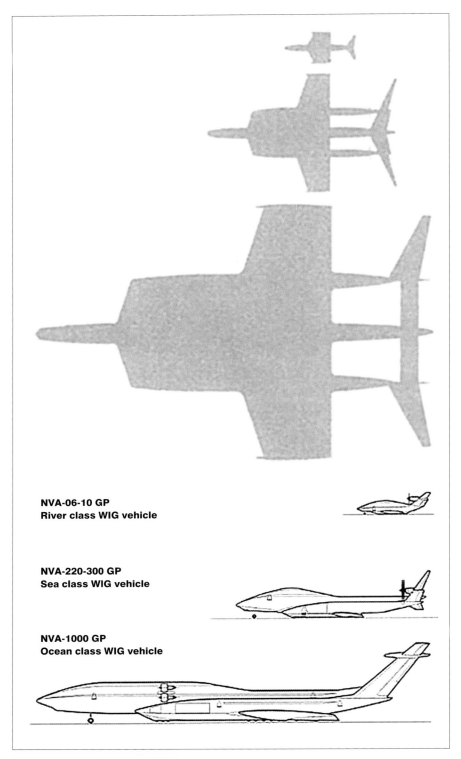

NVA-06-10 GP
River class WIG vehicle

NVA-220-300 GP
Sea class WIG vehicle

NVA-1000 GP
Ocean class WIG vehicle

are located at the junctions of the twin fins and the horizontal tail. A notable feature is the wheel undercarriage – probably for ground handling rather than for take-off and landing on a solid surface.

NVA-1000 GP

This is an ocean-going craft of truly gigantic size, presumably in the 1,000-tonne (3,280,000-lb) class,

looks like a scaled-up version of the NVA-220-300 described above. It is powered by four high-bypass turbofans or propfans mounted in pairs on both sides of the fuselage.

Nazarov's demise might signify the end of the NVA line of projects. However, the Amphibiya JSC continues its work on the NVA projects, having enlisted the support from an organisation which styles itself as Noöspherical industry and transport complex (NPTK). Some of these projects were given a measure of publicity; they were presumably developed while Nazarov was still alive. This series of projects differs from the preceding one in both designations and layout. It encompasses vehicles in the NVA-07 range, such as the NVA-07-025 and NVA-07-530 GP (the last three digits referring to the vehicle's weight in tones).

According to Professor Drachov, one of the leaders of the NPTK company, the WIG vehicles developed by the two companies are to be used primarily in Russia's northern areas, Siberia and the Far East.

NVA-07-025

This vehicle of relatively small size, in the 25-tonne (55,110-lb) weight class, is intended for carrying passengers. It is apparently powered by one turboprop engine with a ducted pusher propeller. A notable feature of the design is the integration of the propeller shroud with the fins placed on the aft parts of the floats flanking the short-span wing centre section. Outer wing panels of small chord and higher aspect ratio are attached to the floats.

NVA-07-530 GP

This is a heavy WIG craft which is intended for operation in the seas encompassed by the so-called Northern Sea Route (the route along Russia's Arctic coast-line). This huge vehicle will be capable of carrying a 200-tonne (441,000-lb) cargo at a speed of 550 km/h (342 mph). An unusual feature of the design is the combination of three fuselages joined together to form an integrated structure. The outboard fuselage sections are very thick airfoils providing lift. At the same time they are provided with endplate floats on their outer sides, thanks to which the space between the floats and beneath the fuselage serves for building up a dynamic air cushion in cruise. The vehicle is provided with low-set wings of composite layout with low-aspect-ratio inboard panels and high-aspect-ratio outer panels.

Plast JSC

PE-201

The NPP Plast (NPP – *naoochno-proizvodstvennoye predpriyatiye*, science and production enterprise) was one of a dozen small enterprises that were founded in Nizhniy Novgorod (Gor'kiy) by former employees of Alekseyev CHDB. Apparently some of the projects developed at CHDB were taken over by the new company and, in a slightly revised configuration, were advertised to prospective customers under new designations. One such project is the PE-201 which bears more than a passing resemblance to the Strizh-3 and Strizh-4 described in Chapter 1. This vehicle, carrying a pilot and seven passengers, is virtually identical to the Strizh-4 and differs primarily in having the winglets moved slightly aft, so that their trailing edge is level with the trailing edge of the wing. The vehicle was to be powered by two 450-hp diesel engines; its specifications include a length of 15.4 m (50 ft 6⅜ in), a wing span of 13,0 m (42 ft 9 in), a height of 4,7

A computer-generated image of the NVA-07-025. The legend on the fuselage reads *Amfibiya* (amphibian).

This computer image depicts a heavy sea-going passenger WIG vehicle in the NVA series (possibly the NVA-07-530). Note the double-deck fuselage with cabins in the thick-profile wing centre section.

An impression of the same vehicle cruising over broken ice. This view shows the composite-layout wings and the four ducted propfans located side by side.

Technoavia JSC

Ronata

The Technoavia company (private design bureau) set up in the 1990s became fairly widely known as the developer and manufacturer of light aircraft, such as the SM-92 Finist, SM-92T Turbo-Finist, SM-94, SM-2000 and others. In late 1990s Technoavia, together with lesser known companies Delta-V and Svetlen and with the assistance from the Moscow Aviation Institute (MAI), developed a light WIG vehicle named Ronata. According to a press report, they *'were ready to conduct flight tests of the Ronata light* ekranoplan *manufactured from aluminium alloys. This vehicle was intended to carry four persons and their luggage over a distance of 1,200 km [745 miles], flying at a speed of 100-200 km/h [62-123 mph] above the water surface with waves not exceeding 0.5 m [1 ft 8 in]'*. The drawing accompanying this report showed a vehicle based on the Lippisch layout and resembling very much the ESKA-1 *ekranoplan* described in the previous chapter. There is no evidence to confirm that this vehicle ws actually tested (or even completed, for that matter).

Tekhaviakompleks JSC

Aeroplane boat

The company Tekhaviakompleks JSC, based in the town of Zhukovskiy near Moscow, was founded in 2004 to handle various questions, such as the

A computer-generated image of a proposed *ekranoplan* **based on the Be-200 amphibian proposed by Tekhnologii SDP.**

The SDP-09 tandem-wing *ekranoplan.* **The vehicle had been awaiting replacement engines for more than two years when this picture was taken.**

Designer V. Surzhik in the open entry hatch of the SDP-09.

The engine bay of the SDP-009 with the cowlings removed, showing the two car engines installed back to back.

in the vertical tail. Two- and four-seat models were under development. The vehicle is expected to be operated in three modes: as a hydroplaning boat; as a WIG craft and as an aircraft; the implementation of this bright idea hinged upon finding investors.

Tekhnologii SDP

(Dynamic Cushion Craft Technologies JSC)

Be-200 conversion into a WIG vehicle (project)

The joint-stock company bearing the name ZAO Tekhnologii SDP (SDP stands for *soodno na dinamicheskoy podooshke* – dynamic cushion craft) was founded in 2002 for the purpose of implementing several projects of WIG vehicles that had been developed in the course of the preceding years by a group of engineers from Kiev who had moved to Irkutsk together with A. Panchenkov (see Chapter 5). The group included V. V. Surzhik, P. A. Skorokhodov, V. V. Taranushenko, S. M. Remizov and G. A. Vzyatkin. Surzhik became the General Designer, with Skorokhodov, a businessman, acting as investor.

The group took out a patent for an invention the essence of which was the description of condition for the autostabilisation of a tail-first WIG vehicle during flight in ground effect. The material submitted together with the application for the patent contained a general arrangement drawing of a WIG vehicle resembling the Be-200 amphibian. In fact, this was a proposal for a conversion of the Be-200 based on the use of the latter's fuselage, engines and equipment. The high-set wings of the amphibian were to be replaced by low-set aft-mounted main wings of low aspect ratio and by smaller-area forward wings, actually a well-developed foreplane. Calculations showed that in the 'WIG version' the Be-200's AUW would reach 80 tonnes (176,400 lb) as against 43 tonne (94,815 lb), and the payload would amount to 40 tonnes (88,200 lb) instead of 12 tonne (26,460 lb). The proposal was brought to the attention of the management of the Irkutsk aircraft plant where the Be-200 amphibian was manufactured, but the response was lukewarm, and the project was not proceeded with.

SDP-09

By mid-2006 the Dynamic Air Cushion Ship Technologies JSC succeeded in building an experimental vehicle designated SDP-09 (SDP = *soodno na dinamicheskoy podooshke* – DAC ship) which was powered by two 220-hp automobile engines. The vehicle was to carry two crew and six to eight passengers. With an AUW of 2,600 kg (5,730 lb), it was expected to cover the distance of 1,500 km (932

upgrading of navigation systems. It did not neglect WIG vehicles either. At the MAKS-2009 show one could sea a curious-looking little craft called *Kater-samolyot* (Aeroplane boat). It was configured as a little boat hull topped by diminutive delta wings with ogival leading edges. The wing trailing edge was provided with elevons. Stabilising floats under the wingtips and a vertical tail at the rear completed the picture. A piston engine hidden inside the hull drove a nose-mounted propeller. The pilot sat well aft, and as the vehicle had a stron nose-up attitude, he had no view forward over the nose. The craft was successfully tested on the Moskva River in August 2009; it was regarded as a proof-of concept vehicle, to be followed by larger craft of the same layout. One of them, shown in model form, featured two propellers, the second one being inserted into a slot

miles) at a speed of 130-150 km/h (81-93 mph), flying at a height of 0.7 – 1.0 m (2-3 ft). It featured a canard layout, with aft-positioned main wings of greater span and forward-mounted wings of lesser span and area, both wings being provided with floats at their tips. The nose part of the hull was 'loaned' from the Sarepta motorboat, and the passenger cabin was an adaptation of the body of the Toyota Caldina stationcar. Two Honda car engines housed in the forward fuselage drove two six-blade tractor propellers via extension shafts with angle drive gearboxes. The vehicle was built with the assistance from A. Vzyatkin who was acting director of Plant No. 403 at that time. However, this machine proved to be ill-fated. Owing to various vicissitudes the vehicle could not begin its test programme in time, and when the tests started in March 2004 (on ice), the Honda engines failed to develop their rated power. Replacement engines proved hard to come by... The ultimate fate of this vehicle is unknown, but some reports suggest that its development was eventually abandoned

Termoplan Design Bureau, JSC

In 2007 this small private design bureau, not previously known to be engaged in WIG craft design, signified its intention to enter this field. The company's Chief Designer Yuriy Ishkov told a journalist that ZAO 'KB Termoplan' (the company's name in Russian) had started work on a WIG craft that probably would be the biggest in the world (a good start, indeed). The vehicle with a cargo-carrying capacity of 2,000 tonnes is visualised as a trimaran based on the use of three fuselages from the An-124 Ruslan super-heavy transport aircraft. The three fuselages are to be topped and joined together by a 'promenade deck' (Ishkov's term, curious enough for a vehicle designed to fly at a speed of up to 500 km/h). They house cargo holds and passenger compartments. The wings imparting WIG qualities to the craft are also borrowed from the An-124. In

the passenger version the vehicle is expected to seat more than 1,500 passengers. In a military version it can be converted into a high-speed aircraft carrier capable of accommodating up to 25 aircraft in its storage holds. Needless to say, the project has slender chances of being implemented, on account of the huge investments that would be required.

WingShips Company

The WingShips Company was established in Moscow in 1998. It is the Russian branch of the WingShips International Corporation which had been established in Delaware, USA, for the purpose of developing new marine transport technology – WIG craft on the basis of Russian expertise in *ekranoplan* construction.

PM-1 and PM-2 *ekranolyots* (projects)

In co-operation with specialists from TsAGI the WingShips Company has developed two projects of multi-purpose C-type WIG vehicles intended to carry up to 8 persons. These are the PM-1 and PM-2, sharing the same fuselage and differing in the powerplant and wing span. Both of them feature a wide fuselage, composite-type wings and a Vee tail unit. The unorthodox hydrodynamic layout of the vehicle features a kind of elongated tunnel under its bottom into which the ram air is fed.

The PM-1 is powered by two 900-kgp (1,980-lbst) AMNTK Soyuz R127-300 turbofans; it has an AUW of 3,500 kg (7,715 lb) and can cover a distance of 1,200 km (745 miles) with a maximum payload of 1,000 kg (2,205 lb) at a cruising speed of up to 270 km/h (168 mph) in ground effect. In aircraft mode it can rise to 4,000 m (13,120 ft).

The PM-2 is powered by two Lycoming TIO-540F piston engines. It can cover a distance of 1,400 km (870 miles) with a maximum payload of 900 kg (1,985 lb) at a cruising speed of up to 210 km/h (131 mph) in ground effect. The maximum altitude of flight out of ground effect is again 4,000 m.

Left: An artist's impression of the AeroMarine Systems AMS-500-4 taking off.

Above: A display model of the AMS-500-4.

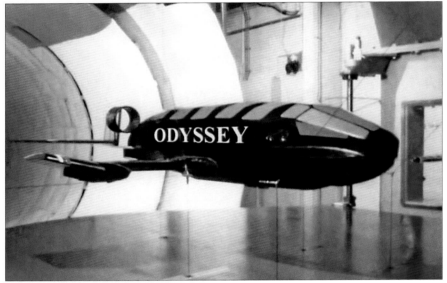

A wind tunnel model of the WingShips PM-2 featuring composite wings and aft-mounted ducted fans; the tail surfaces are just visible behind. The second photo has been altered to advertise the vehicle as the AMS-400 Odyssey; the final configuration of the latter may be rather different.

AeroMarine Systems Company

This company, founded in 2002 as a sister enterprise to the WingShips Company, closely co-operates with the latter and is engaged in the design of WIG craft and light seaplanes as well. It came up with several new designs of WIG vehicles which are listed below. The company advertises these projects in the hope of attracting prospective investors for a joint-venture manufacture of these machines,

AMS-500 Corvette

This multipurpose WIG vehicle is advertised in two versions: the AMS-500-4 patrol version and the AMS-500-30 passenger version. In actual fact, these are two different designs, the AMS-500-30 being a scaled up version of the AMS-500-4. Both of them are intended to be operated in all the three modes included into the IMO classification, that is, as a vehicle of the A. B and C type. In other words,

they can fly in ground effect only sticking to the water surface (Class A); or fly at an altitude of 1 to 4 m above the surface and make occasional hops to a higher altitude (Class B); or rise to an altitude of up to 4,000 m (13,120 ft).

AMS 500-4 Patrol

This craft carrying two crew and eight passengers at a maximum TOW of 4,000 kg (8,820 lb) is based on a machine developed by Yevgeniy P. Groonin way back during his work at the Sukhoi OKB (see Chapter 4). In its present configuration it is has a length of 12 m (39 ft 4 in) and a wing span of 14 m (46 ft) and is to be powered by two M-14 radial engines. Options include two Lycoming piston engines or one M-601 turboprop. With a maximum payload of 1,400 kg (3,087 lb) it can travel at a cruising speed of 240 km/h (149 mph) in ground effect over a distance of 1,600 km (984 miles).

AMS-500-30 Passengers

This is a much larger craft with a maximum TOW of 30,000 kg (66,150 lb), having a length of 21 m and a wing span of 25 m (82 ft). It is intended to carry five crew and 44 passengers at a cruising speed of 360 km/h in ground effect mode over a distance of 3,000 km (1,863 miles). The powerplant comprises two 2,500-shp TV7-117 turboprops.

AMS-400 Odyssey

Again, this is a common name for two different projects – the AMS-400-4 Business and AMS-400-30 Sea Train.

AMS-400-4 Business

This vehicle with an AUW of 4,000 kg (8,820 lb) can carry two crew and up to eight passengers and is powered by one M-601 turboprop. Being closely comparable to the AMS-500-4 in seating capacity and performance, it has lightly smaller dimensions and a different layout looking like a refined version of the WingShips PM-1 described above. The engine is placed above the rear fuselage between the splayed fins of the V-type tail unit. The extensively glazed boat-hull fuselage is to provide comfortable accommodation for four to six VIPs.

AMS-400-30 Sea Train

This larger version with an AUW of 30,000 kg (66,150 lb) is to carry five crew and 50 passengers at a cruising speed of 390 km/h (242 mph) in WIG mode over a distance of 3,200 km (1,989 miles). It is powered by two TV7-117 turboprops.

AMS-410

Two projects bearing this designation differ in function and size and are obviously intended for military

An artist's impression of the
AMS-410-DF Sea hunter in
Russian Navy colours.

duties the exact nature of which is not revealed.
These are the AMS-410 Transporter and the
AMS-410-DF Sea Hunter.

AMS-410 Transporter

This machine with an AUW of 4,500 kg (9,920 lb) is
powered by two RD-127-300 turbofans. With two
pilots and eight crew onboard it can fly at a cruising
speed of up to 270 km/h (168 mph) in WIG mode
and 290 km/h (180 mph) out of ground effect at altitudes up to 4,000 m (13,200 ft)

AMS-410-DF Sea Hunter

The Sea Hunter has an AUW of 9,000 kg (19,840 lb)
an carries a crew of two pilots seated in tandem. It
can be operated in the WIG mode and as an aircraft
at altitudes up to 15,000 m (49,200 ft). The fairly
large machine measuring 14 m (45 ft 11$\frac{3}{16}$ in) in
length and 14.5 m (47 ft 6$\frac{55}{64}$ in) in wingspan is powered by two TRDD-12500 turbojets (thrust rating not
specified). Its maximum payload of 3,500 kg (7,720
lb) presumably comprises various weaponry. The
two versions of the AMS-410 feature an aerohydro-

The ZPKB SM-16 dynamic air
cushion vehicle in cruise
mode. The forward hydrofoil
is barely discernible.

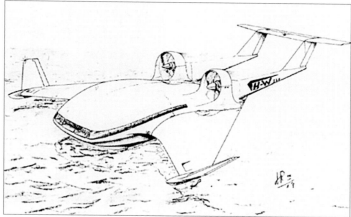

cruise. Two gas turbines drove shrouded propellers mounted on pylons; they also drove compressors which supplied air to the air cushion chamber. The hulls were provided with a flexible skirt. Thus, this

was tasked with research dealing with hydroplanes and amphibious vehicles. The work performed by this institute posed a practical question: who would bear the main responsibility for the design and con-

Artist's impressions of the Hoverwing 80 (left) and the larger Hoverwing 120.

Far left: The TAF VIII-1 two-seat WIG craft in cruise mode.

The TAF VIII-4 features the same tail design, with the propeller rotating in a slit between the fin and the rudder.

The TAF VIII-3 (far left) and the projected TAF VIII-7 both feature tandem wings. The larger TAF VIII-7 has twin engines and twin tails.

was a combination of a WIG vehicle and an ACV. The craft was to be controlled in flight by aerodynamic control surfaces, with special propellers as an option.

The craft was to be equipped with different complements of armament, depending on its mission. For self-defence it was to carry automatic anti-aircraft cannon and surface-to-air missiles. Calculations showed that this vehicle would have a 20-25% greater range in comparison with the conventional ACVs when travelling at a speed of 140-150 km/h (87-93 mph).

struction of WIG vehicles? Was it going to be the shipbuilding industry, or the aircraft industry? Eventually it was decided that the design work would be performed jointly by aviation and shipbuilding spe-

An artist's impression of the ARGE-1 developed by Techno Trans.

China (People's Republic of China)
The beginning of work in China on transport vehicles based on dynamic air cushion dates back to 1961 when China's government established the **Chinese Special Vehicles Research Institute** (CSVRI). Initially based in Beijing, it was transferred to the town of Jingmen, Hubei Province, in 1969. It

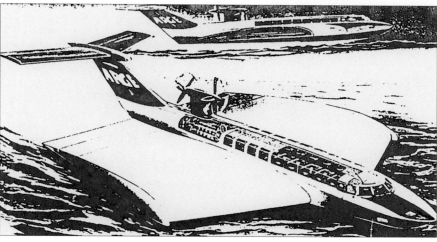

Front view of the Sea Eagle 6 in flight taken from a chase speedboat.

The Australian FlightShip FS-8 in cruise flight; it is identical to the German Airfish-8.

The Chinese Type 750 (aka Swan) research WIG vehicle. Note the false cabin windows painted on the side.

cialists, while the construction of prototypes would be entrusted to aircraft plants.

In due course several other Chinese institutions became involved in the theoretical studies and practical design of WIG craft intended for civil and military purposes. These included the **China Acad-**

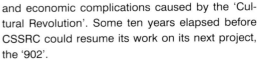

emy of Science and Technology Development (CASTD) and **China Ship Scientific Research Centre** (CSSRC).

In 1967 the latter institution embarked on the design and prototype construction of the first Chinese WIG craft. The project was designated '961'.Some details on this and the subsequent types of WIG vehicles created in China are given below in the chronological succession.

The '961' WIG vehicle was built by CSSRC in Beijing in 1968. This single-seat twin-fuselage machine of purely Chinese design (not patterned on Lippisch designs) was a single-seat experimental craft with twin fuselages. It was tested amid great secrecy in 1968-69.

In the early 1970s the work on *ekranoplans* in China was temporarily halted due to political turmoil

and economic complications caused by the 'Cultural Revolution'. Some ten years elapsed before CSSRC could resume its work on its next project, the '902'.

The '902' vehicle, featuring the Lippisch layout, was built by CSSRC in 1984 and made a series of test flights in the same year. Powered by two HS350A piston engines, it could fly over calm water only. It is regarded as a progenitor of the vehicles in the XTW series.

According to a different source, this single-seat experimental vehicle was under development from 1979 and was developed by CSVRI. It was powered by two HS350A piston engines.

In the early 1990s the Shanghai-based **708 Shipbuilding Institute** joined the work on ground-effect vehicles; it is also known as the ACV section of the **Marine Design and Research Institute of China** (MARIC).

The '750' vehicle was built by MARIC (708 Institute) in 1985 was an experimental two-seat twin-hull vehicle with triple vertical tails, powered by four ducted DT-30 piston engines rated at 22 kW apiece. A pair of engines was placed above aft sections of the hulls, while the other two engines were mounted on outriggers forward of the wing between the hulls, sending their air wash into the space under the wing and thus creating a static air cushion.

In the meantime, three ground-effect vehicles in the XTW series were designed and built by CSSRC.

The XTW-1 was built by CSSRC in 1989. This experimental three-seat machine powered by two Rotax 447 engines, featured the Lippisch layout; it had two vertical tails, canted to the sides and topped by a horizontal stabiliser. The vehicle was capable of flying in surface effect at a height of up to 10 m (33 ft) in sea state 2.

A different sourse states that a single example was built in1988 by CSVRI.

The XTW-2, built by CSSRC in 1990, was a scaled-up version of its predecessor. It carried two pilots and 14 passengers. The cruise flight could be performed in sea state 3, but the take-off was limited to sea state 2. The vehicle was powered by two Lycoming IO-540K1B5 flat-six engines. A different source states that a single example of the XTW-2 was built in 1988 by CSVRI.

The XTW-3, built by CSSRC in 1997, was a reworked version of the XTW-2 with a new tail unit (a T-tail). The photos also show a modified wing without the short 'shoulders' characteristic for the preceding two models.

Again, a different source states that a single example of the XTW-3 was built by CSVRI in 1996.

At about the same time a new organisation made an entry into the WIG craft design field. It was a WIG vehicle development centre (presumed to be the 605 Institute) which is known as a branch of the **China Academy of Science and Technology Development** (CASTD). In 1996 it presented its own project designated DFX-100

The DXF-100 (also referred to as DFXJ-100) was built by CASTD at an aircraft plant in Jingmen in 1998 as an experimental craft weighing 4.8 tonnes and powered by two IO-540K1B5 engines;. DXF = DiXiaoFeiji – WIG vehicle. This vehicle accommodating 12 passengers became the basis for subsequent designs in the TY (*TianYi*) series.

The TY-1 built by CASTD in. 1998 was the first in the TY series. This vehicle accommodating 14 passengers is powered by two engines of 294 kW apiece.

The development programme for the TY-1 has been aimed at commercial operation from the very beginning, so there was no prototype as such. The first craft to be built was put into commercial operation. Design work was started in June 1966 and was completed in August 1997. The TY-1 made its maiden flight on 10th November 1998. Upon completion of a 10-month test programme it was out into commercial operation on a lake in China for touristic purposes on a daily basis, sometimes up to eight runs per day.

The TY-1 was designed to carry 15 passengers, but this number had to be reduced to 11 due to a weight problem. The design weight was exceeded by 100 kg (220 lb); after adding an air conditioning system the weight and drag were such that the payloads decreased by four passengers.

The TY-1 was powered by two 224-kW Lycoming IO-450-K1B5 engines driving variable-pitch ducted propellers. The TY-1 is suitable only for lake or very flat sea operations since its maximum sea state is 3.

A second example was under construction and was to be ready by May 2000. There is no information about the actual progress of the work and the subsequent operational record of the machine.

In 1998 (?) the TY-1 was followed by the next design in the series.

The TY-2 (TianYi-2), built by CASTD, 605 Institute, was an experimental WIG vehicle powered by two ZMZ-4058 piston engines. No further details are available.

In September 1997 the 708 Institute made a come-back with new projects in the 75x series, starting with the '751'.

The '751' (DX-TE), initially known as the AF-1 and later dubbed Swan, was built by MARIC, 708 Institute, in Shanghai in 1998. This is a craft for 15-20 passengers with an AUW of 8.1 tonnes (17,860 lb) powered by three HS6 nine-cylinder radial engines (one 210-kW HS6A as a cruise engine and two 257-kW HS6E-1 forward-mounted engines that

This rather ungainly vehicle is the XTW-1 developed by CSSRC. Note the cranked wings with zero dihedral on the inboard sections; this is due to the need to provide propeller clearance.

could be tilted for take-off boost.) Two years of testing corroborated the soundness of the concept. The machine was built from imported aluminium alloys and composite materials. It could be operated at sea states up to 3.

The year of 1999 saw the beginning of construction of the XTW-4 WIG vehicle – the fourth in the XTW-series.

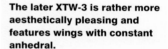

The later XTW-3 is rather more aesthetically pleasing and features wings with constant anhedral.

Front view of the still-unpainted XTW-3 with the engines uncowled.

An artist's impression of an XTW-3 in operation.

The XTW-4 was built by CSSRC (by CSVRI, according to a different source). The XTW-4 reverted to the same tail arrangement as on the XTW-1 and -2, featuring two vertical tails canted outwards and topped by a horizontal tail. Apparently

The almost completed CASTD TY-1 in the assembly shop in primer finish.

The TY-1 taxies on the water during tests.

Here the TY-1 is seen in wing-borne flight, showing the deep endplates/wingtip floats.

the T-tail of the XTW-3 was not successful. On the XTW-4 CSSRC used for the first time turboprop engines in its WIG craft. Two PT-6A-15AG turboprops drove 5-blade propellers. The XTW-4 was launched in September 1999; by early 2000 trials on the Changjiang River were completed. No information is available on the ultimate story of the project.

Unlike the earlier XTW craft, the XTW-4 was intended for sea operation and had a boat-type hull. A new type of high-strength, corrosion-resistant aluminium alloy was used for the structure. All primary control surfaces were operated by push-pull rod and cable systems, hydraulics were used to operate the taxiing gear. This was an experimental vehicle, construction of only one example was planned.

In 1999 MARIC submitted an improved version of the '751' project designated '751G'. Modifications to the original design included relocation of

the engines and of the horizontal tail. The engines were moved from the wings to pylons flanking the forward fuselage; the wing area was increased and the tailplane was raised higher on the vertical tail.

The DY-1 (known in the full-scale mock-up form as the DY-806) was presented by CASTD (605 Institute) in 2000. It was a six-seat craft powered by a single IO-540K1B5 engine The machine featured a composite layout wing with a low-aspect-ratio centre section and high-aspect-ratio outer wing panels. The vehicle could be operated in the surface-effect mode with occasional flight at an altitude of up to 30 m (100 ft).

No prototype was built; the testing was to be conducted on the first production example. This vehicle appears to be near-identical in appearance to the Australian Sea Eagle-6 and its presumed Chinese counterpart, the TSN (see below)

The SDJ-1 WIG vehicle built by CSSRC in 2000 was a six-seat craft of a catamaran layout, with the engine mounted above the wing between the fuselages. The main wing was supplemented by outer wing panels outboard of the hulls.

Two projects were reportedly developed by Hong Kong-based company **Flying Dragon Technology Ltd** by 2000.

The FLHRO-PB ('Orlyonok-2') was a project of a 150-seat passenger WIG craft, developed by the Hong Kong company on the basis of the Soviet (Russian) ekranoplan Orlyonok.

The FLHW from the same company was a project of a 50-seat passenger WIG craft (no details are available.

A notable event took place in 1999. A joint stock company named **Guangzhou Tianxiang WIGcraft Company Ltd** was set up in the Guangzhou Free Zone (the city of Guangzhou, Guangdong Province). The company, created for the purpose of developing and manufacturing ekranoplans of Chinese design, declared its intention to win leading positions on the world market. The company/s designs are described below.

The TX-2 (TY-2) or Tianxiang-2 WIG vehicle was under construction by Guangzhou Tianxiang WIGcraft Co., nearing completion in 2000. According to one source, it was powered by three 150-hp automobile engines and featured a combination of aluminium alloys and composite materials in its structure. A different source states, however, describes this vehicle was as a copy of the Russian twin-engined Volga-2 WIG vehicle, carrying two crew and eight passengers.

Development of this vehicle started in 1999. It had an AUW of 2.85 tonnes (6,280 lb).

The TX-5 (Tianxiang -5) was announced by **Guangzhou Tianxiang WIGcraft Co.** in 2001. It was a 45-seat WIG vehicle powered by two WS-11 turbofans rated at 1,720 kgp (3,790 lbst), and one WJ5E turboprop delivering 2,850 hp. The vehicle had an AUW of 19 tonnes (42,000 lb). According to the project, the vehicle was expected to be able to make short 'hops' to an altitude of up to 150 m (500 ft) for overcoming obstacles.

The TSN (AD-606) six-seat machine with a 300-hp engine was reportedly built by **Yionhe Iongyi Ship Co. Ltd** in Nanjing in 2002. It had composite layout wings and a T-tail.

This craft appears to be identical to a WIG vehicle developed in Australia by the Sea Eagle International company. This company worked with China's CSSRC to develop a civilian range of Class B Wing Effect Craft. The craft, also dubbed Sea Eagle, was reported to be flying in China. Quite recently, in mid-2010, there were reports stating that two Chinese WIG craft had been delivered to United Arab Emirates, and accompanying pictures showed a vehicle closely resembling the TSN (or Sea Eagle).

Development of WIG vehicles in China is conducted by Chinese scientists, designers and engineers with due regard to experience gained in other countries. In particular, China established co-operation with Russian scientific institutions engaged in researching the problems associated with WIG craft. This co-operation included contracts with Russian institutions for studying certain specific problems; Russian specialists came to China to deliver lectures on the subjects of WIG technology. The Guangzhou Tianxiang WIGcraft Co Ltd mentioned above officially acknowledged its close collaboration with the leading Russian organisations in the field of *ekranoplan* design.

In 2005 Vyacheslav Kolganov, the then general director and chief designer of the TREK Science & Production Group (at present the leader of IFPG), told journalists that a number of engineers and designers from the Alekseyev CHDB, from St. Petersburg and from Novosibirsk had been working already for several years at Chinese enterprises engaged in the development of *ekranoplans*. On the basis of the experience accumulated by these specialists, said Kolganov, the Chinese had built and were using operationally close to a hundred ekranoplans with a seating capacity for 7, 14 and 20 passengers; they were engaged in the projecting of WIG vehicles intended to carry 150 passengers and still bigger machines. Kolganov's point obviously was the much greater attention accorded in China to the WIG field as compared to Russia.

The use of Russian experience found its expression, in particular, in the construction of the YH-7, a copy of the Russian Ivolga (EL-7) WIG craft, which was built in Shanghai (in 2002?). However, generally speaking, Chinese designers have demonstrated their ability to produce designs which cannot be classed as mere copies of ekranoplans created in other countries. This was acknowledged by the abovementioned Kolganov, who, while speaking to a journalist in November 2008, said that in China the WIG vehicle construction had become a task of state importance. This subject, he said, was attracting great interest there. He noted that a shipbuilding plant in Shanghai had started the manufacture of the airframe of a 150-seat ekranoplan and there were plans for creating a WIG vehicle capable of carrying 600 passengers. Chinese specialists, noted Kolganov, were very active in studying theoretical aspects of ekranoplan construction and were actively co-operating with their Russian colleagues from SibNIA (the Siberian Aviation Research Institute).

Chinese designers are visualizing new projects of WIG vehicles of ever growing weight and dimensions, some of which are truly gigantic machines.

The '751' (Swan) WIG craft taxies on the water

Two views of the '751' in cruise flight. Note the 'power egg' installation of the centre engine.

A computer-generated image of the DY-1 light WIG vehicle.

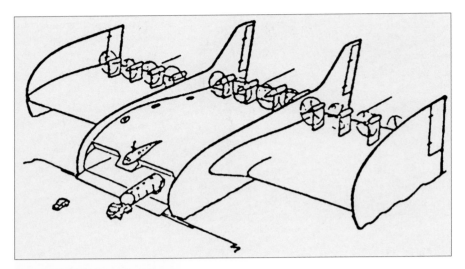

An artist's impression of the French Cygne-14 large transport WIG craft.

For example, there were reports about a vehicle called Xintianwen-4 (project Changhe-6) with a displacement of 500 tonnes (1,102,290 lb). In 2006 publicity was given to a series of projects on military WIG vehicles, including a WIG-type aircraft carrier, a missile-carrying WIG craft and other projects.

In 2007 a WIG vehicle with unknown designation was reported to be under development at Shanghai-based **Tongji University**. With a design payload of 4,000 kg (8,800 lb), top speed of 162 kt (300 km/h) and maximum altitude of 5 m (16 ft), the vehicle was claimed to consume half to two-thirds of fuel of an 'ordinary aircraft' while carrying 'more weight' and being up to six times faster than an ocean-going ship. The report said further development could lead to a 50-seat prototype by 2013, with future versions able to carry up to 400,000 kg (881,850 lb) by 2017.

Representatives of the **Shanghai Engineering and Construction University** announced in 2008 (?) that they would soon complete the work on the projects of several types of WIG vehicles with an all-up weight ranging from 10 to 200 tonnes (22,045-441,000 lb). They claimed that by 2017 more than 200 ekranoplans would be put into operation on regular routes; the machines in question were expected to carry a cargo of more than 400 tonnes (882,000 lb).

The success achieved by the Chinese industry in the design and construction of WIG vehicles was emphasized by recent reports (March 2010) according to which the Chinese company Jiangsu Hengchuan Group from Jiansu province, Eastern China, was going to deliver 17 WIG vehicles to the United Arab Emirates. The value of the contract was estimated at 'several milliard yuan'. The delivery is scheduled to take place in 2011. Unfortunately, no further details concerning this contract are available.

France

In 1977 specialists of French company Bertin developed a project of a transoceanic cargo transport WIG vehicle named Cygne (Swan). The project comprised two versions: the Cygne-10 and the Ceygne-14 with a useful load of 550 and 870 tonnes (1,213,000 and 1,918,000 lb) respectively. The vehicle was configured as a lifting body (thick-section flying wing) with outer wing panels of thinner profile. Both the wing centre section and the outer panels were provided with huge endplates which also acted as vertical tails and accordingly were fitted with rudders. The forward part of the lifting body (wing centre section) incorporated ramps which could be used for the roll-on loading of wheeled vehicles. The powerplant comprised 12 aircraft engines with tractor propellers which were mounted on pylons above the wings. Aerodynamic properties of the craft enabled it to fly at altitudes between 0 and 3,000 m (9,840 ft). It was expected to cross the Atlantic at a speed of 370 km/h (230 mph) in the course of 20 hours.

Great Britain

Several projects of heavy WIG craft for military duties were developed in Great Britain and the USA. In the late 1950s and early 1960s the British engineer **A. Pedrick** developed several projects. One of them was a project of an aircraft-carrier configured as a WIG vehicle. It was intended to provide accommodation for 20-30 light combat aircraft. The vehicle featured a flying wing layout with thin endplates. A streamlined superstructure located on the forward part of the upper deck housed the workstations for the vehicle's control. The empennage comprised two large vertical tails with rudders. The powerplant comprised ten turbojets located in the aft section of the body. The air intakes of the turbojets were located on the wing's upper surface; their suction helped enhance the lifting properties of the wing. A large part of the vehicle's body was occupied by a single-tier hangar with three deck lifts, two of which served for delivering the aircraft from the hangar to the deck and the third one was used for stowing them under the deck after landing. Two catapults were to be used for performing the take-off, while the landing involved the use of arrestor wires.

In 1972 the British designer **W. E. Walledge** took out a patent for an original design of a boat with aerodynamic off-loading. The boat had a streamlined hull with an aft-mounted pylon which carried a lifting wing and an engine with a propeller. The vehicle had an aircraft-type tail unit. Pitch trim was effected with the help of an adjustable stabiliser, ailerons served for lateral control. The vehicle's peculiar feature was the location of the wing far aft of the CG.

In 1976 the **Sunbury** company built the first British WIG boat featuring an aircraft layout. It had a tail unit consisting of twin fins topped by a horizontal tail. The vehicle was powered by a Rolls-Royce Gnome turboshaft engine. The machine underwent flight testing which reportedly was intended to provide a basis for a decision on the feasibility of larger machines of the same kind.

Iran

In 2006 the world mass media carried reports touching upon Iran's work in the field of WIG vehicle construction. This was sparked off by the news item according to which the Iranian military had tested during a 'war game' in the Persian Gulf 'a small propeller-driven aircraft that flies low over water surface at up to 100 knots'. A screenshot from an Iranian video showed a little craft very similar in appearance to the Russian TsLST ESKA-1 *ekranoplan* (see Chapter 5). An interesting comment on this event was published at a Russian web forum in 2009. It stated that 'some ten years ago' (that is, around 1999) a large-scale piloted model of a WIG craft had been built with assistance from TsAGI in the Iranian **Malek-Ashtar University**. It was made of aluminium/magnesium alloy (AMg) and was of riveted construction. The machine performed many successful fights in the Persian Gulf, but eventually fell victim to corrosion, despite all efforts to prevent this. Later (according to the comment) moulds were taken from the basic elements of the airframe, to be used for manufacturing two new airframes in glass fibre which were externally identical but featured internal differences. One of the two airframes was of traditional construction (framework and skin); the other one featured a 'sandwich' construction without the supporting internal framework. Both airframes were completed as piloted vehicles which were tested in the Persian Gulf in late 2005 and early 2006. At the same time the Iranian Navy conducted an exercise in the Persian Gulf, with due attention from the US mass media. It was then that one of these test vehicles was glimpsed in the news. It stands to reason to surmise that there are plans in Iran for further experiments with larger WIG craft intended for practical operation

Japan

At about the same time specialists of the Japanese company **Kawasaki Aircraft Co Ltd** built three WIG craft designated KAG-1, KAG-2 and KAG-3. They were configured as catamarans with large-size floats flanking the short-span wing. In particular, the KAG-3 vehicle had an open cockpit for two pilots located on the forward section of the wing. The vehicle powered by an 80-hp outboard engine attained a speed of 110 km/h (68 mph). The KAG-3 under-

A poor-quality yet interestig photo showing the Iranian WIG craft in flight.

went comprehensive testing during which it was repeatedly subjected to modifications. The design was eventually abandoned because it was inherently unstable.

In 1963 engineer Shigenori from the Kawasaki Company developed a project of a heavy WIG vehicle featuring a 'flying wing' layout. It was to be powered by four aircraft engines mounted on tall pylons located in the aft part of the wing. The pylons doubled as vertical tails topped by a horizontal tail. The wing underside was provided with longitudinal strakes which were intended to prevent transverse airflow under the wings.

Yet another project prepared by Shigenori was a similar WIG craft equipped with a device for power-augmented take-off. It was fitted with a foreplane mounted higher up in relation to the main wing. The foreplane carried two engines with propellers. At the moment of take-off the foreplane was tilted, directing the propeller wash under the wing to create a static air cushion which lifted the vehicle out of the water. When the craft attained the necessary speed for cruise flight, the foreplane was set into the optimum position dictated by hydrodynamic conditions, and the craft's load.

In 1962 a team of Kawasaki designers led by M. Yatsushi developed a project of a WIG craft very similar to the one described above. It differed in having a fixed foreplane carrying booster engines which were given the proper incidence to direct the airflow from their propellers under the wings.

The Kawasaki KAG-3 WIG craft. Note the cockpit offset to port.

In 1972 the Japanese company **Shin Meiwa** built the AF-XS boat with aerodynamic off-loading. This two-seat vehicle with a hull having a length of 4 m (13 ft) was fitted with sweptback wings and with hydroskis. The boat was powered by a 25-hp Yamaha outboard engine. Tests showed an improvement of performance in comparisons with traditional boats in the same power class.

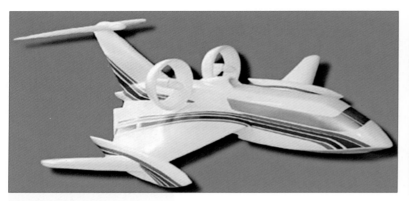

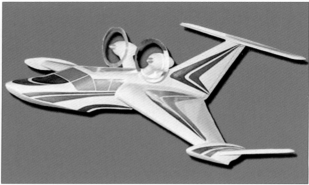

Desktop models of two unnamed WIG vehicles designed by NPO Europa in the Ukraine.

New Zealand

In 2007 it was reported that engineer Rudy Heeman residing in New Zealand had built an air cushion vehicle which, in addition, is capable of flight in ground effect mode thanks to the wings which can be extended after the vehicle has reached a speed of 100 km/h (62 mph). The two-seat vehicle has made some successful flights, but no news is available on the ultimate results of the testing.

South Korea

In September 2007 the government of South Korea announced its plans envisaging the construction of a large WIG craft intended for commercial purposes by 2012. The vehicle is expected to have an all-up weight of 300 tonnes (660,000 lb) and have a load-carrying capacity of up to 100 tonnes (220,400 lb). The machine measuring 77 m (252 ft 7½ in) in length and 65 m (213 ft 3 in) in wing span will have a speed of 250-300 km/h (155-186 mph). The government allotted US$ 91.7 million for the development of this *ekranoplan* for the nearest five years. The work on this project had been started as early as 1995.

Models of South Korean-designed *ekranoplans* were submitted for testing to the Krylov TsNII in St. Petersburg.

Taiwan (Republic of China)

A Taiwanese company built a prototype of a WIG vehicle designated HT-1. It is a medium-sized machine featuring the Lippisch layout. Propulsion is ensured by a single ducted propeller mounted above the fuselage close to the tail unit (a T-tail).

The **Chung-Shan Institute of Science and Technology** (based in Taiwan) conducting research in the field of civil and military aviation, shipbuilding and other branches associated with defence has devoted some attention to WIG technology, too. Detailed information on this score is not available. It is presumed that several relatively small craft intended to carry two to five persons were designed, built and tested.

The Ukraine

As related in the previous chapters, during the Soviet period some research and practical design work related to the WIG vehicles took place in the Ukraine (then a part of the Soviet Union). Moreover, at one time there were plans for starting a series manufacture of large *ekranoplans*, such as the Orlyonok and Loon', at the Feodosiya-based **'More' (Sea) Shipyard** in the Crimea. These plans were not put into practice, but the plant retained a potential for dealing with that sort of production. In the independent Ukraine this plant made an attempt to diversify its product range by starting the manufacture of a light WIG craft on the basis of a Russian design. Kolganov's Ivolga types (the EL-7 and the EK-12) have served as a basis for the development of the Shmel' (Bumblebee) WIG vehicle by the Feodosiya Shipbuilding Company 'More'. It appears to be a cross between the two designs, slightly differing in dimensions from the EK-12 and featuring the straight wings of the EK-7, rather then the swept wings of the EK-12. Its design seating capacity of 7-10 passengers lies between the figures for the EL-7 (seven passengers) and the EK-12 (12-14 passengers). The craft was advertised and offered to prospective customers, but there is no evidence of any results of these plans so far.

Some private companies are offering their own projects of light WIG craft and advertise them in the hope of finding sponsors and investors willing to invest into the construction of prototypes. As an example one may mention the company **NPO Europa** which has spawned two projects of passenger WIG craft, both of them having the same configuration and differing in size. The two machines, carrying four and 20 passengers respectively, have a typical Lippisch layout with the reversed delta wing, T-tail, wingtip floats and small winglets, only moderately upswept. The fuselages are configured as boat hulls. The powerplant comprises an engine buried in the aft fuselage and transmitting its torque via extension shafts to two ducted propellers above the fuselage.